ELECTRONICS

Principles and Applications

Fifth Edition

Charles A. Schuler

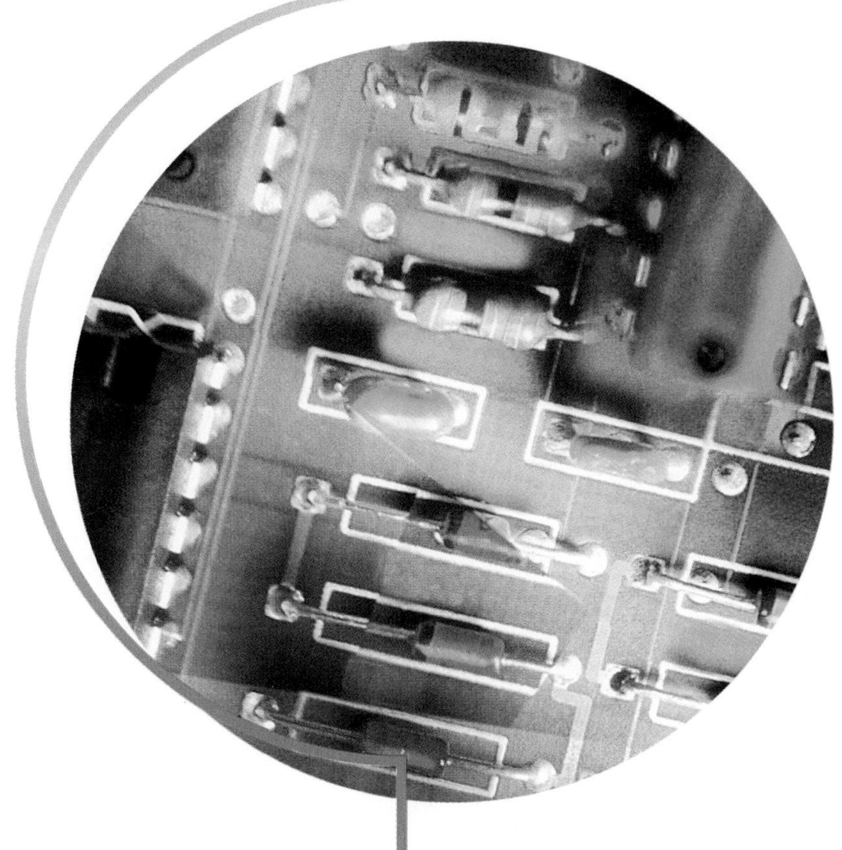

Glencoe
McGraw-Hill

New York, New York Columbus, Ohio Woodland Hills, California Peoria, Illinois

Cover photos: Charles Thatcher/Tony Stone Images; background: David McGlynn/FPG.

Text photos: Page ix: *(top):* Lou Jones/The Image Bank; *(bottom):* © Cindy Lewis. **Page 2:** *(right):* file photo; *(left):* Design Center, Ford Motor Company. **Page 7:** Eric-Leigh Simmons/The Image Bank. **Page 11:** Ron Scott/Tony Stone Worldwide. **Page 12:** Reproduced by permission of Tektronix, Inc. **Page 14:** *(upper right):* Bachmann/© The Stock Solution; *(lower left, both photos):* Chris Fritsch/Life Images Studio. **Page 16:** Courtesy of Motorola. **Page 33:** Mark Steinmetz. **Page 72:** Photo courtesy of Tektronix Corporation. **Page 84:** Toyota. **Page 119:** Courtesy of Tektronix. **Page 143:** Telegraph Colour Library/FPG International. **Page 160:** Reproduced by permission of Tektronix, Inc. **Page 201:** Motorola and Onkyo. **Page 265:** Courtesy of Tektronix. **Page 269:** *(left):* Courtesy of Tektronix; *(second from left):* Motorola and Onkyo; *(right):* Courtesy of Tektronix. **Page 275** (DVD): Courtesy of Sony Electronics Inc. (amplifier): Motorola and Onkyo. **Page 322:** Courtesy of Mercedes Benz. **Page 336:** Courtesy of Motorola. **Page 375:** *(top):* Courtesy of Motorola and Onkyo; *(bottom left and right):* Courtesy of Intel Corporation. **Page 399:** Blair Seitz/Photo Researchers.

Library of Congress Cataloging-in-Publication Data

Schuler, Charles A.
Electronics, principles and applications / Charles A. Schuler.—
5th ed.
p. cm.
Includes index.
ISBN 0-02-804244-1
1. Electronics. I. Title.
TK7816.S355 1999
621.381—dc21 98-25185
CIP
AC

Glencoe/McGraw-Hill
A Division of The ***McGraw-Hill*** *Companies*

Electronics: Principles and Applications, Fifth Edition

Send all inquiries to:
Glencoe/McGraw-Hill
936 Eastwind Drive
Westerville, OH 43081

ISBN 0-02-804244-1 (student edition)
ISBN 0-02-804255-7 (instructor's annotated edition)

2 3 4 5 6 7 8 9 071 06 05 04 03 02 01 00 99

Contents

The Glencoe *Basic Skills in Electricity and Electronics* series has been designed to provide entry-level competencies in a wide range of occupations in the electrical and electronic fields. The series consists of coordinated instructional materials designed especially for the career-oriented student. Each major subject area covered in the series is supported by a textbook, an experiments manual, and an instructor's productivity center. All the materials focus on the theory, practices, applications, and experiences necessary for those preparing to enter technical careers.

There are two fundamental considerations in the preparation of materials for such a series: the needs of the learner and needs of the employer. The materials in this series meet these needs in an expert fashion. The authors and editors have drawn upon their broad teaching and technical experiences to accurately interpret and meet the needs of the student. The needs of business and industry have been identified through personal interviews, industry publications, government occupational trend reports, and reports by industry associations.

The processes used to produce and refine the series have been ongoing. Technological change is rapid and the content has been revised to focus on current trends. Refinements in pedagogy have been defined and implemented based on classroom testing and feedback from students and instructors using the series. Every effort has been made to offer the best possible learning materials.

The widespread acceptance of the *Basic Skills in Electricity and Electronics* series and the positive responses from users confirm the basic soundness in content and design of these materials as well as their effectiveness as learning tools. Instructors will find the texts and manuals in each of the subject areas logically structured, well-paced, and developed around a framework of modern objectives. Students will find the materials to be readable, lucidly illustrated, and interesting. They will also find a generous amount of self-study and review materials and examples to help them determine their own progress.

Both the initial and on-going success of this series are due in large part to the wisdom and vision of Gordon Rockmaker who was a magical combination of editor, writer, teacher, electrical engineer and friend. Gordon has retired but he is still our friend. The publisher and editors welcome comments and suggestions from instructors and students using the materials in this series.

Charles A. Schuler,
Project Editor
and
Brian P. Mackin,
Editorial Director

Basic Skills in Electricity and Electronics

Charles A. Schuler, Project Editor

New Editions in This Series

Electricity: Principles and Applications, Fifth Edition, Richard J. Fowler
Electronics: Principles and Applications, Fifth Edition, Charles A. Schuler
Digital Electronics: Principles and Applications, Fifth Edition, Roger L. Tokheim

Other Series Titles Available:

Communication Electronics, Second Edition, Louis E. Frenzel
Microprocessors: Principles and Applications, Second Edition, Charles M. Gilmore
Industrial Electronics, Frank D. Petruzella
Mathematics for Electronics, Harry Forster, Jr.

Preface

This text introduces the principles and applications of analog electronic devices, circuits, and systems. It is intended for students who have a basic understanding of Ohm's law, Kirchhoff's laws, power, schematic diagrams, and basic components such as resistors, capacitors, and inductors. The only mathematics prerequisite is a command of basic algebra.

The major objective of this text is to provide entry-level knowledge and skills for a wide range of occupations in electricity and electronics. Its purpose is to assist in the education and preparation of technicians who can effectively diagnose, repair, verify, and install electronic circuits and systems. It also provides a solid and practical foundation in analog electronic concepts and device theory for those who may need or want to go on to more advanced study.

The fifth edition, like the earlier ones, combines theory and applications in a logical, evenly paced sequence. It is important that a student's first exposure to analog electronic circuits and devices be based on a smooth integration of theory and practice. This approach helps the student to develop an understanding of how devices such as diodes and transistors function. Then the understanding of these functions can be applied to the solution of practical problems and system applications. Modern electronics is a vital and exciting field, and studying it should be the same.

I have taken a practical approach throughout this text. The devices, circuits, and applications are typical of those used in all phases of electronics. Reference is made to common aids such as parts catalogs and substitution guides, and real-world troubleshooting techniques are applied whenever they are appropriate. The information, theory, and calculations presented here are the same as those used by practicing technicians.

The 15 chapters progress from an introduction to the broad field of electronics through solid-state theory, transistors, and the concepts of gain, amplifiers, oscillators, radio, integrated circuits, control circuitry, and regulated power supplies. As an example of the practicality of the text, an entire chapter is devoted to troubleshooting circuits and systems; in other chapters, entire sections are devoted to this vital topic.

This edition has been significantly improved by giving more attention to student needs. It includes more solved problems and has more lucid and detailed discussions of vital concepts. It introduces computer simulation and modeling because these tools are now available to most students. It reflects changes in technology that have occurred since the last edition. Most notably, it devotes serious attention to the digital revolution that is beginning to take place in analog electronics. More and more solutions are hybrid designs that combine analog and digital approaches. This edition has new information about transistors as switches, switch-mode amplifiers, direct digital synthesis, and digital signal processing.

Students and teachers will find many features designed to make the study of analog electronic devices and circuits interesting and effective. Each chapter starts with objectives to alert the reader to what should be accomplished. The reader is encouraged to take advantage of the self-test items at the end of each chapter section as well as the chapter review problems and tests. All critical facts and principles are reviewed in a summary listing at the end of each chapter. Important items are placed in the margins to call the reader's attention to key terms and concepts. New information has been added to the margins for enrichment of knowledge and to provide practical tips for job seekers. All in all, every effort has been made to optimize this text for learners and to promote their success on the job.

Be sure to visit our website at www.glencoe.com/ps/bsee/schuler. There, you will find valuable information on subjects including careers and skills standards. You will also find practice tests organized by chapter, and links to equipment manufacturers and other related websites.

I am pleased to acknowledge all the instructors and students who have used the earlier editions and especially those who made comments and suggestions. Their feedback has been invaluable, and the fifth edition is markedly better because of their interest and wisdom. I also thank the industry personnel who reviewed the content and provided invaluable guidance. A book such as this is greatly enhanced by this process. As always, I welcome and solicit comments and suggestions from students and instructors using this edition.

The author, project editor, and publisher would like to thank the instructors listed below who provided general and detailed comments, and those who responded to the survey that was sent out while the book was being revised. Their comments and suggestions provided the valuable input necessary to make a good book even better.

Donald D. Barrett Jr.
DeVry Institute of Technology
Dallas, TX

Albert Covington
Chattanooga State Tech Community College
Chattanooga, TN

Bob Crosby
Western Technical Institute
El Paso, TX

Jim Davidson
Southwest School of Electronics
Austin, TX

Ron Dreucci
California University of Pennsylvania
California, PA

Bill Hessmiller
Technical Training Associates
Dunmore, PA

Bill Lamprich
LTC—Minden Campus
Minden, LA

John W. Loney
California University of Pennsylvania
California, PA

James R. McGraw
Tidewater Tech
Virginia Beach, VA

Gregory T. Shepley
Girard, OH

Richard Windley
ECPI College of Technology
Virginia Beach, VA

Charles A. Schuler

Safety

Electric and electronic circuits can be dangerous. Safe practices are necessary to prevent electrical shock, fires, explosions, mechanical damage, and injuries resulting from the improper use of tools.

Perhaps the greatest hazard is electrical shock. A current through the human body in excess of 10 milliamperes can paralyze the victim and make it impossible to let go of a "live" conductor or component. Ten milliamperes is a rather small amount of electrical flow: It is only *ten one-thousandths* of an ampere. An ordinary flashlight uses more than 100 times that amount of current!

Flashlight cells and batteries are safe to handle because the resistance of human skin is normally high enough to keep the current flow very small. For example, touching an ordinary 1.5-V cell produces a current flow in the microampere range (a microampere is one-millionth of an ampere). This amount of current is too small to be noticed.

High voltage, on the other hand, can force enough current through the skin to produce a shock. If the current approaches 100 milliamperes or more, the shock can be fatal. Thus, the danger of shock increases with voltage. Those who work with high voltage must be properly trained and equipped.

When human skin is moist or cut, its resistance to the flow of electricity can drop drastically. When this happens, even moderate voltages may cause a serious shock. Experienced technicians know this, and they also know that so-called low-voltage equipment may have a high-voltage section or two. In other words, they do not practice two methods of working with circuits: one for high voltage and one for low voltage. They follow safe procedures at all times. They do not assume protective devices are working. They do not assume a circuit is off even though the switch is in the OFF position. They know the switch could be defective.

As your knowledge and experience grow, you will learn many specific safe procedures for dealing with electricity and electronics. In the meantime:

1. Always follow procedures.
2. Use service manuals as often as possible. They often contain specific safety information. Read, and comply with, all appropriate material safety data sheets.
3. Investigate before you act.
4. When in doubt, *do not act*. Ask your instructor or supervisor.

General Safety Rules for Electricity and Electronics

Safe practices will protect you and your fellow workers. Study the following rules. Discuss them with others, and ask your instructor about any you do not understand.

1. Do not work when you are tired or taking medicine that makes you drowsy.
2. Do not work in poor light.
3. Do not work in damp areas or with wet shoes or clothing.
4. Use approved tools, equipment, and protective devices.
5. Avoid wearing rings, bracelets, and similar metal items when working around exposed electric circuits.
6. Never assume that a circuit is off. Double-check it with an instrument that you are sure is operational.
7. Some situations require a "buddy system" to guarantee that power will not be turned on while a technician is still working on a circuit.
8. Never tamper with or try to override safety devices such as an interlock (a type of switch that automatically removes power when a door is opened or a panel removed).
9. Keep tools and test equipment clean and in good working condition. Replace insulated probes and leads at the first sign of deterioration.
10. Some devices, such as capacitors, can store a *lethal* charge. They may store this charge for long periods of time. You must be certain these devices are discharged before working around them.
11. Do not remove grounds and do not use adaptors that defeat the equipment ground.
12. Use only an approved fire extinguisher for electrical and electronic equipment. Water can conduct electricity and may severely damage equipment. Carbon dioxide (CO_2) or halogenated-type extinguishers are usually preferred. Foam-type extin-

guishers may also be desired in *some* cases. Commercial fire extinguishers are rated for the type of fires for which they are effective. Use only those rated for the proper working conditions.

13. Follow directions when using solvents and other chemicals. They may be toxic, flammable, or may damage certain materials such as plastics. Always read and follow the appropriate material safety data sheets.
14. A few materials used in electronic equipment are toxic. Examples include tantalum capacitors and beryllium oxide transistor cases. These devices should not be crushed or abraded, and you should wash your hands thoroughly after handling them. Other materials (such as heat shrink tubing) may produce irritating fumes if overheated. Always read and follow the appropriate material safety data sheets.
15. Certain circuit components affect the safe performance of equipment and systems. Use only exact or approved replacement parts.
16. Use protective clothing and safety glasses when handling high-vacuum devices such as picture tubes and cathode-ray tubes.
17. Don't work on equipment before you know proper procedures and are aware of any potential safety hazards.
18. Many accidents have been caused by people rushing and cutting corners. Take the time required to protect yourself and others. Running, horseplay, and practical jokes are strictly forbidden in shops and laboratories.

Circuits and equipment must be treated with respect. Learn how they work and the proper way of working on them. Always practice safety: your health and life depend on it.

Electronics workers use specialized safety knowledge.

Introduction

Chapter Objectives

This chapter will help you to:

1. *Identify* some major events in the history of electronics.
2. *Classify* circuit operation as digital or analog.
3. *Name* major analog circuit functions.
4. *Begin* developing a system viewpoint for troubleshooting.
5. *Analyze* circuits with both dc and ac sources.
6. *List* the current trends in electronics.

Electronics is a relatively recent technology that has undergone explosive growth. It is now so widespread that it touches all of our lives in many ways. This chapter will help you to understand how electronics developed over the years and how it is currently divided into specialty areas. It will help you to understand some basic functions that take place in electronic circuits and systems and will also help you to build upon what you have already learned.

1-1 A Brief History

Electronics is very young. It is hard to place an exact date on its beginning. The year 1899 is one possibility. During that year, J. J. Thompson, at the University of Cambridge in England, discovered the electron. Two important developments at the beginning of the twentieth century made people interested in electronics. The first was in 1901, when Marconi sent a message across the Atlantic Ocean using *wireless* telegraphy. Today we call wireless communication *radio*. The second development came in 1906, when De Forest invented the audion vacuum tube. The term "audion" related to its first use, to make sounds ("audio") louder. It was not long before the wireless inventors used the vacuum tube to improve their equipment.

Another development in 1906 is worth mentioning. Pickard used the first crystal radio detector. This great improvement helped make radio and electronics more popular. It also suggested the use of *semiconductors* (crystals) as materials with future promise for the new field of radio and electronics.

Commercial radio was born in Pittsburgh, Pennsylvania, at station KDKA in 1920. This development marked the beginning of a new era, with electronic devices appearing in the average home. By 1937 more than half the homes in the United States had a radio. Commercial television began around 1946. In 1947 several hundred thousand home radio receivers were manufactured and sold. Complex television receivers and complicated electronic devices made technicians wish for something better than vacuum tubes.

The first vacuum-tube computer was built in 1943 at the University of Pennsylvania. Soon commercial vacuum-tube computers became available. They were large and expensive, and generated much heat because of the many vacuum tubes they used. The numerous vacuum tubes and heat caused many failures. It was not uncommon to have several breakdowns in one day. Television receivers and radio receivers had dozens of vacuum tubes, but computers

Audion

Vacuum tube

Semiconductor

Analog and digital displays. (*Left*) Digital speedometer. (*Right*) Analog speedometer.

had thousands. Something better than the vacuum tube had to be found.

Scientists had known for a long time that many of the jobs done by vacuum tubes could be done more efficiently by semiconducting crystals, but they could not make crystals pure enough to do the job. The breakthrough came in 1947. Three scientists working with Bell Laboratories made the first working transistor.

Solid state

This was such a major contribution to science and technology that the three men—Bardeen, Brittain, and Shockley—were awarded the Nobel Prize.

Improvements in transistors came rapidly, and now they have all but completely replaced the vacuum tube. "Solid state" has become a household term. Many people believe that the transistor is one of the greatest developments ever.

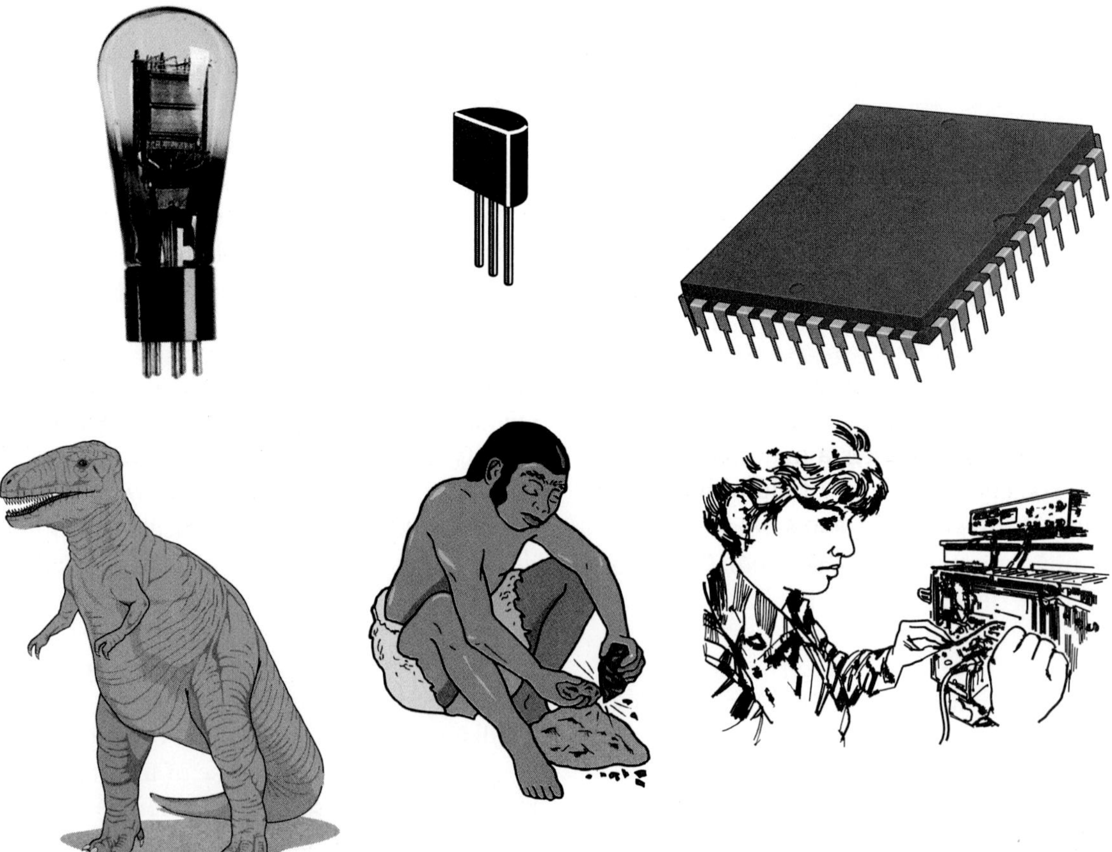

The vacuum tube, the transistor, and then the integrated circuit. The evolution of electronics can be compared to the evolution of life.

Solid-state circuits were small, efficient, and more reliable. But the scientists and engineers still were not satisfied. Work done by Jack Kilby of Texas Instruments led to the development of the integrated circuit in 1958. Integrated circuits are complex combinations of several kinds of devices on a common base, called a *substrate,* or in a tiny piece of silicon. They offer low cost, high performance, good efficiency, small size, and better reliability than an equivalent circuit built from separate parts. The complexity of some integrated circuits allows a single chip of silicon only 0.64 centimeter (cm) [0.25 inch (in.)] square to replace huge pieces of equipment. Although the chip can hold thousands of transistors, it still has diodes, resistors, and capacitors too!

In 1971 Intel Corporation in California announced one of the most sophisticated of all integrated circuits—the microprocessor. A microprocessor is most of the circuitry of a computer reduced to a single integrated circuit. Microprocessors, some containing the equivalent of millions of transistors, have provided billions of dollars worth of growth for the electronics industry and have opened entire new areas of applications.

In 1977 the cellular telephone system entered its testing phase. Since then, the system has experienced immense growth. Its overwhelming success has fostered the development of new technology such as digital communications and linear integrated circuits for communications.

The integrated circuit is producing an electronics explosion. Now electronics is being applied in more ways than ever before. At one time, radio was almost its only application. Today, electronics makes a major contribution to our society and to every field of human endeavor. It affects us in ways we may not be aware of. We are living in the electronic age.

TEST

Determine whether each statement is true or false.

1. Electronics is a young technology that began in the twentieth century.
2. The early histories of radio and of electronics are closely linked.
3. Transistors were invented before vacuum tubes.

About Electronics

Extreme Temperatures and Components/Digital Signal Processing

- Temperature changes can have a greater influence on circuit performance than component tolerance.
- DSP can be used to compress video signals and to identify fingerprints.

4. A modern integrated circuit can contain thousands of transistors.
5. A microprocessor is a small circuit used to replace radio receivers.

Integrated circuit

Substrate

1-2 Digital or Analog

Microprocessor

Today, electronics is such a huge field that it is often necessary to divide it into smaller subfields. You will hear terms such as medical electronics, instrumentation electronics, automotive electronics, avionics, consumer electronics, industrial electronics, and others. One way that electronics can be divided is into digital or analog.

More technicians are fired for lack of people skills than for lack of technical skills.

A digital electronic device or circuit will recognize or produce an output of only several limited states. For example, most digital circuits will respond to only two input conditions: low or high. Digital circuits may also be called *binary* since they are based on a number system with only two digits: 0 and 1.

Digital electronic device

High or Low voltages

Analog circuit

An analog circuit can respond to or produce an output for an infinite number of states. An analog input or output might vary between 0 and 10 volts (V). Its actual value could be 1.5, 2.8, or even 7.653 V. In theory, an *infinite* number of voltages are possible. On the other hand, the typical digital circuit recognizes inputs

Did You Know?

Components to Solid State

- Television receivers with LCDs are 100 percent solid-state, but those with a CRT are not.
- There are two places to find breadboards: electronics labs and kitchens.

Digital circuit

ranging from 0 to 0.4 V as low (binary 0) and those ranging from 2.0 to 5 V as high (binary 1). A digital circuit does not respond any differently for an input of 2 V than it does for one at 4 V. Both of these voltages are in the high range. Input voltages between 0.4 and 2.0 V are not allowed in digital systems because they cause an output that is not predictable.

For a long time, almost all electronic devices and circuits operated in the analog fashion. This seemed to be the most obvious way to do a particular job. After all, most of the things that we measure are analog in nature. Your height, weight, and the speed at which you travel in a car are all analog quantities. Your voice is analog. It contains an infinite number of levels and frequencies. So, if you wanted a circuit to amplify your voice, you would probably think of using an analog circuit.

Telephone switching and computer circuits forced engineers to explore digital electronics. They needed circuits and devices to make logical decisions based on certain input conditions. They needed highly reliable circuits that would always operate the same way. By limiting the number of conditions or states in which the circuits must operate, they could be made more reliable. An infinite number of states—the analog circuit—was not what they needed.

Figure 1-1 gives examples of circuit behavior to help you identify digital or analog operation. The signal going into the circuit is on the left, and the signal coming out is on the right. For now, think of a signal as some electrical quantity, such as voltage, that changes with time. The circuit marked *A* is an example of a digital device. Digital waveforms are rectangular. The output signal is a rectangular wave; the input signal is not exactly a rectangular wave. Rectangular waves have only two voltage levels and are very common in digital devices.

Circuit *B* in Fig. 1-1 is an analog device. The input and the output are sine waves. The output is larger than the input, and it has been shifted above the zero axis. The most important feature is that the output signal is a com-

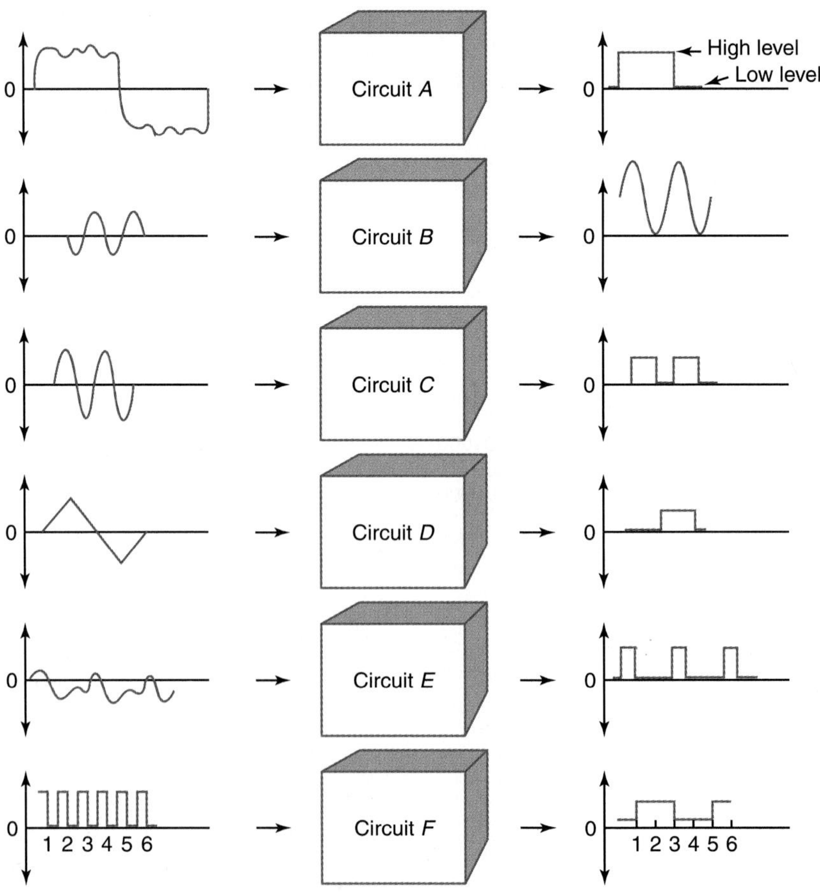

Fig. 1-1 A comparison of digital and analog circuits.

bination of an infinite number of voltages. In a linear circuit, the output is an exact replica of the input. Though circuit *B* is linear, not all analog circuits are linear. For example, a certain audio amplifier could have a distorted sound. This amplifier would still be in the analog category, but would be nonlinear.

Circuits *C* through *F* are all digital. Note that the outputs are all *rectangular* waves (two levels of voltage). Circuit *F* deserves special attention. Its input is a rectangular wave. This could be an analog circuit responding to only two voltage levels except that something has happened to the signal which did not occur in any of the other examples. The output frequency is different from the input frequency. Digital circuits that accomplish this are called *counters,* or *dividers.*

It is now common to convert analog signals to a digital format, which can be stored in computer memory, on magnetic or optical disks, or on magnetic tape. Digital storage has advantages. Everyone who has heard music played from a digital disk knows that it is usually noise free. Digital recordings do not deteriorate with use as analog recordings do.

Another advantage of converting analog signals to digital is that computers can then be used to enhance the signals. Computers are digital machines. They are powerful high-speed number crunchers. A computer can do various things to signals such as eliminate noise and distortion, correct for frequency and phase errors, and identify signal patterns. This area of electronics is known as digital signal processing (DSP). DSP is used in medical electronics to enhance scanned images of the human body, in audio to remove noise from old recordings, and in many other ways. DSP is covered in Chap. 13.

DSP

Linear circuit

Figure 1-2 shows a system that converts an analog signal to digital and then back to analog. An analog-to-digital (A/D) converter is a circuit that produces a binary (only 0s and 1s) output. Note that the numbers stored in memory are binary. A clock (a timing circuit) drives the A/D converter to sample the analog signal on a repetitive basis. Figure 1-3 shows the analog waveform in greater detail. This waveform is sampled by the A/D converter every 20 microseconds (μs). Thus, over a period of 0.8 millisecond (ms), 40 samples are taken. The required sampling rate for any analog signal is a function of the frequency of that signal. The higher the frequency of the signal, the higher the sampling rate.

A/D converter

Refer back to Fig. 1-2. The analog signal can be re-created by sending the binary contents of memory to a digital-to-analog (D/A) converter. The binary information is clocked out of memory at the same rate as the original signal was sampled. Figure 1-4 shows the output of the D/A converter. It can be seen that the waveform is not exactly the same as the original analog signal. It is a series of discrete steps. However, by using more steps, a much closer representation of the original signal can be achieved. Step size is determined by the number of binary digits (bits) used. The number of steps is found by raising 2 to the power of the

D/A converter

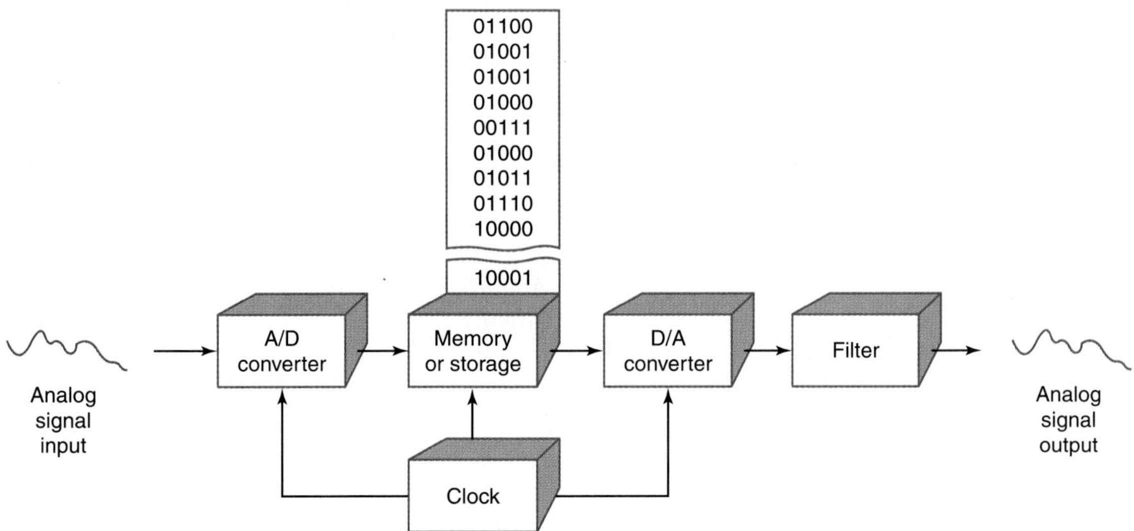

Fig. 1-2 An analog-to-digital-to-analog system.

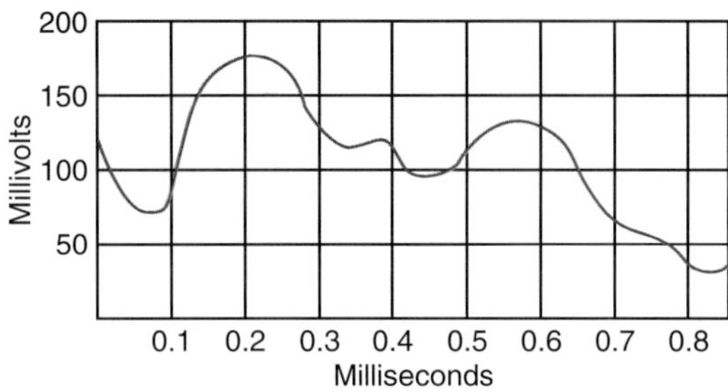

Fig. 1-3 An analog waveform.

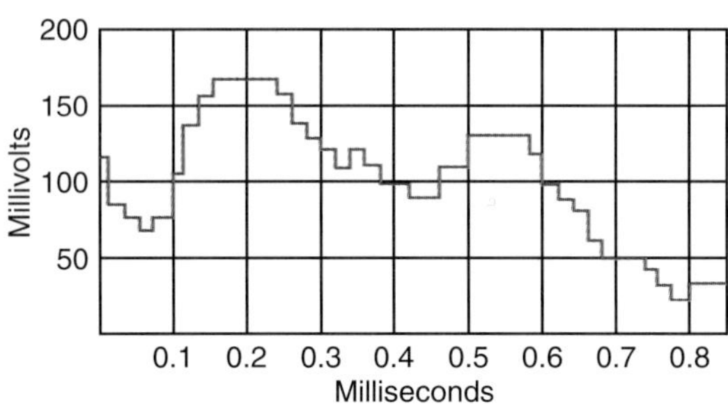

Fig. 1-4 Output of the D/A converter.

interNET CONNECTION

Visit the website for the International Brotherhood of Electrical Workers (IBEW) for career and other information.

number of bits. A 5-bit system provides

$$2^5 = 32 \text{ steps}$$

An 8-bit system would provide

$$2^8 = 256 \text{ steps}$$

Example 1-1

An audio compact disk (CD) uses 16 bits to represent each sample of the signal. How many steps or volume levels are possible? Use the appropriate power of 2:

$$2^{16} = 65{,}536$$

This is easy to solve using a calculator with an x^y key. Press 2, then x^y, and then 16 followed by the = key.

Actually, the filter shown in Fig. 1-2 smooths the steps, and the resulting analog output signal would be quite acceptable for many applications such as speech.

If enough bits and an adequate sampling rate are used, an analog signal can be converted into an accurate digital equivalent. The signal can be converted back into analog form and may not be distinguishable from the original signal. Or it may be noticeably better if DSP were used.

Analog electronics involves techniques and concepts different from those of digital electronics. The rest of this book is devoted mainly to analog electronics. Today most electronic technicians must have skills in both analog and digital circuits and systems.

TEST

Determine whether each statement is true or false.

6. Electronic circuits can be divided into two categories, digital or analog.
7. An analog circuit can produce an infinite number of output conditions.
8. An analog circuit recognizes only two possible input conditions.
9. Rectangular waves are common in digital systems.
10. D/A converters are used to convert analog signals to their digital equivalents.
11. The output of a 2-bit D/A converter can produce eight different voltage levels.

1-3 Analog Functions

This section presents an overview of some functions that analog electronic circuits can provide. Complex electronic systems can be broken down into a collection of individual functions. An ability to recognize individual functions, how they interact, and how each contributes to system operation will make system analysis and troubleshooting easier.

Signals

Analog circuits perform certain operations. These operations are usually performed on *signals.* Signals are electrical quantities, such as voltages or currents, that have some merit or use. For example, a microphone converts a human voice into a small voltage whose frequency and level change with time. This small voltage is called an *audio signal.*

Linear electronic circuits are often named after the function or operation they provide. *Amplification* is the process of making a signal larger or stronger, and circuits that do this are called *amplifiers.* Here is a list of the major types of linear electronic circuits.

1. *Adders:* Circuits that add signals together. Subtracters are also available.
2. *Amplifiers:* Circuits that increase signals.
3. *Attenuators:* Circuits that decrease signals.

Technician inspecting a circuit board.

4. *Clippers:* These prevent signals from exceeding some set amplitude limit or limits.
5. *Comparators:* Compare a signal against some reference, which is usually a voltage.
6. *Controllers:* Regulate signals and load devices. For example, a controller might be used to set and hold the speed of a motor.
7. *Converters:* Change a signal from one form to another (e.g., voltage-to-frequency and frequency-to-voltage converters).
8. *Detectors:* Remove information from a signal (a radio detector removes voice or music from a radio signal).
9. *Dividers:* Perform arithmetic division of a signal.
10. *Filters:* Remove unwanted frequencies from a signal.
11. *Mixers:* Another name for adders. Also, nonlinear circuits that produce the sum and difference frequencies of two input signals.
12. *Multipliers:* Perform arithmetic multiplication of some signal characteristic (there are frequency and amplitude multipliers).
13. *Oscillators:* Change direct current to alternating current.
14. *Rectifiers:* Change alternating current to direct current.
15. *Regulators:* Circuits that hold some value, such as voltage or current, constant.
16. *Switches:* Turn signals on or off or change their routing in an electronic system.

Schematic diagram

Block diagram

A *schematic diagram* shows all of the individual parts of a circuit and how they are interconnected. Schematics use standard symbols to represent circuit components. A *block diagram* shows all of the individual functions of a system and how the signals flow through the system. Schematic diagrams are usually required for what is known as *component-level troubleshooting.* A component is a single part, such as a resistor, capacitor, or an integrated circuit. Component-level repair requires the technician to isolate and replace individual parts that are defective.

System-level repair often requires only a block diagram or a knowledge of the block diagram. The technician observes symptoms and makes measurements to determine which function or functions are improper. Then an entire module, panel, or circuit board is replaced. Component-level troubleshooting usually takes longer than system-level does. Since time is money, it may be economical to replace entire modules or circuit boards.

Troubleshooting

Troubleshooting begins at the system level. Using knowledge of circuit functions, the block diagram, observation of symptoms, and measurements, the technician isolates the difficulty to one or more circuit functions. If replacement boards or modules are on hand, one or more functions can be replaced. However, if component-level troubleshooting is required, the technician continues the isolation process to the component level, using voltmeters and oscilloscopes.

Figure 1-5 shows one block of a block diagram for you to see the process. Troubleshooting is often a series of simple yes or no decisions. For example, is the output signal shown in Fig. 1-5 normal? If so, there is no need to troubleshoot that circuit function. If it is not normal, four possibilities exist: (1) a

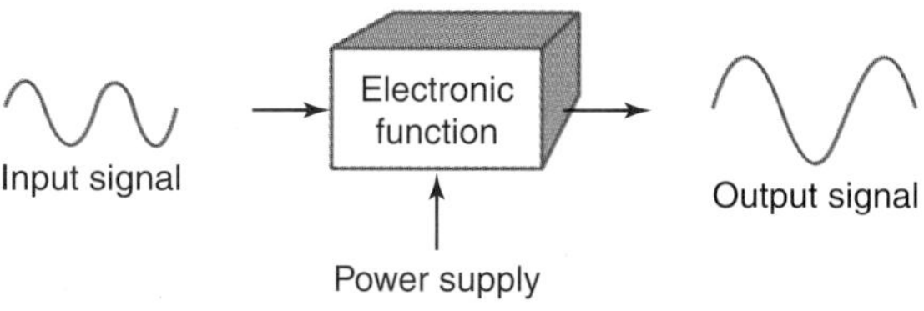

Fig. 1-5 One block of a block diagram.

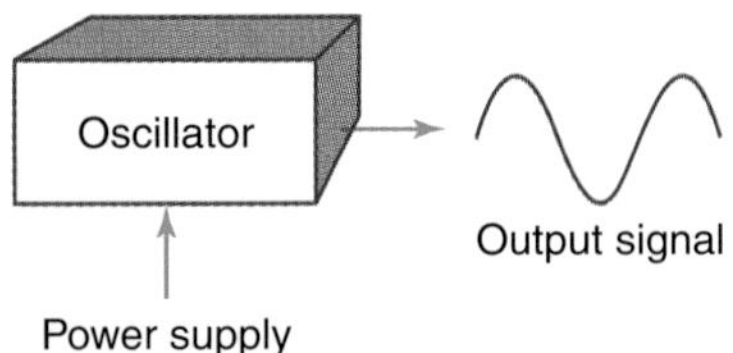

Fig. 1-6 A block with only a power supply input.

power supply problem, (2) an input signal problem, (3) the block (function) is defective, or (4) some combination of these three items. Voltmeters and/or oscilloscopes are generally used to verify the power supply and the input signal to a block. If the supply and input signals are normal, then the block can be replaced or component-level troubleshooting on that circuit function can begin. The following chapters in this book detail how electronic circuits work and cover component-level troubleshooting.

Figure 1-6 shows a block with only one input (power) and one output. Assuming the output signal is missing or incorrect, the possibilities are: (1) the power supply is defective, (2) the oscillator is defective, (3) or both are defective.

Figure 1-7 shows an amplifier that is controlled by a separate input. If its output signal is not correct, the possible causes are: (1) the power supply is defective, (2) the input signal is defective, (3) the control input is faulty, (4) the amplifier has malfunctioned, or (5) some combination of these four items.

Figure 1-8 illustrates a partial block diagram for a radio receiver. It shows how signals flow through the system. A radio signal is amplified, detected, attenuated, amplified again, and then sent to a loudspeaker to produce sound. Knowing how the signal moves from block to block enables a technician to work efficiently. For example, if the signal is missing or weak at point 5, the problem could be caused by a bad signal at point 1, or any of the blocks shown might be defective. The power supply should be checked first, since it affects most of the circuit functions shown. If it checks good, then the signal can be verified at point 1, then point 2, and so on. A defective stage will be quickly located by this orderly process. If the signal is normal at point 3 but not at point 4, then the attenuator block and/or its control input is bad.

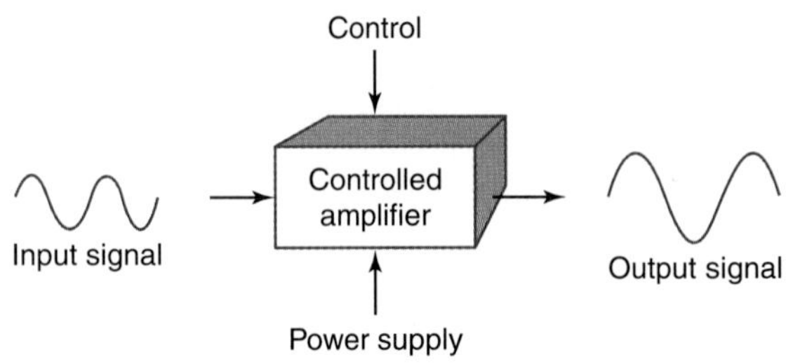

Fig. 1-7 Amplifier with a control input.

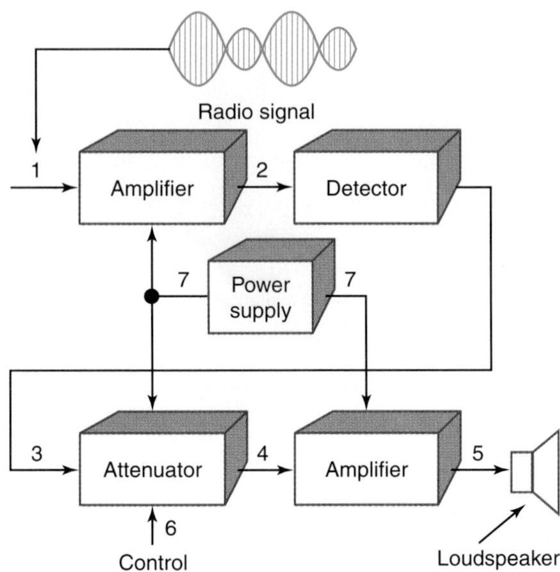

Fig. 1-8 Partial block diagram of a radio receiver.

Much of this book is devoted to the circuit details needed for component-level troubleshooting. However, you should remember that troubleshooting begins at the system level. Always keep a clear picture in your mind of what the individual circuit function is and how that function can be combined with other functions to accomplish system operation.

TEST

Determine whether each statement is true or false.

12. Amplifiers make signals larger.
13. If a signal into an amplifier is normal but the output is not, then the amplifier has to be defective.
14. Component-level troubleshooting requires only a block diagram.
15. A schematic diagram shows how individual parts of a circuit are connected.
16. The first step in troubleshooting is to check individual components for shorts.

1-4 Circuits with Both DC and AC

The transition from the first electricity course to an electronics course can cause some initial confusion. One reason for this is that dc and ac circuit concepts are often treated separately in the first course. Later, students are exposed to electronic circuits that have both dc and ac components. This section will make the transition easier.

Figure 1-9 shows examples of circuits containing both dc and ac components. A battery, a dc source, is connected in series with an ac source. The waveform across the resistor shows that both direct current and alternating current are present. The waveform at the top in Fig. 1-9 shows a sine wave with an average value that is positive. The waveform below this shows a sine wave with a negative average value. The average value in both waveforms is called the *dc component of the waveform,* and it is equal to the battery voltage. Without the batteries, the waveforms would have an average value of 0 V.

Figure 1-10 shows an *RC* (resistor-capacitor) circuit that has both ac and dc sources. This circuit is similar to many linear electronic circuits, which are energized by dc power supplies, such as batteries, and which often process ac signals. Thus, the waveforms in linear electronic circuits often show both ac and dc components.

JOB TIP

Ohm's law and Kirchhoff's laws are the foundations of circuit troubleshooting. Electronic technicians use them on the job, sometimes without even realizing it.

Figure 1-11 on page 10 shows the waveforms that occur at the various nodes in Fig. 1-10. A node is a point at which two or more circuit elements (resistors, inductors, etc.) are connected. These two figures will help you understand some important ideas that you will need in your study of linear electronics.

The waveform for Node A, in Fig. 1-11, shows *pure direct current.* The word "pure" is used because there is no ac component. This is the waveform expected from a dc source such as a battery. Since Node A in Fig. 1-10 is the

DC component

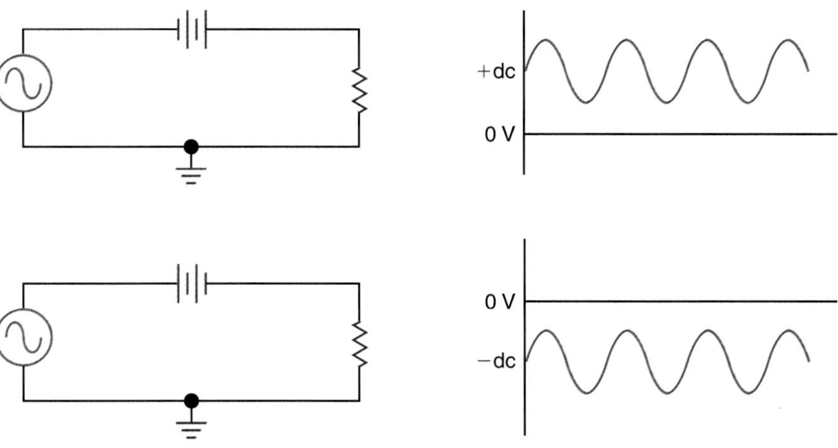

Fig. 1-9 Circuits with dc and ac sources.

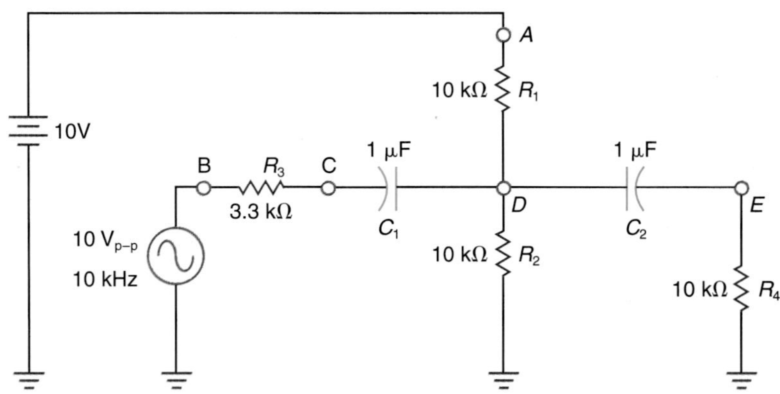

Fig. 1-10 An *RC* circuit with two sources.

positive terminal of the battery, the dc waveform is no surprise.

Node B, in Fig. 1-11, shows *pure alternating current* (there is no dc component). Node B is the ac source terminal in Fig. 1-10, so this waveform is what one would expect it to be.

The other waveforms in Fig. 1-11 require more thought. Starting with Node C, we see a pure ac waveform with about half the amplitude of the ac source. The loss in amplitude is caused by the voltage drop across R_3, discussed later. Node D shows an ac waveform with a 5 V dc component. This dc component is established by R_1 and R_2 in Fig. 1-10, which act as a voltage divider for the 10 V dc battery. Finally, Node E in Fig. 1-11 shows a pure ac waveform. The dc component has been removed by C_2 in Fig. 1-10. A dc component is present at Node D but is missing at Node E because *capacitors block or remove the dc component of signals or waveforms.*

Capacitors block dc component

YOU MAY RECALL . . . that capacitors have infinite reactance (opposition) for direct current and act as open circuits. The formula for capacitive reactance is:

$$X_C = \frac{1}{2\pi fC}$$

As the frequency (f) approaches direct current (0 Hz), the reactance approaches infinity. In capacitors, the relationship between frequency and reactance is *inverse.* As one goes down, the other goes up.

Example 1-2

Determine the reactance of the capacitors in Fig. 1-10 at a frequency of 10 kHz and compare this reactance to the size of the resistors:

$$X_C = \frac{1}{2\pi fC}$$

$$= \frac{1}{6.28 \times 10 \times 10^3 \times 1 \times 10^{-6}}$$

$$= 15.9\ \Omega$$

The reactance 15.9 Ω is low. In fact, we can consider the capacitors to be short circuits at 10 kHz because the resistors in Fig. 1-10 are much larger.

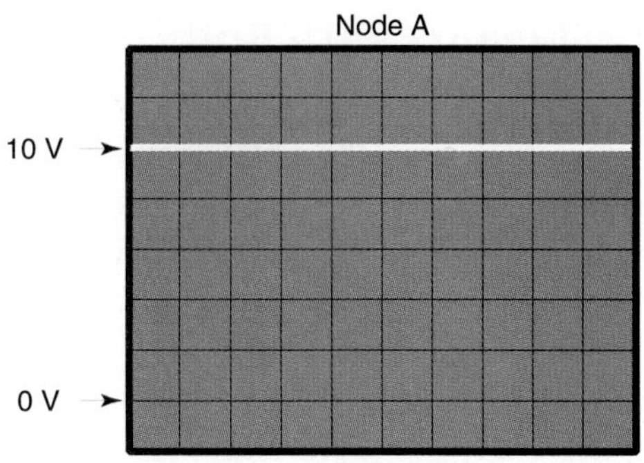

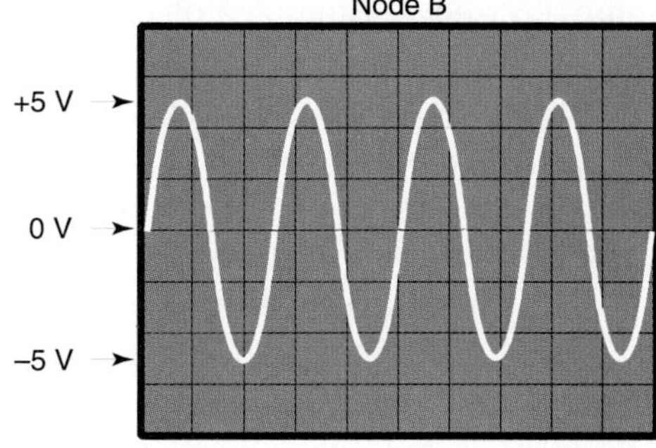

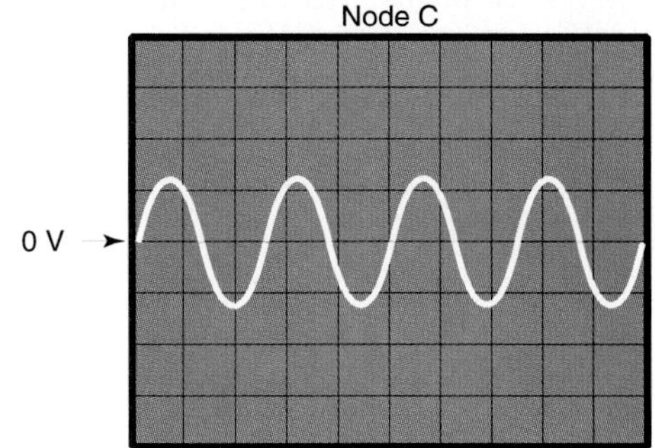

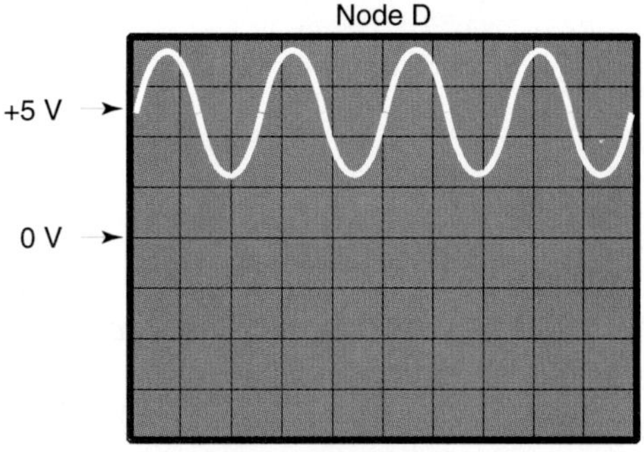

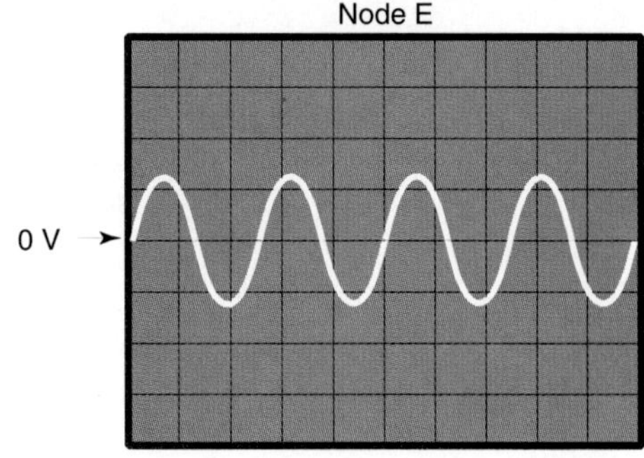

Fig. 1-11 Waveforms for Fig. 1-10.

Let's summarize two points: (1) the capacitors are open circuits for direct current, and (2) the capacitors are short circuits for ac signals when the signal frequency is relatively high. These two concepts are applied over and over again in linear electronic circuits. Please try to remember them.

What happens at other frequencies? At higher frequencies, the capacitive reactance is even lower, so the capacitors can still be viewed as shorts. At lower frequencies, the capacitors show more reactance and the short-circuit viewpoint may no longer be correct. As long as the reactance is less than one-tenth of the effective resistance, the short-circuit viewpoint is generally good enough.

Example 1-3

Determine the reactance of the capacitors in Fig. 1-10 at a frequency of 100 Hz. Will the short-circuit viewpoint be appropriate at this frequency?

$$X_C = \frac{1}{2\pi fC}$$

$$= \frac{1}{6.28 \times 100 \times 1 \times 10^{-6}}$$

$$= 1.59\ \text{k}\Omega$$

This reactance is in the 1000-Ω range, so the capacitors *cannot* be viewed as short circuits at this frequency.

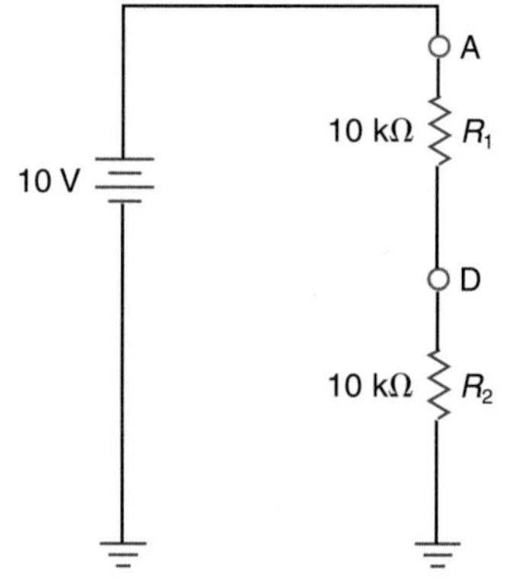

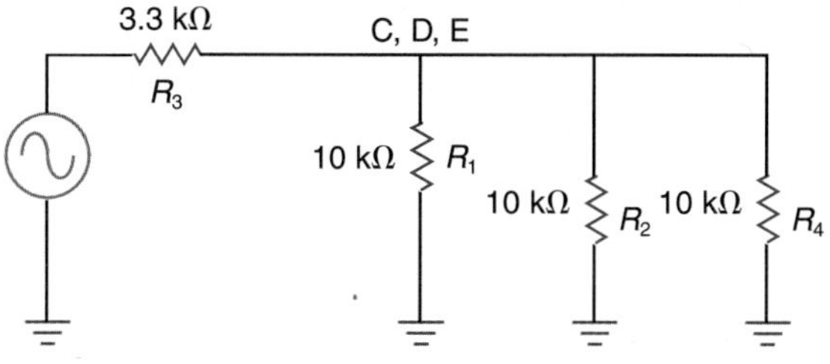

Fig. 1-12 Equivalent circuits for Fig. 1-10.

Figure 1-12 illustrates the equivalent circuits for Fig. 1-10. The dc equivalent circuit shows the battery, R_1, and R_2. Where did the other resistors and the ac source go? They are "disconnected" by the capacitors, which are open circuits for direct current. Since R_1 and R_2 are equal in value, the dc voltage at Node D is half the battery voltage, or 5 V. The ac equivalent circuit is more complicated. Note that resistors R_1, R_2, and R_4 are in parallel. Since R_2 and R_4 are connected by C_2 in Fig. 1-10, they can be joined by a short circuit in the ac equivalent circuit. Remember that the capacitors can be viewed as short circuits for signals at 10 kHz. An equivalent short at C_2 puts R_2 and R_4 in parallel. Resistor R_1 is also in parallel because the internal resistance of a dc voltage source is taken to be 0 Ω. Thus, R_1 in the ac equivalent circuit is effectively grounded at one end and connected to Node D at the other. The equivalent resistance of three 10-kΩ resistors in parallel is one-third of 10 kΩ or 3.33 kΩ—almost equal to the value of R_3. Resistor R_3 and the equivalent resistance of 3.33 kΩ form a voltage divider. So, the ac voltage at Nodes C, D, and E will be about half the value of the ac source, or 5 V_{p-p}.

When the dc and ac equivalent circuits are taken together, the result at Node D is 5 V dc and 5 V_{p-p} alternating current. This explains the waveform at Node D shown in Fig. 1-11. The *superposition theorem,* which you may have studied, provides the explanation for the combining effect.

Superposition theorem

Digital telephone switching system.

About Electronics

Surface-Mount Technology and the Technician Although SMT has reduced the amount of time spent on component-level troubleshooting, technicians with these troubleshooting skills are still in demand.

RFC

Bypassing

There is another very important concept used in electronic circuits, called *bypassing.* Look at Fig. 1-13 and note the C_2 is grounded at its right end. This effectively shorts Node D as far as the ac signal is concerned. The waveform shows that Node D has only 5 V dc, since the ac signal has been *bypassed.* Bypassing is used at nodes in circuits in which the ac signal must be eliminated.

Coupling capacitor

Blocking capacitor

Bypass capacitor

Capacitors are used in many ways. Capacitor C_2 in Fig. 1-10 is often called a *coupling capacitor.* This name serves well since its function is to couple the ac signal from Node D to Node E. However, while it couples the ac signal it *blocks* the dc component. So, it may also be called a *blocking capacitor.* Capacitor C_2 in Fig. 1-13 serves a different function. It eliminates the ac signal at node D and is called a *bypass capacitor.*

Figure 1-14 shows a clever application of the ideas presented here. Suppose there is a problem with weak signals from a television station. An amplifier can be used to boost a weak signal. The best place for one is at the antenna, but the antenna is often on the roof. The amplifier needs power, so one solution would be to run power wires to the roof along with a separate cable for the television signal. The one coaxial cable can serve both needs (power and signal).

The battery in Fig. 1-14 powers an amplifier located at the opposite end of the coaxial cable. The outer conductor of the coaxial cable serves as the ground for both the battery and the remote amplifier. The inner conductor of the coaxial cable serves as the positive connection point for both the battery and the amplifier. Radio-frequency chokes (RFCs) are used to isolate the signal from the power circuit. *RFCs* are coils wound with copper wire. They are inductors and have more reactance for higher frequencies.

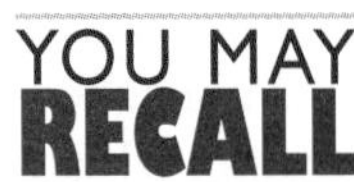

. . . that inductive reactance increases with frequency:

$$X_L = 2\pi fL$$

Frequency and reactance are *directly* related in an inductor. As one increases, so does the other.

At direct current ($f = 0$ Hz) the inductive reactance is zero. The dc power passes through the chokes with no loss. As frequency increases, so does the inductive reactance. The inductive reactance of the choke at the right prevents the battery from shorting the high-frequency signal to ground in Fig. 1-14. The inductive reactance of the other choke keeps the ac signal out of the power wiring to the amplifier.

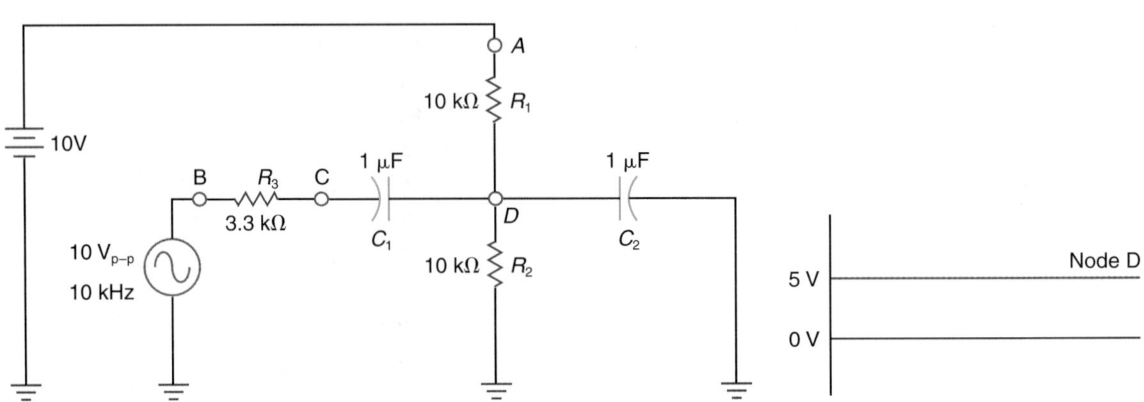

Fig. 1-13 The concept of bypassing.

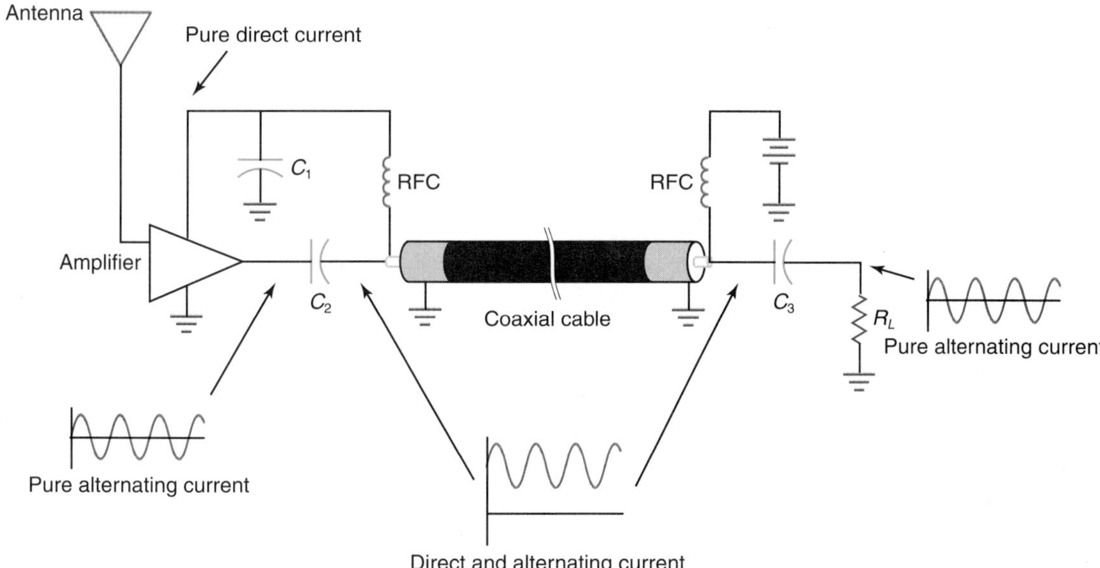

Fig. 1-14 Sending power and signal on the same cable.

Example 1-4

Assume that the RFCs in Fig. 1-14 are 10 μH. The lowest-frequency television channel starts at 54 MHz. Determine the minimum inductive reactance for television signals. Compare the minimum choke reactance to the impedance of the coaxial cable, which is 72 Ω.

$$X_L = 2\pi fL = 6.28 \times 54 \times 10^6 \times 10 \times 10^{-6} = 3.39 \text{ k}\Omega$$

The reactance of the chokes is almost 50 times the cable impedance. This means that the chokes effectively isolate the cable signal from the battery and from the power circuit of the amplifier.

Capacitors C_2 and C_3 in Fig. 1-14 are coupling capacitors. They couple the ac signal into and out of the coaxial cable. These capacitors act as short circuits at the signal frequency, and they are open circuits for the dc signal from the battery. Capacitor C_1 is a bypass capacitor. It ensures that the amplifier is powered by pure direct current. Resistor R_L in Fig. 1-14 is the load for the ac signal. It represents the television receiver.

TEST

Solve problems 17 to 21.

17. Determine the average value of the bottom waveform shown in Fig. 1-9 if the battery develops 7.5 V.
18. Find the average value of the waveform for Node D and for Node E in Fig. 1-10 if the battery provides 25 V.
19. Which components are used in electronics to block direct current, to couple ac signals, and for bypassing?
20. What is the function of C_1 in Fig. 1-14?
21. What is the function of C_2 in Fig. 1-14?

1-5 Trends in Electronics

Trends in electronics are characterized by enormous growth and sophistication. The growth is the result of the "learning curve" and competition. The learning curve simply means that as more experience is gained, more efficiency results. Electronics is maturing as a technology. The yield of integrated circuits is a good example of this. A new integrated circuit (IC), especially a sophisticated one, may yield less than 10 percent. Nine out of ten do not pass the test and are thrown away, making the price of a new device very high. Later, after much is learned about making that part, the yield goes up to 90 percent. The price drops drastically, and many new applications are found for it because of the lower price. Although the new parts are complex and sophisticated, the usual result is a product that is easier to use. In fact, "user friendly" is a term used to describe sophisticated products.

Employers often seek employees with a general background in addition to specific technical skills. It can help to express broad interests when job hunting.

Learning curve

Did You Know?

People Love to Communicate Cellular telephones continue to grow in popularity. With this growth, more and more people are choosing digital phones to fulfill their cellular needs. However, because digital technology is not as widespread as analog, digital cellular telephones are designed to automatically switch to an analog mode when a digital link is not available. People concerned with security need to be aware of this.

Surface-mount technology (SMT)

Microminiaturization

The IC is the key to most electronic trends. These marvels of microminiaturization keep expanding in performance and usually decrease the cost of products. They also require less energy and offer high reliability. One of the most popular ICs, the microprocessor, has created many new products.

Along with ICs, surface-mount technology (SMT) will increase in popularity and expand electronics applications. *SMT* is an alternative to insertion technology for the fabrication of circuit boards. With insertion technology, device leads pass through holes in the circuit board. The insides of the holes are usually

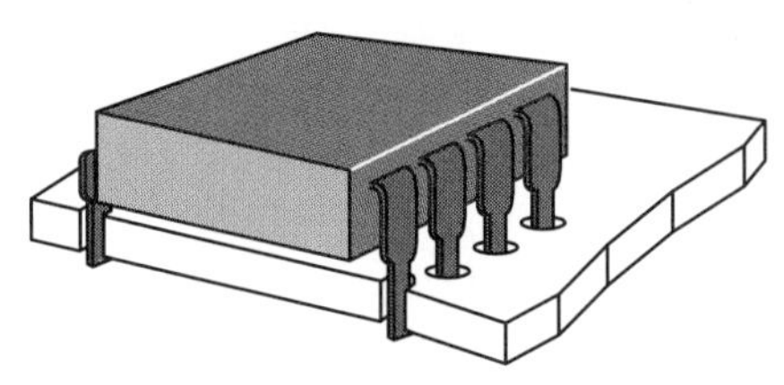

(*a*)

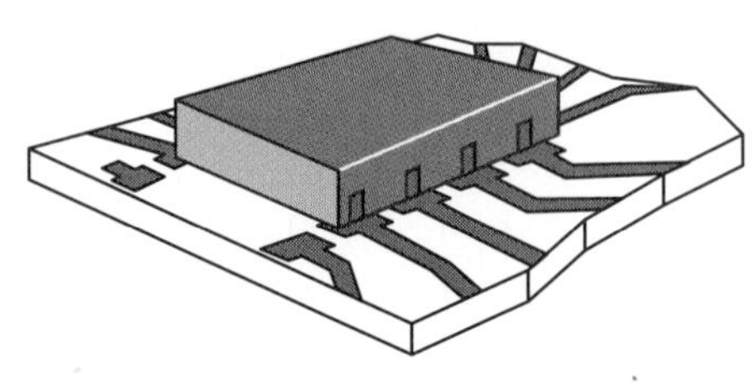

(*b*)

A comparison of conventional-mount and surface-mount technologies. (a) The photo shows both methods, and the drawing shows the conventional method. (b) Photo and drawing of a surface-mount technology (SMT) circuit board.

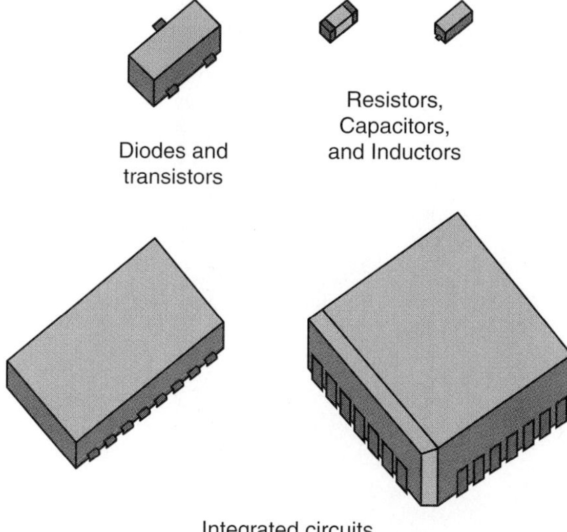

Fig. 1-15 Device packaging for surface-mount technology.

plated with metal to electrically connect the various board layers. Circuit boards designed for insertion technology have more plated-through holes, are larger, and cost more.

The devices intended for SMT have a different appearance. As Fig. 1-15 shows, the device packages have very short leads or just end terminals. These packages are designed to be soldered onto the surface of printed circuit boards. The short leads save material and reduce the stray effects associated with the longer leads used in insertion technology. SMT provides better electrical performance, especially in high-frequency applications.

Two other advantages of SMT are lower circuit assembly cost, since they are easier to automate, and a lower profile. Since more boards can be packed into a given volume, smaller, less expensive products will become available.

A disadvantage of SMT technology is the close spacing of IC leads. Troubleshooting and repair are difficult. Figure 1-16 shows some tools that should be on hand to make measurements on modern circuit boards. The probe allows momentary contact to be safely made at one IC pin. An ordinary probe is uninsulated and will likely slip between two SMT device leads. When this happens, the two leads will be shorted together and damage could result. The single contact test clip in Fig. 1-16 is preferred for making connections that will be used for more than one measurement. The IC test clip in Fig. 1-16 is the best tool for SMT IC measurements. It clips onto an SMT IC and provides larger and widely spaced test contacts for safe probing or test clip connections. Different models are available for the various SMT IC packages.

The uses for electronic devices, products, and systems are expanding. Computer technology finds new applications almost on a daily basis. Electronic communications is expanding rapidly. Thanks to compression and processing breakthroughs, the growth is brisk. Three-dimensional image processing is providing systems for product inspection, automated security monitoring, and even virtual reality for education and entertainment. Computer tech-

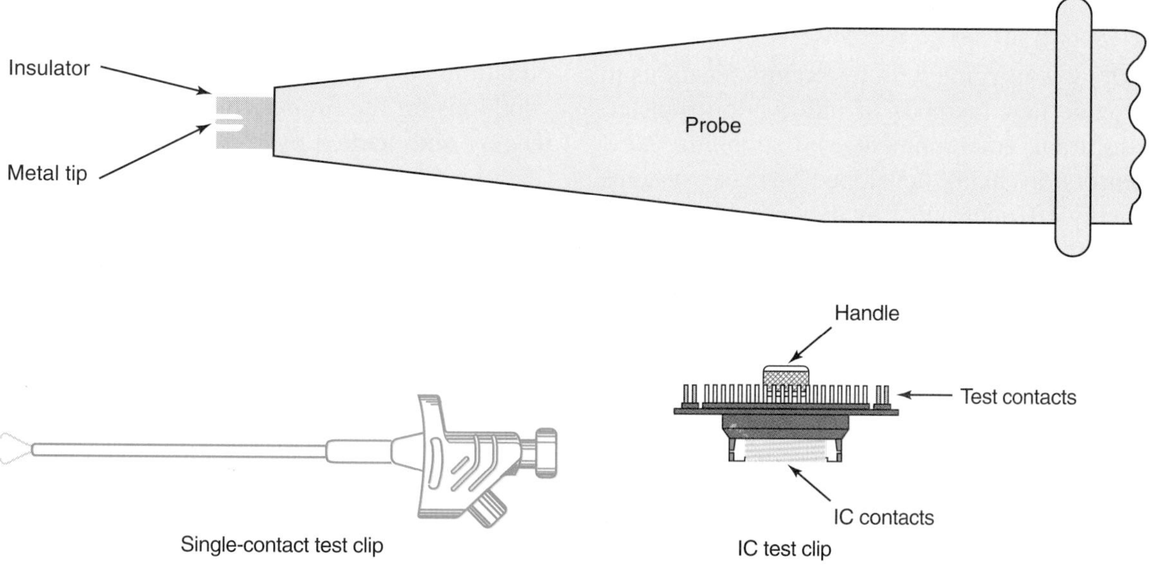

Fig. 1-16 Tools for SMT measurements.

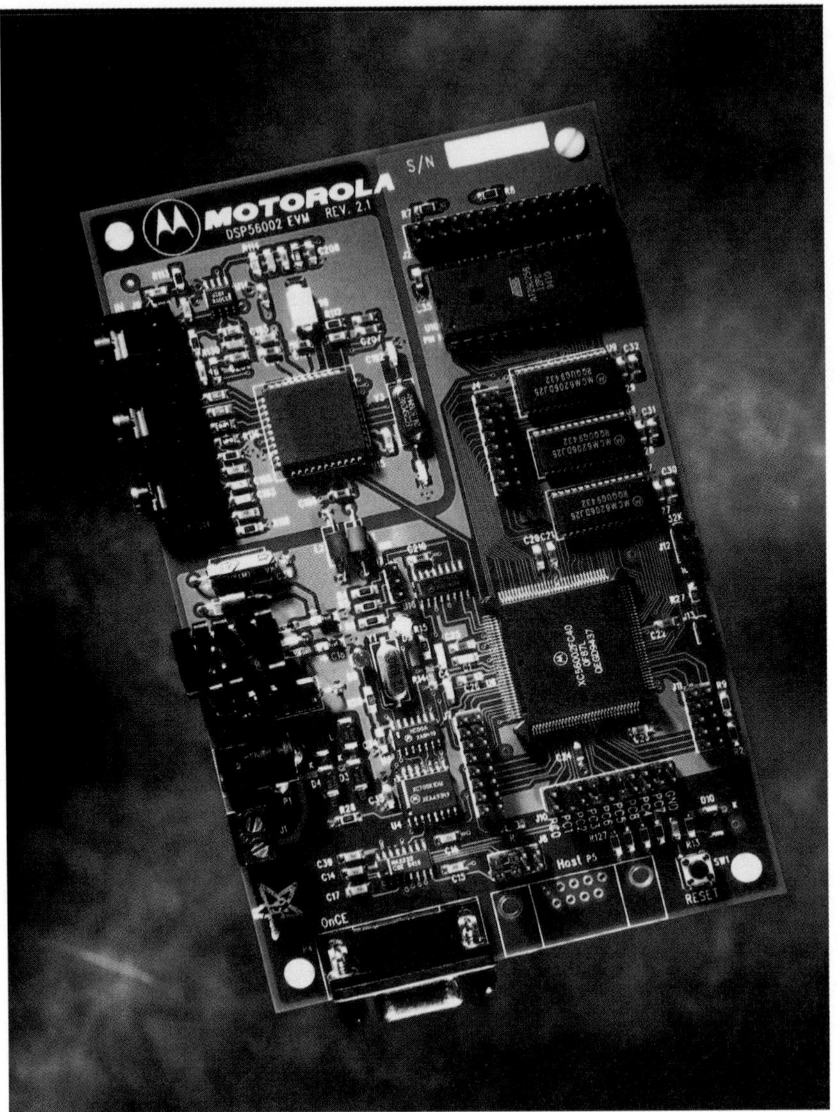

This evaluation module is used to develop digital signal processing ideas and designs.

nology is merging with telecommunications to provide new methods of information transfer, education, entertainment, and shopping. New sensors are being developed to make systems energy-efficient and less damaging to the environment. As an example, heating, ventilating, and air-conditioning systems will use oxygen sensors to direct airflow in buildings on an as-needed basis. It is impossible to do an adequate job of describing the coming changes in any detail. Look at the brief and amazing history of electronics and notice that the rate of change always increases.

The outlook is bright for those with careers in electronics. The new products, the new applications, and the tremendous growth mean good jobs for the future. The jobs will be challenging and marked by constant change.

TEST

Determine whether each statement is true or false.

22. Integrated circuits will be used less in the future.
23. The learning curve makes electronic devices less expensive as time goes on.
24. In the future, more circuits will be fabricated using insertion technology and fewer will be fabricated with SMT.

Chapter 1 Review

Summary

1. Electronics is a relatively young field. Its history began in the twentieth century.
2. Electronic circuits can be classified as digital or analog.
3. The number of states or voltage levels is limited in a digital circuit (usually to two).
4. An analog circuit has an infinite number of voltage levels.
5. In a linear circuit, the output signal is a replica of the input.
6. All linear circuits are analog, but not all analog circuits are linear. Some analog circuits distort signals.
7. Analog signals can be converted to a digital format with an A/D converter.
8. Digital-to-analog converters are used to produce a simulated analog output from a digital system.
9. The quality of a digital representation of an analog signal is determined by the sampling rate and the number of bits used.
10. The number of output levels from a D/A converter is equal to 2 raised to the power of the number of bits used.
11. Digital signal processing uses computers to enhance signals.
12. Block diagrams give an overview of electronic system operation.
13. Schematic diagrams show individual part wiring and are usually required for component-level troubleshooting.
14. Troubleshooting begins at the system level.
15. Ac and dc signals are often combined in electronic circuits.
16. Capacitors can be used to couple ac signals, to block direct current, or to bypass alternating current.
17. SMT is replacing insertion technology.

Chapter Review Questions

Determine whether each statement is true or false.

1-1. Most digital circuits can output only two states, high and low.

1-2. Digital circuit outputs are usually sine waves.

1-3. The output of a linear circuit is an exact replica of the input.

1-4. Linear circuits are classified as analog.

1-5. All analog circuits are linear.

1-6. The output of a 4-bit D/A converter can produce 128 different voltage levels.

1-7. An attenuator is an electronic circuit used to make signals stronger.

1-8. Block diagrams are best for component-level troubleshooting.

1-9. In Fig. 1-8, if the signal at point 4 is faulty, then the signal at point 3 must also be faulty.

1-10. Refer to Fig. 1-8. The power supply should be checked first.

1-11. Refer to Fig. 1-10. Capacitor C_2 would be called a bypass capacitor.

1-12. Node C in Fig. 1-10 has no dc component since C_1 blocks direct current.

1-13. In Fig. 1-11, Node D is the only waveform with dc and ac components.

1-14. Refer to Fig. 1-14. The reactance of the coils is high for dc signals.

Critical Thinking Questions

1-1. Functions now accomplished by using electronics may be accomplished in different ways in the future. Can you think of any examples?

1-2. Can you describe a simple system that uses only two wires but will selectively signal two different people?

1-3. What could go wrong with capacitor C_2 in Fig. 1-10, and how would the fault affect the waveform at Node D?

1-4. What could go wrong with capacitor C_2 in Fig. 1-13, and how would the fault affect the waveform at Node D?

Answers to Tests

1. T
2. T
3. F
4. T
5. F
6. T
7. T
8. F
9. T
10. F
11. F
12. T
13. F
14. F
15. T
16. F
17. −7.5 V
18. 12.5 V, 0 V
19. capacitors
20. bypass
21. coupling (dc block)
22. F
23. T
24. F

Semiconductors

Chapter Objectives

This chapter will help you to:

1. *Identify* some common electronic materials as conductors or semiconductors.
2. *Predict* the effect of temperature on conductors.
3. *Predict* the effect of temperature on semiconductors.
4. *Show* the directions of electron and hole currents in semiconductors.
5. *Identify* the majority and minority carriers in N-type semiconductors.
6. *Identify* the majority and minority carriers in P-type semiconductors.

Electronic circuits used to be based on the flow of electrons in devices called vacuum tubes. Today, almost all electronic circuits are based on current flow in semiconductors. The term "solid state" means that semiconducting crystals are being used to get the job done. The mechanics of current flow in semiconductors is different than in conductors. Some current carriers are not electrons. High temperatures create additional carriers in semiconductors. These are important differences between semiconductors and conductors. The transistor is considered to be one of the most important developments of all time. It is a semiconductor device. Diodes and integrated circuits are also semiconductors. This chapter covers the basic properties of semiconductors.

2-1 Conductors and Insulators

All materials are made from atoms. At the center of any atom is a small, dense core called the *nucleus.* Figure 2-1(*a*) shows that the nucleus of a copper atom is made up of positive (+) particles called *protons* and neutral (N) particles called *neutrons.* Around the nucleus are orbiting electrons that are negative (−) particles. Copper, like all atoms, has an equal number of protons and electrons. Thus, the net atomic charge is zero.

In electronics, the main interest is in the orbit that is farthest away from the nucleus. It is called the *valence orbit.* In the case of copper, there is only one valence electron. A copper atom can be simplified as shown in Fig. 2-1(*b*). Here, the nucleus and the first three orbits are combined into a net positive (+) charge. This is balanced by the single valence electron.

Conductors form the fundamental paths for electronic circuits. Figure 2-2 shows how a copper wire supports the flow of electrons. A copper atom contains a positively charged nucleus and negatively charged electrons that orbit around the nucleus. Figure 2-2 is simplified to show only the outermost orbiting electron, the *valence* electron. The valence electron is very important since it acts as the current carrier.

Even a very small wire contains billions of atoms, each with one valence electron. These electrons are only weakly attracted to the nuclei of the atoms. They are very easy to move. If an electromotive force (a voltage) is applied across the wire, the valence electrons will re-

Nucleus

Proton

Neutron

Valence orbit

Conductor

Electron

Copper atom

Valence electron

Current carrier

Electromotive force (a voltage)

(*a*) Bohr model of the copper atom (not to scale)

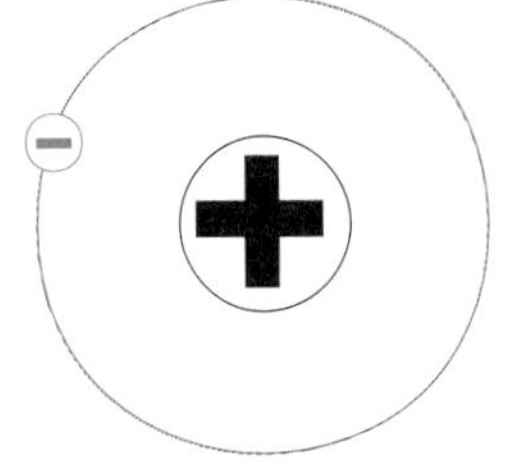

(*b*) Simplified model

Fig. 2-1 Atomic copper.

Low resistance

spond and begin drifting toward the positive end of the source voltage. Since there are so many valence electrons and since they are so easy to move, we can expect tremendous numbers of electrons to be set in motion by even a small voltage. Thus, copper is an excellent electric conductor. It has very low resistance.

Heating a copper wire will change its resistance. As the wire becomes warmer, the valence electrons become more active. They move farther away from their nuclei, and they move more rapidly. This activity increases the chance for collisions as current carrying electrons drift toward the positive end of the wire. These collisions absorb energy and increase the resistance to current flow. The resistance of the wire increases as it is heated.

Positive temperature coefficient

All conductors show this effect. As they become hotter, they conduct less efficiently and their resistance increases. Such materials are said to have a *positive temperature coefficient.* This simply means that the relationship between temperature and resistance is positive—that is, they increase together.

Printed circuit

Copper is the most widely applied conductor in electronics. Most of the wire used in electronics is made from copper. Printed circuits use copper foil to act as circuit conductors. Copper is a good conductor, and it is easy to solder. This makes it very popular.

Aluminum is a good conductor, but not as good as copper. It is used more in power transformers and transmission lines than it is in electronics. Aluminum is less expensive than copper, but it is difficult to solder and tends to corrode rapidly when brought into contact with other metals.

Silver is the best conductor because it has the least resistance. It is also easy to solder. The high cost of silver makes it less widely applied than copper. However, silver-plated conductors are sometimes used in critical electronic circuits to minimize resistance.

Gold is a good conductor. It is very stable and does not corrode as badly as copper and silver. Some sliding and moving electronic contacts are gold-plated. This makes the contacts very reliable.

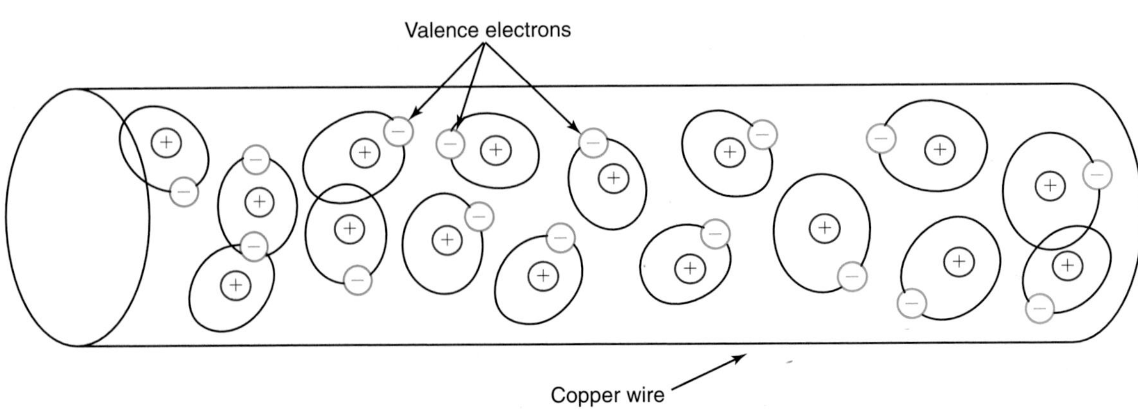

Fig. 2-2 The structure of a copper conductor.

The opposite of a conductor is called an *insulator.* In an insulator, the valence electrons are tightly bound to their parent atoms. They are not free to move, so little or no current flows when a voltage is applied. Practically all insulators used in electronics are based on compounds. A *compound* is a combination of two or more different kinds of atoms. Some of the widely applied insulating materials include rubber, plastic, Mylar, ceramic, Teflon, and polystyrene.

Whether a material will insulate depends on how the atoms are arranged. Carbon is such a material. Figure 2-3(*a*) shows carbon arranged in the diamond structure. With this crystal or diamond structure, the valence electrons cannot move to serve as current carriers. Diamonds are insulators. Figure 2-3(*b*) shows carbon arranged in the graphite structure. Here, the valence electrons are free to move when a voltage is applied. It may seem odd that both diamonds and graphite are made from carbon. One insulates, and the other does not. It is simply a matter of whether the valence electrons are locked into the structure. Carbon in graphite form is used to make resistors and electrodes. So far, the diamond structure of carbon has not been used to make electrical or electronic devices.

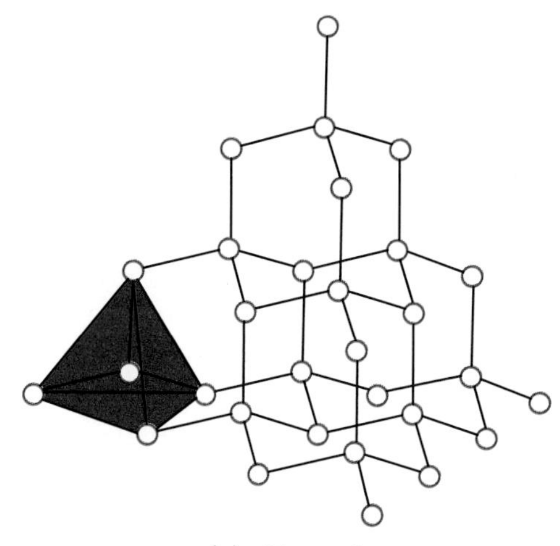

(*a*) Diamond

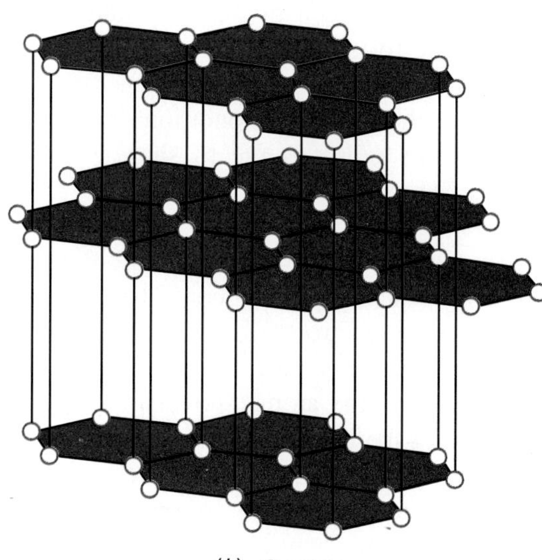

(*b*) Graphite

Fig. 2-3 Structures of diamond and graphite.

About Electronics

Materials Used for Dopants, Semiconductors, and Microwave Devices

- GaAs (gallium arsenide) works better than silicon in microwave devices because it allows faster movement of electrons.
- Materials other than boron and arsenic are used as dopants.
- It is theoretically possible to make semiconductor devices from crystalline carbon.
- Crystal radio receivers were an early application of semiconductors.

Insulator

Compound

TEST

Determine whether each statement is true or false.

1. Valence electrons are located in the nucleus of the atom.
2. Copper has one valence electron.
3. In conductors, the valence electrons are strongly attracted to the nucleus.
4. The current carriers in conductors are the valence electrons.
5. Cooling a conductor will decrease its resistance.
6. Silver is not often used in electronic circuits because of its high resistance.
7. Aluminum is not used as much as copper in electronic circuits because it is difficult to solder.

2-2 Semiconductors

Semiconductor

Silicon

Diode

Transistor

Semiconductors do not allow current to flow as easily as conductors do. Under some conditions semiconductors can conduct so poorly that they behave as insulators.

Silicon is the most widely used semiconductor material. It is used to make diodes, transistors, and integrated circuits. These and other components make modern electronics possible. It is important to understand some of the details about silicon.

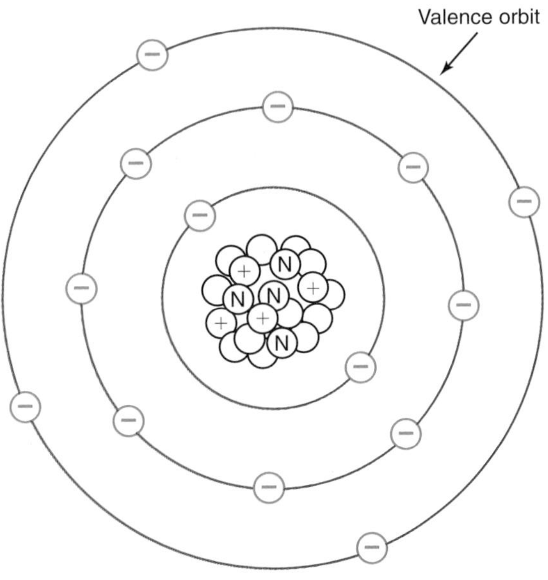

(*a*) The structure of a silicon atom

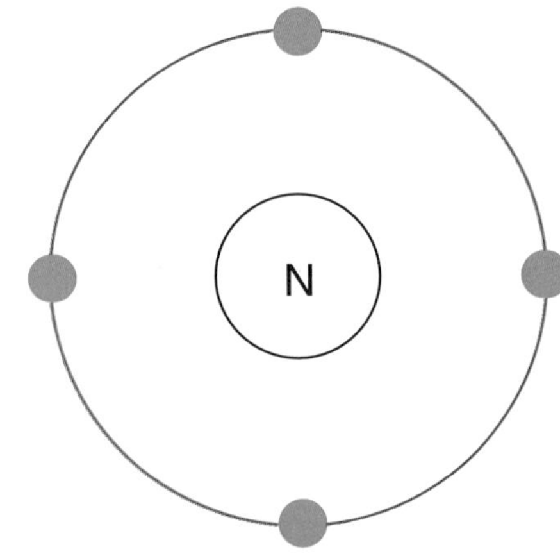

(*b*) A simplified silicon atom

Fig. 2-4 Atomic silicon.

Figure 2-4 shows atomic silicon. The compact bundle of particles in the center of the atom [Fig. 2-4(*a*)] contains protons and neutrons. This bundle is called the *nucleus of the atom.* The protons show a positive (+) electric charge, and the neutrons show no electric charge (N). Negatively charged electrons travel around the nucleus in orbits. The first orbit has two electrons. The second orbit has eight electrons. The last, or outermost, orbit has four electrons. The outermost or valence orbit is the most important atomic feature in the electrical behavior of materials.

> **JOB TIP**
>
> Your physical appearance and personal manner often strongly affect the impression you make on other people. Ask close friends and family for objective suggestions.

Because we are interested mainly in the valence orbit, it is possible to simplify the drawing of the silicon atom. Figure 2-4(*b*) shows only the nucleus and the valence orbit of a silicon atom. Remember that there are four electrons in the valence orbit.

Active material

Materials with four valence electrons are not stable. They tend to combine chemically with other materials. They can be called *active materials.* This activity can lead them to a more stable state. A law of nature makes certain materials tend to form combinations that will make eight electrons available in the valence orbit. Eight is an important number because it gives stability.

Silicon dioxide

Ionic bond

One possibility is for silicon to combine with oxygen. A single silicon atom can join, or link, with two oxygen atoms to form silicon dioxide (SiO_2). This linkage is called an *ionic bond.* The new structure, SiO_2, is much more stable than either silicon or oxygen. It is interesting to consider that chemical, mechanical, and electrical properties often run parallel. Silicon dioxide is stable chemically. It does not react easily with other materials. It is stable mechanically. It is a hard, glasslike material. It is stable electrically. It does not conduct; in fact, it is used as an *insulator* in integrated circuits and other solid-state devices. SiO_2 insulates because all of the valence electrons are tightly locked into the ionic bonds. They are not easy to move and therefore do not support the flow of current.

Covalent bonding

Crystal

Sometimes oxygen or another material is not available for silicon to combine with. The silicon still wants the stability given by eight valence electrons. If the conditions are right, silicon atoms will arrange to share valence electrons. This process of sharing is called *covalent bonding.* The structure that results is called a *crystal.* Figure 2-5 is a symbolic diagram of a crystal of pure silicon. The dots represent valence electrons.

Count the valence electrons around the nucleus of one of the atoms shown in Fig. 2-5. Select one of the internal nuclei as represented by the circled N. You will count eight electrons. Thus the silicon crystal is very stable. At room temperature, pure silicon is a very poor conductor. If a moderate voltage is applied across the crystal, very little current will flow.

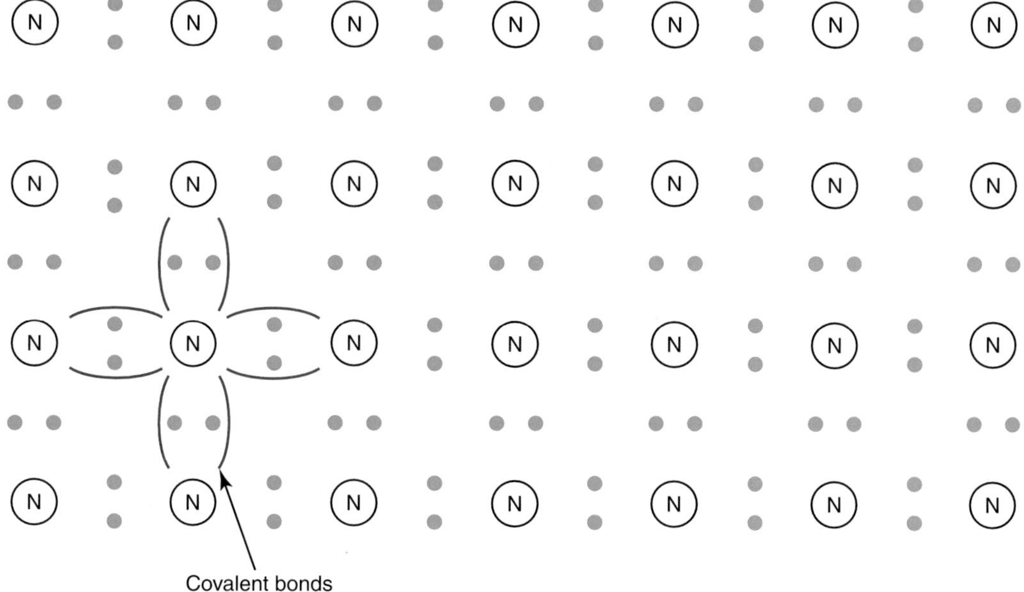

Fig. 2-5 A crystal of pure silicon.

The valence electrons that normally would support current flow are all tightly locked up in covalent bonds.

Pure silicon crystals behave like insulators. Yet silicon itself is classified as a semiconductor. Pure silicon is sometimes called *intrinsic silicon.* Intrinsic silicon contains very few free electrons to support the flow of current and therefore acts as an insulator.

Crystalline silicon can be made to semiconduct. One way to improve its conduction is to heat it. Heat is a form of energy. A valence electron can absorb some of this energy and move to a higher orbit level. The high-energy electron has *broken* its covalent bond. Figure 2-6 shows a high-energy electron in a silicon crystal. This electron may be called a *thermal carrier.* It is free to move, so it can support the flow of current. Now, if a voltage is placed across the crystal, current will flow.

Silicon has a *negative temperature coefficient.* As temperature increases, resistance *decreases* in silicon. It is difficult to predict exactly how much the resistance will change in a given case. One rule of thumb is that the resistance will be cut in half for every 6°C rise in temperature.

The semiconductor material *germanium* is used to make transistors and diodes, too. Germanium has four valence electrons and can form the same type of crystalline structure as silicon. It is interesting to observe that the first transistors were all made of germanium. The first silicon transistor was not developed until 1954. Now silicon has almost entirely replaced germanium. One of the major reasons for this shift from germanium to silicon is the temperature response. Germanium also has a negative temperature coefficient. The rule of thumb for germanium is that the resistance will be cut in half for every 10°C rise in temperature. This would seem to make germanium more stable with temperature change.

The big difference between germanium and silicon is the amount of heat energy needed to

Insulator

Temperature response

Intrinsic silicon

Thermal carrier

Germanium

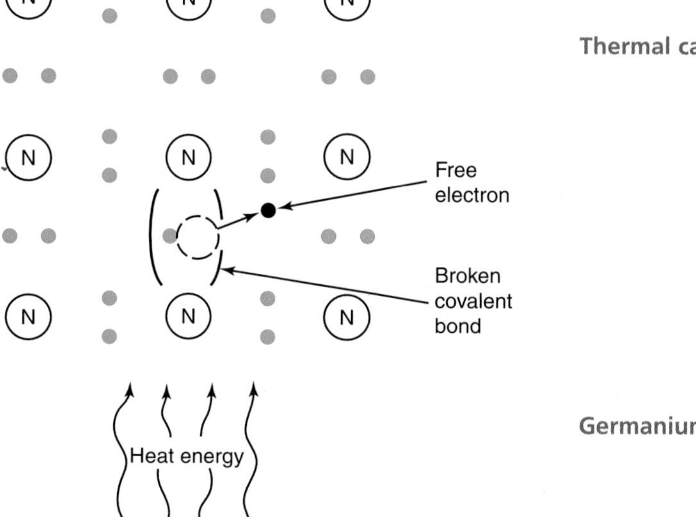

Fig. 2-6 Thermal carrier production.

move one of the valence electrons to a higher orbit level, breaking its covalent bond. This is far easier to do in a germanium crystal. A comparison between two crystals, one germanium and one silicon, of the same size and at room temperature will show about a 1000:1 ratio in resistance. The silicon crystal will actually have 1000 times the resistance of the germanium crystal. So even though the resistance of silicon drops more rapidly with increasing temperature than that of germanium, silicon is still going to show greater resistance than germanium at a given temperature.

Circuit designers prefer silicon devices for most uses. The thermal, or heat, effects are usually a source of trouble. Temperature is not easy to control, and we do not want circuits to be influenced by it. However, all circuits are changed by temperature. Good designs minimize that change.

Sometimes heat-sensitive devices are necessary. A sensor for measuring temperature can take advantage of the temperature coefficient of semiconductors. So the temperature coefficient of semiconductors is not always a disadvantage.

Germanium started the solid-state revolution in electronics, but silicon has taken over. The integrated circuit is a key part of most electronic equipment today. It is not practical to make integrated circuits from germanium, but silicon works well in this application.

interNET CONNECTION

Look for related information on the Texas Instruments web site.

TEST

Determine whether each statement is true or false.

8. Silicon is a conductor.
9. Silicon has four valence electrons.
10. Silicon dioxide is a good conductor.
11. A silicon crystal is formed by covalent bonding.
12. Intrinsic silicon acts as an insulator at room temperature.
13. Heating semiconductor silicon will decrease its resistance.
14. An electron that is freed from its covalent bond by heat is called a thermal carrier.
15. Germanium has less resistance than silicon.
16. Silicon transistors and diodes are not used as often as germanium devices.
17. Integrated circuits are made from germanium.

2-3 N-Type Semiconductors

Thus far we have seen that pure semiconductor crystals are very poor conductors. High temperatures can make them semiconduct because thermal carriers are produced. For most applications, there is a better way to make them semiconduct.

Doping

Arsenic

Doping is a process of adding other materials called *impurities* to the silicon crystal to change its electrical characteristics. One such impurity material is *arsenic*. Arsenic is known as a *donor impurity* because each arsenic atom donates one free electron to the crystal. Figure 2-7 shows a simplified arsenic atom. Arsenic is different from silicon in several ways, but the important difference is in the valence orbit. Arsenic has *five* valence electrons.

When an arsenic atom enters a silicon crystal, a free electron will result. Figure 2-8 shows what happens. The covalent bonds with neighboring silicon atoms will capture four of the arsenic atom's valence electrons, just as if it were another silicon atom. This tightly locks the arsenic atom into the crystal. The fifth valence electron cannot form a bond. It is a *free* electron as far as

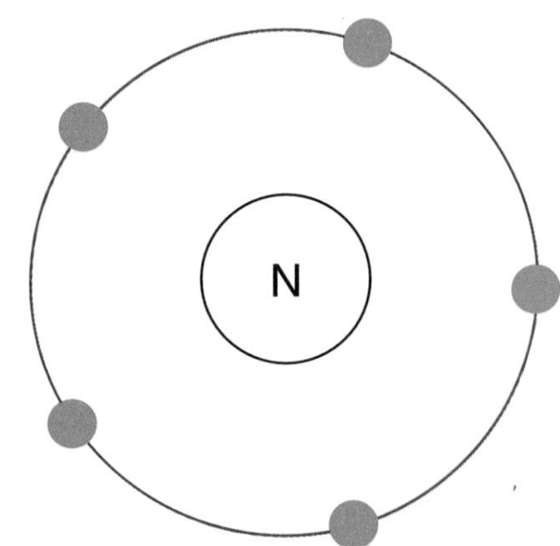

Fig. 2-7 A simplified arsenic atom.

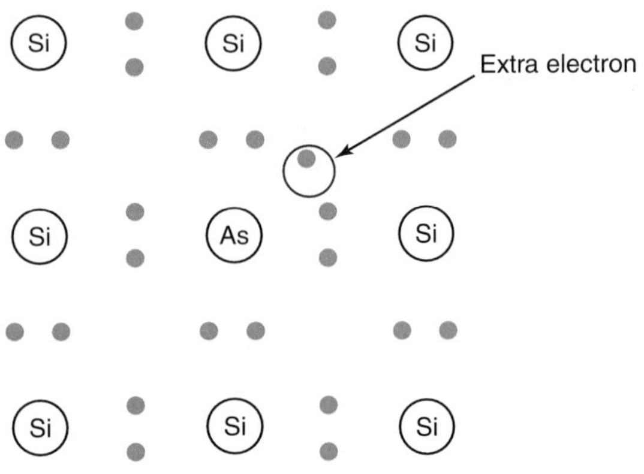

Fig. 2-8 N-type silicon.

the crystal is concerned. This makes the electron very easy to move. It can serve as a current carrier. Silicon with some arsenic atoms will semiconduct even at room temperature.

Doping lowers the resistance of the silicon crystal. When donor impurities with five valence electrons are added, free electrons are produced. Since electrons have a negative charge, we say that an N-type semiconductor material results.

TEST

Supply the missing word in each statement.

18. Arsenic is a DONOR impurity.
19. Arsenic has 5 valence electrons.
20. When silicon is doped with arsenic, each arsenic atom will give the crystal one free ELECTRON.
21. Free electrons in a silicon crystal will serve as current CARRIER.
22. When silicon is doped, its resistance DECREASES.

2-4 P-Type Semiconductors

Doping can involve the use of other kinds of impurity materials. Figure 2-9 shows a simplified *boron atom.* Note that boron has only three valence electrons. If a boron atom enters the silicon crystal, another type of current carrier will result.

Figure 2-10 shows that one of the covalent bonds with neighboring silicon atoms cannot be formed. This produces a *hole,* or *missing electron.* The hole is assigned a *positive charge* since it is capable of attracting, or being filled by, an electron.

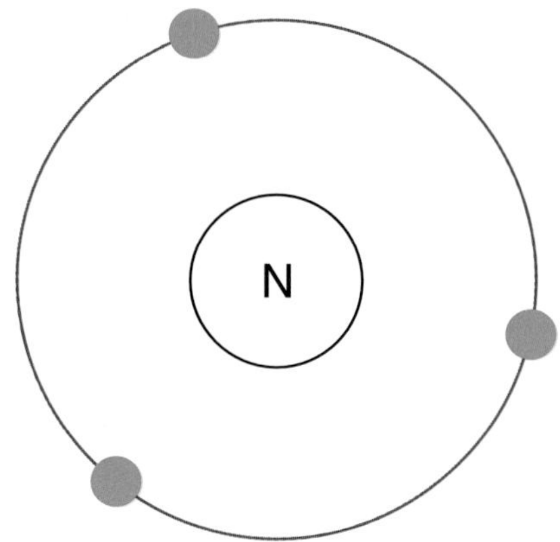

Fig. 2-9 A simplified boron atom.

Boron is known as an *acceptor impurity.* Each boron atom in the crystal will create a hole which is capable of accepting an electron.

Interviewers look for interest in the company and its products. It pays you to learn about the company before the interview.

Holes serve as current carriers. In a conductor or in an N-type semiconductor, the carriers are electrons. The free electrons are set into motion by an applied voltage, and they drift toward the positive terminal. But in a P-type semiconductor, the holes move toward the negative terminal of the voltage source. Hole current is equal to electron current but *opposite* in direction. Figure 2-11 illustrates the difference between N-type and P-type semiconductor ma-

N-type semiconductor material

P-type semiconductor material

Boron atom

Hole, or missing electron

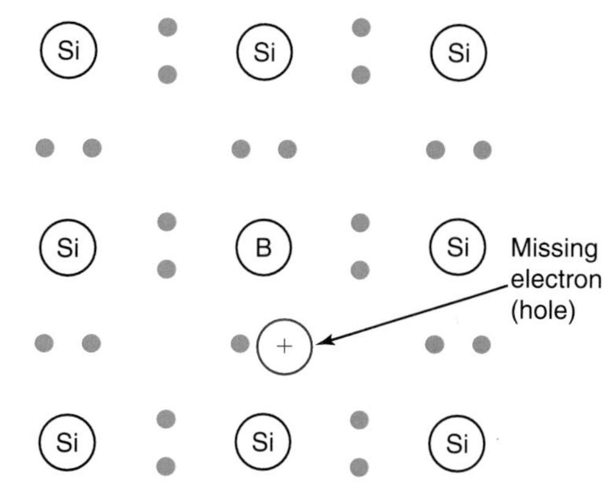
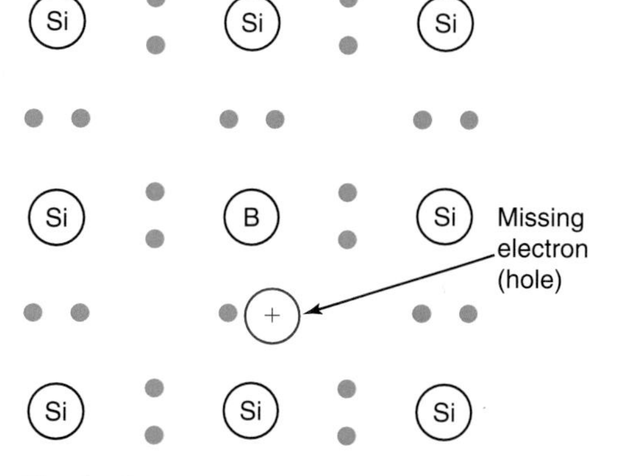

Fig. 2-10 P-type silicon.

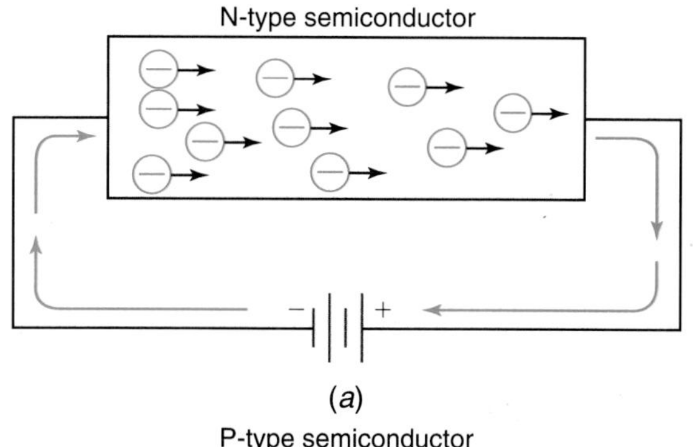

(*a*)

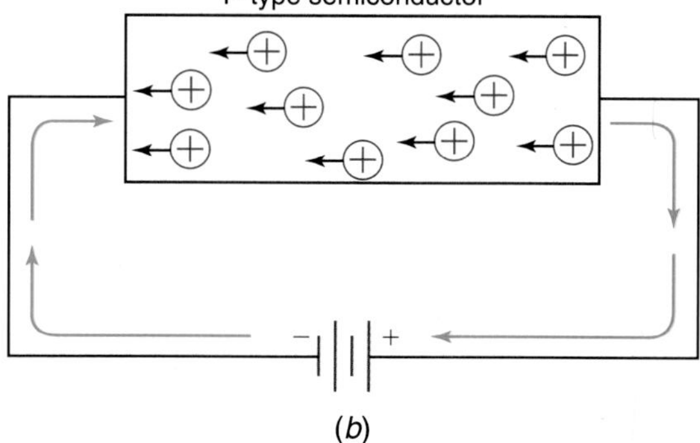

(*b*)

Fig. 2-11 Conduction in N- and P-type silicon.

terials. In Fig. 2-11(*a*) the carriers are electrons, and they drift toward the positive end of the voltage source. In Fig. 2-11(*b*) the carriers are holes, and they drift toward the negative end of the voltage source.

JOB TIP

All companies seek employees who are self-directed and good problem solvers.

Figure 2-12 shows a simple analogy for hole current. Assume that a line of cars is stopped for a red light, but there is space for the first car to move up one position. The driver of that car takes the opportunity to do so, and this makes a space for a car back one position. The driver of the second car also moves up one position. This continues with the third car, the fourth car, and so on down the line. The cars are moving from left to right. Note that the space is moving from right to left. A hole may be considered as a space for an electron. This is why hole current is opposite in direction to electron current.

TEST

Supply the missing word in each statement.

23. Boron is an ________ impurity.
24. Boron has ________ valence electrons.
25. Electrons are assigned a negative charge, and holes are assigned a ________ charge.
26. Doping a semiconductor crystal with boron will produce current carriers called ________.
27. Electrons will drift toward the positive end of the energy source, and holes will drift toward the ________ end.

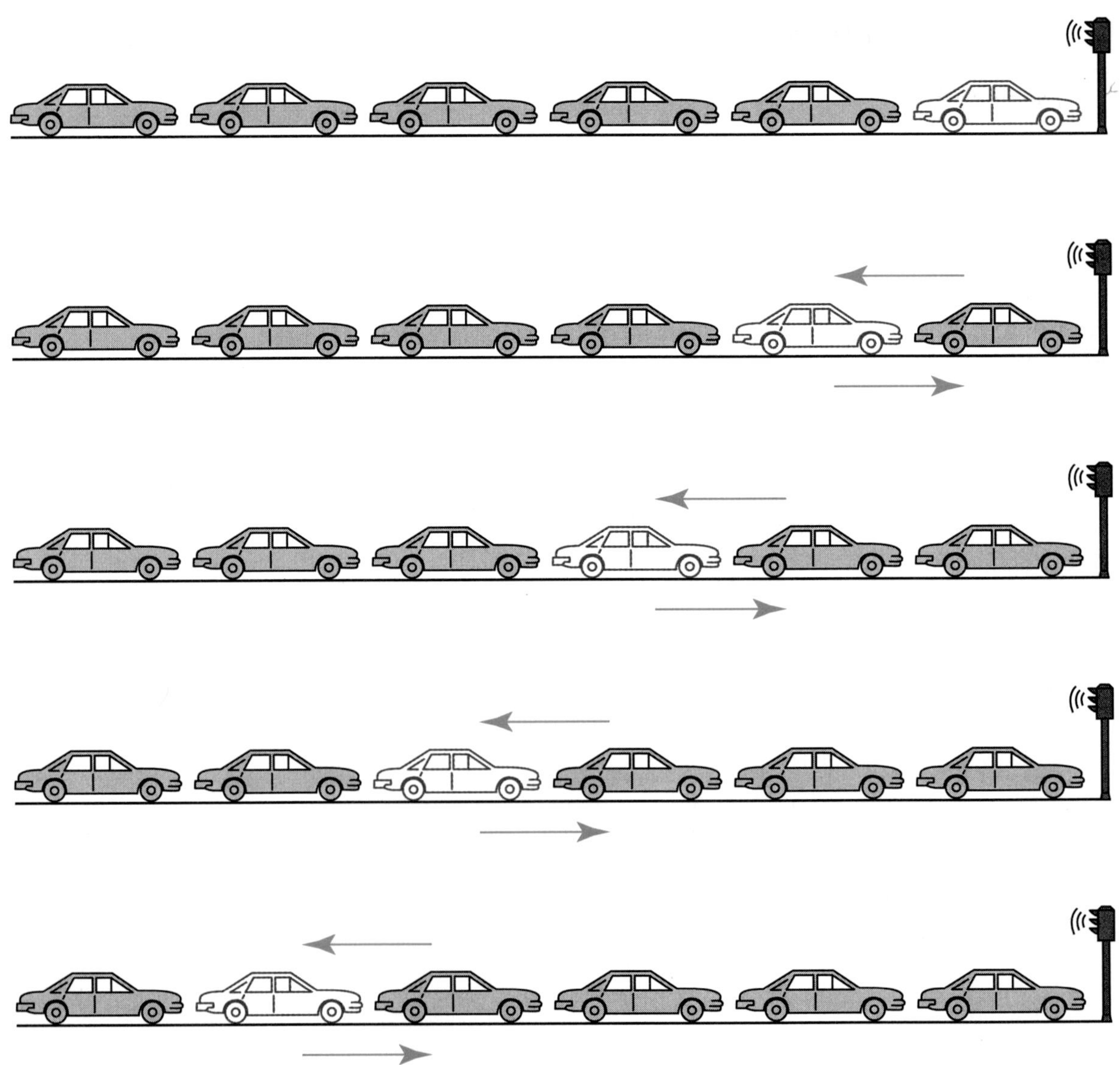

Fig. 2-12 Hole current analogy.

2-5 Majority and Minority Carriers

When N- and P-type semiconductor materials are made, the doping levels can be as small as 1 part per million or 1 part per billion. Only a tiny trace of impurity materials having five or three valence electrons enters the crystal. It is not possible to make the silicon crystal absolutely pure. Thus, it is easy to imagine that an occasional atom with three valence electrons might be present in an N-type semiconductor. An unwanted hole will exist in the crystal. This hole is called a *minority carrier.* The free electrons are the *majority carriers.*

In a P-type semiconductor, one expects holes to be the carriers. They are in the majority. A few free electrons might also be present. They will be the minority carriers in this case.

The majority carriers will be electrons for N-type material and holes for P-type material. Minority carriers will be holes for N-type material and electrons for P-type material.

Today very high-grade silicon can be manufactured. This high-grade material has very few unwanted impurities. Although this keeps the number of minority carriers to a minimum, their numbers are increased by high temperatures. This can be quite a problem in electronic circuits. To understand how heat produces minority carriers, refer to Fig. 2-6. As additional heat energy enters the crystal, more and more electrons will gain enough energy to break their bonds. Each broken bond produces both a free electron and a hole. Heat produces carriers in *pairs.* If the crystal was manufactured to be N-type material, then every thermal hole becomes a minority carrier and the thermal electrons join the other majority carriers. If the crystal was made as P-type material, then the thermal holes join the majority carriers and the thermal electrons become minority carriers.

Minority carrier

Majority carrier

Carrier production by heat decreases the crystal's resistance. The heat also produces minority carriers. Heat and the resulting minority carriers can have an adverse effect on the way semiconductor devices work.

About Electronics

Neils Bohr and the Atom Scientists change the future by improving on the ideas of others. Neils Bohr proposed a model of atomic structure in 1913 that applied energy levels (quantum mechanics) to the Rutherford model of the atom. Bohr also used some of the work of Max Planck.

TEST

Determine whether each statement is true or false.

28. In making N-type semiconductor material, a typical doping level is about 10 arsenic atoms for every 90 silicon atoms.
29. A free electron in a P-type crystal is called a majority carrier.
30. A hole in an N-type crystal is called a minority carrier.
31. As P-type semiconductor material is heated, one can expect the number of minority carriers to increase.
32. As P-type semiconductor material is heated, the number of majority carriers decreases.
33. Heat increases, the number of minority and majority carriers in semiconductors.

Chapter 2 Review

Summary

1. Good conductors, such as copper, contain a large number of current carriers.
2. In a conductor, the valence electrons are weakly attracted to the nuclei of the atoms.
3. Heating a conductor will increase its resistance. This response is called a positive temperature coefficient.
4. Silicon atoms have four valence electrons. They can form covalent bonds which result in a stable crystal structure.
5. Heat energy can break covalent bonds, making free electrons available to conduct current. This gives silicon and other semiconductor materials a negative temperature coefficient.
6. At room temperature, germanium crystals have 1000 times more thermal carriers than silicon crystals do. This makes germanium diodes and transistors less useful than silicon devices for many applications.
7. The process of adding impurities to a semiconductor crystal is called doping.
8. Doping a semiconductor crystal changes its electrical characteristics.
9. Donor impurities have five valence electrons and produce free electrons in the crystal. This forms N-type semiconductor material.
10. Free electrons serve as current carriers.
11. Acceptor impurities have three valence electrons and produce holes in the crystal.
12. Holes in semiconductor materials serve as current carriers.
13. Hole current is opposite in direction to electron current.
14. Semiconductors with free holes are classified as P-type materials.
15. Impurities with five valence electrons produce N-type semiconductors.
16. Impurities with three valence electrons produce P-type semiconductors.
17. Holes drift toward the negative end of a voltage source.
18. Majority carriers are electrons for N-type material. Holes are majority carriers for P-type material.
19. Minority carriers are holes for N-type material. Electrons are minority carriers for P-type material.
20. The number of minority carriers increases with temperature.

Chapter Review Questions

Determine whether each statement is true or false.

2-1. The current carriers in conductors such as copper are holes and electrons.

2-2. It is easy to move the valence electrons in conductors.

2-3. A positive temperature coefficient means the resistance goes up as temperature goes down.

2-4. Conductors have a positive temperature coefficient.

2-5. Silicon does not semiconduct unless it is doped or heated.

2-6. Silicon has five valence electrons.

2-7. A silicon crystal is built by ionic bonding.

2-8. Materials with eight valence electrons tend to be unstable.

2-9. Semiconductors have a negative temperature coefficient.

2-10. Silicon is usually preferred to germanium because it has higher resistance at any given temperature.

2-11. When a semiconductor is doped with arsenic, free electrons are placed in the crystal.

2-12. N-type material has free electrons available to support current flow.

2-13. Doping a crystal increases its resistance.

2-14. Doping with boron produces free electrons in the crystal.

2-15. Hole current is opposite in direction to electron current.
2-16. Holes are current carriers and are assigned a positive charge.
2-17. If a P-type semiconductor shows a few free electrons, the electrons are called minority carriers.
2-18. If an N-type semiconductor shows a few free holes, the holes are called minority carriers.

Critical Thinking Questions

2-1. Suppose that you could perfect a method of inexpensively making ultrapure carbon crystals and then doping them. How could these crystals be used in electronics? (*Hint:* Diamonds are noted for their extreme hardness and ability to withstand high temperatures.)
2-2. Some semiconductors, such as gallium arsenide, show better carrier mobility than silicon. That is, the carriers move faster in the crystal. What kinds of devices could benefit from this?
2-3. Semiconductors respond to temperature by showing decreased resistance leading to problems in many, but not all, electronic products. Can you think of an application where their temperature sensitivity is desired?
2-4. You have learned that conductors and semiconductors have opposite temperature coefficients. How could you use this knowledge to design a circuit that remains stable over a wide temperature range?

Answers to Tests

1. F
2. T
3. F
4. T
5. T
6. F
7. T
8. F
9. T
10. F
11. T
12. T
13. T
14. T
15. T
16. F
17. F
18. donor
19. five
20. electron
21. carriers
22. decreases
23. acceptor
24. three
25. positive
26. holes
27. negative
28. F
29. F
30. T
31. T
32. F
33. T

Junction Diodes

Chapter Objectives

This chapter will help you to:

1. *Predict* the conductivity of junction diodes under the conditions of forward and reverse bias.
2. *Interpret* volt-ampere characteristic curves for diodes.
3. *Identify* the cathode and anode leads of some diodes by visual inspection.
4. *Identify* the cathode and anode leads of diodes by ohmmeter testing.
5. *Identify* diode schematic symbols.
6. *List* several diode types and applications.

This chapter introduces the most basic semiconductor device, the junction diode. Diodes are very important in electronic circuits. Everyone working in electronics must be familiar with them. Your study of diodes will enable you to predict when they will be on and when they will be off. You will be able to read their characteristic curves and identify their symbols and their terminals. This chapter also introduces several important types of diodes and some of the many applications for them.

3-1 The PN Junction

Diode

PN-junction diode

A basic use for P- and N-type semiconductor materials is in *diodes.* Figure 3-1 shows a representation of a *PN-junction diode.* Notice that it contains a P-type region with free holes and an N-type region with free electrons. The diode structure is continuous from one end to the other. It is one complete crystal of silicon or germanium.

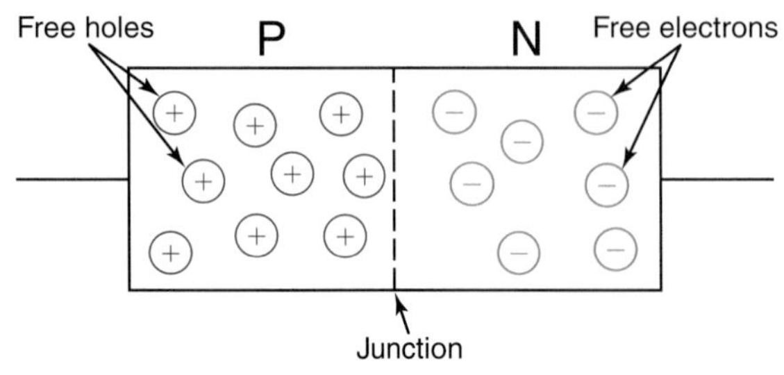

Fig. 3-1 The structure of a junction diode.

The junction shown in Fig. 3-1 is the boundary, or dividing line, that marks the end of one section and the beginning of the other. It does not represent a mechanical joint. In other words, the *junction* of a diode is that part of the crystal where the P-type material ends and the N-type material begins.

Depletion region

Because the diode is a continuous crystal, free electrons can move across the junction. When a diode is manufactured, some of the free electrons cross the junction to fill some of the holes. Figure 3-2 shows this effect. The result is that a *depletion region* is formed. The electrons that have filled holes are effectively captured and are no longer available to support current flow. With the electrons gone and the holes filled, no free carriers are left. The region around the junction has become *depleted.*

The depletion region will not continue to grow for very long. An electric potential, or

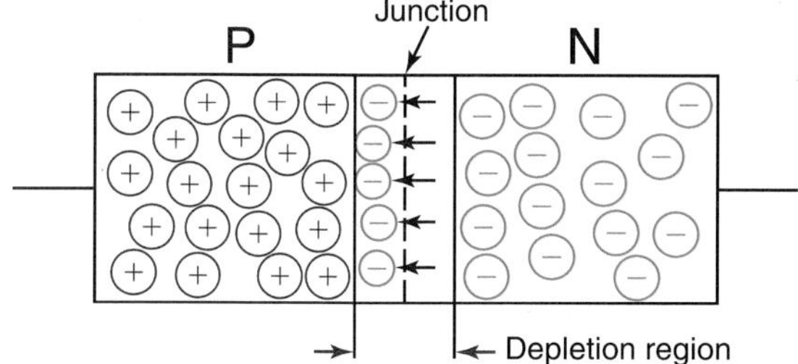

Fig. 3-2 The diode depletion region.

force, forms along with the depletion region and prevents all the electrons from crossing over and filling all the holes in the P-type material.

Figure 3-3 shows why this potential is formed. Any time an atom loses an electron, it becomes unbalanced. It now has more protons in its nucleus than it has electrons in orbit. This gives it an overall positive charge. It is called a *positive ion.* In the same way, if an atom gains an extra electron, it shows an overall negative charge and is called a *negative ion.* When one of the free electrons in the N-type material leaves its parent atom, that atom becomes a positive ion. When the electron joins another atom on the P-type side, that atom becomes a negative ion. The ions form a charge that prevents any more electrons from crossing the junction.

So when a diode is manufactured, some of the electrons cross the junction to fill some of the holes. The action soon stops because a negative charge forms on the P-type side to repel any other electrons that might try to cross over. This negative charge is called the *ionization potential* or the *barrier potential.* "Barrier" is a good name since it does stop additional electrons from crossing the junction.

Now that we know what happens when a PN junction is formed, we can investigate how it will behave electrically. Figure 3-4 shows a

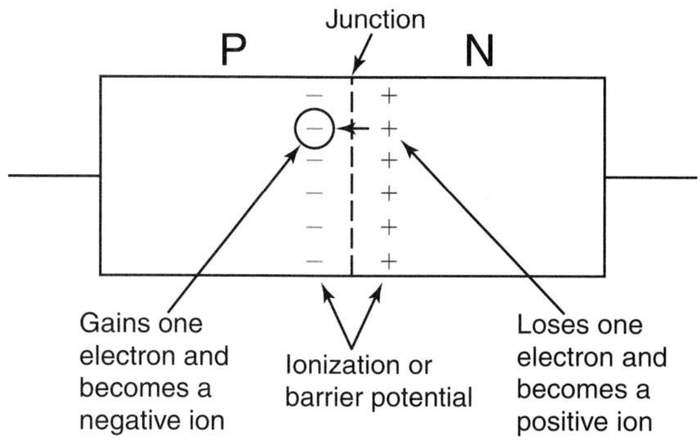

Fig. 3-3 Formation of the barrier potential.

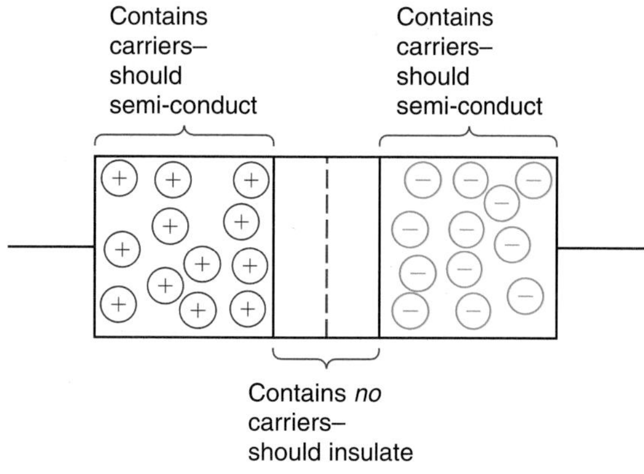

Fig. 3-4 Depletion region as an insulator.

summary of the situation. There are two regions with free carriers. Since there are carriers, we can expect these regions to *semiconduct.* But right in the middle there is a region with no carriers. When there are no carriers, we can expect it to *insulate.*

Positive ion

Negative ion

Any device having an insulator in the middle will not conduct. So we can assume that PN-junction diodes are insulators. However, a depletion region is not the same as a fixed insulator. It was formed in the first place by electrons moving and filling holes. An external voltage can *remove* the depletion region.

In Fig. 3-5 a PN-junction diode is connected to an external battery in such a way that the depletion region is eliminated. The positive terminal of the battery repels the holes on the P-type side and pushes them toward the junction. The negative terminal of the battery repels the electrons and pushes them toward the junction. This *collapses* (removes) the depletion region.

Barrier potential

With the depletion region collapsed, the diode can semiconduct. Figure 3-5 shows elec-

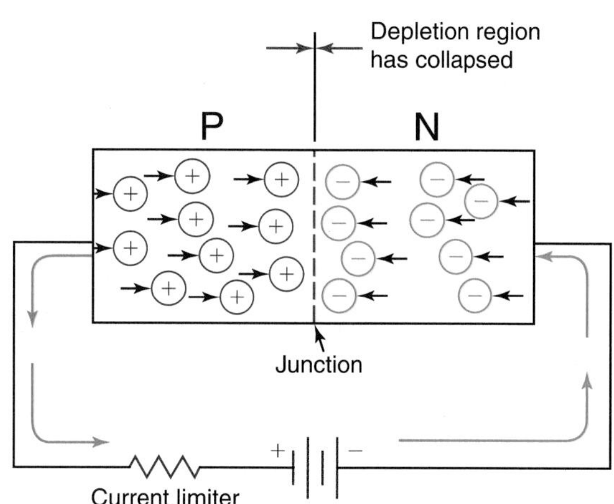

Fig. 3-5 Forward bias.

tron current leaving the negative side of the battery, flowing through the diode, through the current limiter (a resistor), and returning to the positive side of the battery. The current-limiting resistor is needed in some cases to keep the current flow at a safe level. Diodes can be destroyed by excess current. Ohm's law can be used to find current in diode circuits. For example if the battery in Fig. 3-5 is 6 V and the resistor is 1 kilohm (kΩ),

$$I = \frac{V}{R} = \frac{6\ \text{V}}{1\ \text{k}\Omega} = 6\ \text{milliamperes (mA)}$$

The above calculation ignores the diode's resistance and voltage drop. It is only an *approximation* of the circuit current. If we know the drop across the diode, it is possible to accurately predict the current. The diode drop is simply subtracted from the supply voltage:

$$I = \frac{6\text{V} - 0.6\ \text{V}}{1\ \text{k}\Omega} = 5.4\ \text{mA}$$

A typical silicon diode drops about 0.6 V when it is conducting. This is still an approximation, but it is more accurate than our first attempt.

Forward bias

Reverse bias

Example 3-1

Calculate the current in Fig. 3-5 for a 1-V battery and a 1-kΩ resistor. Determine the importance of correcting for the diode voltage drop. First, calculate the current without correcting for the diode drop:

$$I = \frac{1\ \text{V}}{1\ \text{k}\Omega} = 1\ \text{mA}$$

Make a second calculation which includes the correction:

$$I = \frac{1\ \text{V} - 0.6\ \text{V}}{1\ \text{k}\Omega} = 0.4\ \text{mA}$$

It is important to correct for the diode drop when the supply voltage is relatively low.

Example 3-2

Schottky diodes drop about 0.3 V when conducting. Schottky diodes are explained in Sec. 3-4. Calculate the current in Fig. 3-5 for a Schottky diode, a 1-V battery, and a 1-kΩ resistor.

$$I = \frac{1\ \text{V} - 0.3\ \text{V}}{1\ \text{k}\Omega} = 0.7\ \text{mA}$$

The small voltage drop of Schottky diodes makes a significant difference in low-voltage circuits.

Example 3-3

Calculate the current in Fig. 3-5 for a 100-V battery and a 1-kΩ resistor. Determine the importance of correcting for the voltage drop of a silicon diode.

$$I = \frac{100\ \text{V}}{1\ \text{k}\Omega} = 100\ \text{mA}$$

$$I = \frac{100\ \text{V} - 0.6\ \text{V}}{1\ \text{k}\Omega} = 99.4\ \text{mA}$$

It is not as important to correct for the diode drop when the supply voltage is relatively high.

The condition of Fig. 3-5 is called *forward bias.* In electronics, a bias is a voltage or a current applied to a device. Forward bias indicates that the voltage or current is applied so that it turns the device *on.* The diode in Fig. 3-5 has been turned on by the battery, so it is an example of forward bias.

Reverse bias is another possibility. With zero bias connected to the diode, the depletion region is as shown in Fig. 3-6(*a*). When reverse bias is applied to a junction diode, the depletion region does not collapse. In fact, it becomes wider than it was. Figure 3-6(*b*) shows a diode with reverse bias applied. The positive side of the battery is applied to the N-type material. This attracts the free electrons away from the junction. The negative side of the battery attracts the holes in the P-type material away from the junction. This makes the depletion region wider than it was when no voltage was applied.

Because reverse bias widens the depletion region, it can be expected that no current flow will result. The depletion region is an insulator, and it will block the flow of current. Actually, a small current will flow because of *minority carriers.* Figure 3-7 shows why this

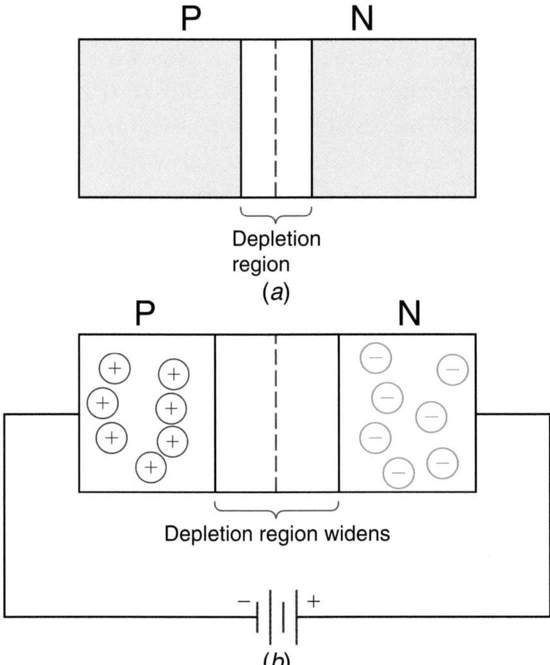

Fig. 3-6 The effect of reverse bias on the depletion region.

happens. The P-type material has a few minority electrons. These are pushed to the junction by the repulsion of the negative side of the battery. The N-type material has a few minority holes. These are also pushed toward the junction. Reverse bias forces the minority carriers together, and a small *leakage current* results. Diodes are not perfect, but modern silicon diodes usually show a leakage current so small that it cannot be measured with ordinary meters. At room temperature there are only a few minority carriers in silicon, so the reverse leakage can be ignored.

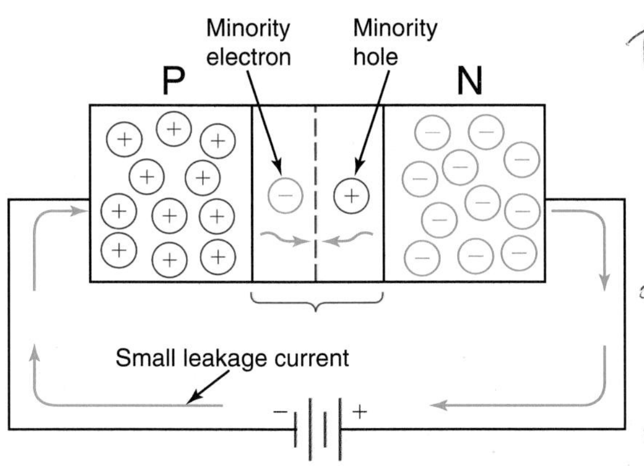

Fig. 3-7 Leakage current due to minority carriers.

About Electronics

Diodes Provide Protection from Reverse Polarity A diode can provide reverse polarity protection. One approach is to use a series protection diode, and the other is to use a shunt protection diode that causes a fuse to blow when polarity is reversed.

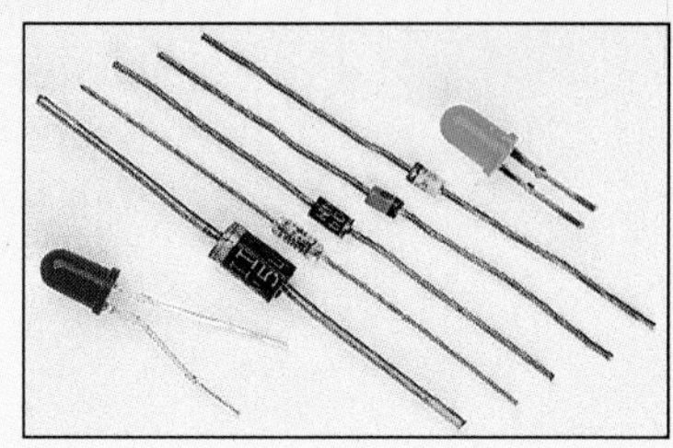

Germanium diodes have more leakage. At room temperature, germanium has about 1000 times as many minority carriers as silicon. Silicon diodes cost less, show very low leakage current, and are better choices for most applications. Germanium diodes do have certain advantages such as low turn-on voltage and low resistance and are therefore still used in a few specific areas.

In summary, the PN-junction diode will conduct readily in one direction and very little in the other. The direction of easy conduction is from the N-type material to the P-type material. If a voltage is applied across the diode to move the current in this direction, it is called forward bias. The diode is very useful because it can steer current in a given direction. It can also be used as a switch and as a means of changing alternating current (ac) to direct current (dc). Other diodes perform many special jobs in electric and electronic circuits.

Leakage current

TEST

Determine whether each statement is true or false.

1. A junction diode is doped with both P- and N-type impurities.
2. The depletion region is formed by electrons crossing over the P-type side of the junction to fill holes on the N-type side of the junction.
3. The barrier potential prevents all the electrons from crossing the junction and filling all the holes.
4. The depletion region is a good conductor.
5. Once the depletion region forms, it cannot be removed.
6. Forward bias expands the depletion region.

7. Reverse bias collapses the depletion region and turns on the diode.
8. A reverse-biased diode may show a little leakage current because of minority-carrier action.
9. High temperatures will increase the number of minority carriers and diode leakage current.

3-2 Characteristic Curves of Diodes

Diodes conduct well in one direction but not in the other. This is the fundamental property of diodes. They have other characteristics too, and some of these must be understood in order to have a working knowledge of electronic circuits.

JOB TIP

Good communication skills are as important as your understanding of circuit theory.

Characteristics of electronic devices can be shown in several ways. One way is to list the amount of current flow for each of several values of voltage. These values could be presented in a table. A better way to do it is to show the values on a graph. Graphs are easier to use than tables of data.

One of the most frequently used graphs in electronics is the volt-ampere characteristic curve. Units of voltage make up the horizontal axis, and units of current make up the vertical axis. Figure 3-8 shows a volt-ampere characteristic curve for a 100-Ω resistor. The origin is the point where the two axes cross. This point indicates zero voltage and zero current. Note that the resistor curve passes through the origin. This means that with zero voltage across a resistor we can expect zero current through it. Ohm's law will verify this:

$$I = \frac{V}{R} = \frac{0}{100} = 0 \text{ A}$$

At 5 V on the horizontal axis, the curve passes through a point exactly opposite 50 mA on the vertical axis. By looking at the curve, we can quickly and easily find the current for any value of voltage. At 10 V the current is 100 mA. We can check this using Ohm's law:

$$I = \frac{V}{R} = \frac{10}{100} = 0.1 \text{ A} = 100 \text{ mA}$$

Moving to the left of the origin in Fig. 3-8, we can obtain current levels for values of reverse voltage. Reverse voltage is indicated by V_R, and V_F indicates the forward voltage. At −5 V the current through the resistor will be −50 mA. The minus signs indicate that when the voltage across a resistor is reversed in polarity, the resistor current will reverse (change direction). Forward current is indicated by I_F, and I_R indicates reverse current.

The characteristic curve for a resistor is a straight line. For this reason, it is said to be a linear device. Resistor curves are not necessary. With Ohm's law to help us, we can easily obtain any data point without a graph.

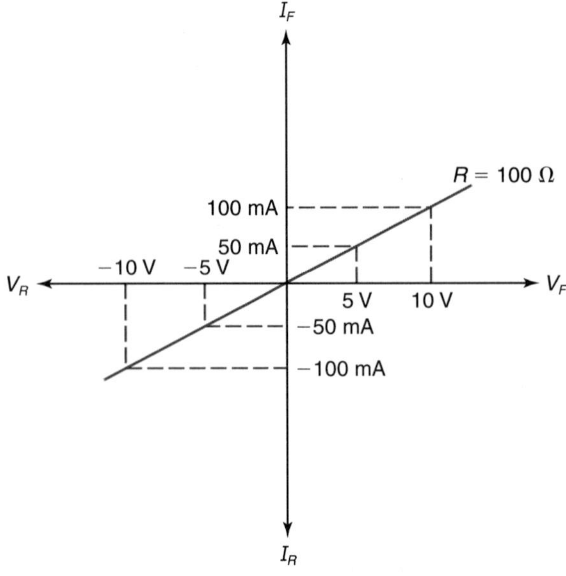

Fig. 3-8 A volt-ampere characteristic curve for a resistor.

Example 3-4

How would Fig. 3-8 appear for a 50-Ω resistor?

$$I = \frac{V}{R} = \frac{10 \text{ V}}{50\ \Omega} = 200 \text{ mA}$$

The curve would be a straight line passing through the origin and through data points at ± 10 V and ± 200 mA. Thus, the 50-Ω curve would be steeper (would have more slope) than the 100-Ω curve.

Diodes are more complicated than resistors. Their volt-ampere characteristic curves give information that cannot be obtained with a sim-

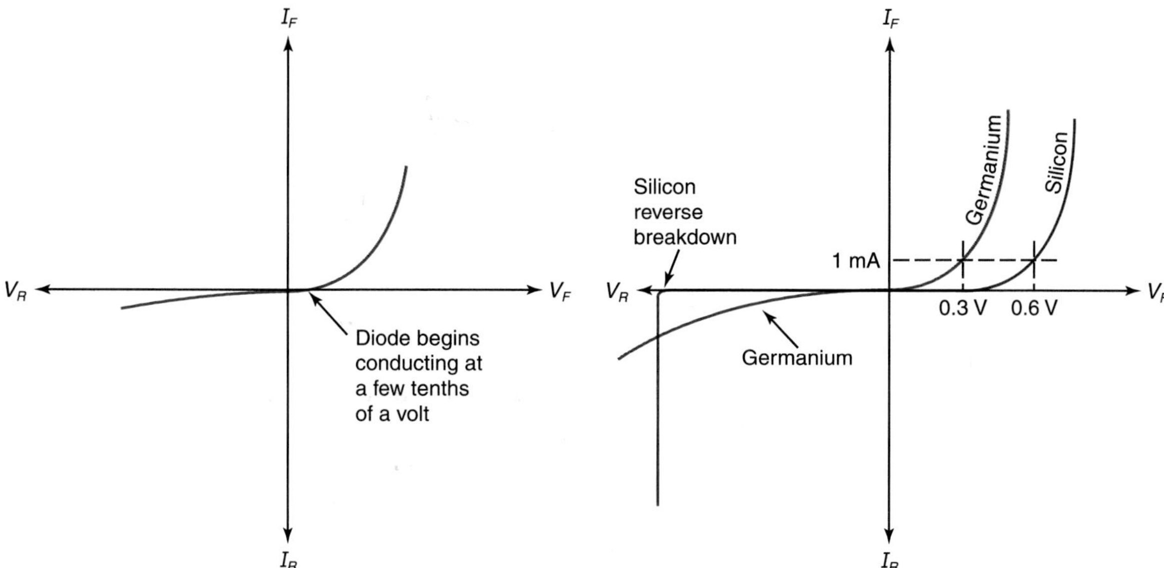

Fig. 3-9 A volt-ampere characteristic curve of a diode.

Fig. 3-10 Comparison of silicon and germanium diodes.

ple linear equation. Figure 3-9 shows the volt-ampere characteristic curve for a typical PN-junction diode. Please notice that the curve is *not* linear. With 0 V across the diode, the diode will not conduct. The diode will not begin to conduct until a few tenths of a volt are applied across it. This is the voltage needed to remove the depletion region. It requires about 0.2 V to turn on a germanium diode and about 0.6 V to turn on a silicon diode.

Nonlinearity

Figure 3-9 also shows what happens when reverse bias is applied to a diode. At increasing levels of V_R, the curve shows some reverse current I_R. This leakage current is caused by minority carriers. It is usually very small. Sometimes the I_R axis is calibrated in microamperes (μA). Reverse current is not significant until there is a large reverse bias across the diode. For that reason, the V_R axis is sometimes calibrated in tens or hundreds of volts.

A comparison of the characteristic curves for a silicon diode and a germanium diode is shown in Fig. 3-10. It is clear that the germanium diode requires much less forward bias to conduct. This can be an advantage in low-voltage circuits. Also, note that the germanium diode will show a lower voltage drop for any given level of current than the silicon diode will. Germanium diodes have less resistance for forward current because germanium is a better conductor. However, the silicon diode is still superior for most applications because of its low cost and lower leakage current.

Figure 3-10 also shows how silicon and germanium diodes compare under conditions of reverse bias. At reasonable levels of V_R, the leakage current of the silicon diode is very low. The germanium diode shows much more leakage. However, if a certain critical value of V_R is reached, the silicon diode will show a rapid increase in reverse current. This is shown as the *reverse breakdown* point. It is also referred to as the *avalanche voltage.* Avalanche breakdown occurs when carriers accelerate and gain enough energy to collide with valence electrons and knock them loose. This causes an "avalanche" of carriers, and the reverse current flow increases tremendously.

Reverse breakdown point

Avalanche voltage

The avalanche voltage for silicon diodes ranges from 50 to over 1000 V, depending on how the diode was manufactured. If the reverse current at avalanche is not limited, the

Did You Know?

Silicon Diodes and the Auto Industry The development of silicon diodes allowed automobile designers to choose alternators rather than generators. This greatly improved the performance and reliability of the charging system.

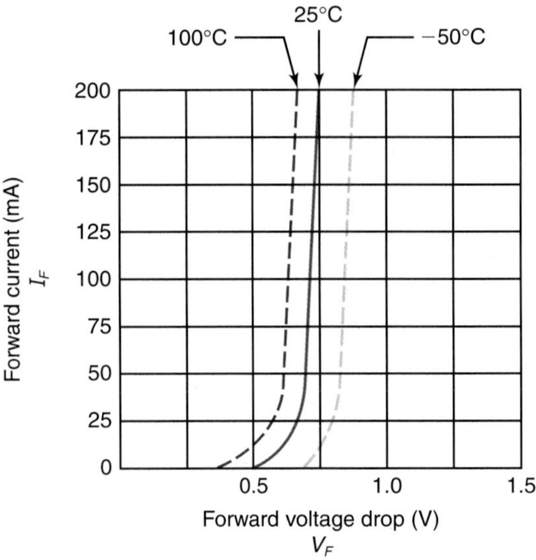

Fig. 3-11 Characteristic curves showing the effect of temperature on a typical silicon diode.

Polarity

diode will be destroyed. Avalanche is avoided by using a diode that can safely withstand circuit voltages.

Schematic symbol

Anode

Cathode

Figure 3-11 shows how volt-ampere characteristic curves can be used to indicate the effects of temperature on diodes. The temperatures are in degrees Celsius (°C). Electronic circuits may have to work over a range of temperatures from −50° to +100°C. At the low end mercury will freeze; at the high end water will boil. The range for military-grade electronic circuitry is −65° to +125°C. For circuits to operate in such a wide temperature range, extreme care must be taken in the selection of materials, the manufacturing processes used, and the handling and testing of the finished product. This is why military-grade devices are more expensive than industrial- and commercial-grade devices.

By examining the curves in Fig. 3-11, you can conclude that silicon conducts better at elevated temperatures. Since the forward voltage drop V_F decreases as temperature goes up, its resistance must be going down. This agrees with silicon's negative temperature coefficient.

TEST

Supply the missing word in each statement.

10. The characteristic curve for a linear device is shaped like a ______.
11. A volt-ampere characteristic curve for a resistor is shaped like a ______.
12. A volt-ampere characteristic curve for a 1000-Ω resistor will, at 10 V on the horizontal axis, pass through a point opposite ______ on the vertical axis.
13. The volt-ampere characteristic curve for an open circuit (∞ Ω) will be a straight line on the ______ axis.
14. The volt-ampere characteristic curve for a short circuit (0 Ω) will be a straight line on the ______ axis.
15. Resistors are linear devices. Diodes are ______ devices.
16. A silicon diode does not begin conducting until ______ V of forward bias is applied.
17. Diode avalanche, or reverse breakdown, is caused by excess reverse ______.

3-3 Diode Lead Identification

Diodes have *polarity.* Components such as resistors can be wired either way into a circuit but diodes must be installed properly. Connecting a diode backward can destroy it and may also damage many other parts of a circuit. A technician must always be absolutely sure that the diodes are correctly connected.

Technicians often refer to schematic diagrams when checking diode polarity. Figure 3-12 shows the *schematic symbol* for a diode. The P-type material makes up the *anode* of the diode. The word "anode" is used to identify the terminal that attracts electrons. The N-type material makes up the *cathode* of the diode. The word "cathode" refers to the terminal that gives off, or emits, electrons. Note that the forward current moves from the cathode to the anode (against the arrowhead).

Diodes are available in many package styles. Some examples are shown in Fig. 3-13. Manufacturers use plastic, glass, metal, ceramic, or a combination of these to package diodes. There are quite a few sizes and shapes available. Generally, the larger devices have

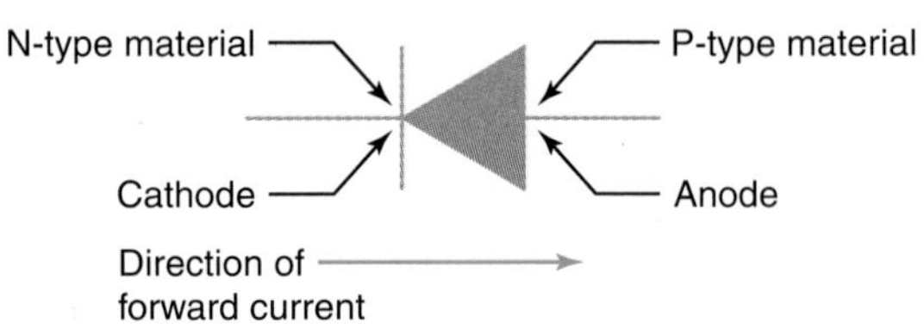

Fig. 3-12 Diode schematic symbol.

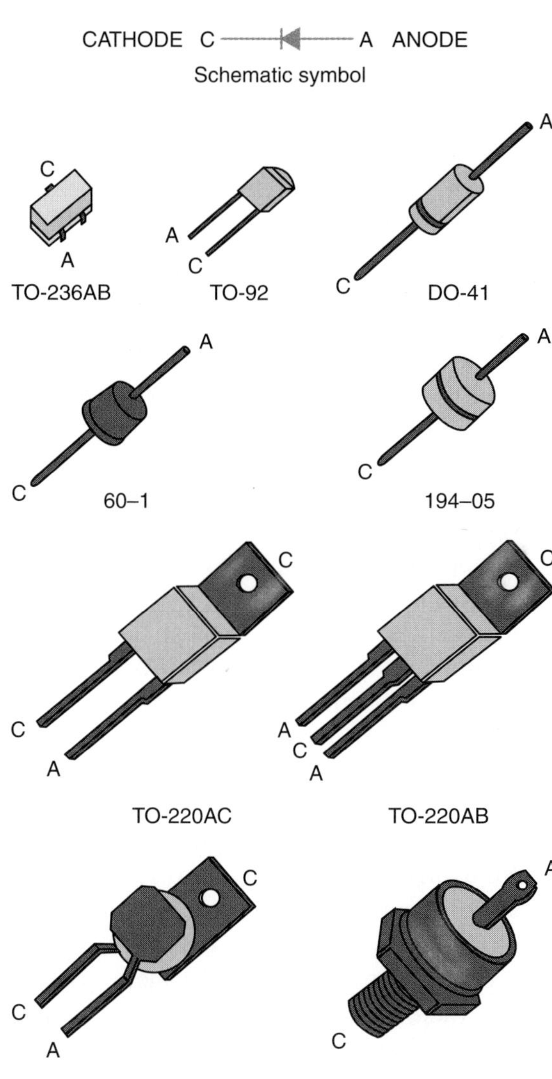

Fig. 3-13 Diode package styles.

higher current ratings. The diode package is often marked to denote the *cathode lead.* This can be done with one or more bands near the cathode lead. An example of this method is shown on the DO-41 package in Fig. 3-13. Some older package styles used a bevel or a plus sign (+) to denote the cathode lead.

Other packages use various schemes for lead identification. A few use an imprint of the diode symbol. This method can be used with the 194-05 package in Fig. 3-13, although the illustration does not show it. The TO-220AC style has both a cathode lead and a metal tab which also serves as a cathode contact. Either the lead or the tab can be used to connect the diode to the rest of the circuit. The TO-220AB case shows two anode leads. This is a different situation because there are two diodes inside the package. The anodes of the two diodes are available as separate terminals, but the cathodes are connected together internally.

Manufacturers can offer diodes in both a normal polarity version and a reverse polarity version. For example, the threaded stud end of the 257-01 package in Fig. 3-13 is the anode in the reverse polarity version. The part number is followed by an "R" to denote the reverse polarity version. However, the part number is rarely marked on the device. Another problem is that manufacturers use the same package to house different devices. Both the TO-236AB package and the TO-220AB package shown in Fig. 3-13 can also be used for transistors. In other words, a casual inspection of an electronic circuit will not always allow you to positively identify components and their leads. You should use schematics or other service literature to be certain.

Since diode package styles can cause confusion, a technician may check a diode and identify the leads using a volt-ohm-milliammeter (VOM), or a digital multimeter (DMM). This check uses the ohmmeter function of the meter. The ohmmeter is connected across the diode, and the resistance is noted as in Fig. 3-14. The $R \times 100$ range is useful for diode checking. (DMMs often have a diode testing function.) It is necessary to note only whether the resistance is high or low. Exact readings are not important. Then the ohmmeter leads are reversed as in Fig. 3-14(*b*). The resistance should change drastically. If it does not, the diode is probably defective. In Fig. 3-14 we can conclude that the diode is good and that the cathode lead is at the left. When the positive lead of the ohmmeter was on the right lead, the diode was turned on. Forward current is from cathode to anode. Making the anode positive is necessary if the anode is going to attract electrons. Remember, in order to turn on the diode, the anode must be positive with respect to the cathode.

Cathode lead

Using ohmmeters to check diodes and identify the leads works well. However, there are two traps that the technician must know about in order to avoid mistakes. First, a few older ohmmeters have reversed polarity. This may occur on some ranges or on all ranges. The only way to be sure is to check the ohmmeter with a separate dc voltmeter. Know your meter. When using an unfamiliar meter, do not assume that the common, or black, lead is nega-

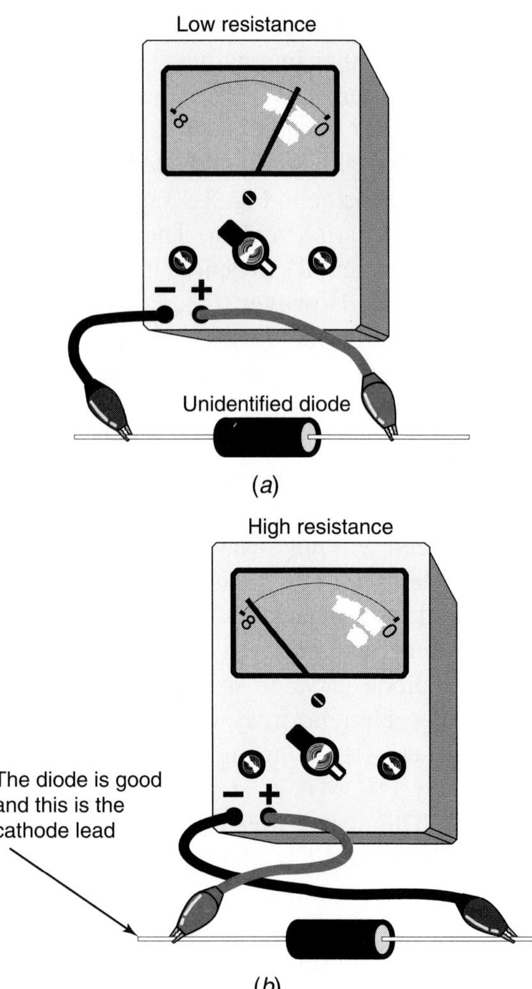

Fig. 3-14 Testing a diode with an ohmmeter.

tive on the ohms function. The second trap is that the ohmmeter supply voltage might not be high enough to turn on the diode. This would indicate that the diode is open (shows a high resistance in both directions). Most newer DMMs have a *low ohms function* that uses less than 0.2 V. Remember, it takes about 0.2 V to turn on a germanium diode and about 0.6 V to turn on a silicon diode. Some special high-voltage diodes may even require much more than this and cannot be tested with ohmmeters.

Schottky diode

A few older ohmmeters can supply enough voltage or current to damage certain diodes. The diodes used in high-frequency detection circuits can be very delicate. Usually, it is not a good idea to check these diodes with an ohmmeter. Technicians must take the time to learn the characteristics of their test equipment.

Characteristic curve

The DMM has decreased in cost and is now less expensive than a good quality analog multimeter. Technicians must be able to use DMMs to test diodes and identify diode leads. Procedures will vary from model to model and it is always good practice to refer to the operator's manual for a particular DMM.

Modern DMMs have an ohms range and a diode range. The diode range is usually marked with the diode schematic symbol. Use the diode range when testing diodes.

Some DMMs have an audible output on the diode range. They beep once when a good diode is forward-biased and beep continuously for a shorted diode. They make no sound when a good diode is reverse-biased.

The diode test function on some DMMs sends approximately 0.6 mA through the component connected to the meter terminals. The digital display reads the voltage drop across the component. A normal, forward-biased junction will read somewhere between 0.250 and 0.700 using this type of meter. A reverse-biased junction will cause the meter display to indicate overrange.

Table 3-1 shows some typical readings obtained using a DMM on its ohms function and on its diode function to test various diode types. In every case, the diode was normal and was forward-biased by the meter. Notice that as the current capacity (size) of the silicon diodes increases, the diode's forward resistance decreases when using the ohms function and the voltage drop across the diode is smaller when using the diode function. Also notice that the Schottky and germanium diodes show the lowest resistances and voltage drops. *Schottky diodes* are explained in the next section.

Diodes are nonlinear devices. They will not show the same resistance when operated at different levels of forward bias. For example, a silicon diode might show 500 Ω of forward resistance when measured on a 2-kΩ range and 5 kΩ of forward resistance when measured on a 20-kΩ range. This is to be expected since the ohmmeter operates the diode at different points on its characteristic curve when different ranges are selected. Figure 3-15 illustrates this idea. Ohm's law is used to calculate diode resistance at two different operating points on the *characteristic curve.* At the upper operating point the diode's resistance is 500 Ω, and it is 5 kΩ at the lower operating point.

Table 3-1 Typical Results of Diode Testing with a Digital Multimeter (DMM)

Device Tested	Results: Ohms Function kΩ	Results: Diode Function
Small silicon diode	19	0.571
1-A silicon diode	17	0.525
5-A silicon diode	14	0.439
100-A silicon diode	8.5	0.394
Small Schottky diode	7	0.339
Small germanium diode	3	0.277

Beginners may be confused by diode polarity. There is a good reason, too. One of the older ways to mark the cathode lead was to use a plus (+) symbol (this is no longer done by diode manufacturers). Yet, we have said that the diode is turned on when its *anode lead* is made positive. This seems to be a contradiction. However, the reason the plus sign was used to indicate the cathode lead is related to how the diode behaves in a *rectifier circuit.* In a rectifier circuit, it is the cathode lead that is in contact with the positive end of the load. So, the plus sign was used to help technicians find the load polarity. Rectifier circuits are covered briefly in the next section and in detail in Chap. 4.

Example 3-5

Find R_D for Fig. 3-15 when $V_D = 0.2$ V. If we attempt to use Ohm's law:

$$R_D = \frac{V_D}{I_D} = \frac{0.2\text{ V}}{0} = \text{undefined}$$

Division by 0 is undefined. However, as the denominator of a fraction *approaches 0,* the value of the fraction approaches *infinity*:

$$R_D \Rightarrow \infty$$

The important idea here: The resistance of a diode is infinite if the voltage drop across the diode is less than its barrier potential.

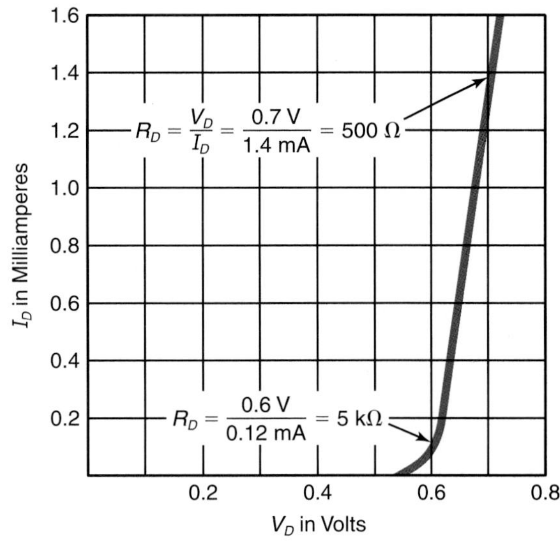

Fig. 3-15 Diode resistance at different operating points.

TEST

Supply the missing word in each statement.

18. Assume that a diode is forward-biased. The diode lead that is connected to the negative side of the source is called the CATHODE.
19. The diode lead near the band or bevel on the package is the CATHODE lead.
20. A plus (+) sign on an older diode indicates the CATHODE lead.
21. An ohmmeter is connected across a diode. A low resistance is shown. The leads are reversed. A low resistance is still shown. The diode is SHORTED.
22. When the positive lead from an ohmmeter is applied to the anode lead of a diode, the diode is turned ON.
23. Diodes show different values of forward resistance on different ohmmeter ranges because they are NON-LINEAR.

3-4 Diode Types and Applications

There are many diode types and applications in electronic circuits. Some of the important ones are presented in this section.

Rectifier diode

Rectifier diodes are widely applied. A rectifier is a device that changes alternating current to direct current. Since a diode will conduct easily in one direction only, just half of the ac cycle will pass through the diode. A diode can be used to supply direct current in a simple battery charger (Fig. 3-16.) A secondary battery can be charged by passing a direct current through it that is opposite in direction to its discharge current. The rectifier will permit only that direction of current that will restore (recharge) the battery.

Notice in Fig. 3-16 that the diode is connected so the current flow during charging is opposite to the current flow during discharging. The cathode of the diode *must* be connected to the positive terminal of the battery. A mistake in this connection would discharge the battery or damage the diode. It is very important to connect diodes correctly.

Hot-carrier diode

An ideal rectifier would turn off at the instant it is reverse-biased. PN-junction diodes cannot turn off instantaneously. There are quite a few holes and electrons around the junction when a diode is conducting. Applying reverse bias will not immediately turn the diode off since it takes time to sweep these carriers away from the junction and establish a depletion region. This effect is not a problem when rectifying low frequencies such as 60 Hz. However, it is a factor in high-frequency circuits.

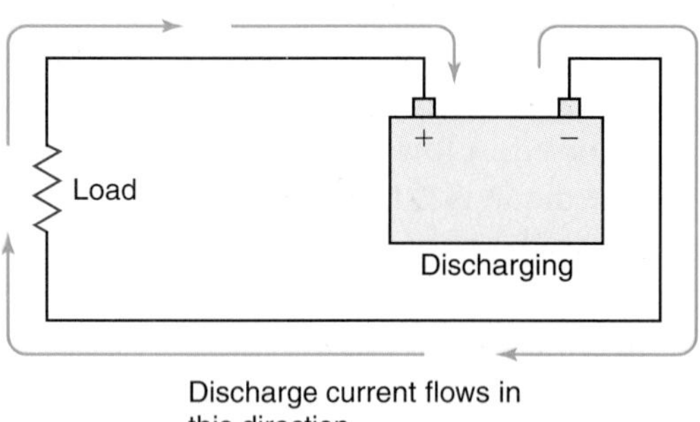

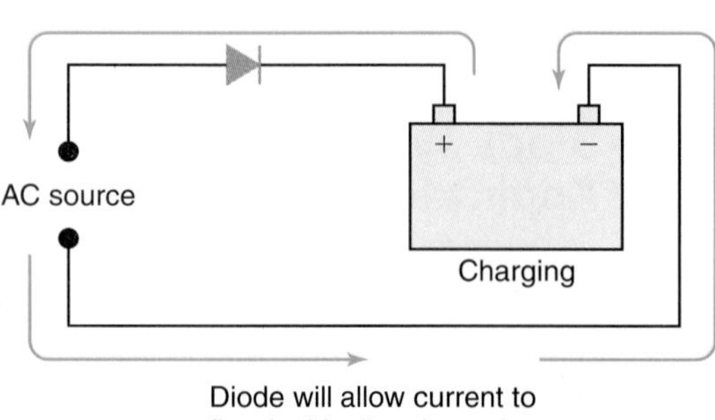

Fig. 3-16 Battery charging with a diode.

Fig. 3-17 Schottky diode schematic symbol.

So far we have looked at an interface of two types of semiconductors to produce diode action. Some metal-to-semiconductor interfaces will also rectify. This type of interface is called a *barrier. Schottky diodes* (or *barrier diodes*) use an N-type chip of silicon bonded to platinum. This semiconductor-to-metal barrier provides diode action and turns off much more quickly than a PN junction. Figure 3-17 shows the schematic symbol for a Schottky diode.

When a Schottky diode is forward-biased, electrons in the N-type cathode must gain energy to cross the barrier to the metal anode. The term "hot carrier diode" is sometimes used because of this fact. Once the "hot carriers" reach the metal, they join the great number of free electrons there and quickly give up their extra energy. When reverse bias is applied, the diode stops conducting almost immediately since a depletion region does not have to be established to block current flow. The electrons cannot cross back over the barrier because they have lost the extra energy required to do so. However, if more than about 50 V of reverse bias is applied, the electrons will gain the required energy and the barrier will break over and conduct. This prevents barrier type devices from being used in high-voltage circuits. Schottky diodes require only about 0.3 V of forward bias to establish forward current. They are well suited for high-frequency, low-voltage applications.

Voltage regulation

Zener diode

A diode can be used to hold a voltage constant. This is called *voltage regulation.* A special type called a *zener diode* is used as a voltage regulator. The characteristic curve and symbol for a zener diode are shown in Fig. 3-18. The symbol is similar to that of a rectifier diode except that the cathode is drawn as a bent line representing the letter Z. Zener diodes are manufactured to regulate voltages from 3.3 to 200 V. As an example, the 1N4733 is a popular 5.1-V zener.

The important difference between zener diodes and rectifier diodes is in how they are used in electronic circuits. As long as zeners are operated within their normal range, their voltage drop will equal their rated voltage plus

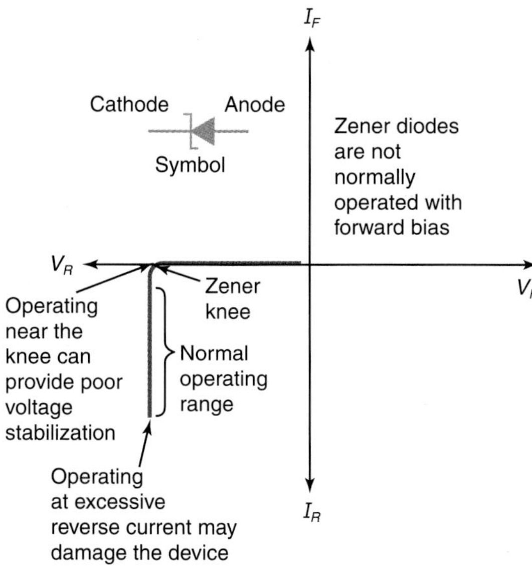

Fig. 3-18 Characteristic curve and symbol of a zener diode.

or minus a small error voltage. They are operated *backward* compared to a rectifier diode. In a rectifier, the normal current is from cathode to anode. Zeners are operated in reverse breakover and conduct from anode to cathode.

A change in zener diode current will cause only a small change in the zener voltage. This can be seen clearly in Fig. 3-19(*a*). Within the normal operating range, the zener voltage is reasonably stable.

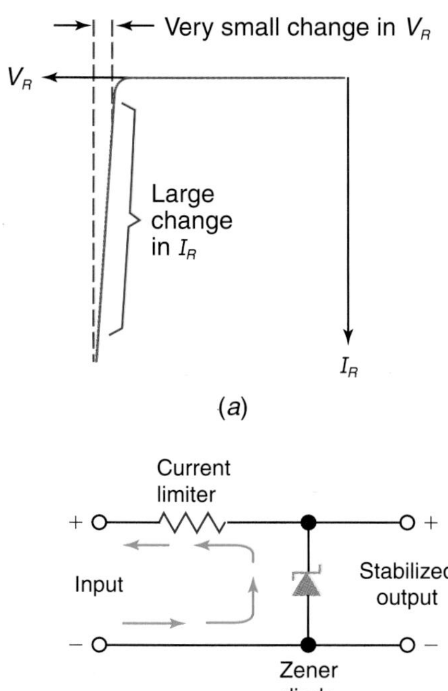

Fig. 3-19 A zener diode used as a voltage regulator.

Figure 3-19(*b*) shows how a zener diode can be used to stabilize a voltage. A current-limiting resistor is included to prevent the zener diode from conducting too much and overheating. The stabilized output is available across the diode itself. Notice that conduction is from anode to cathode. Zener voltage regulators are covered in more detail in the next chapter.

Diodes may be used as *clippers* or *limiters*. Refer to Fig. 3-20. Diode D_1 clips (limits) the input signal at −0.6 V and D_2 clips it at +0.6 V. A signal that is too small to forward-bias either diode will not be affected by the diodes. Diodes have a very high resistance when they are off. However, a large signal will turn the diodes on and they will conduct. When this happens, the excess signal voltage is dropped across R_1. Therefore, the total output swing is limited to 1.2 V peak-to-peak. This kind of limiting action may be used when a signal can get too large. For example, clippers can be used to keep audio signals from exceeding some loudness limit.

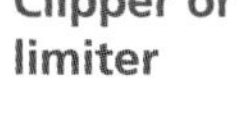

Figure 3-20 shows that the input signal is a sine wave but the output signal is more like a square wave. Sometimes a clipping circuit is used to change the shape of a signal. A third way that clippers can be used is to remove noise pulses riding on a signal. If the noise pulses exceed the clipping points, they will be clipped off or limited. The resulting signal is more noise-free than the original.

Diode D_2 clips the positive part of the signal in Fig. 3-20. As the signal voltage begins increasing from 0 V, nothing happens at first. Then, when the signal voltage reaches 0.6 V, D_2 turns on and begins to conduct. Now its resistance is much less than the resistance of R_1. Resistor R_1 drops the signal source voltage that

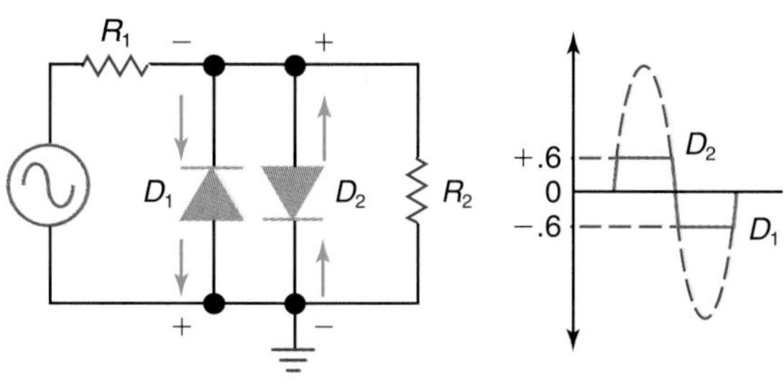

Fig. 3-20 Diode clipper.

About Electronics

Component Applications

- Blue LEDs are difficult to manufacture and are too expensive for most applications.
- Zener diodes are sometimes used to prevent damage from voltage transients.
- The crystal in a radio receiver acts as a diode detector.

is in excess of 0.6 V. Later the negative alternation begins. As the signal first goes negative, nothing happens. When it reaches −0.6 V, D_1 turns on. As D_1 conducts, R_1 drops the signal voltage in excess of −0.6 V. The total output swing is the difference between +0.6 and −0.6 V, or 1.2 V peak-to-peak. Germanium diodes would turn on at 0.2 V and produce a total swing of 0.4 V peak-to-peak if used in a clipper circuit.

The clipping points can be changed to a higher voltage by using series diodes. Examine Fig. 3-21. It will require 0.6 V + 0.6 V, or 1.2 V, to turn on D_3 and D_4. Notice that the positive clipping point is now shown on the graph at +1.2 V. In a similar fashion, D_1 and D_2 will turn on when the signal swings to −1.2 V. The output signal in Fig. 3-21 has been limited to a total swing of 2.4 V peak-to-peak. Higher clipping voltages can be obtained by using zener diodes as shown in Fig. 3-22. Assume that D_2 and D_4 are 4.7-V zeners. The positive-going signal will be clipped at +5.3 V since it takes 4.7 V to turn on D_4 and another +0.6 V to turn on D_3. Diodes D_1 and D_2 clip the negative alternation at −5.3 V. The total peak-to-peak output signal in Fig. 3-22 is limited to 10.6 V.

When a zener diode is *forward-biased,* it drops a bit more than a rectifier diode (about 0.7 V). Therefore, the circuit in Fig. 3-22 can be simplified by using two zeners back to back as shown in Fig. 3-23. If the current is flowing up, then the bottom zener will drop 0.7 V and the top zener will drop its rated voltage. When the current is flowing down, the top zener will drop 0.7 V and the bottom zener its rated voltage. For example, if the circuit uses two 1N4733s (5.1-V devices), the total output swing will be limited to 5.1 + 0.7 = 5.8 V peak voltage or 11.6 V peak-to-peak.

Clamp or dc restorer

Diodes may also be used as *clamps* or *dc restorers.* Refer to Fig. 3-24. The signal source generates an ac waveform. The graph shows that the output signal that appears across the resistor is not ordinary alternating current. It does not have an average value of 0 V. It averages to some positive voltage. Such signals are common in electronic circuits and are said to have both an ac component and a dc com-

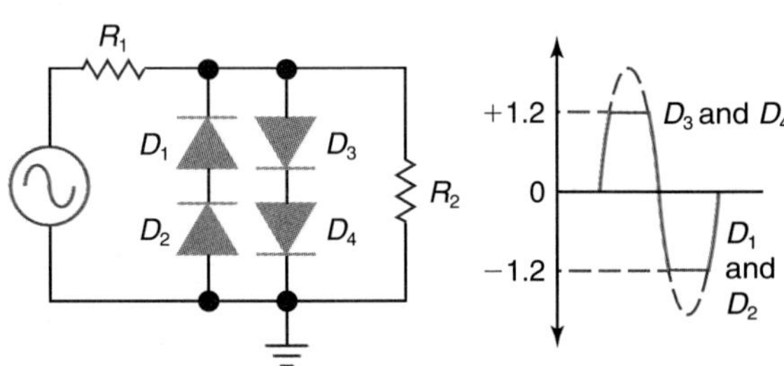

Fig. 3-21 Clipping at a higher threshold.

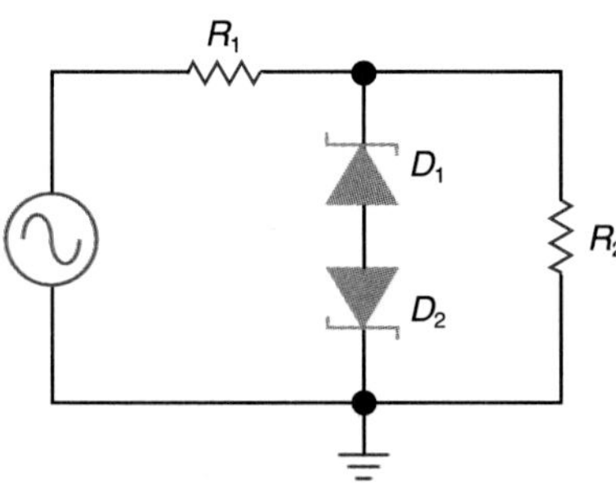

Fig. 3-23 A simplified high-threshold clipper.

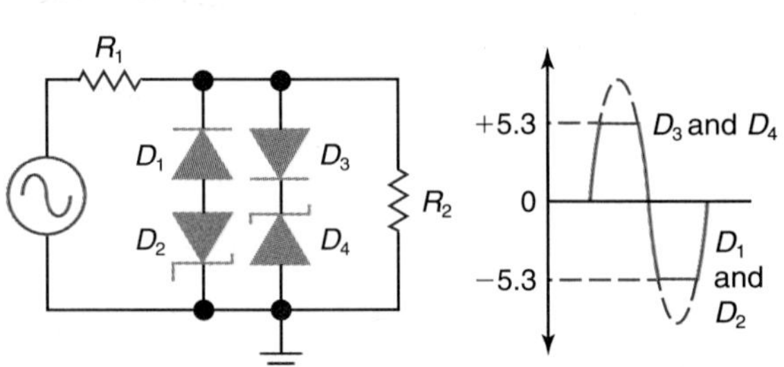

Fig. 3-22 Using zener diodes to set a higher clipping threshold.

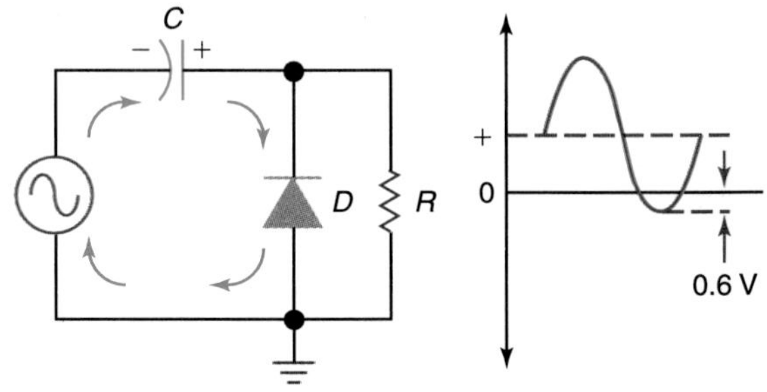

Fig. 3-24 Positive clamp.

ponent. Where does the dc component come from? The diode creates it by charging the capacitor. Note that diode D in Fig. 3-24 will allow a charging current to flow into the left side of capacitor C. This current places extra electrons on the left side of the capacitor and a negative charge results. Electrons flow off the right plate of the capacitor and make it positive. If the discharge time of the circuit ($T = R \times C$) is long compared to the period of the signal, the capacitor will maintain a steady charge from cycle to cycle.

Example 3-6

Evaluate the discharge time for Fig. 3-24 if the capacitor is 1 μF, the resistor is 10 kΩ, and the source develops 1 kHz. Find the RC time constant by:

$$\begin{aligned} T &= R \times C \\ &= 10 \times 10^3\ \Omega \times 1 \times 10^{-6}\ \text{F} \\ &= 0.01\ \text{s} \end{aligned}$$

Find the period of the signal:

$$t = \frac{1}{f} = \frac{1}{1 \times 10^3\ \text{Hz}} = 0.001\ \text{s}$$

The discharge time (T) is 10 times larger than the signal period (t).

Figure 3-25 is the equivalent circuit. It explains the clamp by showing that the charged capacitor acts as a battery in series with the ac signal source. The battery voltage V_{dc} accounts for the upward shift shown in the graph.

Refer again to Fig. 3-24. Note that the graph shows that the output signal goes 0.6 V below the zero axis. This -0.6 V point is when diode D turns on and conducts. The charging current flows briefly once every cycle when the signal source reaches its maximum negative voltage. Figure 3-26 shows what happens if the diode is reversed. The charging current is reversed, and the capacitor develops a negative voltage on its right plate. Notice that the graph shows that the output signal has a negative dc component. This circuit is called a *negative clamp.*

Clamping sometimes happens when we do not want it. For example, a signal generator is often used for circuit testing. Some signal generators use a coupling capacitor between their

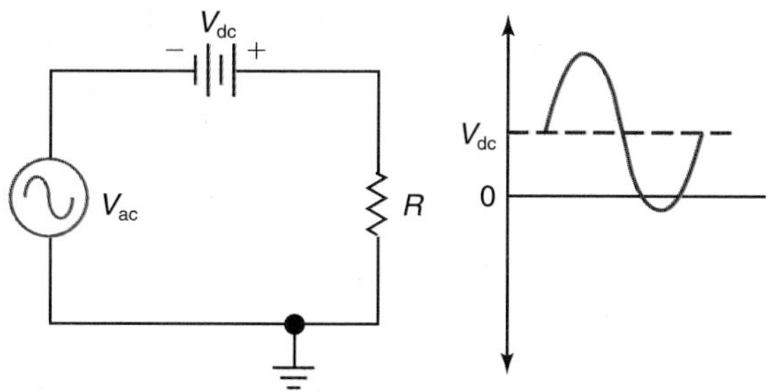

Fig. 3-25 Clamp equivalent circuit.

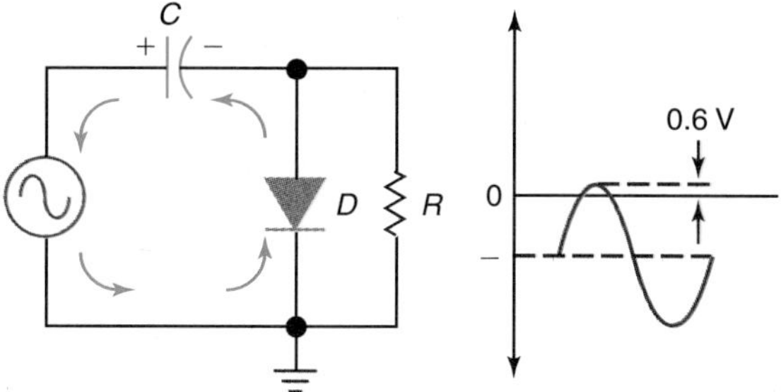

Fig. 3-26 Negative clamp.

output circuitry and their output jack. If you connect such a generator to an unbalanced diode load that allows a charge to build up on the built-in coupling capacitor, confusing results may occur. The resulting dc charge will act in series with the ac signal and may change the way the test circuit works. A dc voltmeter or a dc-coupled oscilloscope can be connected from ground to the output jack to verify that clamping is occurring.

Figure 3-27 shows how diodes are sometimes used to prevent arcing and component damage. When the current is suddenly interrupted in a coil, a large counterelectromotive force (CEMF) is generated across the coil. This high voltage can cause arcing and can also destroy sensitive devices such as integrated circuits and transistors. Note that in Fig. 3-27(a), there is an arc when the switch in series with the relay coil opens. In Fig. 3-27(b), there is a protection diode across the coil. This diode is forward-biased by the CEMF. The diode safely discharges the coil and prevents arcing or damage.

Another important diode type is the *light-emitting diode,* or LED. Its schematic symbol is shown in Fig. 3-28(a). Figure 3-28(b) shows that, as the electrons of the LED cross the junction, they combine with holes. This changes their status from one energy level to a lower

Light-emitting diode (LED)

Negative clamp

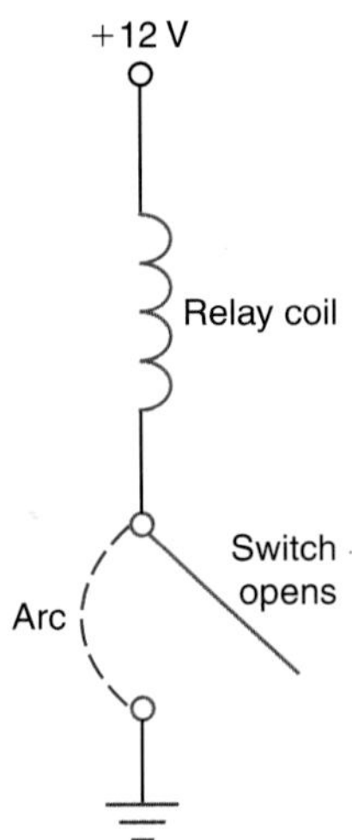

(*a*) Inductive kick causes an arc when switch opens

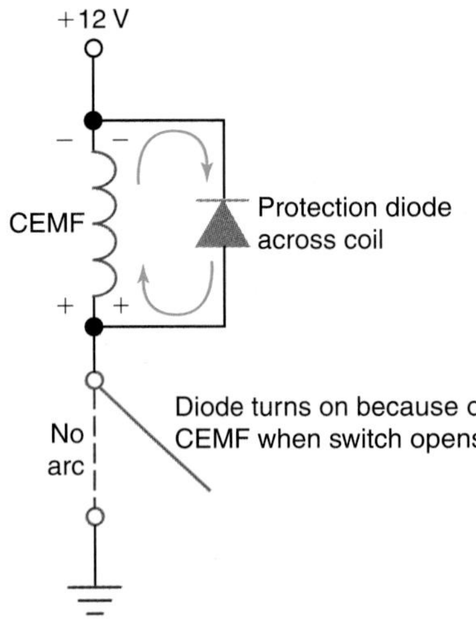

(*b*) No arc

Fig. 3-27 Using a diode to stop "inductive kick."

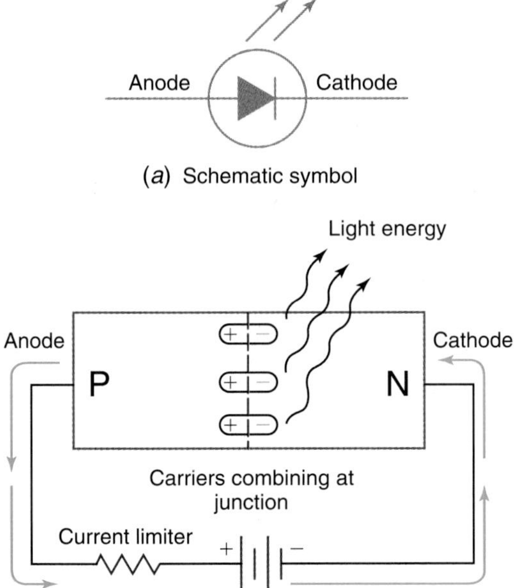

(*a*) Schematic symbol

(*b*) A simple LED circuit

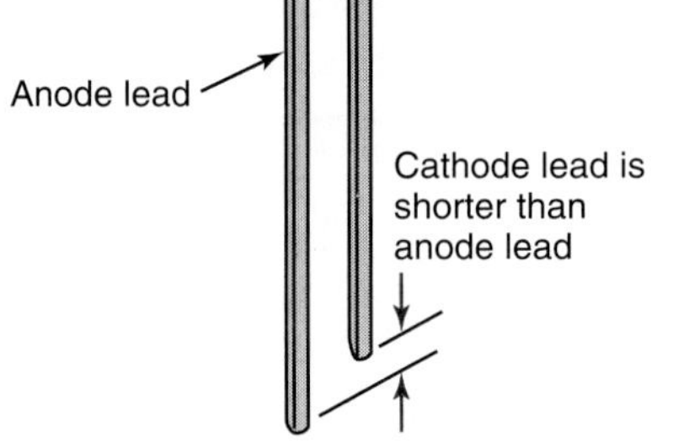

(*c*) Features of a T-1 3/4 plastic LED

Fig. 3-28 Light-emitting diode.

energy level. The extra energy they had as free electrons must be released. Silicon diodes give off this extra energy as heat. *Gallium arsenide diodes* release some of the energy as heat and some as infrared light. This type of diode is called an infrared-emitting diode (IRED). Infrared light is not visible to the human eye. By doping gallium arsenide with various materials, manufacturers can produce diodes with visible outputs of red, green, or yellow light.

The *laser diode* is an LED or IRED with carefully controlled physical dimensions that produce a resonant optical cavity. The resonant cavity provides feedback at one frequency to produce a strong, monochromatic (single-color) output. Laser diodes are used in applications such as fiber-optic communications, interferometry, alignment, and scanning systems.

The LEDs and IREDs have a higher forward voltage drop than do silicon diodes. This drop varies from 1.5 to 2.5 V depending on diode current, the diode type, and its color. If the manufacturer's data is not available, 2 V is a good starting point. Assume that the diode circuit in Fig. 3-28(*b*) is being designed for an LED current of 20 mA and that the supply (battery) produces 5 V. Ohm's law is used to find the value of the current-limiter resistor. The diode drop must be subtracted from the supply to find the voltage across the resistor:

$$R = \frac{V_S - V_D}{I_D} = \frac{5\text{ V} - 2\text{ V}}{20\text{ mA}} = 150\ \Omega$$

Example 3-7

Select a current-limiting resistor for an automotive circuit in which the diode current needs to be 15 mA. Such circuits use 12 V, and we can assume a 2-V diode drop:

$$R = \frac{12\text{ V} - 2\text{ V}}{15\text{ mA}} = 667\ \Omega$$

The power dissipation in a current-limiting resistor can also be important:

$$P = I^2R = (15\text{ mA})^2 \times 667\ \Omega = 150\text{ mW}$$

For better reliability, power dissipation is normally doubled. Since 300 mW is more than ¼ W, a ½-W resistor would be a good choice.

Figure 3-28(*c*) shows the physical appearance of a T-1¾ LED package. The T-1¾ package is 5 millimeters (mm) in diameter and is a common size. Another common size is the T-1 package which is 3 mm in diameter. The figure shows that the cathode lead is shorter than the anode lead and also that the flat side of the dome can be used to identify the cathode lead. As with other diode types, LEDs *must* be installed with the correct polarity.

Light-emitting diodes are rugged, small, and have a very long life. They can be switched rapidly since there is no thermal lag caused by gradual cooling or heating in a filament. They lend themselves to certain photochemical fabrication methods and can be made in various

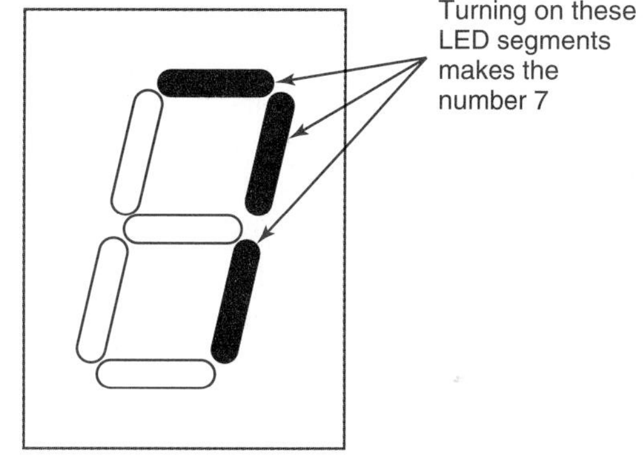

Fig. 3-29 An LED numeric display.

shapes and patterns. They are much more flexible than incandescent lamps. Light-emitting diodes may be used as numeric displays to indicate the numerals 0 through 9. A typical *seven-segment display* is shown in Fig. 3-29. By selecting the correct segments, the desired number is displayed.

Seven-segment display

Photodiodes are silicon devices sensitive to light input. They are normally operated in reverse bias. When light energy enters the depletion region, pairs of holes and electrons are generated and support the flow of current. Thus, a photodiode shows a very high reverse resistance with no light input and less reverse resistance with light input. Figure 3-30 shows an optocoupler circuit. An *optocoupler* is a package containing an LED or IRED and a photodiode or phototransistor. When S_1 is open, the LED is off and no light enters the photodiode. The resistance of the photodiode is high and the output signal will be high. When S_1 is closed, the LED is on. Light enters the photodiode so its resistance drops and the output signal drops to a lower level because of the voltage drop across R_2. Optocouplers are used to electrically isolate one circuit from another. They are also

Photodiode

Optocoupler

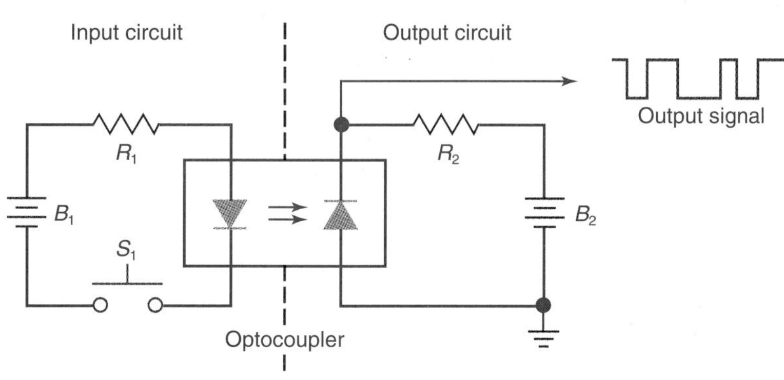

Fig. 3-30 An optocoupler circuit.

Optoisolator

called *optoisolators.* The only thing connecting the input circuit to the output circuit in Fig. 3-30 is light, so they are electrically isolated from each other.

Light-emitting diodes and photodiodes are often used in conjunction with fiber-optic cable for the purpose of data transmission. Compared with wire, fiber-optic cable is more expensive but has several advantages:

1. Elimination of electrical and magnetic field interference.
2. Greater data capacity for long runs.
3. Data security.
4. Safe in explosive environments.
5. Smaller and lighter.

A fiber-optic light source can also be a laser diode. These devices are LEDs that also act like optical cavities. An optical cavity behaves like a parallel *LC*-tuned circuit with very high *Q*. They are both highly responsive at their resonant frequency. A laser diode has both a reflective end and a partly reflective end. When the diode current is right, the light bounces back and forth between the ends and adds in phase. A single phase and frequency of light is released from the partly reflective end. When light is discussed, one phase is called *coherent* and one frequency is called *monochromatic.* Laser diodes are also used in CD-ROM and DVD players.

Allow yourself some extra time when going to job interviews. It is bad form to be late. Also, if you appear flustered and out of breath, you will not create a favorable impression.

Coherent

Monochromatic

Both LEDs and laser diodes can be rapidly pulsed to allow high-speed data transmission. At the other end, a light detector is needed to change the light back into electrical pulses. Photodiodes are used to accomplish this. Figure 3-31 shows that light from a diode enters one end of a cable and leaves the other end where it strikes another diode.

Fiber-optic cables

Figure 3-31 also shows the construction and types of fiber-optic cables. These cables are light pipes. The principle of operation is total internal reflection of light. When light strikes a transparent surface, it is divided into a reflected beam and a refracted (bent) beam. If a ray of light strikes at some angle less than a so-called critical angle, all the light is reflected. If a core material is cladded with a different material having a smaller refractive index, total reflection is achieved for those rays that strike the cladding at shallow angles. Most light cables use various blends of silica glass for the core and for the cladding.

The step-index multimode fiber shown in Fig. 3-31 uses a relatively large core. Thus, some of the light rays that make up a light pulse may travel a direct route, whereas others zig and zag as they bounce off the cladding. Different rays arrive at the detector diode at different times depending on different path lengths. The output pulse is spread in time. Look closely at the relationship between input and output pulses in Fig. 3-31. You can see that the pulse spreading in the multimode cable does not allow high-speed transmission. High-speed pulse transmission requires that the pulses be spaced very close together in time. As the pulses are spaced closer and closer, spreading makes it impossible to separate them into individual pulses. Multimode fibers are not used for long-distance, high-speed communication.

A graded-index multimode fiber is also shown in Fig. 3-31. This cable type suffers less output pulse spreading. Here, the refractive index of a smaller core changes gradually from the center out toward the cladding. Light traveling down the core curves rather than zigzags; this is due to the gradual change in the index. Also, the bent (curved) rays wind up arriving at the diode detector at about the same time as the direct rays because the direct rays must travel more slowly in the core's center.

The single-mode fiber also shown in Fig. 3-31 is capable of the highest speeds. Note that the light travels in a narrow core and only by a direct route. Pulse spreading is minimal, and high speeds can be used. The current speed limit for fiber-optic transmission is about 10 billion bits (pulses) per second. A trillion bits per second is expected to be reached within the next few years.

Fiber-optic cables used for data transmission typically carry light signals at levels of 100 microwatts (μW) or less. Eye damage is not possible at these levels. However, other applications may use much higher power levels. Never look into the end of a fiber-optic cable unless the power level has been verified as absolutely safe. Also, remember that some systems use infrared light. What you can't see can hurt you.

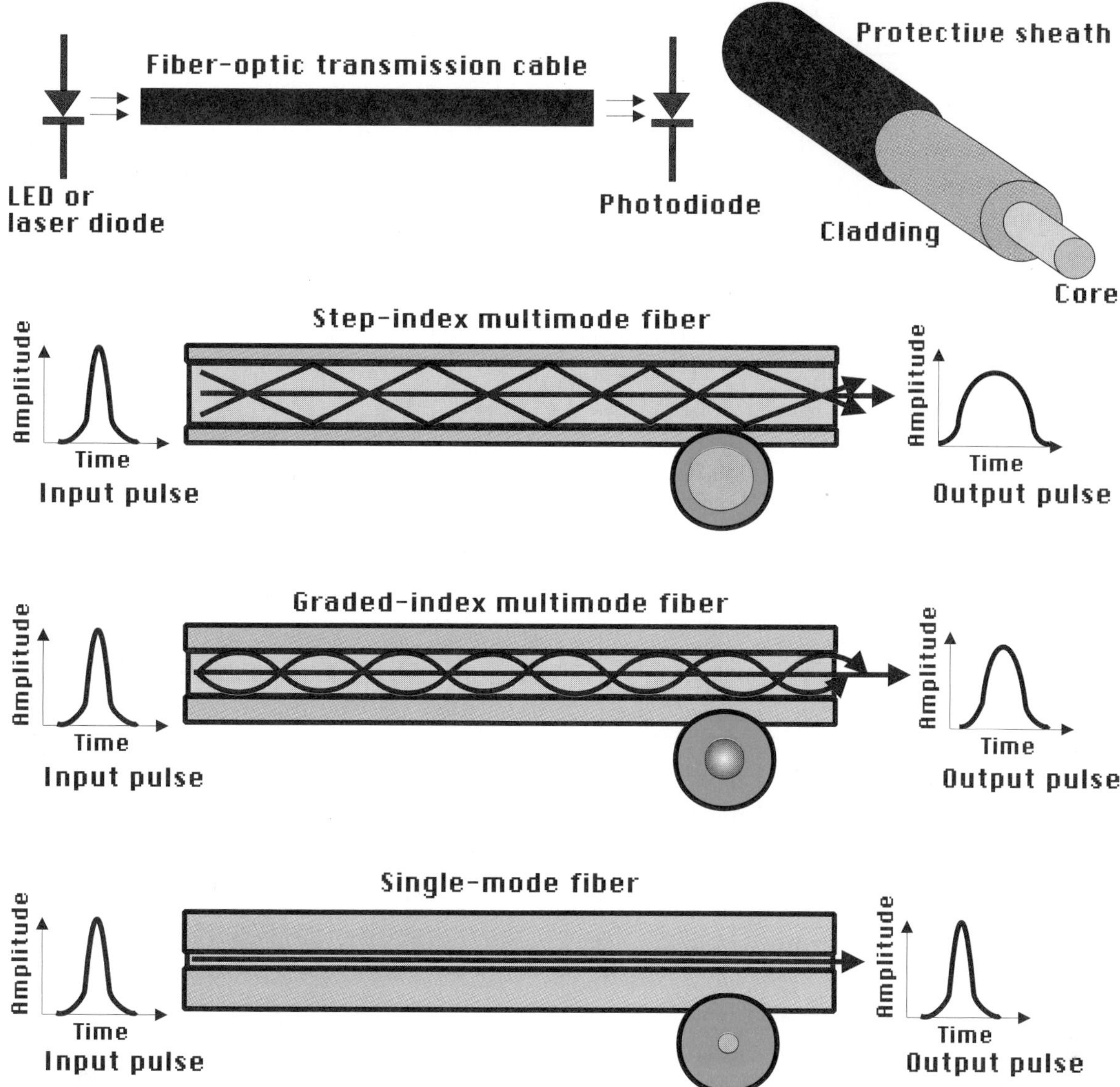

Fig. 3-31 Fiber-optic cables.

Varicap or Varactor

The *varicap* or *varactor* diode is a solid-state replacement for the variable capacitor. Much of the tuning and adjusting of electronic circuits involves changing capacitance. Variable capacitors are often large, delicate, and expensive parts. If the capacitor must be adjusted from the front panel of the equipment, a metal shaft or a complicated mechanical connection must be used. This causes some design problems. The varicap diode can be controlled by voltage. No control shaft or mechanical linkage is needed. The varicap diodes are small, rugged, and inexpensive. They are used instead of variable capacitors in modern electronic equipment.

The capacitor effect of a PN junction is shown in Fig. 3-32. A capacitor consists of two conducting plates separated by a dielectric material or insulator. Its capacitance depends on the area of the plates as well as on their separation. A reverse-biased diode has a similar electrical format. The P-type material semiconducts and forms one plate. The N-type material also semiconducts and forms the other plate. The depletion region is an insulator and forms the dielectric. By adjusting the reverse bias, the width of the depletion region, that is, the dielectric, is changed; and this changes the capacity of the diode. With a high reverse bias, the diode capacitance will be low because the depletion region widens. This is the same effect as moving the plates of a variable capacitor farther apart. With little reverse bias, the depletion region is narrow. This makes the diode capacitance increase.

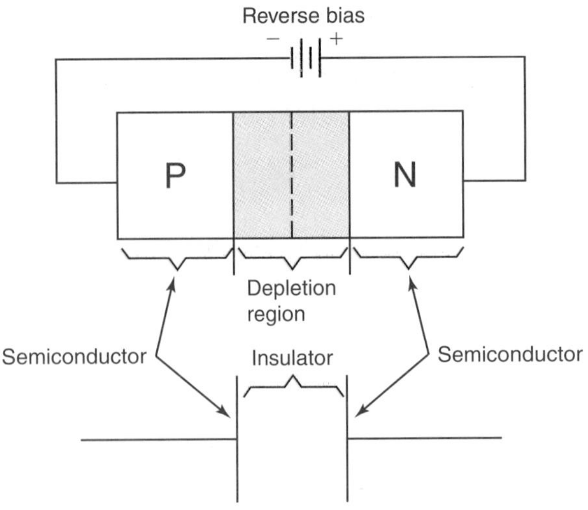

Fig. 3-32 Diode capacitance effect.

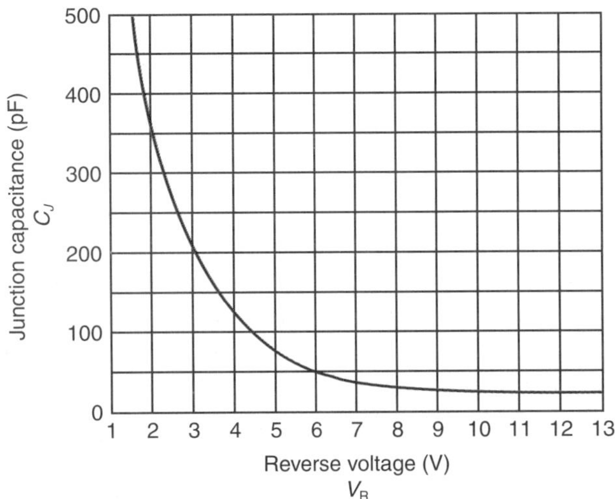

Fig. 3-33 Junction capacitance versus reverse voltage characteristic curve of a varicap diode.

Tuning diode

Figure 3-33 shows the capacitance in picofarads (pF) versus reverse bias for a varicap tuning diode. Capacitance decreases as reverse bias increases. The varicap diode can be used in a simple *LC* tuning circuit, as shown in Fig. 3-34. The tuned circuit is formed by an inductor (*L*) and two capacitors. The top capacitor C_2 is usually much higher in value than the bottom varicap diode capacitor C_1. This makes the resonant frequency of the tuned circuit mainly dependent on the inductor and the varicap capacitor.

YOU MAY RECALL . . . that when capacitors are in series, their total or equivalent capacitance is found with the product over sum formula:

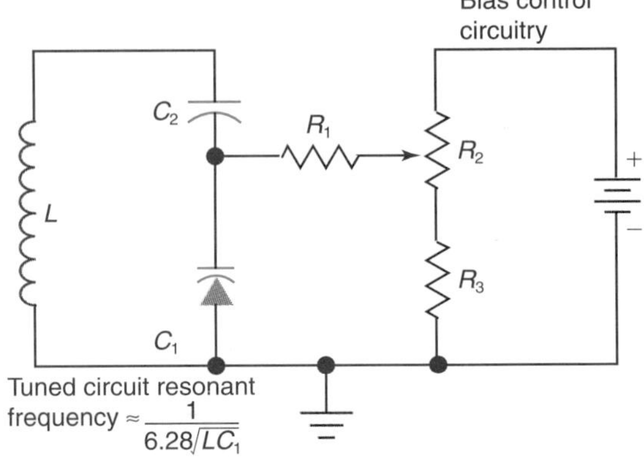

Fig. 3-34 Tuning with a varicap diode.

$$C_s = \frac{C_1 \times C_2}{C_1 + C_2}$$

Example 3-8

Calculate the equivalent series capacitance for Fig. 3-34 if C_2 is 0.005 μF and C_1 varies from 400 to 100 pF as the tuning voltage increases. First, convert 0.005 μF to picofarads:

$$0.005 \times 10^{-6} = 5000 \times 10^{-12}$$

Next, determine the series capacitance for $C_1 = 400$ pF:

$$C_s = \frac{400 \times 5000}{400 + 5000} = 370 \text{ pF}$$

Then, determine the series capacitance for $C_1 = 100$ pF:

$$C_s = \frac{100 \times 5000}{100 + 5000} = 98 \text{ pF}$$

In both cases, the series capacitance is close to the value of C_1 alone.

The series capacitance tunes the inductor in Fig. 3-34. This capacitance is determined by the bias control circuitry, so adjusting R_2 will change the resonant frequency of the *LC* tuning circuit.

. . . that the resonant frequency of an *LC* circuit may be determined with the formula:

$$f_r = \frac{1}{2\pi\sqrt{LC}}$$

Example 3-9

Find the frequency range for Fig. 3-34 for a varicap range of 100 to 400 pF if the coil is 1 μH. Assume that C_2 is large enough so that its value will not have a significant effect. Find the high frequency:

$$f_h = \frac{1}{6.28 \times \sqrt{100 \times 10^{-12} \times 1 \times 10^{-6}}}$$
$$= 15.9 \text{ MHz}$$

Find the low frequency:

$$f_l = \frac{1}{6.28 \times \sqrt{400 \times 10^{-12} \times 1 \times 10^{-6}}}$$
$$= 7.96 \text{ MHz}$$

Subtract to find the frequency range:

$$f_{\text{range}} = f_h - f_l = 15.9 \text{ MHz} - 7.96 \text{ MHz}$$
$$= 7.94 \text{ MHz}$$

Note that the *ratio* of the high frequency to the low frequency is 2:1 for a varicap capacitance range of 4 to 1. This is because frequency varies as the square root of capacitance.

Example 3-10

Find the frequency ratio for Fig. 3-34 if the varicap has a capacitance range of 10 to 1. The frequency ratio is equal to the square root of the capacitance range:

$$f_{\text{ratio}} = \sqrt{10} = 3.16$$

R_1 in Fig. 3-34 is a high value of resistance and isolates the tuned circuit from the bias-control circuit. This prevents the Q of the tuned circuit, that is, the sharpness of the resonance, from being lowered by resistive loading. High resistance gives light loading and better Q. Resistors R_2 and R_3 form the variable-bias divider. As the wiper arm on the resistor is moved up, the reverse bias across the diode will increase. This will decrease the capacitance of the varicap diode and raise the resonant frequency of the tuned circuit. You should inspect the resonant frequency formula and verify this trend. Without R_3, the diode bias could be reduced to zero. In a varicap tuning diode, zero bias is not usually acceptable. An ac signal in the tuned circuit could switch the diode into forward conduction. This would cause undesired effects. A circuit such as the one shown in Fig. 3-34 can be used for many tuning purposes in electronics.

Some diodes are built with an *intrinsic* layer between the P and the N regions. These are called *PIN diodes,* where the I denotes the intrinsic layer between the P material and the N material. The intrinsic layer is pure silicon (not doped). When a PIN diode is forward-biased, carriers are injected into the intrinsic region. Then, when the diode is reverse-biased, it takes a relatively long time to sweep these carriers out of the intrinsic region. This makes PIN diodes useless as high-frequency rectifiers.

PIN diode

The value of PIN diodes is that they can act as variable resistors for radio-frequency (RF) currents. Fig. 3-35 shows how the resistance of a typical PIN diode varies with the direct current flowing through it. As the direct current increases, the diode's resistance drops.

PIN diodes are also used for RF switching. They can be used to replace relays for faster, quieter, and more reliable operation. A typical situation that occurs in two-way radios is shown in Fig. 3-36. The transmitter and receiver share an antenna. The receiver must be isolated from the antenna when the transmitter is on or it may be damaged. This is accomplished by applying a positive voltage to the bias terminal in Fig. 3-36 which turns on both PIN diodes. Direct current will flow from ground, through D_2, through the coil, through D_1, and through the radio frequency choke (RFC) into the bias terminal. Both diodes will

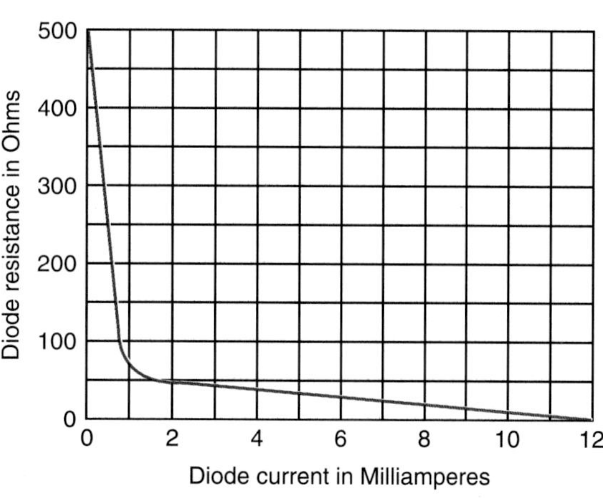

Fig. 3-35 PIN diode resistance versus current.

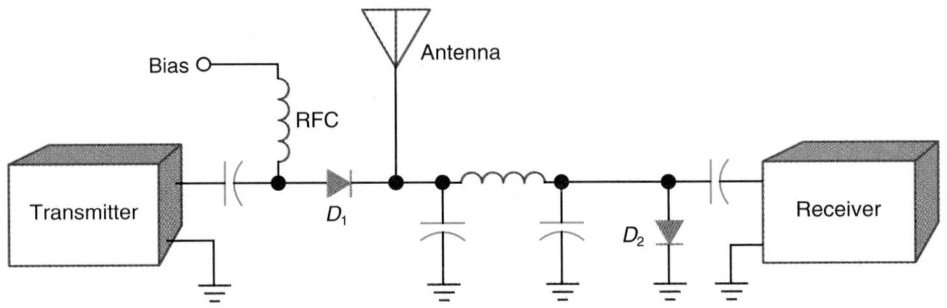

Fig. 3-36 PIN diode transmit-receive switching.

have a low resistance and the radio signal from the transmitter will pass through D_1 and on to the antenna with little loss. D_2 also has low resistance when transmitting and this prevents any significant RF voltage from appearing across the receiver input. The bias voltage is removed when receiving, and both diodes will then show a high resistance. The antenna is effectively disconnected from the transmitter by D_1.

Attenuation

In addition to switching, PIN diodes can also provide *attenuation* of RF signals. Figure 3-37 shows a PIN diode attenuator circuit. When the control point is at 0 V, signals pass through from input to output with little loss. This is because D_1 is forward-biased and in a low-resistance state. D_2 is now reverse-biased and it has almost no effect on the signal. The bias conditions can be determined by solving for the dc voltage drop across the 3000-Ω resistor. With the control point at 0 V, there is 12 V across the series circuit containing the 3000-Ω resistor, D_1, and the 2700-Ω resistor. Follow the blue arrows. The diode resistance is small enough to be ignored. The drop across the 3000-Ω resistor can be found with the voltage divider equation:

$$V = \frac{3000}{3000 + 2700} \times 12 \text{ V} = 6.32 \text{ V}$$

The voltage at the top end of the left-hand 51-Ω resistor is found by subtracting the 6.32-V drop from the 12-V supply:

$$V = 12 \text{ V} - 6.32 \text{ V} = 5.68 \text{ V}$$

Thus, the cathode of D_2 is at +6 V and the anode connects through a 51-Ω resistor to a voltage of 5.68 V. With the cathode more positive than the anode, D_2 is reverse-biased and has a very high resistance.

When the control voltage is changed to +6 V in Fig. 3-37, the situation is opposite. Follow the red arrows. D_2 is now on and D_1 is off. Little of the input signal can reach the output since D_2 is in a low-resistance state. The input signal dissipates in the left-hand 51-Ω resistor. This assumes that the cathode of D_2 is at RF ground (it is usually bypassed to ground with a capacitor that has low reactance at the signal frequency).

To prove that D_1 is off in Fig. 3-37 when the control is at 6 V, we will again use the voltage divider equation. The current is now through D_2, the 51-Ω resistor, and the 3000-Ω resistor

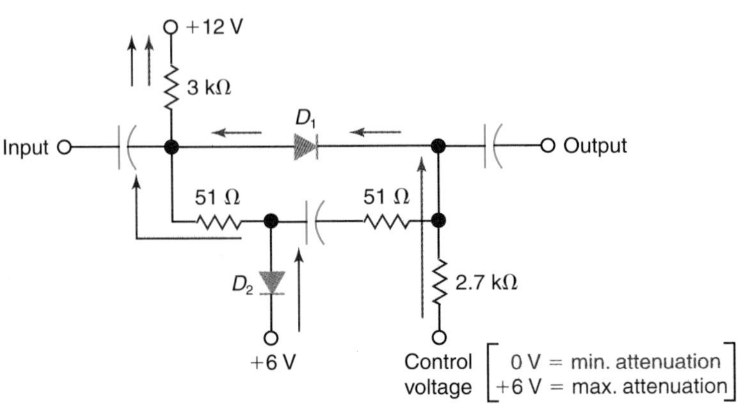

Fig. 3-37 PIN diode attenuator.

(look at the red arrows). The drop across the 3000-Ω resistor is found by

$$V = \frac{3000}{3000 + 51} \times (12\text{ V} - 6\text{ V}) = 5.9\text{ V}$$

The voltage at the anode end of D_1 is found by subtracting the drop from the 12-V supply:

$$V = 12\text{ V} - 5.9\text{ V} = 6.1\text{ V}$$

Thus, the anode end of D_1 is only 0.1 V positive with respect to the cathode end. This is not enough to forward-bias it, so D_1 is off and in a high-resistance state.

TEST

Determine whether each statement is true or false.

24. A rectifier is a device used to change alternating current to direct current.
25. Schottky diodes are used in low-voltage high-frequency applications.
26. A zener diode that is serving as a voltage regulator has electron flow from its anode to its cathode.
27. A normally operating rectifier diode will conduct from its anode to its cathode.
28. A diode clamp is used to limit the peak-to-peak swing of a signal.
29. A diode clamp may also be called a dc restorer.
30. A device containing an LED and a photodiode in the same sealed package is called an optoisolator.
31. Varactor diodes show large inductance change with changing bias.
32. The depletion region serves as the dielectric in a varicap diode capacitor.
33. Increasing the bias (reverse) across a varicap diode will increase its capacitance.
34. Decreasing the capacitance in a tuned circuit will raise its resonant frequency.
35. PIN diodes are used as high-frequency rectifiers.

Chapter 3 Review

Summary

1. One of the most basic and useful electronic components is the PN-junction diode.
2. When the diode is formed, a depletion region appears that acts as an insulator.
3. Forward bias forces the majority carriers to the junction and collapses the depletion region. The diode conducts. (Technically speaking, it semiconducts.)
4. Reverse bias widens the depletion region. The diode does not conduct.
5. Reverse bias forces the minority carriers to the junction. This causes a small leakage current to flow. It can usually be ignored.
6. Volt-ampere characteristic curves are used very often to describe the behavior of electronic devices.
7. The volt-ampere characteristic curve of a resistor is linear (a straight line).
8. The volt-ampere characteristic curve of a diode is nonlinear.
9. It takes about 0.2 V of forward bias to turn on a germanium diode, about 0.6 V for a silicon rectifier, and about 2 V for an LED.
10. A silicon diode will avalanche at some high value of reverse voltage.
11. Diode leads are identified as the cathode lead and the anode lead.
12. The anode must be made positive with respect to the cathode to make a diode conduct.
13. Manufacturers mark the cathode lead with a band, bevel, flange, or plus (+) sign.
14. If there is doubt, the ohmmeter test can identify the cathode lead. It will be connected to the negative terminal. A low resistance reading indicates that the negative terminal of the ohmmeter is connected to the cathode.
15. Caution should be used when applying the ohmmeter test. Some ohmmeters have reversed polarity. The voltage of some ohmmeters is too low to turn on a PN-junction diode. Some ohmmeters' voltages are too high and may damage delicate PN junctions.
16. A diode used to change alternating current to direct current is called a rectifier diode.
17. Schottky diodes do not have a depletion region and turn off much faster than silicon diodes.
18. A diode used to stabilize or regulate voltage is the zener diode.
19. Zener diodes conduct from anode to cathode when they are working as regulators. This is just the opposite from the way rectifier diodes conduct.
20. A diode clipper or limiter can be used to stabilize the peak-to-peak amplitude of a signal. It may also be used to change the shape of a signal or reduce its noise content.
21. Clamps or dc restorers add a dc component to an ac signal.
22. Light-emitting diodes are used as indicators, transmitters, and in optoisolators.
23. Varicap diodes are solid-state variable capacitors. They are operated under conditions of reverse bias.
24. Varicap diodes show minimum capacitance at maximum bias. They show maximum capacitance at minimum bias.
25. PIN diodes are used to switch radio-frequency signals and also to attenuate them.

Chapter Review Questions

Determine whether each statement is true or false.

3-1. A PN-junction diode is made by mechanically joining a P-type crystal to an N-type crystal.

3-2. The depletion region forms only on the P-type side of the PN junction in a solid-state diode.

3-3. The barrier potential prevents all the electrons on the N-type side from crossing the junction to fill all the holes in the P-type side.

3-4. The depletion region acts as an insulator.

3-5. Forward bias tends to collapse the depletion region.

3-6. Reverse bias drives the majority carriers toward the junction.
3-7. It takes 0.6 V of forward bias to collapse the depletion region and turn on a silicon solid-state diode.
3-8. A diode has a linear volt-ampere characteristic curve.
3-9. Excessive reverse bias across a rectifier diode may cause avalanche and damage it.
3-10. Silicon is a better conductor than germanium.
3-11. Less voltage is required to turn on a germanium diode than to turn on a silicon diode.
3-12. The behavior of electronic devices such as diodes changes with temperature.
3-13. The Celsius temperature scale is used in electronics.
3-14. Leakage current in a diode is from the cathode to the anode.
3-15. Forward current in a diode is from the cathode to the anode.
3-16. Diode manufacturers usually mark the package in some way so as to identify the cathode lead.
3-17. Making the diode anode negative with respect to the cathode will turn on the diode.
3-18. It is possible to test most diodes with an ohmmeter and identify the cathode lead.
3-19. Rectifier diodes are used in the same way as zener diodes.
3-20. Zener diodes are normally operated with the cathode positive with respect to the anode.
3-21. Two germanium diodes are connected as shown in Fig. 3-20. With a 10-V peak-to-peak input signal, the signal across R_2 would be 0.4 V peak-to-peak.
3-22. The function of D in Fig. 3-24 is to limit the output signal swing to no more than 0.6 V peak-to-peak.
3-23. Light-emitting diodes emit light by heating a tiny filament red hot.
3-24. The capacitance of a varicap diode is determined by the reverse bias across it.
3-25. Germanium diodes cost less and are therefore more popular than silicon diodes in modern circuitry.
3-26. Diode clippers are also called clamps.
3-27. As the wiper arm of R_2 in Fig. 3-34 is moved up, f_r will increase.

Chapter Review Problems

3-1. Refer to Fig. 3-5. The diode is silicon, the battery is 3 V, and the current-limiter resistor is 150 Ω. Find the current flow in the circuit. (*Hint:* Don't forget to subtract the diode's forward voltage drop.)
3-2. Refer to Fig. 3-11. Calculate the forward resistance of the diode at a temperature of 25°C and a forward current of 25 mA.
3-3. Refer again to Fig. 3-11. Calculate the forward resistance of the diode at a temperature of 25°C and a forward current of 200 mA.
3-4. Refer to Fig. 3-23. Both resistors are 10 kΩ, both zeners are rated at 3.9 V, and the input signal is 2 V peak-to-peak. Calculate the output signal. (*Hint:* Don't forget the voltage divider action of R_1 and R_2.)
3-5. Find the output signal for Fig. 3-23 for the same conditions as given in Prob. 3-4 but with an input signal of 20 V peak-to-peak.
3-6. What value of current-limiter resistor should be used in an LED circuit powered by 8 V if the desired LED current is 15 mA? You may assume an LED forward drop of 2 V.

Critical Thinking Questions

3-1. A nearly *ideal* diode would have, among other characteristics, a very small barrier potential (say a millivolt or so). What would be the advantage of such a tiny barrier potential?
3-2. Can you think of a way to use a diode to measure temperature?
3-3. High-power diodes can get very hot, and heat is a major factor in the failure of electronic devices. Does anything in this chapter suggest a possible solution?
3-4. Infrared remote control units are very popular in products such as television receivers and

VCRs. Can you describe a simple circuit, to be used in conjunction with an oscilloscope, that could help in diagnosing problems with remote control units?

3-5. Can you think of a reason why optocouplers are often used in medical electronics?

3-6. Why is the PIN diode transmit-receive circuit shown in Fig. 3-36 not useful for cellular telephones?

3-7. Can you identify two effects of adding a series rectifier to a string of decorative lights?

Answers to Tests

1. T
2. F
3. T
4. F
5. F
6. F
7. F
8. T
9. T
10. straight line
11. straight line
12. 10 mA (0.01 A)
13. horizontal
14. vertical
15. nonlinear
16. 0.6
17. bias (voltage)
18. cathode
19. cathode
20. cathode
21. shorted
22. on
23. nonlinear
24. T
25. T
26. T
27. F
28. F
29. T
30. T
31. F
32. T
33. F
34. T
35. F

Power Supplies

Chapter Objectives

This chapter will help you to:

1. *Identify* the common rectifier circuits and explain how they work.
2. *Recognize* various filter configurations and list their characteristics.
3. *Measure* and calculate power-supply ripple percentage and voltage regulation.
4. *Predict* and measure dc output voltage for filtered and unfiltered power supplies.
5. *Troubleshoot* common power-supply problems.
6. *Select* replacement parts for power supplies.

Electronic circuits need energy to work. In most cases, this energy is provided by a circuit called the power supply. A power supply failure will affect all of the other circuits. The supply is a key part of any electronic system. Power supplies use rectifier diodes to convert alternating current to direct current. They may also use zener diodes as voltage regulators. This chapter covers the circuits that use diodes in these ways. This chapter also discusses component-level troubleshooting. Knowing what each part of a circuit does and how the circuit functions allows technicians to find faulty components.

4-1 The Power-Supply System

The power supply changes the available electric energy (usually ac) to the form required by the various circuits within the system (usually dc). One of the early steps in the troubleshooting of any electronic system is to check the supply voltages at various stages in the circuitry.

Power supplies range from simple to complex depending on the requirements of the system. A simple power supply may be required to furnish 12 V dc. A more complicated power supply may provide several voltages, some positive and some negative with respect to the chassis ground. A supply that provides voltages at both polarities is called a *bipolar supply.* Some power supplies may have a wide output voltage tolerance. The output may vary ± 20 percent. Another power supply may have to keep its output voltage within ± 0.01 percent. Obviously, a strict tolerance complicates the design of the supply. Such supplies are covered in Chap. 15.

Figure 4-1 shows a block diagram for an electronic system. The power supply is a key part of the system since it energizes the other circuits. If a problem develops in the power supply, the fuse might "blow" (open). In that case, none of the voltages could be supplied to the other circuits. Another type of problem might involve the loss of only one of the outputs of the power supply. Suppose the +12-V dc output drops to zero because of a component failure in the power supply. Circuit *A* and circuit *B* would no longer work.

The second output of the power supply shown in Fig. 4-1 develops both positive and negative dc voltages with respect to the common point (usually the metal chassis). This

Bipolar supply

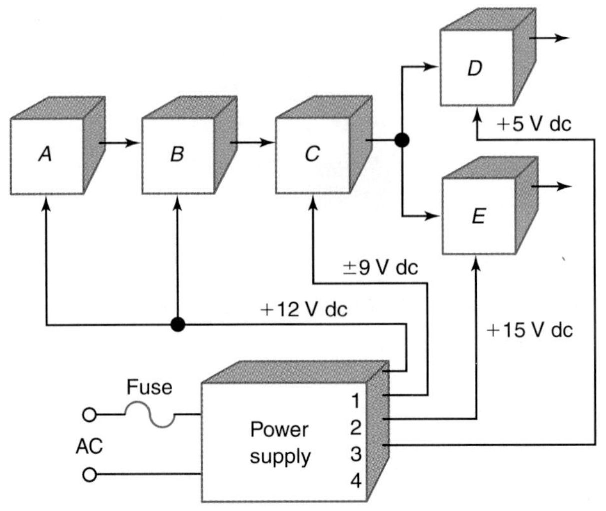

Fig. 4-1 Block diagram of an electronic system.

output could fail, too. It is also possible that only the negative output could fail. In either case, circuit C would not work normally under such conditions.

Troubleshooting electronic systems can be made much easier with block diagrams. If the symptoms indicate the failure of one of the blocks, then the technician will devote special attention to that part of the circuit. Since the power supply energizes most or all the other blocks, it is one of the first things to check when troubleshooting.

TEST

Supply the missing word in each statement.

1. Power supplies will usually change alternating current to ______.
2. Power-supply voltages are usually specified by using the chassis ______ as a reference.
3. Drawings such as Fig. 4-1 are called ______ diagrams.
4. On a block diagram, the circuit that energizes most or all the other blocks is called the ______

4-2 Rectification

Rectification

Most electronic circuits need direct current. Alternating current is supplied by the power companies. The purpose of the power supply is to change alternating current to direct current by *rectification.* Alternating current flows in both directions, and direct current flows in only one direction. Since diodes conduct in only one direction, they serve as rectifiers.

The ac supply available at ordinary wall outlets is 120 V, 60 hertz (Hz). Electronic circuits often require lower voltages. Transformers can be used to step down the voltage to the level needed. Figure 4-2 shows a simple power supply using a step-down transformer and a diode rectifier.

The load for the power supply of Fig. 4-2 could be an electronic circuit, a battery being charged, or some other device. In this chapter, the loads will be shown as resistors designated R_L.

The transformer in Fig. 4-2 has a voltage ratio of 10:1. With 120 V across the primary, 12 V ac is developed across the secondary. If it were not for the diode, there would be 12 V ac across the load resistor. The diode allows current flow only from its cathode to its anode. The diode is in series with the load. Current is the same everywhere in a series circuit, so the diode current and the load current are the same. Since the load current is flowing in only one

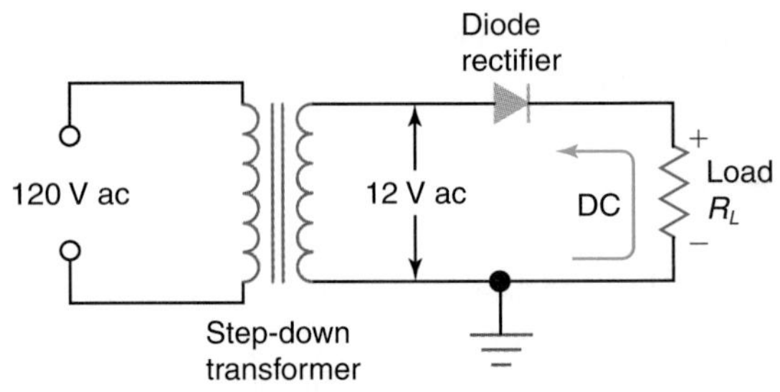

Fig. 4-2 A simple dc power supply.

direction, it is *direct current.* When direct current flows through a load, a dc voltage appears across the load.

Example 4-1

What will the secondary voltage be for Fig. 4-2 if the transformer ratio is 2 : 1? A voltage ratio of 2 : 1 means that we can divide the primary voltage by 2 to find the answer:

$$V_{\text{secondary}} = \frac{V_{\text{primary}}}{2} = \frac{120\text{ V}}{2} = 60\text{ V}$$

Note the polarity across the load in Fig. 4-2. Electrons move from negative to positive through a load. The *positive* end of the load is connected to the *cathode* end of the rectifier. In all rectifier circuits, the positive end of the load will be that end which contacts the cathode of the rectifier. It can also be stated that the *negative* end of the load will be in contact with the *anode* of the rectifier. Figure 4-3 illustrates this point. Compare Fig. 4-2 and Fig. 4-3. Note that the diode polarity determines the load polarity.

In Chap. 3 it was stated that to forward-bias a diode, the anode must be made positive with respect to the cathode. It was also noted that diode manufacturers used to mark the cathode with a plus (+) sign. When the diode acts as a rectifier, the function of the plus sign becomes clear. The plus sign is placed on the cathode end to show the technician which end of the load will be positive. Look again at Fig. 4-2 and verify this.

Figure 4-4 (*a*) shows the input waveform to the rectifier circuits of Figs. 4-2 and 4-3. Two complete cycles are shown. In Fig. 4-4(*b*), the waveform that appears across the load resistor of Fig. 4-2 is shown. The negative half of the cycle is missing since the diode blocks it. This waveform is called *half-wave pulsating direct current.* It represents only the positive half of the ac input to the rectifier.

Half-wave pulsating direct current

In Fig. 4-3 the diode has been reversed. This causes the positive half of the cycle to be blocked [Fig. 4-4(*c*)]. The waveform is also half-wave pulsating direct current. Both circuits, Figs. 4-2 and 4-3, are classified as *half-wave rectifiers.*

Half-wave rectifiers

The ground reference point determines which way the waveform will be shown for a rectifier circuit. For example, in Fig. 4-3 the positive end of the load is grounded. If an oscilloscope is connected across the load, the ground lead of the oscilloscope will be positive

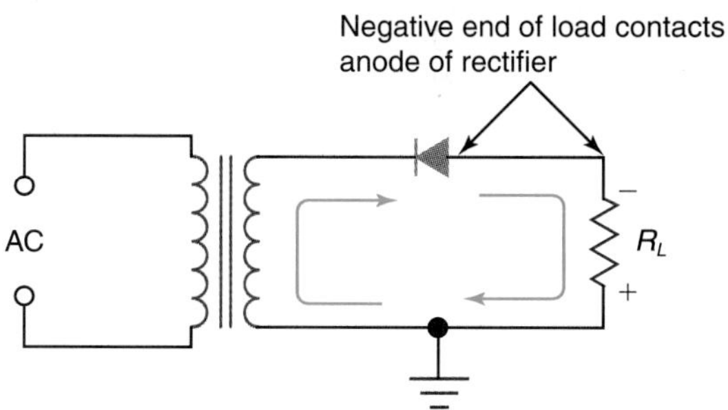

Fig. 4-3 Establishing the polarity in a rectifier circuit.

AC input to rectifier

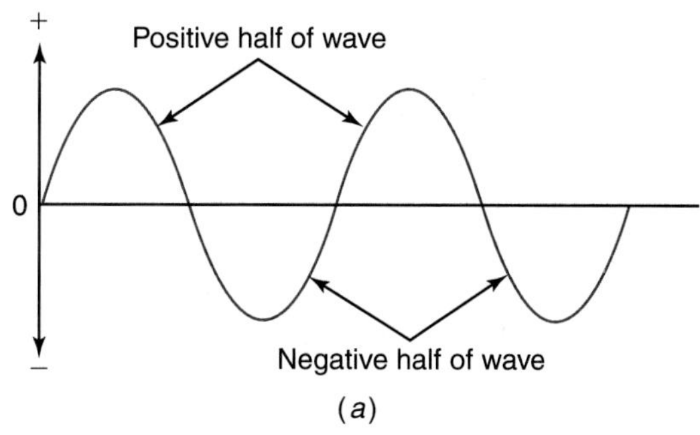

(*a*)

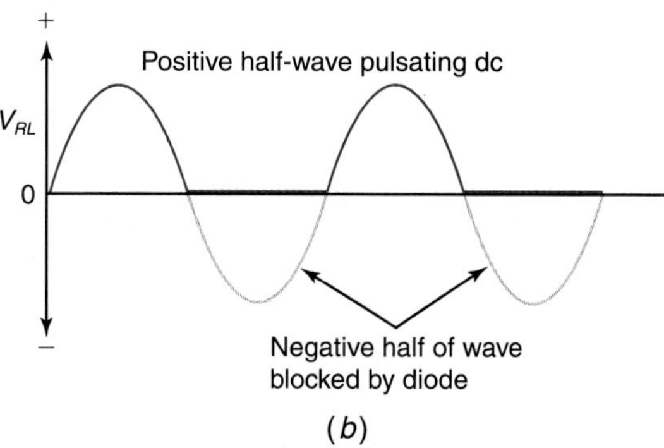

(*b*)

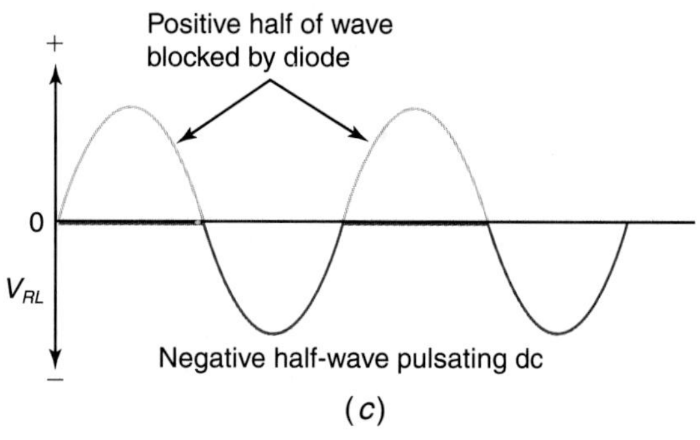

(*c*)

Fig. 4-4 Rectifier circuit waveforms.

Full-wave pulsating direct current

and the probe tip will be negative. Oscilloscopes ordinarily show positive as "up" and negative as "down" on the screen. The actual waveform will appear as that shown in Fig. 4-4(*c*). Waveforms can appear up or down depending on circuit polarity, instrument polarity, and the connection between the instrument and the circuit.

JOB TIP

Most jobs require teamwork. You should work on your interpersonal skills and learn to accept criticism.

Half-wave rectifiers are usually limited to low-power applications. They take useful output from the ac source for only half the input cycle. They are not supplying any load current half the time. This limits the amount of electric energy they can deliver over a given period of time. High power means delivering large amounts of energy in a given time. A half-wave rectifier is a poor choice in high-power applications.

TEST

Determine whether each statement is true or false.

5. Current that flows in both directions is called alternating current.
6. Current that flows in one direction is called direct current.
7. Diodes are used as rectifiers because they conduct in two directions.
8. A rectifier can be used in a power supply to step up voltage.
9. In a rectifier circuit, the positive end of the load will be connected to the cathode of the rectifier.
10. The waveform across the load in a half-wave rectifier circuit is called half-wave pulsating direct current.
11. A half-wave rectifier supplies load current only 50 percent of the time.
12. Half-wave rectifiers are usually used in high-power applications.

4-3 Full-Wave Rectification

Full-wave rectifier

Bridge rectifier

A *full-wave rectifier* is shown in Fig. 4-5(*a*). It uses a center-tapped transformer secondary and two diodes. The transformer center tap is located at the electrical center of the secondary winding. If, for example, the entire secondary winding has 100 turns, then the center tap will be located at the 50th turn. The waveform across the load in Fig. 4-5(*a*) is *full-wave pulsating direct current* with half the peak voltage of the secondary because of the center tap. Both alternations of the ac input are used to energize the load. Thus, a full-wave rectifier can deliver twice the power of a half-wave rectifier.

The ac input cycle is divided into two parts: a *positive alternation* and a *negative alternation.* The positive alternation is shown in Fig. 4-5(*b*). The induced polarity at the secondary is such that D_1 is turned on. Electrons leave the center tap, flow through the load, through D_1, and back into the top of the secondary. Note that the positive end of the load resistor is in contact with the cathode of D_1.

On the negative alternation, the polarity across the secondary is reversed. This is shown in Fig. 4-5(*c*). Electrons leave the center tap, flow through the load, through D_2, and back into the bottom of the secondary. The load current is the same for both alternations: It flows up through the resistor. Since the direction never changes, the load current is *direct current.*

Full-wave rectifiers can be constructed using two separate diodes or by using a package that contains two diodes. An example is shown in Fig. 4-5(*d*).

Figure 4-6 shows a full-wave rectifier with the diodes reversed. This reverses the polarity across the load resistor. Note that the output waveform shows both alternations going in a negative direction. This is what would be seen on an oscilloscope since the output is negative with respect to ground. The diode rule regarding polarity holds true in Fig. 4-6. The negative end of the load is in contact with the anodes of the rectifiers.

Full-wave rectifiers have one slight disadvantage: The transformer must be center-tapped. This may not always be possible. In fact, there are occasions when the use of any transformer is not desirable because of size, weight, or cost restrictions. Figure 4-7(*a*) on page 60 shows a rectifier circuit that gives full-wave performance without the transformer. It is called a *bridge rectifier.* It uses four diodes to give full-wave rectification.

Figure 4-7(*b*) traces the circuit action for the positive alternation of the ac input. The current moves through D_2, through the load, through D_1, and back to the source. The negative alternation is shown in Fig. 4-7(*c*). The current is always moving from left to right through the

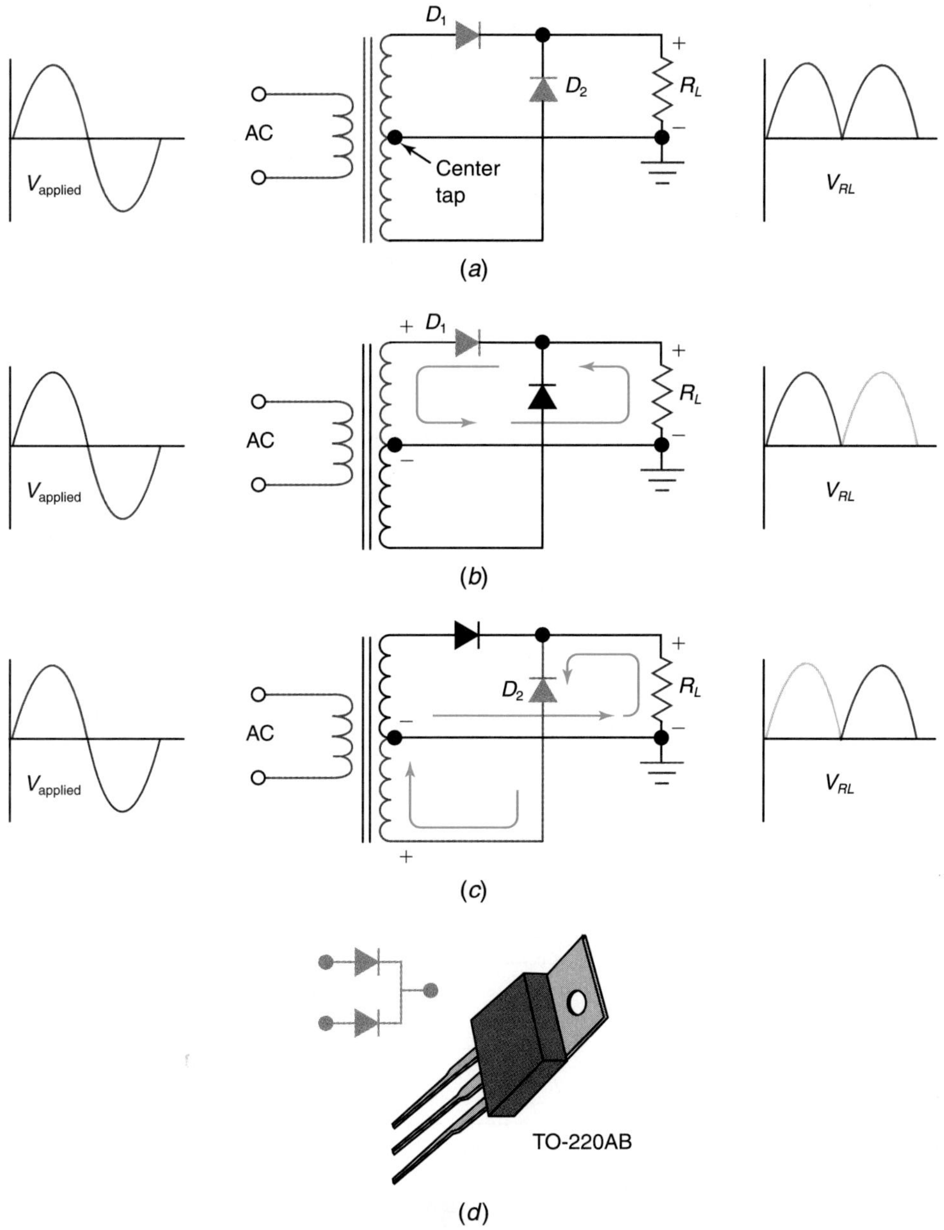

Fig. 4-5 Full-wave rectifiers.

load. Again, the positive end of the load is in contact with the rectifier cathodes. This circuit could be arranged for either ground polarity simply by choosing the left or the right end of the load as the common point.

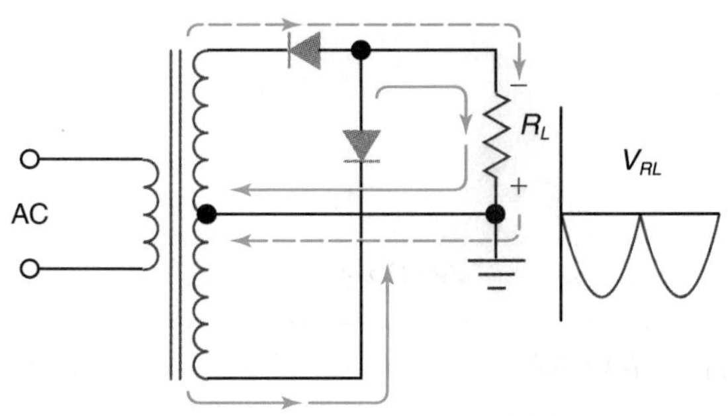

Fig. 4-6 Reversing the rectifier diodes.

A bridge rectifier requires four separate diodes, or a special rectifier package that contains four diodes connected in the bridge configuration. Figure 4-7(*d*) shows three examples of packaged bridge rectifiers.

TEST

Supply the missing word in each statement.

13. A transformer secondary is center-tapped. If 50 V is developed across the entire secondary, the voltage from either end to the center tap will be 25 ✓.

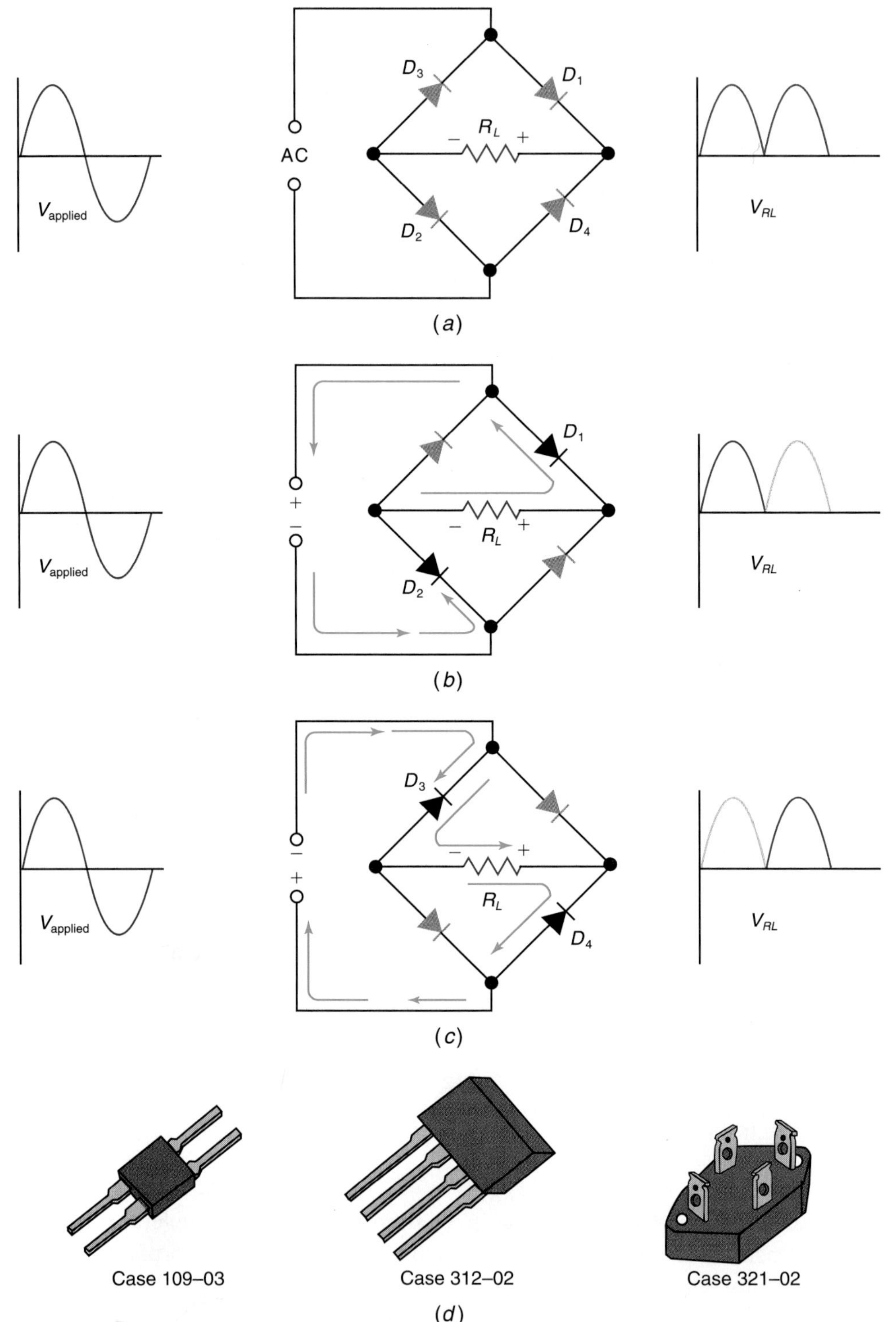

Fig. 4-7 Bridge rectifier circuit and case styles.

14. A half-wave rectifier uses ________ diode(s).
15. A full-wave rectifier using a center-tapped transformer requires ________ diodes.
16. Each cycle of the ac input has two ________.
17. In rectifier circuits, the load current never changes ________.
18. A bridge rectifier eliminates the need for a ________.
19. A bridge rectifier requires ________ diodes.

4-4 Conversion of RMS Values to Average Values

There is a significant difference between *pure direct current* and *pulsating direct current* (rectified alternating current). Meter readings taken in rectifier circuits can be confusing if the difference is not understood. Figure 4-8 compares a pure dc waveform with a pulsating dc waveform. A meter used to make measurements in a pure dc circuit can respond to the steady value of the direct current. In the case of pulsating dc, the meter will try to follow the pulsating waveform. At one instant in time, the meter tries to read zero. At another instant in time, the meter tries to read the peak value. Meter movements cannot react to the rapid changes because of the *damping* in their mechanism. Damping in a meter limits the speed with which the pointer can change position. The meter settles on the *average value* of the waveform.

Average value

Digital meters do not have damping, but they produce the same results. The display is not updated often enough to follow the pulsating dc waveform. If it did rapidly follow the waveform, the display would be a useless blur of constantly changing numbers. For this reason, digital meters also display the average value of a pulsating dc waveform.

Alternating current supply voltages are typically specified by their *root-mean-square (rms) values.* It would be convenient to have a way of converting rms values to average values when working with rectifier circuits.

RMS value

. . . that sinusoidal alternating current can be measured in several ways and that it is pos-

Damping

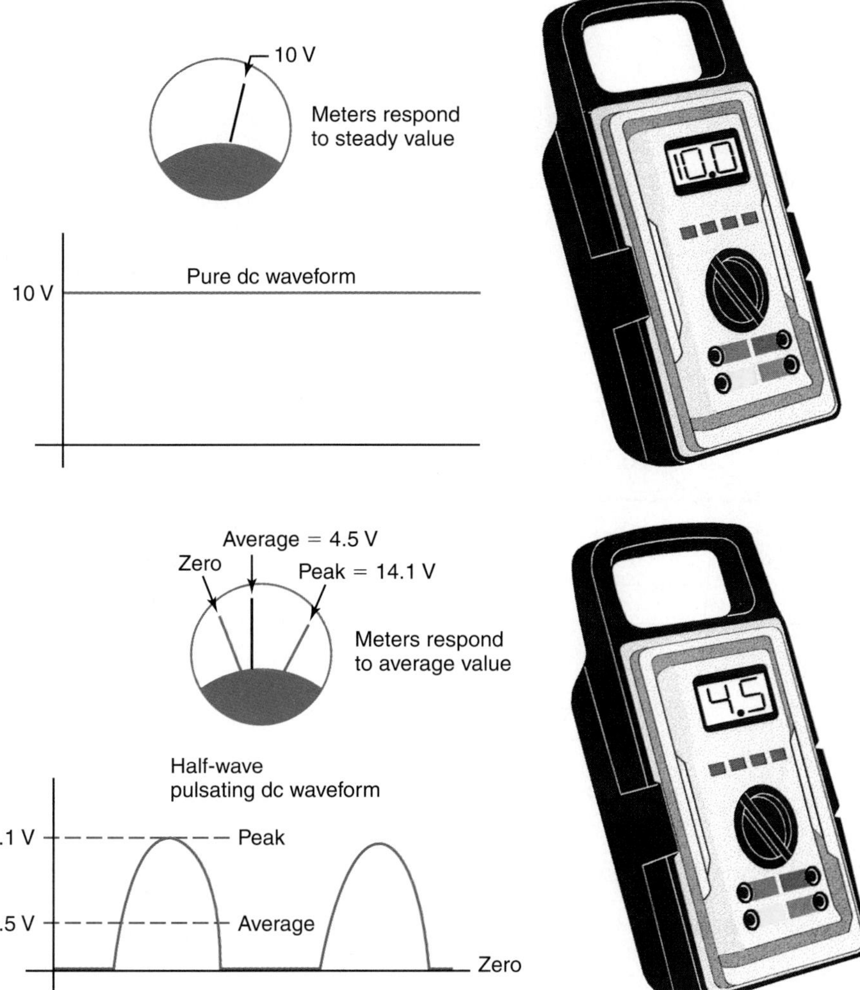

Fig. 4-8 Meters measure steady or average values.

About Electronics

Practical Power Supply for Portables "Wall transformer" power supplies are a good choice for products like notebook computers because they reduce product weight and size.

sible to convert from one to another. Figure 4-9 shows some measurements and conversion factors. If you have access to a calculator, it might be easier to calculate the conversion factors than to remember them:

$$0.707 = \frac{1}{\sqrt{2}} \text{ (to go from peak to rms values)}$$

$$0.637 = \frac{2}{\pi} \text{ (to go from peak to average values)}$$

As another aid, remember that rms means *root mean square* and you will know which one to use when converting peak to rms.

Example 4-2

Suppose the peak-to-peak value of the sine wave shown in Fig. 4-9 is 340 V. Find the average and the rms values. First, divide by 2 to find the peak value:

$$V_p = \frac{V_{p\text{-}p}}{2} = \frac{340 \text{ V}}{2} = 170 \text{ V}$$

$$V_{av} = V_p \times 0.637 = 170 \text{ V} \times 0.637 = 108 \text{ V}$$

$$V_{rms} = V_p \times 0.707 = 170 \text{ V} \times 0.707 = 120 \text{ V}$$

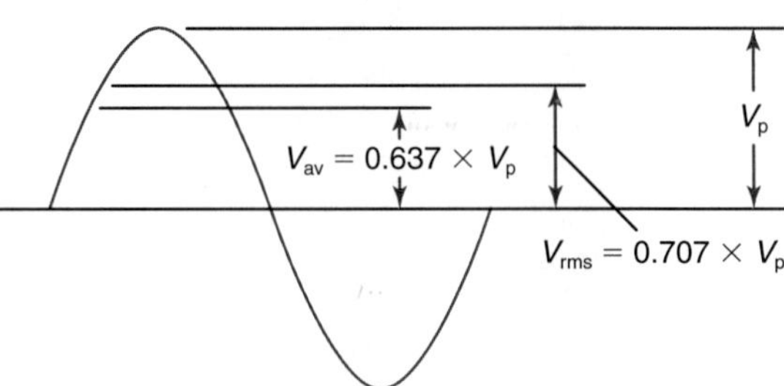

Alternating current

Fig. 4-9 Measuring sinusoidal alternating current.

A little algebra can be used to relate rms values to average values:

$$V_{av} = 0.637 \times V_p$$

$$V_{rms} = 0.707 \times V_p$$

Rearranging the second equation gives

$$V_p = \frac{V_{rms}}{0.707}$$

Substituting the right-hand side into the first equation gives

$$V_{av} = 0.637 \times \frac{V_{rms}}{0.707} = 0.9 \times V_{rms}$$

Thus, the *average* value of a rectified sine wave is 0.9 or 90 percent of the rms value. This means that a dc voltmeter connected across the output of a rectifier should indicate 90 percent of the rms voltage input to the rectifier. This would be true for the entire waveform. But, as Fig. 4-10 shows, the average value must be less for half the waveform. Half-wave pulsating direct current has one-half of the average value compared to full-wave pulsating direct current. So, for a half-wave rectifier, the average value of the waveform is 0.9/2 = 0.45 or 45 percent of the rms value.

Example 4-3

What should the dc voltmeter shown in Fig. 4-11 read? Taking the step-down action of the transformer into account first:

$$V_{secondary} = \frac{120}{10} = 12 \text{ V}$$

Next, note that Fig. 4-11 shows a half-wave rectifier. The appropriate conversion factor is 0.45:

$$V_{av} = V_{rms} \times 0.45$$
$$= 12 \times 0.45$$
$$= 5.4 \text{ V}$$

The meter should read 5.4 V.

If the circuit in Fig. 4-11 were constructed, how close could we expect the actual reading to be? The actual reading would be influenced by several factors: (1) the actual line voltage, (2) transformer winding tolerance, (3) meter accuracy, (4) rectifier loss, and (5) transformer losses. The actual line voltage and the actual transformer secondary voltage can be accounted for by accurate measurements. The

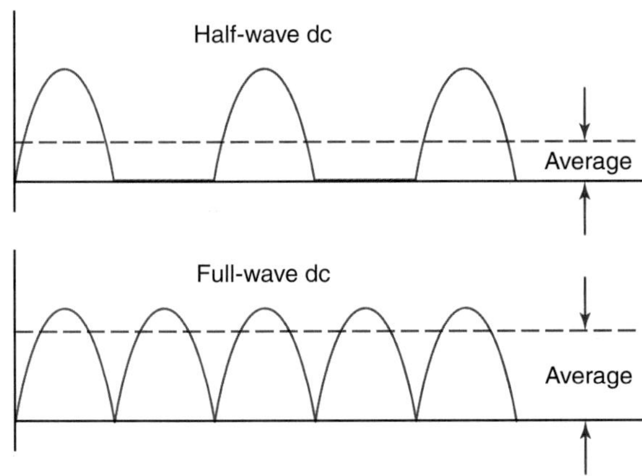

Fig. 4-10 Comparing half-wave and full-wave direct current.

meter accuracy can be high with a quality meter that has been checked against a standard. The rectifier loss is caused by the 0.6-V forward drop needed for conduction in a silicon diode. At high current levels, the drop will be greater. For example, if the rectifier current is several amperes, the diode loss will be close to 1 V. Transformer losses also increase at high current levels. Thus, the actual readings can be expected to be a little on the low side, especially at high-load current levels.

Example 4-4

What should the dc voltmeter shown in Fig. 4-12 read? Taking the step-down action of the transformer into account first:

$$V_{\text{secondary}} = \frac{120}{2} = 60\text{ V}$$

Since Fig. 4-12 shows a full-wave circuit, the appropriate conversion factor is 0.9. However, we *must* take into account that only *half* of the secondary is conducting at any given time. Please review Fig. 4-5 if you are confused.

$$V_{\text{av}} = \frac{V_{\text{secondary}}}{2} \times 0.9$$
$$= \frac{60}{2} \times 0.9 = 27\text{ V}$$

The meter should read 27 V. If the load demands a high current, then the actual voltage will be less. What would happen if one of the diodes should "burn out" (open)? This would

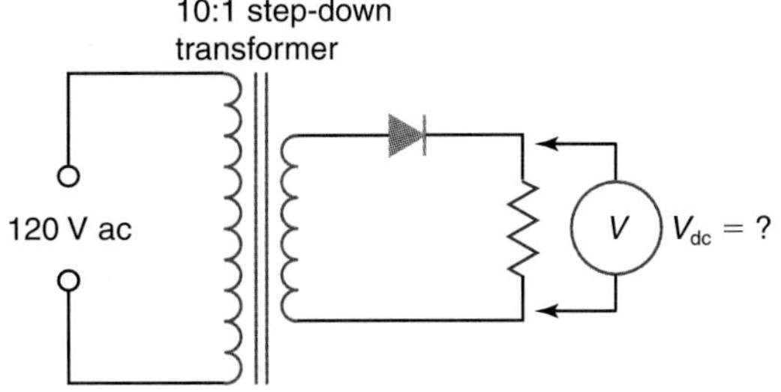

Fig. 4-11 Calculating dc voltage in a half-wave rectifier circuit.

change the circuit from *full-wave* to *half-wave.* The dc voltmeter could then be expected to read

$$V_{\text{av}} = V_{\text{rms}} \times 0.45$$
$$= 30 \times 0.45$$
$$= 13.5\text{ V}$$

Example 4-5

Calculate the average dc voltage for Fig. 4-12 assuming that a bridge rectifier will be used with the entire secondary (in other words, the center tap will not be connected).

$$V_{\text{av}} = V_{\text{secondary}} \times 0.9 = 60\text{ V} \times 0.9 = 54\text{ V}$$

Example 4-6

What should the dc voltmeter shown in Fig. 4-13 on page 64 read? Taking the step-down action of the transformer into account first:

$$V_{\text{secondary}} = \frac{120}{4} = 30\text{ V}$$

Since Fig. 4-13 shows a full-wave bridge circuit, the appropriate conversion factor is 0.9:

$$V_{\text{av}} = 30 \times 0.9 = 27\text{ V}$$

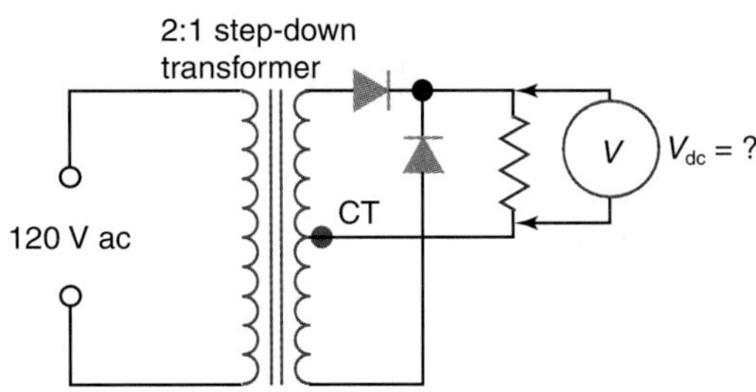

Fig. 4-12 Calculating dc voltage in a full-wave rectifier circuit.

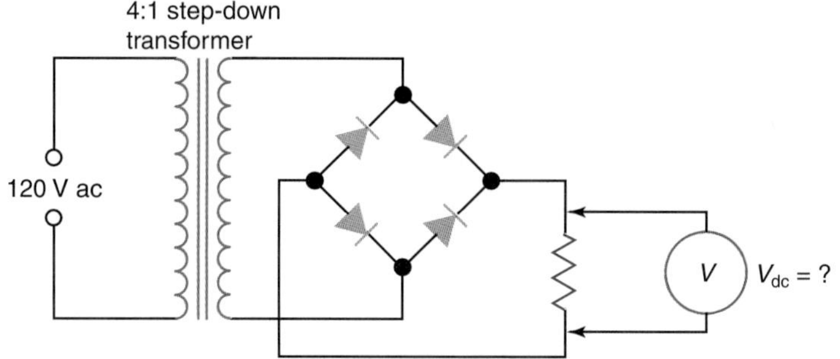

Fig. 4-13 Calculating dc voltage in a bridge rectifier circuit.

The diode loss in a bridge rectifier is *twice* that of the other circuits. A review of Fig. 4-7 will show that *two* diodes are always conducting in *series.* The 0.6-V drop will be doubled to 1.2 V. In low-voltage rectifier circuits, this can be significant. If the current demand is high, each diode may drop about 1 V, giving a total loss of 2 V. For the purposes of this chapter, you may ignore diode loss when performing calculations for dc output voltage.

TEST

Supply the missing word or number in each statement

20. A transformer has five times as many primary turns as secondary turns. If 120 V ac is across the primary, the secondary voltage should be 24 V AC.
21. Suppose the transformer in question 20 is center-tapped and connected to a full-wave rectifier. The average dc voltage across the load should be 10.8 V DC.
22. The average dc load voltage for the data in question 21 will change to 5.4 V DC if one of the rectifiers burns out (develops an open).
23. The ac input to a half-wave rectifier is 32 V. A dc voltmeter connected across the load should read 14.4 DC.
24. The ac input to a bridge rectifier is 20 V. A dc voltmeter connected across the load should read 18 V DC.
25. In rectifier circuits, one can expect the output voltage to drop as load current INCREASES.
26. Rectifier loss is more significant in LOW voltage rectifier circuits.
27. If each diode in a high-current bridge rectifier drops 1 V, then the total rectifier loss is 2 V.

4-5 Filters

Pulsating direct current is not directly usable in most electronic circuits. Something closer to pure direct current is required. Batteries produce pure direct current. Battery operation is usually limited to low-power and portable types of equipment. Figure 4-14 shows a battery connected to a load resistor. The voltage waveform across the load resistor is a straight line. There are no pulsations.

AC component

Pulsating direct current is not pure because it contains an *ac component.* Figure 4-15 shows how both direct current and alternating current can appear across one load. An ac generator and a battery are series-connected. The voltage waveform across the load shows both ac and dc content. This situation is similar to the output of a rectifier. There is dc output because of the rectification and there is also an ac component (the pulsations).

Ripple

Filter

The ac component in a dc power supply is called *ripple.* Much of the ripple must be removed for most applications. The circuit used to remove the ripple is called a *filter.* Filters can produce a very smooth waveform that will

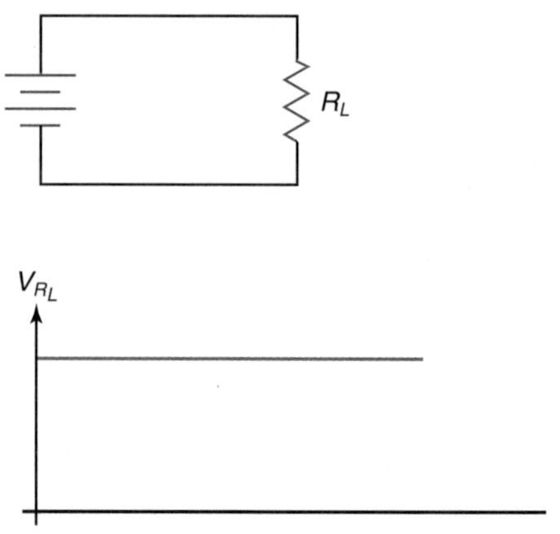

Fig. 4-14 Pure direct current.

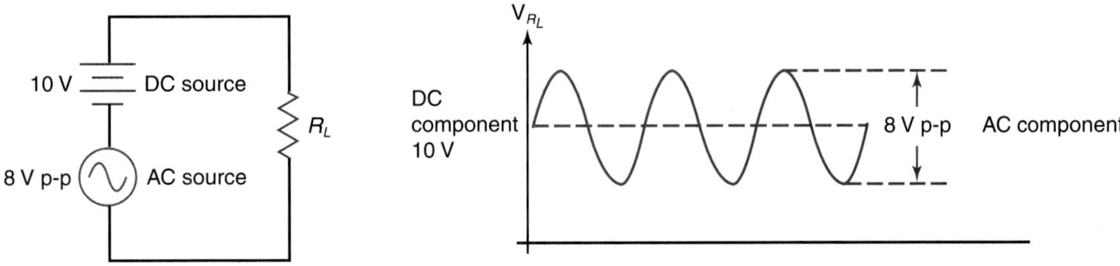

Fig. 4-15 An ac source in series with a battery.

approach the waveform produced by a battery.

The most common technique used for filtering is a capacitor connected across the output. Figure 4-16 shows a simple *capacitive filter* that has been added to a full-wave rectifier circuit. The voltage waveform across the load resistor shows that the ripple has been greatly reduced by the addition of the capacitor.

Capacitors are energy storage devices. They can take a charge and then later deliver that charge to a load. In Fig. 4-17(*a*) the rectifiers are producing peak output, load current is flowing, and the capacitor is *charging*. Later, when the rectifier output drops off, the capacitor *discharges* and furnishes the load current [Fig. 4-17(*b*)]. Since the current through the load has been maintained, the voltage across the load will be maintained also. This is why the output voltage waveform shows less ripple.

The effectiveness of a capacitive filter is determined by three factors:

1. The size of the capacitor
2. The value of the load
3. The time between pulsations

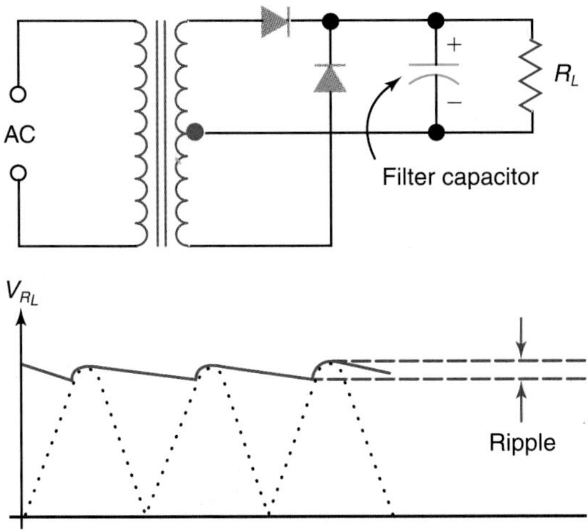

Fig. 4-16 A full-wave rectifier with a capacitive filter.

These three factors are related by the formula

$$T = R \times C$$

where T = time in seconds (s)
R = resistance in ohms (Ω)
C = capacitance in farads (F)

Capacitive filter

The product RC is called the *time constant of the circuit.* A charged capacitor will lose 63.2 percent of its voltage in T seconds. It takes approximately $5 \times T$ seconds to completely discharge the capacitor.

To be effective, a filter capacitor should be only *slightly* discharged between peaks. This will mean a small voltage change across the load and thus little ripple. The time constant will have to be long when compared to the time between peaks. This makes it interesting to compare half-wave and full-wave filtering. The time between peaks for full-wave and half-

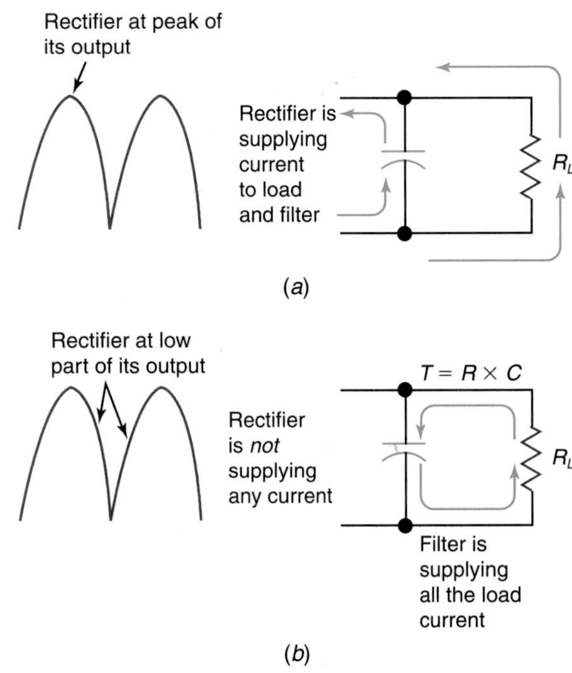

Fig. 4-17 Filter capacitor action.

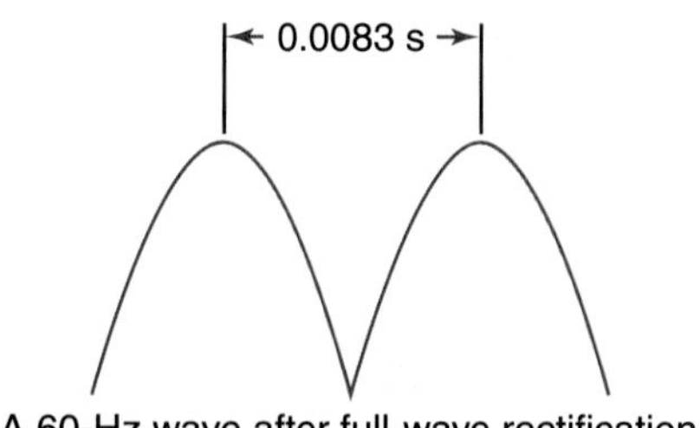

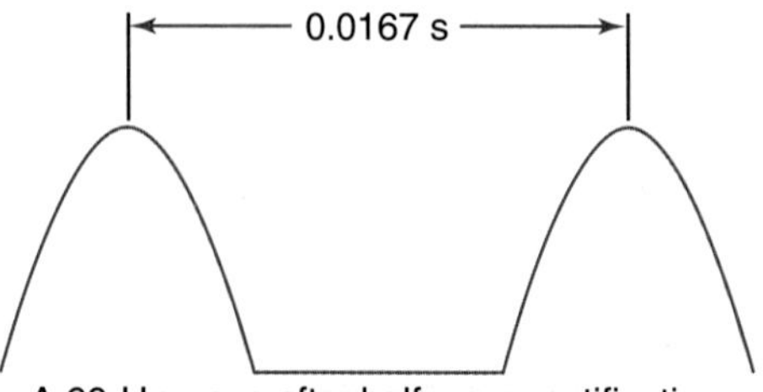

Fig. 4-18 A rectified 60-Hz wave.

wave rectifiers is shown in Fig. 4-18. Obviously, in a half-wave circuit the capacitor has twice the time to discharge, and the ripple will be greater. Full-wave rectifiers are desirable when most of the ripple must be removed. This is because it is easier to filter a wave whose peaks are closer together. Looking at it another way, it will take a capacitor twice the size to adequately filter a half-wave rectifier, if all other factors are equal.

Example 4-7

Estimate the relative effectiveness of 100-μF and 1000-μF capacitors when they will be used to filter a 60-Hz half-wave rectifier loaded by 100 Ω. First, find both time constants:

$$T = R \times C$$
$$T_1 = 100\ \Omega \times 100\ \mu\text{F} = 0.01\ \text{s}$$
$$T_2 = 100\ \Omega \times 1000\ \mu\text{F} = 0.1\ \text{s}$$

If we look at Fig. 4-18, we see that the discharge time for 60-Hz half-wave circuits is in the vicinity of 0.01 s. This means that the smaller filter will discharge for about one time constant, losing about 60 percent, for a significant amount of ripple. The larger capacitor has a 0.1-s time constant, which is long compared to the discharge time. The 1000-μF capacitor will be a much more effective filter (a lot less ripple).

The choice of a filter capacitor can be based on the following equation:

$$C = \frac{I}{V_{\text{p-p}}} \times T$$

where C = the capacitance in farads (F)
I = the load current in amperes (A)
$V_{\text{p-p}}$ = the peak-to-peak ripple in volts (V)
T = the period in seconds (s)

Example 4-8

Choose a filter capacitor for a full-wave, 60-Hz power supply when the load current is 5 A and the allowable ripple is 1 $V_{\text{p-p}}$. The power supply operates at 60 Hz, but as Fig. 4-18 shows, the ripple frequency is twice the input frequency for full-wave rectifiers:

$$T = \frac{1}{f} = \frac{1}{2 \times 60} = 8.33\ \text{ms}$$

This agrees with Fig. 4-18. Find the filter size next:

$$C = \frac{I}{V_{\text{p-p}}} \times T = \frac{5}{1} \times 8.33 \times 10^{-3}$$
$$= 41.7\ \text{mF}$$
$$= 41{,}700\ \mu\text{F}$$

The size of filter capacitors is often expressed in microfarads.

Example 4-9

Choose a filter capacitor for a full-wave, 100-kHz power supply when the load current is 5 A and the allowable ripple is 1 $V_{\text{p-p}}$. Compare the capacitor to that found in the previous example.

$$T = \frac{1}{2 \times 100 \times 10^3} = 5\ \mu\text{s}$$

$$C = \frac{5}{1} \times 5 \times 10^{-6} = 25\ \mu\text{F}$$

The size of the capacitor is much smaller when compared to the previous example.

Figure 4-18 is based on the 60-Hz power-line frequency. If a much higher frequency were used, the job of the filter could be even easier. For example, if the frequency were 1 kilohertz (kHz), the time between peaks in a full-wave rectifier output would be only 0.0005 s.

In this short period of time, the filter capacitor would be only slightly discharged. Another interesting point about high frequencies is that transformers can be made much smaller. Some power supplies convert the power-line frequency to a much high frequency to gain these advantages. Power supplies of this type are called *switch-mode supplies.* They are covered in Chap. 15.

One way to get good filtering is to use a large filter capacitor. This means that it will take longer for the capacitor to discharge. If the load resistance is low, the capacitance will have to be very high to give good filtering. Inspect the time constant formula, and you will see that if R is made lower, then C must be made higher if T is to remain the same. So, with heavy current demand (a low value of R), the capacitor value must be quite high.

Electrolytic capacitors are available with very high values of capacitance. However, a very high value in a capacitive-input filter can cause problems. Figure 4-19 shows waveforms that might be found in a capacitively filtered power supply. The unfiltered waveform is shown in Fig. 4-19(*a*). In Fig. 4-19(*b*) the capacitor supplies energy between peaks. Note that the rectifiers do not conduct until their peak output exceeds the capacitive voltage. The rectifier turns off when the peak output ends. The rectifiers conduct *for only a short time.* Figure 4-19(*d*) shows the rectifier current waveform. Notice the high peak-to-average ratio.

One of the common characteristics of all valuable employees is pride in their work.

In some power supplies, the peak-to-average current ratio in the rectifiers may exceed 100:1. This causes the rms rectifier current to be greater than eight times the current delivered to the load. The rms current determines the actual heating effect in the rectifiers. This is why diodes may be rated at 10 amperes (A) when the power supply is designed to deliver only 2 A.

The dc output voltage of a filtered power supply tends to be *higher* than the output of a nonfiltered supply. Figure 4-20 shows a bridge rectifier circuit with a switchable filter capacitor. Before the switch is closed, the meter will read the average value of the waveform:

$$\begin{aligned} V_{av} &= 0.9 \times V_{rms} \\ &= 0.9 \times 10 \\ &= 9 \text{ V} \end{aligned}$$

After the switch is closed, the capacitor charges to the *peak* value of the waveform:

$$\begin{aligned} V_{p} &= 1.414 \times V_{rms} \\ &= 1.414 \times 10 \\ &= 14.14 \text{ V} \end{aligned}$$

This represents a *significant* change in output voltage. However, as the supply is loaded, the capacitor will not be able to maintain the peak voltage, and the output voltage will drop. The

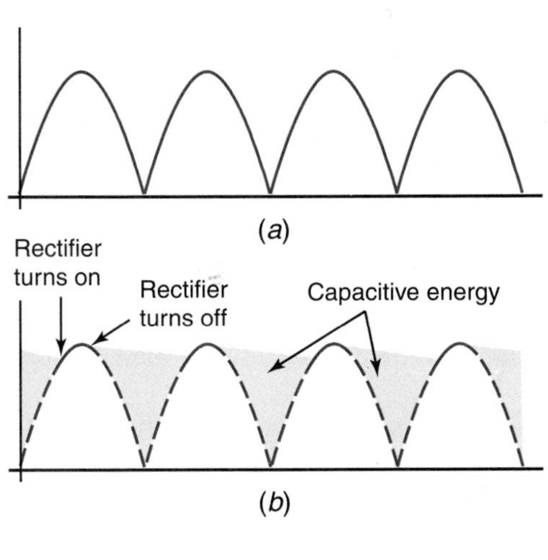

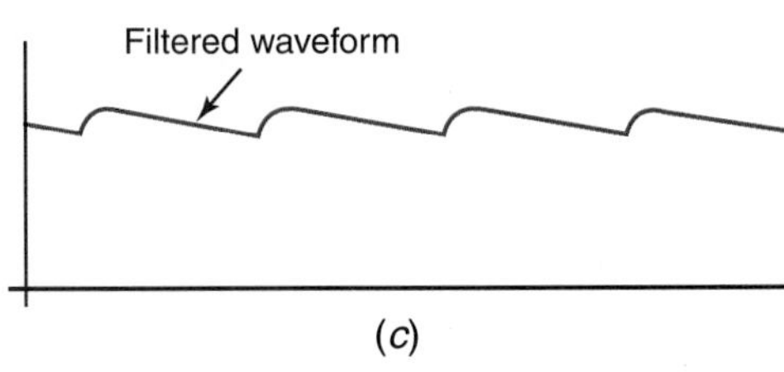

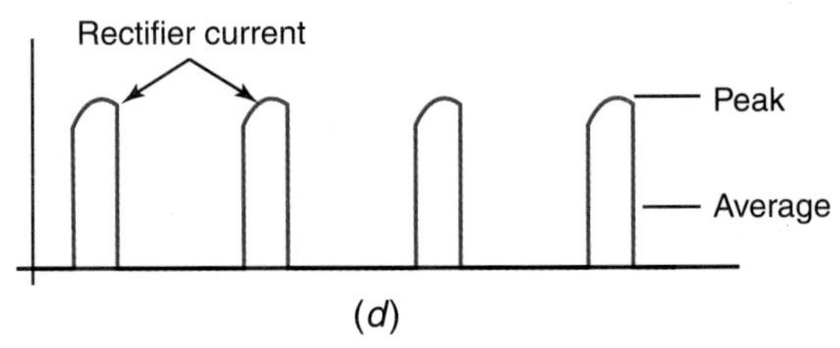

Fig. 4-19 Capacitor filter circuit waveforms.

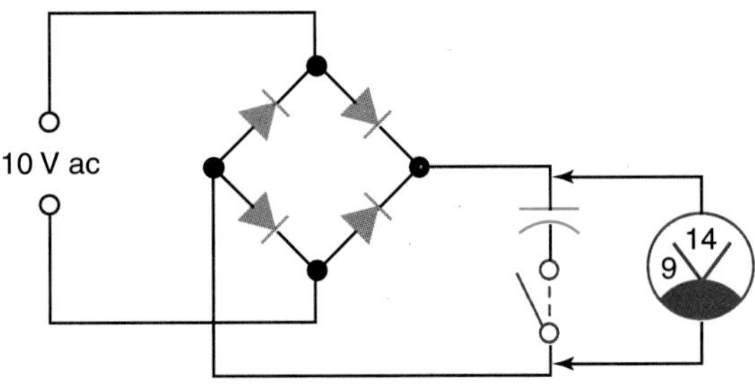

Fig. 4-20 Calculating dc output voltage with a capacitive filter.

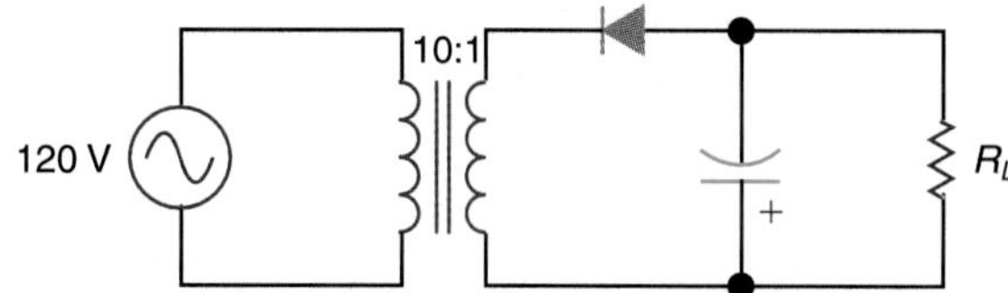

Fig. 4-21 A filtered half-wave rectifier circuit.

more heavily it is loaded (the more current there is), the lower the output voltage will be. Therefore, you can assume that the dc output voltage in a capacitively filtered supply is equal to the peak value of the ac input *when the supply is lightly loaded* or not loaded at all as in Fig. 4-20.

Figure 4-21 shows a filtered half-wave rectifier circuit. What is the procedure for predicting the voltage across R_L? When filters are used, do not use the 0.9 or the 0.45 conversion constants. Remember, the filter charges to the *peak* value of the input.

Referring to Fig. 4-21, the input is 120 V ac and is stepped down by the transformer:

$$\frac{120 \text{ V rms}}{10} = 12 \text{ V rms}$$

The peak value is found next:

$$V_p = 1.414 \times 12 \text{ V} = 16.97 \text{ V}$$

Assuming a light load, the dc voltage across the load resistor in Fig. 4-21 is nearly 17 V. If the filter capacitor would *open,* the dc output voltage would drop quite a bit. Its average value would be

$$V_{av} = 0.45 \times 12 \text{ V} = 5.4 \text{ V}$$

So, a good capacitor in Fig. 4-21 makes the output nearly 17 V and an open capacitor means that the output will be only 5.4 V. Understanding this can be quite important when troubleshooting power supplies.

Figure 4-22 shows the same transformer and input but the half-wave rectifier has been replaced with a bridge rectifier. Since the circuit is filtered, the dc output will again be equal to the peak value, or 16.97 V. If the capacitor opens in this circuit, the output will be

$$V_{av} = 0.9 \times 12 \text{ V} = 10.8 \text{ V}$$

Obviously, the failure (open type) of a filter capacitor in full-wave circuits will have a less drastic effect on the dc output voltage than it does in half-wave circuits.

The fact that a filter capacitor charges to the *peak* value of the ac waveform is important. Filter capacitors must be rated for this higher voltage. Another important point is capacitor *polarity.* If you will check Figs. 4-21 and 4-22, you will notice that the + lead is at the bottom. Verify that this is correct by checking the rectifier connections. Most filter capacitors are of the electrolytic type. These can *explode* if connected backward. This includes many tantalum capacitors.

Example 4-10

What voltage rating will be required for the filter capacitor in Fig. 4-22 if the transformer ratio is 1:1? The secondary voltage will be equal to the primary voltage, so the capacitor will charge to the peak value of the ac line:

$$V_p = 1.414 \times V_{rms} = 1.414 \times 120 = 170 \text{ V}$$

The capacitor will charge to 170 V. A margin of safety is required, so a capacitor rated at 250 V or more would likely be used in this case.

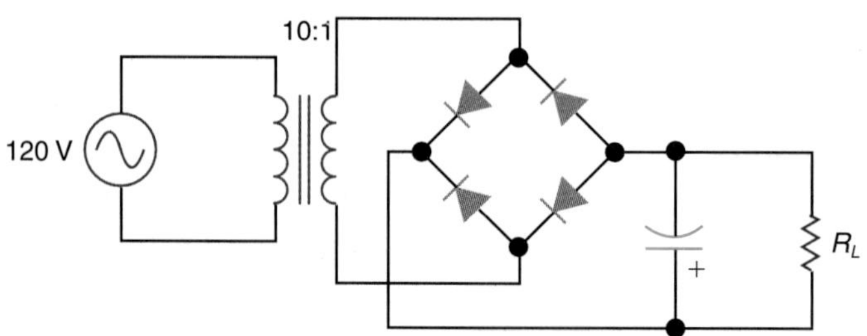

Fig. 4-22 A filtered bridge rectifier circuit.

Figure 4-23 shows a *choke-input filter.* A *choke* is another name for an inductor. The term "choke" is used in power supplies because of its use to "choke off" the ripple. You may recall that inductance was defined as the circuit property that *opposes any change in current.* The choke in fig. 4-23 is in series with R_L. It will therefore oppose changes in load current. This opposition will reduce the ripple current in the load and the ripple voltage across the load.

Choke-input filter

Chokes are not applied as often in 60-Hz power supplies as they once were. The change to solid-state circuits and the improvements in electrolytic capacitors have made it less expensive to remove ripple using only capacitors. Chokes for 60-Hz supplies tend to be large, heavy, and expensive components. Chokes are used more often in switch-mode supplies. Here the frequencies are so high that physically small inductors can be used to advantage.

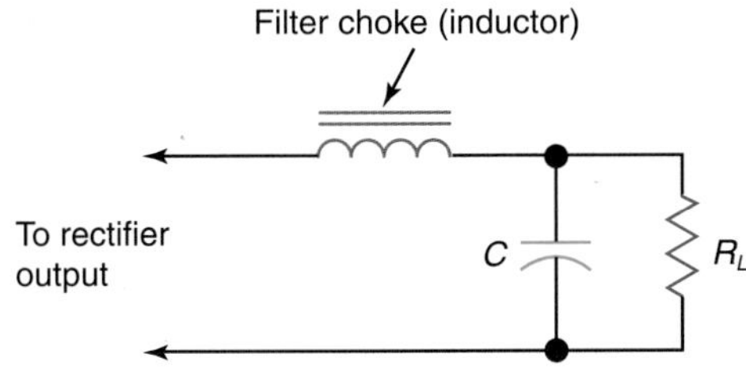

Fig. 4-23 Choke-input filter.

TEST

Supply the missing word or number in each statement.

28. Pure dc contains no ______.
29. Rectifiers provide ______ dc.
30. Power supplies use filters to reduce ______.
31. Capacitors are useful in filter circuits because they store electric ______.
32. In a power supply with a capacitor filter, the effectiveness of the filter is determined by the size of a capacitor, the ac frequency, and the ______
33. Half-wave rectifiers are more difficult to filter because the filter has more time to ______
34. Heating effect is determined by the ______ value of a current.
35. In a filtered power supply, the dc output voltage can be as high as ______ times the rms input voltage.
36. The conversion factor that is useful when predicting the dc output voltage of a filtered supply is ______.
37. The conversion factors of 0.45 and 0.90 are useful for predicting the dc output of ______ supplies.
38. A filter capacitor's voltage rating must be greater than the ______ value of the pulsating waveform.
39. The dc output from a lightly loaded supply using a bridge rectifier with 15 V ac input and a filter capacitor at the output will be ______.

4-6 Voltage Multipliers

The typical general-purpose line voltage in this country is about 115 to 120 V ac. Usually, solid-state circuits require lower voltage for operation. Sometimes, higher voltages are required. One way to obtain a higher voltage is to use a step-up transformer. Unfortunately, transformers are expensive devices. They are also relatively large and heavy. For these reasons, designers may not want to use them to obtain high voltages.

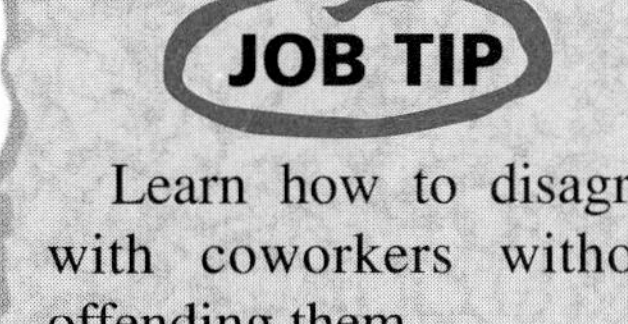

Voltage multipliers

Voltage multipliers can be used to produce higher voltages and eliminate the need for a transformer. Figure 4-24(*a*) on page 70 shows the diagram for a full-wave voltage doubler. This circuit can produce an output voltage as high as 2.8 times the rms input voltage. The output will be a dc voltage with some ripple.

Figure 4-24(*b*) shows the operation of the full-wave doubler. It shows how C_1 charges through D_1 when the ac line is on its positive alternation. Capacitor C_1 can be expected to charge to the peak value of the ac line. Assuming the input voltage is 120 V, we have

$$\begin{aligned} V_p &= 1.414 \times V_{rms} \\ &= 1.414 \times 120 \text{ V} \\ &= 169.68 \text{ V} \end{aligned}$$

On the negative alternation of ac line voltage 4-24(*c*), C_2 charges through D_2 to the peak value of 169.68 V. Now both C_1 and C_2 are charged. In Fig. 4-24(*d*) it can be seen that C_1 and C_2 are in series. Their polarities are series-aiding, and they will produce double the peak line voltage across the load:

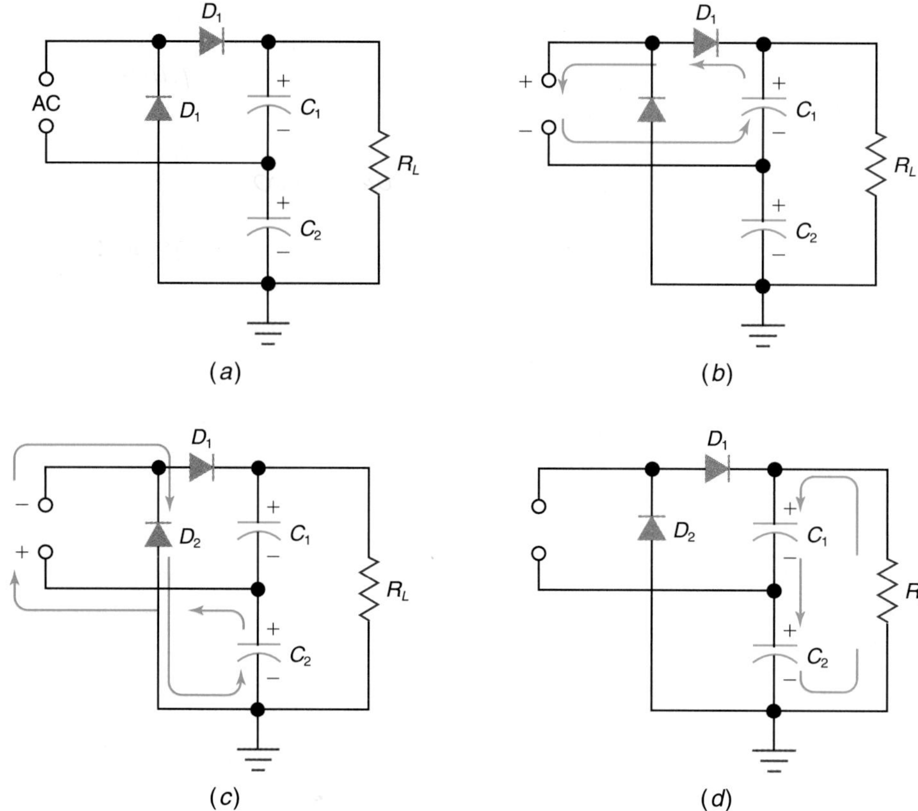

Fig. 4-24 Full-wave voltage doubler.

$$\begin{aligned} V_{RL} &= V_{C1} + V_{C2} \\ &= 169.68 \text{ V} + 169.68 \text{ V} \\ &= 339 \text{ V} \end{aligned}$$

Shock hazard

Voltage doublers can come close to producing three times the line voltage. As they are loaded, their output voltage tends to drop rapidly. Thus, a voltage doubler energized by a 120-V ac line might produce a voltage near 240 V dc when delivering current to a load. A voltage multiplier is a poor choice when stable output voltages are required.

Example 4-11

Find the voltage across R_L for Fig. 4-24 assuming a light load and an ac input of 230 V. Rather than calculate each capacitor voltage, it is easier to use double the 1.414 factor:

$$V_{RL} = 2.82 \times V_{\text{rms}} = 2.82 \times 230 \text{ V} = 649 \text{ V}$$

Ground loop

Lack of line isolation is the greatest problem with transformerless power supplies. Most electronic equipment is fabricated on a metal framework or chassis. Often, this chassis is the common conductor for the various circuits. If the chassis is not isolated from the ac line, it can present an extreme *shock hazard.* The chassis is usually inside a nonconducting cabinet. The control knobs and shafts are made of nonconducting materials such as plastic. This gives some protection. However, a technician working on the equipment may be exposed to a shock hazard.

Figure 4-25(*a*) shows a situation that has surprised more than one technician. Most bench-type test equipment is wired with a three-conductor power cord that automatically grounds its chassis, its case, and the shield on the test lead. If the shield, which is usually terminated with a black alligator clip, comes into contact with a "hot" chassis, there is a *ground loop* or short circuit across the ac line. Trace the short circuit in Fig. 4-25(*a*). The path is from the hot wire, through the polarized outlet, through the power cord, through the metal chassis, through the alligator clip lead of the test equipment, and through the power cord of the test equipment to ground. Since ground and

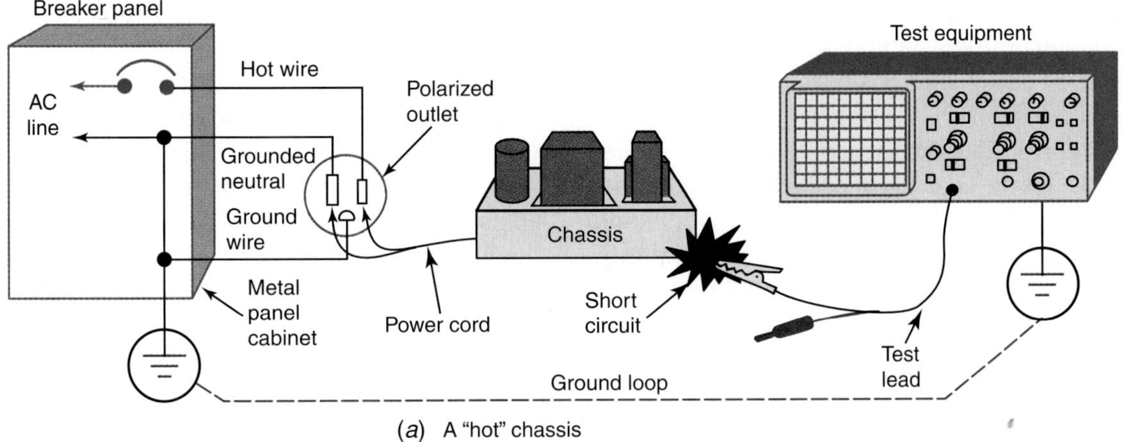

(*a*) A "hot" chassis

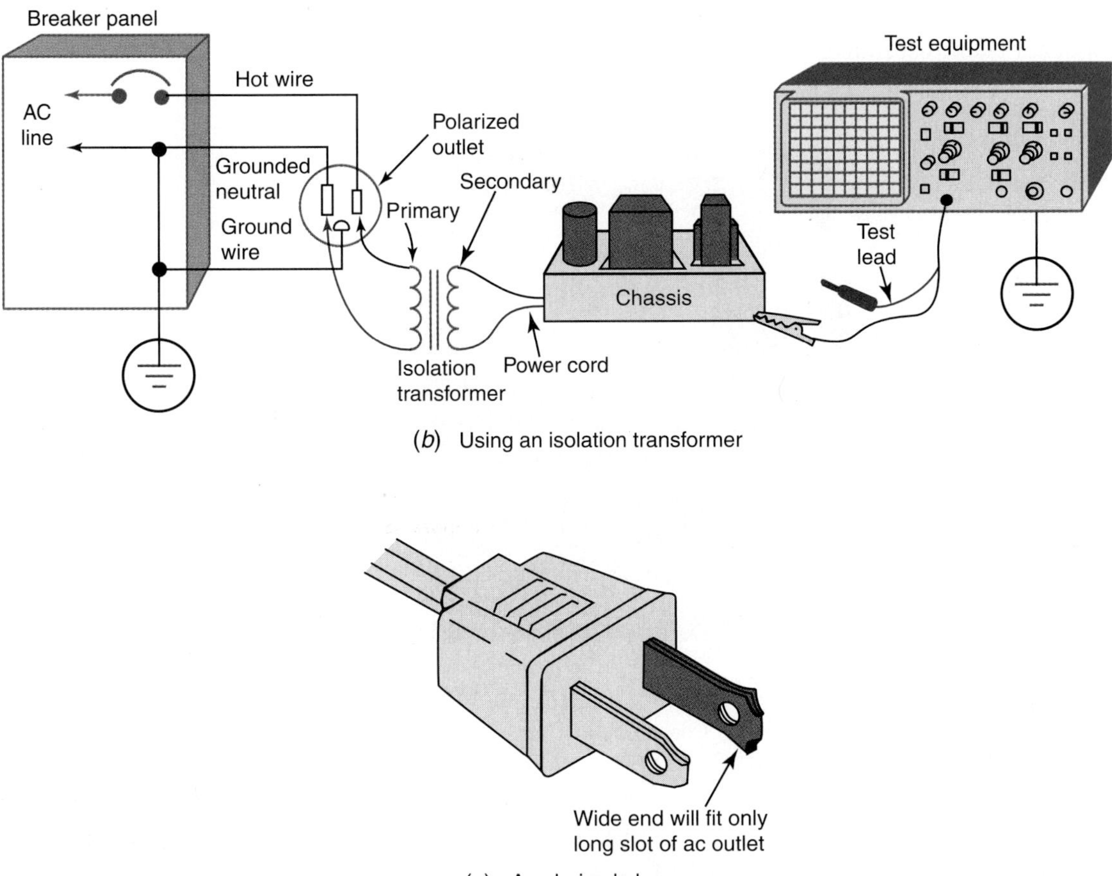

(*b*) Using an isolation transformer

(*c*) A polarized plug.

Fig. 4-25 The "hot" chassis problem and two solutions.

the grounded neutral wire are tied together in the breaker panel, this traced path is a short circuit across the ac line. Thus, connecting test equipment to "hot" equipment can open (trip) circuit breakers, blow fuses, damage test leads, and damage circuits. Worse than this, a technician's body may become a part of the ground loop and a serious electric shock can result. Working on equipment that is not isolated from the ac line is dangerous.

Figure 4-25(*b*) shows how an *isolation transformer* can be used to solve the hot-chassis problem. The transformer is plugged into the polarized outlet, and the chassis is energized from the secondary. There is very high electrical resistance from the primary of the

Isolation transformer

transformer to the secondary. Now a fault current cannot flow from the hot wire to the metal chassis. The chassis has been *isolated* from the ac line.

Figure 4-25(*c*) shows a polarized power plug that keeps the chassis connected to the grounded neutral side of the ac line. However, some equipment and some buildings may be improperly wired so that the chassis would still be hot.

Floating Measurements

Battery-operated, portable oscilloscopes are becoming more popular. With these, it is sometimes possible to safely make what are known as *floating measurements.* Floating measurements are taken with the common connections of the test equipment at some potential other than ground. For example, you might need a reading across a resistor that has no ground connection. Figure 4-26 shows the architecture used in some Tektronix® handheld oscilloscope/DMMs. It is important to realize that the channel 1, channel 2, and the digital multimeter inputs are isolated from the main chassis and from each other. This architecture allows floating measurements with channel 1, channel 2, and the DMM.

Floating measurements

When using an instrument with the type of architecture shown in Fig. 4-26, you must follow these precautions:

- You must attach the probe reference lead for each channel directly to your circuit (use both black leads).

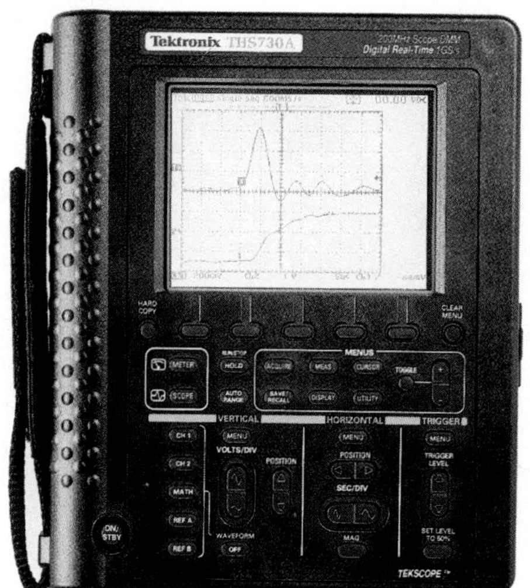

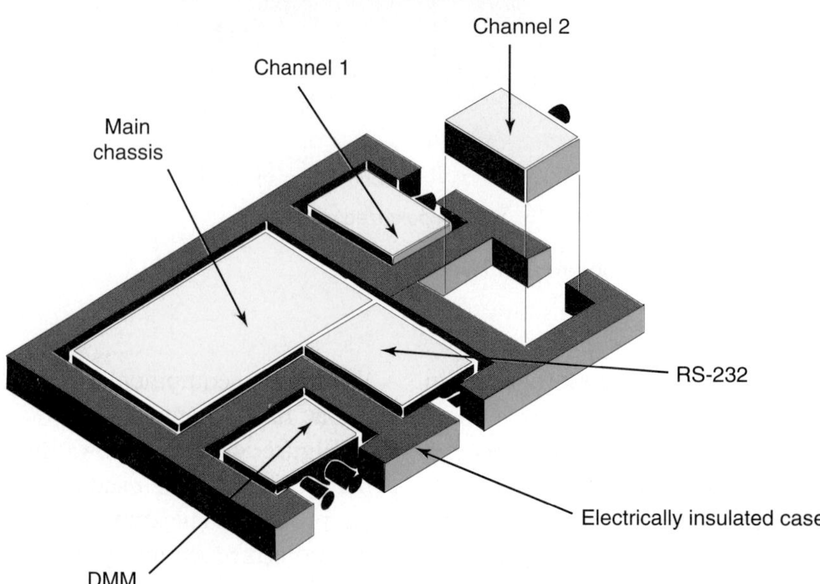

Fig. 4-26 TekScope® instrument architecture.

- If you are using the DMM, you must also connect the DMM common lead to your circuit.
- *Do not* exceed the voltage ratings for your instrument and its probes:

  Typical maximum voltage ratings are:

  From probe tip to reference lead: 300 V_{rms}

  From reference lead to earth ground: 30 V_{rms}

  From reference lead to reference lead: 30 V_{rms}

It is critical that you take the time to learn your test equipment. Your safety depends on it.

Many handheld oscilloscope/DMM products use a different architecture than that shown in Fig. 4-26. They have in common the reference for the oscilloscope channels and the DMM. With this arrangement, all input signals must have the same reference voltage when multichannel measurements are taken. Finally, remember that bench-top oscilloscopes have a common reference and *do not have an insulated case*. Bench-top oscilloscopes are not suited to floating measurements.

The *half-wave voltage doubler* shown in Fig. 4-27(*a*) offers some improvement in safety over the full-wave voltage doubler. Compare Figs. 4-24(*a*) and 4-27(*a*). The chassis is always hot in the full-wave doubler. In the half-wave doubler, the chassis is hot only if the connection to the ac outlet is wrong.

The half-wave doubler works a little differently from the full-wave doubler. On the negative alternation, C_1 will be charged [Fig. 4-27(*b*)]. Then in Fig. 4-27(*c*), C_1 adds in series with the ac line's positive alternation, and C_2 will be charged to twice the peak line voltage. Load resistance R_L is in parallel with C_2 and will see a peak voltage of about 340 V with a line voltage of 120 V ac. The key differences are the capacitor voltage ratings and the ripple frequency across the load. Full-wave doublers use two identical capacitors. Each would have to be rated at least equal to the peak line voltage. Half-wave doublers require the load capacitor to be rated at least equal to twice the peak line voltage. The ripple frequency in a full-wave doubler will be twice the line frequency. Half-wave doublers will show a ripple frequency equal to the line frequency.

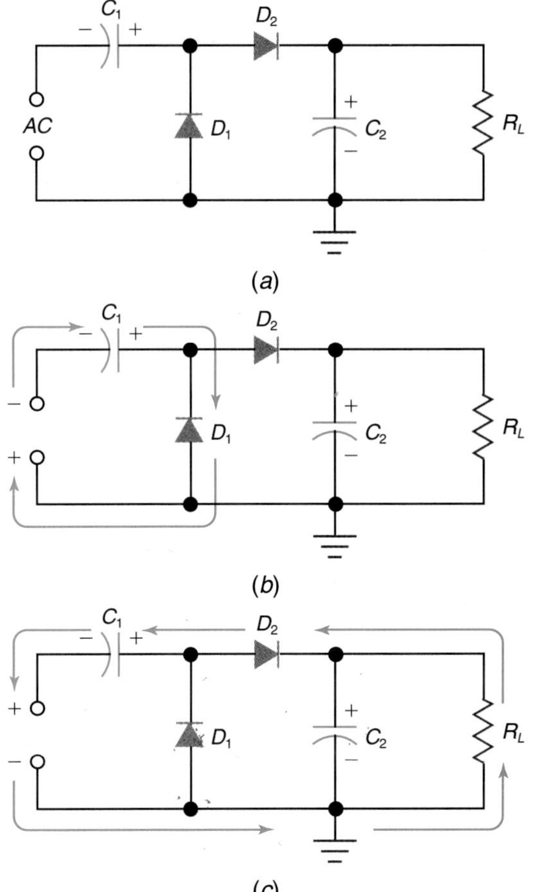

Fig. 4-27 Half-wave voltage doubler.

Half-wave voltage doubler

It is possible to build voltage multipliers that triple, quadruple, and multiply even more. Figure 4-28(*a*) on page 74 shows a voltage tripler. On the first positive alternation, C_1 is charged through D_1. On the next alternation, C_2 is charged to twice the peak line voltage through D_2 and C_1. Finally, C_3 is charged to three times the peak line voltage through D_3 and C_2. With 120-V ac input, the load would see a peak voltage of 509 V. A voltage quadrupler is shown in Fig. 4-28(*b*). This circuit is actually two half-wave doublers connected back to back and sharing a common input. The voltages across C_2 and C_4 will add to produce four times the peak ac line voltage. Assuming a line input of 120 V, R_L would see a peak voltage of 679 V.

All voltage multipliers tend to produce an output voltage that *drops* quite a bit when the load is increased. This drop can be offset somewhat by using large values of filter capacitors. Using large capacitors may cause the peak rectifier current to be very high. The surge is usually worst when the supply is first turned on. If the power supply should happen to be switched

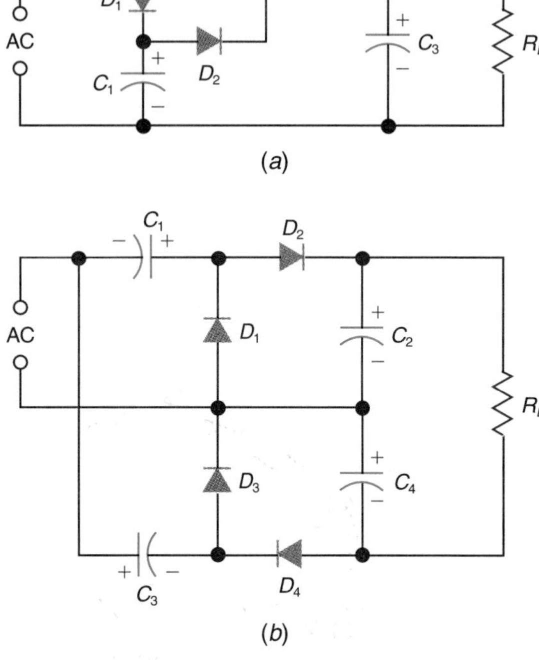

Fig. 4-28 Voltage multipliers.

Surge limiter

Percentage of ripple

on just as the ac line is at its peak, the surge may damage the diodes. Surge limiting must be added to some multiplier circuits. A *surge limiter* is usually a low-value resistor connected into the circuit so that it can limit the surge current to a value safe for the rectifiers. Figure 4-29 shows a surge limiter in a full-wave doubler circuit.

TEST

Supply the missing word in each statement.

40. Connecting grounded test equipment to a hot chassis will result in a ground LOOP.

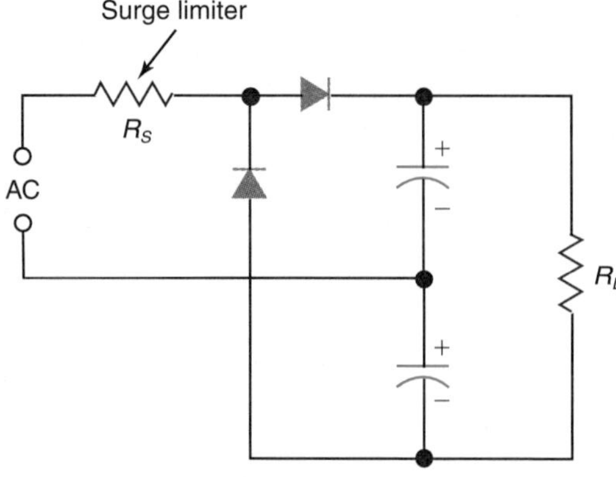

Fig. 4-29 A voltage multiplier with a surge-limiting resistor.

41. Voltage doublers may be used to obtain higher voltages and eliminate the need for a(n) XFMR.
42. A lightly loaded voltage doubler will give a dc output voltage that is 2.82 times the rms input.
43. The output of voltage multipliers tends to DECREASE quite a bit when the load is increased.
44. To reduce shock hazard and equipment damage, a technician should use a(n) ISOLATION transformer.
45. The ripple frequency in a 60-Hz half-wave doubler supply will be 60 Hz.
46. The ripple frequency in a 60-Hz full-wave doubler supply will be 120 Hz.
47. Voltage multipliers may use surge-limiting resistors to protect the RECTIFIERS.

4-7 Ripple and Regulation

A power-supply filter reduces ripple to a low level. The actual effectiveness of the filter can be checked with a measurement and then a simple calculation. The formula for calculating the *percentage of ripple* is

$$\text{Ripple} = \frac{\text{ac}}{\text{dc}} \times 100\%$$

For example, assume the ac ripple remaining after filtering is measured and found to be 1 V in a 20-V dc power supply. The percentage of ripple is

$$\text{Ripple} = \frac{\text{ac}}{\text{dc}} \times 100\%$$
$$= \frac{1}{20} \times 100\%$$
$$= 5\%$$

Example 4-12

Find the percentage of ac ripple if the ac content is 0.5 V in a 20-V supply.

$$\text{Ripple} = \frac{\text{ac}}{\text{dc}} \times 100\% = \frac{0.5\text{ V}}{20\text{ V}} \times 100\%$$
$$= 2.5\%$$

Notice that the percentage is smaller when the ac content is less.

Ripple should be measured only when the supply is delivering its *full* rated output. At zero load current, even a poor filter will reduce the ripple to almost zero. Ripple can be measured with an oscilloscope or a voltmeter. The oscilloscope will easily give the peak-to-peak value of the ac ripple. Many meters will indicate the approximate value of the rms ripple content. It will not be exact since the ripple waveform is *nonsinusoidal.* In a capacitive filter, the ripple is similar to a sawtooth waveform. This causes an error with most meters since they are calibrated to indicate rms values for sine waves. There are meters that will read the true rms value of nonsinusoidal alternating current. True rms meters are becoming more popular as prices drop.

To measure the ac ripple riding on a dc waveform, an older meter may have to be switched to a special function, or one of the test leads may have to be moved to a special jack. The special function or jack may be labeled *output.* The output jack is connected to the meter circuitry through a *coupling capacitor.* This capacitor is selected to have a low reactance at 60 Hz. Thus, 60- or 120-Hz ripple will reach the meter circuits with little loss. Capacitors have infinite reactance for direct current (0 Hz). This means that the dc content of the waveform will be blocked and will not interfere with the measurement. If an unusually high ripple content is measured, the meter circuit should be checked to be certain the dc component is not upsetting the reading.

The *regulation* of a power supply is its ability to hold the output steady under conditions of changing input or changing load. As power supplies are loaded, the output voltage tends to drop to a lower value. The quality of the voltage regulation can be checked with two measurements and then a simple calculation. The formula for calculating the *percentage of voltage regulation* is

$$\text{Regulation} = \frac{\Delta V}{V_{F_L}} \times 100\%$$

where ΔV = voltage change from no load to full load

V_{F_L} = output voltage at full load

For example, a power supply is checked with a dc voltmeter and shows an output of 14 V when no (0) load current is supplied. When the power supply is loaded to its rated maximum current, the meter reading drops to 12 V. The percentage of voltage regulation is

$$\text{Regulation} = \frac{\Delta V}{V_{F_L}} \times 100\%$$

$$= \frac{2\text{ V}}{12\text{ V}} \times 100\%$$

$$= 16.7\%$$

Example 4-13

Find the percentage of voltage regulation when the output drops from 14.5 V to 14.0 V as the supply is loaded.

$$\text{Regulation} = \frac{\Delta V}{V_{F_L}} \times 100\%$$

$$= \frac{0.5\text{ V}}{14\text{ V}} \times 100\% = 3.57\%$$

Notice that the percentage is smaller when the voltage change is less.

Coupling capacitor

The output voltage of some power supplies can increase quite a bit when there is a *no-load condition.* The no-load condition can be avoided by connecting a fixed load called a *bleeder* to the output of a power supply. Figure 4-30 on page 76 shows the use of a bleeder resistor. If R_L is disconnected, the bleeder will continue to load the output of the supply. Thus, some minimum output current will always flow. This fixed load can reduce the fluctuations in output voltage with changes in R_L. So one function of a bleeder is to improve supply regulation.

Bleeder resistor

Bleeder resistors perform another important function. They drain the filter capacitors after the power is turned off. Some filter capacitors can store a charge for months. Charged capacitors can present a shock hazard. It is *not* safe to assume that the capacitors have been drained even if there is a bleeder resistor across them. The bleeder could be open. Technicians who work on high-voltage supplies use a shorting rod or a shorting stick to be certain that all the filters are drained before working on the equipment. High-energy capacitors can discharge violently, so it is important that the shorting rod contain a high-wattage resistor of about

Percentage of voltage regulation

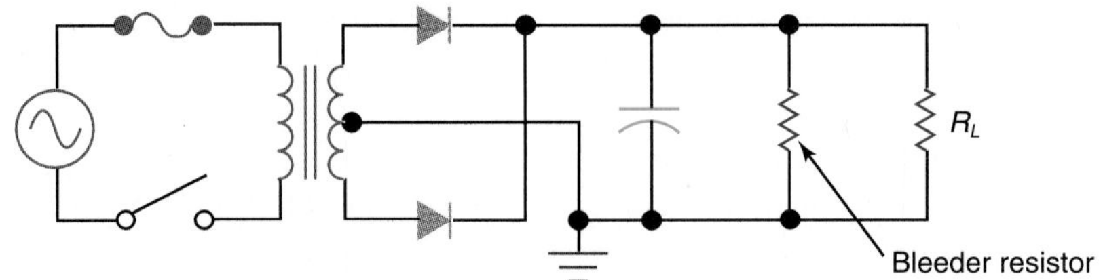

Fig. 4-30 Power supply with a bleeder resistor.

100 Ω to keep the discharge current reasonable. Figure 4-31 shows such a device.

TEST

Supply the missing word or number in each statement.

48. As the load current increases, the ac ripple tends to _____.
49. As the load current increases, the dc output voltage tends to _____.
50. A power supply develops 13 V dc with 1-V ac ripple. Its percentage of ripple is _____.
51. A power supply develops 28 V under no-load conditions and drops to 24 V when loaded. Its percentage of regulation is _____.
52. A bleeder resistor may improve supply regulation and help to ensure that the capacitors are _____ after the supply is turned off.

4-8 Zener Regulators

Regulator

Zener shunt regulator

Zener diode

Power-supply output voltage tends to change as the load on the power supply changes. The output also tends to change as the ac input voltage changes. This can cause some electronic circuits to operate improperly. When a stable voltage is required, the power supply must be *regulated.* The block diagram of a power supply (Fig. 4-32) shows where the regulator is often located in the system.

Regulators can be elaborate circuits using integrated circuits and transistors. Such circuits are covered in Chap. 15. For some applications, however, a simple *zener shunt regulator* does the job (Fig. 4-33). The regulator is a *zener diode,* and it is connected in shunt (parallel) with the load. If the voltage across the diode is constant, then the load voltage must also be constant.

The design of a shunt regulator using a zener diode is based on a few simple calculations. For example, suppose a power supply develops 16 V and that a regulated 12 V is required for the load. A simple calculation shows the need to drop 4 V (16 V − 12 V = 4 V). This voltage will drop across R_Z in Fig. 4-33. Assume that the load current is 100 mA. Also assume that we want the zener current to be 50 mA. Now we can calculate a value for R_Z using Ohm's law:

$$R_Z = \frac{V}{I_{\text{total}}} = \frac{4\text{ V}}{0.100\text{ A} + 0.050\text{ A}} = 26.67\ \Omega$$

The nearest standard value of a resistor is 27 Ω, which is very close to the calculated value. The power dissipation in the resistor can be calculated:

$$P = V \times I = 4\text{ V} \times 0.150\text{ A} = 0.6\text{ watt (W)}$$

We can use a 1-W resistor, although a 2-W resistor may be required for better reliability. Next, the power dissipation in the diode is

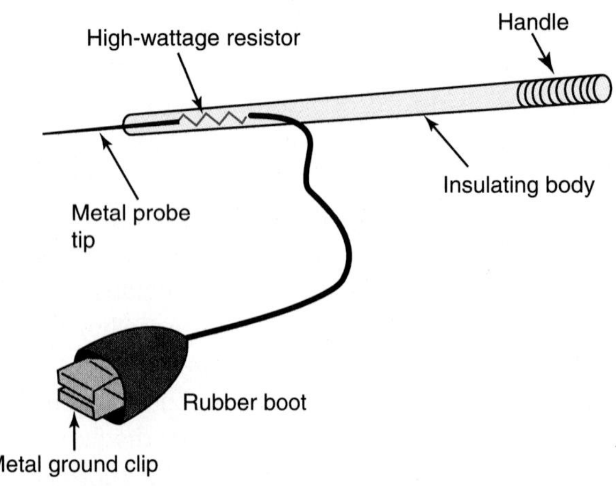

Fig. 4-31 A shorting rod.

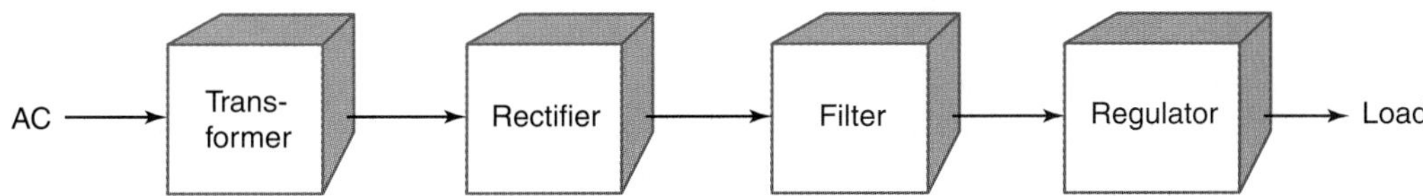

Fig. 4-32 Location of a regulator in a power supply.

$$P = V \times I$$
$$= 12 \text{ V} \times 0.050 \text{ A}$$
$$= 0.6 \text{ W}$$

A 1-W zener diode may be adequate. However, if the load is disconnected, the zener has to dissipate quite a bit more power. All the current (150 mA) flows through the diode. The diode dissipation increases to

$$P = V \times I$$
$$= 12 \text{ V} \times 0.150 \text{ A}$$
$$= 1.8 \text{ W}$$

Obviously, the zener must be capable of handling more power if there is a possibility of the load being disconnected.

Example 4-14

Determine R_Z for a 12-V zener regulator with a dc input of 20 V, a load current of 65 mA, and a zener current of 20 mA.

$$R_Z = \frac{V_{R_Z}}{I_{\text{total}}} = \frac{20 \text{ V} - 12 \text{ V}}{65 \text{ mA} + 20 \text{ mA}} = 94.1\ \Omega$$

Use 91 Ω, which is the closest standard value. Find the dissipation in R_Z next:

$$P_{R_Z} = V \times I = 8 \text{ V} \times 85 \text{ mA} = 0.68 \text{ W}$$

Use a 2-W resistor for good reliability.

Example 4-15

Find the zener diode dissipation for Example 4-14 if the load is disconnected from the regulator.

$$P = V \times I = 12 \text{ V} \times 85 \text{ mA} = 1.02 \text{ W}$$

Use a 2-W zener for good reliability.

Another possibility is that the load might demand more current. Suppose that the load current increases to 200 mA. Resistor R_Z would drop

$$V = I \times R$$
$$= 0.200 \text{ A} \times 27\ \Omega$$
$$= 5.4 \text{ V}$$

This would cause a decrease in voltage across the load:

$$V_{\text{load}} = V_{\text{supply}} - V_{R_Z}$$
$$= 16 \text{ V} - 5.4 \text{ V}$$
$$= 10.6 \text{ V}$$

The regulator is no longer working. Shunt regulators work only up to the point at which the zener stops conducting. The zener current should not be allowed to approach zero. As shown in Fig. 4-34 on page 78, the region of the characteristic curve near the zener knee shows poor regulation.

Zener regulators *reduce* ac ripple. This is because zener diodes have a low impedance when biased properly. For example, one manufacturer of the 1N4733 zener diode rates its typical dynamic impedance (Z_Z) at 5 Ω when it is biased at 10 mA. Let's determine what this characteristic can mean in terms of ripple performance.

Fig. 4-35(*a*) shows a regulator circuit based on the 1N4733 zener. This diode regulates at 5.1 V. This will establish a 5.1-V drop across the 470-Ω load resistor and a load current of:

$$I = \frac{V}{R}$$
$$= \frac{5.1 \text{ V}}{470\ \Omega}$$
$$= 10.9 \text{ mA}$$

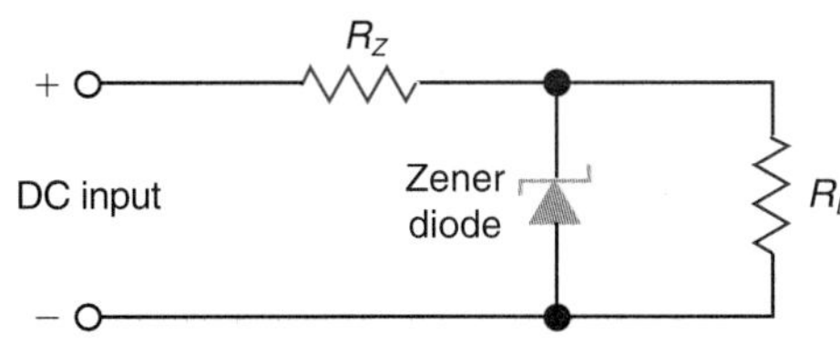

Fig. 4-33 Zener-diode shunt regulator.

If we assume a zener current of 10 mA, then the total current through the series resistor is

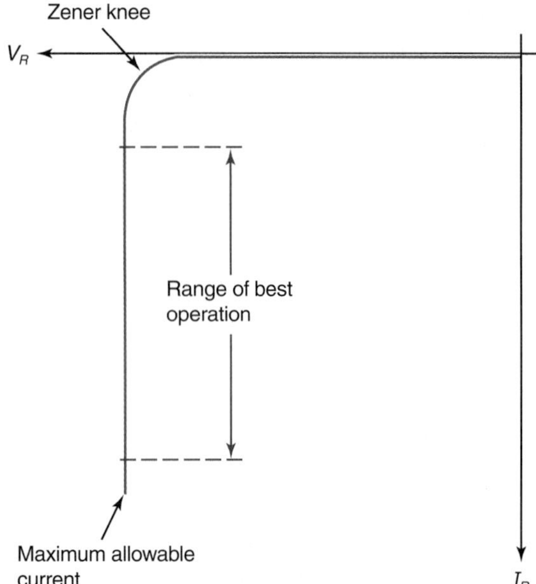

Fig. 4-34 Characteristic curve of a zener diode.

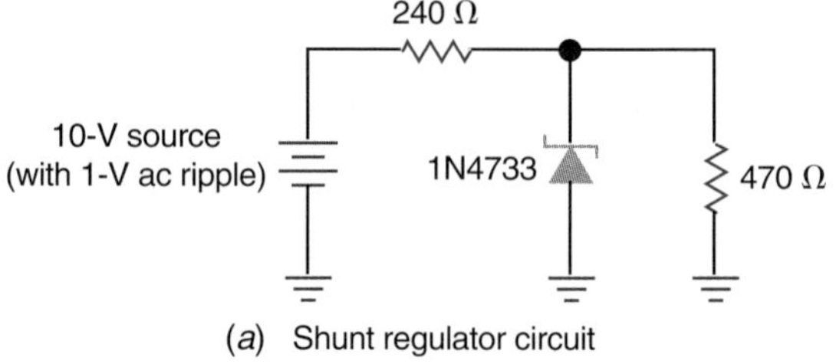

(*a*) Shunt regulator circuit

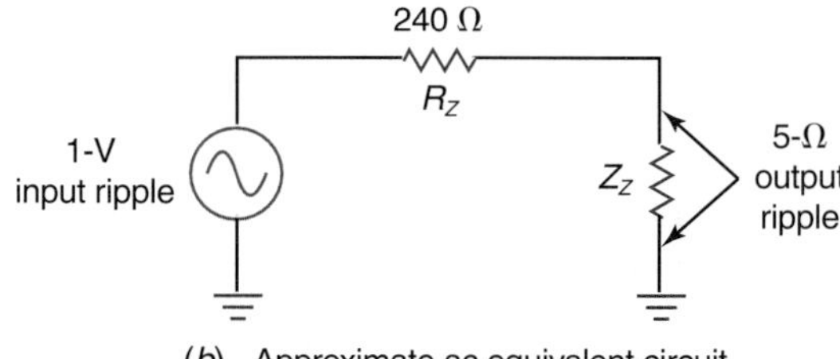

(*b*) Approximate ac equivalent circuit

Fig. 4-35 Determining zener regulator output ripple.

$$I_T = 10.9 \text{ mA} + 10 \text{ mA}$$
$$= 20.9 \text{ mA}$$

The series resistor drops the difference between the supply voltage and the load voltage:

$$V_{RZ} = 10 \text{ V} - 5.1 \text{ V}$$
$$= 4.9 \text{ V}$$

Ohm's law gives us the value for the series resistor:

$$R_Z = \frac{4.9 \text{ V}}{20.9 \text{ mA}}$$
$$= 234 \ \Omega$$

The closest standard value is 240 Ω, and this is shown in Fig. 4-35(*a*).

Figure 4-35(*b*) shows the approximate *ac equivalent circuit* for the regulator. The 470-Ω load resistor has been ignored because its resistance is much greater than the zener impedance. The 1 V of ac ripple will be divided by R_Z and Z_Z. The voltage divider equation will predict the ac ripple at the output of the regulator:

$$\text{Ripple} = \frac{5 \ \Omega}{240 \ \Omega + 5 \ \Omega} \times 1 \text{ V}$$
$$= 20.4 \text{ mV}$$

This very small voltage shows that zener shunt regulators are effective in reducing ac ripple.

Component derating

Solid-state devices such as zener diodes have to be *derated* in some applications. The power rating of zener diodes and other solid-state devices must be *decreased* as the device temperature goes up. The temperature inside the cabinet of an electronic system might increase from 25° to 50°C after hours of continuous operation. This increase in temperature decreases the safe dissipation levels of the devices in the cabinet. Figure 4-36 shows a typical power derating curve for a zener diode.

Cabinet temperatures are only a part of the problem. A zener diode that is dissipating a watt or so will be self-heating. So depending on dissipation levels and the environment components like zeners may have to be *derated* for reliable operation.

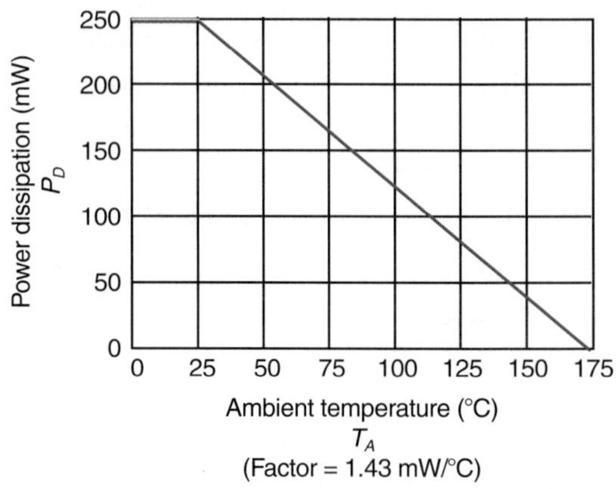

Fig. 4-36 Derating curve of a zener diode.

TEST

Supply the missing word or number in each statement.

53. A zener diode shunt regulator uses the zener connected in PARALLEL with the load.
54. A power supply develops 8 V. Regulated 5 V is required at a load current of 500 mA. A zener diode shunt regulator will be used. The zener current should be 200 mA. The value of R_Z should be 4.29 Ω.
55. The dissipation in R_Z in question 54 is 2.1 W.
56. The zener dissipation in question 54 is ________.
57. If the load current were interrupted in question 54, the zener would dissipate ________.
58. A zener shunt regulator can provide voltage regulation and reduce ________.

4-9 Troubleshooting

One of the major skills of an electronic technician is *troubleshooting.* The process involves the following steps:

1. Observing the symptoms
2. Analyzing the possible causes
3. Limiting the possibilities by tests and measurements

Good troubleshooting is an orderly process. To help keep things in order, remember the word *"GOAL."* GOAL stands for *G*ood, *O*bserve, *A*nalyze, and *L*imit.

Electronic equipment that is broken usually shows very definite *symptoms.* These are extremely important. Technicians should try to note all the symptoms before proceeding. This demands a knowledge of the equipment. You must know what the normal performance of a piece of equipment is in order to be able to identify what is abnormal. It is often necessary to make some adjustments or run some checks to be sure that the symptoms are clearly identified. For example, if a radio receiver has a hum or whistle on one station, several other stations should be tuned in to determine whether the symptom persists. Another example might involve checking the sound from a tape recorder with a tape recorded on another machine known to be working properly. These kinds of adjustments and checks will help the technician to properly observe the symptoms.

Analyzing possible causes comes after the symptoms are identified. This part of the process involves a general knowledge of the block diagram of the equipment. Certain symptoms are closely tied to certain blocks on the diagram. Experienced technicians "think" the block diagram. They do not need a drawing in front of them. Their experience tells them how the major sections of the circuit work, how signals flow from stage to stage, and what happens when one section is not working properly. For example, suppose a technician is troubleshooting a radio receiver. There is only one major symptom: No sound of any kind is coming from the speaker. Experience and knowledge of the block diagram will tell the technician that two major parts of the circuit can cause this symptom: the power supply and/or the audio output section.

After the possibilities are established, it is time to limit them by tests and measurements. A few voltmeter checks generally will tell the technician if the power-supply voltages are correct. If they are not correct, then the technician must further limit the possibilities by making more checks. Circuit breakdown is usually limited to one component. Of course, one component failure may damage several others because of the way they interact. A resistor that has burned black is almost always a sure sign that another part has shorted.

Troubleshooting

Power-supply troubleshooting

GOAL

Symptom

Power-supply troubleshooting follows the general process. The symptoms that can be observed are

1. No output voltage
2. Low output voltage
3. Excessive ripple voltage
4. High output voltage

Note that the symptoms are all limited to voltages. This is the way technicians work. Voltages are easy to measure. Current analysis is rarely used because it is necessary to break into the circuit and insert an ammeter. It is also worth mentioning that two of the power-supply symptoms might appear at the same time: low output voltage and excessive ripple voltage.

Once the symptoms are clearly identified, it

is time to analyze possible causes. For *no output voltage,* the possibilities include

1. Open fuse or circuit breaker
2. Defective switch, line cord, or outlet
3. Defective transformer
4. Open surge-limiting resistor
5. Open diode or diodes (rare)
6. Open filter choke or doubler capacitor

The last step is to limit the list of possibilities. This step is accomplished by making some measurements. Figure 4-37 is the schematic diagram for a half-wave doubler power supply. The technician can make ac voltage measurements as shown at *A*, *B*, *C*, and *D* to find the cause of no output voltage. For example, suppose the measurement as *A* is 120 V alternating current but 0 V at *B*. This indicates a blown fuse. Suppose *A* and *B* show line voltage and *C* shows zero. This would indicate an open surge-limiting resistor. If measurements *A*, *B*, and *C* are 120 V alternating current and if measurement *D* is zero, then capacitor C_1 is open.

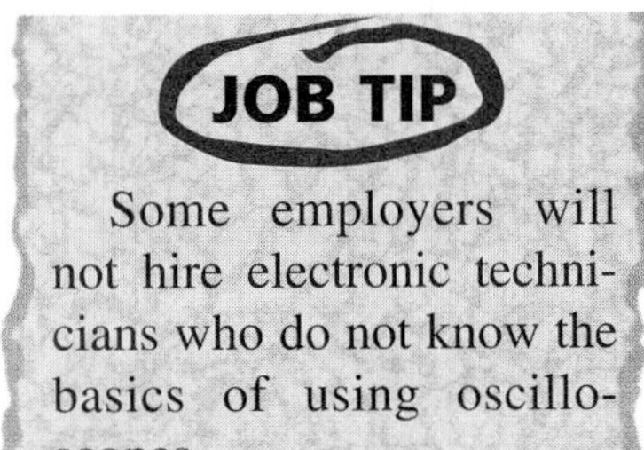

Some defects show the need for more checking. Again referring to Fig. 4-37, if the surge-limiting resistor R_S is open, it may be because another component is defective. Simply replacing R_S may result in the new part burning out. It is a good idea to check the diodes and the capacitors when a surge limiter opens or a fuse blows. One of the capacitors or diodes could be shorted.

Solid-state rectifier diodes usually do not open (show a very high resistance in both directions). There are exceptions, of course. Their typical failure mode is to short-circuit. Diodes can be checked with an ohmmeter, but this requires disconnecting at least one side of the diode. Sometimes it is possible to obtain a rough check with the diode still in the circuit. *Always remove power* before making ohmmeter tests and make sure the filter capacitors are *discharged.*

Figure 4-38 shows the schematic for a full-wave power supply. An ohmmeter test (with the supply off) across the diodes will show a low resistance when the diode is forward-biased and a higher resistance roughly equal to the total load resistance when the diode is reverse-biased. This will prove that the diode is not shorted but it may have excessive leakage. The sure method is to disconnect one end of the diode from the circuit. Bridge rectifier diodes can also be checked in circuit with similar results and limitations.

Figure 4-38 also shows some possible waveforms. Many technicians prefer troubleshooting with an oscilloscope. The ac waveform at the transformer secondary proves that the supply is plugged into an ac source, that it is turned on, and that the fuse is not blown. The full-wave pulsating dc waveform at the unregulated output is *not normal*: It indicates that C_1 is open. With a normal filter capacitor, some ripple might be seen, as also shown in Fig. 4-38. The waveform at the regulated output shows no ripple and a lower dc voltage, which is to be expected.

Electrolytic capacitor

Many of the filter capacitors used in modern power supplies are of the *electrolytic* type. These capacitors can short-circuit, open, develop leakage, or lose much of their capacity. They can be tested on a capacitor tester, or a rough check can be made with an ohmmeter. Be *sure* the supply is off and that all capacitors are discharged. Disconnect one lead and observe polarity when testing them. A good electrolytic capacitor will show a momentary low resistance as it draws a charging current from the ohmmeter. The larger the capacitor, the longer the low resistance will be shown. After some time, the ohmmeter should show a high

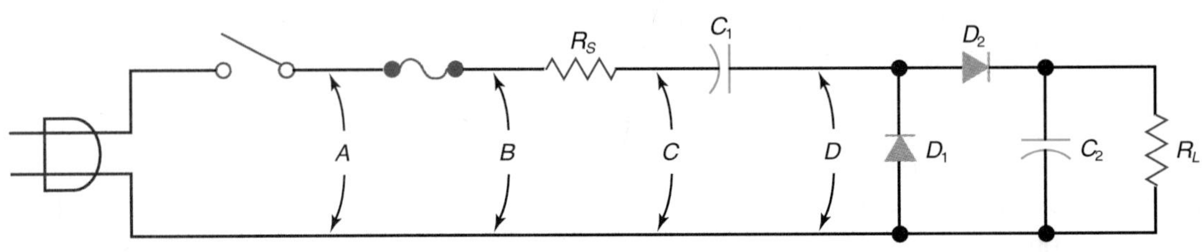

Fig. 4-37 A half-wave doubler schematic.

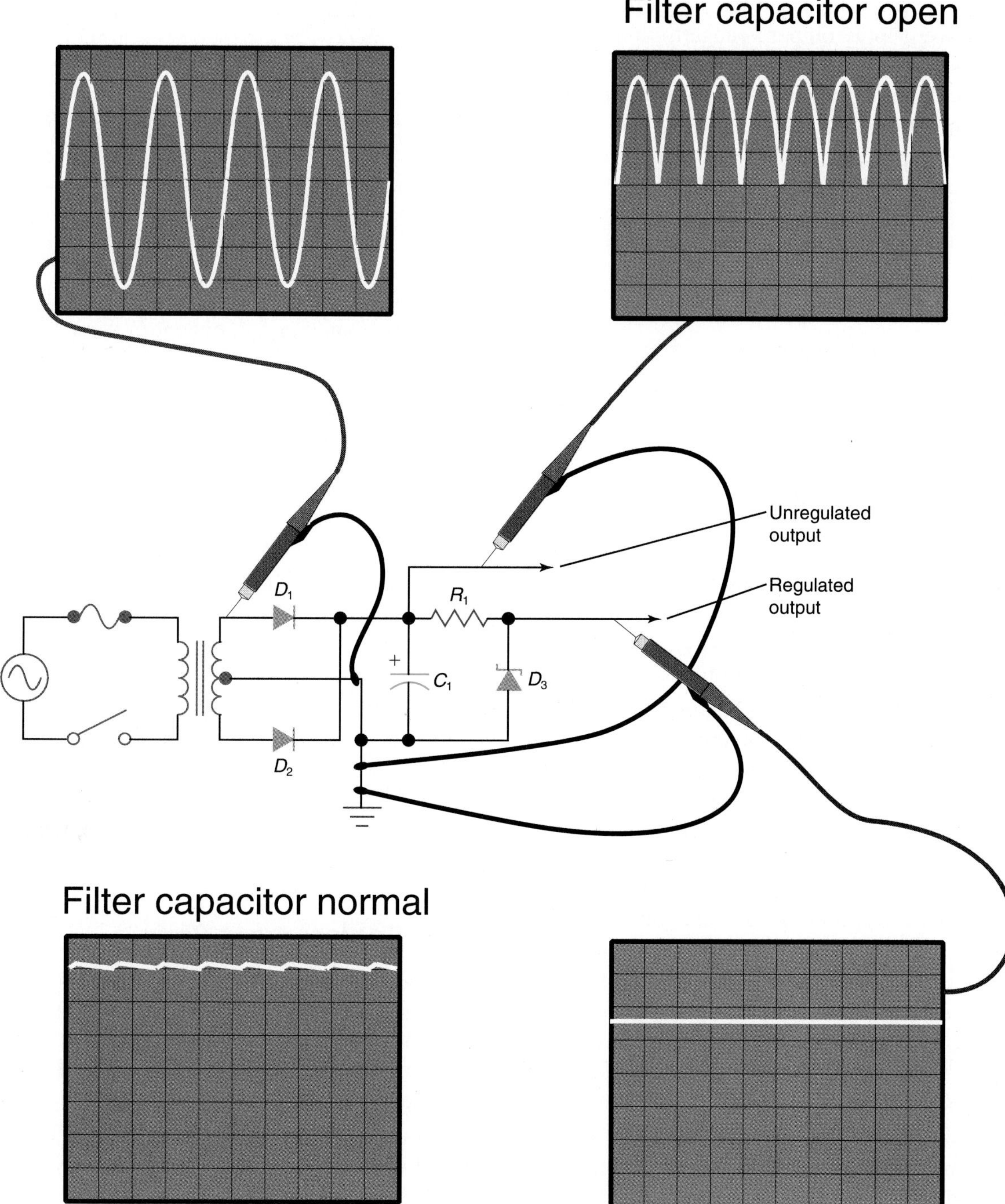

Fig. 4-38 Full-wave supply and waveforms.

resistance. It may not be infinite. All electrolytic capacitors have some leakage, and it is more pronounced in those with very high values. A large capacitor may show a leakage resistance of 100,000 Ω. Usually this is not significant in a power supply. This same leakage in a smaller capacitor used elsewhere in an electronic circuit could cause trouble.

The symptom of *low output voltage* in a power supply can be caused by

1. Excessive load current (overload)
2. Low input (line) voltage
3. Defective surge-limiting resistor
4. Defective filter capacitors
5. Defective rectifiers

Power supplies are often one part of an electronic system. Some other part of the system can fault and demand excess current from the

Low output voltage

power supply. This *overload* will often cause the power-supply output voltage to drop. There may not be anything wrong with the power supply itself. It is a good idea to first make sure that the current demand is normal when the power-supply output is low. This is one case when a current measurement may be required.

If the load is normal, then the supply itself must be checked. Some of the defects that might cause the half-wave doubler of Fig. 4-37 to produce low output voltage are

1. R_S has increased in value.
2. C_1 has lost much of its capacity.
3. C_2 has lost much of its capacity.
4. Defective rectifiers.
5. Low line voltage.

Low output voltage may be accompanied by excessive ripple. For example, suppose C_1 in Fig. 4-38 has lost much of its capacity. This will cause a drop in the unregulated output, and it will also cause the ripple voltage to increase. The regulated output may or may not show symptoms. It depends on the zener voltage, the regulated load, how bad C_1 is, and so on. Excessive loading on the power supply will also increase the ripple. Again, a current measurement may be required.

Excessive ripple is often caused by defective filter capacitors. Some technicians use clip leads to connect a test capacitor in parallel with the one they suspect. This will restore the circuit to normal operation in those cases where the original capacitor is open or low in capacity. *Be very careful* when making this kind of test. Remember, the power supply can store quite a charge. Be *sure* to observe the correct polarity with the test capacitor. If the test proves the capacitor is defective, it should be removed from the circuit. It is poor practice to leave the original capacitor in the circuit with a new one soldered across it.

The last power-supply symptom is high output voltage. Usually this is caused by low load current (underload). The trouble is not in the power supply but somewhere else in the circuit. It may be that a bleeder resistor is open. This decreases the load on the power supply, and the output voltage goes up. High output in a regulated power supply would indicate a defect in the regulator. Regulator troubleshooting is covered in Chap. 15.

TEST

Determine whether each statement is true or false.

59. A skilled troubleshooter uses a random trial-and-error technique to find circuit faults.
60. In troubleshooting, it is often possible to limit the problem to one area of the block diagram by observing the symptoms.
61. A resistor that is burned black may indicate that another component in the circuit has failed.
62. A supply that is overloaded will often show low output voltage.
63. Refer to Fig. 4-29. Resistor R_S burns out (opens). The symptom will be zero dc output voltage.
64. Refer to Fig. 4-37. The fuse blows repeatedly. Rectifier D_1 is probably open.
65. Refer to Fig. 4-37. The output voltage is low. Capacitor C_2 could be defective.
66. Refer to Fig. 4-38. The zener diode burns out. Regulated output voltage will be high.
67. Refer to Fig. 4-38. The zener diode is shorted. Both outputs will be zero.
68. Refer to Fig. 4-38. R_1 is open. Both outputs will be zero.

4-10 Replacement Parts

Exact replacement

After the defective parts are located, it is time to choose replacement parts. *Exact replacements* are the safest choice. If exact replacements are not available, it may be possible to make substitutions. A substitution should have ratings at least equal to those of the original. It would never do to replace a 2-W resistor with a 1-W resistor. The replacement resistor would probably fail in a short time. It may *not* be a good idea to replace a resistor with one having a higher power rating. In some circuits, the resistor may protect another more expensive part by increasing in value under overload conditions. Also, a fire hazard can result in some circuits if a carbon-composition resistor is substituted for a film resistor. It is easy to see why exact replacements are the safest.

Rectifier diodes have several important ratings. They are rated for average current and for surge current. The current peaks can be much higher than the average current with capacitive filters. However, the current peaks caused by filter capacitors are *repetitive*. Therefore, the average current rating of a rectifier diode is often *greater* than the actual circuit load current. Table 4-1 lists some of the maximum ratings for several common rectifiers.

TABLE 4-1 Common Rectifier Diode Ratings

Device	Peak Inverse Voltage, V	Average Rectified Output, Current, A (Resistive Load)	Nonrepetitive Peak Surge Current, A (1 Cycle)
1N4001	50	1	30
1N4002	100	1	30
1N4003	200	1	30
1N4004	400	1	30
1N4005	600	1	30
1N4006	800	1	30
1N4007	1000	1	30
1N5400	50	3	200
1N5401	100	3	200
1N5402	200	3	200
1N5404	400	3	200
1N5406	600	3	200

The maximum reverse-bias voltage that the diode can withstand is another important rating. In a half-wave power supply with a capacitive-input filter or in a full-wave power supply with a center-tapped transformer, the diodes are subjected to a reverse voltage equal to two times the peak value of the ac input. This is because the charged capacitor adds in series with the input when the diodes are off. Thus, the rectifier diodes must block twice the peak input. Figure 4-39 shows the diode ratings for various power supply circuits.

Electrolytic capacitors are rated for a dc working voltage (dcWV or VdcW or WVdc). This voltage must not be exceeded. Filter capacitors charge to the peak value of the rectified wave. Such a capacitor's dcWV rating should be greater than the peak voltage value.

The capacity of the electrolytic filters is also very important. Substituting a lower value may result in low output voltage and excessive ripple. Substituting a much higher value may cause the rectifiers to run hot and be damaged. A value close to the original is the best choice.

Transformers and filter chokes may also have to be replaced. The replacements should have the same voltage ratings, the same current ratings, and the same taps.

Sometimes, the physical characteristics of the parts are just as important as the electrical

Schematic	Name	PIV per diode	PIV per diode with capacitive filter	Diode current
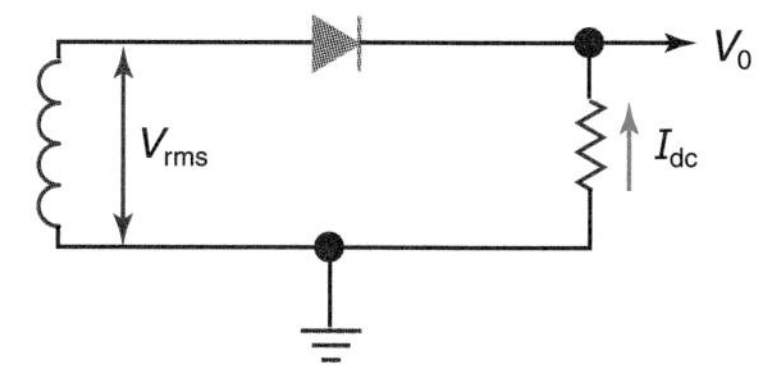 V_{rms}, V_o, I_{dc}	Half-wave	1.41 V_{rms}	2.82 V_{rms}	I_{dc}
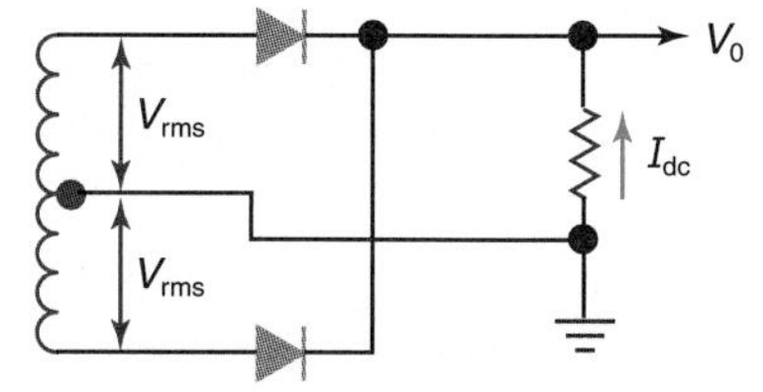 V_{rms}, V_{rms}, V_o, I_{dc}	Full-wave	2.82 V_{rms}	2.82 V_{rms}	0.5 I_{dc}
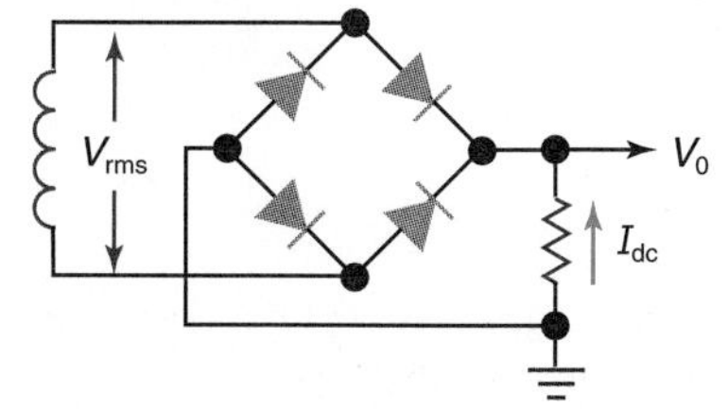 V_{rms}, V_o, I_{dc}	Bridge (full-wave)	1.41 V_{rms}	1.41 V_{rms}	0.5 I_{dc}

Fig. 4-39 Diode ratings for various supply circuits.

characteristics. A replacement transformer may be too large to fit in the same place on the chassis, or the mounting bolt pattern may be different. A replacement filter capacitor may not fit in the space taken by the old one. The stud on a power rectifier may be too large for the hole in the heat sink. It pays to check into the mechanical details when choosing replacement parts.

Substitution guide

Technicians use *substitution guides* to help them choose replacement parts. These are especially helpful for finding replacements for solid-state devices. The guides list many device numbers and the numbers for the replacement parts. The guides often include some of the ratings and physical characteristics for the replacement parts. Even though the guides are generally very good, at times the recommended part will not work properly. Some circuits are critical, and the recommended replacement part may be just different enough to cause trouble. There may also be some physical differences between the original and the replacement recommended by the guide.

JEDEC

PRO-ELECTRON

House number

Registered EIA number

Solid-state devices have two types of part numbers: registered and nonregistered. There are two major groups of registered devices: *JEDEC* and *PRO-ELECTRON. JEDEC* stands for the Joint Electronic Device Engineering Council and accounts for all devices registered in the United States. Some books and manuals refer to *EIA registration.* The letters *EIA* stand for the Electronic Industries Association. It is an association of American electronic manufacturers. EIAJ is a related association of Japanese electronic manufacturers. The PRO-ELECTRON devices have been registered in Europe with Association International.

When a solid-state device manufacturer uses a registered number for a part, that device must conform to registered specifications. This means that a diode or a transistor could be purchased from any of several manufacturers and its registered number will guarantee its similarity to the original part. The EIA or JEDEC part numbers for solid-state devices have the prefix 1N, 2N, 3N, or 4N. Examples of JEDEC registered numbers for solid-state devices are 1N4002, 2N3904, and 3N128. JEDEC also registers case numbers. This ensures that the physical size and characteristics of the case or package will conform to standards. Examples of JEDEC registered numbers for cases are DO-4, TO-9, and TO-92. The PRO-ELECTRON devices are preceded by a two- or three-letter prefix. Examples of PRO-ELECTRON registered numbers for solid-state devices are BAX13, BC531, and BSX29.

Military parts use JAN (Joint Army-Navy) numbers. A military electronic part is similar to a commercial- or industrial-rated part. It may have better specifications, and it will always have passed a far more rigorous and thorough testing procedure.

The second type of part number is the nonregistered or so-called *house number.* These part numbers are "invented" by the manufacturers. They do not conform to any agreed-upon standards. They do not indicate who the manufacturer was. However, most manufacturers do use a consistent pattern when numbering their parts. Thus, with experience, it may

Did You Know?

Inverted Electric Automobiles
Electric automobiles with ac motors use inverters to convert battery direct current to alternating current.

be possible to identify the manufacturer and the type of part just from the part number alone. For example, Motorola uses the following pattern of prefixes for some of their house numbers: MZ for zener diodes, MR for rectifier diodes, MRA for power rectifiers, and MLED for light-emitting diodes.

Nonregistered device numbers can cause problems for technicians who do not have years of experience. Luckily, substitution guides include nonregistered, as well as registered, part numbers. A good assortment of up-to-date substitution guides is a valuable part of the technician's library. Also, manufacturers' data manuals are extremely valuable for identifying and classifying nonregistered parts. Some examples of nonregistered part numbers for solid-state devices are MR1816, MCB5405F, CA200, and 2000287-28.

Another category of nonregistered numbers that can present difficulties is the so-called proprietary system. Suppose that company *X* buys 1000 parts from ABC Devices Inc. ABC has an established house number system, but customer *X* wants a special proprietary number placed on the parts. ABC wants the order and uses the customer's number. Later, a technician attempts to find a replacement for one of these parts. Usually, the technician must have company *X*'s service literature and obtain the part from them. Substitution guides and data manuals are of little help in this type of situation.

TEST

Determine whether each statement is true or false.

69. It may not be good practice to replace a 1-W resistor with a 2-W resistor.
70. It may not be safe to replace a film resistor with a carbon-composition resistor.
71. It may not be good practice to replace a 1000-μF filter capacitor with a 2000-μF capacitor.
72. A transistor is marked 2N3904. This is a house number.
73. The safest replacement part is the exact replacement.
74. The 1N914 is an example of a JEDEC registered part.

Chapter 4 Review

Summary

1. The power supply provides the various voltages for the circuits in an electronic system.
2. Bipolar power supplies develop both polarities with respect to the chassis ground.
3. Diagrams that show the major sections of electronic systems and how they are related are called block diagrams.
4. Power supplies usually change voltage levels and change alternating current to direct current.
5. In a diode rectifier circuit, the positive end of the load will be in contact with the cathode of the rectifier. The negative end of the load will be in contact with the anode of the rectifier.
6. A single diode forms a half-wave rectifier.
7. Half-wave rectification is generally limited to low-power applications.
8. A full-wave rectifier utilizes both alternations of the ac input.
9. One way to achieve full-wave rectification is to use a center-tapped transformer secondary and two diodes.
10. It is possible to achieve full-wave rectification without a center-tapped transformer by using four diodes in a bridge circuit.
11. A dc voltmeter or a dc ammeter will read the average value of a pulsating waveform.
12. The average value of half-wave, pulsating direct current is 45 percent of the rms value.
13. The average value of full-wave, pulsating direct current is 90 percent of the rms value.
14. Pulsating direct current contains an ac component called ripple.
15. Ripple can be reduced in a power supply by adding filter circuits after the rectifiers.
16. Filters for 60-Hz supplies are usually capacitive.
17. Filter chokes are more likely to be used in high-frequency supplies.
18. Capacitive filters cause a heating effect in the rectifiers that requires them to have ratings greater than the dc load current.
19. The factors for predicting dc output voltage are 0.45 for half-wave, 0.90 for full-wave, and 1.414 for any supply with a capacitive filter.
20. Full-wave rectifiers are easier to filter than half-wave rectifiers.
21. Line-operated equipment should always be operated with an isolation transformer to protect the technician and the equipment being serviced.
22. A surge-limiting resistor may be included in power supplies to protect the rectifiers from damaging current peaks.
23. Ripple should be measured when the power supply is delivering its rated full-load current.
24. Ripple is usually nonsinusoidal.
25. The percent regulation is a comparison of the no-load voltage and the full-load voltage.
26. Bleeder resistors can improve voltage regulation and drain the filter capacitors when the power supply is off.
27. A voltage regulator can be added to a power supply to keep the output voltage constant.
28. Zener diodes are useful as shunt regulators.
29. Limiting the possible causes to one or two defects usually involves making tests with meters and other equipment. The schematic diagram is very helpful in this phase of the troubleshooting process.
30. Defects may come in groups. One part shorting out could damage several others.
31. In troubleshooting power supplies, no output voltage usually is caused by open components.
32. Open components can be isolated by voltage measurements or resistance checks with the circuit turned off and the filters drained.
33. Electrolytic capacitors can short, develop excess leakage, open, or lose much of their capacity.
34. Power-supply voltages are affected by load current.
35. Excessive ripple is usually caused by defective filter capacitors.
36. Maximum ratings of parts must never be exceeded. A substitute part should be at least equal to the original.
37. Substitution guides are very helpful in choosing replacement parts.

Chapter Review Questions

Determine whether each statement is true or false.

4-1. A schematic shows only the major sections of an electronic system in block form.

4-2. In troubleshooting, one of the first checks that should be made is power-supply voltages.

4-3. Rectification is the same as filtering.

4-4. Diodes make good rectifiers.

4-5. A transformer has 120 V alternating current across its primary and 40 V ac across its secondary. It is a step-down transformer.

4-6. The positive end of the load will be in contact with the cathode of the rectifier.

4-7. A single diode can give full-wave rectification.

4-8. Half-wave rectifiers are limited to low-power applications.

4-9. A full-wave rectifier uses two diodes and a center-tapped transformer.

4-10. A bridge rectifier can provide full-wave rectification without a center-tapped transformer.

4-11. A bridge rectifier uses three diodes.

4-12. The average value of a sine wave is 0.637 times its rms value.

4-13. With pulsating direct current, a dc voltmeter will read the rms value of the waveform.

4-14. The ac input to a half-wave rectifier is 20 V. A dc voltmeter connected across the load should read 10 V.

4-15. Increasing the load current taken from a power supply will tend to make the output voltage drop.

4-16. Diode losses can always be ignored when they are used as rectifiers.

4-17. With light loads, power supply filter capacitors hold the dc output near the peak value of the input.

4-18. A filter capacitor loses much of its capacity. The symptoms could be excess ripple and low output voltage.

4-19. Capacitive filters increase the heating effect in the rectifiers.

4-20. Filter chokes are widely applied in 60-Hz power supplies.

4-21. The conversion factors 0.45 and 0.90 are not used to predict the dc output voltage of filtered power supplies.

4-22. Pure direct current means that no ac ripple is present.

4-23. A lightly loaded voltage doubler may give a dc output voltage nearly 4 times the ac input voltage.

4-24. An isolation transformer eliminates all shock hazards for an electronics technician.

4-25. The ripple frequency for a half-wave doubler will be twice the ac line frequency.

4-26. A 5-V dc power supply shows 0.2 V of ac ripple. The ripple percentage is 4.

4-27. From no load to full load, the output of a supply drops from 5.2 to 4.8 V. The regulation is 7.69 percent.

4-28. Alternating current ripple can be measured with a dc voltmeter.

4-29. It is necessary to load a power supply to measure its ripple and regulation.

4-30. The main function of a bleeder resistor is to protect the rectifiers from surges of current.

4-31. A zener diode shunt regulator is generally used to filter out ac ripple.

4-32. The dissipation in a shunt regulator goes up as the load current goes down.

4-33. A power supply blows fuses. The trouble could be a shorted filter capacitor.

4-34. A power supply develops too much output voltage. The problem might be high load current.

4-35. A burned-out surge resistor is found in a voltage doubler circuit. It might be a good idea to check the diodes and filter capacitors before replacing the resistor.

4-36. A shorted capacitor can be found with an ohmmeter check.

4-37. A shorted diode can be found with an ohmmeter check.

4-38. There is no way to locate data on parts using house numbers.

4-39. The EIA is a European association of electronics manufacturers.

Chapter Review Problems

4-1. Refer to Fig. 4-3. The ac line is 120 V and the transformer is 3:1 step-down. What would a dc voltmeter read if connected across R_L?

4-2. Refer to Fig. 4-5. The ac line is 120 V and the primary turns equal the secondary turns. What would a dc voltmeter read if connected across R_L?

4-3. Refer to Fig. 4-7. The ac input is 120 V. What would a voltmeter read if connected across the load resistor?

4-4. Refer to Fig. 4-16. The ac input is 120 V and the primary turns equal the secondary turns. What would a dc voltmeter read if connected across R_L?

4-5. Refer to Fig. 4-16. Assume a light load and a source voltage of 240 V ac. What would a dc voltmeter read if connected across R_L?

4-6. Refer to Fig. 4-27. Assume a light load and an ac source of 240 V. What is the dc voltage across R_L?

4-7. Refer to Fig. 4-33. The dc input is 24 V and the zener is rated at 9.1 V. Assume a zener current of 100 mA and a load current of 50 mA. Calculate the value for R_Z.

4-8. What is the dissipation in R_Z in problem 4-7?

4-9. What is the dissipation in the zener diode in problem 4-7?

4-10. What is the zener dissipation in problem 4-7 if R_L burns out (opens)?

4-11. A power supply output drops from 14 to 12.5 V dc when it is loaded. Find its regulation.

4-12. The output of the supply in problem 4-11 shows 500 mV ac ripple when it is loaded. Find its ripple percentage.

Critical Thinking Questions

4-1. Referring to Fig. 4-1, we see that stage *A* and stage *B* are both energized by the +12 V dc output of the power supply. Is it likely that stage *A* would have a power supply problem that stage *B* would not have?

4-2. Diode manufacturers package two diodes in one case for use in full-wave rectifier circuits. These packages have a metal tab that contacts both cathodes. They also offer a *reverse polarity* version in which the tab contacts both anodes. Why are reverse polarity versions offered?

4-3. Is there ever a situation when there is ac ripple in a circuit that is powered by a battery?

4-4. If an isolation transformer has a short circuit from its primary winding to its secondary winding, will it still work?

4-5. How would you check an isolation transformer to make sure that it does not have a problem such as the one mentioned in question 4?

4-6. A friend asks you to help troubleshoot an electronic gadget that she built. You agree, and when you look at the components, you notice a capacitor with a pronounced bulge. What would you do?

4-7. A story in the newspaper relates an incident when a ham radio operator was electrocuted in his basement *during a prolonged power outage.* Does the story make any sense?

Answers to Tests

1. direct current
2. ground (common)
3. block
4. power supply
5. T
6. T
7. F
8. F
9. T
10. T
11. T
12. F
13. 25 V
14. 1
15. 2
16. alternations
17. direction
18. center-tapped transformer
19. 4
20. 24 V ac
21. 10.8 V dc
22. 5.4 V dc
23. 14.4 V dc
24. 18 V dc
25. increases
26. low
27. 2 V
28. ripple (ac)
29. pulsating
30. ripple
31. energy
32. load resistance (current)
33. discharge
34. rms
35. 1.414
36. 1.414
37. unfiltered
38. peak

39. 21.2 V dc
40. loop
41. transformer
42. 2.82
43. decrease
44. isolation
45. 60 Hz
46. 120 Hz
47. diodes (rectifiers)
48. increase
49. decrease
50. 7.69 percent
51. 16.7 percent
52. discharged
53. parallel (shunt)
54. 4.29 Ω
55. 2.1 W
56. 1 W
57. 3.5 W
58. ripple
59. F
60. T
61. T
62. T
63. T
64. F
65. T
66. T
67. F
68. F
69. T
70. T
71. T
72. F
73. T
74. T

Junction Transistors

Chapter Objectives

This chapter will help you to:

1. *Identify* the schematic symbols for several types of transistors.
2. *Define* the meaning of amplification and power gain.
3. *Predict* the correct bias polarity for several types of transistors.
4. *Calculate* current gain from data and from characteristic curves.
5. *Calculate* collector dissipation from data and from characteristic curves.
6. *Test* bipolar transistors with an ohmmeter.

This chapter introduces the transistor. Transistors are solid-state devices similar in some ways to the diodes you have studied. Transistors are more complex and can be used in many more ways. The most important feature of transistors is their ability to amplify signals. Amplification can make a weak signal strong enough to be useful in an electronic application. For example, an audio amplifier can be used to supply a strong signal to a loudspeaker.

5-1 Amplification

Power gain

Amplifier

Amplification is one of the most basic ideas in electronics. Amplifiers make sounds louder and signal levels greater and, in general, provide a function called *gain.* Figure 5-1 shows the general function of an amplifier. Note that the amplifier must be provided two things: *dc power* and the *input signal.* The signal is the electrical quantity that is too small in its present form to be usable. With gain, it becomes usable. As shown in Fig. 5-1, the output signal is greater because of the gain provided by the amplifier.

Voltage gain

Current gain

Gain can be measured in several ways. If an oscilloscope is used to measure the amplifier input signal voltage and the output signal voltage, then the *voltage gain* can be determined. A certain amplifier may provide an output voltage that is 10 times greater than the input voltage. The voltage gain of the amplifier is 10. If an ammeter is used to measure amplifier input and output currents, then the *current gain* can be obtained. With a 0.1-A input signal, an amplifier might produce a 0.5-A output signal for a current gain of 5. If the voltage gain and the current gain are both known, then the *power gain* can be established. An amplifier that produces a voltage gain of 10 and a current gain of 5 will give the following power gain:

$$P = V \times I$$

or

$$\begin{aligned} P_{gain} &= V_{gain} \times I_{gain} \\ &= 10 \times 5 \\ &= 50 \end{aligned}$$

Example 5-1

Calculate the power gain of an amplifier that has a voltage gain of 0.5 and a current gain of 100.

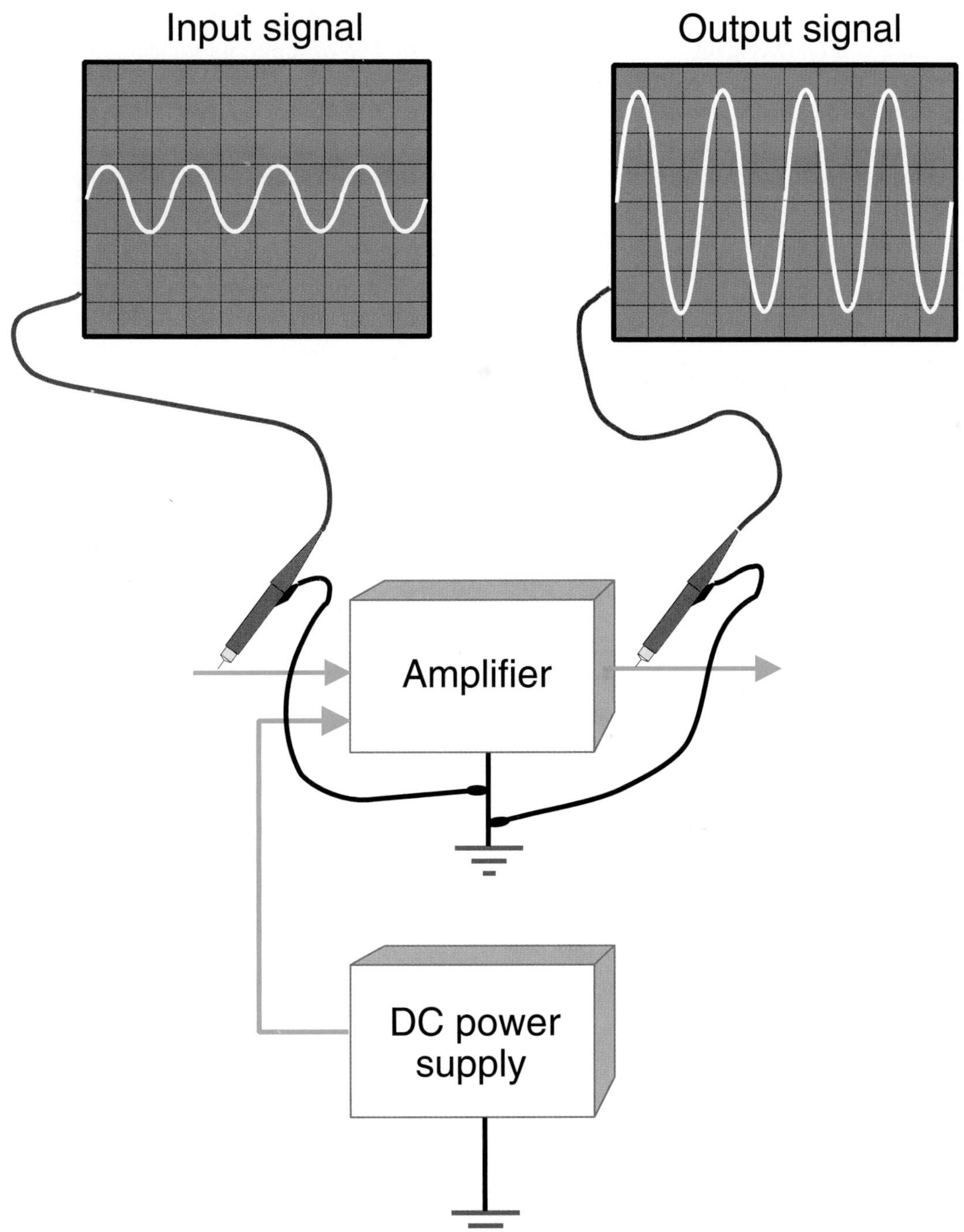

Fig. 5-1 Amplifiers provide gain.

$$P_{gain} = V_{gain} \times I_{gain} = 0.5 \times 100 = 50$$

Note that an amplifier can show a voltage loss and yet still have a significant power gain. Likewise, another amplifier might have a current loss and still have power gain.

Only amplifiers provide a power gain. Other devices might give a voltage gain or a current gain, but not both. A step-up transformer provides voltage gain but is *not* an amplifier. A transformer does not provide any power gain. If the transformer steps up the voltage 10 times, then it steps down the current 10 times. The power gain, ignoring loss in the transformer, will be

$$\begin{aligned} P_{gain} &= V_{gain} \times I_{gain} \\ &= 10 \times 0.1 \\ &= 1 \end{aligned}$$

A step-down transformer provides a current gain. It cannot be considered an amplifier. The current gain is offset by a voltage loss, and thus, there is no power gain.

Even though power gain seems to be the important idea, some amplifiers are classified as *voltage amplifiers.* In some circuits, only the voltage gain is mentioned. This is especially true in amplifiers designed to handle *small signals.* You will run across many voltage amplifiers or *small-signal amplifiers* in electronic devices. You should remember that they provide power gain, too.

Small-signal amplifier

Power amplifier

The term *power amplifier* is generally used to refer to amplifiers that develop a *large signal.* A signal can be large in terms of its voltage level, its current level, or both. In the electronic system of Fig. 5-2, the speaker requires several watts for good volume. The signal from the tape head is a fraction of a milliwatt (mW). A total power gain of thousands is needed. However, only the final large-signal amplifier is called a power amplifier.

In electronics, *gain* has no unit. If voltage gain is being discussed, it will be a pure number. Gain is the ratio of some output to some input. The letter A is often used as the symbol for gain or amplification. For voltage gain:

$$A_V = \frac{V_{out}}{V_{in}}$$

The units cancel. So if an amplifier outputs 10 V for 1 V of input, its voltage gain equals 10. It does not equal 10 V.

Example 5-2

Calculate the voltage gain of an amplifier if it has an input signal of 15 mV and an output signal of 1 V.

$$A_V = \frac{V_{out}}{V_{in}} = \frac{1\text{ V}}{15 \times 10^{-3}\text{ V}} = 66.7$$

TEST

Determine whether each statement is true or false.

1. An amplifier must be powered and have an input signal to develop a normal output signal.
2. An amplifier has a voltage gain of 50. If the input signal is 2 millivolts (mV), the output signal should be 50 mV.
3. The input signal to an amplifier is 1 mA. The output signal is 10 mA. The amplifier has a current gain of 10 W.

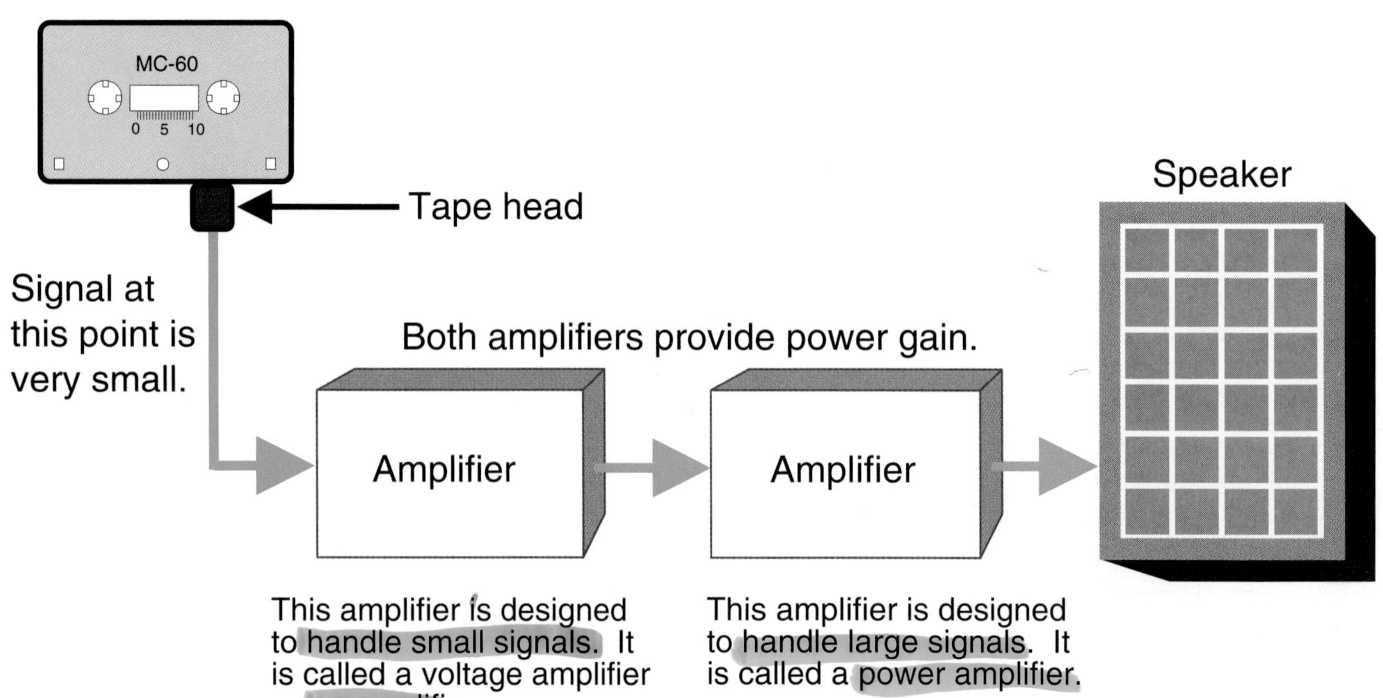

Fig. 5-2 Small-signal and large-signal amplifiers.

4. The input signal to an amplifier is 100 microvolts (μV) and its output signal is 50 mV so its voltage gain is 500.
5. A step-up transformer has voltage gain so it may be considered an amplifier.
6. All amplifiers have power gain.

5-2 Transistors

Transistors provide the power gain that is needed for most electronic applications. They also can provide voltage gain and current gain. There are several important types of transistors. This chapter will be mainly concerned with the *bipolar junction transistor* (BJT).

Bipolar junction transistors are similar to junction diodes, but one more junction is included. Figure 5-3 shows one way to make a transistor. A P-type semiconductor region is located between two N-type regions. The polarity of these regions is controlled by the valence of the materials used in the doping process. If you have forgotten this process and how it works, review the information in Chap. 2.

The transistor regions shown in Fig. 5-3 are named *emitter, base,* and *collector*. The *emitter* is very rich in current carriers. Its job is to send its carriers into the base region and then on to the collector. The *collector* collects the carriers. The emitter emits the carriers. The base acts as the control region. The *base* can allow none, some, or many carriers to flow from the emitter to the collector.

The transistor of Fig. 5-3 is *bipolar* because both holes and electrons will take part in the current flow through the device. The N-type regions contain free electrons which are negative carriers. The P-type region contains free holes which are positive carriers. Two (bi) polarities of carriers are present. Note that there are also two PN junctions in the transistor. It is a BJT.

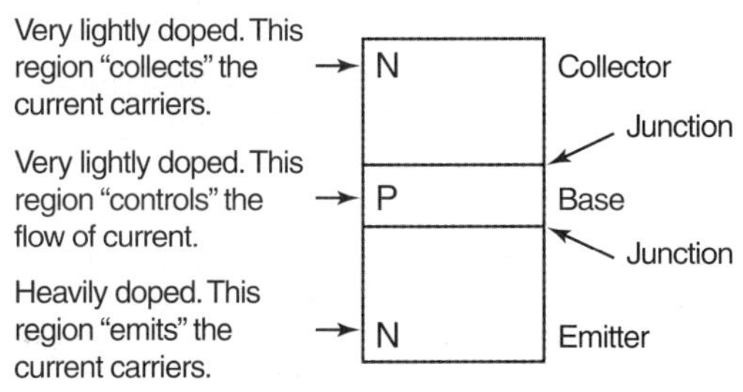

Fig. 5-3 NPN transistor structure.

The transistor shown in Fig. 5-3 would be classified as an *NPN transistor.* Another way to make a bipolar junction transistor is to make the emitter and collector of P-type material and the base of N-type material. This type would be classified as a *PNP transistor.* Figure 5-4 shows both possibilities and the schematic symbols for each. You should memorize the symbols. Remember that the emitter lead is always the one with the arrow. Also remember that if the arrow is Not Pointing iN, the transistor is an NPN type.

NPN transistor

PNP transistor

The two transistor junctions must be biased properly. This is why you cannot replace an NPN transistor with a PNP transistor. The polarities would be wrong. Transistor bias is shown in Fig. 5-5 on page 94. The *collector-base junction* must be *reverse-biased* for proper operation. In an NPN transistor, the collector will have to be positive with respect to the base. In a PNP transistor, the collector will have to be negative with respect to the base. PNP and NPN transistors are not interchangeable.

Bipolar junction transistor

Collector-base junction

The *base-emitter junction* must be *forward-biased,* as shown in Fig. 5-5. This makes the resistance of the base-emitter junction very low as compared with the resistance of the collector-base junction. A forward-biased semiconductor junction has low resistance. A reverse-biased junction has high resistance. Figure 5-6 on page 94 compares the two junction resistances.

Base-emitter junction

Emitter

Collector

Base

The large difference in junction resistance makes the transistor capable of power gain.

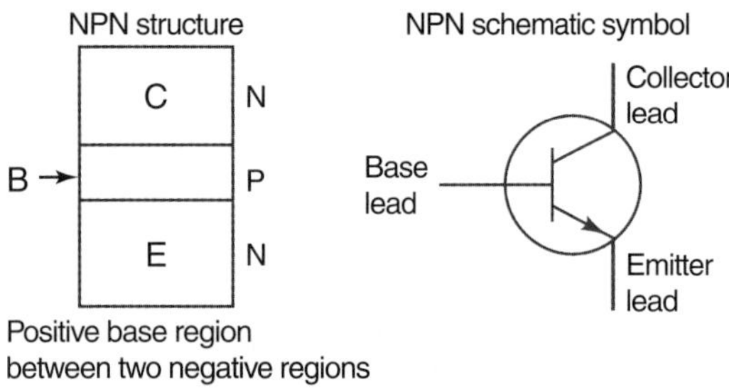

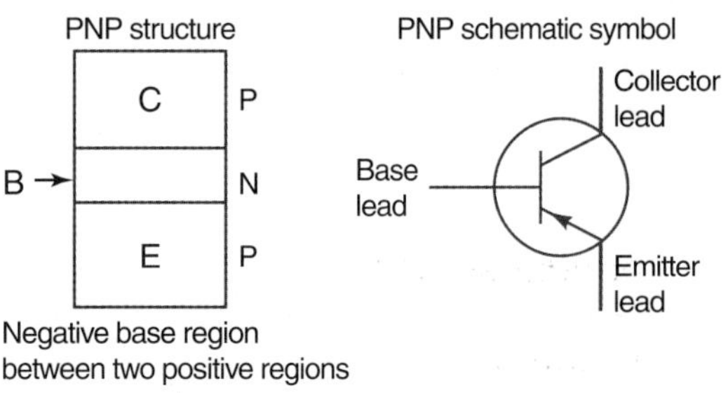

Fig. 5-4 Transistor structures and symbols.

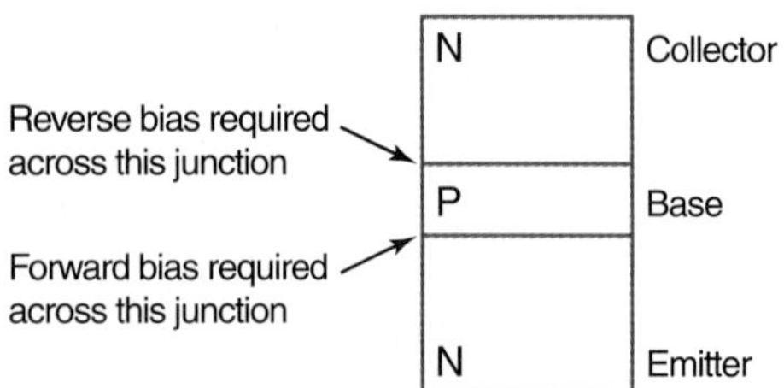

Fig. 5-5 Biasing the transistor junctions.

Assume that a current is flowing through the two resistances shown in Fig. 5-6. Power can be calculated using:

$$P = I^2 \times R$$

The power gain from R_{BE} to R_{CB} could be established by calculating the power in each and dividing:

$$P_{gain} = \frac{I^2 \times R_{CB}}{I^2 \times R_{BE}}$$

If the current through R_{CB} happened to be equal to the current through R_{BE}, I^2 would cancel out and the power gain would be

$$P_{gain} = \frac{R_{CB}}{R_{BE}}$$

The currents are not equal in transistors, but they are very close. A typical value for R_{CB} might be 10 kΩ. It is high since the collector-base junction is reverse-biased. A typical value for R_{BE} might be 100 Ω. It is low because the base-emitter junction is forward-biased. The power gain for this typical transistor would be

$$P_{gain} = \frac{R_{CB}}{R_{BE}} = \frac{10 \times 10^3\ \Omega}{100\ \Omega}$$

$$= 100$$

Note: The units (Ω) cancel and the gain is a pure number.

Perhaps the biggest puzzle is why the current through the reverse-biased junction is as high as the current through the forward-biased junction. Diode theory tells us to expect almost no current through a reverse-biased junction. This is true in a diode but not true in the collector-base junction of a transistor.

When job hunting, keep in mind that some companies offer better growth potential than others.

Figure 5-7 shows why the collector-base junction current is high. The collector-base voltage V_{CB} produces a reverse bias across the collector-base junction. The base-emitter voltage V_{BE} produces a forward bias across the base-emitter junction. If the transistor were simply two diode junctions, the results would be:

- I_B and I_E would be high
- I_C would be zero

The base region of the transistor is very narrow (about 0.0025 cm, or 0.001 in.). The base region is lightly doped. It has only a few free holes. It is not likely that an electron coming from the emitter will find a hole in the base with which to combine. With so few electron-hole combinations in the base region, the base current is *very small.* The collector is an N-type region but is charged positively by V_{CB}. Since the base is such a narrow region, the positive field of the collector is quite strong and the great majority of the electrons coming from the emitter are attracted and collected by the collector. Thus,

- I_E and I_C are high
- I_B is low

The emitter current of Fig. 5-7 is the highest current in the circuit. The collector current is just a bit less. Typically, about 99 percent of the emitter carriers go on to the collector. About 1 percent of the emitter carriers combine with carriers in the base and become base current. The current equation for Fig. 5-7 is

$$I_E = I_C + I_B$$

By using typical percentages, it can be stated as

$$100\% = 99\% + 1\%$$

The base current is quite small but *very* important. Suppose, for example, that the base lead of the transistor in Fig. 5-7 is opened. With the lead open, there can be no base cur-

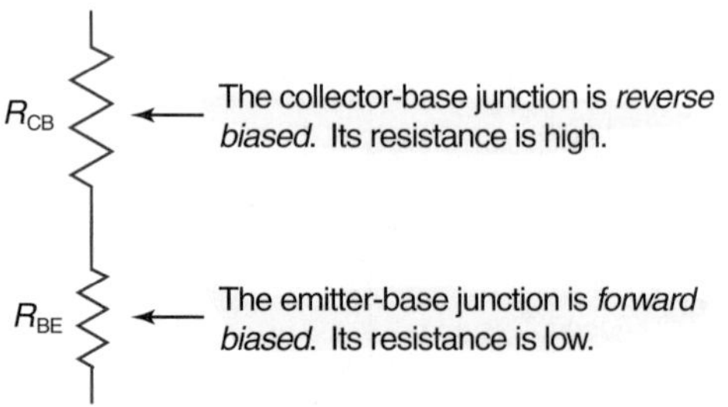

Fig. 5-6 Comparing junction resistances.

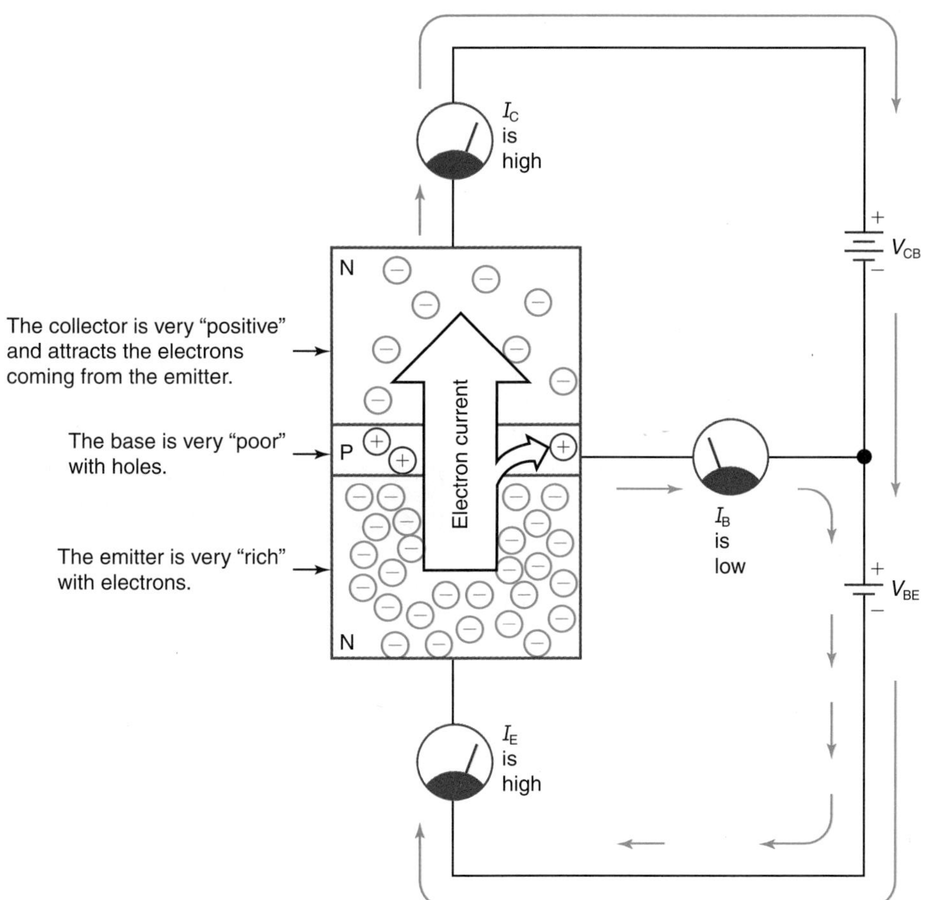

Fig. 5-7 NPN transistor currents.

rent. The two voltages V_{CB} and V_{BE} would add in series to make the collector positive with respect to the emitter. You might guess that current would continue to flow from the emitter to the collector, but *it does not. With no base current, there will be no emitter current and no collector current.* The base-emitter junction *must* be forward-biased for the emitter to emit. Opening the base lead removes this forward bias. If the emitter is not emitting, there is nothing for the collector to collect. Even though the base current is very low, it must be present for the transistor to conduct from emitter to collector.

Example 5-3

Determine the emitter current in a transistor when the base current is 1 mA and the collector current is 150 mA.

$I_E = I_C + I_B = 150 \text{ mA} + 1 \text{ mA} = 151 \text{ mA}$

Example 5-4

What is the base current in a transistor when the emitter current is 58 mA and the collector current is 56 mA? The equation is rearranged:

$I_B = I_E - I_C = 58 \text{ mA} - 56 \text{ mA} = 2 \text{ mA}$

Collector current

Base current

The fact that a low base current controls much higher currents in the emitter and collector is very important. This shows how the transistor is capable of good current gain. Quite often, the current gain from the base terminal to the collector terminal will be specified. This is one of the most important transistor characteristics. The characteristic is called β (Greek beta), or h_{FE}:

β (Greek beta)

$$\beta = \frac{I_C}{I_B} \quad \text{or} \quad h_{FE} = \frac{I_C}{I_B}$$

What is the β of a typical transistor? If the base current is 1 percent and the collector current is 99 percent, then

$$\beta = \frac{99\%}{1\%} = 99$$

Note that the percent symbol cancels since it appears in both the numerator and the denominator. This is also the case if actual current readings are used. The unit of current will cancel, leaving β as a pure number.

Example 5-5

Find β for a transistor with a base current of 0.3 mA and a collector current of 60 mA.

$$\beta = \frac{I_C}{I_B} = \frac{60 \text{ mA}}{0.3 \text{ mA}} = 200$$

Don't forget to take *prefixes* such as milli and micro into account when using the β equation. For example, if a transistor has a collector current of 5 mA and a base current of 25 μA, its β is found by

$$\beta = \frac{I_C}{I_B} = \frac{5 \times 10^{-3} \text{ A}}{25 \times 10^{-6} \text{ A}} = 200$$

The ampere units cancel. β is a pure number. Sometimes β is known and must be used to find either base current or collector current. If a transistor has a β of 150 and a collector current of 10 mA, how much base current is flowing? Rearranging the β equation and solving for I_B gives

$$I_B = \frac{I_C}{\beta} = \frac{10 \times 10^{-3} \text{ A}}{150} = 66.7\ \mu\text{A}$$

About Electronics

Transistor Applications—Then and Now

- The earliest commercial transistors worked only at frequencies below 1 MHz.
- Transistors intended for harsh environments use metal, glass, and ceramic packages.
- Pseudomorphic high electron mobility transistors (pHEMTs) provide new application areas in radar and microwave communications.

As another example, let's find the collector current in a transistor circuit with a β of 40 and a base current of 85 mA:

$$I_C = \beta \times I_B = 40 \times 85 \text{ mA} = 3.4 \text{ A}$$

Occasionally a current must be calculated first before the current gain can be determined. Don't forget that the emitter current is the sum of the collector and base currents.

Example 5-6

A transistor has an emitter current of 12.1 mA and a collector current of 12.0 mA. What is the β of this transistor? First, rearrange the current equation to find the base current:

$$I_B = I_E - I_C = 12.1 \text{ mA} - 12.0 \text{ mA} = 0.1 \text{ mA}$$

Then find β:

$$\beta = \frac{I_C}{I_B} = \frac{12 \text{ mA}}{0.1 \text{ mA}} = 120$$

The β of actual transistors varies greatly. Certain power transistors can have a β as low as 20. Small-signal transistors can have a β as high as 400. If you have to guess, a β of 150 can be used for small transistors and a β of 50 for power transistors.

The value of β varies among transistors with the same part number. A 2N2222 is a registered transistor. One manufacturer of this particular device lists a typical β range of 100 to 300. Thus, if three seemingly identical 2N2222 transistors are checked for β, values of 108, 167, and 256 could be obtained. It is *very* unlikely that they would check the same.

The value of β is important but unpredictable. Luckily, there are ways to use transistors that make the actual value of β less important than other, more predictable circuit characteristics. This will become clear in a later chapter. For now, focus on the idea that the current gain from the base terminal to the collector terminal tends to be high. Also, remember that the base current is small and controls the collector current.

Figure 5-8 shows what happens in a PNP transistor. Again, the base-emitter junction must be forward-biased. Note that V_{BE} is reversed in polarity when compared to Fig. 5-7. The collector-base junction of the PNP transistor must be reverse-biased. Note also that V_{CB} has been reversed in polarity. This is why PNP and NPN transistors are not interchangeable. If one were substituted for the other, both the collector-base and the base-emitter junction would be biased incorrectly.

Figure 5-8 shows the flow from emitter to collector as *hole current.* In an NPN transistor, it is *electron* current. The two transistor structures operate about the same in most ways. The emitter is very rich with carriers. The base is quite narrow and has only a few carriers. The collector is charged by the external bias source and attracts the carriers coming from the emitter. The major difference between PNP and NPN transistors is polarity.

The NPN transistor is more widely applied than the PNP transistor. Electrons have better *mobility* than holes; that is, they can move more quickly through the crystal structure. This gives NPN transistors an advantage in high-frequency circuits where things have to happen quickly. Transistor manufacturers have more NPN types in their line. This makes it easier for circuit designers to choose the exact characteristics they need from the NPN group. Finally, it is often more convenient to use NPN devices in negative ground systems. Negative ground systems are more prevalent than positive ground systems.

You will find both types of transistors in use. Many electronic systems use both PNP and NPN transistors in the same circuit. It is very convenient to have both polarities available. This adds flexibility to circuit design.

For more information on transistors, visit the web site for Motorola.

Hole current

TEST

Determine whether each statement is true or false.

7. The emitter region of a junction transistor is heavily doped to have many current carriers.

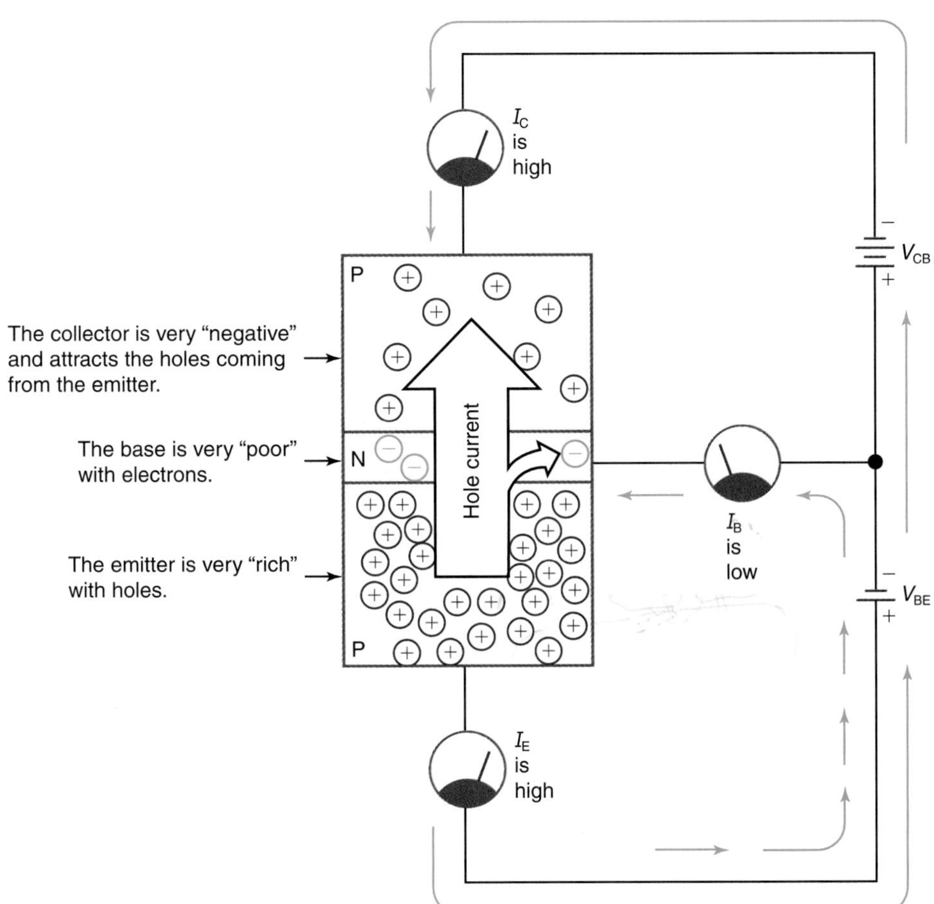

Fig. 5-8 PNP transistor currents.

8. A bipolar device may be connected in either direction and still give proper operation.
9. The collector-base junction must be forward-biased for proper transistor action.
10. A defective NPN transistor can be replaced with a PNP type.
11. Even though the collector-base junction is reverse-biased, considerable current can flow in this part of the circuit.
12. The base of BJTs is thin and lightly doped with impurities.
13. When I_B is equal to zero a BJT is off and I_C will also be close or equal to zero.
14. Base current controls collector current.
15. Base current is greater than emitter current.
16. Transistor β is measured in milliamperes.
17. 2N2222 transistors are manufactured to have a current gain of 222 from the base to the collector.
18. In a PNP transistor, the emitter emits holes and the collector collects them.
19. A PNP transistor is turned on by forward biasing its base-emitter junction.

Solve the following problems.

20. A transistor has a base current of 500 μA and a β of 85. Find the collector current.
21. A transistor has a collector current of 1 mA and a β of 150. Find its base current.
22. A transistor has a base current of 200 μA and a collector current of 50 mA. Find its β.
23. A transistor has a collector current of 1 A and an emitter current of 1.01 A. Find its base current.
24. Find β for the transistor described in problem 23.

5-3 Characteristic Curves

Collector family of curves

As with diodes, transistor characteristic curves can provide much information. There are many types of transistor characteristic curves. One of the more popular types is the *collector family of curves*. An example of this type is shown in Fig. 5-9. The vertical axis shows collector current (I_C) and is calibrated in milliamperes. The horizontal axis shows collector-emitter bias (V_{CE}) and is calibrated in volts. Figure 5-9 is called a collector *family* since several volt-ampere characteristic curves are presented for the same transistor.

Figure 5-10 shows a circuit that can be used to measure the data points for a collector family of curves. Three meters are used to monitor base current I_B, collector current I_C, and collector-emitter voltage V_{CE}. To develop a graph of three values, one value can be held constant as the other two vary. This produces

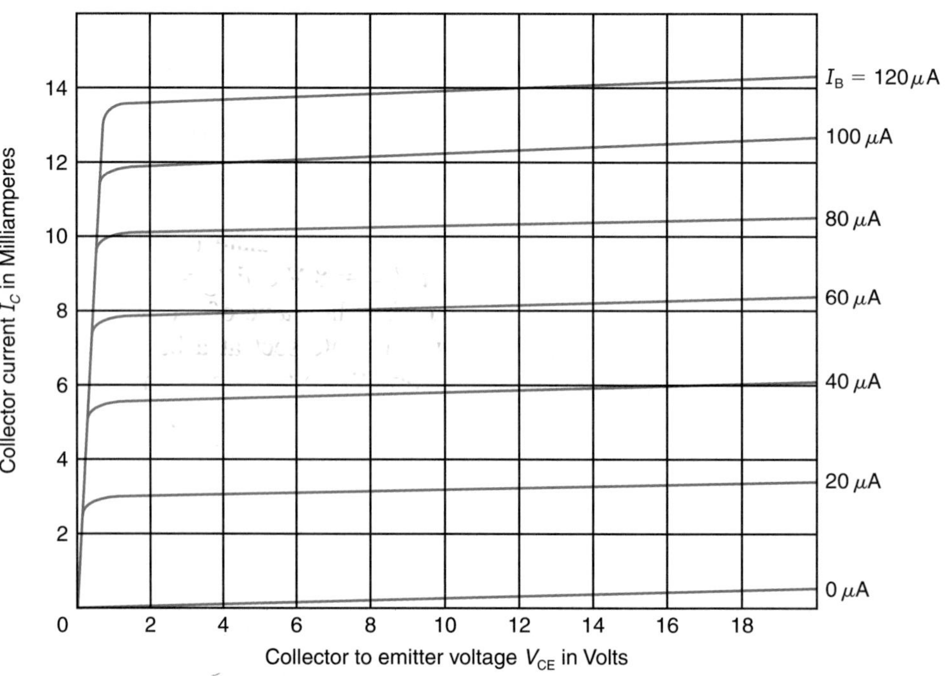

Fig. 5-9 A collector family of curves for an NPN transistor.

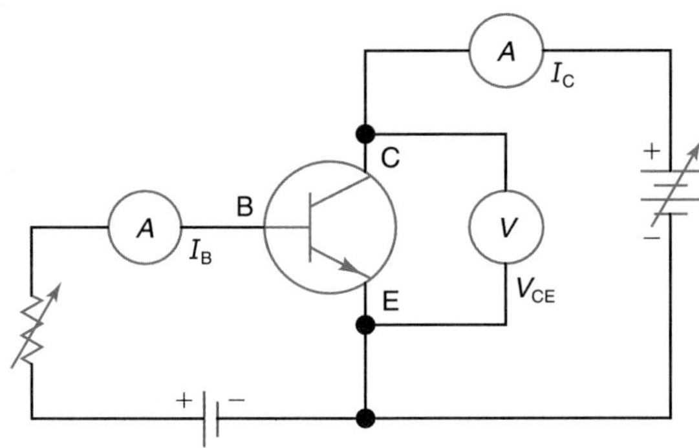

Fig. 5-10 Circuit for collecting transistor data.

one curve. Then the constant value is set to a new level. Again, the other two values are changed and recorded. This produces the second curve. The process can be repeated as many times as required. For a collector family of curves, the constant value is the *base current.* The variable resistor in Fig. 5-10 is adjusted to produce the desired level of base current. Then the adjustable source will be set to some value of V_{CE}. The collector current is recorded. Next, V_{CE} is changed to a new value. Again, I_C is recorded. These data points are plotted on a graph to produce a volt-ampere characteristic curve of I_C versus V_{CE}. A very accurate curve can be produced by recording many data points. The next curve in the family is produced in exactly the same way but at a new level of base current.

The curves of Fig. 5-9 show some of the important characteristics of junction transistors. Notice that over most of the graph the collector-emitter voltage has *little* effect on the *collector current.* Examine the curve for $I_B = 20$ μA. How much change in collector current can you see over the range from 2 to 18 V? This is a ninefold increase in voltage. Ohm's law tells us to expect the current to increase 9 times. It would increase 9 times if the transistor were a simple resistor. In a transistor, the base current has the major effect on collector current. Notice that the collector voltage affects current *only when it is very low* (below 1 V in Fig. 5-9).

It is important to be able to convert the curves back into data points. For example, can you read the value of I_C when $V_{CE} = 6$ V and $I_B = 20$ μA? Refer to Fig. 5-9. First, locate 6 V on the horizontal axis. Project up from this point until you reach the 20-μA curve. Now, project from this point to the left and read the value of I_C on the vertical axis. You should obtain a value of 3 mA. Try another: Find the value of I_B when $I_C = 10$ mA and $V_{CE} = 4$ V. These two data points cross on the 80-μA curve. The answer is 80 μA. It may be necessary to estimate a value. For example, what is the value of base current when $V_{CE} =$ 10 V and $I_C = 7$ mA? The crossing of these two values occurs well away from any of the curves in the family. It is about halfway between the 40-μA curve and the 60-μA curve, so 50 μA is a good estimate.

Example 5-7

Base current

Use Fig. 5-9 to determine the collector current when $V_{CE} = 8$ V and $I_B = 20$ μA. Using estimation gives a value of I_C of about 3.1 mA.

Example 5-8

Use the curves of Fig. 5-9 to find the emitter current when V_{CE} is 6 V and I_B is 100 μA. The collector curves do not show any emitter data, but emitter current can be found from base current and collector current. We already know the base current, so we inspect the curves to find the collector current. Figure 5-9 shows that $V_{CE} = 6$ V and $I_B =$ 100 μA intersect at $I_C = 12$ mA. Thus:

Collector current

$$I_E = I_C + I_B = 12 \text{ mA} + 100\ \mu\text{A}$$
$$= 12.1 \text{ mA}$$

The curves of Fig. 5-9 give enough information to calculate β. What is the value of β at $V_{CE} = 8$ V and $I_C = 8$ mA? The first step is to find the value of the base current. The two values intersect at a base current of 60 μA. Now, β can be calculated:

$$\beta = \frac{I_C}{I_B}$$
$$= \frac{8 \text{ mA}}{60\ \mu\text{A}}$$
$$= 133$$

Calculate β for the conditions of $V_{CE} = 12$ V and $I_C = 14$ mA. These values intersect at $I_B =$ 120 μA:

h_{fe}

$$\beta = \frac{14 \text{ mA}}{120 \ \mu\text{A}} = 117$$

The two prior calculations reveal another fact about transistors. Not only does β vary from transistor to transistor, but *it also varies with* I_C. Later, it will be shown that temperature also affects β.

What happens to a transistor when its base current is relatively large? For example, what about Fig. 5-11 if $I_B = 1$ mA? This is off the graph! But, it can be interpreted. First, this will not damage the transistor unless its maximum collector dissipation rating is exceeded. This subject will be covered a little later in this section. Transistors can be damaged by excess base current, but something important happens before that extreme is reached. So, let's return to the graph. The transistor will operate somewhere along the *steep vertical part* of the characteristic curves. Look closely at Fig. 5-11, and you can see that the transistor voltage V_{CE} is always small for base currents greater than 35 μA. When a transistor is operated this way, it is turned on hard (saturated). It acts like a closed switch from collector to emitter. Ideally, a closed switch drops zero volts since it has no resistance. Real switches have a little resistance and drop a small voltage. Operating bipolar transistors with large base currents makes them act like closed switches. Switching applications are covered in the last section of this chapter.

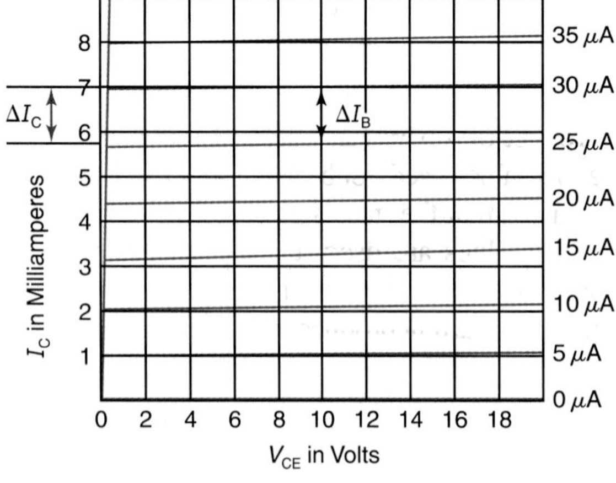

Fig. 5-11 Calculating β_{ac} with characteristic curves.

There is another form of current gain from base to collector called β_{ac} or h_{fe}. Study the following equations to see how β_{ac} differs from what has already been discussed:

$$\beta_{dc} = h_{FE} = \frac{I_C}{I_B}$$

$$\beta_{ac} = h_{fe} = \frac{\Delta I_C}{\Delta I_B} \Big| V_{CE}$$

The symbol Δ means "change in" and the symbol $|$ means that V_{CE} is to be held constant. Figure 5-11 shows the process. The collector-to-emitter voltage is constant at 10 V. The base current changes from 30 to 25 μA, for a ΔI_B value of 5 μA. Projecting to the left shows a corresponding change in collector current from 7.0 to 5.7 mA. This represents a ΔI_C value of 1.3 mA. Dividing gives a β_{ac} of 260.

There is no significant difference between β_{dc} and β_{ac} at low frequencies. This book emphasizes β_{dc}. The beta symbol with no subscript will designate *dc current gain.* Alternating current gain will be designated by β_{ac}.

Example 5-9

Use Fig. 5-9 to obtain the data needed to calculate β_{ac} when $V_{CE} = 4$ V and I_B is changing from 20 to 40 μA.

$$\beta_{ac} = \frac{\Delta I_C}{\Delta I_B} = \frac{2.6 \text{ mA}}{20 \ \mu\text{A}} = 130$$

Gain-bandwidth product

At high frequencies the ac current gain of bipolar junction transistors starts to fall off. This effect limits the useful frequency range of transistors. The *gain-bandwidth product* is the frequency at which the ac current drain drops to 1. The symbol for gain-bandwidth product is f_T. This transistor specification is important in high-frequency applications.

It is standard practice to plot positive values to the right on the horizontal axis and up on the vertical axis. Negative values go to the left and down. A family of curves for a PNP transistor may be plotted on a graph as shown in Fig. 5-12. The collector voltage must be negative in

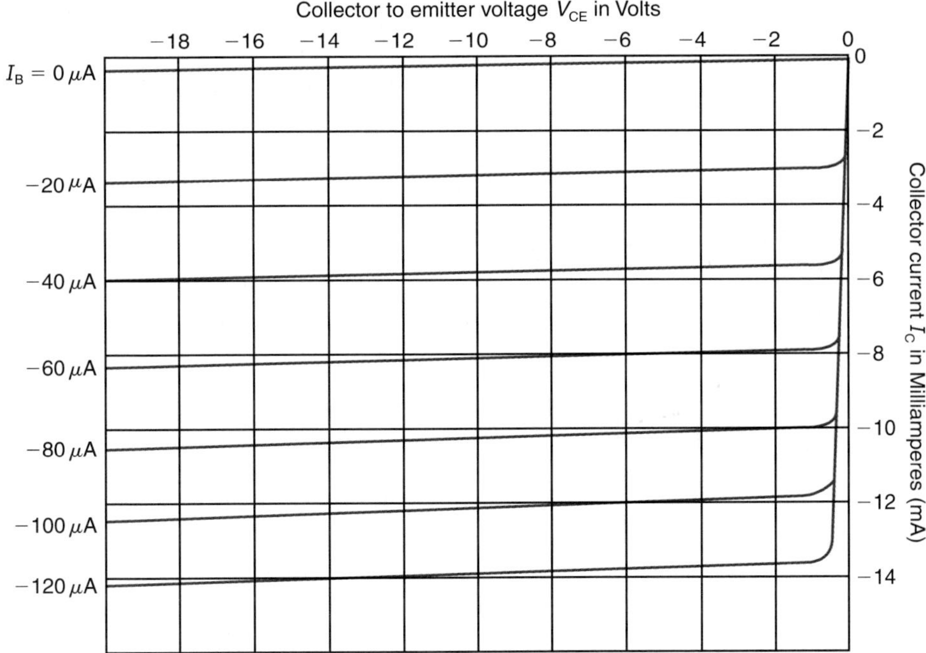

Fig. 5-12 A collector family of curves for a PNP transistor.

a PNP transistor. Thus, the curves go to the left. The collector current is in the opposite direction, compared to an NPN transistor. Thus, the curves go down. However, curves for PNP transistors are sometimes drawn up and to the right. Either method is equally useful for presenting the collector characteristics.

Some shops and laboratories are equipped with a device called a *curve tracer.* This device draws the characteristic curves on a cathode-ray tube or picture tube. This is far more convenient than collecting many data points and plotting the curves by hand. Curve tracers show NPN curves in the first quadrant (as in Fig. 5-9) and PNP curves in the third quadrant (as in Fig. 5-12).

The collector family of curves can be used to show the *safe operating area* for a transistor. Figure 5-13 on page 102 shows an example of this. A *constant-power curve* has been added to the graph that clearly divides the curves into those operating points below 7.5 W and those above 7.5 W. This makes it very easy to find safe areas of operation for the transistor. For example, if the maximum safe transistor dissipation is 7.5 W, then no operating point that falls to the right of the power curve would be safe.

Transistor dissipation is usually calculated for the collector circuit. It is based on this power formula:

$$P = V \times I$$

Thus, *collector dissipation* is calculated by

$$P_C = V_{CE} \times I_C$$

Now, the power curve on Fig. 5-13 can be verified. At $V_{CE} = 4$ V, the power curve crosses at a little less than 1.9 A on the I_C axis:

$$P_C = 4\text{ V} \times 1.9\text{ A} = 7.6\text{ W}$$

At $V_{CE} = 8$ V, the power curve crosses a bit above 0.9 A on the I_C axis:

$$P_C = 8\text{ V} \times 0.9\text{ A} = 7.2\text{ W}$$

All points along the power curve represent a product of 7.5 W (± graphical error). The negative values need not be taken into account. They indicate that the transistor is a PNP type. If negative values are used, the answers remain the same since multiplying a negative voltage by a negative current produces a positive power value.

If the collector characteristic curves are extended to include higher voltages, *collector breakdown* can be shown. Like diodes, transistors have limits as to the amount of reverse bias that can be applied. Transistors have two junctions, and their breakdown ratings are complicated. Figure 5-14 on page 102 shows a collector family of curves where the

Collector dissipation

Curve tracer

Constant-power curve

Collector breakdown

Transistor dissipation

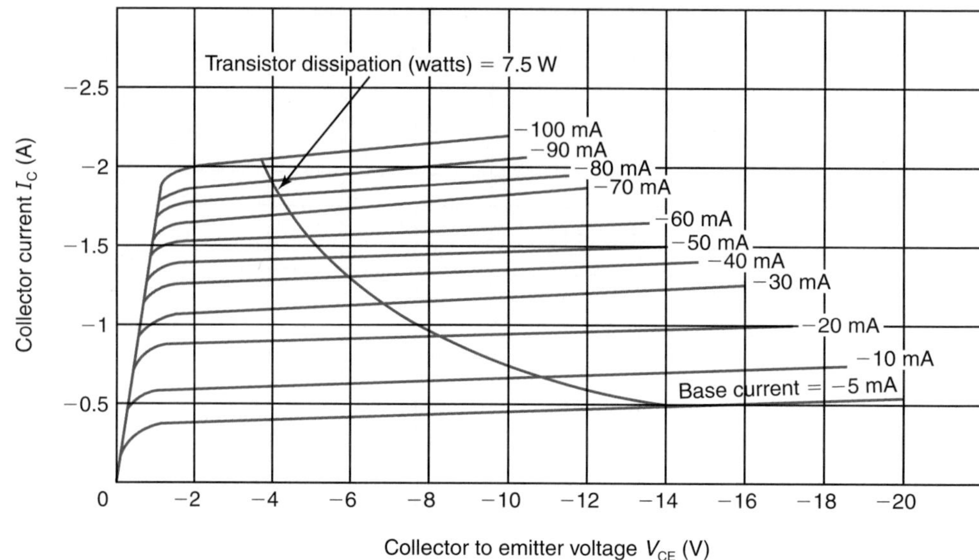

Fig. 5-13 Constant-power curve.

horizontal axis is extended to 140 V. If collector voltage becomes very high, it begins to control collector current. This is undesirable. The base current is supposed to control the collector current. Transistors should not be operated near or over their maximum voltage ratings. As can be seen from Fig. 5-14, collector breakdown is not a fixed point as it is with diodes. It varies with the amount of base current. At 15 μA, the collector breakdown point is around 110 V. At 0 μA, it occurs near 130 V.

Example 5-10

Calculate the power dissipation for Fig. 5-13 when $V_{CE} = -10$ V and $I_B = -70$ mA.

$$P_C = 10 \text{ V} \times 1.8 \text{ A} = 18 \text{ W}$$

Note: This exceeds the safe limit for this device.

Transfer characteristic curve

The *transfer characteristic curves* shown in Fig. 5-15 are another example of how curves

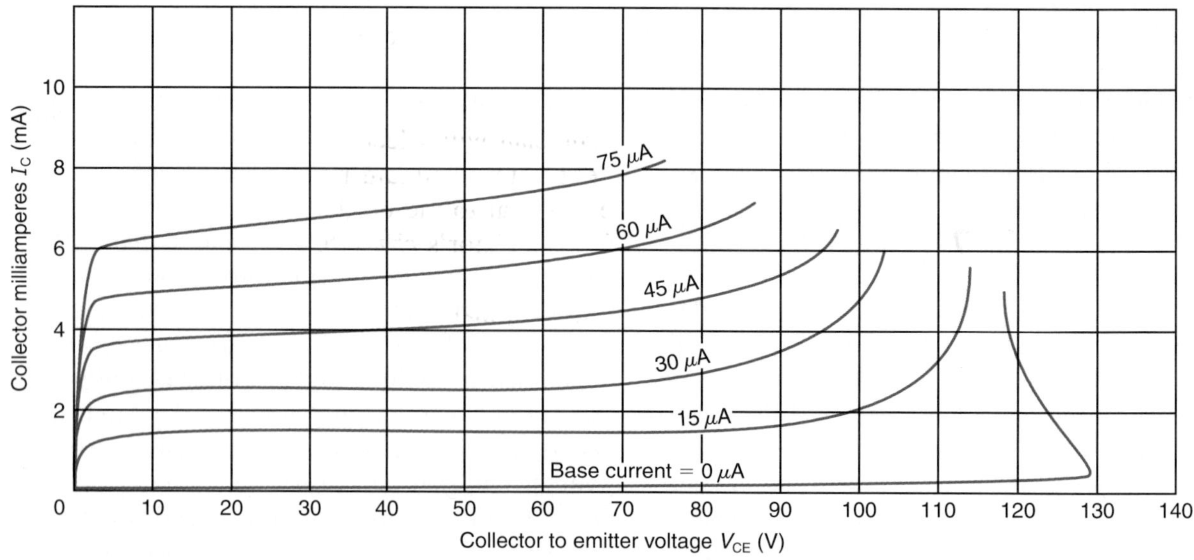

Fig. 5-14 Collector breakdown.

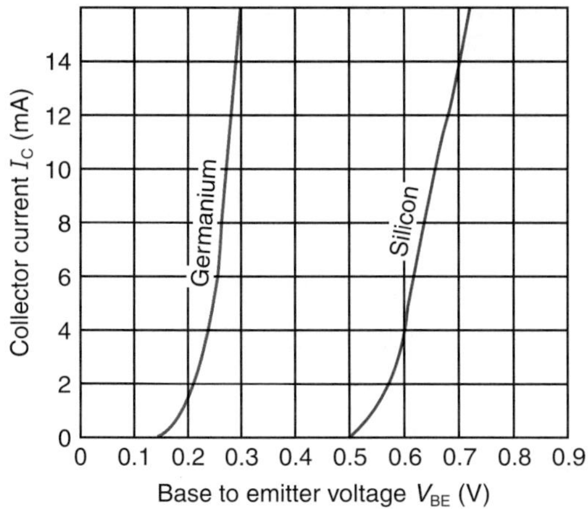

Fig. 5-15 Comparing silicon and germanium transistors.

can be used to show the electrical characteristics of a transistor. Curves of this type show how one transistor terminal (the base) affects another (the collector). This is why they are called transfer curves. We know that base current controls collector current. Figure 5-15 shows how base-emitter voltage controls collector current. This is because the base-emitter bias sets the level of base current.

Figure 5-15 also shows one of the important differences between silicon transistors and germanium transistors. Like diodes, germanium transistors turn on at a much lower voltage (approximately 0.2 V). The silicon device turns on near 0.6 V. These voltages are important to remember. They are reasonably constant and can be of great help in troubleshooting transistor circuits. They can also help a technician determine whether a transistor is made of silicon or germanium.

Germanium transistors are not widely applied in modern systems. Germanium does offer a few advantages for some applications, however. A few high-power transistors use germanium since it is a better conductor than silicon. The low turn-on voltage of germanium is also an advantage in some circuits. Silicon transistors are inexpensive and show much better high-temperature performance. These two reasons make them the logical choice for most applications.

TEST

Supply the missing word or number in each statement.

25. Refer to Fig. 5-9. Voltage V_{CE} = 4 V and current I_C = 3 mA. I_B = ________.
26. Refer to Fig. 5-9. Current I_B = 90 μA and voltage V_{CE} = 4 V. I_C = ________.
27. Refer to Fig. 5-9. Voltage V_{CE} = 6 V and current I_C = 8 mA. β = ________.
28. Refer to Fig. 5-9. Current I_B = 100 μA and voltage V_{CE} = 8 V. P_C = ________.
29. Refer to Fig. 5-9. V_{CE} is held constant at 4 V. I_B changes from 60 to 80 μA. β_{ac} = ________.
30. Germanium transistors turn on when V_{BE} reaches __0.2__ V.
31. Silicon transistors turn on when V_{BE} reaches __0.6__ V.
32. Of the two popular semiconductor materials, __Germanium is__ the better conductor.

5-4 Transistor Data

Transistor manufacturers prepare data sheets that detail the mechanical, thermal, and electrical characteristics of the parts they make. These data sheets are often bound into volumes called *data manuals.* Table 5-1 on page 104 is a sample from a data manual. It shows the *maximum ratings* and some of the *characteristics* for 2N2222A transistors. Data manuals also contain characteristic curves such as those discussed in the previous section of this chapter.

Data manuals

Technicians usually try to replace a defective transistor with one having the same part number. This is considered an "exact replacement" even when the manufacturer is different. Sometimes it is impossible to find an exact replacement. Data, such as that shown in Table 5-1, is very useful in these cases. The technician will select a replacement with maximum ratings at least equal to the original part. The transistor's characteristics must also be examined and matched as closely as possible to the original.

A good question to ask during a job interview concerns company-sponsored training and continuing education for employees.

One way for a technician to learn something about a particular transistor is to use *substitution* guides. These guides are not totally accurate, but they do provide a good, general idea about the device of interest. Another good source of information is a *parts catalog.* Figure 5-16 on page 104 is a sample of transistor listings from a parts catalog. The prices have been

Parts catalog

Table 5-1 Selected Specifications for the 2N2222A Bipolar Junction Transistor

Parameter	Symbol	Value
	Maximum ratings	
Collector-emitter voltage	V_{CEO}	40 V dc
Collector-base voltage	V_{CB}	75 V dc
Emitter-base voltage	V_{EB}	6 V dc
Collector current	I_C	800 mA dc
Total device dissipation		1.8 W
(derate above 25°C)	P_D	12 mW/°C
	Characteristics	
DC current gain	h_{FE}	100 to 300
AC current gain	h_{fe}	50 to 375
Gain-bandwidth product	f_T	300 MHz
Collector-emitter saturation	$V_{CE(sat)}$	0.3 V dc
Noise figure	NF	4 dB

deleted. These listings may even include some of the nonregistered device numbers. Notice that quite a bit of information is listed in the catalog for each transistor number. For example, a 2N5179 transistor is seen to use a TO-72 *case style,* to be a silicon NPN type, to be used as an ultra-high-frequency (UHF) amplifier, to dissipate 0.2 W, and so on. Parts catalogs are available for a small cost or often are free. It is a good idea to gather a collection of these catalogs and obtain new ones as they become available.

Case style

Figure 5-17 shows another example of the information that can be found in substitution guides and parts catalogs. Transistors are made in many case styles. The physical characteristics can be just as important as the electrical details. Figure 5-17 is only a sample of the many case styles used today. Notice that this material is also valuable because it can help you to identify the emitter, base, and collector leads. This information is not usually marked on the transistor case.

In some cases the part number cannot be found in any of the available guides or on the original transistor. It may be possible to use a general type of unit in these situations. For example, the 2N2222A (or the 2N3904) is a good, general-purpose replacement for small-signal silicon NPN BJTs. Likewise, the 2N2905A (or the 2N3906) is a general-purpose PNP replacement. General-purpose replacements should be avoided in these cases:

- VHF or UHF applications
- High-power applications
- High-voltage applications

Substitute transistors must be of the same material and the same polarity. Also, be sure that size and lead arrangements are compatible. They must be based on the same technol-

Type	Case	Material Function	Maximum Ratings: Dissipation Watts	Maximum Ratings: Coll. To Base Volts	Maximum Ratings: Coll. Curr. mA	Beta H_{FE}@Ic: Min. Max.	Beta H_{FE}@Ic: mA	f_T MHz
2N2870/ 2N301	TO-3	GP AP	30C	80	3A	50-165	1A	.200
2N2876	TO-60	SN AV	17.5C	80	2.5A			.200
2N2894	TO-18	SP SH	1.2C	12	200	40-150	30	400
2N2895	TO-18	SN GP	.500	120	1A	60-150	1	120
2N5070	TO-60	SN AP	70C	65	3.3A	10-100	3A	100
2N5071	TO-60	SN AP	70C	65	3.3A	10-100	3A	100
2N5086	TO-92	SP GP	.310	50	50	150-	1	40
2N5087	TO-92	SP GP	.310	50	50	250-	1	40
2N5088	TO-92	SN GP	.310	35	50	350-	1	
2N5172	TO-98	SN GP	.200	25	100	100-	10	
2N5179	TO-72	SN AU	.200	20	50	25-	20	900
2N5180	TO-104	SN AU	.180	30		20-	2	650
2N5183	TO-104	SN GP	.500	18	1A	70-	10	62
2N5184	TO-104	SN GP	.500	120	50	10-	50	50

Material code:
GP .... Germanium, PNP
SN .... Silicon, NPN
SP .... Silicon, PNP

Function code:
AP .... Amplifier, power
AV .... Amplifier, VHF
SH .... Switch, high speed
GP .... General purpose
AU .... Amplifier, UHF

Fig. 5-16 Transistor catalog listings.

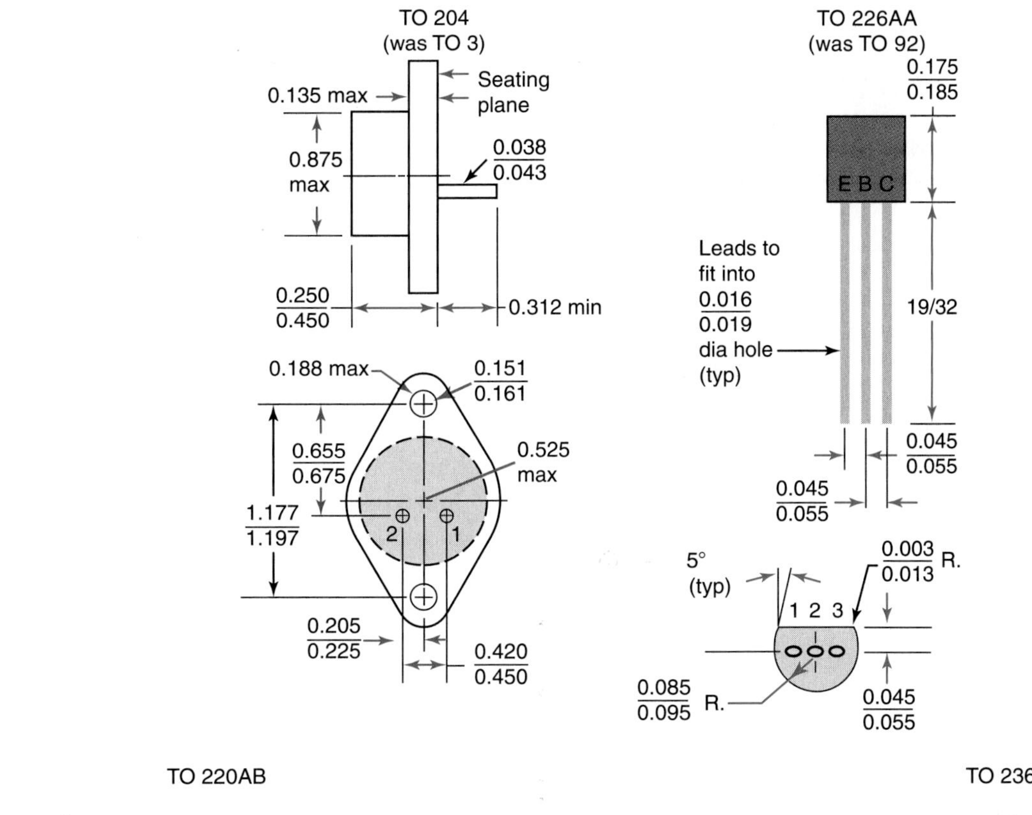

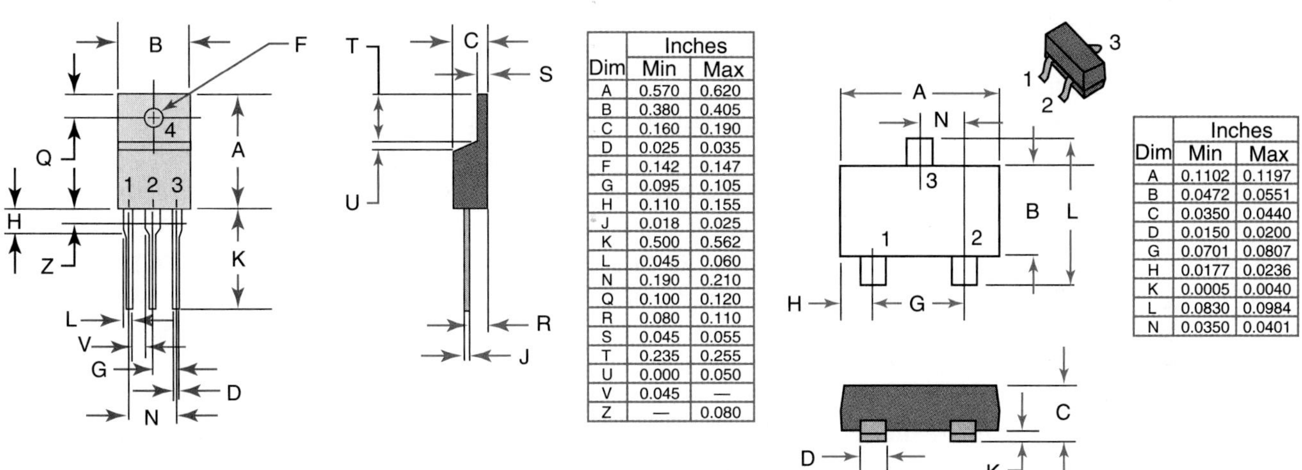

TO 220AB

Dim	Inches Min	Max
A	0.570	0.620
B	0.380	0.405
C	0.160	0.190
D	0.025	0.035
F	0.142	0.147
G	0.095	0.105
H	0.110	0.155
J	0.018	0.025
K	0.500	0.562
L	0.045	0.060
N	0.190	0.210
Q	0.100	0.120
R	0.080	0.110
S	0.045	0.055
T	0.235	0.255
U	0.000	0.050
V	0.045	—
Z	—	0.080

TO 236AB

Dim	Inches Min	Max
A	0.1102	0.1197
B	0.0472	0.0551
C	0.0350	0.0440
D	0.0150	0.0200
G	0.0701	0.0807
H	0.0177	0.0236
K	0.0005	0.0040
L	0.0830	0.0984
N	0.0350	0.0401

Note: All dimensions in inches.

Fig. 5-17 Transistor case styles.

ogy. For example, the last section of this chapter covers some transistors that *will not* interchange with BJTs.

Circuit voltages can be inspected to give some idea of the voltage ratings that the new transistor should have. The power ratings can be established by inspecting the circuit current levels and voltages. Of course, the physical characteristics should also be similar. Finally, knowing the function of the original unit is helpful in picking a substitute. Substitution guides and catalogs often list transistors as audio types, very-high-frequency (VHF) types, switching types, and in other descriptive ways.

TEST

Determine whether each statement is true or false.

33. Device manufacturers publish data sheets and data manuals for solid-state devices.
34. Almost all solid-state devices have the leads marked on the case.

Did You Know?

Much Is Learned on the Job With most companies, an *entry-level position* means that related work experience is not required.

Overlay-type transistor

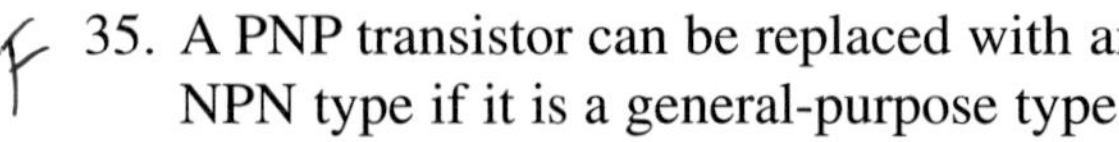

35. A PNP transistor can be replaced with an NPN type if it is a general-purpose type.

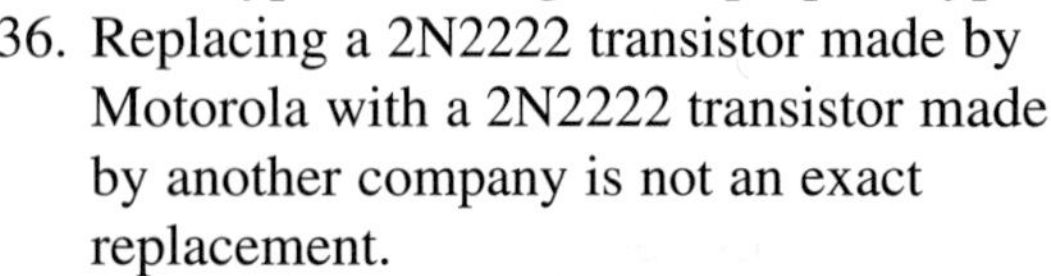

36. Replacing a 2N2222 transistor made by Motorola with a 2N2222 transistor made by another company is not an exact replacement.

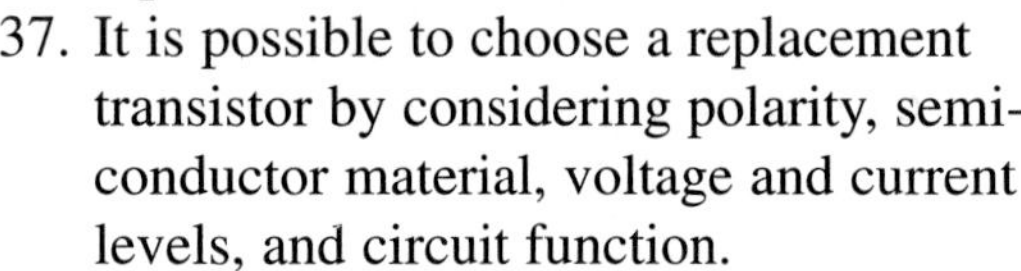

37. It is possible to choose a replacement transistor by considering polarity, semiconductor material, voltage and current levels, and circuit function.

5-5 Transistor Testing

One way to test transistors is to use a curve tracer. This technique is used by semiconductor manufacturers and by equipment makers to test incoming parts. Curve tracers are also used in design labs. Most technicians do not have access to a curve tracer.

Dynamic test

Noise figure

Another technique used at manufacturing and design centers is to place the transistor in a special fixture or test circuit. This is a *dynamic test* because it makes the device operate with real voltages and signals. This method of testing is often used for VHF and UHF transistors. Dynamic testing reveals power gain and *noise figure* under signal conditions. Noise figure is a measure of a transistor's ability to amplify weak signals. Some transistors make enough electrical noise to overpower a weak signal. These transistors are said to have a poor noise figure.

A few transistor types may show a gradual loss of power gain. Radio-frequency power amplifiers, for example, may use *overlay-type transistors.* These transistors can have over 100 separate emitters. They can suffer base-emitter changes which can gradually degrade power gain. Another problem is moisture, which can enter the transistor package and gradually degrade performance. Even though gradual failures are possible in transistors, they are *not* typical.

For the most part, *transistors fail suddenly and completely.* One or both junctions may short-circuit. An internal connection can break loose or burn out from an overload. This type of failure is easy to check. Most bad transistors can be identified with a few ohmmeter tests out-of-circuit or with voltmeter checks in-circuit.

A good transistor has two PN junctions. Both can be checked with an ohmmeter. As shown in Fig. 5-18, a PNP transistor can be compared to two diodes with a common cathode connection. The base lead acts as the common cathode. Figure 5-19 shows an NPN transistor as two diodes with a common anode connection. If two good diodes can be verified by ohmmeter tests, the transistor is probably good.

The ohmmeter can also be used to identify the polarity (NPN or PNP) of a transistor and the three leads. This can be helpful when data are not available. Analog ohmmeters should be set to the $R \times 100$ range for testing most transistors. Germanium power transistors may be easiest to check by using the $R \times 1$ range. Power transistors are easy to recognize because they are physically larger than small-signal transistors. When using a DMM, use the diode function.

The first step in testing transistors is to connect the ohmmeter leads across two of the tran-

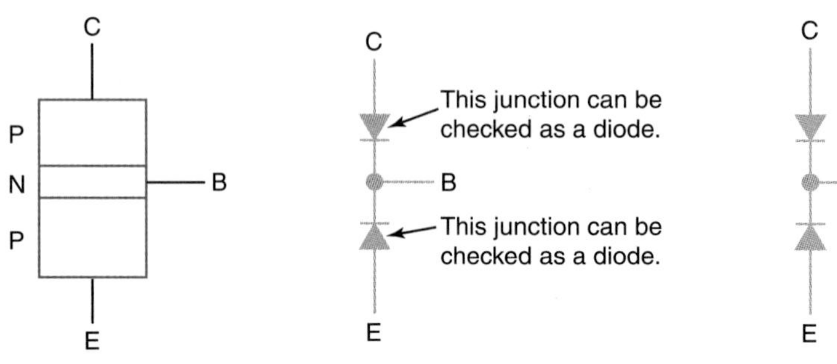

Fig. 5-18 PNP junction polarity.

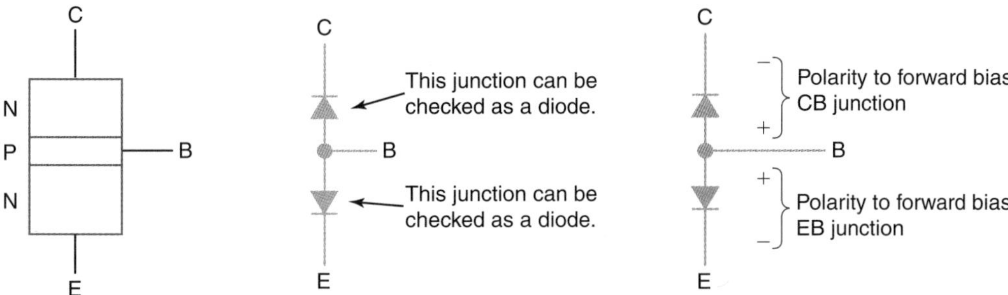

Fig. 5-19 NPN junction polarity.

sistor leads, as shown in Fig. 5-20. If a lower resistance is indicated, the leads are across one of the diodes or else the transistor is shorted. To decide which is the case, reverse the ohmmeter leads. If the *transistor junction* is good, the ohmmeter will show a high resistance, as seen in Fig. 5-21. If you happen to connect across the emitter and collector leads of a good transistor, the ohmmeter will show high resistance in both directions. The reason is that two junctions are in the ohmmeter circuit. Study Figs. 5-18 and 5-19 and verify that with either polarity applied from emitter to collector, one of the diodes will be reverse-biased.

Once the emitter-collector connection is found, the base has been identified by the process of elimination. Now connect the negative lead of the ohmmeter to the base lead. Touch the positive lead to one and then the other of the two remaining leads. If a low resistance is shown, the transistor is a PNP type. Connect the positive lead to the base lead. Touch the negative lead to one and then the other of the two remaining leads. If a low resistance is shown, then the transistor is an NPN type.

Transistor junction

Checking junction

Thus far, you have identified the base lead and the polarity of the transistor. Now it is possible to check the transistor for gain and to identify the collector and emitter leads. All that is needed is a 100,000-Ω resistor and an analog ohmmeter. If you are checking a germanium power transistor on the $R \times 1$ range, use a 1000-Ω resistor. DMMs are usually *not* useful for gain checking (some do have a gain function).

The resistor will be used to provide the transistor with a small amount of base current. If

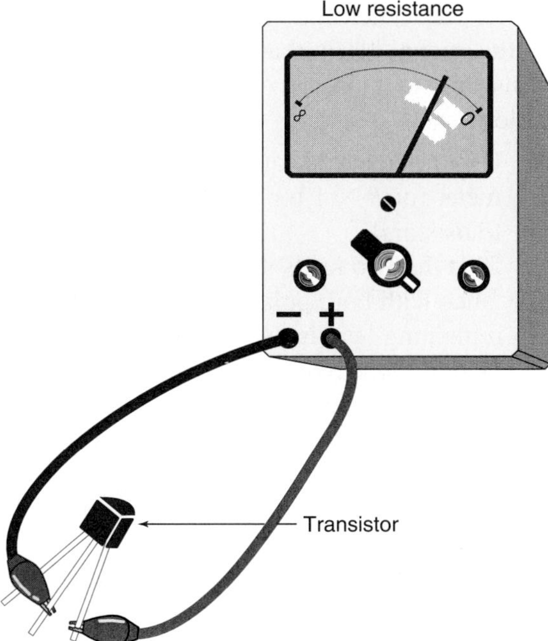

Fig. 5-20 Establishing a forward-biased junction.

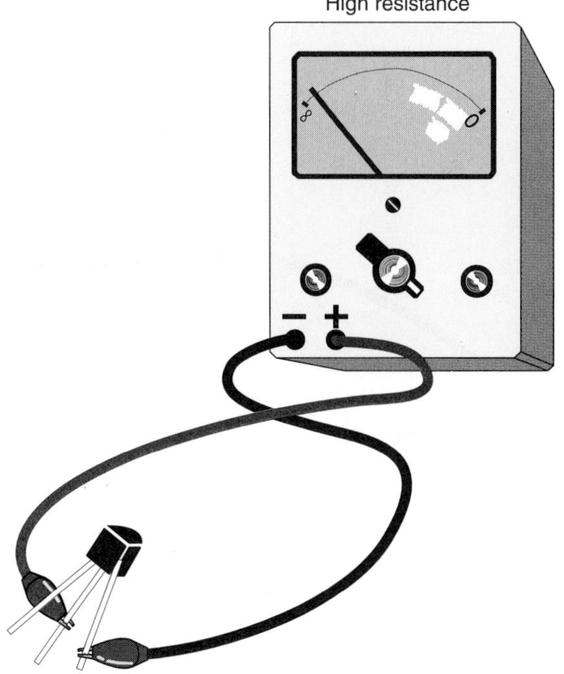

Fig. 5-21 Establishing a reverse-biased junction.

the transistor has good current gain, the collector current will be much greater. The ohmmeter will indicate a resistance much lower than 100,000 Ω, and this proves that the transistor is capable of current gain. This check is made by connecting the ohmmeter across the emitter and collector leads at the same time that the resistor is connected across the collector and base leads. For an NPN transistor, the technique is shown in Fig. 5-22. If you guess wrong and have the positive lead to the emitter and the negative lead to the collector, a low resistance reading will not be seen. Just remember that the resistor must be connected from the positive lead to the base when testing for gain in an NPN transistor. The emitter-collector combination showing the most gain (lowest resistance) is the correct connection. When this is obtained, the ohmmeter leads will identify the collector and the emitter, as shown in Fig. 5-22 for NPN transistors. You will also be sure that the transistor has gain because of the low resistance reading. Usually it will be much less than 100,000 Ω.

Checking gain

In *checking* a PNP transistor for *gain,* the connection for the lowest resistance reading is shown in Fig. 5-23. Remember that the resistor must be connected from the negative lead to the base when a PNP transistor is tested for gain. The combination that shows the best gain (lowest resistance) is as shown in Fig. 5-23.

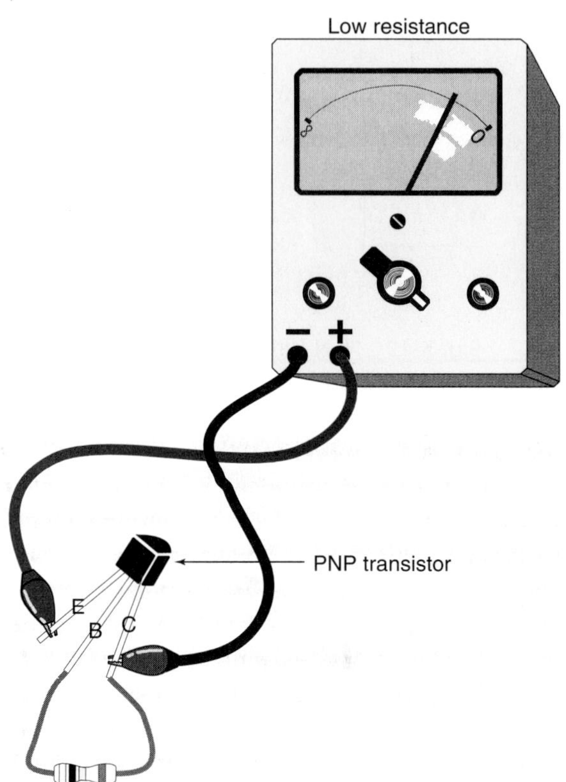

Fig. 5-23 Checking PNP gain.

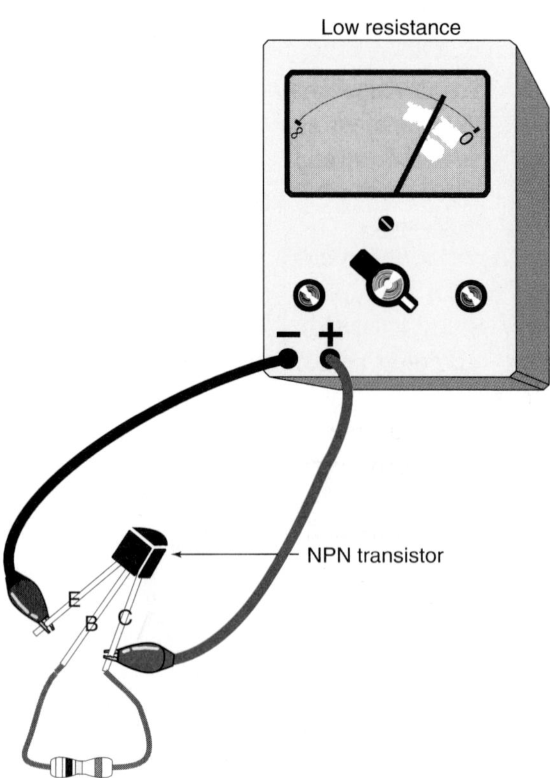

Fig. 5-22 Checking NPN gain.

The entire process is more difficult to describe than to do. With some practice, it becomes very quick and easy. The only drawback to this technique is that it cannot be used on transistors in a circuit. A summary of the steps follows:

1. Use the $R \times 100$ range of an analog ohmmeter (or $R \times 1$ for germanium power transistors).
2. Find the two leads showing high resistance with both polarities applied. The remaining lead is the base lead.
3. With the positive lead on the base, a low resistance should be found by connecting to either of the two remaining leads if the transistor is NPN. For a PNP transistor, the negative lead will have to be on the base to obtain a low resistance.
4. With the ohmmeter across the emitter-collector combination, connect the resistor (100 kΩ or 1 kΩ) from the positive lead to the base terminal for an NPN unit. Reverse the emitter-collector combination.

The lowest resistance is obtained when the positive lead is on the collector.

5. In checking a PNP transistor, the resistor goes from the negative lead to the base. The correct combination (lowest resistance) is when the negative lead is on the collector.

The process is easier to remember and less confusing if you know why it works. Figure 5-24 shows what happens when an NPN transistor is checked for gain. The positive lead of the ohmmeter is applied directly to the collector. This reverse-biases the collector, as it should be. The positive lead of the ohmmeter is also connected to the base but through a high resistance. This forward-biases the base, as it should be. However, the high value of the resistor keeps the base current very low. If the transistor has gain, the emitter-collector current will be greater. The current is supplied by the ohmmeter, and the ohmmeter shows a low resistance because of the increased current from emitter to collector.

Some ohmmeters may have reversed polarity, and some use a very low supply voltage to avoid turning on PN junctions. These characteristics of the ohmmeter must be known.

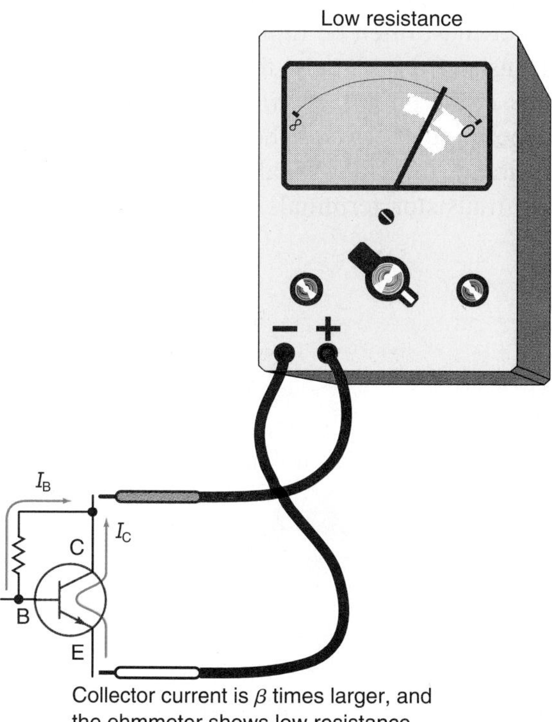

Fig. 5-24 How the ohmmeter test works.

Leakage current

Transistors have some *leakage current.* This is due to minority carrier action. One leakage current in a transistor is called I_{CBO}. (The symbol I stands for current, CB stands for the collector-base junction, and O tells us the emitter is open.) This is the current that flows across the collector-base junction under conditions of reverse bias and with the emitter lead open. Another transistor leakage current is I_{CEO}. (The symbol I stands for current, CE stands for the collector-emitter terminals, and O tells us that the base terminal is open.) I_{CEO} is the largest leakage current. It is an amplified form of I_{CBO}:

$$I_{CEO} = \beta \times I_{CBO}$$

With the base terminal open, any current leaking across the reverse-biased collector-base junction will have the same effect on the base-emitter junction as an externally applied base current. With the base terminal open, there is no other place for the leakage current to go. The transistor amplifies this leakage just as it would any base current:

$$I_C = \beta \times I_B$$

Silicon transistors have very low leakage currents. When ohmmeter tests are made, the ohmmeter should show an infinite resistance when the junctions are reverse-biased. Anything less may mean the transistor is defective. Germanium transistors have much greater leakage currents. This will probably show up as a high, but not infinite, reverse resistance. It will be most noticeable when checking from the emitter to the collector terminal. This is because I_{CEO} is an amplified version of I_{CBO}. Some technicians use this test to tell the difference between a silicon transistor and a germanium transistor. It works, but remember that you could be confused by a leaky silicon transistor.

Digital multimeters have decreased in cost and are readily available. Testing transistors with a DMM is likely to produce different readings than testing with an analog meter. As discussed in Chap. 3, a forward-biased PN junction will show a different amount of resistance from one ohmmeter range to the next. The typical digital ohmmeter does not apply as much current to the device being tested as its analog counterpart. This makes the measured values much higher when testing junctions. For example, an analog ohmmeter might show a for-

Darlington transistor

In-circuit testing

ward resistance of 20 Ω for the emitter-base junction of a transistor. The same junction will typically measure greater than 200 kΩ when a DMM is used. In fact, it is not unusual to measure several megohms with a digital meter.

In general, you can expect the following results when testing good transistors with a DMM:

1. The reading will be greater than 20 MΩ (overrange) when a junction is reverse-biased.
2. The reading will be greater than 20 MΩ when the collector-emitter circuit is tested, regardless of polarity.
3. Readings will be lower for power transistors. A forward-biased junction in a small-signal transistor may measure several megohms. A power transistor junction may test at several hundred kilohms.
4. The gain tests, as shown in Figs. 5-22 and 5-23, will not work with a DMM.

Of course, the actual results will vary from one model of DMM to another. Some DMMs have autoranging. The use of this feature tends to make readings higher because the meter will automatically step to its highest range when the leads are open (not connected). Then, when a junction is connected to the meter the test current is quite low. With this low current bias, the junction resistance is very high and the meter stays on its highest range.

Most DMMs have a diode test. This function can be used for testing transistors. Both the emitter-base and the collector-base should test as diode junctions. The emitter-collector should *not.* Technicians must take the time to read operator's manuals and learn the various features of their equipment.

There is no way to determine what is actually inside a solid-state device by just looking at the package. Figure 5-25 shows a *Darlington transistor* that is packaged in the popular TO-204 (formerly TO-3) case. A Darlington is really two transistors connected for high current gain. The MJ3000 is rated for a minimum h_{FE} of 1000 but is usually closer to 4000. Note in Fig. 5-25 that the emitter of the left-hand transistor controls the base of the right-hand transistor. The current gain from the *B* terminal to the *C* terminal is approximately equal to the product of both transistor gains. If each transistor has a current gain of 50:

$$h_{FE(BOTH)} = h_{FE(1)} \times h_{FE(2)} = 50 \times 50 = 2500$$

An ohmmeter test of a device like that of Fig. 5-25 might be misleading. If the ohmmeter does not turn on the series base-emitter junctions, then about 2 kΩ would be measured in both directions. There would be no indication of a good junction. Also, note that the collector and emitter leads would not check normally, because of the added internal diode. Ohmmeter tests are OK if you remember their limitations. Part numbers, schematics, and data manuals are also required.

Quite a bit of information can be learned from ohmmeter tests. Unfortunately, the transistor must often be removed from the circuit. Transistor testers exist that will check transistors in the circuit. *In-circuit testing* is usually done in other ways. When a transistor fails in a circuit, there are usually voltage changes at the transistor terminals. These can be found

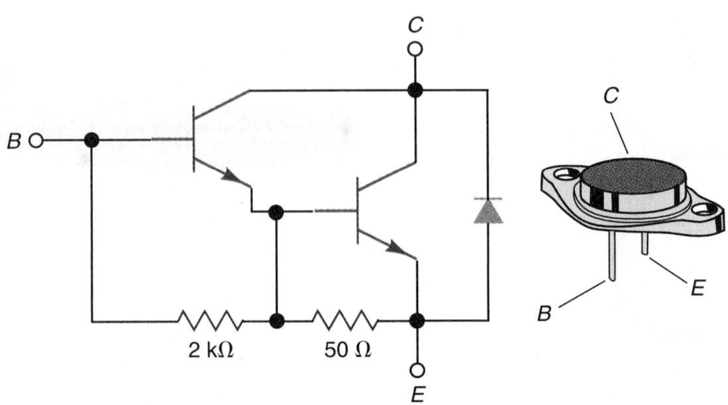

Fig. 5-25 The MJ3000 Darlington transistor.

with a voltmeter. This is called *voltage analysis*. Another in-circuit test uses an oscilloscope to check transistor input and output signals (*signal tracing*). A bad transistor may have an input signal but no output signal.

Voltage analysis

In-circuit checking can also be accomplished by *signal injection.* The technician applies a test signal from a generator. If the signal goes through the rest of the circuit when applied at the output but not when applied at the input of an amplifier, it is fairly certain that something is wrong with that amplifier.

Signal injection

In summary, in-circuit transistor testing can be done in four ways (there are others, too):

1. Using an in-circuit transistor tester
2. Using a voltmeter (voltage analysis)
3. Using an oscilloscope (signal tracing)
4. Using a signal generator (signal injection)

Technicians use any or all of these techniques. One technique may prove to be quicker or better in a given situation. Voltage analysis, signal tracing, and signal injection are covered in more detail in later chapters.

TEST

Determine whether each statement is true or false.

38. Transistor junctions can be checked with an ohmmeter.
39. Junction failures account for most bad transistors.
40. A good transistor should show a low resistance from emitter to collector, regardless of the ohmmeter polarity.
41. It is not possible to locate the base lead of a transistor with an ohmmeter.
42. Suppose that the positive lead of an ohmmeter is connected to the base of a good transistor. Also assume that touching either of the remaining transistor leads with the negative lead shows a moderate resistance. The transistor must be an NPN type.
43. It is possible to verify transistor gain with an analog ohmmeter.
44. Transistor testing with an ohmmeter is limited to in-circuit checks.
45. It is not possible to check transistors that are soldered into a circuit.

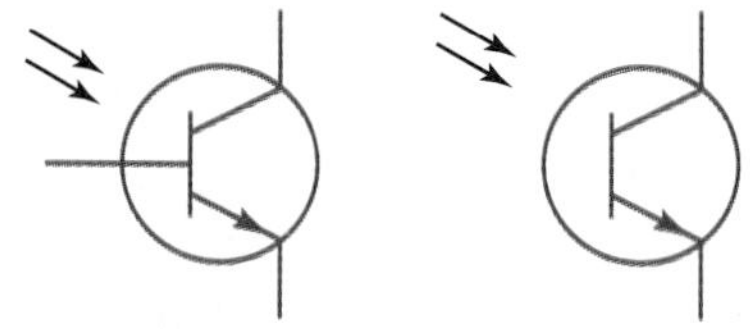

(*a*) Schematic symbols

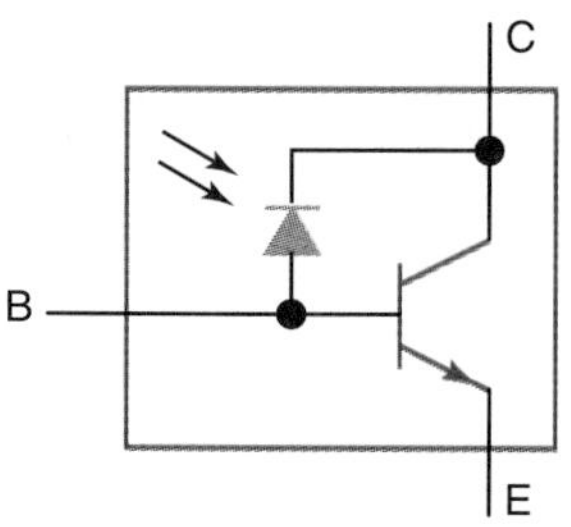

(*b*) Equivalent circuit

Fig. 5-26 Phototransistors.

5-6 Other Transistor Types

Bipolar junction transistors are *light-sensitive,* and their packages are designed to eliminate this effect. *Phototransistors* are packaged to allow light to enter the crystal. Light energy will create hole-electron pairs in the base region and turn the transistor on. Thus, phototransistors can be controlled by light instead of base current. In fact, some phototransistors are manufactured without a base lead as shown by the schematic symbol at the right in Fig. 5-26(*a*).

Phototransistor

Figure 5-26(*b*) shows the equivalent circuit for a phototransistor. You may assume that the collector is several volts positive with respect to the emitter. With no light entering the package, only a small current flows. It is typically on the order of 10 nanoamperes (nA) at room temperature (data sheets call this *flow dark current*). When light does enter, it penetrates the diode depletion region and generates carriers. The diode conducts and provides base current for the phototransistor. The transistor has current gain, so the collector current can be expected to be a great deal larger than the current flow in the diode. A typical phototransistor might show 5 mA of collector current with a light input of 3 mW per square centimeter.

One application for a phototransistor is shown in Fig. 5-27 on page 112. This circuit provides automatic lighting. With daylight conditions, the transistor conducts and holds the normally closed contacts of the relay open.

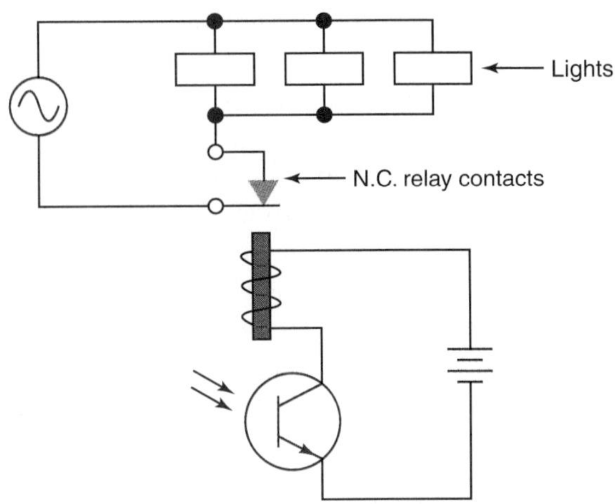

Fig. 5-27 Phototransistor-controlled lighting.

N-channel JFET

P-channel JFET

Optoisolator

Depletion mode

This keeps the lights turned off. When night falls, the phototransistor dark current is too small to hold the relay in and the contacts close and turn on the lights.

Phototransistors can be used in *optoisolators* (also called optocouplers). Fig. 5-28 shows the 4N35 optoisolator package which houses a gallium arsenide infrared-emitting diode and an NPN silicon phototransistor. The diode and transistor are optically coupled. Applying forward bias to the diode will cause it to produce infrared light and turn on the transistor. The 4N35 can safely withstand as much as 2.5 kV of voltage difference from its input terminals (1 and 2) to its output terminals (4, 5, and 6) for up to one minute. This high rating gives an indication of its ability to isolate one circuit from another.

Bipolar junction transistors are used in most circuits. However, another transistor type is also popular. This type is classified as *unipolar.* A unipolar (one-polarity) transistor uses only one type of current carrier. The *junction field-effect transistor* (JFET) is an example of a unipolar transistor. Figure 5-29 shows the structure and schematic symbol for an *N-channel JFET.* Notice that the leads are named *source, gate,* and *drain.*

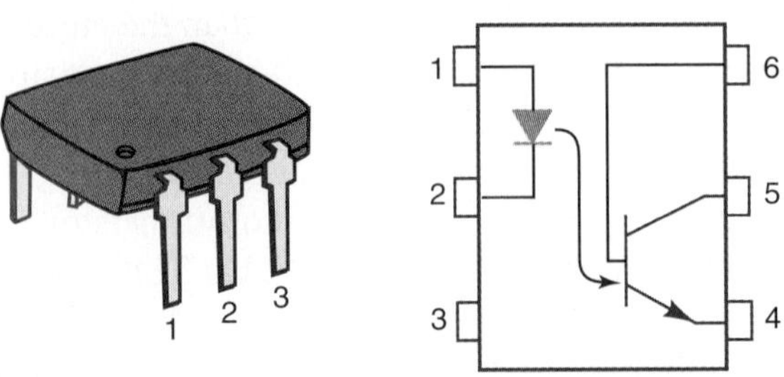

Fig. 5-28 4N35 optoisolator.

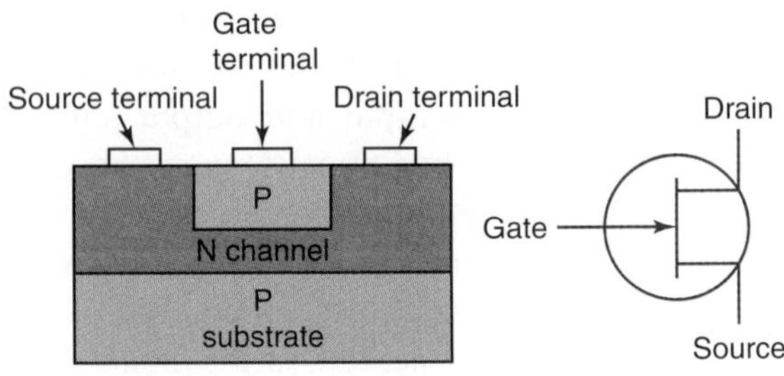

Fig. 5-29 An N-channel JFET.

The JFET can be made in two ways. The channel can be N-type material or P-type material. The schematic symbol of Fig. 5-29 is for an N-channel device. The symbol for a *P-channel device* will show the arrow on the gate lead pointing out. Remember, pointing i*N* indicates an *N-channel device.*

In a BJT, both holes and electrons are used to support conduction. In an *N-channel JFET, only electrons* are used. In a *P-channel JFET, only holes* are used.

The JFET operates in the *depletion mode.* A control voltage at the gate terminal can deplete (remove) the carriers in the channel. For example, the transistor of Fig. 5-29 will normally conduct from the source terminal to the drain terminal. The N channel contains enough free electrons to support the flow of current. If the gate is made negative, the free electrons can be *pushed out of the channel.* Like charges repel. This leaves the channel with fewer free carriers. The resistance of the channel is now much higher, and this tends to decrease the source and drain currents. In fact, if the gate is made negative enough, the device can be turned off and no current will flow.

Examine the curves of Fig. 5-30. Notice that as the voltage from gate to source ($-V_{GS}$) increases, the drain current I_D decreases. Compare this operation with BJTs:

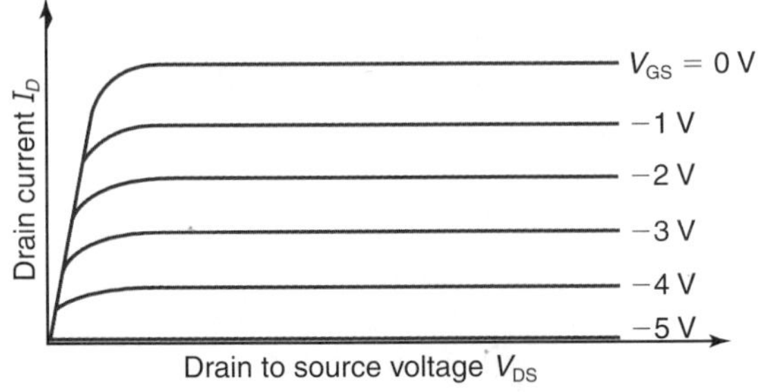

Fig. 5-30 Characteristic curves of a JFET.

- A BJT is off (there is no collector current) until base current is provided.
- A JFET is on (drain current is flowing) until the gate voltage becomes high enough to remove the carriers from the channel.

These are *important* differences: (1) The bipolar device is current-controlled. (2) The unipolar device is voltage-controlled. (3) The bipolar transistor is normally off. (4) The JFET is normally on.

Will there be any *gate current* in the JFET? Check Fig. 5-29. The gate is made of P-type material. To control channel conduction, the gate is made negative. This reverse-biases the gate-channel diode. The gate current should be *zero* (there may be a very small leakage current).

There are also P-channel JFETs. They use P-type material for the channel and N-type material for the gate. The gate will be made positive to repel the holes in the channel. Again, this reverse-biases the gate-channel diode, and the gate current will be zero. Since the polarities are opposite, N-channel JFETs, and P-channel JFETs are not interchangeable.

Field-effect transistors (FETs) do not require any gate current for operation. This means the gate structure can be completely insulated from the channel. Thus any slight leakage current resulting from minority carrier action is blocked. The gate can be made of metal. The insulation used is an oxide of silicon. This structure is shown in Fig. 5-31. It is called a *metal oxide semiconductor field-effect transistor* (MOSFET). The MOSFET can be made with a P channel or an N channel. Again, the arrow pointing i*N* tells us that the channel is N-type material.

Early MOSFETs were very delicate. The thin oxide insulator was easily damaged by excess voltage. The static charge on a technician's body could easily break down the gate insulator. These devices had to be handled very carefully. Their leads were kept shorted together until the device was soldered into the circuit. Special precautions were needed to safely make measurements in some MOSFET circuits. Today most MOSFET devices have built-in diodes to protect the gate insulator. If the gate voltage goes too high, the diodes turn on and safely discharge the potential. However,

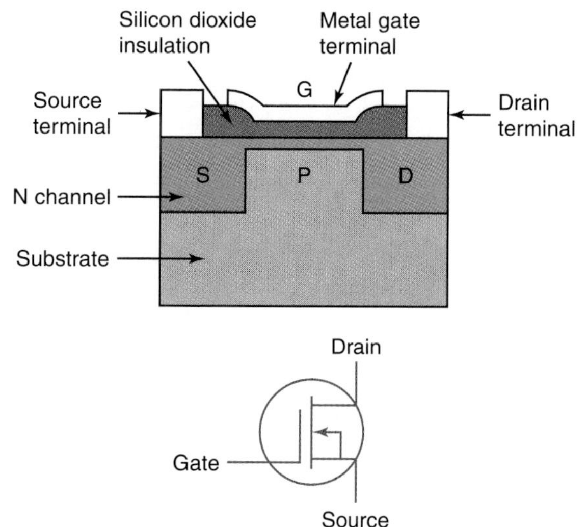

Fig. 5-31 An N-channel MOSFET.

manufacturers still advise careful handling of MOSFET devices.

The gate voltage in a MOSFET circuit can be of either polarity since a diode junction is not used. This makes another mode of operation possible—the *enhancement mode.* An enhancement-mode device normally has no conductive channel from the source to the drain. It is a normally off device. The proper gate voltage will attract carriers to the gate region and form a conductive channel. The channel is *enhanced* (aided by gate voltage). Figure 5-32 on page 114 shows the schematic symbols used for enhancement-mode MOSFETs. Note that the line from source to drain is broken. This implies that the channel is not always present.

Enhancement mode

Figure 5-33 on page 114 shows a family of curves for an N-channel enhancement-mode device. As the gate is made more positive, more electrons are attracted into the channel area. This enhancement improves channel conduction, and the drain current increases. A JFET should not be operated in the enhancement mode because the gate diode would become forward-biased and gate current would flow. Gate current is not desired in any type of FET. Field-effect transistors are normally *voltage-controlled.*

MOSFET

Field-effect transistors have some advantages over bipolar transistors that make them attractive for certain applications. Their gate terminal does not re-

Be careful with all materials supplied to a prospective employer. Have others check for accuracy, clarity, spelling, and grammar.

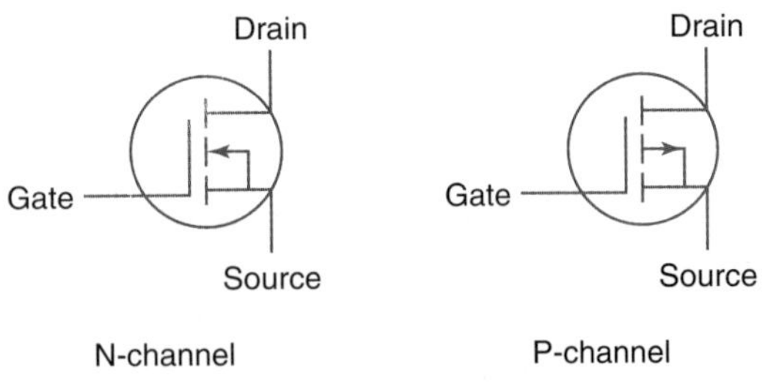

Fig. 5-32 Enhancement-mode MOSFETs.

quire any current. This is a good feature when an amplifier with high input resistance is needed. This is easy to understand by inspecting Ohm's law:

$$R = \frac{V}{I}$$

Consider V to be a signal voltage supplied to an amplifier and I the current taken by the amplifier. In this equation, as I decreases, R increases. This means that an amplifier which draws very little current from a signal source has a high input resistance. Bipolar transistors are current-controlled. A bipolar amplifier must take a great deal more current from the signal source. As I increases, R decreases. Bipolar-junction transistor amplifiers have a low input resistance compared to FET amplifiers.

Vertical metal oxide semiconductor (VMOS)

Thermal runaway

Secondary breakdown

Another type of field-effect transistor has advantages over bipolar transistors in certain high-power applications. This device is the *vertical metal oxide semiconductor* (VMOS) and is usually referred to as a VMOS or VFET transistor. VMOS transistors are similar to enhancement-mode MOSFETs. In fact, they are represented with the same schematic symbols as shown in Fig. 5-32. The difference is that the current flow is vertical rather than lateral. Refer to Fig. 5-31. The source terminal is at the left, and the drain terminal is at the right. Current flows laterally from left to right.

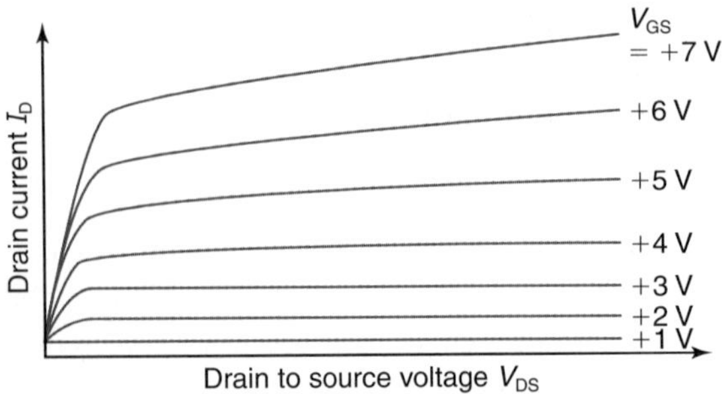

Fig. 5-33 Enhancement-mode characteristic curves.

Now look at Fig. 5-34. The current in the VMOS transistor flows vertically from the metal source contacts down to the drain contact on the bottom of the structure. Placing the drain at the bottom of the structure allows a wide, short channel for current flow. Such a channel can have very low resistance which permits high current flow. The vertical structure is an advantage in high-power applications.

Bipolar power transistor

Minority carrier storage

Bipolar junction transistors are also available in larger, high-power models that can safely conduct higher current flow. These are known as *bipolar power transistors.* However, they suffer from several limitations. First, they are current-driven. They must be supplied with larger and larger controlling currents as the power level increases. The base drive power is lost and detracts from circuit efficiency. Second, they exhibit *minority carrier storage,* which limits their speed. An NPN transistor, for example, when conducting a large current flow, will have many electrons moving through the P-type base region. These electrons are minority carriers in the P-type material. When the forward bias is removed from the base-emitter junction, the transistor will not turn off until all the minority carriers are cleared from the crystal. This takes time and limits the switching speed of bipolar transistors. Third, they tend to conduct more current when they get hot. This increase in current makes them even hotter so they conduct still more current. This undesirable effect is called *thermal runaway* and can continue until the transistor is damaged or destroyed. Finally, bipolar power transistors sometimes fail because of a phenomenon called *secondary breakdown.* This takes place when hot spots due to current crowding occur inside the crystal. The crowding is caused by the electric fields set up by the flow of current. At high currents, the fields can become intense enough to squeeze the flow into a small area which overheats, and the transistor fails.

Vertical field-effect transistors do not have the limitations associated with power bipolar devices. They are well-suited to certain high-power applications. However, it must be emphasized that bipolar power transistors are still widely applied. They are inexpensive, and there are ways to design around some of their limitations.

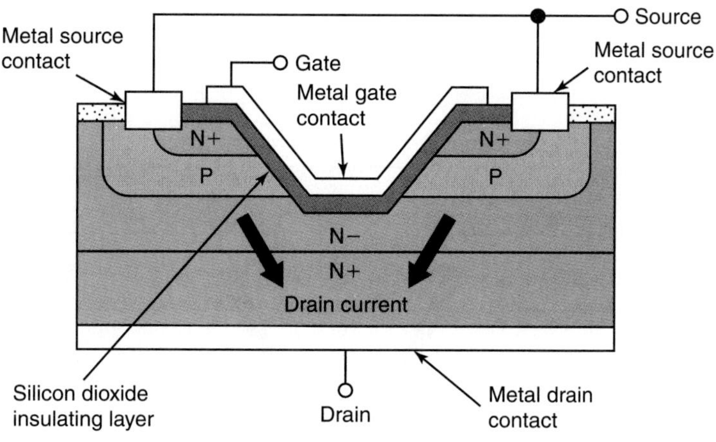

Fig. 5-34 VMOS structure.

The VFET structure shown in Fig. 5-34 operates in the enhancement mode. A positive gate voltage will set up an electric field which will attract electrons into the P-type regions on each side of the V-shaped gate channel. This enhanced channel will support the flow of electron current from the source contacts down to the drain contact. The N− and N+ areas are indicated to show the level of doping. Lightly doped regions (less than 10^{15} impurity atoms per cubic centimeter) are marked with a minus sign, and heavily doped regions (greater than 10^{19} impurity atoms per cubic centimeter) are marked with a plus sign.

Figure 5-35 shows the characteristic curves for a typical VMOS transistor. These curves are similar to those shown in Fig. 5-33 for the enhancement-mode MOSFET. However, the high current capability must be emphasized. Note that the VMOS device is capable of drain currents in the ampere region, whereas the MOSFET is capable of only milliamperes. Also note that the VMOS curves are flat rather than sloping and that they are evenly spaced above 200 mA. The flatness of the curves tells us that the drain-to-source voltage has little effect on drain current. The even spacing means that a given change in gate-to-source voltage will produce the same change in drain current over most of the operating area. Both flatness and even spacing are highly desirable for many transistor applications.

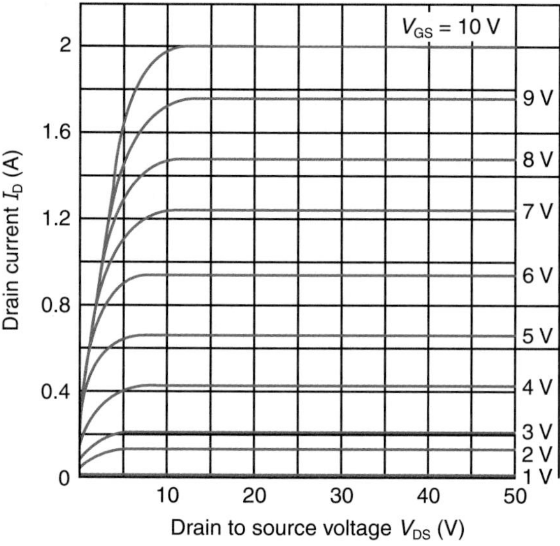

Fig. 5-35 VMOS characteristic curves.

Example 5-11

What is the condition of the VMOS transistor represented by Fig. 5-35 when $V_{GS} = 0$ V? The curves show zero drain current. The transistor is off.

Figure 5-36 shows the structure of another power field-effect transistor. It is also a type of vertical field-effect transistor. It uses a double-diffused MOS design and is called a DMOS device. The polarity of semiconducting crystals can be reversed by diffusing impurity atoms into the crystal. A double-diffusion process is used to form the N+ source structures in the P-type wells shown in Fig. 5-36. This DMOS transistor is also an N-channel enhancement-mode device. A positive gate voltage will attract electrons into the P-type wells and enhance their ability to act as N channels

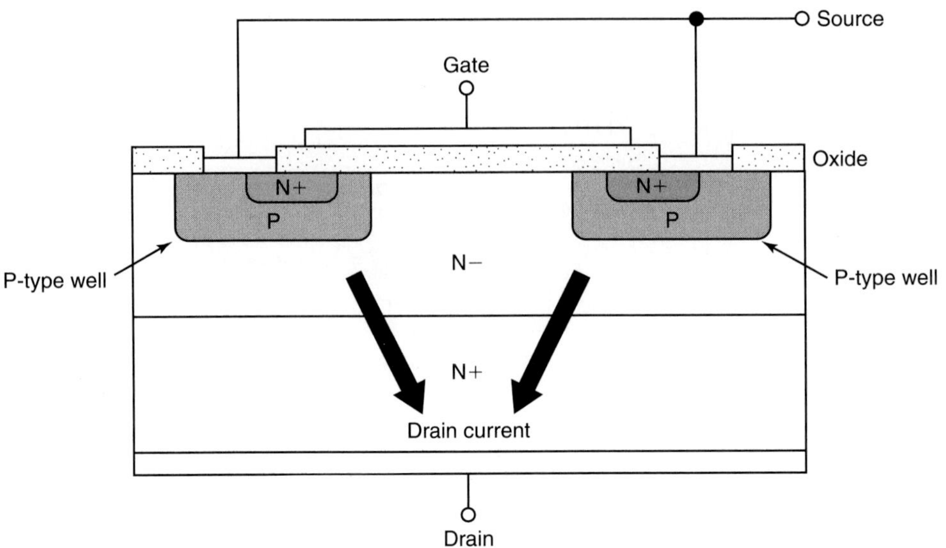

Fig. 5-36 DMOS structure.

Unijunction transistor (UJT)

Negative resistance

to support the flow of current from the source terminals down to the drain terminal. Note that once again the flow of current is vertical. Various manufacturers use variations in the structure of the power field-effect transistors that they build. They may also choose to use a unique name for their devices.

Figure 5-37 shows a gallium arsenide (GaAs) VFET. It has as little as one-eighth of the on resistance of an equivalent DMOS device. This makes it more efficient in power switching applications because the power lost in a switch is proportional to its on resistance:

$$P_{\text{switch}} = I^2 R_{\text{switch(on)}}$$

When the on resistance is very low in a switch, the switch is efficient. In fact, as $R_{\text{switch(on)}}$ approaches 0 Ω, the dissipation approaches 0 W. This is true even at high current flows. Also, thanks to the high mobility of electrons in GaAs, this VFET is an extremely fast device, with switching speeds approaching 10 nanoseconds (ns).

The final transistor type that will be covered is the *unijunction transistor (UJT).* This transistor is *not* used as an amplifier. It is used in timing and control applications. The structure can be seen in Fig. 5-38. The device has an N-type silicon structure with a tiny P-type zone near the center. This produces only one PN junction in the device ("uni" means *one*).

The characteristic curve for the UJT is shown in Fig. 5-39. This curve has a unique feature called the *negative-resistance region.* When a device shows decreasing voltage drop with increasing current, it can be said to have *negative resistance.* According to Ohm's law, $V = I \times R$. Thus, as current increases, the volt-

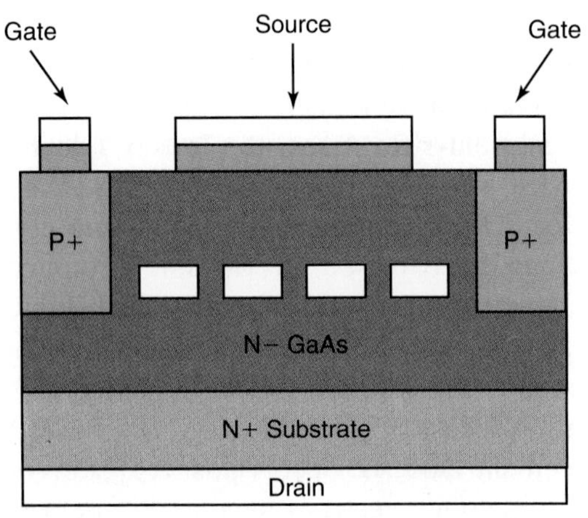

Fig. 5-37 Texas Instrument GaAs VFET.

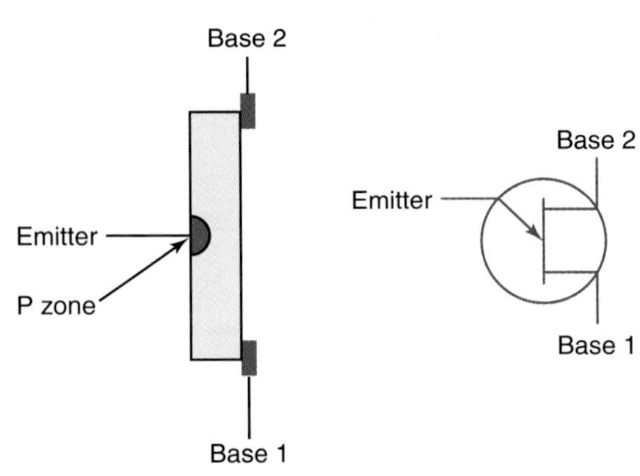

Fig. 5-38 The unijunction transistor.

age drop is expected to increase also. If the reverse occurs, the resistance must be changing. It is easy to explain by using some numbers. Suppose that the current through a device is 1 A and the resistance is 10 Ω. The voltage drop would be

$$V = I \times R$$
$$= 1\ \text{A} \times 10\ \Omega$$
$$= 10\ \text{V}$$

Now, if the current increases to 2 A, we can expect the voltage drop to increase:

$$V = 2\ \text{A} \times 10\ \Omega$$
$$= 20\ \text{V}$$

But, if the resistance drops to 2 Ω, then

$$V = 2\ \text{A} \times 2\ \Omega$$
$$= 4\ \text{V}$$

Strictly speaking, negative resistance does not exist. It is simply a property that has been applied to a family of devices which show a sudden drop in resistance at some point on their characteristic curve.

The UJT is a member of the negative-resistance family. When the emitter voltage reaches a certain point (V_P) on the curve of Fig. 5-38, the emitter diode becomes forward-biased. This causes holes to cross over from the P-type zone into the N-type silicon. These holes are injected into the region between the emitter and the base 1 connection of the transistor. The injected holes greatly improve the conductivity of this part of the N-type material. Greater conductivity means lower resistance. This sudden drop in resistance occurs with an increase of current. It can be used to trigger or turn on other devices. The *trigger point* V_P is predictable. This makes the UJT very useful in timing and control circuits. Remember, it is not used as an amplifier.

Trigger point

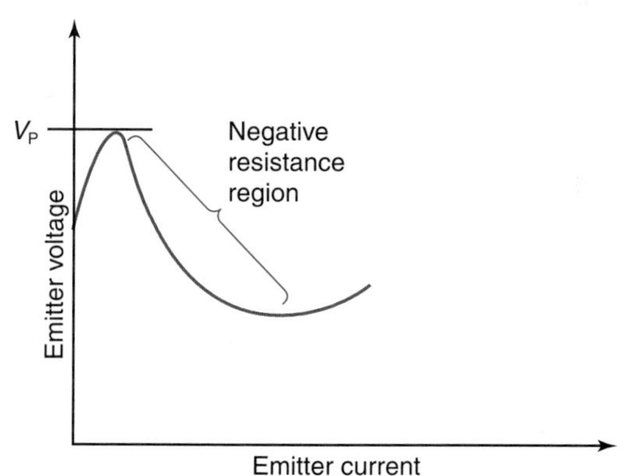

Fig. 5-39 Characteristic curve of a UJT.

TEST

Determine whether each statement is true or false.

46. The schematic symbol for a phototransistor may or may not show a base lead.
47. Refer to Fig. 5-27. The current in the relay coil will increase as more light enters the transistor.
48. Refer to Fig. 5-28. Applying forward bias across pins 1 and 2 will allow current to flow from pin 4 to pin 5.
49. Bipolar junction transistors are bipolar devices.
50. The JFET is a unipolar device.
51. A depletion-mode transistor uses gate voltage in order to increase the number of carriers in the channel.
52. Bipolar transistors are current amplifiers while unipolar transistors are voltage amplifiers.
53. It is possible to turn off an N-channel JFET with negative gate voltage.
54. In P-channel JFET circuits, the gate diode is normally forward-biased.
55. A MOSFET must be handled carefully to prevent breakdown of the gate insulator.
56. It is possible to operate a MOSFET in the enhancement mode.
57. The enhancement mode means that carriers are being pushed out of the channel by gate voltage.
58. The FET makes a better high-input-resistance amplifier than the bipolar type.
59. Power field-effect transistors are often called VMOS or VFETs.
60. Vertical metal oxide semiconductor transistors operate in the depletion mode.
61. Vertical metal oxide semiconductor transistors have the same schematic symbols as JFETs.
62. The UJT is a negative-resistance device.
63. The UJT makes a good amplifier.

5-7 Transistors as Switches

The term "solid-state switch" refers to a switch that has no moving parts. Such switches are usually bipolar or field-effect transistors. They lend themselves for use as switches because they can be turned on with a base current or a gate voltage to produce a low resistance path (the switch is on), or they can be turned off by removing the base current or the gate voltage to produce a high resistance (the switch is off). They are very widely applied because they are small, quiet, inexpensive, reliable, capable of high-speed operation, easy to control, and relatively efficient.

Figure 5-40 shows a typical application. It is a computer-controlled battery conditioner. It is used to determine the condition of rechargeable batteries. It automatically cycles a battery from charge to discharge to charge while it monitors both battery voltage and temperature. Modern rechargeable batteries are often expensive. Proper use will extend their life.

Q_2 is on in Fig. 5-40 when the computer sends a signal to the charge/discharge control block to forward-bias its base-emitter junction. A typical circuit would apply about +5 V to the 1-kΩ base resistor. In switching applications, a transistor is either off or is turned on very hard. In other words, it is expected to drop little voltage from collector to emitter when it is on. So, in Fig. 5-40, when Q_2 is on, only the 10-Ω resistor limits current flow. We can use Ohm's law to find the discharge current:

$$I_{\text{discharge}} = \frac{12\text{ V}}{10\ \Omega} = 1.2\text{ A}$$

When Q_2 is on in Fig. 5-40, Q_1 must be off since we can't charge and discharge at the same time. The computer will reverse the conditions when it is time to charge the battery. For charging, Q_1 must be on so that the 20-V power source is effectively connected to the battery through the 40-Ω resistor. Q_1 will be turned on hard, so once again the current can be calculated by viewing the transistor as a closed switch. This time, however, the battery voltage must be subtracted from the source voltage:

$$I_{\text{charge}} = \frac{20\text{ V} - 12\text{ V}}{40\ \Omega} = 200\text{ mA}$$

The computer turns on Q_3 when it is time to charge the battery. When Q_3 is switched on, it grounds the voltage divider formed by the two 1-kΩ resistors. Ignoring Q_1's base current, we would expect half the 20-V source at the center of the divider. This means that the base of Q_1

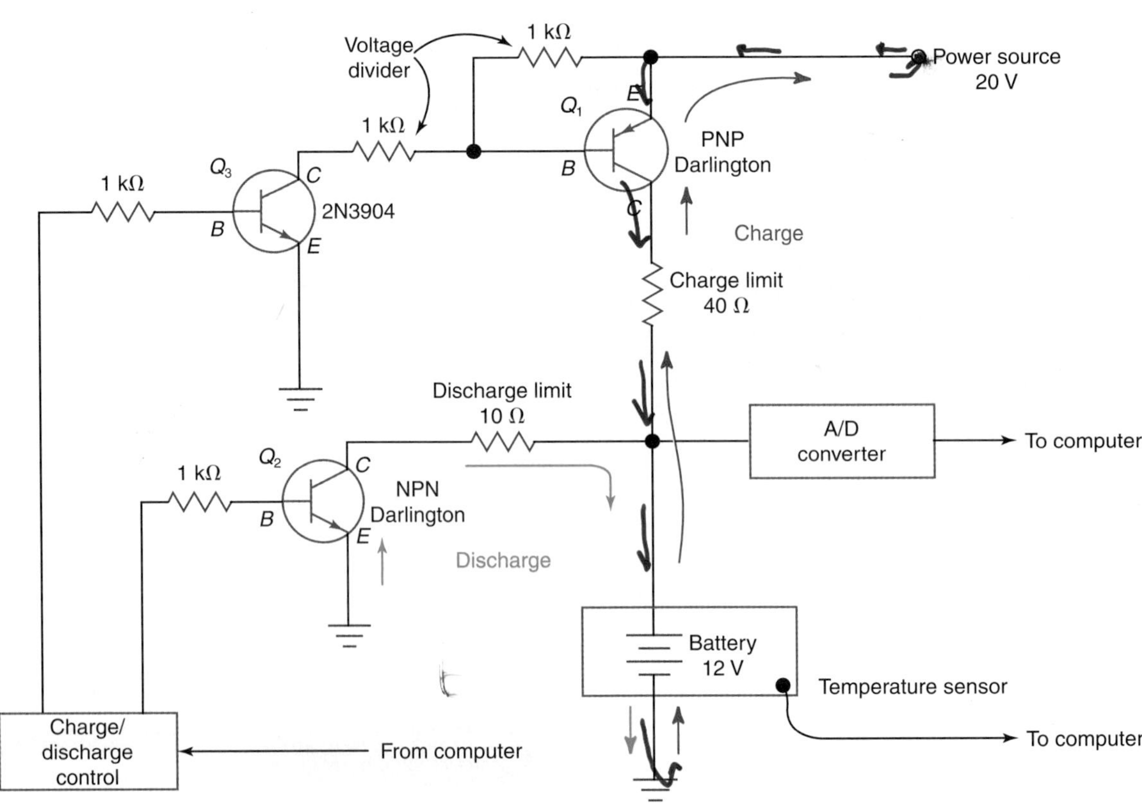

Fig. 5-40 Computer-controlled battery conditioner.

would be at +10 V (referenced to ground) while its emitter is at +20 V. Looking at this another way, we can say that the base of Q_1 is less positive than its emitter by an amount equal to 10 V. When the base of a PNP transistor is negative with respect to its emitter by this amount, the PNP transistor is turned on very hard. Actually, the base-emitter voltage will not be that high in Fig. 5-40 (10 V ignores Q_1). However, the general reasoning is correct, and the PNP transistor is turned on hard when Q_3 is turned on by the computer.

With Q_3 off in Fig. 5-40, there is no current flow in the voltage divider and no drop across either of the 1-kΩ divider resistors. So, Q_1's emitter and base are then at the same potential. This is zero bias, so there is no base current, there is no collector current, and Q_1 is off.

Figure 5-41 on page 120 shows another application for transistor switches. Stepper motors can be used in applications where tight control of speed and position is required. Such applications include computer disk drives, numerically controlled (automated) lathes and milling machines, and automated surface mount assembly lines. The shaft of a stepper motor moves in defined increments such as 1, 2, or 5 degrees per step. If a motor is a 1 degree type, then 180 pulses will move the shaft exactly one-half turn. As Fig. 5-41(*b*) shows, four groups of pulses are required with this particular motor.

A computer or a microprocessor sends precisely timed waveforms to the switching transistors to control the four motor leads: *W*, *X*, *Y*, and *Z* in Fig. 5-41(*a*). The control waveforms are shown as phases *A*, *B*, *C*, and *D* in Fig. 5-41(*b*); note that they are *rectangular.* This is typical when transistors are used as switches. They are either on or off (high or low). Figure 5-41(*d*) shows that power MOSFETs can also be used to control stepper motors.

Motors are inductive loads. They generate a large CEMF when they are switched off. Notice the protection diodes in Fig. 5-41(*a*). They can be separate diodes or built into the transistor, as was shown in Fig. 5-25. When the transistor turns off, the CEMF across the associated motor coil will turn on the diode. Without this diode, the collector of the switching transistor would break down from high voltage and the transistor would be damaged. Figure 5-41(*d*) shows that power MOSFETs have an integral body diode associated with their structure. This body diode is useful in handling the CEMF generated by an inductive load.

Stepper motors are expensive and are not available for high-power applications. Variable-speed induction motors are often better choices, and they can also be controlled by solid-state switches. They are more like the ordinary motors found in home appliances. They are good choices when efficient control of motor speed is the main issue, especially in multiple horsepower applications. An electric vehicle can use solid-state switch control for efficient control of vehicle speed.

You might have guessed by now that this section of the chapter has been devoted to digital applications of transistors. When transistors are used as switches, they are on, or they are off, or they are quickly changing from one state to the next. The next chapter deals with analog applications. There, the transistor will be modeled as a variable resistance and not as a switch.

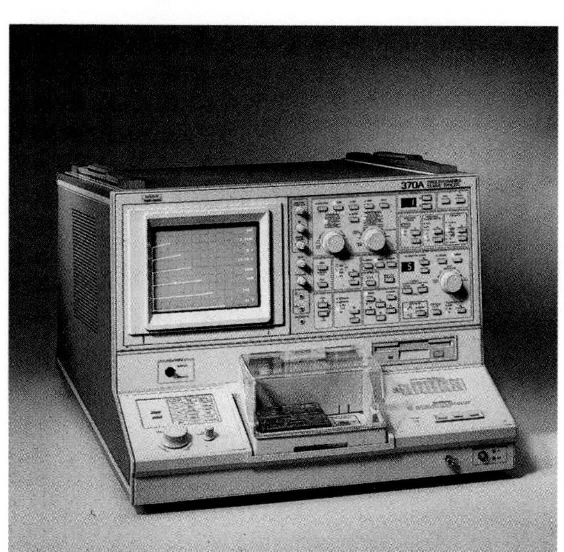

A Tektronix curve tracer.

TEST

Answer the following questions with a short phrase or word.

64. The ideal switch has infinite off resistance. What about its on resistance?
65. An ideal switch dissipates no power when it is on. Why?
66. An ideal switch dissipates no power when it is off. Why?

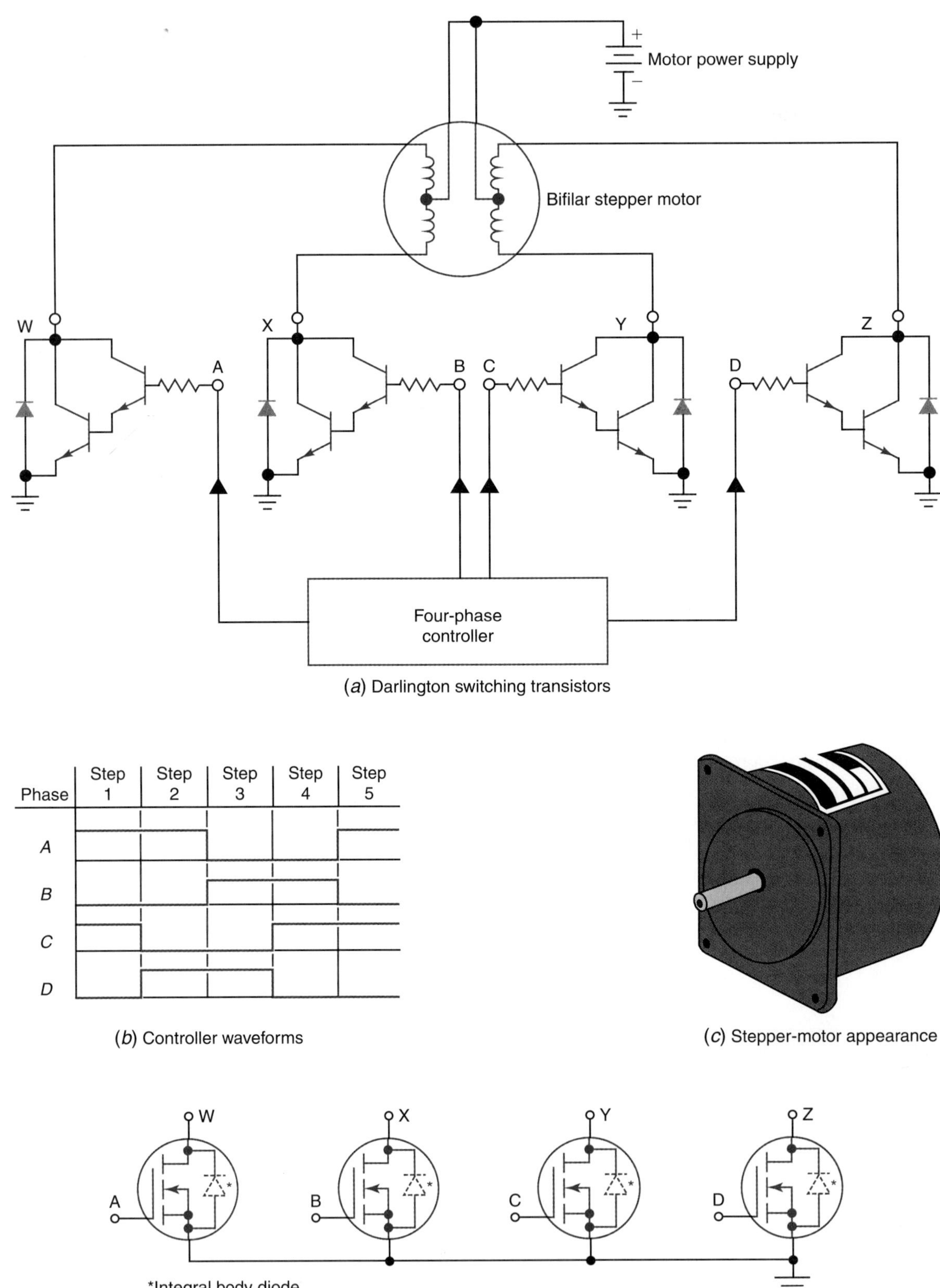

(*a*) Darlington switching transistors

(*b*) Controller waveforms

(*c*) Stepper-motor appearance

(*d*) Power MOSFET switching transistors

Fig. 5-41 Control of stepper motors.

67. What is happening to the battery in Fig. 5-40 when Q_3 is turned on by the computer?
68. What is happening to the battery in Fig. 5-40 when Q_2 is turned on by the computer?
69. What is happening to the battery in Fig. 5-40 when both lines coming out of the charge/discharge control block are at 0 V?
70. Assuming normal operation, are both Q_3 and Q_2 in Fig. 5-40 ever turned on at the same time?
71. Refer to Fig. 5-41(*b*). Ignoring transitions, how many coils are active at any given time?
72. Stepper motors have poor efficiency since they use power when they are standing still. Which section of Fig. 5-41 proves this?

Solve the following problems.

73. Determine a new value for the charge limit resistor in Fig. 5-40 to set the charging current to 1 A.
74. Determine a new value for the discharge limit resistor in Fig. 5-40 to set the discharge current to 0.5 A.

Chapter 5 Review

Summary

1. Gain is the basic function of any amplifier.
2. Gain can be calculated using voltage, current, or power. In all cases, the units cancel and gain is simply a number.
3. Power gain is the product of voltage gain and current gain.
4. The term "voltage amplifier" is often used to describe a small-signal amplifier.
5. The term "power amplifier" is often used to describe a large-signal amplifier.
6. Bipolar junction transistors are manufactured in two polarities: NPN and PNP. The NPN types are more widely applied.
7. In a BJT, the emitter emits the carriers, the base is the control region, and the collector collects the carriers.
8. The schematic symbol of an NPN transistor shows the emitter lead arrow *N*ot *P*ointing i*N*.
9. Normal operation of a BJT requires that the collector-base junction be reverse-biased and the base-emitter junction be forward-biased.
10. Most of the current carriers coming from the emitter cannot find carriers in the base region to combine with. This tends to make the base current much less than the other currents.
11. The base is very narrow, and the collector bias attracts the carriers coming from the emitter. This tends to make the collector current almost as high as the emitter current.
12. Beta (β), or h_{FE}, is the current gain from the base terminal to the collector terminal. The value of β varies considerably, even among devices with the same part number.
13. Base current controls collector current and emitter current.
14. Emitters of PNP transistors produce holes. Emitters of NPN transistors produce electrons.
15. A collector characteristic curve is produced by plotting a graph of I_C versus V_{CE} with I_B at some fixed value.
16. Collector voltage has only a small effect on collector current over most of the operating range.
17. A power curve can be plotted on the graph of the collector family to show the safe area of operation.
18. Collector dissipation is the product of collector-emitter voltage and collector current.
19. Germanium transistors require a base-emitter bias of about 0.2 V to turn on. Silicon units need about 0.6 V.
20. Silicon transistors are much more widely used than germanium transistors.
21. Substitution guides provide the technician with needed information about solid-state devices.
22. The physical characteristics of a part can be just as important as the electrical characteristics.
23. Transistors can be tested with curve tracers, dynamic testers, and ohmmeters and with various in-circuit checks.
24. Most transistors fail suddenly and completely. One or both PN junctions may short or open.
25. An analog ohmmeter can check both junctions, identify polarity, identify leads, check gain, indicate leakage, and may even identify the transistor material.
26. Leakage current I_{CEO} is β times larger than I_{CBO}.
27. Phototransistors are biased on with light.
28. Phototransistors can be packaged with LEDs to form devices called optoisolators or optocouplers.
29. Bipolar transistors (NPN and PNP) use both holes and electrons for conduction.
30. Unipolar transistors (N-channel and P-channel types) use either electrons or holes for conduction.
31. A BJT is a normally off device. It is turned on with base current.
32. A JFET is a normally on device. It is turned off with gate voltage. This is called the depletion mode.
33. A MOSFET uses an insulated gate structure. Manufacturers make both depletion-type and enhancement-type MOSFETs.
34. An enhancement-mode MOSFET is a normally off device. It is turned on by gate voltage.

35. Field-effect transistors have a very high input resistance.
36. The abbreviations VFET and VMOS are used to refer to power field-effect transistors that have a vertical flow of current from source to drain.
37. Power FETs do not have some of the limitations of power bipolar transistors. The FETs are voltage-controlled, they are faster (no minority-carrier storage), they do not exhibit thermal runaway, and they are not prone to secondary breakdown.
38. Power FETs operate in the enhancement mode.
39. Unijunction transistors have one junction. They are not used as amplifiers.
40. Unijunction transistors are negative-resistance devices, and they are useful in timing and control applications.
41. When a bipolar junction transistor is operated as a switch, there is going to be either no base current or a lot of base current.
42. A switching transistor is either turned on hard or is off.
43. Ideally, switching is very efficient since an open switch shows no current for zero power dissipation and a closed switch shows no voltage drop, which is another case of zero power dissipation.
44. In switching circuits, the control waveforms are often rectangular.
45. When switching inductive loads, some sort of protection device or circuit is needed because of the CEMF generated by the inductance.

Chapter Review Questions

Supply the missing word in each statement.

5-1. Small-signal amplifiers are usually called __V__ amplifiers.

5-2. Large-signal amplifiers are usually called __P__ amplifiers.

5-3. Bipolar junction transistors are made in two basic polarities: NPN and __PNP__.

5-4. Current flow in bipolar transistors involves two types of carriers: electrons and __holes__.

5-5. The base-emitter junction is normally __forward__ biased.

5-6. The collector-base junction is normally __reversed__ biased.

5-7. The smallest current in a BJT is normally the __base__ current.

5-8. In a normally operating BJT, the collector current is controlled mainly by the __base__ current.

5-9. Turning on an NPN BJT requires that the base be made __positive__ with respect to the emitter terminal.

5-10. For proper operation, the base terminal of a PNP BJT should be __negative__ with respect to the emitter terminal.

5-11. The emitter of a PNP transistor produces __holes__ current.

5-12. The emitter of an NPN transistor produces __electrons__ current.

5-13. The symbol h_{FE} represents the __DC__ current gain of a transistor.

5-14. The symbol h_{fe} represents the __AC__ current gain of a transistor.

5-15. The equivalent symbol for h_{FE} is __Bdc__.

5-16. The equivalent symbol for h_{fe} is __Bac__.

5-17. In testing bipolar transistors with an ohmmeter, a good diode indication should be noted at the collector-base and __emitter__ junctions.

5-18. In an ohmmeter test of a good BJT, the collector and emitter leads should check __high resistance__ regardless of meter polarity.

5-19. A phototransistor's current is usually controlled by __light__.

5-20. Optocoupler is another name for __optoisolator__.

5-21. Refer to Fig. 5-30. As V_{GS} becomes more positive, drain current __~~decrease~~ increase__.

5-22. An N-channel JFET uses __electrons__ to support the flow of current.

5-23. A P-channel JFET uses __holes__ to support the flow of current.

5-24. Gate voltage in a JFET can remove carriers from the channel. This is known as the __depletion__ mode.

5-25. Gate voltage in a MOSFET can produce carriers in the channel. This is known as the __enhancement__ mode.

5-26. A JFET is not normally operated in the enhancement mode because the gate diode may become __forward__ biased.

5-27. The current flow in power field-effect transistors is __vertical__ rather than lateral.

5-28. Power bipolar transistors may be damaged by hot spots in the crystal caused by current crowding. This phenomenon is known as ________.

5-29. Vertical metal oxide semiconductor transistors operate in the __enhancement__ mode.

5-30. Once the firing voltage V_P is reached in a UJT, the resistance is expected to __decrease__.

Chapter Review Problems

5-1. An amplifier provides a voltage gain of 20 and a current gain of 35. Find its power gain.

5-2. An amplifier must give an output signal of 5 V peak-to-peak. If its voltage gain is 25, determine its input signal.

5-3. If an amplifier develops an output signal of 8 V with an input signal of 150 mV, what is its voltage gain?

5-4. A BJT has a base current of 25 μA and its $\beta = 200$. Determine its collector current.

5-5. A BJT has a collector current of 4 mA and a base current of 20 μA. Find its β.

5-6. A BJT has a $\beta = 250$ and a collector current of 3 mA. What is its base current?

5-7. A bipolar transistor has a base current of 200 μA and an emitter current of 20 mA. What is the collector current?

5-8. Find β for problem 5-7.

5-9. Refer to Fig. 5-11. $V_{CE} = 10$ V and $I_B = 20$ μA. Find β.

5-10. Refer to Fig. 5-12. $V_{CE} = -16$ V and $I_C = -7$ mA. Find I_B.

5-11. Refer to Fig. 5-12. $I_B = -100$ μA and $V_{CE} = -10$ V. Find P_C.

5-12. Refer to Fig. 5-15. The transistor is silicon and $V_{BE} = 0.65$ V. What is I_C?

Critical Thinking Questions

5-1. If a transistor has a current gain of 100, how much current gain would be available by using three transistors? How would they be arranged?

5-2. You are looking at a collector family of characteristic curves on a curve tracer and you notice that the curves appear to be spreading (moving apart). What is happening?

5-3. Transistor heating is a big problem in some circuits. Today, it is becoming popular to operate transistors in a digital mode to alleviate the heat problem. Why?

5-4. Transistors are very popular, but an older technology based on *vacuum tubes* is still in use in very-high-power applications such as large radio and television transmitters. Why? (*Hint:* Vacuum tubes can operate at thousands of volts.)

5-5. FETs are unipolar devices, and BJTs are bipolar devices. Will the future bring a new category of electronic devices that are tripolar?

5-6. When examining a piece of electronic equipment, why can't you assume that the transistors will all have three leads and the diodes will all have two leads?

Answers to Tests

1. T
2. F
3. F
4. T
5. F
6. T
7. T
8. F
9. F
10. F
11. T
12. T
13. T
14. T
15. F
16. F
17. F
18. T
19. T
20. 42.5 mA
21. 6.67 μA
22. 250
23. 10 mA
24. 100
25. 20 μA
26. 11 mA
27. 133
28. 96 mW
29. 100
30. 0.2
31. 0.6
32. germanium
33. T
34. F
35. F
36. F

37. T
38. T
39. T
40. F
41. F
42. T
43. T
44. F
45. F
46. T
47. T
48. T
49. T
50. T
51. F
52. T
53. T
54. F
55. T
56. T
57. F
58. T
59. T
60. F
61. F
62. T
63. F
64. ideally, zero
65. ideally, it drops zero volts
66. ideally, the current flow is zero
67. it is charging
68. it is discharging
69. nothing (it is not charging or discharging)
70. no
71. two
72. (*b*)
73. 8 Ω
74. 24 Ω

Introduction to Small-Signal Amplifiers

Chapter Objectives

This chapter will help you to:

1. *Calculate* decibel gain and loss.
2. *Draw* a load line for a basic common-emitter amplifier.
3. *Define* clipping in a linear amplifier.
4. *Find* the operating point for a basic common-emitter amplifier.
5. *Determine* common-emitter amplifier voltage gain.
6. *Identify* common-base and common-collector amplifiers.
7. *Explain* the importance of impedance matching.

This chapter deals with gain. Gain is the ability of an electronic circuit to increase the level of a signal. As you will see, gain can be expressed as a ratio or as a logarithm of a ratio. Transistors provide gain. This chapter will show you how they can be used with other components to make amplifier circuits. You will learn how to evaluate a few amplifiers using some simple calculations. This chapter is limited to small-signal amplifiers. As mentioned before, these are often called voltage amplifiers.

6-1 Measuring Gain

Gain

Gain is the basic function of all amplifiers. It is a comparison of the signal fed into the amplifier with the signal coming out of the amplifier. Because of gain, we can expect the output signal to be greater than the input signal. Figure 6-1 shows how measurements are used to calculate the voltage gain of an amplifier. For example, if the input signal is 1 V and the output signal is 10 V, the gain is

$$\text{Gain} = \frac{\text{signal out}}{\text{signal in}} = \frac{10\text{ V}}{1\text{ V}} = 10$$

Note that the units of voltage cancel and gain is a pure number. It is *not* correct to say that the gain of the amplifier is 10 V.

Example 6-1

Calculate the gain of an amplifier if the output signal is 4 V and the input signal is 50 mV.

$$\text{Gain} = \frac{\text{out}}{\text{in}} = \frac{4\text{ V}}{50 \times 10^{-3}\text{ V}} = 80$$

Example 6-2

If an amplifier has a gain of 50, find its output signal amplitude when its input is 20 mV. Rearranging:

$$\text{Out} = \text{gain} \times \text{in} = 50 \times 20\text{ mV} = 1\text{ V}$$

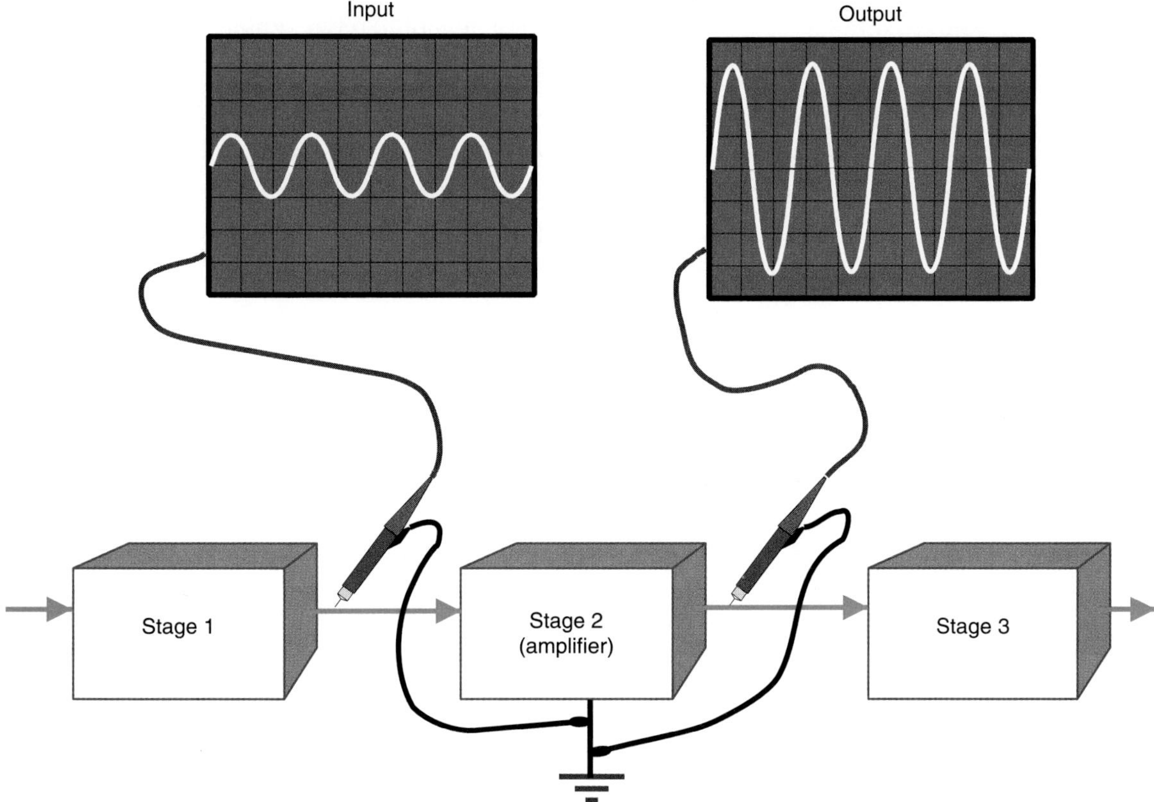

Fig. 6-1 Measuring gain.

A circuit that has gain *amplifies.* The letter A is the general symbol for gain or amplification in electronics. A subscript can be added to specify the type of gain:

$$A_V = \frac{V_{out}}{V_{in}} = \text{voltage gain}$$

$$A_I = \frac{I_{out}}{I_{in}} = \text{current gain}$$

$$A_P = \frac{P_{out}}{P_{in}} = \text{power gain}$$

Voltage gain A_V is used to describe the operation of small-signal amplifiers. Power gain A_P is used to describe the operation of large-signal amplifiers. If the amplifier of Fig. 6-1 were a power or large-signal amplifier, the gain would be based on watts rather than on volts. For example, if the input signal is 0.5 W and the output signal is 8 W, the power gain is

$$A_P = \frac{P_{out}}{P_{in}} = \frac{8\text{ W}}{0.5\text{ W}} = 16$$

Early work in electronics was in the communications area. The useful output of most circuits was audio for headphones or speakers. Thus, engineers and technicians needed a way to align circuit performance with human hearing. The human ear is *not linear* for audio power. It does not recognize intensity or loudness in the way a linear device does. For example, if you are listening to a speaker with 0.1-W input and the power suddenly increases to 1 W, you will notice that the sound has become louder. Then assume that the power suddenly increases again to 10 W. You will notice a second increase in loudness. The interesting thing is that you will probably rate the second increase in loudness as about *equal* to the first increase in loudness.

Voltage gain

Example 6-3

Find P_{in} for an amplifier with a power gain of 20 and an output power of 1 W. Rearranging:

$$P_{in} = \frac{P_{out}}{A_p} = \frac{1\text{ W}}{20} = 50\text{ mW}$$

A linear detector would rate the second increase to be 10 times greater than the first. Let us see why:

- First increase from 0.1 to 1 W, which is a *0.9-W* linear change.
- Second increase from 1 to 10 W, which is a *9-W* linear change.

The second change is 10 times greater than the first:

Decibel (dB)

$$\frac{9\ \text{W}}{0.9\ \text{W}} = 10$$

The loudness response of human hearing is *logarithmic*. Logarithms are therefore often used to describe the performance of audio systems. We are often more interested in the *logarithmic gain* of an amplifier than in its *linear gain*. Logarithmic gain is very convenient and widely applied. What started out as a convenience in audio work has now become the universal standard for amplifier performance. It is used in radio-frequency systems, video systems, and just about anywhere there is electronic gain.

Logarithmic gain

Common logarithm

Common logarithms are *powers (exponents) of 10*. For example,

$$\begin{aligned}10^{-3} &= 0.001\\ 10^{-2} &= 0.01\\ 10^{-1} &= 0.1\\ 10^{0} &= 1\\ 10^{1} &= 10\\ 10^{2} &= 100\\ 10^{3} &= 1000\end{aligned}$$

Power gain

The logarithm of 10 is 1. The logarithm of 100 is 2. The logarithm of 1000 is 3. The logarithm of 0.01 is −2. Any positive number can be converted to a common logarithm. Logarithms can be found with a scientific calculator. Enter the number and then press the "log" key to obtain the common logarithm for the number.

Example 6-4

Use a scientific calculator to find the common log of 2138. Enter 2138 and then press the log key:

$$\text{Display} \approx 3.33$$

Example 6-5

Use a scientific calculator to find the common log of 0.0316. Enter .0316 and then press the log key:

$$\text{Display} \approx -1.5$$

Or, enter 31.6, then press the EXP key, then enter 3, then press the ± key, and finally, press the log key.

Power gain is very often measured in *decibels (dB)*. The decibel is a logarithmic unit. Decibels can be found with this formula:

$$\text{dB power gain} = 10 \times \log_{10} \frac{P_{out}}{P_{in}}$$

Gain in decibels is based on *common logarithms*. Common logarithms are based on 10. This is shown in the above equation as $\log_{10}$ (the base is 10). Hereafter the base 10 will be dropped, and log will be understood to mean $\log_{10}$.

Logarithms for numbers less than 1 are *negative*. This means that any part of an electronic system that produces *less* output than input will have a negative gain (−dB) when the above formula is used.

Let's apply the formula to the example given previously. The first loudness increase:

$$\begin{aligned}\text{dB } \textit{power gain} &= 10 \times \log \frac{1\ \text{W}}{0.1\ \text{W}}\\ &= 10 \times \log 10\end{aligned}$$

The logarithm of 10 (log 10) is 1, so:

$$\text{dB power gain} = 10 \times 1 = 10$$

Thus, the first increase in level or loudness was equal to 10 dB. The second loudness increase:

$$\begin{aligned}\text{dB power gain} &= 10 \times \log \frac{10\ \text{W}}{1\ \text{W}}\\ &= 10 \times \log 10\\ &= 10 \times 1 = 10\end{aligned}$$

The second increase was also equal to 10 dB. Since the decibel is a logarithmic unit and because your hearing is logarithmic, the two 10-dB increases sound about the same. The average person can detect a change as small as 1 dB. Any change smaller than 1 dB would be very difficult for most people to hear.

Why has the decibel, which was developed for audio, come to be used in all areas of electronics where gain is important? The answer is

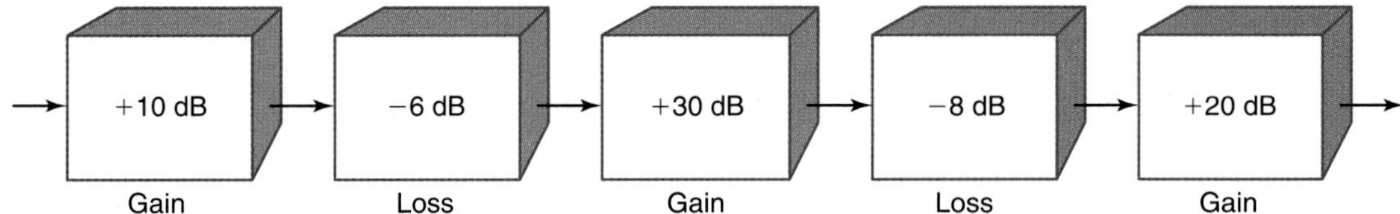

Fig. 6-2 Gain and loss in decibels.

that it is so convenient to work with. Figure 6-2 shows why. Five stages, or parts, of an electronic system are shown. Three of the stages show gain (+dB), and two show loss (−dB). To evaluate the overall performance of the system shown in Fig. 6-2, all that is required is to add the numbers:

$$\text{Overall gain} = +10 - 6 + 30 - 8 + 20 = +46 \text{ dB}$$

When the gain or loss of individual parts of a system is given in decibels, it is very easy to evaluate the overall performance. This is why the decibel has come to be so widely used in electronics.

Figure 6-3 shows the same system with the individual stage performance stated as *ratios.* Now, it is not so easy to evaluate the overall performance. The overall performance will be given by

$$\text{Overall gain} = \frac{10}{4} \times \frac{1000}{6.31} \times 100 = 39{,}619.65$$

Notice that it is necessary to *multiply* for the gain stages and *divide* for the loss stages. When stage performance is given in decibels, gains are added and losses subtracted. The overall system performance is easier to determine using the dB system.

Figures 6-2 and 6-3 describe the same system. One has an overall gain of +46 dB, and the other has an overall gain of 39,619.65. The dB gain and the ratio gain should be the same:

$$\text{dB} = 10 \times \log 39{,}619.65 = 10 \times 4.60 = 46$$

The decibel is based on the ratio of the power output to the power input. It can also be used to describe the ratio of two voltages. The equation for finding *dB voltage gain* is slightly different from the one used for finding dB power gain:

Decibel voltage gain

$$\text{dB voltage gain} = 20 \times \log \frac{V_{out}}{V_{in}}$$

Notice that the logarithm is multiplied by 20 in the above equation. This is because power varies as the *square* of the voltage:

$$\text{Power} = \frac{V^2}{R}$$

Power gain can therefore be written as

$$A_P = \frac{(V_{out})^2/R_{out}}{(V_{in})^2/R_{in}}$$

If R_{out} and R_{in} happen to be equal, they will cancel. Now power gain reduces to

$$A_P = \frac{(V_{out})^2}{(V_{in})^2} = \left(\frac{V_{out}}{V_{in}}\right)^2$$

Since the log of $V^2 = 2 \times \log$ of V, the logarithm can be multiplied by 2 to eliminate the need for squaring the voltage ratio:

$$\text{dB voltage gain} = 10 \times 2 \times \log \frac{V_{out}}{V_{in}} = 20 \times \log \frac{V_{out}}{V_{in}}$$

The requirement that R_{in} and R_{out} be equal is often set aside for voltage amplifiers. It is important to remember that if the resistances are not equal, then the dB voltage gain will *not* be equal to the dB power gain. For example,

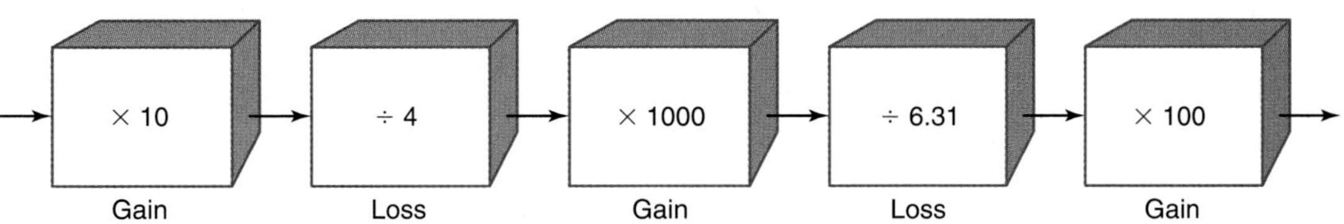

Fig. 6-3 Gain and loss in ratios.

suppose that an amplifier has a voltage gain of 50, an input resistance of 1 kΩ, and an output resistance of 150 Ω. Its dB voltage gain will be

$$A_V = 20 \times \log 50 = 34 \text{ dB}$$

The power gain of this same amplifier can be found by assigning some input voltage (the value does not matter). If we assign the input to be 1 V, then the output will be 50 V because of the stated voltage ratio. Now, power in and power out can be calculated:

$$P_{in} = \frac{V^2}{R} = \frac{1^2}{1000} = 1 \text{ mW}$$

$$P_{out} = \frac{50^2}{150} = 16.7 \text{ W}$$

The power gain of the amplifier in decibels is

$$A_P = 10 \times \log \frac{16.7 \text{ W}}{1 \text{ mW}} = 42.2 \text{ dB}$$

Note that this does *not* equal the dB voltage gain.

As another example, suppose an amplifier has a voltage gain of 1, an input resistance of 50,000 Ω, and an output resistance of 100 Ω. The dB voltage gain of this amplifier is

$$A_V = 20 \times \log 1 = 0 \text{ dB}$$

Assuming a 1-V input signal,

$$P_{in} = \frac{1^2}{50{,}000} = 20 \ \mu\text{W}$$

$$P_{out} = \frac{1^2}{100} = 10 \text{ mW}$$

The power gain of this amplifier in dB is

$$A_P = 10 \times \log \frac{10 \times 10^{-3}}{20 \times 10^{-6}} = 27 \text{ dB}$$

Once again, we see that the dB power gain is *not* equal to the dB voltage gain because R_{in} does not equal R_{out}. We see another interesting fact: *An amplifier may have no voltage gain yet offer a significant power gain.*

TABLE 6-1 Common Values for Estimating dB Gain and Loss

Change	Power	Voltage
Multiplied by 2	+3 dB	+6 dB
Divided by 2	−3 dB	−6 dB
Multiplied by 10	+10 dB	+20 dB
Divided by 10	−10 dB	−20 dB

Technicians should have a feeling for gain and loss expressed in decibels. Often, a quick estimate is all that is required. Table 6-1 contains the common values used by technicians for making estimates.

Example 6-6

A 100-W amplifier has a power gain of 10 dB. What input signal power is required to drive the amplifier to full output? Table 6-1 shows the multiplication (ratio) to be 10 for a power gain of 10 dB. The required input power is therefore one-tenth the desired output power:

$$P_{in} = \frac{100 \text{ W}}{10} = 10 \text{ W}$$

Example 6-7

A transmitter feeds an antenna through a long run of coaxial cable. The transmitter develops 1 kW of output power and only 500 W reaches the antenna. What is the performance of the coaxial cable in dB? The 500 W is one-half of the input power. Therefore, the power has been divided by 2. Table 6-1 shows that this equals −3 dB. The performance of this cable can be verbalized in different ways:

1. The cable gain is −3 dB.
2. The cable loss is 3 dB.
3. The cable loss is −3 dB.

The first statement is technically correct. A negative dB gain means that there is actually a loss. The second statement is also technically correct. The word "loss" means that the value of 3 dB is to be preceded by a minus sign when used in system calculations. The third statement is *not* technically correct. Since the word "loss" means to precede the dB value with a minus sign, the result would be −(−3 dB) = +3 dB, which is a gain. A coaxial cable cannot produce a power gain. Double negatives should be avoided when describing dB losses.

Example 6-8

The response of a low-pass filter is specified to be −6 dB at 5 kHz. A technician measures the filter output and finds 1 V at 1 kHz and notes that it drops to 0.5 V at 5 kHz. Is the filter working properly? Table 6-1 shows that a voltage division of 2 is equal to −6 dB. The filter is working properly.

Example 6-9

An amplifier develops a 2-W output signal when its input signal is 100 mW. What is the power gain of this amplifier in dB? Find the ratio first:

$$\frac{2\text{ W}}{0.1\text{ W}} = 20$$

The value 20 is not in Table 6-1. However, it may be possible to *factor* a gain value into values that are in the table. A power gain of 20 can be broken down into a power gain of 10 (+10 dB) times a power gain of 2 (+3 dB). Add the dB gains:

$$\text{Gain} = 10\text{ dB} + 3\text{ dB} = 13\text{ dB}$$

Example 6-10

An amplifier has a voltage gain of 60 dB. If its input signal is 10 μV, what output signal can be expected? The table shows that a voltage gain of 20 dB produces a multiplication of 10; 60 dB = 3 × 20 dB, thus a gain of 60 dB will multiply the signal by 10 three times:

$$\begin{aligned} V_{out} &= V_{in} \times 10 \times 10 \times 10 \\ &= 10\ \mu\text{V} \times 1000 = 10\text{ mV} \end{aligned}$$

Calculators with logarithms are inexpensive. Technicians are expected to be able to use calculators to find dB gain and loss. The following example is easy to work using a calculator.

Example 6-11

If the input signal to a voltage amplifier is 350 mV and the output signal is 15 V, what is the performance of this amplifier in decibels?

$$\begin{aligned} \text{dB} &= 20 \times \log \frac{V_{out}}{V_{in}} \\ &= 20 \times \log \frac{15\text{ V}}{0.35\text{ V}} \\ &= 20 \times \log 42.9 \\ &= 20 \times 1.63 \\ &= 32.6 \end{aligned}$$

The amplifier shows a voltage gain of 32.6 dB. The calculator manipulation is straightforward. First, divide the input signal into the output signal. Next, press the log key. Finally, multiply by 20.

A little algebraic manipulation will be required to solve some problems. The following example demonstrates this.

Example 6-12

You are using an oscilloscope to measure a high-frequency waveform. The manufacturer of your oscilloscope specifies that its response at −3 dB is at the frequency of measurement. If the screen shows a peak-to-peak value of 7 V, what is the actual value of the signal? Begin by plugging the known information into the dB equation for voltage:

$$-3 = 20 \times \log \frac{7\text{ V}_{\text{p-p}}}{V_{in}}$$

Divide both sides of the equation by 20:

$$-0.15 = \log \frac{7\text{ V}_{\text{p-p}}}{V_{in}}$$

Take the *inverse* log of both sides of the equation. This *removes* the log term from the right-hand side of the equation. For the left-hand side, you must find the inverse log of −0.15 using your calculator. On some calculators, you must use an INV key in conjunction with the log key. Press INV and then press log. On other calculators, you will find a key marked 10^X. Simply press this key. Perform the operation with −0.15 showing in the calculator display. The calculator should respond with 0.708. We are now at this point:

$$0.708 = \frac{7 \text{ V}_{\text{p-p}}}{V_{\text{in}}}$$

Rearrange and solve:

$$V_{\text{in}} = \frac{7 \text{ V}_{\text{p-p}}}{0.708} = 9.89 \text{ V peak-to-peak}$$

---

dBm scale

The dB system is sometimes misused. Absolute values are often given in decibel form. For example, you may have heard that the sound level of a musical group is 90 dB. This provides no information at all unless there is an agreed-upon reference level. One reference level used in sound is a pressure of 0.0002 dynes per square centimeter (dyn/cm^2) or 2×10^{-5} newtons per square meter (N/m^2). This reference pressure is equated to 0 dB, the *threshold of human hearing*. Now, if a second pressure is compared to the reference pressure, the dB level of the second pressure can be found. For example, a jet engine produces a sound pressure of 2000 dyn/cm^2:

$$\text{Sound level} = 20 \times \log \frac{2000}{0.0002} = 140 \text{ dB}$$

(The log is multiplied by 20 since sound power varies as the square of sound pressure.) In the average home there is a sound pressure of 0.063 dyn/cm^2, which can be compared to the reference level:

$$\text{Sound level} = 20 \times \log \frac{0.063}{0.0002} = 50 \text{ dB}$$

It is interesting to note that the decibel scale that places the threshold of human hearing at 0 dB places the threshold of *feeling* at 120 dB. You may have noticed that a very loud sound can be felt in the ear in addition to hearing it. An even louder sound will produce pain. The total dynamic range of hearing is 140 dB. Any sound louder than 140 dB (a jet engine) will not sound any louder to a person (although it would cause more pain).

dBA scale

Loudness is often measured using the *dBA scale*. This scale also places the threshold of hearing at 0 dB. The *A* refers to the *weighting* used when making measurements. A filter tailors the frequency response to match how people hear. Tests have confirmed that *A weighting* closely matches what the instruments report to what people hear. The dBA scale is often used to determine if workers need hearing protection. For example, 90 dBA is the maximum safe work week (40 hours) exposure level. Some common levels are

Whisper	30 dBA
Conversation	60 dBA
Busy city street	80 dBA
Nearby auto horn	100 dBA
Nearby thunder	120 dBA

The *dBm scale* is widely applied in electronic communications. This scale places the 0 dB reference level at a power level of 1 mW. Signals and signal sources for radio frequencies and microwaves are often calibrated in dBm. With this reference level, a signal of 0.25 W is

$$\text{Power level} = 10 \times \log \frac{0.25}{0.001} = +24 \text{ dBm}$$

And a 40-μW signal is

$$\text{Power level} = 10 \times \log \frac{40 \times 10^{-6}}{1 \times 10^{-3}}$$
$$= -14 \text{ dBm}$$

TEST

Determine whether each statement is true or false.

1. The ratio of output to input is called gain.
2. The symbol for voltage gain is A_V.
3. Human hearing is linear for loudness.
4. The dB gain or loss of a system is proportional to the common logarithm of the gain ratio.
5. The overall performance of a system is found by multiplying the individual dB gains.
6. The overall performance of a system is found by adding the ratio gains.
7. If the output signal is less than the input signal the dB gain will be negative.
8. The voltage gain of an amplifier in decibels will be equal to the power gain in decibels only if $R_{\text{in}} = R_{\text{out}}$.
9. The dBm scale uses 1 μW as the reference level.

Solve the following problems.

10. A two-stage amplifier has a voltage ratio of 35 in the first stage and a voltage ratio of 80 in the second stage. What is the overall voltage ratio of the amplifier?
11. A two-stage amplifier has a voltage gain of 26 dB in the first stage and 38 dB in the second stage. What is the overall dB gain?
12. A two-way radio needs about 3 V of audio input to the speaker for good volume. If the receiver sensitivity is specified at 1 μV, what will the overall gain of the receiver have to be in decibels?
13. A 100-W audio amplifier is specified at -3 dB at 20 Hz. What power output can be expected at 20 Hz?
14. A transmitter produces 5 W of output power. A 12-dB power amplifier is added. What is the output power from the amplifier?
15. A transmitting station feeds 1000 W of power into an antenna with an 8-dB gain. What is the effective radiated power of this station?
16. The manufacturer of an RF generator specifies its maximum output as $+10$ dBm. What is the maximum output power available from the generator in watts?

6-2 Common-Emitter Amplifier

Common-emitter amplifier

Figure 6-4 shows a *common-emitter amplifier.* It is so named because the emitter of the transistor is *common* to both the input circuit and the output circuit. The input signal is applied across ground and the base circuit of the transistor. The output signal appears across ground and the collector of the transistor. Since the emitter is connected to ground, it is common to both signals, input and output.

The *configuration* of an amplifier is determined by which transistor terminal is used for signal input and which is used for signal output. The common emitter configuration is one of three possibilities. The last section of this chapter discusses the other two configurations.

People who seek technical jobs must also seek lifelong learning.

There are two resistors in the circuit of Fig. 6-4. One is a *base bias resistor* R_B, and the other is a *collector load resistor* R_L. The base bias resistor is selected to limit the base current to some low value. The collector load resistor makes it possible to develop a voltage swing across the transistor (from collector to emitter). This voltage swing becomes the *output signal.*

Coupling capacitor

C_C in Fig. 6-4 is called a *coupling capacitor.* Coupling capacitors are often used in amplifiers where only ac signals are important. For example, in an audio amplifier there is no need to amplify frequencies lower than 20 Hz because people cannot hear them. A capacitor *blocks direct current.* Coupling capacitors may also be called *dc blocking capacitors. Capacitive reactance* is infinite at 0 Hz:

$$X_C = \frac{1}{2\pi fC}$$

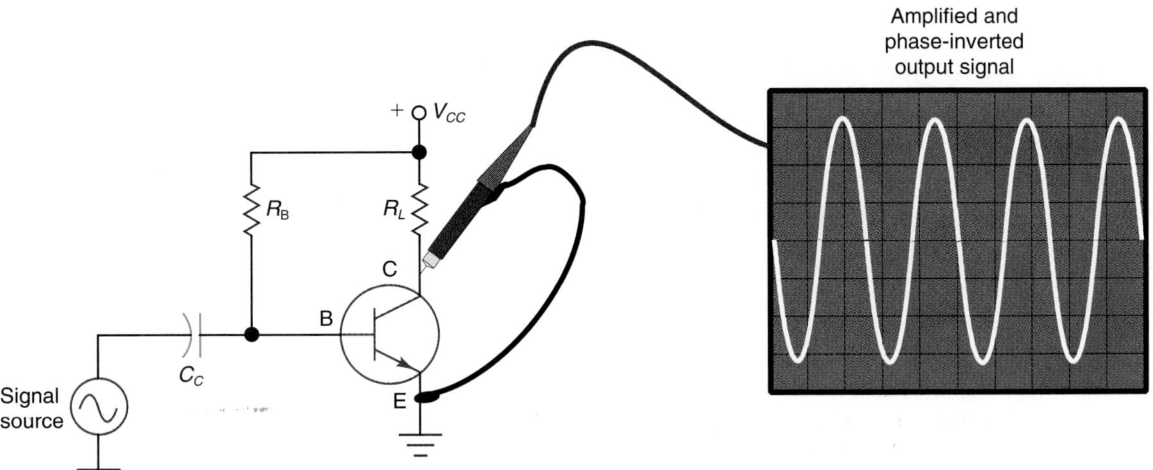

Fig. 6-4 A common-emitter amplifier.

As frequency f approaches that of direct current (0 Hz), capacitive reactance X_C approaches infinity.

A coupling capacitor may be required if the signal source provides a dc path. For example, the signal source could be a pickup coil in a microphone. This coil can have low resistance. Current flow takes the path of least resistance, and without a blocking capacitor the direct current will flow in the coil instead of the transistor base circuit. Figure 6-5 shows this. The direct current flow has been diverted from the transistor base by the signal source.

The direct current through R_B is supposed to come from the base of the transistor, as shown in Fig. 6-6. There *must* be base current for the transistor to be on.

Figure 6-6 has enough information to show how the amplifier operates. We will begin by finding the base current. Two parts can limit base current: R_B and the base-emitter junction. Resistor R_B has high resistance. The base-emitter junction is forward-biased so its resistance is low. Thus, R_B and the supply voltage are the major factors determining base current. By Ohm's law:

$$I_B = \frac{V_{CC}}{R_B} = \frac{12 \text{ V}}{100 \times 10^3 \ \Omega}$$
$$= 120 \times 10^{-6} \text{ A}$$

It is possible to make a better approximation of base current by taking into account the drop across the base-emitter junction of the transistor. This drop is about 0.6 V for a silicon transistor. It is subtracted from the collector supply:

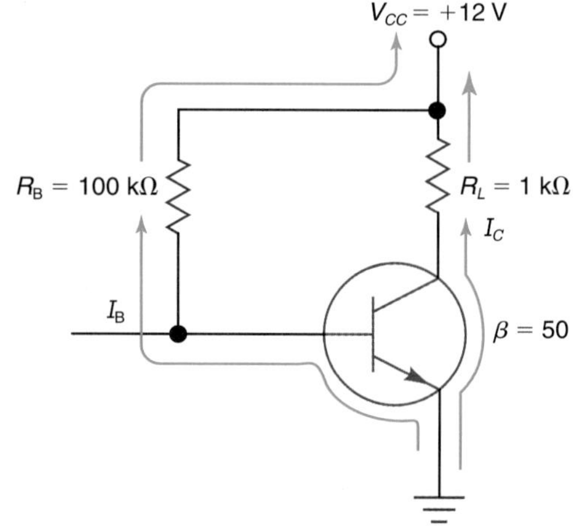

Fig. 6-6 Transistor circuit currents.

$$I_B = \frac{V_{CC} - 0.6 \text{ V}}{R_B}$$

Applying this to the circuit of Fig. 6-6:

$$I_B = \frac{12 \text{ V} - 0.6 \text{ V}}{100 \text{ k}\Omega} = 114 \times 10^{-6} \text{ A}$$

This shows that ignoring the transistor does not produce a large error.

The base current in Fig. 6-6 is small. Since β is given, the collector current can now be found. We will use the first approximation of base current (120 μA) and β to find I_C:

$$I_C = \beta \times I_B = 50 \times 120 \times 10^{-6} \text{ A}$$
$$= 6 \times 10^{-3} \text{ A}$$

The collector current will be 6 mA. This current flows through load resistor R_L. The voltage drop across R_L will be

$$V_{RL} = I_C \times R_L = 6 \times 10^{-3} \text{ A} \times 1 \times 10^3 \ \Omega$$
$$= 6 \text{ V}$$

With a 6-V drop across R_L, the drop across the transistor will be:

$$V_{CE} = V_{CC} - V_{RL} = 12 \text{ V} - 6 \text{ V} = 6 \text{ V}$$

Static condition

The calculations show the condition of the amplifier at its *static,* or resting, state. An input signal will cause the static conditions to change. Figure 6-7 shows why. As the signal source goes positive with respect to ground, the base current increases. The positive-going signal causes additional base current to flow onto the plate of the coupling capacitor. This is shown in Fig. 6-7(*a*). Figure 6-7(*b*) shows the

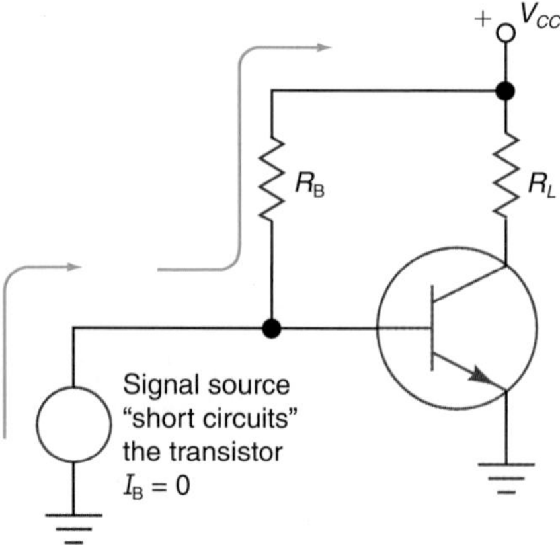

Fig. 6-5 The need for a coupling capacitor.

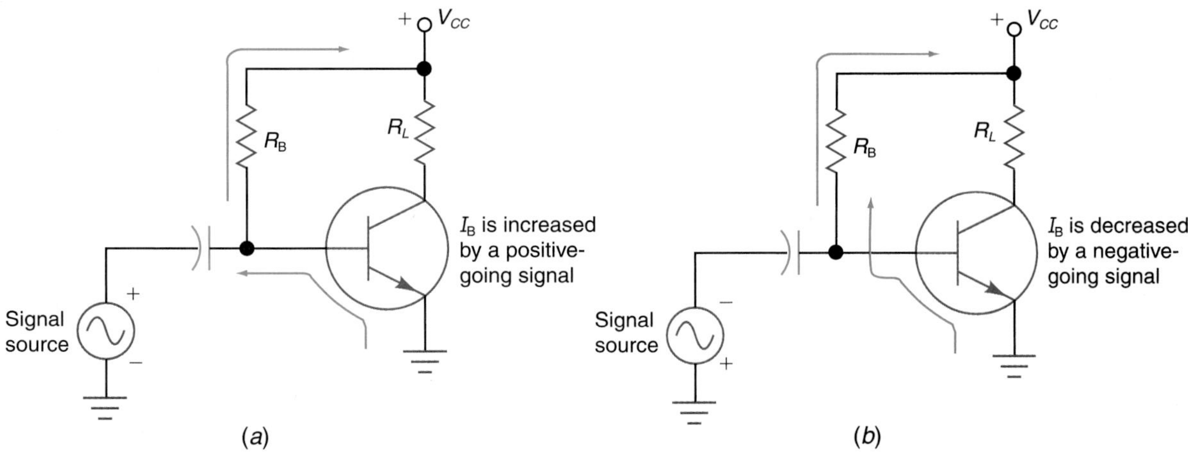

Fig. 6-7 The effect of the input signal on I_B.

input signal going negative. Current flows off the capacitor plate and up through R_B. This decreases the base current.

As the base current increases and decreases, so does the collector current. This is because base current controls collector current. As the collector current increases and decreases, the voltage drop across the load resistor also increases and decreases. This means that the voltage drop across the transistor must also be changing. It does not remain constant at 6 V.

Figure 6-8 on page 136 shows how the output signal is produced. A transistor can be thought of as a resistor from its collector terminal to its emitter terminal. The better the transistor conducts, the lower this resistor is in value. The poorer it conducts, the higher this resistor is in value. Transistor conduction does change as base current changes. So we can assume that an input signal will change the collector-emitter resistance of the transistor.

In Fig. 6-8(*a*) the amplifier is at its static state. The supply voltage is divided equally between R_L and R_{CE}. Resistor R_L is the load, and R_{CE} represents the resistance of the transistor. The time graph shows that the output V_{CE} is a steady 6 V.

Figure 6-8(*b*) shows the input signal going negative. This decreases the base current and, in turn, decreases the collector current. The transistor is now offering more resistance to current flow. Resistance R_{CE} has increased to 2 kΩ. The voltages do not divide equally:

$$V_{CE} = \frac{R_{CE}}{R_{CE} + R_L} \times V_{CC}$$

$$= \frac{2\text{ k}\Omega}{2\text{ k}\Omega + 1\text{ k}\Omega} \times 12\text{ V}$$

$$= 8\text{ V}$$

Thus, the output signal has increased to 8 V. The time graph in Fig. 6-8(*b*) shows this change. Another way to solve for the voltage drop across the transistor would be to solve for the current:

$$I = \frac{V_{CC}}{R_L + R_{CE}}$$

$$= \frac{12\text{ V}}{1\text{ k}\Omega + 2\text{ k}\Omega}$$

$$= 4 \times 10^{-3}\text{ A}$$

Now, this current can be used to calculate the voltage across the transistor:

$$V_{CE} = I \times R_{CE}$$

$$= 4 \times 10^{-3}\text{ A} \times 2 \times 10^{3}\ \Omega$$

$$= 8\text{ V}$$

This agrees with the voltage found by using the ratio technique.

Figure 6-8(*c*) shows what happens in the amplifier circuit when the input signal goes positive. The base current increases. This makes the collector current increase. The tran-

Did You Know?

Not Entirely Obsolete Some currently manufactured audio equipment uses vacuum tubes. This market is tiny compared to the market for solid-state amplifiers.

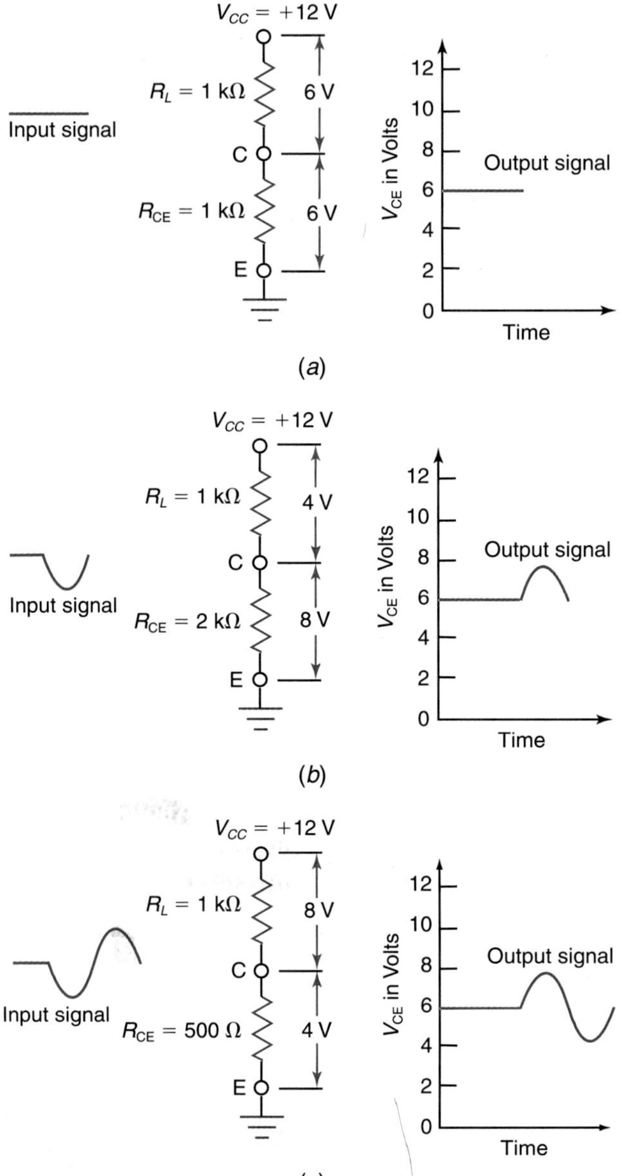

Fig. 6-8 How the output signal is created.

sistor is conducting better, so its resistance has decreased. The output voltage V_{CE} is now:

$$V_{CE} = \frac{0.5\ \text{k}\Omega}{0.5\ \text{k}\Omega + 1\ \text{k}\Omega} \times 12\ \text{V} = 4\ \text{V}$$

The time graph shows this change in output voltage.

Distortion

Note that in Fig. 6-8 that the output signal is *180° out of phase* with the input signal. When the input goes negative [Fig. 6-8(*b*)], the output goes in a positive direction. When the input goes positive [Fig. 6-8(*c*)], the output goes in a negative direction (less positive). This is called *phase inversion.* It is an important characteristic of the common-emitter amplifier.

Phase inversion

The output signal should be a good replica of the input signal. If the input is a sine wave, the output should be a sine wave. When this is achieved, the amplifier is *linear.*

One thing that can make an amplifier nonlinear is too much input signal. When this occurs the amplifier is *overdriven.* This will cause the output signal to show *distortion,* as shown in Fig. 6-9. The waveform is clipped. V_{CE} cannot exceed 12 V. This means that the output signal will approach this limit and then suddenly stop increasing. Note that the positive-going part of the sine wave has been clipped off at 12 V. V_{CE} cannot go below 0 V. Note that the negative-going part of the signal clips at

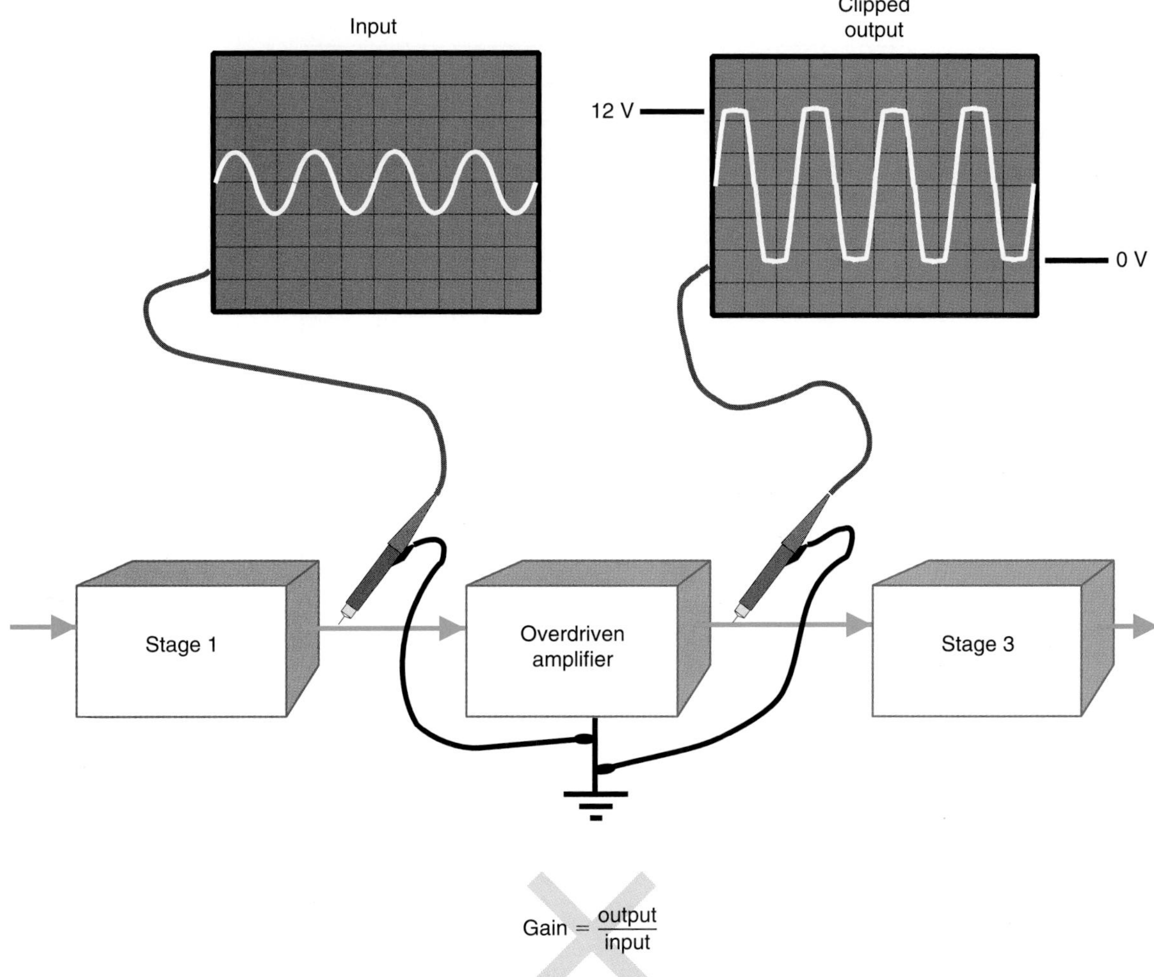

Fig. 6-9 An overdriven amplifier has a clipped output.

0 V. 12 V and 0 V are the *limits* for this particular amplifier. It is not appropriate to calculate gain when the output is clipped.

Clipping is a form of *distortion.* Such distortion in an audio amplifier will cause speech or music to sound bad. This is what happens when the volume control on a radio or stereo is turned up too high. One or more stages are overdriven and distortion results.

Clipping can be avoided by controlling the amplitude of the input and by operating the amplifier at the proper static point. This is best shown by drawing a *load line.* Figure 6-10 on page 138 shows a load line drawn on the collector family of characteristic curves. To draw a load line, you must know the supply voltage (V_{CC}) and the value of the load resistor (R_L). V_{CC} sets the lower end of the load line. If V_{CC} is equal to 12 V, one end of the load line is found at 12 V on the horizontal axis. The other end of the load line is set by the *saturation current.* This is the current that will flow if the collector resistance drops to zero. With this condition, only R_L will limit the flow. Ohm's law is used to find the saturation current:

Clipping

$$I_{sat} = \frac{V_{CC}}{R_L} = \frac{12\text{ V}}{1\text{ k}\Omega}$$
$$= 12 \times 10^{-3}\text{ A, or } 12\text{ mA}$$

This value of current is found on the vertical axis. It is the other end of the load line. As shown in Fig. 6-10, the load line for the amplifier runs between 12 mA and 12 V. These two values are the circuit *limits.* One limit is called *saturation* (12 mA in the example) and the other is called *cutoff* (12 V in the example). No matter what the input signal does, the collector current cannot exceed 12 mA and the output voltage cannot exceed 12 V. If the input

Load line

Saturation

Cutoff

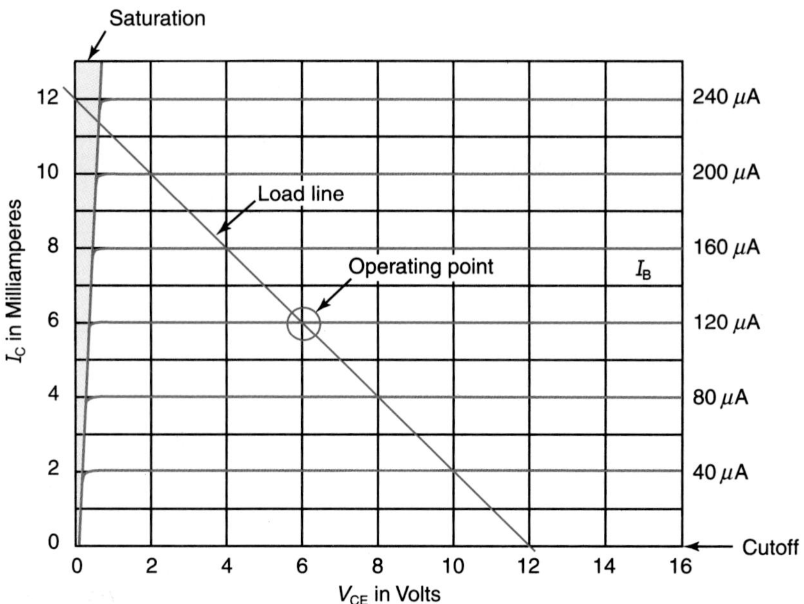

Fig. 6-10 Transistor amplifier load line.

signal is too large, the output will be clipped at these points.

It is possible to operate the amplifier at any point along the load line. The *best* operating point is usually in the *center* of the load line. This point is circled in Fig. 6-10. Notice that the operating point is the *intersection* of the *load line* and the *120-μA curve.* This is the same value of base current we calculated before. Project straight down from the operating point and verify that V_{CE} is 6 V. This also agrees with the previous calculation. Project to the left from the operating point to verify that I_C is 6 mA. Again, there is agreement with the previous calculations.

JOB TIP

It is wise to use polite and ethical behavior when quitting a job.

The load line does not really provide new information. It is helpful because it provides a visual or graphical view of circuit operation. For example, examine Fig. 6-11(*a*). The operating point is in the center of the load line. Notice that as the input goes positive and negative, the output signal follows with no clipping. Look at Fig. 6-11(*b*). The operating point is near the saturation end of the load line. The output signal is now clipped on the negative-going portion. Figure 6-11(*c*) shows the operating point near the cutoff end of the load line. The signal is clipped on the positive-going portion. It should be obvious now why the best operating point is near the center of the load line. It allows the most output before clipping occurs.

Saturation and cutoff must be avoided in linear amplifiers. They cause signal distortion. Three possible conditions for an amplifier are shown in Fig. 6-12. Figure 6-12(*a*) indicates that a *saturated* amplifier is similar to a *closed switch. Saturation* is caused by *high* base current. The collector current is maximum because the transistor is at its minimum resistance. No voltage drops across a closed switch, so $V_{CE} = 0$. Figure 6-12(*b*) shows the amplifier at cutoff. *Cutoff* is caused by *no* base current. The transistor is turned off, and no current flows. All the voltage drops across the open switch, so V_{CE} is equal to the supply voltage. An active transistor amplifier lies between the two extremes. An *active* transistor has some *moderate* value of base current. The transistor is partly on. It can be represented as a resistor. The current is about half of the saturation value, and V_{CE} is about half of the supply voltage.

The conditions of Fig. 6-12 should be memorized. They are very useful when trou-

About Electronics

Quiet Amplifiers Amplifiers that process very small signals are designed to produce as little noise as possible. They are often called *low-noise amplifiers.*

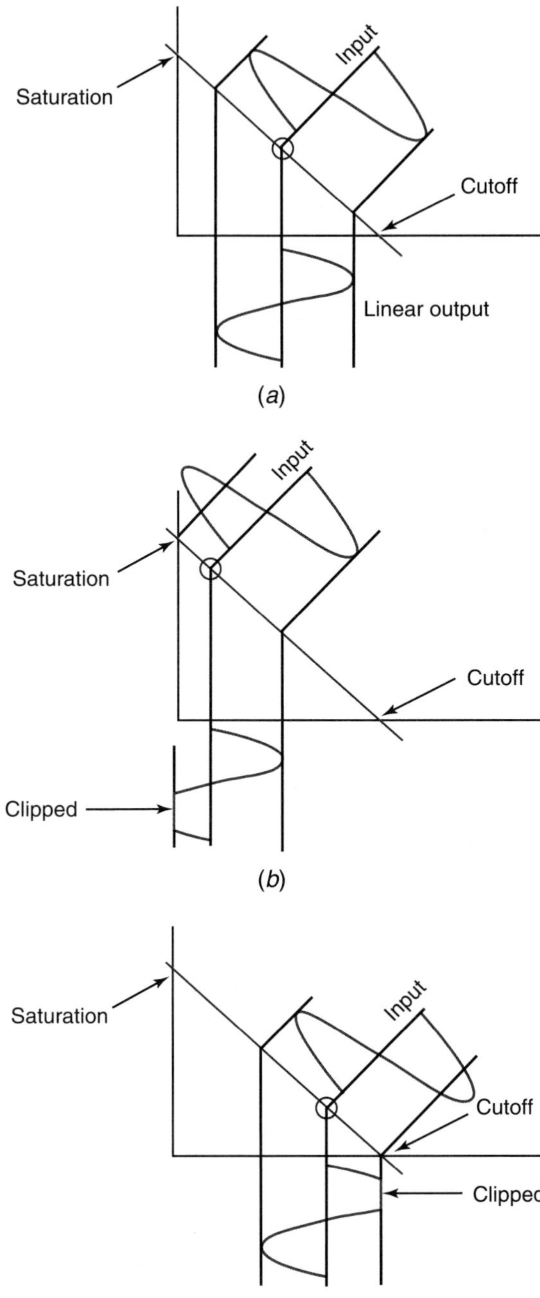

Fig. 6-11 Comparing amplifier operating points.

bleshooting. Also, try to remember that it is the base current that determines whether a transistor will be saturated, in cutoff, or active:

Saturation	High base current
Cutoff	No base current
Active	Moderate base current

This is easy to verify by referring to Fig. 6-10. The operating point can be anywhere along the load line, depending on the base current. When the base current is moderate (120 μA) the operating point is active and in the center of the load line. When the base current is high (240 μA or more), the operating point is at saturation. When the base current is 0, the operating point is at cutoff. Transistor amplifiers that are saturated or in cutoff *cannot* provide linear amplification.

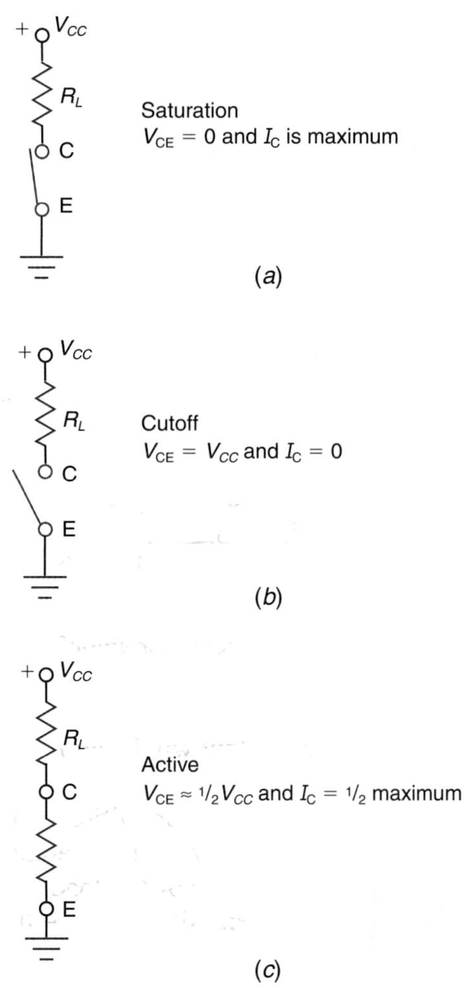

Fig. 6-12 Three possible amplifier conditions.

YOU MAY RECALL . . . that digital or switching circuits use transistors only at cutoff or at saturation. Switching applications were covered in the preceding chapter.

Example 6-13

Use Fig. 6-10 to select a base resistor for a linear amplifier that uses a 12-V power supply and a 2-kΩ collector load resistor. The first step is to draw a new load line. The saturation current is:

$$I_{sat} = \frac{V_{CC}}{R_L} = \frac{12 \text{ V}}{2 \text{ k}\Omega} = 6 \text{ mA}$$

The new load line runs from 12 V on the horizontal axis to 6 mA on the vertical axis. The center of this load line is in between the 40- and 80-μA curves on Fig. 6-10. For linear operation, 60 μA of base current will be about right. Use Ohm's law to find the base resistor:

$$R_B = \frac{V_{CC}}{I_B} = \frac{12 \text{ V}}{60 \ \mu\text{A}} = 200 \text{ k}\Omega$$

TEST

Determine whether each statement is true or false.

17. In a common-emitter amplifier, the input signal is applied to the collector.
18. In a common-emitter amplifier, the output signal is taken from the emitter terminal.
19. A coupling capacitor allows ac signals to be amplified but blocks direct current.
20. Common-emitter amplifiers show a 180° phase inversion.
21. Overdriving an amplifier causes the output to be clipped.
22. The best operating point for a linear amplifier is at saturation.

Solve the following problems.

23. Refer to Fig. 6-6. Change the value of R_B to 75 kΩ. Do not take V_{BE} into account. Find I_B.
24. With R_B changed to 75 kΩ in Fig. 6-6, what is the new value of collector current?
25. With R_B changed to 75 kΩ in Fig. 6-6, what is V_{RL}?
26. With R_B changed to 75 kΩ in Fig. 6-6, what is V_{CE}?
27. Refer to Fig. 6-10. Find the new operating point on the load line using your answer from problem 23, and project down to the voltage axis to find V_{CE}.
28. Refer to Fig. 6-10. Project to the left from the new operating point and find I_C.
29. Refer to Fig. 6-6. Change the value of R_B to 50 kΩ. Do not correct for V_{BE} and determine the base current, the collector current, the voltage drop across the load resistor, and the voltage drop across the transistor. Is the transistor in saturation, in cutoff, or in the linear range?
30. Refer to Figs. 6-6 and 6-10. If V_{CC} is changed to 10 V, determine both end points for the new load line.

6-3 Stabilizing the Amplifier

Figure 6-13 shows a common-emitter amplifier that is the same as the one shown in Fig. 6-6 with one important exception. The transistor has a β of 100. If we analyze the circuit for base current, we get

$$I_B = \frac{V_{CC}}{R_B} = \frac{12 \text{ V}}{100 \text{ k}\Omega} = 120 \ \mu\text{A}$$

This is the same value of base current that was calculated before. The collector current, however, is greater:

$$I_C = \beta \times I_B = 100 \times 120 \ \mu\text{A} = 12 \text{ mA}$$

This is twice the collector current of Fig. 6-6. Now, we can solve for the voltage drop across R_L:

$$V_{RL} = I_C \times R_L = 12 \text{ mA} \times 1 \text{ k}\Omega = 12 \text{ V}$$

And the voltage drop across the transistor is

$$V_{CE} = V_{CC} - V_{RL} = 12 \text{ V} - 12 \text{ V} = 0 \text{ V}$$

There is no voltage across the transistor. The transistor amplifier is in *saturation.* A saturated transistor is *not* capable of linear amplification. The circuit of Fig. 6-13 would produce severe clipping and distortion.

The only change from Fig. 6-6 to Fig. 6-13 is in the β of the transistor. The value of β does vary widely among transistors with the same

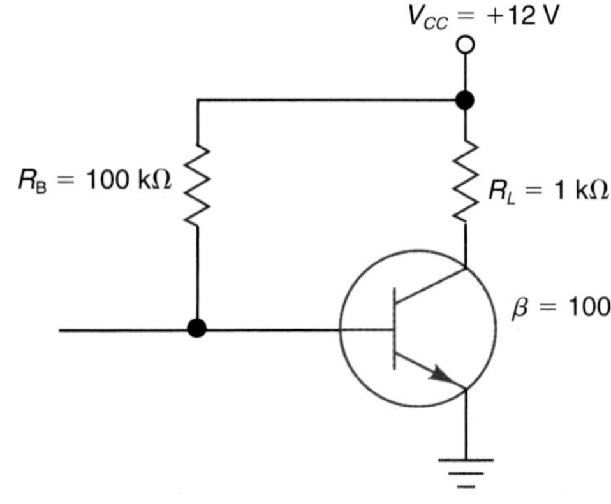

Fig. 6-13 Amplifier with a higher β transistor.

part number. This means that the amplifier circuit shown in Figs. 6-6 and 6-13 is not practical. It is too sensitive to β, which varies from transistor to transistor and with temperature. We need a circuit that is not so sensitive to β.

Sensitivity to β

Example 6-14

Find a better estimate of the conditions for the amplifier shown in Fig. 6-13 by taking the base-emitter junction voltage into account:

$$I_B = \frac{V_{CC} - V_{BE}}{R_B} = \frac{12 \text{ V} - 0.7 \text{ V}}{100 \text{ k}\Omega} = 113\ \mu\text{A}$$

$$I_C = \beta \times I_B = 100 \times 113\ \mu\text{A} = 11.3 \text{ mA}$$

$$V_{R_L} = I_C \times R_L = 11.3 \text{ mA} \times 1 \text{ k}\Omega = 11.3 \text{ V}$$

$$V_{CE} = V_{CC} - V_{R_L} = 12 \text{ V} - 11.3 \text{ V} = 0.7 \text{ V}$$

Note: The better estimate does not make a large difference. V_{CE} is very close to saturation, and the amplifier is still in trouble.

Example 6-15

Solve Fig. 6-13 assuming a current gain of 200. The base current remains the same. Using the value of 120 μA found earlier:

$$I_C = \beta \times I_B = 200 \times 120\ \mu\text{A} = 24 \text{ mA}$$

$$V_{R_L} = I_C \times R_L = 24 \text{ mA} \times 1 \text{ k}\Omega = 24 \text{ V}$$

Note: V_{R_L} can't be larger than V_{CC}. When this happens, the amplifier is in hard saturation and the actual collector current is limited by Ohm's law:

$$I_C = \frac{V_{CC}}{R_L} = \frac{12 \text{ V}}{1 \text{ k}\Omega} = 12 \text{ mA}$$

The collector current in Fig. 6-13 cannot exceed 12 mA, regardless of β.

It is possible to significantly improve a common emitter amplifier by adding two resistors: one in the emitter circuit and another from base to ground. These resistors can make an amplifier less sensitive to β and to temperature changes. The additional resistors create some additional voltage drops. Figure 6-14 shows the drops most commonly used for analyzing and troubleshooting transistor amplifiers.

Beginning at the left in Fig. 6-14, V_B refers to the drop from the base terminal of the transistor to ground. This is an example of *single-subscription notation. When single subscripts (the B in V_B)* are used, ground is the reference point. V_B is measured from the base to ground. V_{BE} is an example of double subscript notation. When double subscripts (B and E in V_{BE}) are used, the voltage drop is measured from one subscript point to the other subscript point. Another possibility is the drop across a part with two leads, such as a resistor. An example is V_{R_L} in Fig. 6-14, which specifies the drop across the

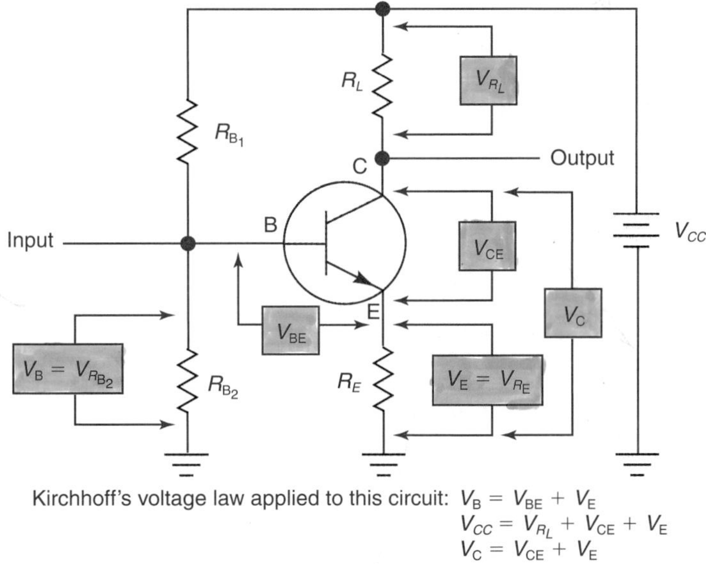

Fig. 6-14 Common labels for voltage drops in a transistor amplifier.

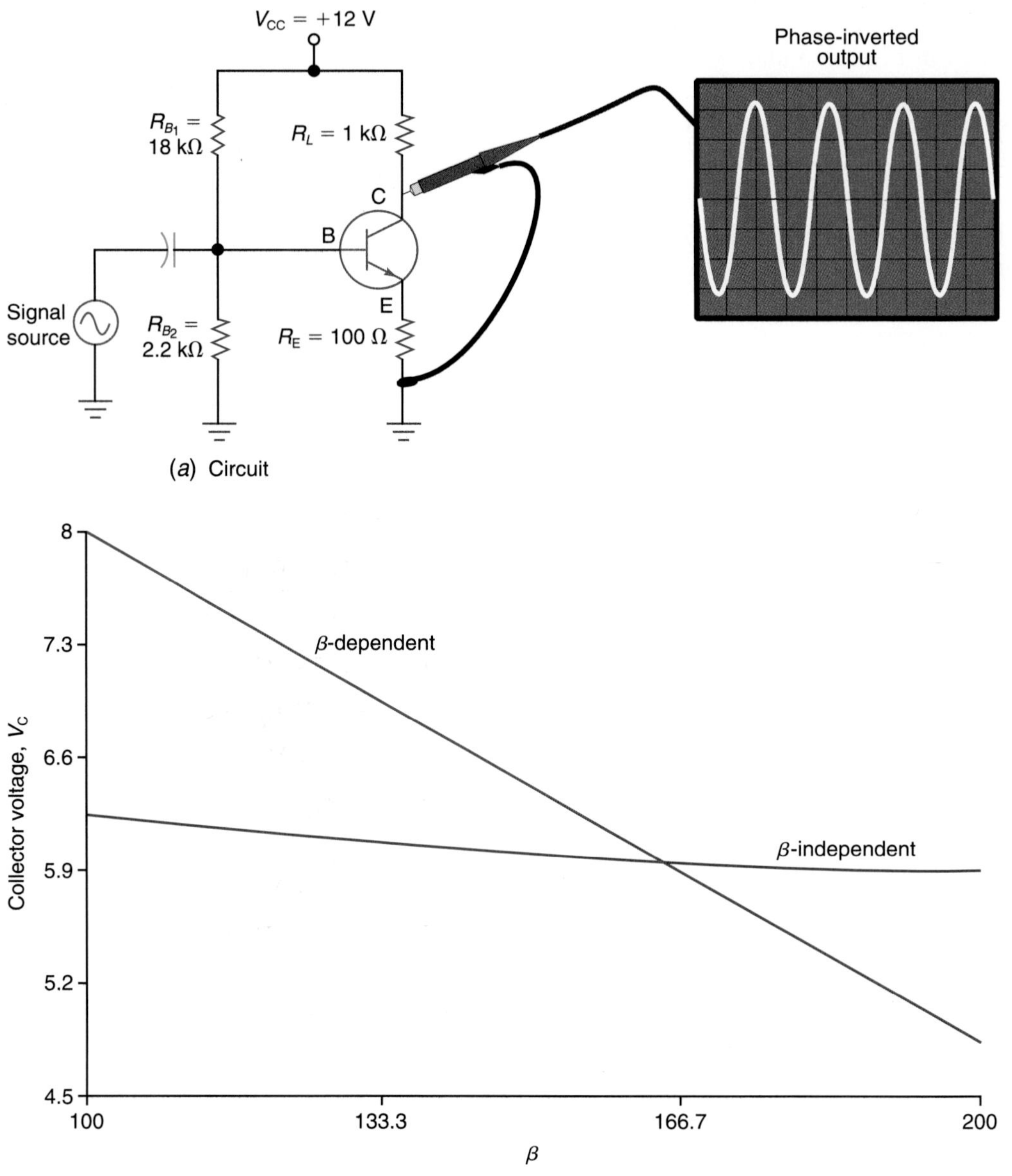

Fig. 6-15 A practical (β-independent) common-emitter amplifier.

load resistor. Study this system of notation for voltage drops carefully. It will help you analyze and troubleshoot solid-state circuits.

Figure 6-15 shows an amplifier with all of the resistor values and the supply voltage specified. This information will allow us to analyze the circuit using the following steps:

1. Calculate the voltage drop across R_{B_2}. This is also called the base voltage, or V_B. The two base resistors form a voltage divider across the supply V_{CC}. The voltage divider equation is:

$$V_B = \frac{R_{B_2}}{R_{B_1} + R_{B_2}} \times V_{CC}$$

2. Assume a 0.7-V drop from base to emitter for silicon transistors and a 0.2-V drop for germanium transistors. Calculate V_E by subtracting this drop from V_B:

$$V_E = V_B - 0.7 \text{ (for silicon transistors)}$$

3. Calculate the emitter current using Ohm's law:

$$I_E = \frac{V_E}{R_E}$$

4. Assume that the collector current equals the emitter current:

$$I_C = I_E$$

5. Calculate the voltage drop across the load resistor using Ohm's law.

$$V_{RL} = I_C \times R_L$$

6. Calculate the collector-to-emitter drop using Kirchhoff's law:

$$V_{CE} = V_{CC} - V_{RL} - V_E$$

7. Calculate V_C using:

$$V_C = V_{CE} + V_E$$

or

$$V_C = V_{CC} - V_{RL}$$

This seven-step process is not exact, but it is accurate enough for practical work. The first step ignores the base current that flows through R_{B1}. This current is usually about one-tenth the divider current. Thus, a small error is made by using only the resistor values to compute V_B. The second step is based on what we have already learned about forward-biased silicon and germanium junctions. However, to get better accuracy, 0.7 V, rather than 0.6 V, is used for silicon transistors. Making V_{BE} a little high tends to reduce the error that was caused by ignoring the base current in the first step. The third step is Ohm's law, and no error is introduced here. The fourth step introduces a small error. We know that the emitter current is slightly larger than the collector current. The last two steps are circuit laws, and no error is introduced. All in all, the procedure can provide very useful answers. Notice that β is not used anywhere in the seven steps.

Let's apply the procedure to the circuit of Fig. 6-15:

1. $V_B = \dfrac{R_{B2}}{R_{B1} + R_{B2}} \times V_{CC} = \dfrac{2.2\ \text{k}\Omega}{18\ \text{k}\Omega + 2.2\ \text{k}\Omega} \times 12\ \text{V} = 1.307\ \text{V}$

2. $V_E = V_B - 0.7\ \text{V} = 1.307\ \text{V} - 0.7\ \text{V} = 0.607\ \text{V}$

3. $I_E = \dfrac{V_E}{R_E} = \dfrac{0.607\ \text{V}}{100\ \Omega} = 6.07\ \text{mA}$

4. $I_C = I_E = 6.07\ \text{mA}$

5. $V_{RL} = I_C \times R_L = 6.07\ \text{mA} \times 1\ \text{k}\Omega = 6.07\ \text{V}$

6. $V_{CE} = V_{CC} - V_{RL} - V_E = 12\ \text{V} - 6.07\ \text{V} - 0.607\ \text{V} = 5.32\ \text{V}$

7. $V_C = V_{CE} + V_E = 5.32\ \text{V} + 0.607\ \text{V} = 5.93\ \text{V}$

Did You Know?

Products Come and Products Go . . . The basics of electricity and electronics will never become obsolete.

Since the collector-emitter voltage is nearly half the supply voltage, we can assume the circuit will make a good linear amplifier. The circuit will work well with any reasonable value of β and will be stable over a wide temperature range.

The graph shown in Fig. 6-15(*b*) shows the performance of the amplifier with transistor gain ranging from 100 to 200. Note that there is only a minor change in the performance of the β-independent amplifier. For contrast, the performance of the β-dependent circuit of Fig. 6-13 is also shown. The graph was produced using a circuit simulator.

Example 6-16

Modify the circuit shown in Fig. 6-15 so that V_{CE} is half the supply voltage. There are many ways to accomplish this. Perhaps the easiest approach is to change the value of R_L, since doing so will not affect the collector current or the emitter voltage. Begin by rearranging the Kirchhoff voltage equation to find the value of V_{RL} with V_{CE} equal to half the supply:

$$V_{RL} = V_{CC} - V_{\mathrm{CE}} - V_{\mathrm{E}}$$
$$V_{RL} = 12\text{ V} - 6\text{ V} - 0.607\text{ V} = 5.39\text{ V}$$

Find the new value for R_L with Ohm's law:

$$R_L = \frac{V_{RL}}{I_{\mathrm{C}}} = \frac{5.39\text{ V}}{6.07\text{ mA}} = 888\ \Omega$$

DC analysis

AC analysis

Voltage gain

So far the discussions and examples have been concerned with the *dc analysis* of transistor amplifiers. The dc conditions include all of the static currents and voltage drops. An *ac analysis* of an amplifier will allow us to determine its *voltage gain.*

Voltage gain is an *ac* amplifier characteristic and one of the most important for small-signal amplifiers. It is the easiest form of gain to measure. For example, an oscilloscope can be used to look at the input signal and then the output signal. Dividing the input into the output will give the gain. Current gain and power gain are not as easy to measure.

Bipolar junction transistors are current amplifiers. A changing base current will produce an output signal. However, as the input signal *voltage* changes, the input signal *current* will also change. In other words, it is the input signal voltage that controls the input signal current. This makes it possible to discuss, calculate, and measure signal voltage gain even though BJTs are current-controlled.

The first step in calculating voltage gain is to estimate the *ac resistance* of the *emitter* of the transistor. This resistance is determined by the dc emitter current. Its symbol is r_{E}, and it can be estimated with:

$$r_{\mathrm{E}} = \frac{25\text{ mV}}{I_{\mathrm{E}}}$$

Since the numerator is in millivolts, the formula can be applied directly when the emitter current is in milliamperes. However, if the emitter current is in amperes, change the numerator to volts (25 mV = 0.025 V) or change the emitter current to milliamperes.

We have already solved the circuit of Fig. 6-15 for the dc emitter current, so we can estimate r_{E}:

$$r_{\mathrm{E}} = \frac{25\text{ mV}}{6.07\text{ mA}} = 4.12\ \Omega$$

In actual circuits, and with higher temperatures, r_{E} tends to be higher. It can be estimated with:

$$r_{\mathrm{E}} = \frac{50\text{ mV}}{I_{\mathrm{E}}}$$

So, for the circuit of Fig. 6-15, r_{E} could be as high as:

$$r_{\mathrm{E}} = \frac{50\text{ mV}}{6.07\text{ mA}} = 8.24\ \Omega$$

Knowing r_{E} allows the voltage gain to be found from:

$$A_V = \frac{R_L}{R_{\mathrm{E}} + r_{\mathrm{E}}}$$

Let's use this formula to find the voltage gain for the circuit of Fig. 6-15:

$$A_V = \frac{1000\ \Omega}{100\ \Omega + 4.12\ \Omega} = 9.6$$

The amplifier will have a voltage gain of 9.6. If the input signal is 1 V peak-to-peak, then the output signal should be 9.6 V peak-to-peak. If the input signal is 2 V peak-to-peak, then the output will be clipped. It is *not possible* to exceed the supply voltage in peak-to-peak output in this type of amplifier. The calculated gain will hold true *only* if the amplifier is operating in a linear fashion.

Emitter bypass capacitor

Sometimes, a much higher gain is needed. It is possible to improve the gain quite a bit by adding an *emitter bypass capacitor.* Figure

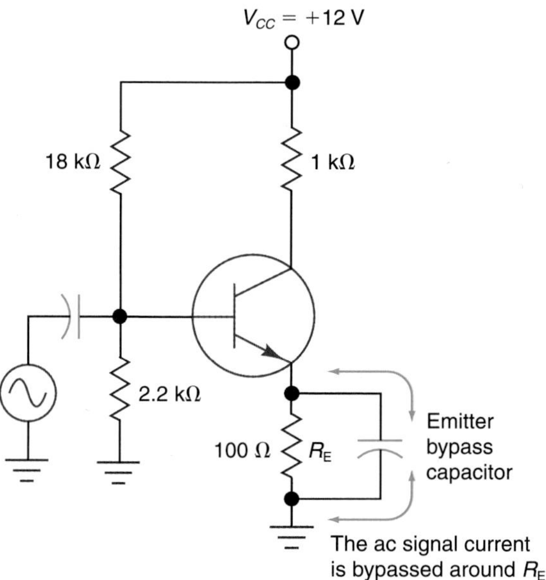

Fig. 6-16 Adding an emitter bypass capacitor.

6-16 shows this capacitor added to the amplifier. The *capacitor* is *chosen* to have a low reactance at the frequency of operation. It acts as a short circuit for the ac signal. This means that the ac signal is bypassed around R_E. Current will take the path of least opposition. Since R_E has been *bypassed,* the voltage gain is set by R_L and r_E:

$$A_V = \frac{R_L}{r_E}$$

The voltage gain for the circuit of Fig. 6-16 is

$$A_V = \frac{1000\ \Omega}{4.12\ \Omega} = 243$$

But, remember, r_E can be as high as 8.24 Ω. The gain might be

$$A_V = \frac{1000\ \Omega}{8.24\ \Omega} = 121$$

The voltage gain for Fig. 6-16 will be between 121 and 243. In practice, the *lower* estimate is safer. A conservative design has extra gain. It is possible to make more precise estimates of amplifier gain. A circuit simulator, using a 2N4401 transistor model, predicts a voltage gain of 200 for the emitter bypass amplifier. For practical work, the procedure used here is good.

Without the bypass capacitor, the voltage gain was calculated at 9.6. With the bypass capacitor, the voltage gain will be at least 121. This is almost an amazing increase in gain for simply adding a capacitor. The increased gain is not without its drawbacks, however. The bypass capacitor may be a relatively costly item. It must have a low reactance at the lowest signal frequency. If the amplifier is to operate at very low frequencies, the capacitor may have to be quite large. As a rough guess, the capacitor can be selected to have one-tenth the reactance of the emitter resistor, in ohms. In Fig. 6-16, the emitter resistor is 100 Ω. This means that the capacitor must not have a reactance greater than 10 Ω at the lowest operating frequency. An audio amplifier may have to operate down to 15 Hz. The capacitive reactance equation can be used to select the bypass capacitor:

$$X_C = \frac{1}{2\pi f C}$$

Rearranging the equation and solving for C gives

Capacitor selection

$$C = \frac{1}{2\pi f X_C} = \frac{1}{6.28 \times 15\ \text{Hz} \times 10\ \Omega}$$
$$= 1061\ \mu\text{F}$$

This is a large capacitor. Electrolytic types are typically used for bypassing in audio amplifiers. The voltage rating can be very low since the dc emitter voltage is only 0.61 V in this circuit. However, the capacitor will still be relatively costly and large.

There are other considerations for emitter bypassing in addition to the size and cost of the capacitor. Emitter bypassing affects the input impedance of the amplifier, its frequency range, and its distortion. These effects can make bypassing a poor choice. Therefore, circuit designers may choose to develop a gain of 100 by using two amplifier stages, each having a gain of 10. These ideas are expanded in the next chapter.

TEST

Solve the following problems.

31. Refer to Fig. 6-14. If V_B = 1.5 V and V_{BE} = 0.7 V, what is the voltage drop across R_E?
32. Refer to Fig. 6-14. If V_{CC} = 10 V, V_{RL} = 4.4 V, and V_{RE} = 1.2 V, what is V_{CE}?
33. Using the data from problem 32, find V_C.

Problems 34 through 40 refer to Fig. 6-15 with these changes: R_{B_2} = 1.5 kΩ and R_L = 2700 Ω.

34. Calculate V_B.
35. Calculate I_E.

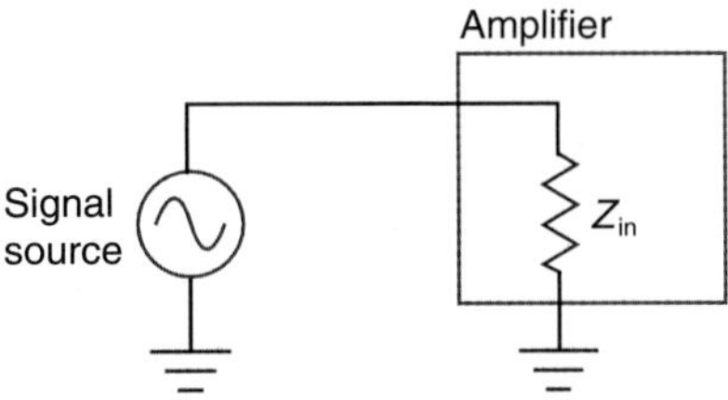

Fig. 6-17 Amplifier loading effect.

36. Calculate V_{R_L}.
37. Calculate V_{CE}.
38. Is the amplifier operating in the center of the load line?
39. Calculate A_V.
40. Calculate the range for A_V if an emitter bypass capacitor is added to the circuit.

6-4 Other Configurations

The common-emitter configuration is a very popular circuit. It serves as the basis for most linear amplifiers. However, for some circuit conditions, one of two other configurations may be a better choice.

Input impedance

Amplifiers have many characteristics. Among these is *input impedance.* The input impedance of an amplifier is the loading effect it will present to a signal source. Figure 6-17 shows that when a signal source is connected to an amplifier, that source sees a load, not an amplifier. The load seen by the source is the input impedance of the amplifier.

Characteristic impedance

Signal sources vary widely. An antenna is the signal source for a radio receiver. An antenna might have an impedance of 50 Ω. A microphone is the signal source for a public address system. A microphone might have an impedance of 100,000 Ω. Every signal source has a *characteristic impedance.*

Impedance matching

The situation can be stated simply: For the best power transfer, the source impedance should *equal* the amplifier input impedance. This is called *impedance matching.* Figure 6-18 shows why impedance matching gives the best power transfer. In Fig. 6-18(*a*) a 60-V signal source has an impedance of 15 Ω. This impedance (Z_G) is drawn as an external resistor in series with the generator (signal source). Since the generator impedance does act in series, the circuit is a good model, as shown. The load impedance in Fig. 6-18(*a*) is also 15 Ω. Thus, we have an *impedance match.* Let us see how much power is transferred to the load. We begin by finding the current flow:

$$I = \frac{V}{Z} = \frac{60\text{ V}}{15\ \Omega + 15\ \Omega} = 2\text{ A}$$

The power dissipated in the load is

$$P = I^2 \times Z_L = (2\text{ A})^2 \times 15\ \Omega = 60\text{ W}$$

A 15-Ω load will dissipate 60 W. Figure 6-18(*b*) shows the same source with a load of 5 Ω. Solving this circuit gives

$$I = \frac{60\text{ V}}{15\ \Omega + 5\ \Omega} = 3\text{ A}$$
$$P = (3\text{ A})^2 \times 5\ \Omega = 45\text{ W}$$

Note that the dissipation is less than maximum when the load impedance is less than the source impedance. Figure 6-18(*c*) shows the load impedance at 45Ω. Solving this circuit gives

$$I = \frac{60\text{ V}}{15\ \Omega + 45\ \Omega} = 1\text{ A}$$
$$P = (1\text{ A})^2 \times 45\ \Omega = 45\text{ W}$$

The dissipation is *less* than maximum when the load impedance is greater than the source impedance. Maximum load power will occur *only* when the impedances are matched.

The common-emitter configuration typically has an input impedance of around 1000 Ω. The actual value depends on both the transistor used and the other parts in the amplifier. This may

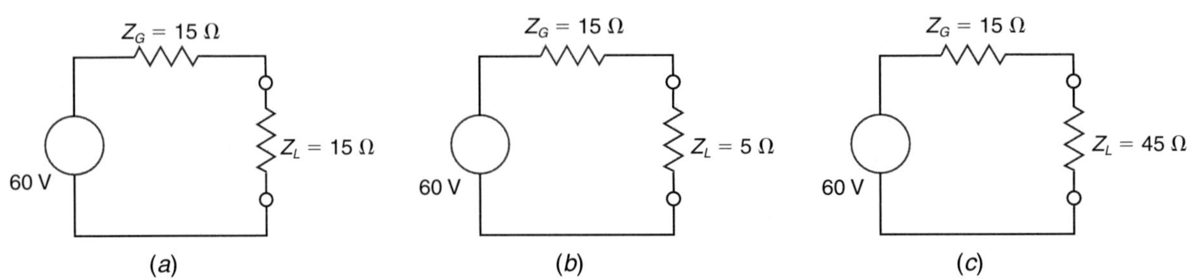

Fig. 6-18 The need for impedance matching.

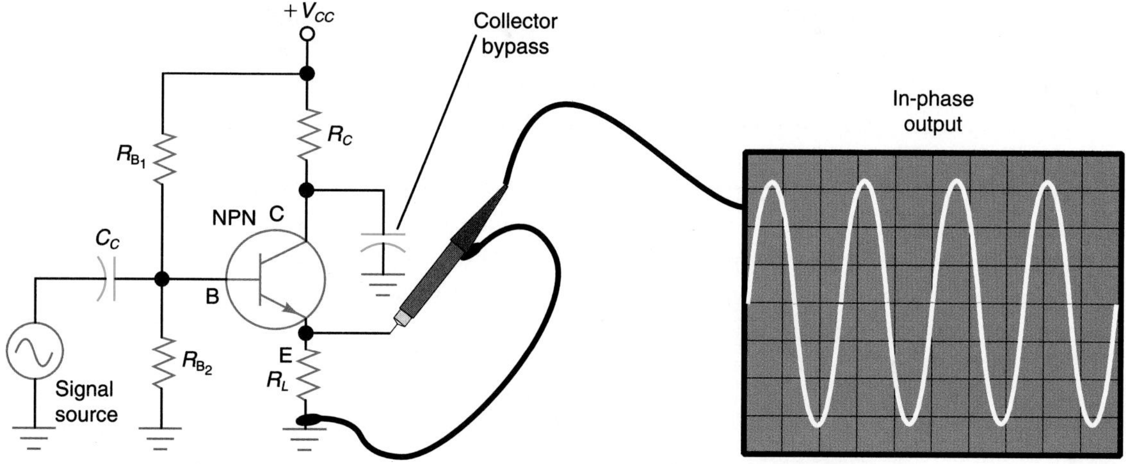

Fig. 6-19 A common-collector amplifier.

or may not be a desirable input impedance. It depends on the signal source.

Figure 6-19 shows a *common-collector amplifier.* It is so named because the collector terminal is common to the input and the output circuits. At first glance, this circuit may seem to be the same as a common-emitter amplifier. There are two very important differences, however:

1. The collector is bypassed to ground with a capacitor. This capacitor has very low reactance at the signal frequency. As far as signals are concerned, the *collector is grounded.*
2. The load resistor is in the emitter circuit. The output signal is across this load. Thus, the *emitter* is now the output terminal. The collector is the output terminal for the common-emitter configuration.

Common-collector amplifier

The common-collector amplifier can have a very high input impedance, as much as several hundred thousand ohms. If the signal source has a very high characteristic impedance, the common-collector amplifier may prove to be the best choice. The stage following the common collector could be a common-emitter configuration. Figure 6-20 shows this arrangement.

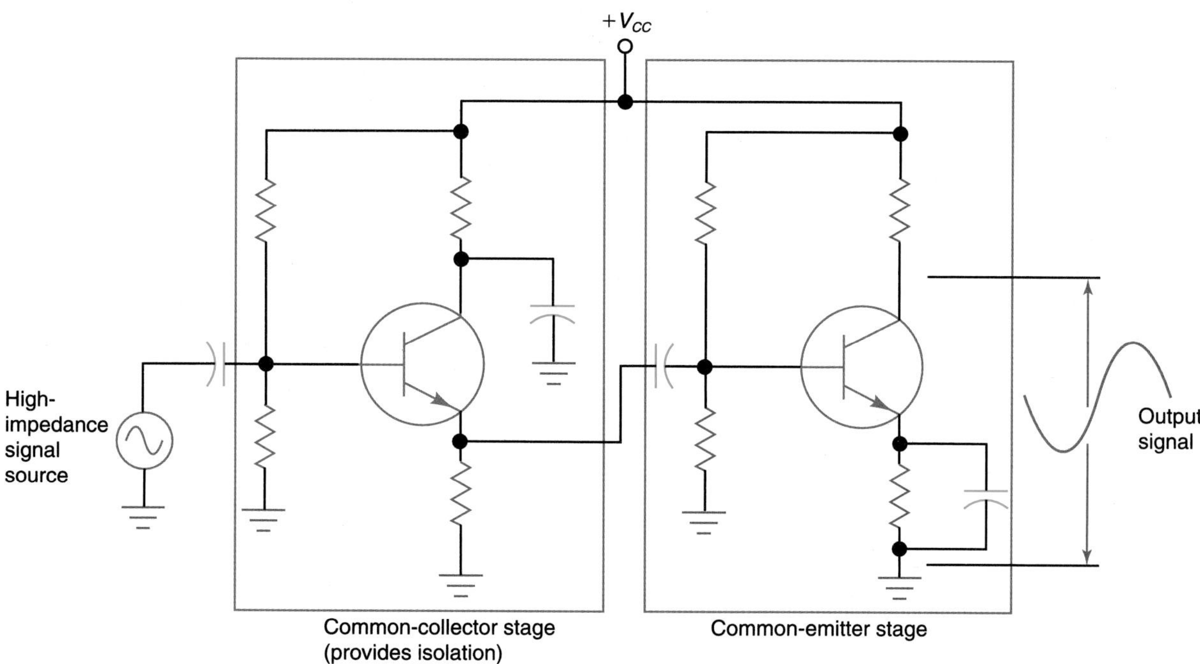

Fig. 6-20 A two-stage amplifier.

Isolation amplifier

The common-collector stage is sometimes referred to as an *isolation amplifier* or *buffer amplifier.* Its high input impedance loads the signal source very lightly. Only a very small signal current will flow. Thus, the signal source has been isolated (buffered) from the loading effects of the rest of the circuit.

In addition to a high input impedance, the common-collector amplifier has some other important characteristics. It is not capable of giving any voltage gain. The output signal will always be less than the input signal as far as voltage is concerned. The current gain is very high. There is also a moderate power gain. There is no phase inversion in the common-collector amplifier. As the signal source drives the base terminal in a positive direction, the output (emitter terminal) also goes in a positive direction. The fact that the output *follows* the input has led to a second name for this amplifier. It is frequently called an *emitter follower.*

Emitter follower

Emitter followers are also noted for their low *output impedance.* This is an advantage when a signal must be supplied to a low-impedance load. For example, a speaker typically has an impedance of 4 to 8 Ω. An emitter follower can drive a speaker reasonably well, whereas the basic common-emitter amplifier cannot.

Common-base amplifier

The last circuit to be discussed is the *common base amplifier.* This configuration has its base terminal common to both the input and output signals. It has a very low *input impedance,* on the order of 50 Ω. Therefore, it is useful only with low-impedance signal sources. It is a good performer at radio frequencies.

Figure 6-21 is a schematic diagram of a common-base RF amplifier. It is designed to amplify weak radio signals from the antenna circuit. The antenna impedance is low, on the order of 50 Ω. This makes a good impedance match from the antenna to the amplifier. The base is grounded at the signal frequency by C_4. The signal is fed into the emitter terminal of the transistor, and the amplified output signal is taken from the collector. Circuits L_1C_2 and L_2C_5 are for *tuning.* They are *resonant* at the desired frequency of operation. This allows the amplifier to reject other frequencies that could cause interference. Coil L_2 and capacitor C_5 form the collector load for the amplifier. This load will be a high impedance at the resonant frequency. This makes the voltage gain high for this frequency. Other frequencies will have less gain through the amplifier. Figure 6-22 shows the gain performance for a tuned RF amplifier of this type.

New hires should remember that some workers tend to be disgruntled even when the job environment is far above average.

The common-base amplifier is not capable of providing any current gain. The input current will always be more than the output current. This is because the emitter current is always highest in BJTs. The amplifier is capable of a large voltage gain. It is also capable of power gain. As with the emitter-follower configuration, it does not invert the signal (no phase inversion).

Example 6-17

Find the emitter current and V_{CE} for Fig. 6-21 if $R_{B_1} = 10\ k\Omega$, $R_{B_2} = 2.2\ k\Omega$, and $R_E = 470\ \Omega$. Although this circuit looks very different, it can be analyzed with the same approach that was used for Fig. 6-15. As before, find the base voltage:

$$V_B = \frac{R_{B_2}}{R_{B_1} + R_{B_2}} \times V_{CC}$$

$$= \frac{2.2\ k\Omega}{10\ k\Omega + 2.2\ k\Omega} \times 12\ V = 2.16\ V$$

Subtract for the base-emitter drop:

$$V_E = V_B - 0.7\ V = 2.16\ V - 0.7\ V = 1.46\ V$$

Note: The dc resistance of coil L_1 is very low and will have no effect on the dc emitter current. Calculate the emitter current:

$$I_E = \frac{V_E}{R_E} = \frac{1.46\ V}{470\ \Omega} = 3.11\ mA$$

Note: The dc resistance of coil L_2 is very low and will have no effect on the dc voltage drop across the transistor. Find the drop across the transistor:

$$V_{CE} = V_{CC} - V_E = 12\ V - 1.46\ V = 10.5\ V$$

Only the common-emitter amplifier provides all three forms of gain: voltage, current, and power. It provides the best power gain of any of the three configurations. It is the most useful of the three. Table 6-2 sum-

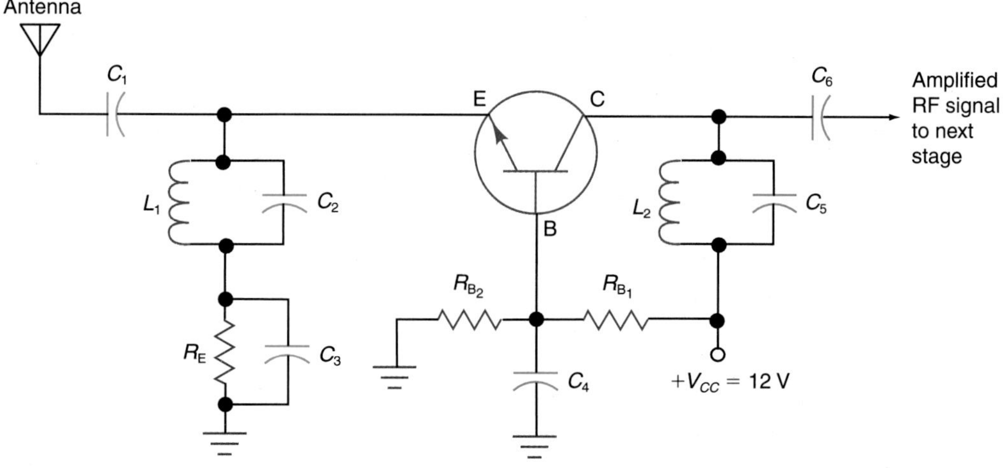

Fig. 6-21 A common-base RF amplifier.

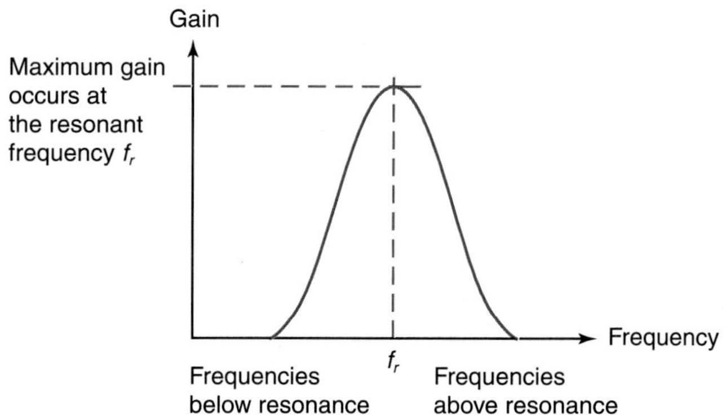

Fig. 6-22 Frequency response of the tuned RF amplifier.

Table 6-2 Summary of Amplifier Configurations

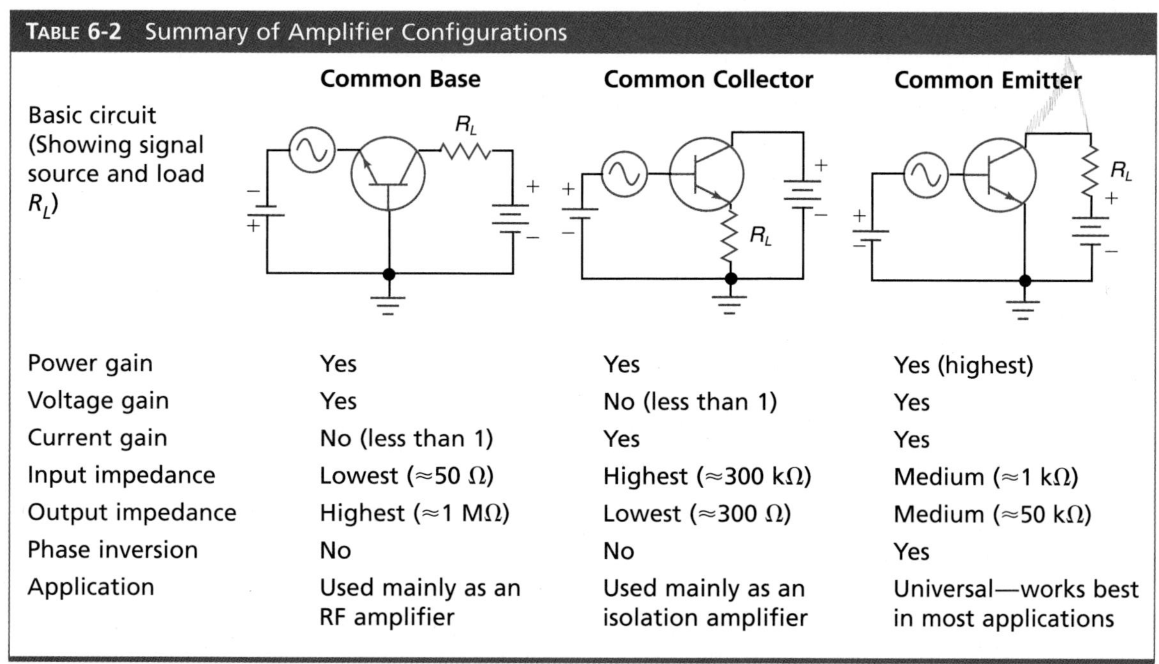

	Common Base	Common Collector	Common Emitter
Basic circuit (Showing signal source and load R_L)			
Power gain	Yes	Yes	Yes (highest)
Voltage gain	Yes	No (less than 1)	Yes
Current gain	No (less than 1)	Yes	Yes
Input impedance	Lowest (≈50 Ω)	Highest (≈300 kΩ)	Medium (≈1 kΩ)
Output impedance	Highest (≈1 MΩ)	Lowest (≈300 Ω)	Medium (≈50 kΩ)
Phase inversion	No	No	Yes
Application	Used mainly as an RF amplifier	Used mainly as an isolation amplifier	Universal—works best in most applications

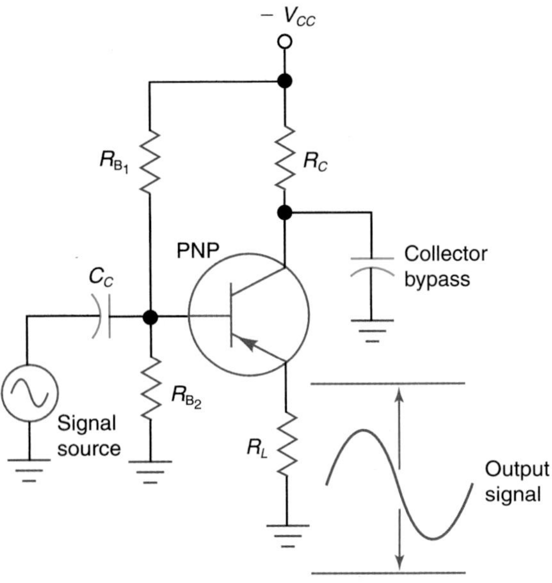

Fig. 6-23 A PNP transistor amplifier.

marizes the important details for the three configurations.

So far, only NPN transistor amplifiers have been shown. Everything that has been discussed for NPN circuits applies to PNP circuits, with the exception of *polarity.* Figure 6-23 shows a PNP amplifier. Note that the supply V_{CC} is *negative.* If you will compare this with the NPN amplifiers shown in this chapter, you will find that they are energized with a positive supply.

NPN circuits *can* be energized with a negative supply. When this is done, the supply terminal is named V_{EE} as opposed to V_{CC}. This is shown in Fig. 6-24. Study this circuit and compare it with Fig. 6-15 to verify that both transistors are properly biased. You should determine in both circuits that the collector is reverse-biased and that the base-emitter junction is forward-biased. Remember, these bias conditions must be met for a transistor to serve as a linear amplifier.

Figure 6-24 is a printout from a circuit simulator. Note how closely the meter readings agree with the values that were found earlier for Fig. 6-15. Some of the meter labels are dif-

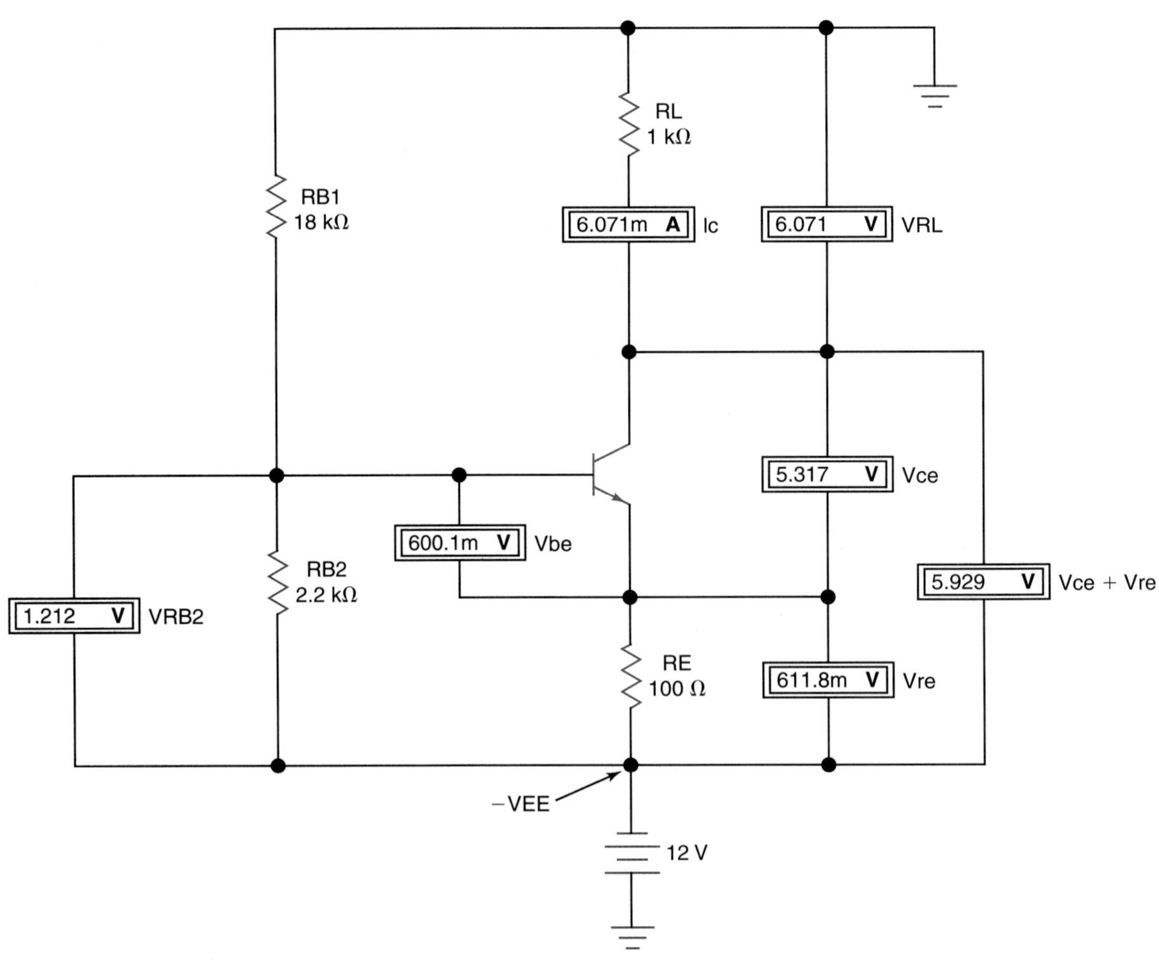

Fig. 6-24 An NPN transistor amplifier with an emitter supply.

ferent, to reflect the change in ground reference for this circuit.

TEST

Determine whether each statement is true or false.

41. The configuration of an amplifier can be determined by inspecting which transistor terminals are used for input and output.
42. For maximum power transfer, the source resistance must be equal to the load resistance.
43. The term "emitter follower" is applied to common-emitter amplifiers.
44. The common collector configuration is the best choice for matching a high impedance source to a low impedance load.
45. The amplifier shown in Fig. 6-23 is in the common-emitter configuration.

Solve the following problems.

46. A signal source has an impedance of 300 Ω. It develops an output of 1 V. Calculate the power transfer from this source into each of the following amplifiers:
    a. Amplifier A has an input impedance of 100 Ω.
    b. Amplifier B has an input impedance of 300 Ω.
    c. Amplifier C has an input impedance of 900 Ω.
47. Refer to Fig. 6-19. V_{CC} = 12 V. The transistor is silicon; R_{B_1} = 47 kΩ, and R_{B_2} = 68 kΩ. If R_L is a 470-Ω resistor, what is the dc emitter current I_E?
48. Refer to Fig. 6-21. The transistor is silicon; R_{B_1} = 5.6 kΩ, R_{B_2} = 2.2 kΩ, R_E = 270 Ω. Assume both coils have zero resistance. What is the collector current I_C?

6-5 Simulation and Models

Models have been used for thousands of years. Designers and artisans learned long ago that it is often easier to work with a model than with the real thing. If a model is a good one, then what is learned from it transfers to reality. Models can be used to study things such as bridges, buildings, molecules, weather, drug interactions, vehicles, and circuits. In engineering, models are usually mathematical. The following is a prime example:

$$I_D = I_S[e^{V_D/V_T} - 1]$$

where I_D = current flow in a PN junction diode
I_S = reverse bias saturation current (perhaps 1×10^{-14} A)
$e \cong 2.718$
V_D = voltage across the PN junction
V_T = thermal voltage (0.026 V at 27°C)

Technicians seldom use models such as this. Here is why: Look at the circuit in Fig. 6-25. It can be solved easily with a simplified model that fixes the drop across a forward-biased diode at 0.7 V. The sophisticated model is not easy to use because we don't know V_D until the circuit is solved. We could attempt to use circuit laws as a strategy. We know by Kirchhoff's voltage law that the two voltage drops in Fig. 6-25 must add up to equal V_S. We also know that the drop across the resistor can be expressed as I_DR (Ohm's law). Thus:

$$V_S = I_DR + V_D$$

Now, combining this with the sophisticated diode model gives:

$$3\text{ V} = [1 \times 10^{-14}][e^{\frac{V_D}{0.026}} - 1][1 \times 10^3] + V_D$$

Our strategy produced an equation with only one unknown (V_D). But, it happens to be a *transcendental equation* and cannot be directly solved. The equation is transcendental because of the exponent (V_D/0.026). Computers (and people) can use a process of repeated guessing called *iteration* to solve such equations. One keeps plugging values for V_D into the right-hand side of the equation and solving until the result is close to 3 V. Iteration doesn't always work. If

Transcendental equation

Iteration

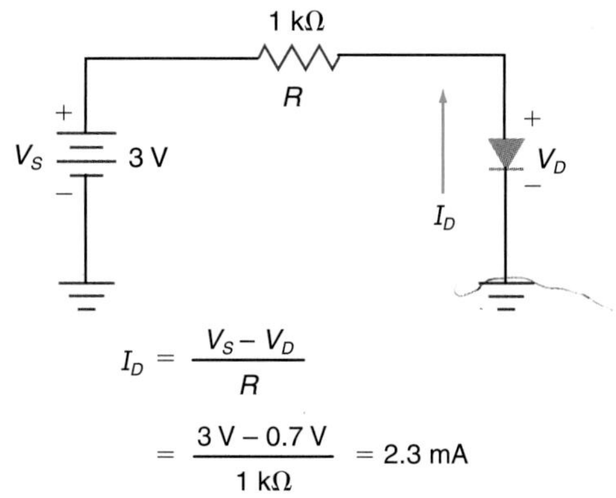

Fig. 6-25 A diode circuit solved using a simple model.

you have ever used software that reported "failed to converge," the equation did not balance after a number of attempts. Yes, computers can be programmed to "know" when to give up. When software iteration fails, it may be possible to change the simulation tolerance or the iteration limits to obtain a solution.

Parameter sweep

By the way, if you want to try iteration for the diode problem, use a technical calculator with an e^x key. Solve the fractional exponent first by using a guess value of 0.7 V for V_D, divide by 0.026, then press the e^x key, subtract 1, multiply by the saturation current, multiply by the resistor value, and then add 0.7 V. *Hint:* 0.68 V is a better guess value for V_D in this case.

Computers are tireless, and people are not. Computers can do millions of calculations in a second. Computers rarely make mistakes. Computers are very good at simulation using complex models. IBM developed one of the first circuit simulators about 1960 and called it *ECAP* (Electronic Circuit Analysis Program). After ECAP came *SPICE* (Simulation Program with Integrated Circuit Emphasis). SPICE was developed at the University of California at Berkeley in the early 1970s. This program is in the public domain and has become a standard for electronic simulation software. Like ECAP, SPICE originally ran on mainframe computers that were available only in governmental agencies, large companies, and some universities. Today, relatively inexpensive personal computers have more computing power than the old mainframes did.

SPICE development continues to evolve, with more models, more features, and more versions adapted to personal computers. It's very likely that most electronic workers will use some type of circuit simulator. Even some electronic hobbyists use simulators. It is worth knowing whether a given simulator is based on SPICE. If it is, chances are that it will be compatible with other simulators and with device models that have been developed for SPICE. Device models are important. As an example, the reverse-bias saturation current for a real diode varies according to its physical size and how it is doped. The range of practical values for I_S are from 10^{-15} to 10^{-13} A. I_S doubles for approximately every 5°C rise in temperature (this is one way that circuit simulators deal with temperature change). Thus, accurate SPICE models can predict how various real parts perform in circuits and what effects temperature will have.

Simulation software allows things to be investigated that would take too long or cost too much by using other means. For example, a *parameter sweep* can allow a circuit to be solved for a range of values. This is how the graph of Fig. 6-15(*b*) was prepared. Transistor current gain β was swept over a range of 100 to 200. Temperature sweeps are another example. It's easy to solve a diode circuit for 10 different temperatures. Imagine doing this using a real circuit. Circuit simulation offers the following advantages:

- Savings in time and money
- A dynamic learning environment
- An easy way to try ideas (called "what if?")
- A safe way of working with high-energy circuits
- Allowing investigations that would not be practical using other methods
- Permitting an independent check on circuit designs (detect and correct mistakes)
- Improving designs (reliability, cost, etc.)

Modeling is not perfect. If a model is not appropriate, it does not represent reality. Also, simulators must use guessing and iteration to solve certain kinds of problems. Circuit simulators have some limitations, which may include:

- Inaccuracy in high-frequency simulator behavior
- Failure to check for failure modes (voltage breakdown, temperature breakdown, etc.)
- Inability to simulate some circuits (a solution is not reached)
- Inability to simulate some circuits accurately (solution is not realistic)
- Inability to provide a particular type of analysis
- Unavailability of needed device models

Let's examine some things that occur at high frequencies. Figure 6-26 shows that, at high frequencies, basic components are more complicated. Why does this happen? There are so-called *stray* inductances and *stray* capacitances in all real devices. If a device has wire leads, they produce stray inductance. If a resistor has a film deposited on an insulating body, there is

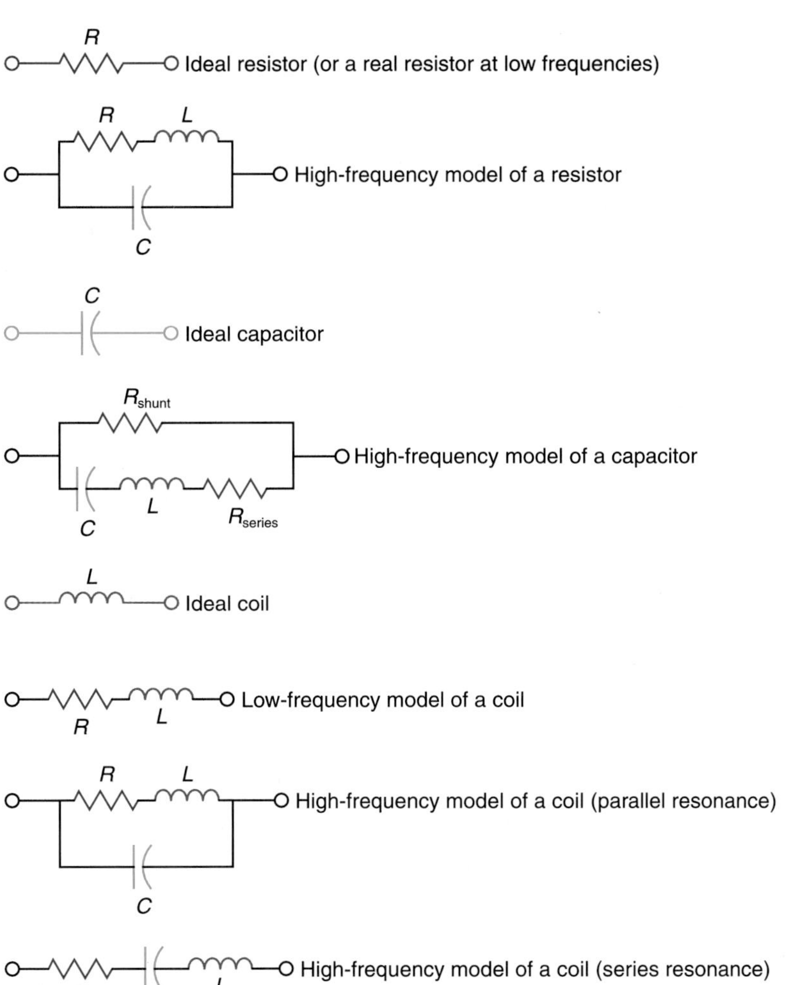

Fig. 6-26 Device models.

a stray capacitance between the film and the body. These stray effects are very small. They are in the nanohenry and picofarad ranges. But at very high frequencies, these stray effects should not be ignored.

What about capacitors? They often have wire leads. The leads have a small inductance, but it becomes significant at high frequencies. High-frequency designs often use leadless parts (like chip capacitors and resistors) to reduce stray inductance.

Consider an inductor or coil. It has turns of wire that are close to each other. There is a small capacitance from turn to turn. The overall effect is called the *distributed capacitance* of the coil. Distributed capacitance is small and can be ignored at low frequencies. At high frequencies, it becomes significant and can make a coil act like a resonant circuit. The wire used to wind a coil has both a low-frequency resistance and a different high-frequency resistance caused by *skin effect.* You may recall that skin effect means that most of the high-frequency current flow is confined to the skin of a conductor.

Distributed capacitance

Skin effect

People who work with radio frequencies (RF) often use specialized RF circuit simulators that are designed to be more realistic at high frequencies. RF design software may also use different models for devices such as transistors.

TEST

Answer the following with a short phrase or a word.

49. List some things that can be simulated with computers.
50. Give an example of a mathematical model.
51. What does the acronym SPICE represent?
52. Give an example of a device model.
53. What is another name for repeated guessing?
54. What does failure to converge mean?
55. What is a parameter sweep good for?
56. Why are ideal models OK at low frequencies?

Chapter 6 Review

Summary

1. Amplifier gain is determined by dividing the output by the input.
2. Gain is specified as a voltage ratio, a current ratio, a power ratio, or as the logarithm of a ratio.
3. When each part of a system is specified in decibel gain or loss, the overall performance can be obtained by simply adding all gains and subtracting all losses.
4. When each part of a system is specified in ratios, the overall performance is obtained by multiplying all gains and dividing by all losses.
5. The decibel is based on power gain or loss. It can be adapted to voltage gain or loss by assuming the input and output resistances to be equal. When they are not equal, the dB voltage gain does not equal the dB power gain.
6. In a common-emitter amplifier, the emitter of the transistor is common to both the input signal and the output signal.
7. The collector load resistor in a common-emitter amplifier allows the output voltage swing to be developed.
8. The base bias resistor limits base current to the desired steady or static level.
9. The input signal causes the base current, the collector current, and the output voltage to change.
10. The common-emitter amplifier produces a 180° phase inversion.
11. One way to show amplifier limits is to draw a load line. Linear amplifiers are operated in the center of the load line.
12. A saturated transistor can be compared with a closed switch. The voltage drop across it will be zero (or very low).
13. A cutoff transistor can be compared with an open switch. The voltage drop across it will be equal to the supply voltage.
14. A transistor set up for linear operation should be between saturation and cutoff. The voltage drop across it should be about half the supply voltage.
15. For a transistor amplifier to be practical, it cannot be too sensitive to β.
16. A practical and stable amplifier circuit uses a voltage divider to set the base voltage and a resistor in the emitter lead.
17. The voltage gain of a common-emitter amplifier is set by the load resistance and the emitter resistance.
18. The common-emitter amplifier is the most popular of the three possible configurations.
19. The best transfer of signal power into an amplifier occurs when the source impedance matches the amplifier input impedance.
20. The common-collector, or emitter-follower, amplifier has a very high input impedance and a low output impedance.
21. The common-collector amplifier has a voltage gain of less than 1.
22. Because of its high input impedance, the common-collector amplifier makes a good isolation amplifier.
23. The common-base amplifier has a very low input impedance.
24. The common-base amplifier is used mainly as an RF amplifier.
25. The common-emitter amplifier is the only configuration that gives both voltage and current gain. It has the best power gain.
26. Any of the three amplifier configurations can use either NPN or PNP transistors. The major difference is in polarity.
27. When the collector circuit of an amplifier is powered, the supply point is called V_{CC}.
28. When the emitter circuit of an amplifier is powered, the supply point is called V_{EE}.
29. A lot can be learned by working with realistic models of real things like circuits.
30. In engineering and technology, mathematical models are popular.
31. Although some equations cannot be directly solved, many of them yield to iteration, which is a process of repeated attempts using guess values.
32. Computers are excellent tools for working with mathematical models and iterative solutions.
33. SPICE, developed by the University of

California at Berkeley, is the most popular electronic simulation software.

34. SPICE models are available for many electronic devices.

35. Although circuit simulators are very useful and powerful, they have limitations.

Chapter Review Questions

Supply the missing word in each statement.

6-1. A_V is the symbol for ________.

6-2. A_P is the symbol for ________.

6-3. Common logarithms are powers of ________.

6-4. If the signal out is less than the signal in, the dB gain will be a ________ number.

6-5. The sensitivity of human hearing to loudness is not linear but ________.

6-6. In a common-emitter amplifier, the signal is fed to the base circuit of the transistor, and the output is taken from the ________.

6-7. Refer to Fig. 6-4. The component that prevents the signal source from bypassing the base-emitter direct current flow is ________.

6-8. Refer to Fig. 6-4. The component that allows the amplifier to develop an output voltage signal is ________.

6-9. Refer to Fig. 6-4. If R_B opens (infinite resistance), then the transistor will operate in ________.

6-10. Refer to Fig. 6-6. As an input signal drives the base in a positive direction, the collector will change in a ________ direction.

6-11. Refer to Fig. 6-6. As an input signal drives the base in a positive direction, the collector current should ________.

6-12. Clipping can be avoided by controlling the input signal and by operating the amplifier at the ________ of the load line.

6-13. Refer to Fig. 6-10. The base current is zero. The amplifier will be in ________.

6-14. Refer to Fig. 6-10. The base current is 300 μA. The amplifier will be in ________.

6-15. A technician is troubleshooting an amplifier and measures V_{CE} to be near 0 V. Voltage V_{CC} is normal. The transistor is operating in ________.

6-16. Refer to Fig. 6-13. This amplifier circuit is not practical because it is too sensitive to temperature and to ________.

6-17. A signal source has an impedance of 50 Ω. For best power transfer, an amplifier designed for this signal source should have an input impedance of ________.

6-18. Refer to Fig. 6-19. As the base is driven in a positive direction, the emitter will go in a ________ direction.

6-19. Refer to Fig. 6-19. This configuration is noted for a high input impedance and a ________ output impedance.

6-20. An amplifier is needed with a low input impedance for a radio-frequency application. The best choice is probably the common ________ configuration.

6-21. The only amplifier that produces a 180° phase inversion is the ________ configuration.

6-22. An amplifier is needed with a moderate input impedance and the best possible power gain. The best choice is probably the common ________ configuration.

6-23. An amplifier is needed to isolate a signal source from any loading effects. The best choice is probably the common ________ configuration.

6-24. Refer to Fig. 6-24. If this circuit would be designed for a PNP transistor, then V_{EE} would have to be ________ with respect to ground.

Chapter Review Problems

6-1. The signal fed into an amplifier is 100 mV, and the output signal is 8.5 V. What is A_V?
6-2. What is the dB voltage gain in problem 6-1?
6-3. If $R_{in} = R_{out}$ in problem 6-1, what is the dB power gain?
6-4. An amplifier with a power gain of 6 dB develops an output signal of 20 W. What is the signal input power?
6-5. A 1000-W transmitter is fed into a coaxial cable with a 2-dB loss. How much power reaches the antenna?
6-6. A two-stage amplifier has a gain of 40 in the first stage and a gain of 18 in the second stage. What is the overall ratio gain?
6-7. A two-stage amplifier has a gain of 18 dB in the first stage and 22 dB in the second stage. What is the overall dB gain?
6-8. An oscilloscope has a frequency response that is −3 dB at 50 MHz. A 10-V peak-to-peak, 50-MHz signal is fed into the oscilloscope. What voltage will the oscilloscope display?
6-9. The signal coming from a microwave antenna is rated at −90 dBm. What is the level of this signal in watts?
6-10. Refer to Fig. 6-4. Assume $R_B = 100$ kΩ and $V_{CC} = 10$ V. Do not correct for V_{BE}, and find I_B.
6-11. Refer to Fig. 6-6. Do not correct for V_{BE}. Assume that $\beta = 80$. Determine V_{CE}.
6-12. Refer to Fig. 6-10. Assume the base current is 180 μA. Find I_C.
6-13. Refer to Fig. 6-10. If the base current is 200 μA, what is V_{CE}?
6-14. Refer to Fig. 6-15(*a*). Assume that R_L is 1500 Ω. Calculate V_{CE}.
6-15. Refer to Fig. 6-15(*a*). Assume the transistor to be germanium and calculate I_E.
6-16. Refer to Fig. 6-16. Assume the emitter current to be 5 mA. The voltage gain could be as high as ________.
6-17. Find A_V for the data in problem 6-16 if the emitter bypass capacitor is open (a common defect in electrolytics).
6-18. Refer to Fig. 6-23. The transistor is silicon and $V_{CC} = -20$ V, $R_{B_1} = R_{B_2} = 10$ kΩ, $R_L = 1$ kΩ, and $R_C = 100$ Ω. Find V_B, V_E, I_E, V_{RC}, and V_{CE}.

Critical Thinking Questions

6-1. Is there any advantage to human hearing being logarithmic?
6-2. Suppose an amplifier is defective and no matter what the input signal is, the output is always zero. Can the performance of this amplifier be expressed using decibels?
6-3. You are approached by an inventor who wants you to invest money in a new development called an *energy amplifier.* Why should you be extremely cautious?
6-4. We know that amplifiers can make sounds louder. Can they also improve the quality of sound?
6-5. A transistor has an operating point at the center of the load line. Assuming no clipping, will this transistor run at a different temperature when it is amplifying signals as compared to when it is not fed any input signal?
6-6. In some cases, gain is needed but a phase inversion is not acceptable. Can the common-emitter configuration be used in these cases?

Answers to Tests

1. T
2. T
3. F
4. T
5. F
6. F
7. T
8. T
9. F
10. 2800
11. 64 dB
12. 130 dB
13. 50 W
14. 79.2 W
15. 6.31 kW
16. 10×10^{-3} W
17. F
18. F
19. T
20. T
21. T
22. F
23. 160 μA
24. 8 mA

25. 8 V
26. 4 V
27. $V_{CE} = 4$ V
28. $I_C = 8$ mA
29. $I_B = 240\ \mu$A
    $I_C = 12$ mA
    $V_{R_L} = 12$ V
    $V_{CE} = 0$ V
    saturation
30. $V_{CE(cutoff)} = 10$ V
    $I_{sat} = 10$ mA
31. 0.8 V
32. 4.4 V
33. 5.6 V
34. 0.923 V
35. 2.23 mA
36. 6.02 V
37. 5.75 V
38. very close to it
39. 24
40. 120 to 241
41. T
42. T
43. F
44. T
45. F
46. a. 0.625 mW
    b. 0.833 mW
    c. 0.625 mW
47. 13.6 mA
48. 9.94 mA
49. weather, circuits, physical systems, etc.
50. Ohm's law, the diode equation, etc.
51. simulation program with integrated circuit emphasis
52. data for an actual device, including its temperature performance, its saturation current, etc.
53. iteration
54. iteration failed to produce an answer close enough to a required value
55. investigating the effect of a device parameter or temperature
56. the stray inductance and capacitance are too small to be significant

More About Small-Signal Amplifiers

Chapter Objectives

This chapter will help you to:

1. *Identify* the standard methods of signal coupling and list their characteristics.
2. *Calculate* the input impedance of common-emitter amplifiers.
3. *Find* voltage gain in cascade amplifiers.
4. *Draw* a signal load line for a common-emitter amplifier.
5. *Solve* FET amplifier circuits.
6. *Identify* negative feedback and list its effects.
7. *Determine* the frequency response of a common-emitter amplifier.

A single stage of amplification is often not enough. This chapter covers multistage amplifiers and the methods used to transfer signals from one stage to the next. It also covers field-effect transistor amplifiers. The FET has certain advantages that make it attractive for some amplifier applications. This chapter also discusses negative feedback and frequency response.

7-1 Amplifier Coupling

Coupling refers to the method used to transfer the signal from one stage to the next. There are three basic types of amplifier coupling (capacitive, direct, and transformer).

Capacitive coupling

Capacitive coupling is useful when the signals are *alternating current.* Coupling capacitors are selected to have a low reactance at the lowest signal frequency. This gives good performance over the frequency range of the amplifier. Any dc component will be blocked by a coupling capacitor.

Figure 7-1 shows why it is important to block the dc component in a multistage amplifier. Transistor Q_1 is the first gain stage. Its static collector voltage is 7 V. This is measured from ground to the collector terminal. Transistor Q_2 in Fig. 7-1 has a static base potential of 3 V. This is measured from ground to the base terminal. Because the grounds are common, it is easy to calculate the voltage across the coupling capacitor from the collector of transistor Q_1 to the base of Q_2:

$$V = 7\text{ V} - 3\text{ V} = 4\text{ V}$$

There is 4 V dc across the capacitor.

What would happen in Fig. 7-1 if the coupling capacitor shorted? The collector of Q_1 and the base of Q_2 would show the *same* voltage with respect to ground. This would greatly change the operating point of Q_2. The base voltage of Q_2 would be higher than 3 V. The increase in base voltage would drive Q_2 into saturation. It would no longer be capable of linear operation.

Polarity

Coupling capacitors used in transistor circuits are often of the electrolytic type. This is especially true in low-frequency amplifiers. High values of capacitance are needed to pass the signals with little loss. *Polarity* is an important factor when working with electrolytic capacitors. Again, refer to Fig. 7-1. The collector of Q_1 is 4 V more positive than the base of Q_2. The capacitor *must* be installed with the polarity shown.

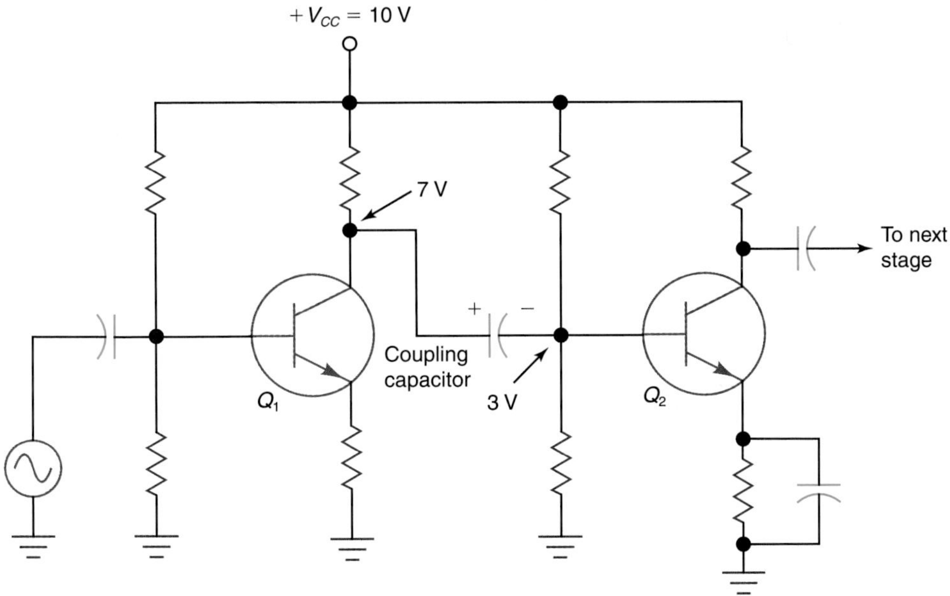

Fig. 7-1 A capacitively coupled amplifier.

Capacitive coupling is widely applied in electronic amplifiers that process ac signals. Some applications, however, require operation down to dc (0 Hz). Electronic instruments, such as oscilloscopes and meters, often have to respond to direct current. The amplifiers in these instruments *cannot* use capacitive coupling.

Direct coupling does work at 0 Hz (direct current). A direct-coupled amplifier uses wire or some other dc path between stages. Figure 7-2 shows a direct-coupled amplifier. Notice that the emitter of Q_1 is directly connected to the base of Q_2. An amplifier of this type will have to be designed so that the static terminal voltages are compatible with each other. In Fig. 7-2, the emitter voltage of Q_1 will be the same as the base voltage of Q_2.

Temperature sensitivity can be a problem in direct-coupled amplifiers. As temperature goes up, β and leakage current increase. This tends to shift the static operating point of an amplifier. When this happens in an early stage of a dc amplifier, all of the following stages will *amplify* the temperature drift. In Fig. 7-2, assume the temperature has gone up. This will make Q_1 conduct more current. More current will flow through Q_1's emitter resistor, increasing its voltage drop. The base of Q_2 now sees more voltage, so it is turned on harder. If a third and fourth stage follow, even a slight shift in the operating point of Q_1 may cause the fourth stage to be driven out of the linear range of operation.

Direct coupling a few stages is not difficult. It may be the least expensive way to get the gain needed. Direct coupling may be used in sections of an audio amplifier where the lowest frequency is around 20 Hz. Direct coupling provides good low-frequency response, and it is used in audio work when it is less expensive than another coupling method.

Direct coupling

A *Darlington circuit* is another example of direct coupling. You should recall that Darlington pairs may be packaged as a single device or connected with two devices, as shown

Darlington circuit

Temperature sensitivity

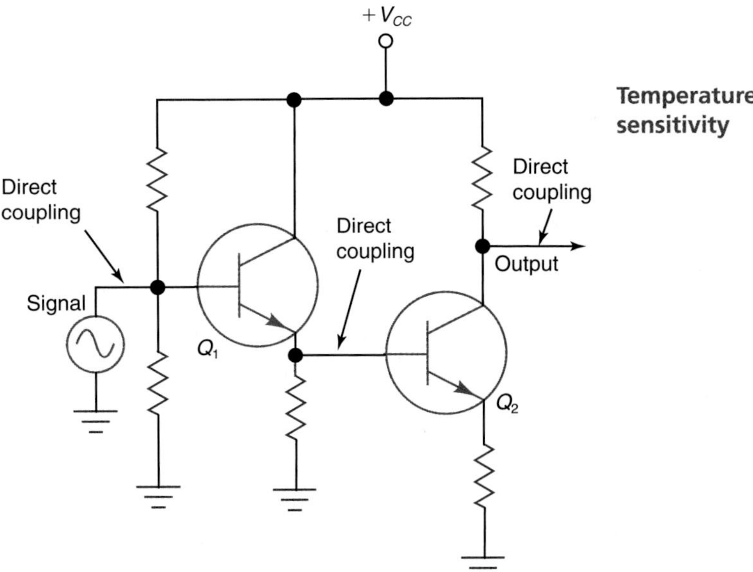

Fig. 7-2 A direct-coupled amplifier.

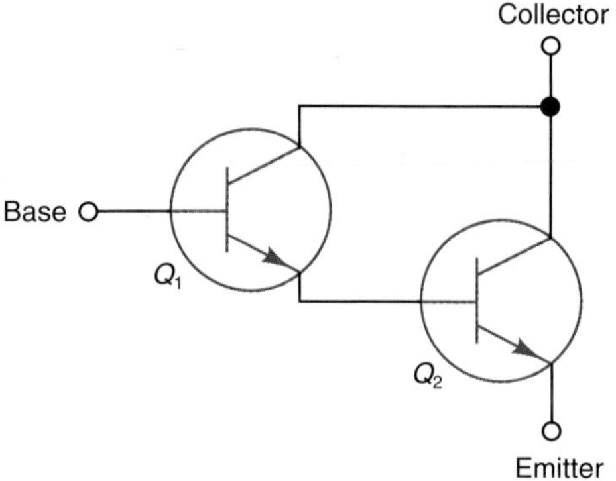

Fig. 7-3 A Darlington amplifier.

in Fig. 7-3. The current gain for Fig. 7-3 is approximately equal to the product of the individual betas:

$$\text{Current gain} = A_I \approx \beta_1 \times \beta_2$$

If each transistor has a β of 100, then

$$A_I \approx 100 \times 100 = 10{,}000$$

The Darlington circuit is a good choice when a high current gain or a high input impedance is required. Since the circuit has such high current gain, it requires little signal current. Low signal current means that the source is lightly loaded by the Darlington amplifier. This is especially true in a Darlington emitter follower such as the one shown in Fig. 7-4. Let's find the static conditions for this circuit. Q_1's base voltage is set by the divider:

$$V_{B1} = \frac{220\ \text{k}\Omega}{220\ \text{k}\Omega + 470\ \text{k}\Omega} \times 12\ \text{V} = 3.83\ \text{V}$$

Transformer coupling

About Electronics

Troubleshooting and the Phase of a Signal When a dual-trace oscilloscope is used to inspect the phase of a signal at various points in a circuit, triggering on the input signal works well.

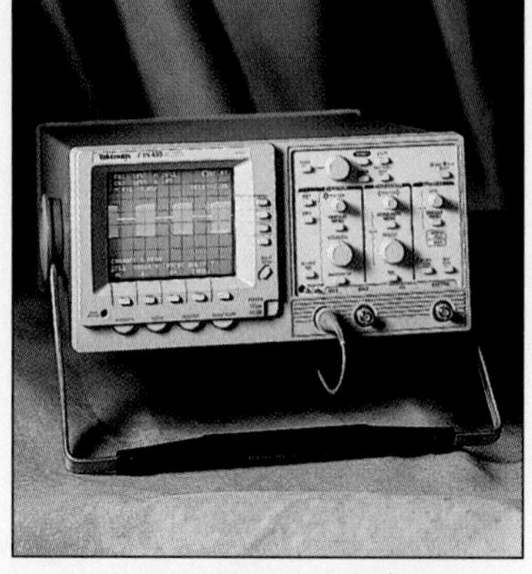

The emitter resistor of Q_2 will see this voltage *minus two base-emitter drops:*

$$V_{E(Q_2)} = 3.83\ \text{V} - 0.7\ \text{V} - 0.7\ \text{V} = 2.43\ \text{V}$$

Ohm's law will give Q_2's emitter current:

$$I_{E(Q_2)} = \frac{2.43\ \text{V}}{1\ \text{k}\Omega} = 2.43\ \text{mA}$$

And Kirchhoff's voltage law will give the drop across Q_2:

$$V_{CE(Q_2)} = 12\ \text{V} - 2.43\ \text{V} = 9.57\ \text{V}$$

Example 7-1

Modify the circuit shown in Fig. 7-4 so that $V_{CE(Q_2)}$ is half the supply voltage. This means that the emitter of Q_2 should be at 6 V. We can change the voltage divider to produce a voltage that is two base-emitter drops above 6 V:

$$V_{\text{divider}} = 6\ \text{V} + 0.7\ \text{V} + 0.7\ \text{V} = 7.4\ \text{V}$$

A new value for the 470-kΩ resistor of the divider can be found by solving the following equation for R:

$$7.4\ \text{V} = \frac{220\ \text{k}\Omega}{220\ \text{k}\Omega + R} \times 12\ \text{V}$$

Multiply both sides by the denominator:

$$7.4\ \text{V}(220\ \text{k}\Omega + R) = 220\ \text{k}\Omega \times 12\ \text{V}$$

Divide both sides by 12 V:

$$0.617(220\ \text{k}\Omega + R) = 220\ \text{k}\Omega$$

Multiply:

$$136\ \text{k}\Omega + 0.617R = 220\ \text{k}\Omega$$

Subtract 136 kΩ from both sides:

$$0.617R = 84\ \text{k}\Omega$$

Divide both sides by 0.617:

$$R = 136\ \text{k}\Omega$$

Replace the 470-kΩ resistor in Fig. 7-4 with a 130-kΩ resistor, which is the closest standard value.

Figure 7-5 shows a *transformer-coupled* amplifier. The transformer serves as the collector load for the transistor and as the coupling device to the amplifier load. The advantage of

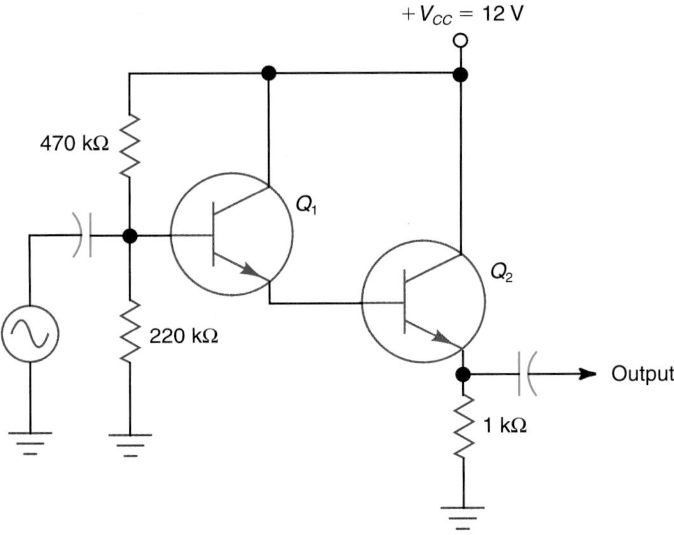

Fig. 7-4 A Darlington emitter follower.

transformer coupling is easy to understand if we examine its *impedance-matching* property. The *turns ratio* of a transformer is given by:

$$\text{Turns ratio} = \frac{N_P}{N_S}$$

where N_P = number of primary turns and
N_S = number of secondary turns

The *impedance ratio* is given by:

$$\text{Impedance ratio} = (\text{turns ratio})^2$$

If the transformer in Fig. 7-5 has 100 primary turns and 10 secondary turns, its turns ratio is

$$\text{Turns ratio} = \frac{100}{10} = 10$$

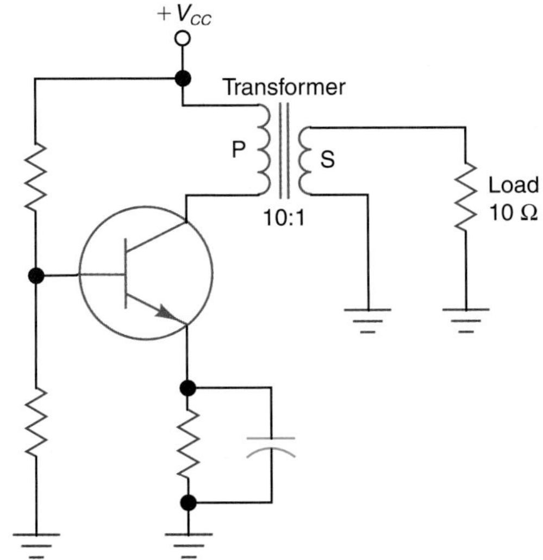

Fig. 7-5 A transformer-coupled amplifier.

and its impedance ratio is

Impedance matching

$$\text{Impedance ratio} = 10^2 = 100$$

This means that the load seen by the collector of the transistor will be 100 times the impedance of the actual load. If the load is 10 Ω, the collector will see 100 × 10 Ω = 1 kΩ.

The output impedance of common-emitter amplifiers is much higher than 10 Ω. When the amplifier must deliver signal energy to such a low impedance, a matching transformer will greatly improve the power transfer to the load. The collector load of 1000 Ω in Fig. 7-5 is high enough to provide good voltage gain. If we assume that the emitter current is 5 mA, we have enough information to calculate the gain. First, we must estimate the ac resistance of the emitter as discussed in the preceding chapter:

$$r_E = \frac{25\text{ mV}}{5\text{ mA}} = 5\ \Omega$$

The emitter resistor is bypassed, so the voltage gain will be given by

$$A_V = \frac{R_L}{r_E}$$

There is no load resistor in Fig. 7-5, but there is a *transformer-coupled 10-Ω load.* This load is *transformed* to 1 kΩ by the transformer and it will set the voltage gain along with r_E:

$$A_V = \frac{1000\ \Omega}{5\ \Omega} = 200$$

Does the external 10-Ω load see 200 times the signal voltage sent to the base of the transistor? *No,* because the transformer gives a 10:1 voltage *step-down.* Therefore, the voltage gain from the base circuit to the 10-Ω load is

Tuned transformer

$$A_V = \frac{200}{10} = 20$$

This is still better than we can do without the transformer. If the 10-Ω load were connected directly in the collector circuit as a load resistor, the gain would be

$$A_V = \frac{10\ \Omega}{5\ \Omega} = 2$$

Obviously, the transformer does quite a bit to improve the voltage gain of the circuit.

Selectivity

Example 7-2

Calculate the voltage gain for Fig. 7-5 using the conservative estimate for the ac emitter resistance. Assume that the dc emitter current is 5 mA. The conservative estimate for ac emitter resistance uses 50 mV, so the voltage gain will be 10, which is half that calculated when using 25 mV. The emitter resistance using 50 mV:

$$r_E = \frac{50\ \text{mV}}{5\ \text{mA}} = 10\ \Omega$$

The transformed collector load is 1000 Ω, and the gain to the collector circuit is:

$$A_V = \frac{1000\ \Omega}{10\ \Omega} = 100$$

The signal voltage across the 10-Ω load is stepped down and so is the gain:

$$A_V = \frac{100}{10} = 10$$

Audio-coupling transformers present problems. If high quality is needed, the transformers will be expensive. They will need good cores, and this means that the size and weight of the transformers will be another problem. Transformer coupling is not often used in high-power audio work. It is more popular in low- and medium-quality audio devices in which some shortcuts are acceptable.

At radio frequencies the transformers become less expensive. The cores are much smaller and lighter. Transformer coupling is more popular in RF amplifiers.

Tuned transformers are used often in RF amplifiers to provide *selective gain.* Selective gain means that some frequencies are amplified more than others. Figure 7-6 shows a tuned, transformer-coupled, RF amplifier. When T_1 is at or near resonance, it will present a high impedance load to the collector circuit of Q_1. A high collector load impedance gives high voltage gain. Thus, frequencies at or near the resonant point of the tuned circuit get the most gain. Transistor Q_2 is tuned by T_2 in Fig. 7-6. Using additional tuned stages improves the *selectivity* of amplifiers of this type. Selectivity is the ability to reject unwanted frequencies.

The transformers in Fig. 7-6 also provide an impedance match. Transformer T_1 normally will have more primary turns than secondary turns. This matches the high collector impedance of Q_1 to the lower input impedance of Q_2.

The secondary of T_1 delivers the ac signal to the base of Q_2. It also provides the dc base voltage for Q_2. Figure 7-6 shows that the base divider network for Q_2 is connected to the bottom of T_1's secondary. The secondary winding has a low dc resistance, so the voltage at the junction of the divider resistors will also appear at the base of Q_2. Note that a *bypass* capacitor provides a *signal ground* for the bottom of the secondary winding of T_1. Without this capacitor, signal current would flow in the divider network and much of the signal energy would be dissipated (lost).

Knowing the function of circuit components is important for component-level troubleshooting. For example, in Fig. 7-6 if the bypass capacitor just discussed *opens,* the symptom is loss of gain because much of the signal will be dissipated in the divider network. On the other hand, if the capacitor *shorts,* the amplifier may pass no signal because there won't be any base bias for Q_2 and it will be in *cutoff.* Also, if the capacitor opens, the fault *cannot* be found by checking dc voltages but if the capacitor shorts the fault *can* be found by checking dc voltages.

Three methods of coupling have been presented. Table 7-1 summarizes some of the important points for each method discussed.

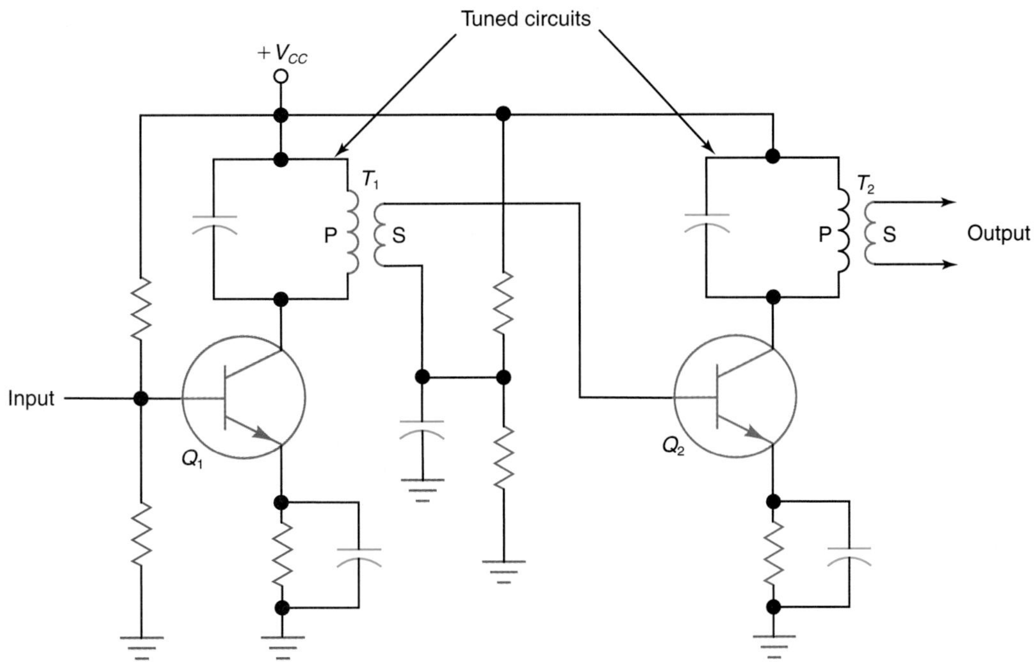

Fig. 7-6 A tuned RF amplifier.

TEST

Determine whether each statement is true or false.

1. Capacitive coupling cannot be used in dc amplifiers.
2. Transformer coupling cannot be used in dc amplifiers.
3. A shorted coupling capacitor cannot be found by making dc voltage checks.
4. An open coupling capacitor can be found by making dc voltage checks.
5. A shorted bypass capacitor can be found by making dc voltage checks.
6. If a signal source and a load have two different impedances, transformer coupling can be used to achieve an impedance match.

Solve the following problems.

7. Refer to Fig. 7-1. A coupling capacitor should present no more than one-tenth the impedance of the load it is working into. If the second stage has an input impedance of 2 kΩ and the circuit must amplify frequencies as low as 20 Hz, what is the minimum value for the coupling capacitor?
8. Refer to Fig. 7-1. Assume the coupling capacitor shorts and the base voltage of Q_2 increases to 6 V. Also assume that Q_2 has a load resistor of 1200 Ω and an emit-

Table 7-1 Summary of Coupling Methods

	Capacitor Coupling	Direct Coupling	Transformer Coupling
Response to direct current	No	Yes	No
Provides impedance match	No	No	Yes
Advantages	Easy to use. Terminals at different dc levels can be coupled.	Simplicity when a few stages are used.	High efficiency. Can be tuned to make a selective amplifier.
Disadvantages	May require high values of capacity for low-frequency work.	Difficult to design for many stages. Temperature sensitivity.	Cost, size, and weight can be a problem.

ter resistor of 1000 Ω. Solve the circuit and prove that Q_2 goes into saturation.

9. Refer to Fig. 7-3. If Q_1 has a β of 50 and Q_2 has a β of 100, what is the current gain from the base terminal to the emitter terminal?
10. Refer to Fig. 7-4. Assume that the 220-kΩ resistor is changed to a 330-kΩ resistor. Find the current flow in the 1-kΩ resistor.
11. Find the voltage drop from collector to emitter for Q_2 for the data given in problem 10.
12. Refer to Fig. 7-5. Assume the turns ratio is 14:1 (primary to secondary). What load does the collector of the transistor see?

7-2 Voltage Gain in Coupled Stages

Figure 7-7(*a*) shows a common-emitter amplifier driven by a 100-mV signal source. This particular signal source has an internal impedance of 10 kΩ. Signal sources with high internal impedances deliver only a fraction of their output capability when connected to amplifiers with moderate input impedances. As Fig. 7-7(*b*) shows, the internal impedance of the source and the input impedance of the amplifier form a *voltage divider.* To find the actual signal voltage delivered to the transistor amplifier, the voltage divider equation is used:

$$V = \frac{6.48\ \text{k}\Omega}{10\ \text{k}\Omega + 6.48\ \text{k}\Omega} \times 100\ \text{mV}$$
$$= 39.3\ \text{mV}$$

This calculation demonstrates why it is sometimes important to know the input impedance of an amplifier.

Example 7-3

How much signal would be delivered to the amplifier of Fig. 7-7 if the signal source had an internal impedance of only 50 Ω?

$$V = \frac{6.48\ \text{k}\Omega}{50\ \Omega + 6.48\ \text{k}\Omega} \times 100\ \text{mV}$$
$$= 99.2\ \text{mV}$$

This demonstrates that amplifier loading has little effect when the signal source impedance is relatively low.

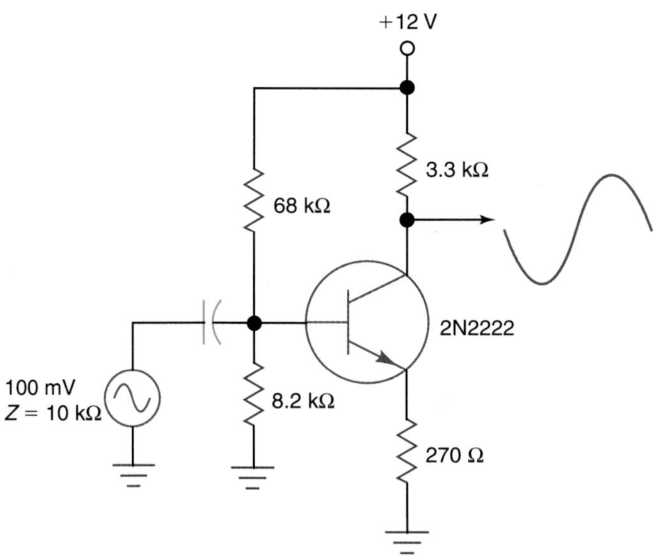

(*a*) A common-emitter amplifier

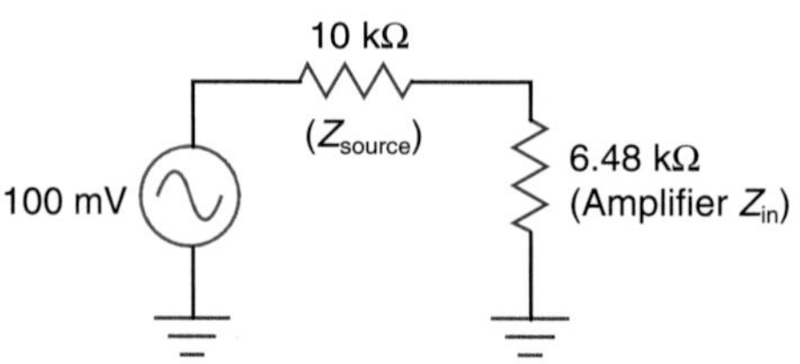

(*b*) The equivalent input circuit

Fig. 7-7 An amplifier input loads a signal source.

Input impedance

Finding the *input impedance* of a common-emitter amplifier is detailed in Fig. 7-8. As Fig. 7-8(*a*) shows, the total ac signal current divides into *three paths*. The power supply point is marked + and is at *ground* potential as far as ac signals are concerned. Power supplies normally have a very low impedance for ac signals. The top of R_{B1} is effectively at signal ground. So both base bias resistors plus the transistor itself load the ac signal source. Figure 7-8(*b*) shows the equivalent circuit. If all three of these loads are known, the input impedance of the amplifier can be found by using the reciprocal equation normally used for parallel resistors.

Figure 7-7 will be used as an example for determining the input impedance of a common-emitter amplifier. The dc conditions are solved first by the approach from the previous chapter:

$$V_B = \frac{8.2\text{ k}\Omega}{68\text{ k}\Omega + 8.2\text{ k}\Omega} \times 12\text{ V} = 1.29\text{ V}$$

$$V_E = 1.29\text{ V} - 0.7\text{ V} = 0.591\text{ V}$$

$$I_E = \frac{0.591\text{ V}}{270\ \Omega} = 2.19\text{ mA}$$

$$r_E = \frac{25}{2.19} = 11.4\ \Omega$$

The input resistance of the transistor itself can now be determined. To avoid confusion, we will use the symbol r_{in} for this resistance and later the symbol Z_{in} to represent the input impedance of the overall amplifier. Input resistance r_{in} is found by multiplying β times the sum of the unbypassed emitter resistances. R_E is *not* bypassed in Fig. 7-7, so it must be used:

$$\begin{aligned} r_{in} &= \beta(R_E + r_E) \\ &= 200(270\ \Omega + 11.4\ \Omega) \\ &= 56.3\text{ k}\Omega \end{aligned}$$

Note: 200 is a good estimate of β for a 2N2222 transistor. Also note that r_E could have been ignored without significantly affecting the result:

$$r_{in} = 200 \times 270\ \Omega = 54\text{ k}\Omega$$

The result of 56.3 kΩ is more accurate. The more accurate approach requires that r_E be

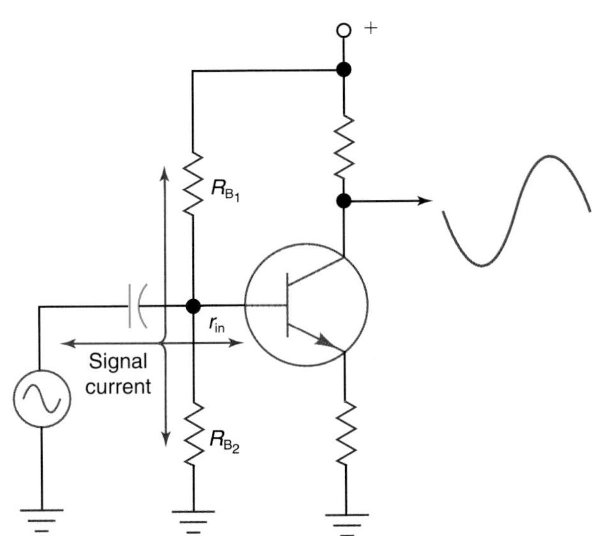

(*a*) The signal input current flows in three paths

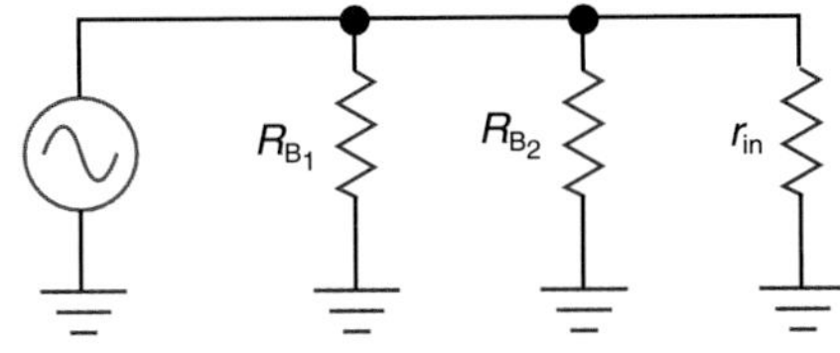

(*b*) The equivalent circuit

Fig. 7-8 Amplifier input impedance.

Parallel equivalent

known. When the emitter resistor is bypassed with a capacitor, r_E *must* be used because the emitter resistor (R_E) is eliminated from the calculation just as it is when solving for voltage gain. If a bypass capacitor were connected across the 270-Ω emitter resistor in Fig. 7-7(*a*), then the result would be

$$r_{in} = 200 \times 11.4\ \Omega = 2.28\ \text{k}\Omega$$

We are now prepared to find the input impedance of the amplifier shown in Fig. 7-7 using the standard reciprocal equation and the more accurate value of r_{in} for the unbypassed condition:

$$Z_{in} = \frac{1}{1/R_{B_1} + 1/R_{B_2} + 1/r_{in}}$$
$$= \frac{1}{1/68\ \text{k}\Omega + 1/8.2\ \text{k}\Omega + 1/56.3\ \text{k}\Omega}$$
$$= 6.48\ \text{k}\Omega$$

If the 270-Ω emitter resistor in Fig. 7-7 *is* bypassed, the input impedance *drops* to 1.74 kΩ. You are encouraged to verify this by combining 2.28 kΩ with the two base resistors. Because it may be desirable to have the input impedance of amplifiers as high as possible, it can be seen that emitter bypassing must be avoided in some applications.

Cascade

Let's apply what we have learned to the cascade amplifier shown in Fig. 7-9. *Cascade* means that the output of one stage is connected to the input of the next. Knowing how to calculate the input impedances of the amplifiers will allow us to find the overall gain of this circuit from the source to the 680-Ω load resistor and the amplitude of the output signal. We begin by solving the *second* stage for its dc conditions:

$$V_B = \frac{3.9\ \text{k}\Omega}{27\ \text{k}\Omega + 3.9\ \text{k}\Omega} \times 12\ \text{V} = 1.51\ \text{V}$$
$$V_E = 1.51\ \text{V} - 0.7\ \text{V} = 0.815\ \text{V}$$
$$I_E = \frac{0.815\ \text{V}}{100\ \Omega} = 8.15\ \text{mA}$$
$$r_E = \frac{25}{8.15} = 3.07\ \Omega$$

Knowing r_E allows us to find the voltage gain for the second stage. The voltage gain in common-emitter amplifiers is found by dividing the collector load by the resistance in the emitter circuit. However, when the output signal is loaded by another resistor such as the 680-Ω resistor in Fig. 7-9, then the total collector load is the *parallel equivalent* of the collector resistor and the other load resistor. Once again, we see that the power supply point is a ground as far as the ac signal is concerned. So the second transistor in Fig. 7-9 supplies signal current to both the 1-kΩ collector resistor and the 680-Ω resistor. Using the product-over-sum technique to find the parallel equivalent of these two resistors gives

$$R_P = \frac{1\ \text{k}\Omega \times 680\ \Omega}{1\ \text{k}\Omega + 680\ \Omega} = 405\ \Omega$$

The voltage gain is found next:

$$A_V = \frac{R_P}{R_E + r_E}$$
$$= \frac{405\ \Omega}{100\ \Omega + 3.07\ \Omega} = 3.93$$

Example 7-4

Calculate the gain for the second stage of Fig. 7-9 using the conservative estimate for ac emitter resistance (use 50 mV). The calculated gain won't change much because the emitter resistor is not bypassed. Find the ac emitter resistance:

$$r_E = \frac{50\ \text{mV}}{8.15\ \text{mA}} = 6.13\ \Omega$$

Determine the voltage gain:

$$A_V = \frac{R_P}{R_E + r_E} = \frac{405\ \Omega}{100\ \Omega + 6.13\ \Omega} = 3.82$$

This is not much different than 3.93.

Once again, we can determine that ignoring r_E will not significantly change the result. The voltage gain calculates to 4.05 when this approach is taken ($^{405}/_{100}$). If the 100-Ω emitter resistor is bypassed, then r_E *must* be used since R_E is eliminated:

$$A_{V(\text{bypassed})} = \frac{405}{3.07} = 132$$

Adding a load to a transistor amplifier changes the way its gain is calculated. Amplifier gain is *always decreased* by loading. The question now is what does the second stage in Fig. 7-9 do to the first stage. The answer is that

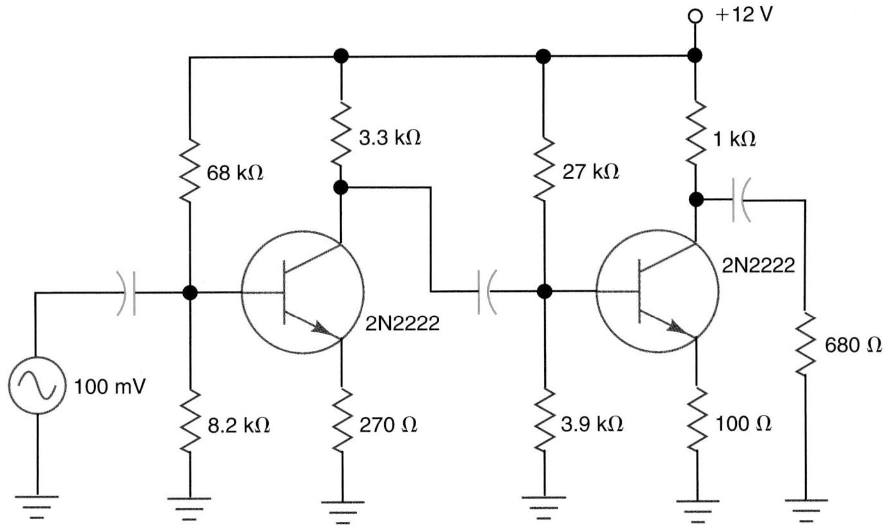

Fig. 7-9 A cascade amplifier.

it loads it. Therefore to find the first-stage voltage gain we must first find the input impedance of the second stage. We begin by finding the input resistance of the second transistor itself:

$$r_{in} = 200\ (100\ \Omega + 3.07\ \Omega) = 20.6\ \text{k}\Omega$$

The input impedance is calculated next:

$$Z_{in} = \frac{1}{1/27\ \text{k}\Omega + 1/3.9\ \text{k}\Omega + 1/20.6\ \text{k}\Omega}$$
$$= 2.92\ \text{k}\Omega$$

This 2.92-kΩ input impedance of the second stage acts in *parallel* with the 3.3-kΩ collector resistor of the first stage in Fig. 7-9:

$$R_P = \frac{3.3\ \text{k}\Omega \times 2.92\ \text{k}\Omega}{3.3\ \text{k}\Omega + 2.92\ \text{k}\Omega} = 1.55\ \text{k}\Omega$$

Therefore, the gain of the first stage is

$$A_V = \frac{1550}{270 + 11.4} = 5.51$$

The *overall gain* of the two-stage amplifier shown in Fig. 7-9 is found by multiplying the individual gains:

$$A_{V(\text{total})} = A_{V_1} \times A_{V_2}$$
$$= 5.51 \times 3.93 = 21.7$$

Since no output impedance is specified for the signal source, we will assume that it is an ideal voltage source (has zero internal impedance). In these cases, all of the signal is delivered to the first stage and the output signal is

$$V_{out} = 100\ \text{mV} \times 21.7 = 2.17\ \text{V}$$

You may have become used to the idea that the quiescent (static) collector-to-emitter voltage should be about half the supply voltage in a linear amplifier. This is *not* the case when *RC*-coupled amplifiers are loaded, as the second stage in Fig. 7-9 is loaded by the 680-Ω resistor.

Some interviewers look for hobby interests in related areas such as computers, radio control, and ham radio.

We have already solved for most of the dc conditions for the second stage of Fig. 7-9. We found that the emitter voltage (V_E) is 0.815 V and that the emitter current is 8.15 mA. Assuming that the collector current is equal to the emitter current. Ohm's law will give us the drop across the collector resistor:

$$V_{RL} = 8.15\ \text{mA} \times 1\ \text{k}\Omega = 8.15\ \text{V}$$

Kirchhoff's voltage law will give us the transistor static drop:

$$V_{CE} = V_{CC} - V_{RL} - V_E$$
$$= 12\ \text{V} - 8.15\ \text{V} - 0.815\ \text{V}$$
$$= 3.04\ \text{V}$$

Half the supply is 6 V. A graphical approach will be used to investigate whether the amplifier is biased properly for linear work.

Signal load line

Figure 7-10 on page 168 shows the development of a *signal load line* for the loaded amplifier. The dc load line is drawn first. It extends from the supply voltage value (12 V) on the horizontal axis to the dc saturation current value (10.9 mA) on the vertical axis. The *quiescent* (Q) point is located on the dc load line

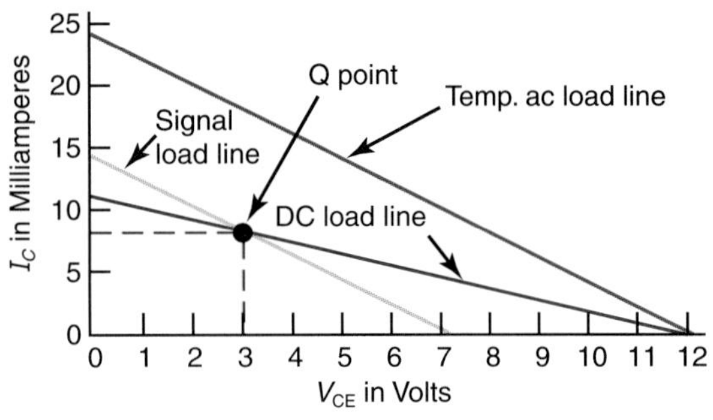

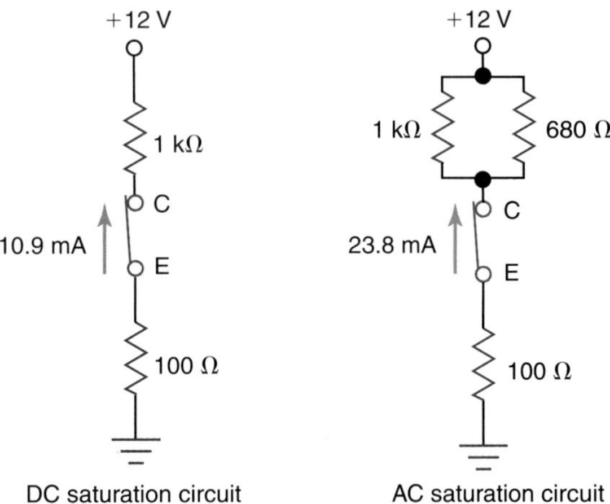

Fig. 7-10 Development of the signal load line.

by projecting the transistor static current or the transistor static voltage drop from the appropriate axis. Quiescent is another word for *static* in electronics. Note that the Q point is *not* in the center of the dc load line.

When the dc load line has been drawn and the Q point has been located on it, it is time to draw a *temporary ac load line.* Figure 7-10 shows that the ac saturation circuit is different from the dc saturation circuit. Once again, we find that the collector resistor and the 680-Ω resistor act in parallel. This makes the ac saturation current higher than the dc saturation current. A temporary ac load line is drawn from the supply voltage value (12 V) to the ac saturation current value (23.8 mA). This temporary line has the correct slope, but it does not pass through the Q point. The last step is to construct a *signal load line* that is parallel (same slope) to the temporary ac load line and that passes through the Q point.

The signal load line shown in Fig. 7-10 determines the clipping points for the amplifier. Since the Q point is near the center of the signal load line, the clipping will be approximately symmetrical. In other words, the amplifier is biased properly for linear operation. When clipping does occur, it will affect the positive and negative peaks of the output signal to about the same extent.

The signal load line shows whether an amplifier is biased properly for linear operation. It is often a requirement to have the Q point in the center of the signal load line. The signal load line shows the maximum peak-to-peak swing of V_{CE}. The output swing across the 680-Ω load will be less for Fig. 7-9 because of signal voltage drop across the 100-Ω emitter resistor. Figure 7-11 shows the performance of the amplifier when its output is driven into clipping. The larger of the two waveforms shows the

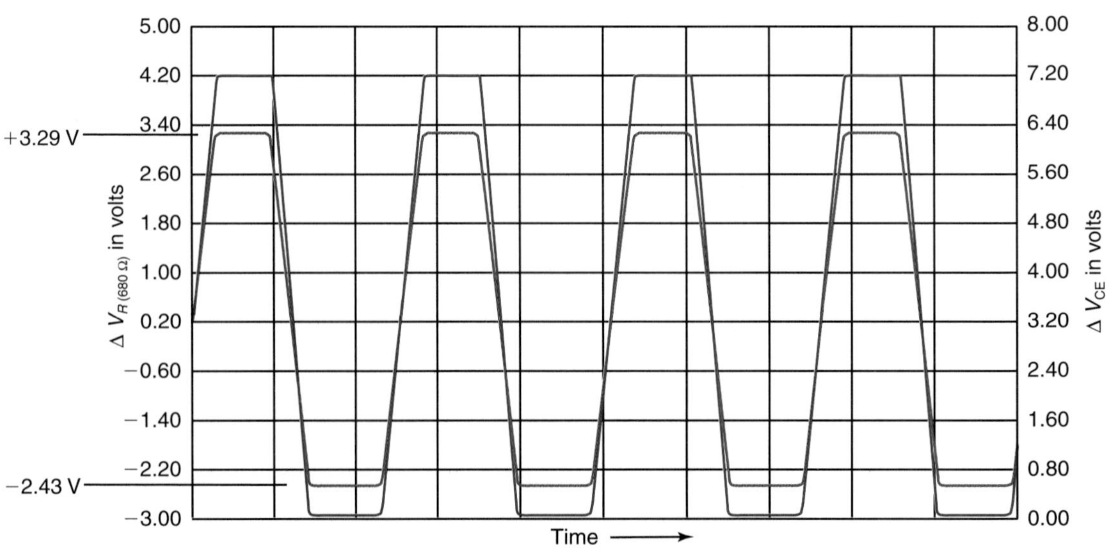

Fig. 7-11 Driving the amplifier output into clipping.

swing of V_{CE} and is a bit more than 7 V_{p-p}. This agrees with the signal load line in Fig. 7-10.

An analysis of the clipping points of the amplifier will provide more insight as to how the circuit operates. There are two clipping points: cutoff and saturation. Figure 7-12(*a*) shows what the output circuit looks like at cutoff. The transistor is off, so all that must be considered is the supply voltage R_L, the 680-Ω load, and the output coupling capacitor, which has a charge of 3.86 V. This charge is equal to the quiescent collector voltage. We have already solved the output circuit for the quiescent values of V_E and V_{CE}. The quiescent collector voltage is found by adding them (Kirchhoff's voltage law):

$$V_C = V_E + V_{CE} = 0.815\text{ V} + 3.04\text{ V} = 3.86\text{ V}$$

If the time constant of the output circuit is relatively long compared to the signal period, the capacitor will maintain a constant voltage. This is why the cutoff circuit shown in Fig. 7-12(*a*) shows a 3.86-V battery. In many cases, a charged capacitor can be viewed as a battery. Note that the battery voltage opposes the + 12-V supply. The voltage drop across the 680-Ω resistor can be found with the voltage divider equation:

$$V = (+12\text{ V} - 3.86\text{ V}) \times \frac{680\ \Omega}{1\text{ k}\Omega + 680\ \Omega}$$
$$= +3.29\text{ V}$$

About Electronics

Television Amplifier Tuning Some of the amplifiers used in television receivers are tuned with piezoelectric devices.

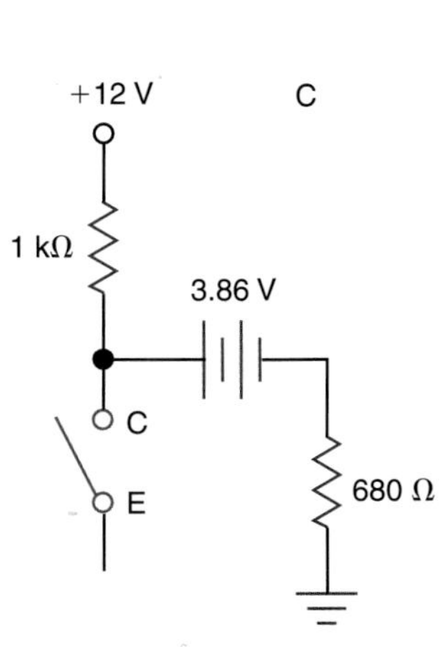

(*a*) Amplifier circuit at cutoff

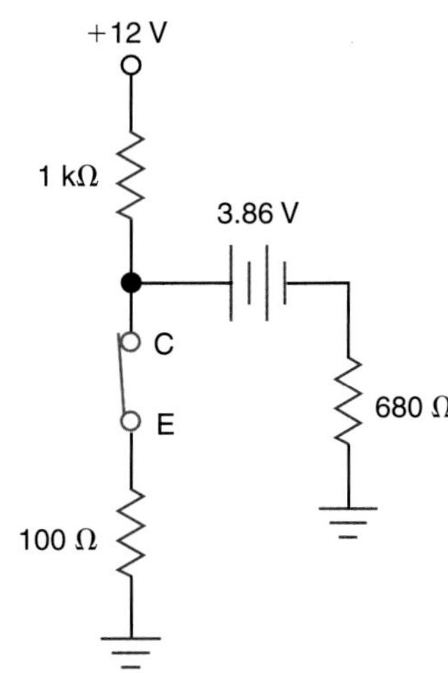

(*b*) Amplifier circuit at saturation

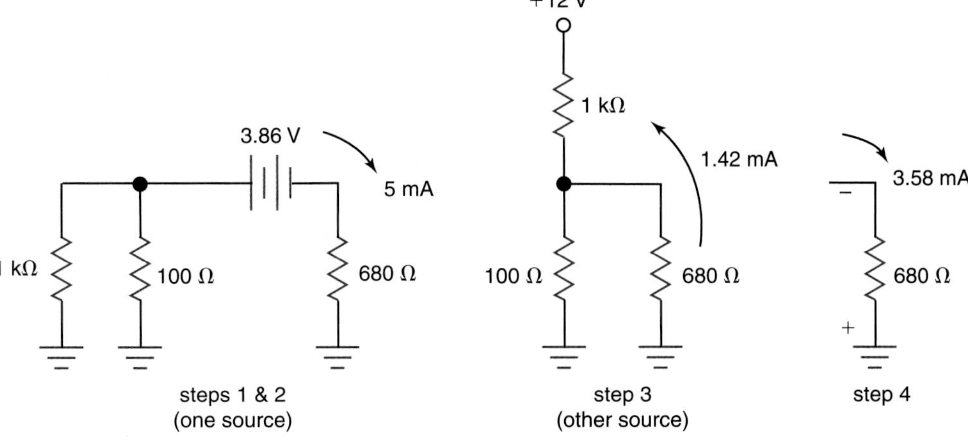

(*c*) Solving the saturation circuit using superposition

Fig. 7-12 Verifying the load waveform of Fig. 7-11.

This verifies the positive clipping point shown in Fig. 7-11.

Figure 7-12(*b*) shows the output circuit at saturation. It is more complicated because it is a multiple-source circuit.

Superposition theorem

. . . that one way to solve multiple-source circuits is to use the *superposition theorem.* The steps are:

1. Replace every voltage source but one with a short circuit.
2. Calculate the magnitude and direction of the current through each resistor in the temporary circuit produced by step 1.
3. Repeat steps 1 and 2 until each voltage source has been used as an active source.
4. Algebraically add all the currents from step 2.

Figure 7-12(*c*) shows the steps. The result is a current *down* through the 680-Ω resistor of 3.58 mA. The top of the resistor is now negative with respect to ground:

$$V = -3.58 \text{ mA} \times 680\ \Omega = -2.43 \text{ V}$$

This verifies the negative limit of the output swing across the 680-Ω load resistor as shown in Fig. 7-11. The maximum signal swing across the 680-Ω resistor is the difference between both limits:

$$V_{\text{out, max}} = 3.29 - (-2.43) = 5.72\ \text{V}_{\text{p-p}}$$

This is quite a bit less than the V_{CE} swing due to the loss across the 100-Ω emitter resistor.

Amplifier clipping can cause problems. As Fig. 7-11 shows, the clipped output starts to look like a square wave. Square waves have high-frequency components called *harmonics.* Harmonics can damage components. For example, it is possible to burn out the high-frequency speakers in some stereo systems by playing them too loudly. This results in clipping and the generation of high-frequency harmonics. Another problem caused by clipping and harmonics is interference. A transmitter that is driven into clipping will generate harmonic frequencies that will interfere with other communications channels.

Harmonics

Harmonics can be investigated using test equipment (such as a spectrum analyzer) or with *Fourier analysis.* Fourier analysis is a mathematical approach used to find the frequency content of a waveform. Figure 7-13 shows the result of a circuit simulator using Fourier analysis for both a sine wave and a square wave. Note that the content of a sine wave is at only one frequency but that the content of a square wave is spread over a range of *odd harmonic frequencies.* In this case, 1 kHz is the fundamental frequency, and there is also content at the third, fifth, seventh, and other *odd harmonic frequencies.*

Fourier analysis

Harmonics are not always a bad thing. They can be intentional and applied for special effects in music, or they can be used to generate higher frequencies in a circuit called a *frequency multiplier.* The important idea is that when the shape of a waveform is changed, so is its frequency content.

TEST

Determine whether each statement is true or false.

13. The open-circuit output voltage of a signal source with a moderate characteristic impedance will not change when connected to an amplifier that has a moderate input impedance.
14. Emitter bypassing in a common-emitter stage increases the amplifier's gain and input impedance.
15. Loading an amplifier always decreases its voltage gain.
16. An amplifier's quiescent current is the same as its static current.
17. An amplifier will provide the most undistorted peak-to-peak output swing when it is biased for the center of the signal load line.
18. Checking to see whether V_{CE} is half of the supply will not confirm that a loaded linear amplifier is properly biased.
19. The maximum output swing from a loaded amplifier is less than the supply voltage.

Solve the following problems.

20. A microphone has a characteristic impedance of 100 kΩ and an open-circuit output of 200 mV. How much signal voltage will this microphone deliver to an amplifier with an input impedance of 2 kΩ?
21. It was determined that the overall voltage gain for the two-stage amplifier shown in Fig. 7-9 is 21.7. Find the maximum input signal for this amplifier that will not cause

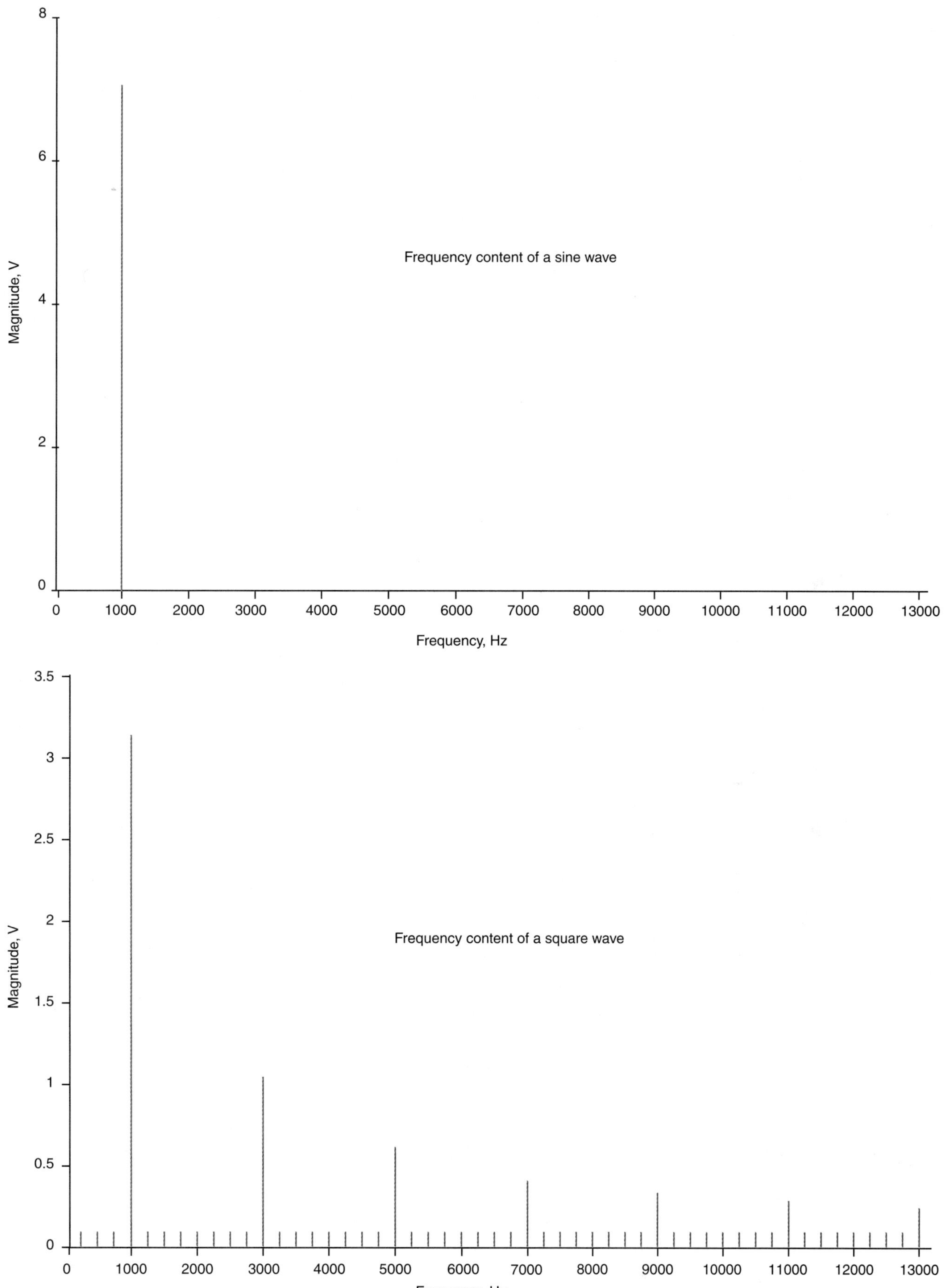

Fig. 7-13 Frequency content of sine and square waves.

clipping. (*Hint:* Use Fig. 7-11 to determine the maximum output first.)

22. Find the input impedance for the first stage in Fig. 7-9 if the 270-Ω emitter resistor is bypassed. Assume that the current in the 270-Ω resistor is 5 mA and that $\beta = 100$. Use 50 mV when estimating r_E.
23. Find the voltage gain of the second stage in Fig. 7-9 assuming that the 100-Ω emitter resistor is bypassed and the emitter current is 10 mA. Use 50 mV when estimating r_E.

7-3 Field-Effect Transistor Amplifiers

The silicon BJT is the workhorse of modern electronic circuitry. Its low cost and high performance make it the best choice for most applications. Field-effect transistors do, however, offer certain advantages that make them a better choice in some circuits. Some of these advantages are as follows:

1. They are voltage-controlled amplifiers. Because of this, their input impedance is very high.
2. They have a low noise output. This makes them useful as *preamplifiers* when noise must be very low because of high gain in following stages.
3. They have better linearity. This makes them attractive when distortion must be minimized.
4. They have low interelectrode capacity. At very high frequencies, interelectrode capacitance can make an amplifier work poorly. This makes the FET desirable in some RF stages.
5. They can be manufactured with two gates. The second gate is useful for gain control or the application of a second signal.

Common-source amplifier

Figure 7-14(*a*) shows an FET *common-source amplifier*. The source terminal is common to both the input and the output signals. This circuit is similar to the BJT common-emitter configuration. The supply voltage V_{DD} is positive with respect to ground. The current will flow from ground, through the N channel, through the load resistor, and into the positive end of the power supply. Note that a bias supply V_{GS} is applied across the gate-source junction. The polarity of this bias supply is arranged to *reverse-bias* the junction. Therefore, we may expect the gate current to be zero.

The gate resistor R_G in Fig. 7-14(*a*) will normally be a very high value [around 1 megohm (MΩ)]. It will not drop any dc voltage because there is no gate current. Using a large value of R_G keeps the input impedance high. If $R_G = 1$ MΩ, the signal source sees an impedance of 1 MΩ. This high input impedance is ideal for amplifying high-impedance signal sources. At very high frequencies, other effects can lower this impedance. At low frequencies, the amplifier input impedance is simply equal to the value of R_G.

The load resistor in Fig. 7-14(*a*) allows the circuit to produce a voltage gain. Figure 7-14(*b*) shows the characteristic curves for the transistor and the load line. As before, one end of the load line is equal to the supply voltage, and the other end of the load line is set by Ohm's law:

$$I_{sat} = \frac{V_{DD}}{R_L} = \frac{20\text{ V}}{5\text{ k}\Omega} = 4\text{ mA}$$

Thus, the load line runs from 20 V on the voltage axis to 4 mA on the current axis. This is shown in Fig. 7-14(*b*). In linear work, the operating point should be near the center of the load line. The operating point is set by the gate-source bias voltage V_{GS} in an FET amplifier. The bias voltage is equal to −1.5 V in Fig. 7-14(*a*). In Fig. 7-14(*b*) the operating point is shown on the load line. By projecting down from the operating point, it is seen that the resting or static voltage across the transistor will be about 11 V. This is roughly half the supply voltage, so the transistor is biased properly for linear work.

An input signal will drive the amplifier above and below its operating point. As shown in Fig. 7-14(*b*), a 1-V peak-to-peak input signal will swing the output about 8 V peak-to-peak. The voltage gain is

$$A_V = \frac{V_{out}}{V_{in}} = \frac{8\text{ V}_{p-p}}{1\text{ V}_{p-p}} = 8$$

As with the common-emitter circuit, the common-source configuration produces a 180° phase shift. Look at Fig. 7-14(*b*). As the input signal shifts the operating point from −1.5 to −1.0 V (positive direction), the output signal swings from about 11 to 7 V (negative direc-

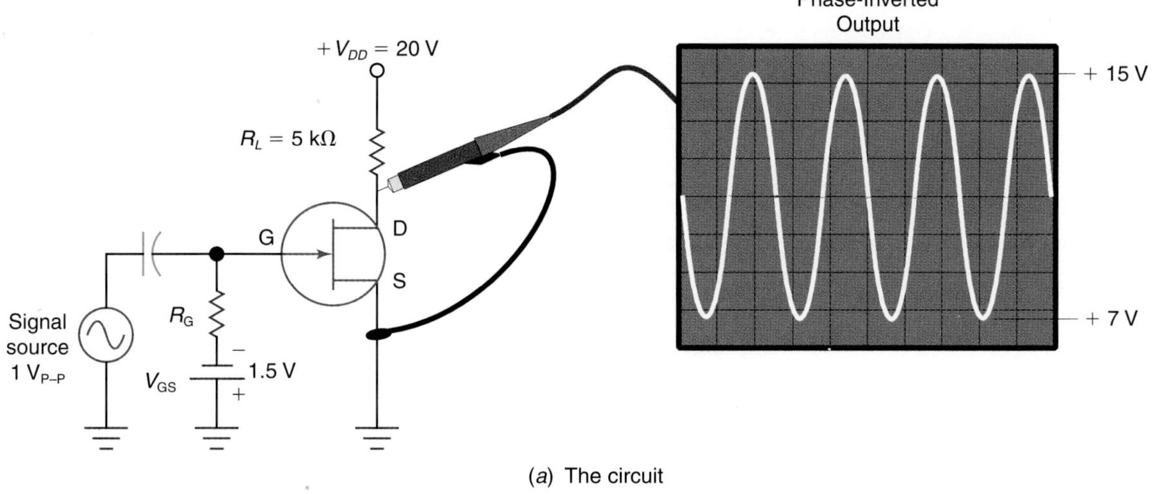

(*a*) The circuit

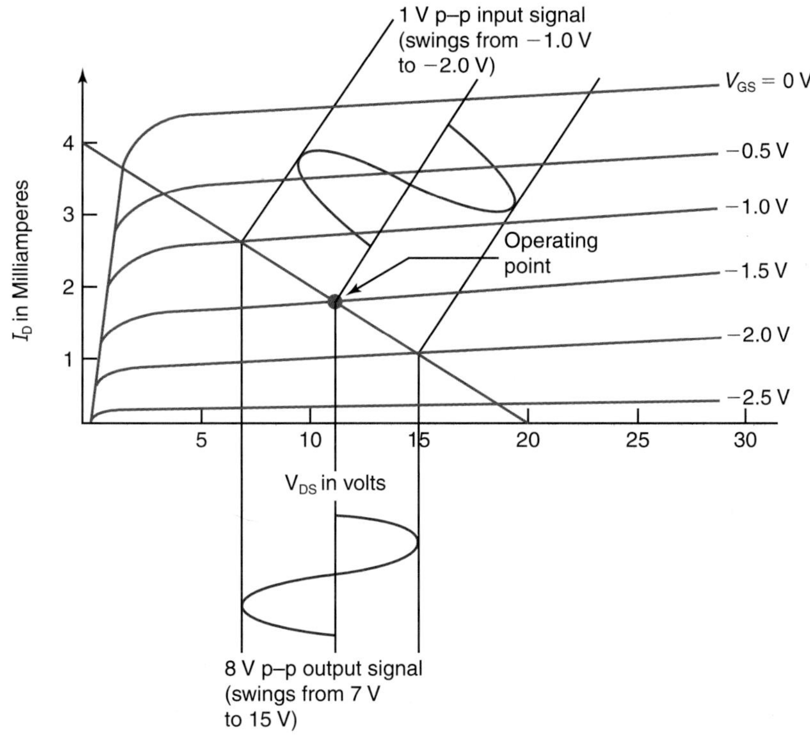

(*b*) The drain family of characteristic curves

Fig. 7-14 An N-channel FET amplifier with fixed bias.

tion). Common-drain and common-gate FET amplifiers do not produce this phase inversion.

There is a second way to calculate voltage gain for the common-source amplifier. It is based on a characteristic of the transistor called the *forward transfer admittance* Y_{fs}.

$$Y_{fs} = \frac{\Delta I_D}{\Delta V_{GS}} \Big| V_{DS}$$

where ΔV_{GS} = change in gate-source voltage ΔI_D = change in drain current, and $| V_{DS}$ means that the drain-source voltage is held constant.

The drain family of curves can be used to calculate Y_{fs} for the FET. In Fig. 7-15 the drain-source voltage V_{DS} is held constant at 11 V. The change in gate-source voltage V_{GS} is from −1.0 to −2.0 V. The change is (−1.0 V) − (−2.0 V) = 1 V. The change in drain current

Forward transfer admittance Y_{fs}

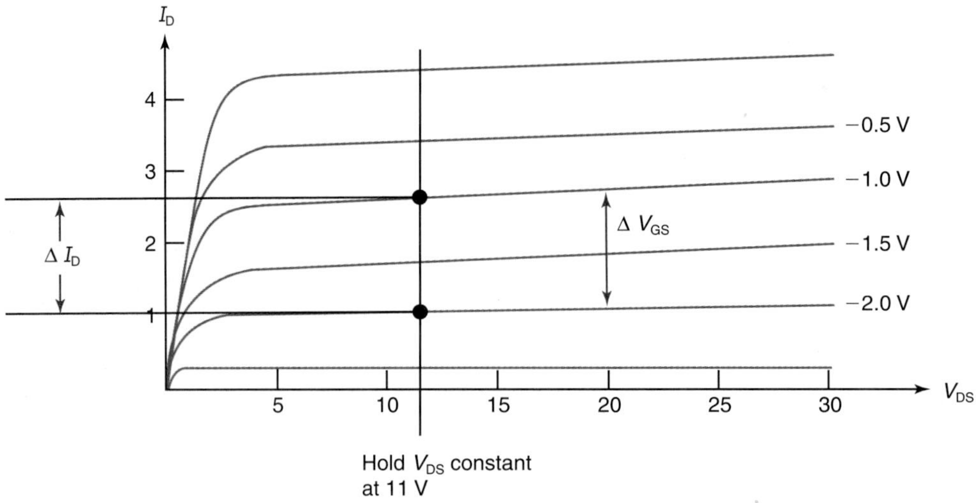

Fig. 7-15 Calculating forward transfer admittance.

ΔI_D is from 2.6 to 1 mA for a change of 1.6 mA. We can now calculate the forward transfer admittance of the transistor:

$$Y_{fs} = \frac{1.6 \times 10^{-3}\text{ A}}{1\text{ V}}$$
$$= 1.6 \times 10^{-3}\text{ siemens (S)}$$

The *siemen* is the unit for *conductance* (although some older references may still use the former unit, the mho). Its abbreviation is the letter *S*. Conductance (letter symbol *G*) is the *reciprocal* of resistance:

$$G = \frac{1}{R}$$

Conductance is a dc characteristic. Admittance is an ac characteristic equal to the reciprocal of impedance. They both use the siemen unit.

The voltage gain of a common-source FET amplifier is given by

$$A_V \approx Y_{fs} \times R_L$$

For the circuit of Fig. 7-14, the voltage gain will be

$$A_V = 1.6 \times 10^{-3}\text{ S} \times 5 \times 10^3\ \Omega = 8$$

Note: Since the siemen and ohm units have a *reciprocal* relationship, the units *cancel* and the gain is just a number as always.

A voltage gain of 8 agrees with the graphical solution of Fig. 7-14(*b*). One advantage of using the gain equation is that it makes it easy to calculate the voltage gain for different values of load resistance. It will not be necessary to draw additional load lines. If the load resistance is changed to 8.2 kΩ, the voltage gain will be

$$A_V = 1.6 \times 10^{-3}\text{ S} \times 8.2 \times 10^3\ \Omega$$
$$= 13.12$$

With a load resistance of 5 kΩ, the circuit gives a voltage gain of 8. With a load resistance of 8.2 kΩ, the circuit gives a voltage gain slightly over 13. This shows that gain is *directly* related to load resistance. This was also the case in the *common-emitter* BJT amplifier circuits. Remember this concept because it is valuable for understanding and troubleshooting amplifiers.

Figure 7-16 shows an improvement in the common-source amplifier. The bias supply V_{GS} has been eliminated. Instead, we find resistor

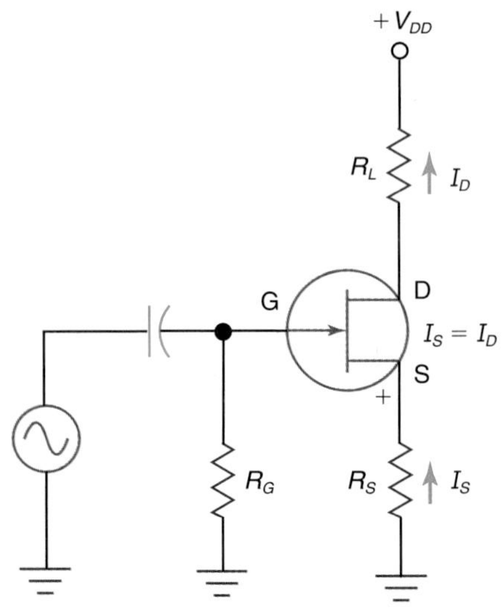

Fig. 7-16 An N-channel FET amplifier using source bias.

R_S in the source circuit. As the source current flows through this resistor, voltage will drop across it. This voltage drop will serve to bias the gate-source junction of the transistor. If the desired bias voltage and current are known, it is an easy matter to calculate the value of the source resistor. Since the drain current and the source current are equal,

$$R_S = \frac{V_{GS}}{I_D}$$

If we assume the same operating conditions as in the circuit of Fig. 7-14(*a*), the gate bias voltage should be −1.5 V (the sign is not used in the calculation). The source resistor should be

$$R_S = \frac{1.5 \text{ V}}{1.9 \text{ mA}} = 790\ \Omega$$

Note that the value of current used in the calculation is about half the saturation current. Check Fig. 7-14(*b*) and verify that this value of drain current is near the center of the load line.

The circuit of Fig. 7-16 is called a *source bias* circuit. The bias voltage is produced by source current flowing through the source resistor. The drop across the resistor makes the source terminal *positive* with respect to ground. The gate terminal is at ground potential. There is no gate current and therefore no drop across the gate resistor R_G. Thus, the source is positive with respect to the gate. To say it another way, the gate is negative with respect to the source. This accomplishes the same purpose as the separate supply V_{GS} in Fig. 7-14(*a*).

Source bias is much simpler than using a separate bias supply. The voltage gain does suffer, however. To see why, examine Fig. 7-17. As the input signal drives the gate in a positive direction, the source current increases. This makes the voltage drop across the source resistor increase. The source terminal is now more positive with respect to ground. This makes the gate more negative with respect to the source. The overall effect is that *some of the input signal is canceled.*

JOB TIP

Don't overlook related areas when assembling your portfolio. If your hobby is building models, some photographs could be helpful.

When an amplifier develops a signal that interacts with the input signal, the amplifier is said to have *feedback.* Figure 7-17 shows one way feedback can affect an amplifier. In this example, the feedback is acting to *cancel* part of the effect of the input signal. When this happens, the feedback is said to be *negative.*

Negative feedback

Negative feedback decreases the amplifier's gain. It is also capable of increasing the frequency range of an amplifier. Negative feedback may be used to decrease an amplifier's distortion. So, negative feedback is not good or bad—it is a mixture. If maximum voltage gain is required, the negative feedback will have to be eliminated. In Fig. 7-18 the source bias circuit has a *source bypass capacitor.* This capacitor will eliminate the negative feedback and increase the gain. It is selected to have low reactance at the signal frequency. It will prevent the source terminal voltage from swinging with the increases and decreases in source current. It has pretty much the same effect as the emitter bypass capacitor in the common-emitter amplifier studied before.

Source bias

When we studied BJTs, we found them to have an unpredictable *β*. This made it necessary to investigate a circuit that was not as sensitive to *β*. Field-effect transistors also have characteristics that vary widely from unit to unit. It is necessary to have circuits that are not as sensitive to certain device characteristics.

Fixed bias

The circuit of Fig. 7-14(*a*) is called *fixed bias.* This circuit will work well only if the

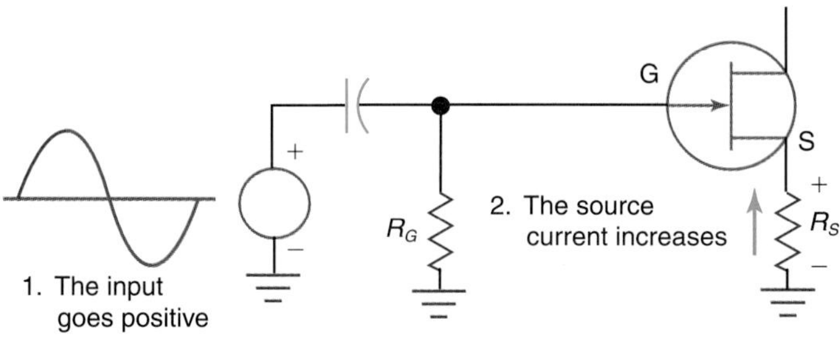

Fig. 7-17 Source feedback.

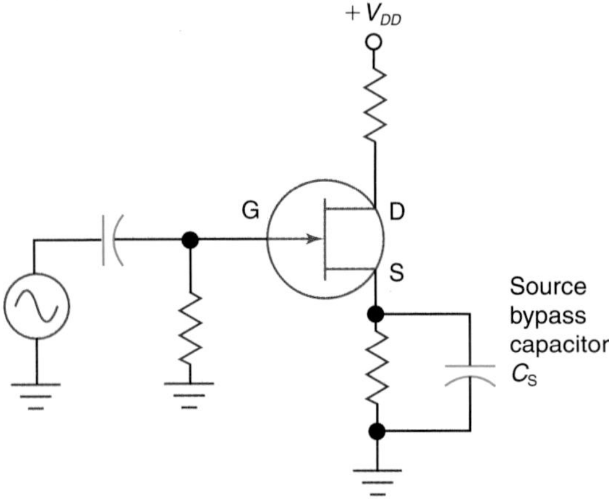

Fig. 7-18 Adding the source bypass capacitor.

transistor has predictable characteristics. The fixed-bias circuit usually is not a good choice. The circuit of Fig. 7-16 uses source bias. It is much better and allows the transistor characteristics to vary. If, for example, the transistor tended toward more current, the source bias would automatically increase. More bias would reduce the current. Thus, it can be seen that the source resistor stabilizes the circuit.

The greater the source resistance, the more stability we can expect in the operating point. But too much source resistance could create too much bias, and the circuit will operate too close to cutoff. If there were some way to offset this effect, a better circuit would result. Figure 7-19 shows a way. This circuit uses *combination bias.* The bias is a combination of a fixed positive voltage applied to the gate terminal and source bias. The positive voltage is set by a voltage divider. The divider network is made up of R_{G_1} and R_{G_2}. These resistors will usually be high in value to maintain a high input impedance for the amplifier.

Combination bias

The combination-bias circuit can use a larger value for R_S, the source resistor. The bias voltage V_{GS} will not be excessive because a positive, fixed voltage is applied to the gate. This fixed, positive voltage will reduce the effect of the voltage drop across the source resistor.

A few calculations will show how the combination-bias circuit works. Assume in Fig. 7-19 that the desired source current is to be 1.9 mA. The voltage drop across R_S is

$$V_{R_S} = 1.9 \text{ mA} \times 2.2 \text{ k}\Omega = 4.18 \text{ V}$$

Next, the voltage divider drop across R_{G_2} sets V_G at

$$V_G = \frac{2.2 \text{ M}\Omega}{2.2 \text{ M}\Omega + 15 \text{ M}\Omega} \times 20 \text{ V} = 2.56 \text{ V}$$

Both of the above voltages are positive with respect to ground. The source voltage is *more* positive. The gate is therefore *negative* with respect to the source by an amount of

$$V_{GS} = 2.56 \text{ V} - 4.18 \text{ V} = -1.62 \text{ V}$$

In measuring bias voltage in circuits such as Fig. 7-19, remember that V_G and V_{GS} are different. The gate voltage V_G is measured with respect to ground. The gate-source voltage V_{GS} is measured with respect to the source. Also, don't forget to take the loading effect of your meter into account. When measuring V_G in a high-impedance circuit like Fig. 7-19, a meter

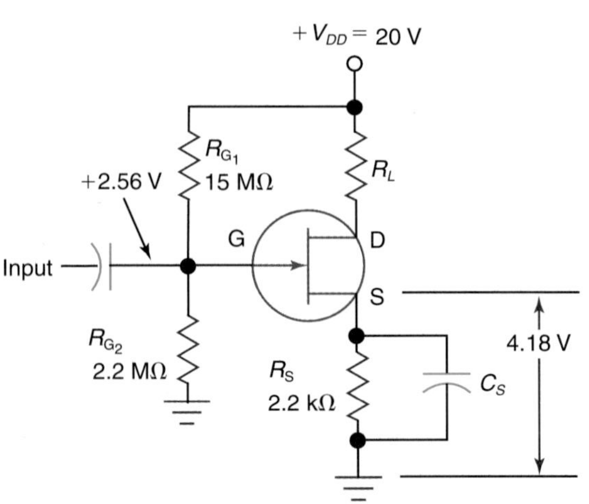

Fig. 7-19 An FET amplifier using combination bias.

with a high input impedance is required for reasonable results.

Figure 7-20 shows a P-channel JFET amplifier. The supply voltage is *negative* with respect to ground. Note the direction of the source current I_S. The voltage drop produced by this current will reverse-bias the gate-source diode. This is proper, and no gate current will flow through R_G.

Example 7-5

Determine the gate-source bias for Fig. 7-20 if the drain current is 2 mA and the source resistor is 860 Ω. Is there any difference when this bias is compared to Fig. 7-16? The bias is found using Ohm's law, and it is understood that the source current is equal to the drain current:

$$V_{GS} = I_S \times R_S = 2 \text{ mA} \times 860\ \Omega = 1.72 \text{ V}$$

V_{GS} is positive in Fig. 7-20 and negative in Fig. 7-16.

It is possible to have linear operation with *zero bias,* as shown in Fig. 7-21. You should recognize the transistor as a MOSFET. The gate is *insulated* from the source in this type of transistor. This prevents gate current regardless of the gate-source polarity. As the signal goes positive, the drain current will increase. As the signal goes negative, the drain current will decrease. This type of transistor can operate in both the *enhancement* and the *depletion* modes. This is *not* true with junction FETs. The zero-bias circuit of Fig. 7-21 is restricted to depletion-mode MOSFETs for linear work.

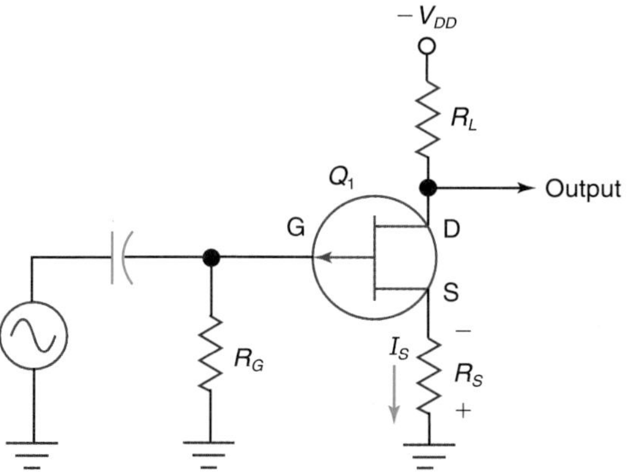

Fig. 7-20 A P-channel JFET amplifier.

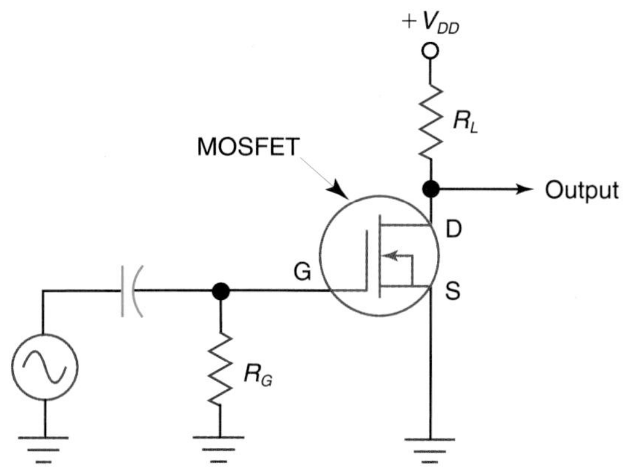

Fig. 7-21 A zero-bias MOSFET amplifier.

Dual-gate MOSFET

Figure 7-22 shows the schematic for a *dual-gate MOSFET* amplifier. The circuit uses tuned transformer coupling. Good gain is possible at frequencies near the resonant point of the transformers. The signal is fed to gate 1 (G_1) of the MOSFET. The output signal appears in the drain circuit. The gain of this amplifier can be controlled over a large range by gate 2 (G_2).

The graph of Fig. 7-23 on page 178 shows a typical gain range for this type of amplifier. Note that maximum power gain, 20 dB, occurs when gate 2 is positive with respect to the source by about 3 V. At zero bias, the gain is only about 5 dB. As gate 2 goes negative with respect to the source, the gain continues to drop. The minimum gain is about −28 dB. Of course, −28 dB represents a large loss.

The total range of gain for the amplifier is from +20 to −28 dB, or 48 dB. This means a power ratio of about 63,000:1. Thus, with the proper control voltage applied to G_2, the circuit of Fig. 7-22 can keep a constant output over a tremendous range of input levels. Amplifiers of this type are used in applications such as communications where a wide range of signal levels is expected.

Field-effect transistors have some very good characteristics. However, bipolar transistors usually cost less and give much better voltage gains. This makes the bipolar transistor the best choice for most applications.

Enhancement mode

Depletion mode

Field-effect transistors are used when some special feature is needed. For example, they are a good choice if a very high input impedance is required. When used, FETs are generally found in the first stage or two of a linear system.

Examine Fig. 7-24 on page 179. Transistor

Did You Know?

"Electronics" and the Human Body Without negative feedback supplied by your nerves to your brain, you could not walk.

Q_1 is a JFET, and Q_2 is a BJT. Thanks to the JFET, the input impedance is high. Thanks to the bipolar device, the power gain is good and the cost is reasonable. This is typical of the way circuits are designed to have the best performance for the least cost.

TEST

Determine whether each statement is true or false.

24. Voltage-controlled amplifiers usually have a lower input impedance than current-controlled amplifiers.
25. Gate current is avoided in JFET amplifiers by keeping the gate-source junction reverse-biased.
26. A separate gate supply, such as shown in Fig. 7-14(*a*), is the best way to keep the gate junction reverse-biased.
27. The voltage gain of an FET amplifier is given in siemens.

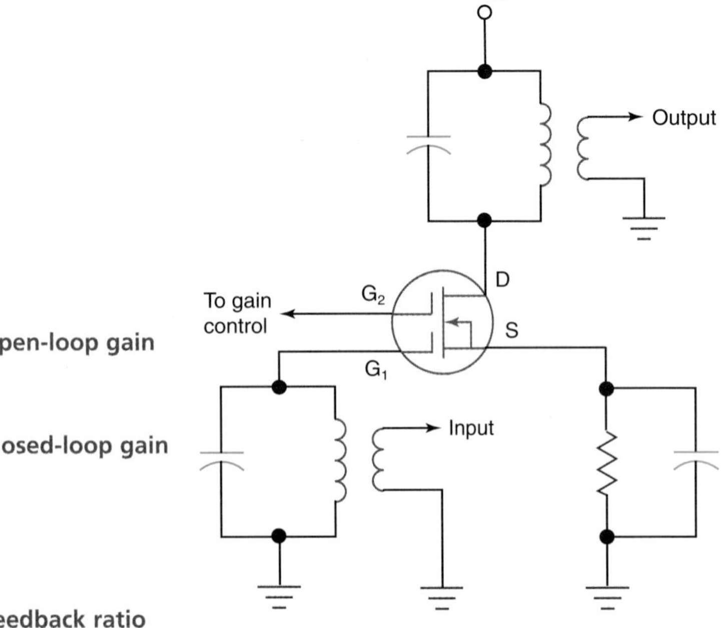

Fig. 7-22 A dual-gate MOSFET amplifier.

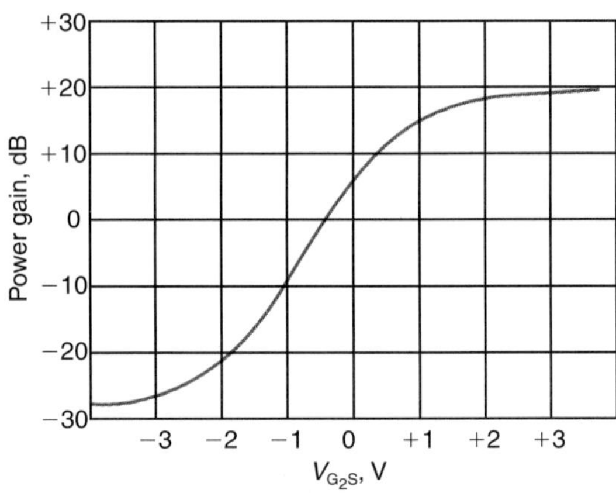

Fig. 7-23 The effect of V_{G_2} on gain.

28. Source bias tends to stabilize FET amplifiers.
29. Dual-gate MOSFETs are not used as linear amplifiers.

Solve the following problems.

30. Refer to Fig. 7-14(*a*). Assume I_D = 3 mA. Find V_{DS}.
31. Refer to Fig. 7-14(*b*). Assume V_{GS} = 0 V. Where would the transistor be operating?
32. Refer to Fig. 7-14(*b*). Assume V_{GS} = −3.0 V. Where would the transistor be operating?
33. Refer to Fig. 7-16. If I_D = 1.5 mA and R_S = 1000 Ω, what is the value of V_{GS}?
34. Refer to Fig. 7-19. Assume V_{DD} = 15 V and I_D = 2 mA. What is the value of V_{GS}? What is the gate polarity with respect to the source?
35. Refer to Fig. 7-24. In what circuit configuration is Q_1 connected? Q_2?
36. Refer to Fig. 7-24. If the input is going in a positive direction, in what direction will the output go?

7-4 Negative Feedback

Open-loop gain

Closed-loop gain

Feedback ratio

Figure 7-25 shows two block diagrams. In Fig. 7-25(*a*), block *A* represents the *open-loop gain* of the amplifier. This is the gain with *no negative feedback.* As an example, perhaps *A* = 50. With negative feedback, it is the *closed-loop gain* that determines V_{out}, and this gain is always less than *A*. So if *A* = 50, the closed-loop gain must be less than 50. Block *B* represents the feedback circuit that returns some of the output signal back to the input. The *feedback ratio* tells us how much of the output is

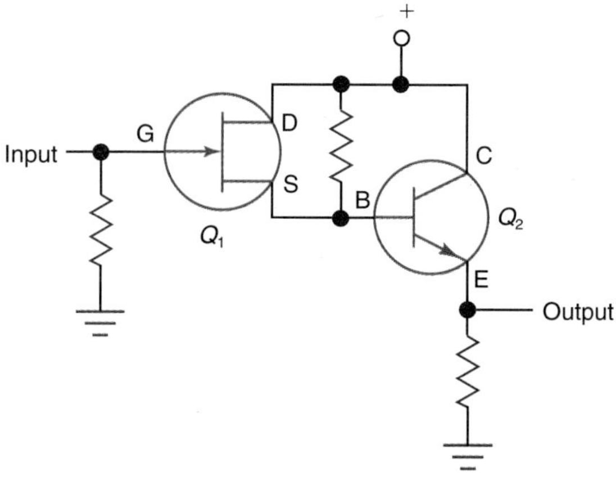

Fig. 7-24 Combining FET and bipolar devices.

returned. If block B is a voltage divider formed with two equal resistors, then $B = 0.5$. The summing junction in Fig. 7-25(a) is where the feedback and V_{in} are put together (combined).

Figure 7-25(b) shows that it is possible to simplify the circuit. The simplification makes it easy to find the closed-loop gain.

Negative feedback must lower the gain of an amplifier, so it doesn't seem to provide any advantage. The fact is that it is an excellent trade-off in many cases. Here are the reasons why negative feedback is used:

1. *Stabilize an amplifier:* Make the gain and/or the operating point independent of device characteristics and temperature.
2. *Increase the bandwidth of an amplifier:* Make it provide useful gain over a broader range of frequencies.

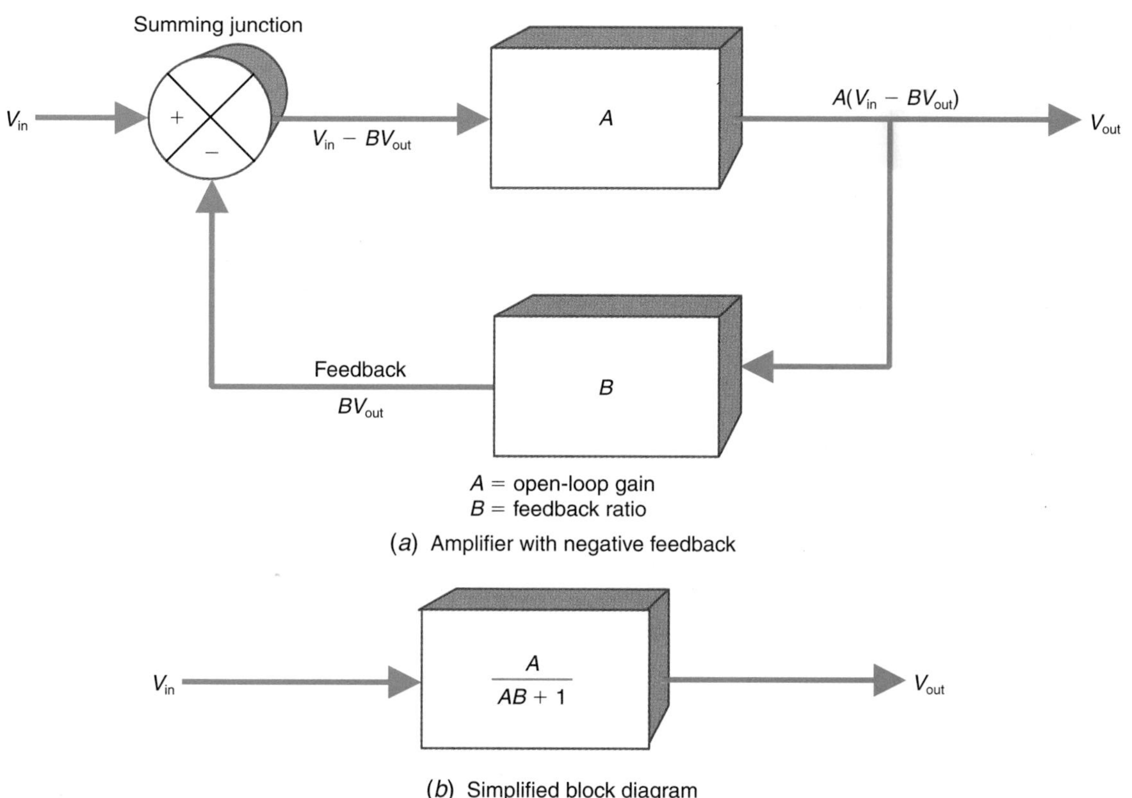

Fig. 7-25 Negative feedback block diagrams.

3. *Improve the linearity of an amplifier:* Decrease the amount of signal distortion.
4. *Improve the noise performance of an amplifier:* Make the amplifier quieter.
5. *Change amplifier impedances:* Raise or lower the input or output impedance.

Example 7-6

What is the closed-loop gain for a negative feedback amplifier with an open-loop gain of 50 and a feedback ratio of 0.5? Use the simplified model:

$$\frac{V_{out}}{V_{in}} = A_{CL} = \frac{A}{AB + 1}$$

$$= \frac{50}{50 \times 0.5 + 1} = 1.92$$

This shows that negative feedback can reduce the open-loop gain by quite a bit.

The simplified closed-loop gain equation can be derived using algebra. We begin by looking at Fig. 7-25(*a*) and noting that the output signal can be written in these terms:

$$V_{out} = A(V_{in} - BV_{out})$$

Multiply:

$$V_{out} = AV_{in} - ABV_{out}$$

Divide each term by V_{out}:

$$1 = \frac{AV_{in}}{V_{out}} - AB$$

Add AB to both sides:

$$AB + 1 = \frac{AV_{in}}{V_{out}}$$

Divide both sides by A:

$$\frac{AB + 1}{A} = \frac{V_{in}}{V_{out}}$$

Invert both sides:

$$\frac{A}{AB + 1} = \frac{V_{out}}{V_{in}} = A_{CL}$$

Collector feedback

DC feedback

AC feedback

Figure 7-26 shows a common-emitter amplifier with *collector feedback.* Resistor R_F provides feedback from the collector terminal (the output) back to the base terminal (the input). Feedback resistor R_F provides *both dc and ac feedback.* The dc feedback stabilizes the amplifier's operating point. If temperature increases, the gain of the transistor increases and more collector current will flow. However, in Fig. 7-26, R_F is connected to the collector and not to the V_{CC} supply point. Thus, when the collector current tries to increase, the collector voltage starts to drop. This effectively decreases the supply voltage for the base current. With less supply voltage, the base current decreases. This decrease in base current makes the collector current decrease, which tends to offset the original effect of a collector current increase caused by temperature rise. If the transistor is replaced by one with higher current gain, once again R_F would help to stabilize the circuit. Direct current feedback in an amplifier is helpful for keeping the operating point near its desired value. The emitter resistor in Fig. 7-26 provides additional feedback and improves stability of the operating point even more.

In Fig. 7-26 R_F also provides ac feedback. When the amplifier is driven with a signal, a larger out-of-phase signal appears at the collector of the transistor. This signal feeds back and reduces the ac current gain of the amplifier and decreases its input impedance. Suppose, for example, that the voltage gain of the amplifier is 50 and that R_F is a 100-kΩ resistor. Any voltage change at the signal source would cause the collector end of R_F to go 50 times in the opposite direction. This means that the *ac signal current* flow in R_F would be *50 times greater* than it would be if the resistor were connected to V_{CC}. The signal current in

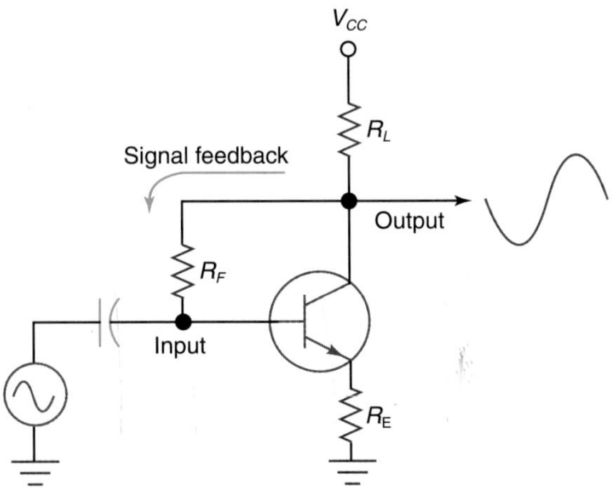

Fig. 7-26 Common-emitter amplifier with collector feedback.

R_F is proportional to the signal voltage across it. The gain of the amplifier is 50; therefore the signal current must be 50 times the value that would be predicted by the signal voltage and the value of R_F. It also means that R_F loads the signal source as if it were 1⁄50 of its actual value or 2 kΩ in this example. Therefore we see that the ac feedback in Fig. 7-26 causes extra signal current to flow in the input circuit. The current gain of the amplifier has been decreased, and its input impedance has been decreased. The voltage gain has not been changed by R_F.

The ac feedback may not be desirable. Figure 7-27 shows how it can be eliminated. The feedback resistor has been replaced with two resistors, R_{F_1} and R_{F_2}. The junction of these two resistors is bypassed to ground with capacitor C_B. This capacitor is chosen to have a very low reactance at the signal frequencies. It acts as a short circuit and prevents any ac feedback from reaching the base of the transistor. Now the input impedance and the current gain of the amplifier are both higher than in Fig. 7-26. The bypass capacitor in Fig. 7-27 has *no effect* on the dc feedback. Capacitive reactance is infinite at 0 Hz; therefore, the amplifier has the same operating point stability as discussed for Fig. 7-26.

In many cases, an emitter resistor provides enough feedback. In Fig. 7-28 on page 182, the emitter resistor R_E provides both dc and ac feedback. The dc feedback acts to stabilize the operating point of the amplifier. Emitter resistor R_E is not bypassed, and a signal appears across it when the amplifier is driven. This signal is in phase with the input signal. If the signal source drives the base in a positive direction, the signal across R_E will cause the emitter to also go in a positive direction. This action reduces the base-emitter signal voltage and decreases the voltage gain of the amplifier. It also increases the input impedance of the amplifier as discussed earlier in this chapter. We know that R_E can be bypassed for improved voltage gain at the cost of decreased input impedance.

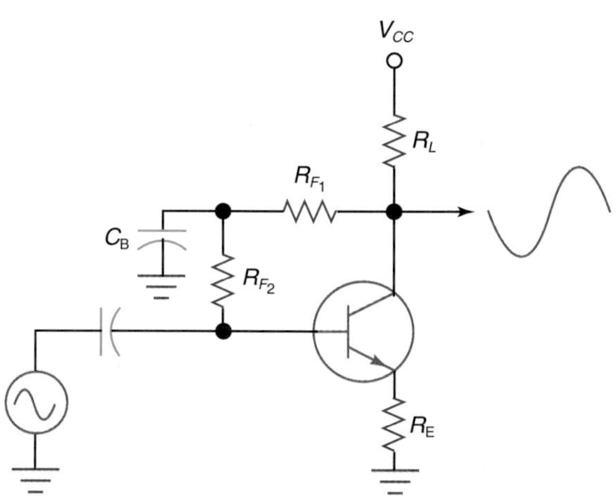

Fig. 7-27 Eliminating the ac feedback.

Bootstrap circuit

When very high input impedances are required, the circuit of Fig. 7-29 on page 183 may be used. It is often called a *bootstrap circuit.* Capacitor C_B and resistor R_B provide a feedback path that works to increase the input impedance of the amplifier. An in-phase signal is developed across the unbypassed part of the emitter resistor. This signal is coupled by C_B to the right end of R_B. Since it is in phase with the input signal, it decreases the signal current flowing in R_B. For example, if the feedback signal at the right end of R_B was equal to the signal supplied by the source, there would be no signal voltage difference across R_B and no signal current could flow in it. This would make R_B appear as an *infinite impedance* as far as signal currents are concerned. In actual circuitry, the feedback signal is lower in amplitude than the input signal; therefore, there is some signal current flow in R_B. However, for the input signal, R_B appears to be many times greater in impedance than its value in ohms. Since R_B is in series with the dc bias divider, it effectively isolates the signal source from the two bias resistors. A bootstrap amplifier, such as the one shown in Fig. 7-29, will have an input impedance of several hundred thousand ohms, whereas the circuit shown in Fig. 7-28 would be on the order of several thousand ohms.

Once hired, continue to study the company. The extra knowledge that you gain will play a role in your promotion.

So far we have seen that negative feedback in an amplifier can decrease current gain or voltage gain. We have seen that it can lower or raise the input impedance of the amplifier. Why use feedback if gain (current or voltage) must suffer? Sometimes it is required to achieve the proper input impedance. The lost gain can be offset by using another stage of amplification. Another reason for using feedback is to obtain *better bandwidth.* Look at Fig. 7-30 on page 183. The amplifier gain is best without negative feedback. At higher frequencies, the gain begins to drop off. This begins to occur at f_1 in Fig. 7-30. The

Improve bandwidth

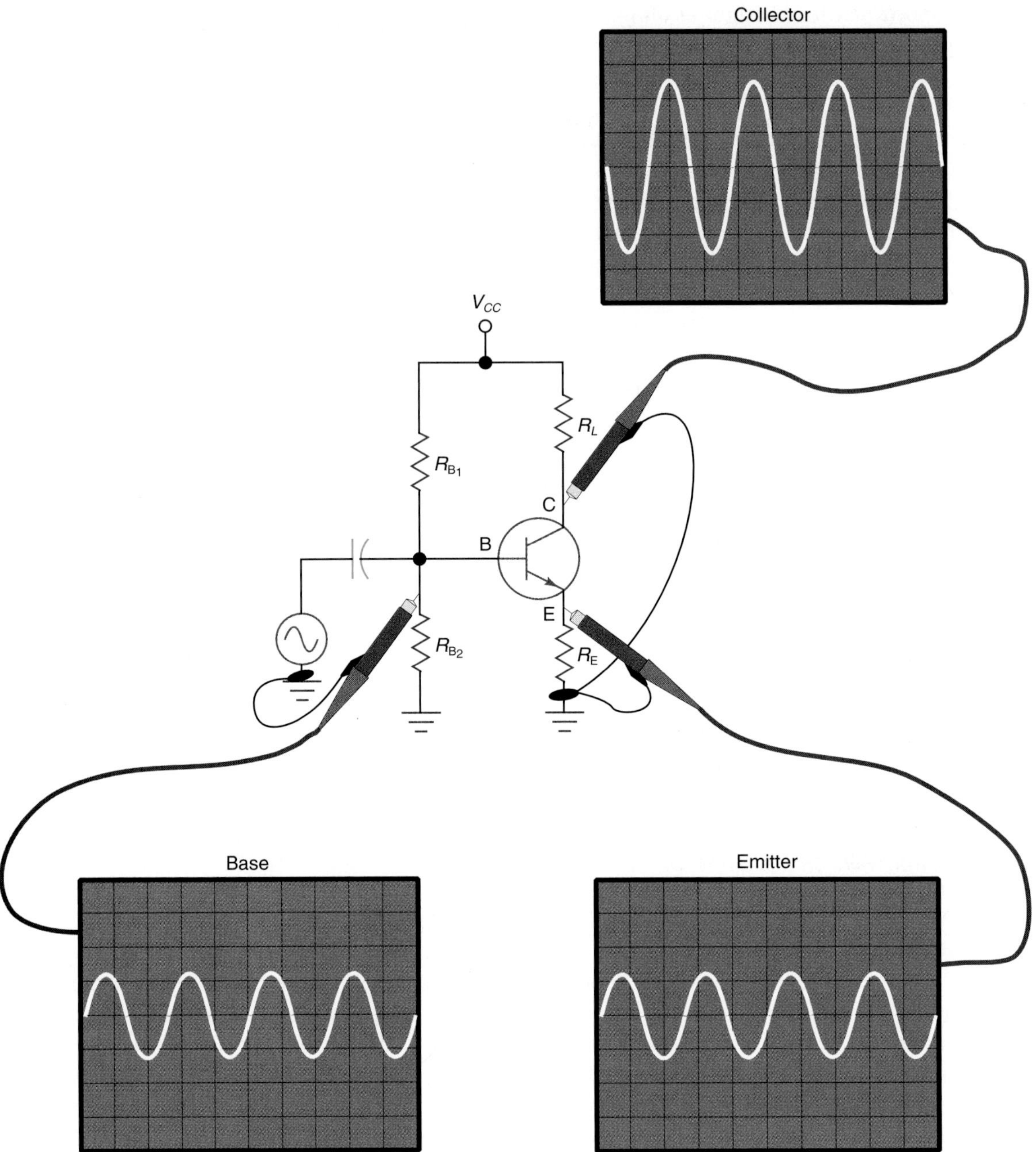

Fig. 7-28 Common-emitter amplifier with emitter feedback.

decrease in gain is a result of the performance of the transistor and circuit capacitance. All transistors show less gain as the frequency increases. The reactance of a capacitor decreases as the frequency increases. This decreasing reactance loads the amplifier, and the gain drops.

Now, look at the performance of the amplifier in Fig. 7-30 with negative feedback. The gain is much less at the lower frequencies, but it does not start dropping off until f_2 is reached. Frequency f_2 is much higher than f_1. With negative feedback, the amplifier provides less gain, but the gain is constant over a wider frequency range. The loss of gain in one stage can be easily overcome by adding another stage of amplification.

Noise and distortion

Negative feedback also *reduces noise and distortion.* Suppose the signal picks up some noise or distortion in the amplifier. This appears on the output signal. Some of the output signal is fed back to the input. The noise or distortion will be placed on the input signal in an opposite way. Remember, the feedback is out of phase with, or opposite to, the input.

Much of the noise and distortion is canceled in this way. By intentionally distorting the input signal, opposite to the way the amplifier distorts it, the circuit becomes more linear.

Figure 7-31 on page 184 shows a cascade amplifier with switchable negative feedback. The feedback is for alternating current only since the 25-μF coupling capacitor blocks direct current. The top oscilloscope display shows the performance with negative feedback. Notice that both waveforms are triangular, with no apparent distortion. The bottom display shows performance without negative feedback. The output triangle wave is distorted. Table 7-2 compares performance with and without feedback. Notice that the *bandwidth* is also much better with negative feedback: bandwidth = $f_H - f_L$.

Triangle waves are often used when checking for clipping or other forms of distortion. Sine waves are not as useful for distortion checks. It is much easier to notice deviations from straight lines.

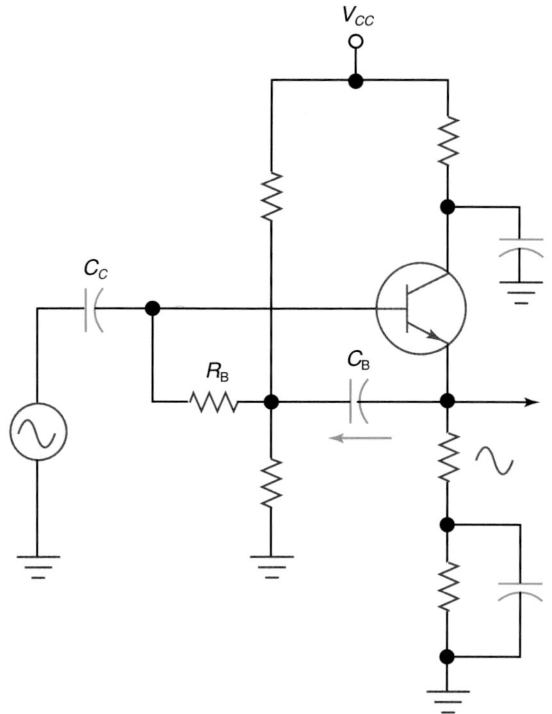

Fig. 7-29 Bootstrapping for high input impedance.

Example 7-7

Verify the negative feedback gain in Table 7-2. A look at Fig. 7-31 shows that the negative feedback is applied to Q_1's source resistor, which forms a voltage divider with the 5.1-kΩ resistor. The feedback ratio is found with the familiar voltage divider equation:

$$B = \frac{180\ \Omega}{180\ \Omega + 5.1\ \text{k}\Omega} = 0.0341$$

Now, we can apply the closed-loop gain equation that was developed earlier. The open-loop gain value is found in Table 7-2 ($A_{OL} = 240$):

$$A_{CL} = \frac{A}{AB + 1}$$

$$= \frac{240}{(240)(0.0341) + 1} = 26.1$$

This agrees with the value in the table. The decrease in gain is the price paid for low distortion and wide bandwidth.

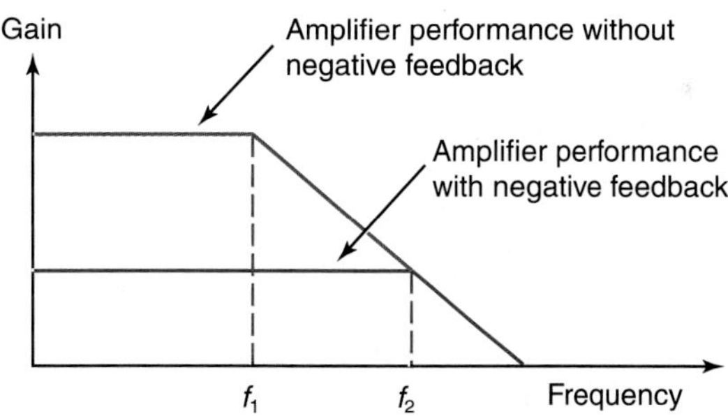

Fig. 7-30 The effect of feedback on gain and bandwidth.

Figure 7-32 on page 185 shows an NPN-PNP amplifier with negative feedback. The PNP transistor shares its collector circuit with the emitter circuit of the NPN transistor. There is both dc and ac feedback in this amplifier.

TABLE 7-2 Performance with and without feedback

	With Feedback	Without Feedback
A_V	26 (28 dB)	240 (48 dB)
f_L	150 Hz	1.2 kHz
f_H	1.9 MHz	190 kHz
Distortion	None	High

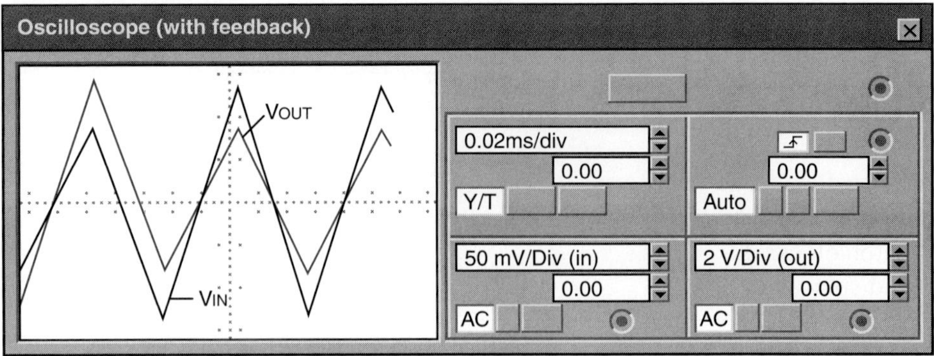

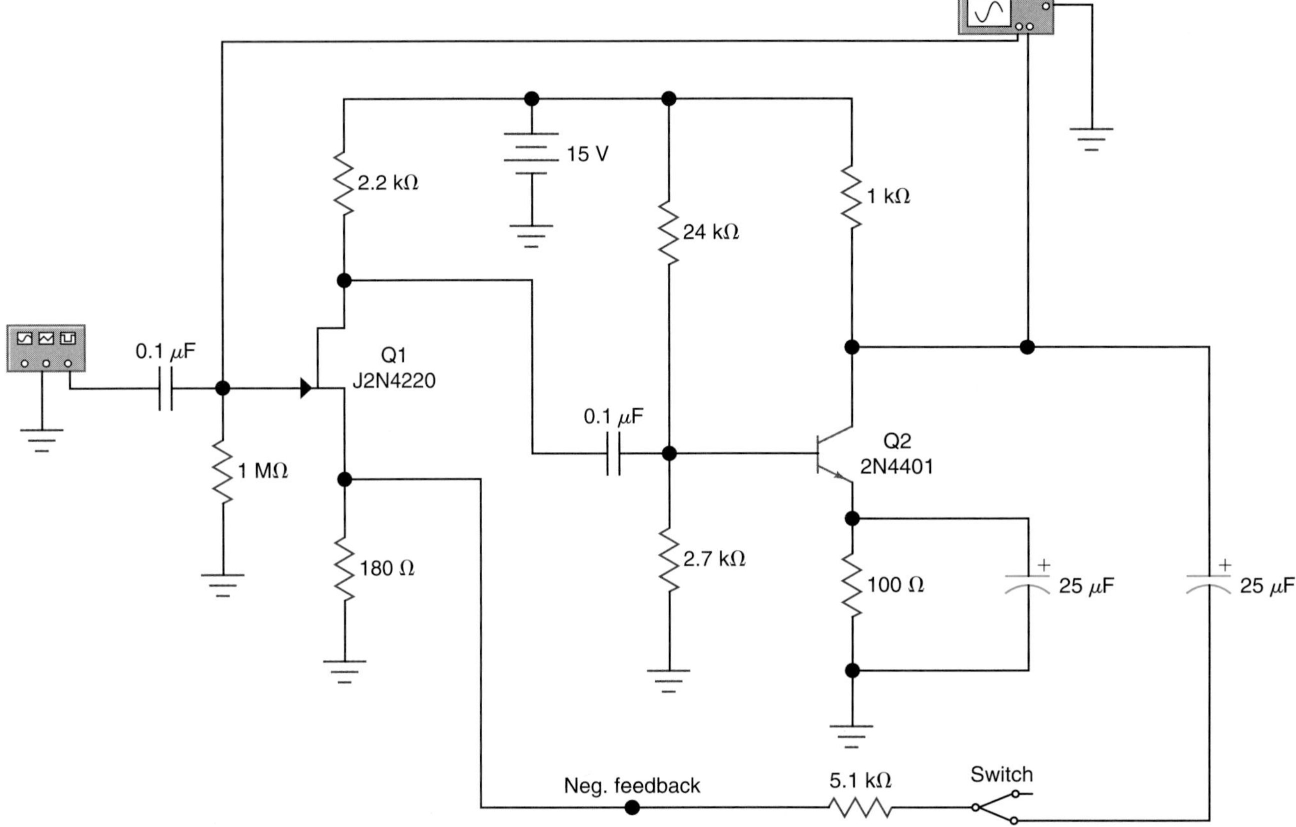

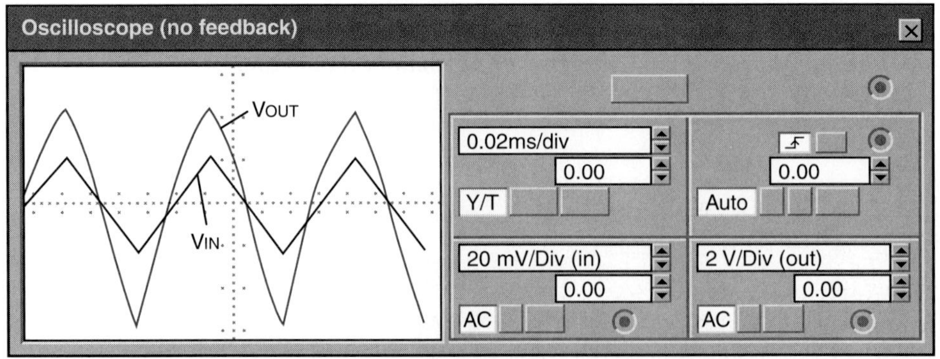

Fig. 7-31 A two-stage amplifier with switchable negative feedback.

Example 7-8

Find the dc conditions for the amplifier shown in Fig. 7-32. Beginning with the base voltage at the input:

$$V_{B(NPN)} = \frac{5.1\ k\Omega}{5.1\ k\Omega + 27\ k\Omega} \times 12\ V$$
$$= 1.91\ V$$
$$I_{(560\ \Omega)} = \frac{1.91\ V - 0.7\ V}{560\ \Omega} = 2.16\ mA$$

This is *not* the current flow in the NPN transistor because both transistors share this resistor. Notice in Fig. 7-32 that the base-emitter junction of the PNP transistor is in *parallel* with a 680-Ω resistor. As always, it is reasonable to assume a 0.7-V drop, so:

$$I_{680\ \Omega} = \frac{0.7\ V}{680\ \Omega} = 1.03\ mA$$

We usually assume that $I_E = I_C$, so it's possible to subtract this current from 2.16 mA to determine the PNP transistor current:

$$I_{C(PNP)} = I_{E(PNP)} = 2.16\ mA - 1.03\ mA$$
$$= 1.13\ mA$$

Finish by finding the dc output voltage of the amplifier. First, the drop across the 4.7-kΩ resistor is:

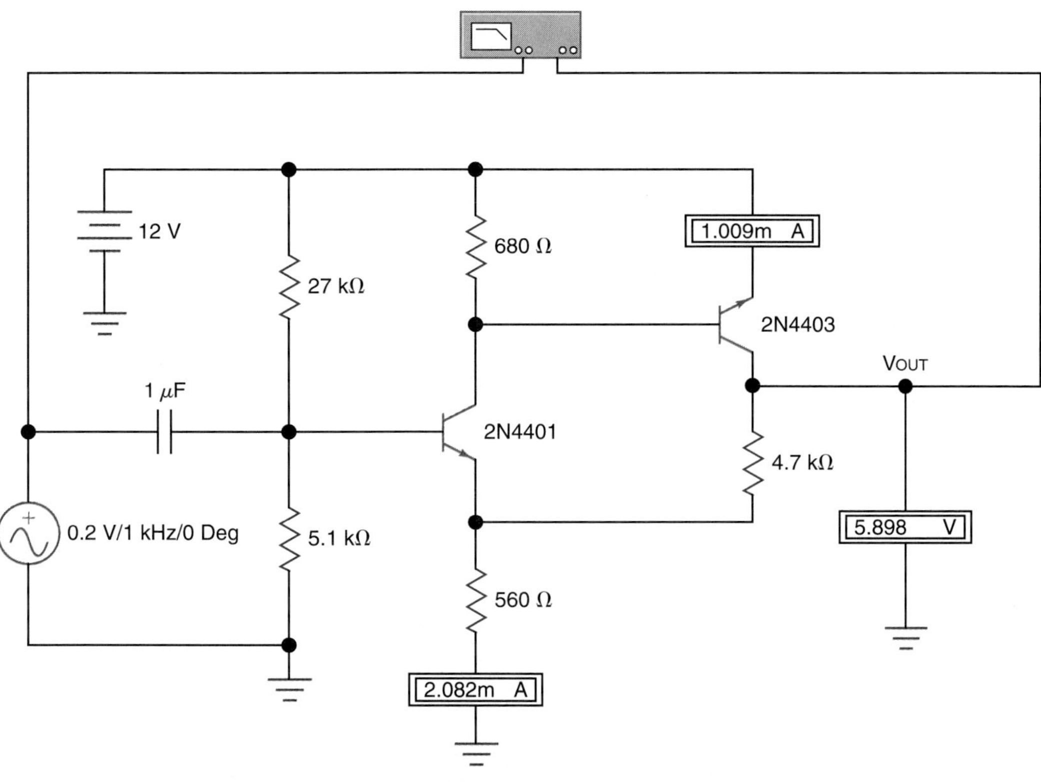

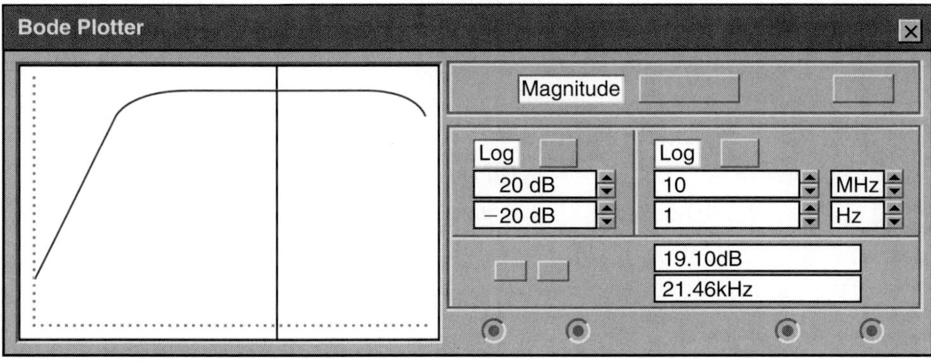

Fig. 7-32 NPN-PNP amplifier with series feedback.

$$V_{4.7\ k\Omega} = 1.13\ mA \times 4.7\ k\Omega = 5.31\ V$$

$$V_{out} = V_{560\ \Omega} + V_{4.7\ k\Omega}$$

$$= 1.21\ V + 5.31\ V = 6.52\ V$$

The fact that this voltage is close to half the supply voltage tells us that the circuit is biased for linear operation. The method used here is based on several assumptions, and there is some error. As Fig. 7-32 shows, a circuit simulator reports different dc conditions—but not drastically different.

Example 7-9

Find the dB closed-loop gain for Fig. 7-32 and compare it to the simulation. Begin by finding the gain of each stage. The emitter resistor for the NPN transistor is large and is not bypassed, so the gain of this stage is:

$$A_{V(NPN)} = \frac{R_L}{R_E} = \frac{680\ \Omega}{560\ \Omega} = 1.21$$

The PNP transistor has no emitter resistor, so to find the gain, we need r_E first:

$$r_E = \frac{25}{I_E} = \frac{25}{1.13} = 22.1\ \Omega$$

$$A_{V(PNP)} = \frac{R_L}{r_E} = \frac{4.7\ k\Omega}{22.1\ \Omega} = 213$$

The open-loop gain for Fig. 7-32 is the product of the gains:

$$A_{OL} = A_{V(NPN)} \times A_{V(PNP)} = 1.21 \times 213 = 258$$

The closed-loop gain is less because of the feedback. The feedback ratio is set by the voltage division produced by the 560-Ω and 4.7-kΩ resistors:

$$B = \frac{560\ \Omega}{560\ \Omega + 4.7\ k\Omega} = 0.106$$

The closed-loop gain is:

$$A_{CL} = \frac{A}{AB + 1}$$

$$= \frac{258}{(258)(0.106) + 1} = 9.11$$

Convert to dB:

$$= 20 \times \log 9.11 = 19.2\ dB$$

This agrees closely with the simulation of the circuit in Fig. 7-32 (the gain is shown on the Bode plotter. Bode plots are covered in Chap. 9).

TEST

Determine whether each statement is true or false.

37. Negative feedback always decreases the voltage or current gain of an amplifier.
38. Feedback can be used to raise or lower the input impedance of an amplifier.
39. Negative feedback decreases the frequency range of amplifiers.
40. Negative dc feedback in an amplifier will make it less temperature-sensitive.
41. The second amplifier stage in Fig. 7-31 is operating in the common-collector configuration.

Solve the following problems.

42. Refer to Fig. 7-26. Assume the voltage gain of the amplifier is 50 and the collector feedback resistor is 100 kΩ. What signal loading effect will R_F present to the signal source?
43. Refer to Fig. 7-28. Assume that the current gain of the transistor is 100 and R_E a 220-Ω resistor. What is the base-to-ground impedance r_{in} for the input signal, ignoring R_{B1} and R_{B2}?
44. Refer to Fig. 7-28. What will happen to the input impedance of the amplifier if R_E is bypassed?
45. Refer to Fig. 7-32. Assume the signal at the base of the 2N4401 is negative-going. What will be the signal at its collector?
46. What is the configuration of the 2N4403 in Fig. 7-31?

7-5 Frequency Response

Figure 7-33 shows a common-emitter amplifier. A dc analysis of this circuit determines that the emitter current is 8.38 mA. The ac resistance of the emitter is:

$$r_E = \frac{50\ mV}{8.38\ mA} = 5.96\ \Omega$$

With the emitter-bypass switch open, the voltage gain of the amplifier is:

$$A_V = \frac{1\ k\Omega \parallel 680\ \Omega}{100\ \Omega + 5.96\ \Omega} = 3.82$$

To get a more precise gain estimate, the internal resistance of the signal source must be

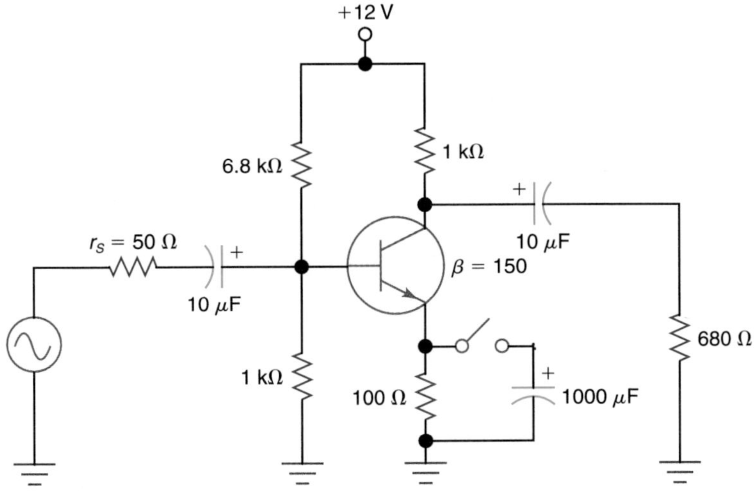

Fig. 7-33 A common-emitter amplifier.

taken into account. The amplifier's input impedance causes a loading effect and some loss of signal voltage across r_S in Fig. 7-33. For this loss to be found, the input impedance of the amplifier must be known. The first step is to find the input resistance of the transistor:

$$r_{\text{in}} = \beta(R_E + r_E)$$
$$= 150\ (100\ \Omega + 5.96\ \Omega) = 15.9\ \text{k}\Omega$$

The input impedance of the amplifier is determined next:

$$Z_{\text{in}} = \frac{1}{\dfrac{1}{R_{B_1}} + \dfrac{1}{R_{B_2}} + \dfrac{1}{r_{\text{in}}}}$$
$$= \frac{1}{\dfrac{1}{6.8\ \text{k}\Omega} + \dfrac{1}{1\ \text{k}\Omega} + \dfrac{1}{15.9\ \text{k}\Omega}} = 826\ \Omega$$

Some fraction of the signal source voltage in Fig. 7-33 will reach the base of the transistor. Using the voltage divider equation gives:

$$V_{\text{fraction}} = \frac{826}{826 + 50} = 0.943$$

So, the unbypassed voltage gain of the common-emitter amplifier in Fig. 7-33 is:

$$A_V = 0.943 \times 3.82 = 3.60$$

And the dB gain is:

$$A_V = 20 \times \log_{10} 3.60 = 11.1\ \text{dB}$$

Figure 7-34 on page 188 shows the gain versus frequency response for the amplifier. The lower curve is for the unbypassed condition. In the *midband* of the amplifier the gain is 11.1 dB. Note that the gain curve starts to drop at frequencies above and below the midband.

The curves of Fig. 7-34 are plotted on a *semilog graph.* The frequency axis is logarithmic, and the vertical axis is linear. If both axes were logarithmic, the graph would be called *log-log.* One reason for using a logarithmic axis is to obtain good resolution over a large range.

Frequency response graphs show the range over which an amplifier is useful. Generally, the *bandwidth* of an amplifier is the range of frequencies where the gain of the amplifier is within 3 dB of its midband gain. An examination of the lower curve of Fig. 7-34 shows that the gain is down 3 dB around frequencies of 20 Hz and 40 MHz. So the amplifier is useful from 20 Hz to 40 MHz, and the bandwidth is just a little less than 40 MHz.

Break frequency

The point at which an amplifier's gain drops 3 dB from its best gain is sometimes called a *break frequency.* The lower break frequency is caused by the capacitors in Fig. 7-33. In the midband of the amplifier, the capacitors can be viewed as ac short circuits. At a break frequency, a capacitor has a reactance *equal* to the equivalent resistance that it is coupling or bypassing. In other words, at the break frequency:

$$X_C = R_{\text{eq}}$$

The capacitive reactance at the break frequency (f_b) is:

$$X_C = \frac{1}{2\pi f_b C}$$

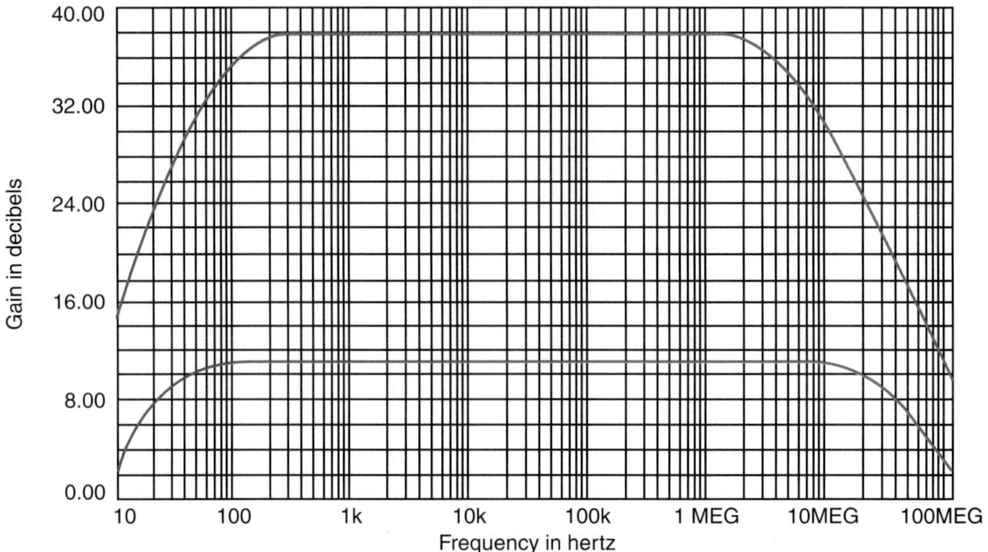

Fig. 7-34 Frequency response of common-emitter amplifier.

Substituting:

$$R_{eq} = \frac{1}{2\pi f_b C}$$

Solving for f_b:

$$f_b = \frac{1}{2\pi R_{eq} C}$$

The preceding equation shows that it is not difficult to calculate a break frequency if the equivalent resistance and capacitance are known.

The 10-μF output coupling capacitor in Fig. 7-33 will cause a 3-dB drop in gain at some frequency. This capacitor couples into 680 Ω. It is fed by the collector circuit of the amplifier, which has an output impedance of 1 kΩ (it is equal to R_L). The equivalent resistance loading the output capacitor is:

$$R_{eq} = 1\ \text{k}\Omega + 680\ \Omega = 1.68\ \text{k}\Omega$$

The break frequency is:

$$f_b = \frac{1}{2\pi R_{eq} C}$$

$$= \frac{1}{6.28 \times 1.68\ \text{k}\Omega \times 10 \times 10^{-6}\ \text{F}}$$

$$= 9.47\ \text{Hz}$$

The other 10-μF capacitor in Fig. 7-33 couples into the amplifier. The input impedance has already been calculated to be 826 Ω. The internal resistance of the signal source is 50 Ω. The equivalent resistance for the input circuit is:

$$R_{eq} = 50\ \Omega + 826\ \Omega = 876\ \Omega$$

The break frequency is:

$$f_b = \frac{1}{6.28 \times 876\ \Omega \times 10 \times 10^{-6}\ \text{F}} = 18.2\ \text{Hz}$$

This is higher than the break frequency for the output circuit. The highest number is the one used to determine the low-frequency response. In this case, the break frequencies are close (9.47 and 18.2 Hz). When this happens, the actual breakpoint of the amplifier will be higher than the highest individual break frequency. The circuit of Fig. 7-33 has a break frequency around 24 Hz.

The upper break frequency of a common-emitter amplifier such as the one shown in Fig. 7-33 is partly determined by capacitances that do not show on the schematic. Transistors have internal junction capacitances which act to bypass high frequencies. This topic is covered in some detail in Chap. 9.

The performance of the common-emitter amplifier in Fig. 7-33 changes quite a bit when the switch is closed. The switch connects an emitter bypass capacitor. This capacitor raises the gain, decreases the input impedance of the amplifier, and decreases the bandwidth of the amplifier. The gain with R_E bypassed is:

$$A_V = \frac{1\ \text{k}\Omega \parallel 680\ \Omega}{5.96\ \Omega} = 67.9$$

The input impedance with R_E bypassed:

$$r_{in} = \beta \times r_E = 150 \times 5.96\ \Omega = 894\ \Omega$$

$$Z_{in} = \frac{1}{1/R_{B_1} + 1/R_{B_2} + 1/r_{in}}$$

$$= \frac{1}{1/6.8\ \text{k}\Omega + 1/1\ \text{k}\Omega + 1/894\ \Omega} = 441\ \Omega$$

As before, using the voltage divider equation:

$$V_{\text{fraction}} = \frac{441}{441 + 50} = 0.898$$

So, the bypassed voltage gain of the CE amplifier in Fig. 7-33 is:

$$A_V = 0.898 \times 67.9 = 61.0$$

And the dB gain is:

$$A_V = 20 \times \log_{10} 61.0 = 35.7\ \text{dB}$$

The higher curve in Fig. 7-34 shows the midband gain to be 38 dB. We have used the conservative value of 50 mV to find the ac emitter resistance. The actual gain of the amplifier tends to be somewhat higher.

The breakpoint for the output coupling capacitor remains the same (9.47 Hz). The breakpoint changes for the input coupling capacitor because the input impedance of the amplifier is lower:

$$R_{\text{eq}} = 50\ \Omega + 441\ \Omega = 491\ \Omega$$

The break frequency is now:

$$f_b = \frac{1}{6.28 \times 491\ \Omega \times 10 \times 10^{-6}\ \text{F}} = 32.4\ \text{Hz}$$

Another lower break frequency is caused by the emitter bypass capacitor. We must find the equivalent resistance bypassed by this capacitor. This resistance is partly determined by the base circuit, as viewed from the emitter terminal. This is sort of a backward view through the amplifier, and we find that resistors appear smaller by a factor equal to β:

$$r_{\text{EB}} = \frac{r_S \parallel R_{\text{B}_1} \parallel R_{\text{B}_2}}{\beta}$$
$$= \frac{50\ \Omega \parallel 6.8\ \text{k}\Omega \parallel 1\ \text{k}\Omega}{150}$$
$$= 0.315\ \Omega$$

This resistance is in series with r_{E} and that combination is in parallel with R_{E}:

$$R_{\text{eq}} = (r_{\text{EB}} + r_{\text{E}}) \parallel R_{\text{E}}$$
$$= (0.315\ \Omega + 5.96\ \Omega) \parallel 100\ \Omega$$
$$= 5.90\ \Omega$$

$$f_b = \frac{1}{6.28 \times 5.9\ \Omega \times 1000 \times 10^{-6}\ \text{F}}$$
$$= 27.0\ \text{Hz}$$

As Fig. 7-34 shows, the lower break frequency is about 80 Hz. Because the three break frequencies are so close to each other (9.47, 32.4, and 27.0 Hz), the actual breakpoint is greater than the highest of the three.

Please notice that the bandwidth suffers in Fig. 7-34 when the gain is increased by the addition of the emitter bypass capacitor. As discussed in the preceding section, the negative feedback provided by using emitter feedback makes the amplifier useful over a wider range of frequencies.

The upper break frequency in Fig. 7-34 drops from 40 to 4 MHz when the capacitor is added. This is because the increased gain multiplies the effect of the internal capacitance of the transistor. When designers need very-wide-band amplifiers, they often keep the gain in any one stage at a moderate value and add an additional stage or two to obtain the overall gain required.

Cascode amplifier

Figure 7-35 on page 190 shows a *cascode amplifier*. This is a special cascade arrangement in which a common-emitter stage is direct-coupled to a common-base stage. Cascode amplifiers extend the frequency limitations of common-emitter amplifiers by decreasing the effect of one of the internal capacitances in transistors. This allows extended bandwidth. Cascode amplifiers are used in radio-frequency (RF) applications.

Common-base amplifiers have extended bandwidth, but they also have a very low input impedance. The cascode arrangement is suited to situations in which the extended frequency performance of a common-base amplifier is desired along with the higher input impedance of a common-emitter amplifier. In Fig. 7-35, Q_1 is the common-emitter input amplifier, and Q_2 is the common-base output amplifier. The cascode arrangement provides high-gain (almost 33 dB), wide bandwidth (almost 300 MHz), and an input impedance in the kilohm range.

TEST

Solve the following problems.

47. Suppose the amplifier in Fig. 7-33 has an input impedance of 600 Ω and a midband voltage gain of 10. Find the midband output voltage from the amplifier if the sig-

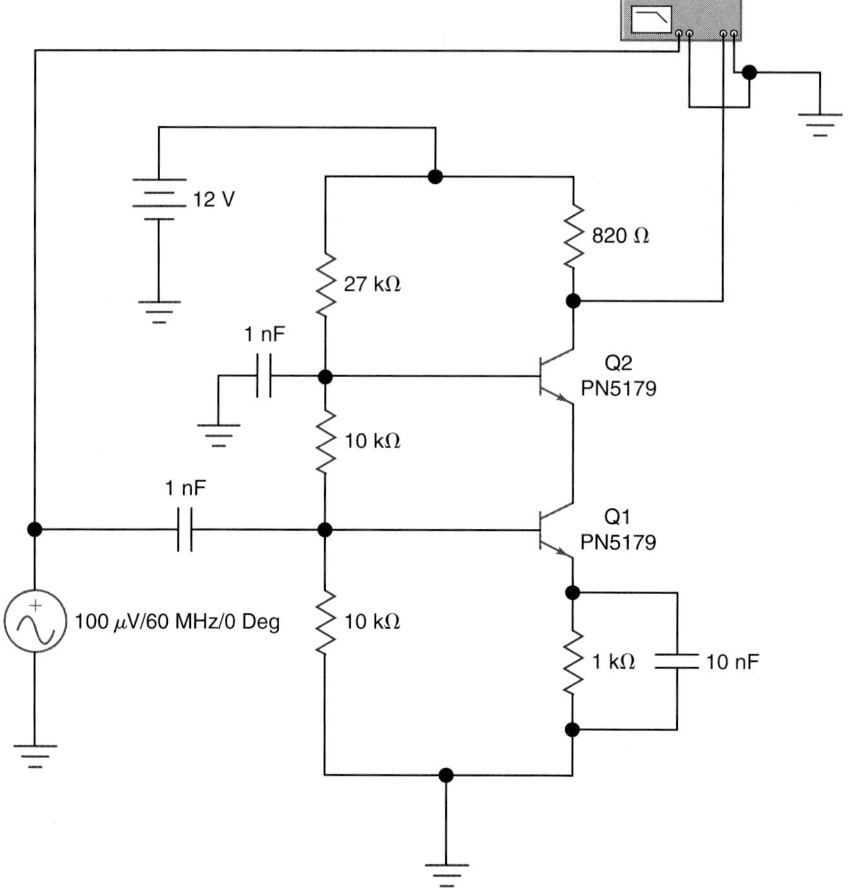

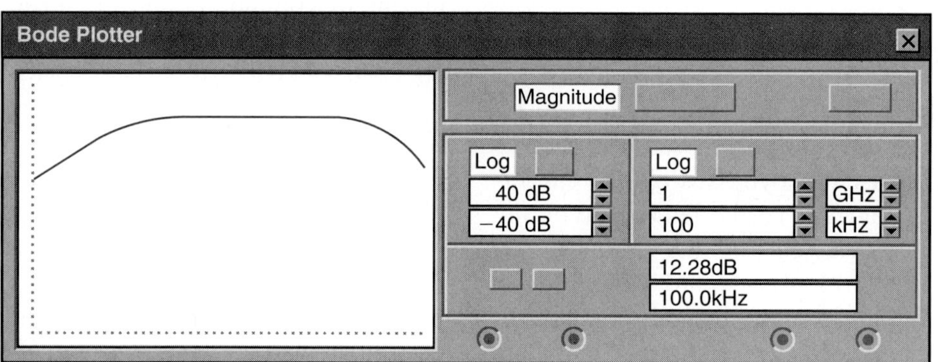

Fig. 7-35 A cascode amplifier.

nal source has an impedance of 600 Ω and develops 100 mV_{p-p}.

48. Find the lower break frequency for the input circuit in question 47 if the input coupling capacitor is 0.1 μF.
49. What is the midband dB gain of the amplifier described in question 47?
50. What is the gain of the amplifier in question 47 at its break frequency?
51. Select an input coupling capacitor for the amplifier described in question 47 that will change its break frequency to 10 Hz.
52. An amplifier has three capacitors. Calculations reveal break frequencies of 10, 15, and 150 Hz. What is the lower break frequency for the entire amplifier?
53. An amplifier has three capacitors. Calculations reveal break frequencies of 135, 140, and 150 Hz. What is the lower break frequency for the entire amplifier?
54. Determine the dc conditions for Fig. 7-35.

Chapter 7 Review

Summary

1. There are three basic types of amplifier coupling: capacitive, direct, and transformer.
2. Capacitor coupling is useful in ac amplifiers because a capacitor will block direct current and allow the ac signal to be coupled.
3. Electrolytic coupling capacitors must be installed with the correct polarity.
4. Direct coupling provides dc gain. It can be used only when the dc terminal voltages are compatible.
5. A Darlington amplifier uses direct coupling. Darlington transistors have high current gain.
6. Transformer coupling gives the advantage of impedance matching.
7. The impedance ratio of a transformer is the square of its turns ratio.
8. Radio-frequency transformers can be tuned to give selectivity. Those frequencies near the resonant frequency will receive the most gain.
9. Loading a signal source will reduce its voltage output. This effect is often most noticeable with high-impedance signal sources.
10. In multistage amplifiers, each stage is loaded by the input impedance of the next stage.
11. When an amplifier is loaded, its voltage gain decreases.
12. Emitter bypassing in a common-emitter amplifier increases voltage gain but lowers the amplifier's input impedance.
13. Loaded amplifiers should be biased at the center of the signal load line for best linear output swing.
14. The maximum output swing from a loaded amplifier is less than the supply voltage. Its value can be found by drawing a signal load line or analyzing the output circuit.
15. It is necessary to reverse-bias the gate-source junction for linear operation with junction field-effect transistors.
16. Because there is no gate current, the input impedance of FET amplifiers is very high.
17. The voltage gain in an FET amplifier is approximately equal to the product of the load resistance and the forward transfer admittance of the transistor.
18. More load resistance means more voltage gain.
19. Fixed bias, in FET amplifiers, is not desirable because the characteristics of the transistor vary quite a bit from unit to unit.
20. Source bias tends to stabilize an FET amplifier and make it more immune to the characteristics of the transistor.
21. Combination bias uses fixed and source bias to make the circuit even more stable and predictable.
22. A common-source FET amplifier using source bias must use a source bypass capacitor to realize maximum voltage gain.
23. The dual-gate MOSFET amplifier is capable of a tremendous range of gain by applying a control voltage to the second gate.
24. When the feedback tends to cancel the effect of the input to an amplifier, that feedback is negative. Another way to identify negative feedback is that it is out of phase with the input.
25. Direct current negative feedback can be used to stabilize the operating point of an amplifier.
26. Negative signal feedback may be used to decrease the current gain or the voltage gain of an amplifier.
27. Negative signal feedback may be used to decrease or increase the input impedance of an amplifier.
28. Negative signal feedback increases the bandwidth of an amplifier. It also reduces noise and distortion.

Chapter Review Questions

Determine whether each statement is true or false.

7-1. When electrolytic capacitors are used as coupling capacitors, they may be installed without checking polarity.

7-2. Capacitors couple alternating current and block direct current.

7-3. If an early stage in a direct-coupled amplifier drifts with temperature, the drift is amplified by following stages.

7-4. A Darlington transistor has three leads, but contains two BJTs.

7-5. A transformer can match a high-impedance collector circuit to a low-impedance load.

7-6. Refer to Fig. 7-6. This amplifier will provide a little gain at 0 Hz.

7-7. Amplifier input impedance can be ignored when the signal source is ideal (0 internal impedance).

7-8. Refer to Fig. 7-9. It can be determined that the input impedance of the first amplifier cannot be greater than 8.2 kΩ by inspection.

7-9. Refer to Fig. 7-10. The Q point is also called the operating point.

7-10. Refer to Fig. 7-10. The amplifier can develop a maximum output swing of 12 V peak-to-peak.

7-11. Refer to Fig. 7-14(*b*). If $V_{GS} = -2.5$ V, the positive-going portion of the output signal will be severely clipped.

7-12. Refer to Fig. 7-16. Increasing the value of R_L should increase the voltage gain of the amplifier.

7-13. Refer to Fig. 7-18. The effect of capacitor C_S is to increase the voltage gain.

7-14. Refer to Fig. 7-20. Transistor Q_1 is in the source follower configuration.

7-15. Refer to Fig. 7-21. The bias V_{GS} is -1.5 V.

7-16. Refer to Figs. 7-22 and 7-23. To decrease the gain of the amplifier, G_2 must be made more negative with respect to the source.

7-17. Refer to Fig. 7-24. The input terminal and the output terminal should be 180° out of phase.

7-18. Negative feedback tends to cancel the input signal.

7-19. Negative feedback increases the voltage gain of the amplifier but at the expense of reduced bandwidth.

7-20. Negative feedback improves amplifier linearity.

7-21. Negative dc feedback can be used to stabilize the amplifier operating point.

7-22. Refer to Fig. 7-33. There is more amplifier distortion when the switch is closed.

7-23. Refer to Fig. 7-33. There is wider frequency response when the switch is open.

Chapter Review Problems

7-1. Refer to Fig. 7-4. Assume that the 220-kΩ resistor is changed to 470 kΩ. Find the base voltage for Q_1.

7-2. For the data of problem 7-1, find the emitter voltage of Q_1.

7-3. For the data of problem 7-1, find the emitter voltage of Q_2.

7-4. For the data of problem 7-1, find the emitter current for Q_2.

7-5. For the data of problem 7-1, find V_{CE} for Q_2.

7-6. For the data of problem 7-1, find Z_{in} for Q_1. (*Hint:* Because of the high current gain and the 1-kΩ emitter resistor, r_{in} for Q_1 is so high in circuits of this type that it can be ignored.)

7-7. Refer to Fig. 7-3. Each transistor has a β of 80. What is the overall current gain of the pair?

7-8. Refer to Fig. 7-5. The secondary has 5 turns, and the primary has 200 turns. What is the turns ratio of the transformer?

7-9. Use the data from problem 7-8 and find the peak-to-peak signal across the load if the collector signal is 40 V peak-to-peak.

7-10. Use the data from problem 7-8 and find the collector load if the load resistor at the transformer secondary is 4 Ω.

7-11. Refer to Fig. 7-6. The inductance of the transformer primaries is 100 μH. The capacitors across the primaries are both 680 pF. At what frequency will the gain of the amplifier be the greatest?

7-12. Refer to Fig. 7-14. If $R_L = 1000$ Ω, where will the load line terminate on the vertical axis?

7-13. Refer to Fig. 7-14. If $V_{DD} = 12$ V, where will the load line terminate on the horizontal axis?

7-14. An FET drain swings 2 mA with a gate swing

of 1 V. What is the forward transfer admittance for this FET?

7-15. An FET has a forward transfer admittance of 4×10^{-3} S. It is to be used in the common-source configuration with a load resistor of 4700 Ω. What voltage gain can be expected?

7-16. Refer to Fig. 7-16. Assume a source current of 10 mA and a source resistor of 100 Ω. What is the value of V_{GS}?

7-17. Refer to Fig. 7-18. It is desired that $V_{GS} = -2.0$ V at $I_D = 8$ mA. What should be the value of the source resistor?

7-18. Refer to Fig. 7-28. V_{CC} is 15 V, R_L is 1.2 kΩ, R_{B1} is 22 kΩ, R_{B2} is 4.7 kΩ, R_E is 470 Ω, and β is 150. Find Z_{in}.

7-19. Using the data from problem 7-18, find A_V.

7-20. Using the data from problem 7-18, find the amplifier ouput signal assuming that the signal source develops an open-circuit output of 1 V peak-to-peak and has a characteristic impedance of 10 kΩ.

Critical Thinking Questions

7-1. List some advantages and disadvantages for a direct-coupled audio amplifier.

7-2. A transformer can match impedances for best power transfer. Are there any mechanical analogies for this fact?

7-3. Can you think of any other methods of signal coupling that are different than the ones discussed in this chapter?

7-4. The gain of an amplifier tends to drop when it is loaded. Are there any analogies in the mechanical world?

7-5. Based on what you have learned about negative feedback, what do you think positive feedback would do to an amplifier?

7-6. Why is it not possible for any amplifier to have infinite bandwidth?

Answers to Tests

1. T
2. T
3. F
4. F
5. T
6. T
7. 39.8 μF
8. $I_E = 5.3$ mA
   $V_{RL} = 6.36$ V
   $V_{RE} + V_{RL} > V_{CC}$
   Amplifier is in saturation
9. 5000
10. 3.55 mA
11. 8.45 V
12. 1.96 kΩ
13. F
14. F
15. T
16. T
17. T
18. T
19. T
20. 3.92 mV
21. 0.264 V peak-to-peak
22. 880 Ω
23. 81
24. F
25. T
26. F
27. F
28. T
29. F
30. 5 V
31. saturation
32. cutoff
33. −1.5 V
34. −2.48 V; negative
35. common drain (source follower) common collector (emitter follower)
36. positive
37. T
38. T
39. F
40. T
41. F
42. 2 kΩ
43. 22 kΩ
44. It will decrease.
45. positive-going
46. common emitter
47. 500 mV_{p-p}
48. 1.33 kHz
49. 20 dB
50. 17 dB (7.07)
51. 13.3 μF
52. 150 Hz
53. greater than 150 Hz
54. For Q_1: $V_B = 2.55$, $V_E = 1.85$, $I_E = 1.85$ mA, $V_C = 4.41$, $V_{CE} = 2.56$; for Q_2: $V_C = 10.5$, $V_E = 4.41$, $V_{CE} = 6.09$

Chapter 8

Large-Signal Amplifiers

Chapter Objectives

This chapter will help you to:

1. *Calculate* amplifier efficiency.
2. *Identify* the class of amplifier operation.
3. *Recognize* crossover distortion in push-pull amplifiers.
4. *Explain* the operation of complementary symmetry amplifiers.
5. *Describe* tank circuit action in class C amplifiers.

This chapter introduces the idea of efficiency in an amplifier. An efficient amplifier delivers a large part of the power it receives from the supply as a useful output signal. Efficiency is most important when large amounts of signal power are required. It will be shown that amplifier efficiency is related to how the amplifier is biased. It is possible to make large improvements in efficiency by moving the operating point away from the center of the load line.

8-1 Amplifier Class

All amplifiers are power amplifiers. However, those operating in the early stages of the signal processing system deal with small signals. These early stages are designed to give good voltage gain. Since voltage gain is the most important function of these amplifiers, they are called voltage amplifiers. Figure 8-1 is a block diagram of a simple audio amplifier. The tape head produces a very small signal, in the millivolt range. The first stage amplifies this audio signal, and it becomes larger. The last stage produces a much larger signal. It is called a power amplifier.

A power amplifier is designed for good power gain. It must handle large voltage and current swings. These high voltages and currents make the power high. It is very important to have good *efficiency* in a power amplifier. An efficient power amplifier delivers the most signal power for the dc power it takes from the supply. Look at Fig. 8-2 on page 196. Note that the job of the power amplifier is to change dc power into signal power. Its efficiency is given by:

Efficiency

$$\text{Efficiency} = \frac{\text{signal power output}}{\text{dc power input}} \times 100\%$$

The power amplifier of Fig. 8-2 produces 8 W of signal power output. Its power supply develops 16 V and the amplifier draws 1 A. The dc power input to the amplifier is

$$P = V \times I = 16 \text{ V} \times 1 \text{ A} = 16 \text{ W}$$

The efficiency of the amplifier is

$$\text{Efficiency} = \frac{8 \text{ W}}{16 \text{ W}} \times 100\% = 50\%$$

Efficiency is very important in high-power systems. For example, assume that 100 W of audio power is required in a music amplifier. Also assume that the power amplifier is only 10 percent efficient. What kind of a power supply would be required? The power supply would have to deliver 1000 W to the amplifier! A 1-kilowatt (kW) power supply is a large, heavy, and expensive item. Heat would be another problem in this music amplifier. Of the 1-kW

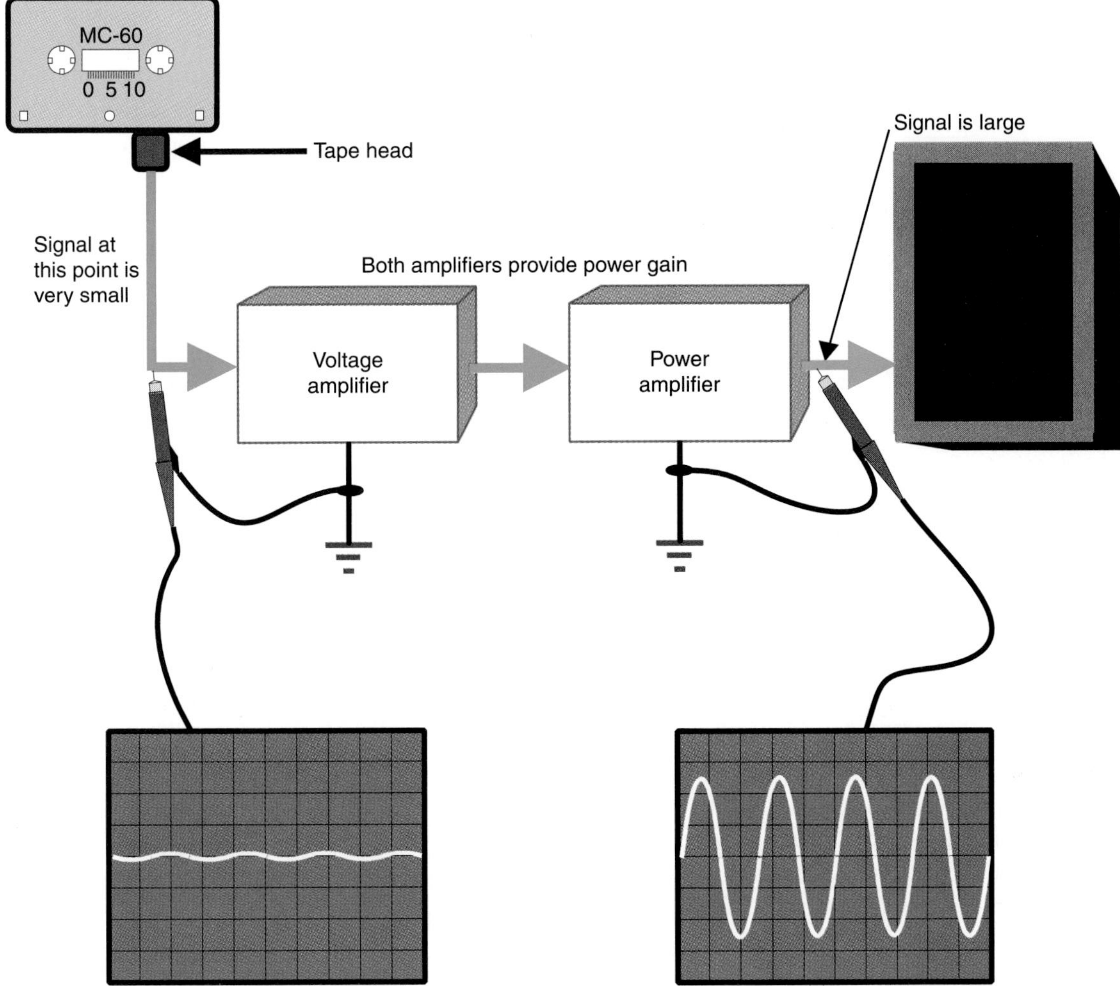

Fig. 8-1 A typical amplifier.

input, 900 W would become heat. This system would probably need a cooling fan.

Example 8-1

What is the efficiency of an amplifier that draws 2 A from a 40-V supply when delivering 52 W of signal power? How many watts are dissipated in this amplifier? Find the dc input power first:

$$P_{in} = V \times I = 40 \text{ V} \times 2 \text{ A} = 80 \text{ W}$$

Find the efficiency:

$$\text{Efficiency} = \frac{P_{out}}{P_{in}} \times 100\% = \frac{52 \text{ W}}{80 \text{ W}} \times 100\% = 65\%$$

The dissipation (heat) is the difference between input and output:

$$\text{Dissipation} = P_{in} - P_{out} = 80 \text{ W} - 52 \text{ W} = 28 \text{ W}$$

Example 8-2

A stereo automotive audio amplifier is rated at 70 W of output per channel at an efficiency of 60 percent. How much current will this amplifier require when it is delivering rated output? Automotive electrical systems operate on 12 V. The first step is to find the input power, and then the current can be determined. Rearranging the efficiency formula gives:

$$P_{in} = \frac{P_{out}}{\text{Efficiency}} = \frac{140 \text{ W}}{0.6} = 233 \text{ W}$$

Rearranging the power formula gives:

$$I = \frac{P}{V} = \frac{233 \text{ W}}{12 \text{ V}} = 19.4 \text{ A}$$

The amplifier circuits covered in previous chapters have been *class A*. *Class A* amplifiers **Class A**

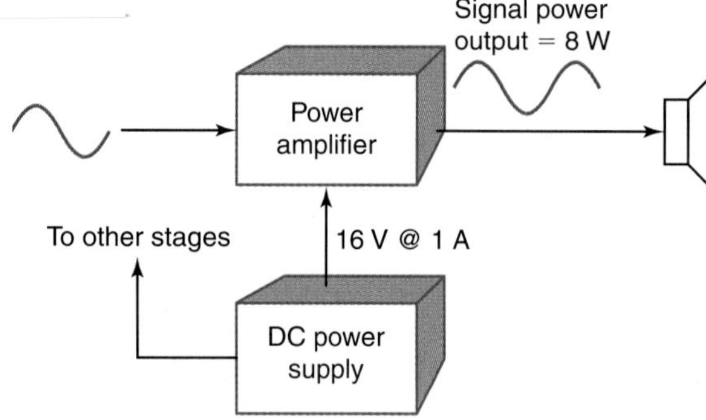

Fig. 8-2 Comparing the output signal to the dc power input.

Conduction angle

operate at the center of the load line. Refer to Fig. 8-3. The operating point is class A. This gives the maximum possible output swing without clipping. The output signal is a good replica of the input signal. This means that distortion is low. This is the greatest advantage of class A operation.

Figure 8-4 shows another class of operation. The operating point is at *cutoff* on the load line. This is done by running the base-emitter junction of the transistor with *zero bias.* Zero bias means that only half the input signal will be amplified. Only that half of the signal which can turn on the base-emitter diode will produce any output signal. The transistor conducts for *half* of the input cycle. A *class B* amplifier is said to have a *conduction angle of 180°*. Class A amplifiers conduct for the entire input cycle. They have a conduction angle of 360°.

Class B

What is to be gained by operating in class B? Obviously, we have a distortion problem that was not present in class A. In spite of the distortion, class B is useful because it gives better efficiency. Biasing an amplifier at cutoff saves power.

Read your employee handbook and all other materials supplied by your employer.

Class A wastes power. This is especially true at very low signal levels. The class A operating point is in the center of the load line. This means that about half the supply voltage is dropped across the transistor. The transistor is conducting half the saturation current. This voltage drop and current produce a power loss in the transistor. This power loss is *constant* in class A. There is a drain on the power supply even if no signal is being amplified.

The class B amplifier operates at *cutoff.* The transistor current is zero. *Zero* current means 0 W. There is no drain on the supply *until* a signal is being amplified. The larger the amplitude of the signal, the larger the drain on the supply. The class B amplifier eliminates the fixed drain from the power supply and is therefore more efficient.

The better efficiency of class B is very important in high-power applications. Much of the distortion can be eliminated by using two transistors: each to amplify one-half of the signal. Such circuits are a bit more complicated, but the improved efficiency is worth the effort.

There are also *class AB* and *class C* amplifiers. Again, it is a question of *bias. Bias controls the operating point, the conduction angle, and the class of operation.* Table 8-1 summarizes the important features of the amplifier classes. Study this table now and refer to it after completing later sections in this chapter.

It is easy to become confused when studying amplifiers for the first time. There are so many categories and descriptive terms. Table 8-2 has been prepared to help you organize your thinking.

TEST

Determine whether each statement is true or false.

1. A voltage amplifier or small-signal amplifier gives no power gain.

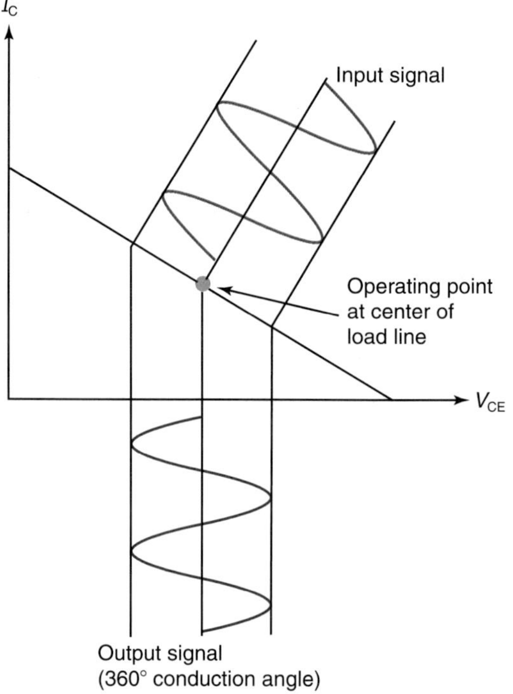

Fig. 8-3 Class A operating point.

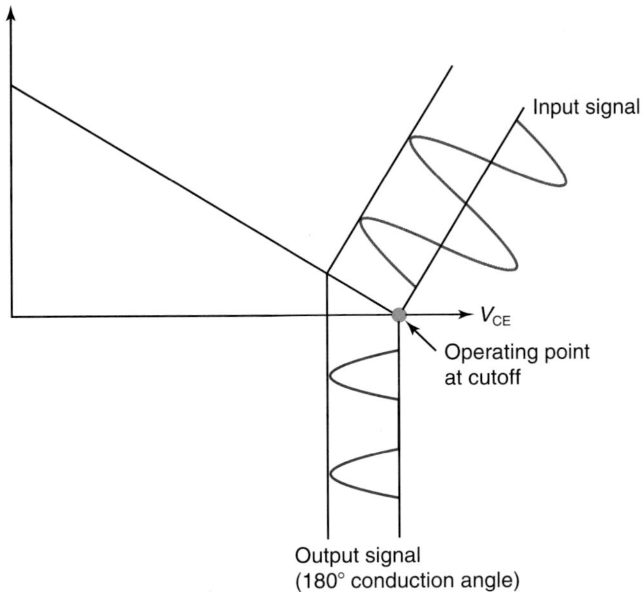

Fig. 8-4 Class B operating point.

2. The efficiency of a class A amplifier is less than that of a class B amplifier.
3. The conduction angle for a class A power amplifier is 180°.
4. Refer to Fig. 8-3. With no input signal, the power taken from the supply will be 0 W.
5. Refer to Fig. 8-4. With no input signal, the power taken from the supply will be 0 W.
6. Bias controls an amplifier's operating point, conduction angle, class of operation, and efficiency.

Solve the following problems.

7. Refer to Fig. 8-2. Suppose the power supply is rated at 20 V. What is the efficiency of the power amplifier?
8. A certain amplifier is producing an output power of 100 W. Its efficiency is 60 percent. How much power will the amplifier take from the supply?

8-2 Class A Power Amplifiers

The class A power amplifier operates near the center of the load line. It is not highly efficient,

Table 8-1 Summary of Amplifier Classes

	Class A	Class AB	Class B	Class C
Efficiency	50%*	Between classes A and B	78.5%*	100%*
Conduction angle	360°	Between classes A and B	180°	small (approx. 90°)
Distortion	Low	Moderate	High	Extreme
Bias (emitter-base)	Forward (center of load line)	Forward (near cutoff)	Zero (at cutoff)	Reverse (beyond cutoff)
Applications	Practically all small-signal amplifiers. A few moderate power amplifiers in audio applications.	High-power stages in both audio and radio-frequency applications.	High-power stages—generally not used in audio applications due to distortion.	Generally limited to radio-frequency applications. Tuned circuits remove much of the extreme distortion.

*Theoretical maximums. Cannot be achieved in practice.

TABLE 8-2 Amplifier Characteristics	
	Explanations and Examples
Voltage amplifiers	Voltage amplifiers are small-signal amplifiers. They can be found in early stages in the signal system. They are often designed for good voltage gain. An audio preamplifier would be a good example of a voltage amplifier.
Power amplifiers	Power amplifiers are large-signal amplifiers. They can be found late in the signal system. They are designed to give power gain and reasonable efficiency. The output stage of an audio amplifier would be a good example of a power amplifier.
Configuration	The configuration of an amplifier tells how the signal is fed to and taken from the amplifying device. For bipolar transistors, the configurations are common-emitter, common-collector, and common-base. For field-effect transistors, the configurations are common-source, common-drain, and common-gate.
Coupling	How the signal is transferred from stage to stage. Coupling can be capacitive, direct, or transformer.
Applications	Amplifiers may be categorized according to their use. Examples are audio amplifiers, video amplifiers, RF amplifiers, dc amplifiers, band-pass amplifiers, and wide-band amplifiers.
Classes	This category refers to how the amplifying device is biased. Amplifiers can be biased for class A, B, AB, or C operation. Voltage amplifiers are usually biased for class A operation. For improved efficiency, power amplifiers may use class B, AB, or C operation.

but it does offer low distortion. It is also the most simple design.

Figure 8-5 shows a class A power amplifier. We will use a load line to see how much signal power can be produced. The load line will be set by the supply voltage V_{CC} and the saturation current:

$$I_{sat} = \frac{V_{CC}}{R_{load}} = \frac{16\text{ V}}{80\ \Omega} = 0.2\text{ A or } 200\text{ mA}$$

The load line will run from 16 V on the horizontal axis to 200 mA on the vertical axis.

Next, we must find the operating point for the amplifier. Solving for the base current, we get

$$I_B = \frac{V_{CC}}{R_B} = \frac{16\text{ V}}{16\text{ k}\Omega} = 1\text{ mA}$$

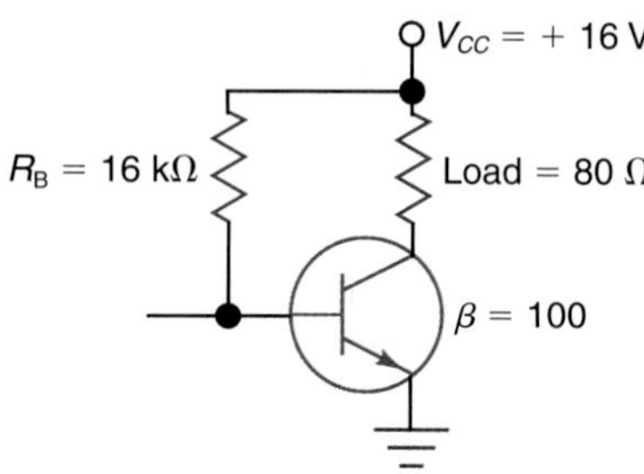

Fig. 8-5 Class A power amplifier.

The transistor has a β of 100. The collector current will be

$$\begin{aligned} I_C &= \beta \times I_B \\ &= 100 \times 1\text{ mA} = 100\text{ mA} \end{aligned}$$

The load line can be seen in Fig. 8-6 with the 100-mA operating point.

The amplifier can be driven to the load-line limits before clipping occurs. The maximum voltage swing will be 16 V peak-to-peak. The maximum current swing will be 200 mA peak-to-peak. Both of these maximums are shown in Fig. 8-6.

We now have enough information to calculate the *signal power.* The peak-to-peak values must be converted to rms values. This is done by:

$$\begin{aligned} V_{rms} &= \frac{V_{p\text{-}p}}{2} \times 0.707 \\ &= \frac{16\text{ V}}{2} \times 0.707 = 5.66\text{ V} \end{aligned}$$

Next, the rms current is

$$\begin{aligned} I_{rms} &= \frac{I_{p\text{-}p}}{2} \times 0.707 \\ &= \frac{200\text{ mA}}{2} \times 0.707 = 70.7\text{ mA} \end{aligned}$$

Finally, the signal power will be given by:

$$\begin{aligned} P &= V \times I \\ &= 5.66\text{ V} \times 70.7\text{ mA} = 0.4\text{ W} \end{aligned}$$

The *maximum power* (sine wave) is 0.4 W.

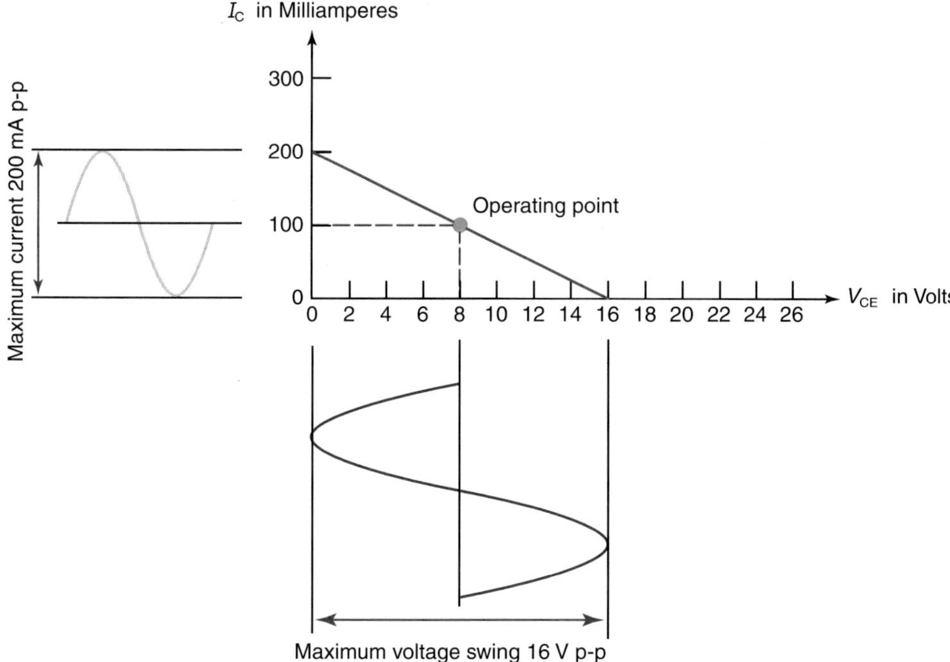

Fig. 8-6 Class A load line.

How much dc power is involved in producing this signal power? The answer is found by looking at the power supply. The supply develops 16 V. The current taken from the supply must also be known. The base current is small enough to ignore. The *average* collector current is 100 mA. Therefore, the *average* power is

$$P = V \times I$$
$$= 16 \text{ V} \times 100 \text{ mA} = 1.6 \text{ W}$$

The amplifier takes 1.6 W from the power supply to produce a signal power of 0.4 W. The efficiency of the amplifier is

$$\text{Efficiency} = \frac{P_{ac}}{P_{dc}} \times 100\%$$
$$= \frac{0.4 \text{ W}}{1.6 \text{ W}} \times 100\% = 25\%$$

The class A amplifier shows a maximum efficiency of *25 percent.* This occurs *only* when the amplifier is driven to its maximum output. The efficiency is *less* when the amplifier is not driven hard. The 1.6 W in the above equation is *fixed.* As the drive decreases, the efficiency drops. With *no drive,* the efficiency drops to *zero.* An amplifier of this type would be a poor choice for high-power applications. The power supply would have to produce *4 times* the required signal power. Three-fourths of this power would be wasted as *heat* in the load and the transistor. The transistor would probably require a large heat sink.

Example 8-3

The amplifier shown in Fig. 8-5 is producing a peak-to-peak sine wave output of 8 V. Determine its efficiency. First, find the rms signal voltage:

$$V_{rms} = \frac{V_{p-p}}{2} \times 0.707 = \frac{8}{2} \times 0.707 = 2.83 \text{ V}$$

We could inspect Fig. 8-6 to determine the peak-to-peak signal current and convert it to rms to find the signal power. However, it is easier to use the power formula to calculate the output power directly:

$$P = \frac{V^2}{R} = \frac{2.83^2}{80} = 0.1 \text{ W}$$

The dc input power doesn't change in a class A amplifier, so the efficiency is:

$$\text{Efficiency} = \frac{P_{ac}}{P_{dc}} \times 100\%$$
$$= \frac{0.1 \text{ W}}{1.6 \text{ W}} \times 100\% = 6.25\%$$

Example 8-4

What is the dissipation in the amplifier of example 8-3? What is its dissipation with no input signal? Dissipation is the difference between input and output:

$$\text{Dissipation} = P_{in} - P_{out} = 1.6\text{ W} - 0.1\text{ W} = 1.5\text{ W}$$

With no input signal, there is no useful output:

$$\text{Dissipation} = P_{in} - P_{out} = 1.6\text{ W} - 0\text{ W} = 1.6\text{ W}$$

One reason why the class A amplifier is so wasteful is that dc power is dissipated in the load. A big improvement is possible by removing the load from the dc circuit. Figure 8-7 shows how to do this. The *transformer* will couple the signal power to the load. Now, there is no direct current flow in the load. *Transformer coupling* allows the amplifier to produce *twice* as much signal power.

Transformer coupling

Figure 8-7 shows the same supply voltage, the same bias resistor, the same transistor, and the same load as in Fig. 8-5. The only difference is the coupling transformer. The dc conditions are now quite different. The transformer primary will have very low resistance. This means that all the supply voltage will drop across the transistor at the operating point.

The dc load line for the transformer-coupled amplifier is shown in Fig. 8-8. It is *vertical.* The operating point is still at 100 mA. This is because the base current and β have not changed. The change is the absence of the 80-Ω dc resistance in series with the collector. All the supply voltage now drops across the transistor.

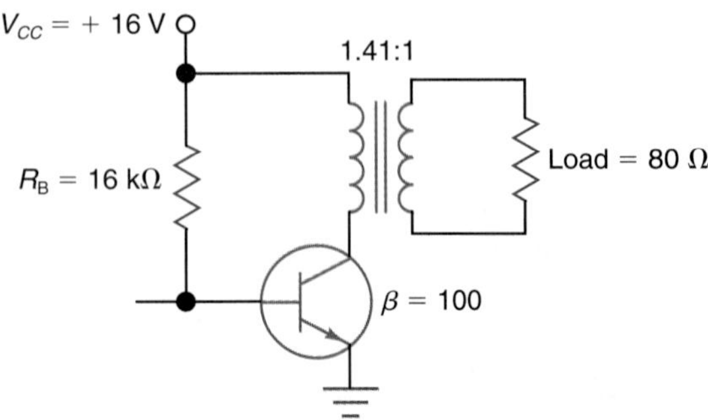

Fig. 8-7 Class A power amplifier with transformer coupling.

Actually, the load line will not be perfectly vertical. The transformer and even the power supply always have a little resistance. However, the dc load line is very steep. We cannot show any output swing from this load line.

There is a second load line in a transformer-coupled amplifier. It is the result of the ac load in the collector circuit and is called the ac load line. The ac load is not 80 Ω in the collector circuit of Fig. 8-7. The transformer is a step-down type. Remember, transformer impedance ratio is equal to the *square of the turns ratio.* Therefore, the ac load in the collector circuit will be:

$$\text{Load}_{ac} = (1.41)^2 \times 80\ \Omega = 160\ \Omega$$

Notice that the ac load line in Fig. 8-8 runs from 32 V to 200 mA. This satisfies an impedance of

$$Z = \frac{V}{I} = \frac{32\text{ V}}{200\text{ mA}} = 160\ \Omega$$

Also notice that the ac load line passes through the operating point. The dc load line and the ac load line must always pass through the same operating point.

How does the ac load line extend to 32 V? This is *twice* the supply voltage! There are two ways to explain this. First, it *must* extend to 32 V if it is to pass through the operating point and satisfy a slope of 160 Ω. Second, a transformer is a type of inductor. When the field collapses, a voltage is generated. This voltage adds in series with the supply voltage. Thus, V_{CE} can swing to *twice* the supply voltage in a transformer-coupled amplifier.

Compare Fig. 8-8 with Fig. 8-6. The output swing doubles with transformer coupling. It is safe to assume the output power also doubles. The *dc power input to the amplifier has not changed.* The supply voltage is still 16 V, and the average current is still 100 mA. Transformer coupling the class A amplifier provides twice as much signal power for the same dc power input. The maximum efficiency of the transformer-coupled class A amplifier is

$$\text{Efficiency} = \frac{P_{ac}}{P_{dc}} \times 100\% = \frac{0.8\text{ W}}{1.6\text{ W}} \times 100\% = 50\%$$

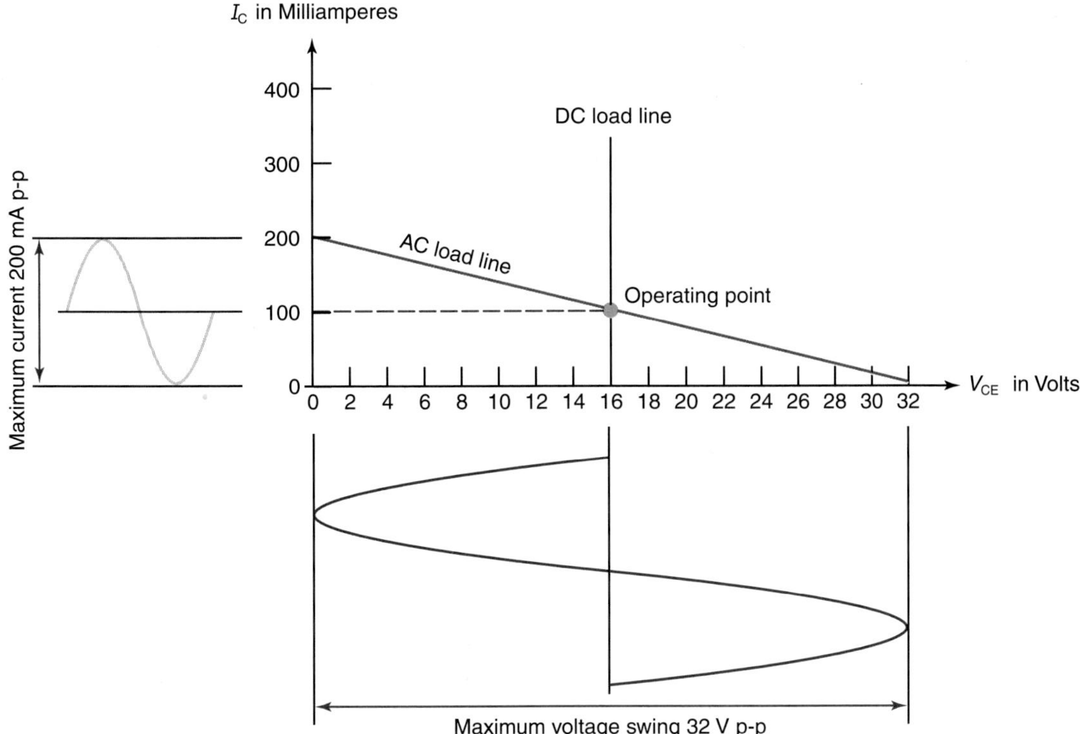

Fig. 8-8 Load lines for the transformer-coupled amplifier.

Remember, however, that this efficiency is reached only at maximum signal level. The efficiency is less for smaller signals and drops to zero when the amplifier is not driven with a signal.

An efficiency of 50 percent may be acceptable in some applications. Class A power amplifiers are sometimes used in medium-power applications (up to 5 W or so). However, the transformer can be an expensive component. For example, in a high-quality audio amplifier, the output transformer can cost more than all the other amplifier parts combined! So for high-power and high-quality amplifiers, something other than class A is usually a better choice.

Our efficiency calculations have ignored some losses. First, we have ignored the *saturation voltage* of the transistor. In practice, V_{CE} cannot drop to 0 V. A power transistor might show a saturation of 0.7 V. This would have to be subtracted from the output swing. Second, we ignored transformer loss in the transformer-coupled amplifier. Transformers are not 100 percent efficient. Small, inexpensive transformers may be only 75 percent efficient at low audio frequencies. The calculated efficien-

Did You Know?

Audio Equipment Facts. Audio listening tests are highly subjective. Graphic equalizers compensate for room acoustics and speakers which do not project well. The physical placement of speakers is important, but subwoofers can go anywhere in a room.

cies of 25 and 50 percent are *theoretical maximums.* They are not realized in actual circuits.

Example 8-5

What is the best efficiency for the amplifier shown in Fig. 8-7 if the efficiency of the transformer is 80 percent? The overall efficiency of a system is the product of the individual efficiencies:

$$\text{Efficiency} = 0.5 \times 0.8 \times 100\% = 40\%$$

Another problem with the class A circuit is the fixed drain on the power supply. Even when no signal is being amplified, the drain on the supply in our example was fixed at 1.6 W. Most power amplifiers must handle signals that change in level. An audio amplifier, for example, will handle a broad range of volume levels. When the volume is low, the efficiency of class A is quite poor.

TEST

Solve the following problems.

9. Refer to Fig. 8-5. The current gain of the transistor is 120. Calculate the power dissipated in the transistor with no input signal.
10. Refer to Fig. 8-6. The operating point is at $V_{CE} = 12$ V. Calculate the power dissipated in the transistor with no input signal.
11. Refer to Fig. 8-7. The transformer has a turns ratio from primary to secondary of 3:1. What load does the collector of the transistor see?
12. Refer to Fig. 8-7. The transformer has a turns ratio of 4:1. An oscilloscope shows a collector sinusoidal signal of 30 V peak-to-peak. What will the amplitude of the signal be across the 80-Ω load? What will be the rms signal power delivered to the load?

Determine whether each statement is true or false.

13. Transformer coupling the output does not improve the efficiency of a class A amplifier.
14. Refer to Fig. 8-8. The dc load line is very steep because the dc resistance of the output transformer primary winding is so low.
15. In practice, it is possible to achieve 50 percent efficiency in class A by using transformer coupling.

8-3 Class B Power Amplifiers

The class B amplifier is biased at *cutoff.* No current will flow until an input signal provides the bias necessary to turn on the transistor. This eliminates the large fixed drain on the power supply. The efficiency is much better. Only *half the input* signal is amplified, however. This produces extreme distortion. A single class B transistor would not be useful in audio work. The sound would be horrible.

Two transistors can be operated in class B. One can be arranged to amplify the positive-going portion of the input and the other to amplify the negative-going portion. Combining the two halves, or portions, will reduce much of the distortion. Two transistors operating in this way are said to be in *push-pull.*

Push-pull

Figure 8-9 shows a class B push-pull power amplifier. Two transformers are used. Transformer T_1 is called the *driver transformer.* It provides Q_1 and Q_2 with signal drive. Transformer T_2 is called the *output transformer.* It combines the two signals and supplies the output to the load. Notice that both transformers have one winding that is center-tapped.

With no signal input, there will not be any current flow in Fig. 8-9. Both Q_1 and Q_2 are *cut off.* There is no dc supply to turn on the base-emitter junctions. When the input signal produces the secondary polarity in T_1 as shown, Q_1 is turned on. Current will flow through half of the primary of T_2. Since the primary current is changing in the output transformer, a signal will appear across the secondary. The positive-going portion of the input signal has been amplified and appears across the load.

When the signal reverses polarity, Q_1 is cut off and Q_2 turns on. This is shown in Fig. 8-10. Current will flow through the other half of the primary of T_2. This time the current is flowing up through the primary. When Q_1 was on, the current was flowing down through the primary winding. This current *reversal* produces the negative alternation across the load. By operating two transistors in push-pull, much

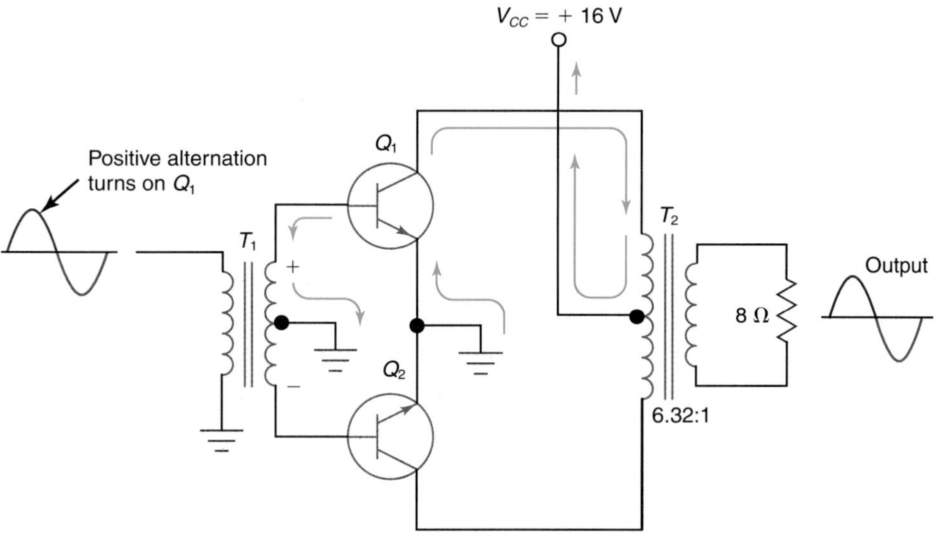

Fig. 8-9 Class B push-pull power amplifier with Q_1 turned on.

of the distortion has been eliminated. The circuit amplifies almost the entire input signal.

We can use graphs to show the output swing and efficiency for the class B push-pull amplifier. Figure 8-11 on page 204 shows the dc and ac load lines for the push-pull circuit. The dc load line is vertical. There is very little resistance in the collector circuit. The ac load line slope is set by the *transformed load* in the collector circuit.

Transformer T_2 shows a turns ratio of 6.32:1. This turns ratio determines the ac load that will be seen in the collector circuit. Only half the transformer primary is conducting at any time. Therefore, only half the turns ratio will be used to calculate the impedance ratio:

$$\frac{6.32}{2} = 3.16$$

Now, the collector load will be equal to the square of half the turns ratio times the load resistance:

$$\text{Load}_{ac} = (3.16)^2 \times 8\ \Omega = 80\ \Omega$$

Each transistor sees an ac load of 80 Ω. The load line of Fig. 8-11 runs from 16 V to 200 mA. This satisfies a slope of 80 Ω:

$$R = \frac{V}{I} = \frac{16\text{ V}}{0.2\text{ A}} = 80\ \Omega$$

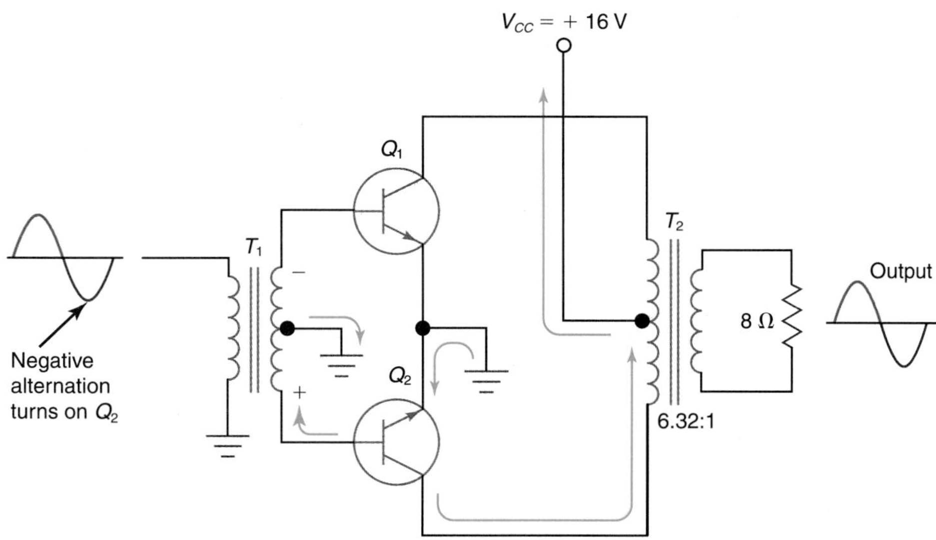

Fig. 8-10 Class B push-pull power amplifier with Q_2 turned on.

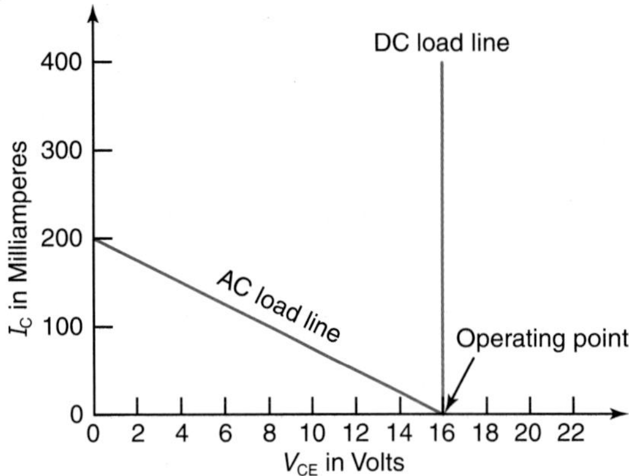

Fig. 8-11 Load lines for the class B amplifier.

Figure 8-11 is correct but shows only one transistor.

There is another way to graph a push-pull circuit, as shown in Fig. 8-12. This graph allows the entire output swing to be shown. The output voltage swings 32 V peak-to-peak. This must be converted to an rms value:

$$V_{\text{rms}} = \frac{V_{\text{p-p}}}{2} \times 0.707$$
$$= \frac{32\text{ V}}{2} \times 0.707 = 11.31\text{ V}$$

Next, the rms current is found:

$$I_{\text{rms}} = \frac{I_{\text{p-p}}}{2} \times 0.707$$
$$= \frac{400\text{ mA}}{2} \times 0.707 = 141.4\text{ mA}$$

Finally, the signal power is

$$P = V \times I$$
$$= 11.31\text{ V} \times 141.4\text{ mA} = 1.6\text{ W}$$

To find the efficiency of the class B push-pull circuit, we will need the dc input power. The supply voltage is 16 V. The supply current varies from 0 to 200 mA. As in class A, the average collector current is what we need:

$$I_{\text{av}} = I_P \times 0.637 = 200\text{ mA} \times 0.637$$
$$= 127.4\text{ mA}$$

The average input power is

$$P = V \times I = 16\text{ V} \times 127.4\text{ mA} = 2.04\text{ W}$$

The class B push-pull amplifier takes 2.04 W from the supply to give a signal output of 1.6 W. The efficiency is

$$\text{Efficiency} = \frac{P_{\text{ac}}}{P_{\text{dc}}} \times 100\%$$
$$= \frac{1.6\text{ W}}{2.04\text{ W}} \times 100\% = 78.5\%$$

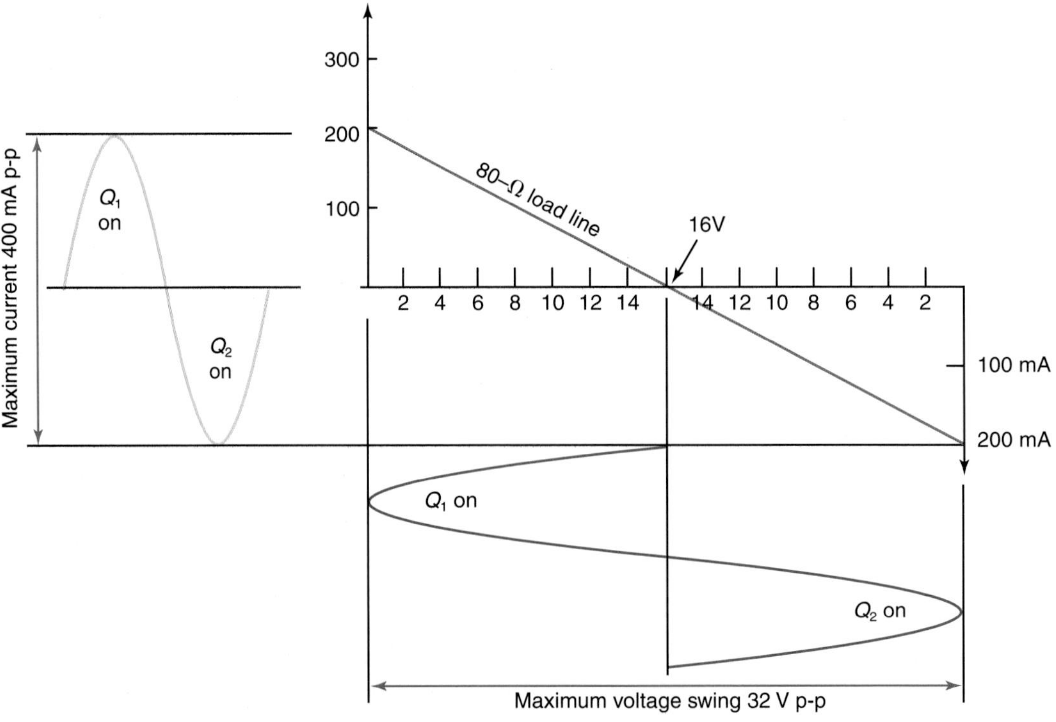

Fig. 8-12 Load line for push-pull operation (both transistors).

The best efficiency for class A is 50 percent. The best efficiency for class B is 78.5 percent. This improved efficiency makes the class B push-pull circuit attractive for high-power applications. For smaller signals, the class B amplifier takes less from the power supply. The 2.04-W factor is not fixed in the above calculation. As the input signal decreases, the power demand on the supply also decreases.

Example 8-6

Find the efficiency for the push-pull amplifier in Fig. 8-10 when it is driven to half of its maximum voltage swing. The power output will decrease to one-fourth of maximum, or 0.4 W, because power varies as the square of voltage. However, let's be certain and do the calculations. Half voltage swing is 16 V_{p-p} for Fig. 8-10 and:

$$V_{rms} = \frac{V_{p-p}}{2} \times 0.707 = \frac{16}{2} \times 0.707 = 5.66 \text{ V}$$

The peak-to-peak current is now 200 mA, and the rms signal current is:

$$I_{rms} = \frac{I_{p-p}}{2} \times 0.707$$
$$= \frac{200 \text{ mA}}{2} \times 0.707 = 70.7 \text{ mA}$$
$$P_{ac} = V_{rms} \times I_{rms}$$
$$= 5.66 \text{ V} \times 70.7 \text{ mA} = 0.4 \text{ W}$$

This verifies our expectation for the signal power. The average dc current is:

$$I_{av} = I_P \times 0.637 = 100 \text{ mA} \times 0.637$$
$$= 63.7 \text{ mA}$$

The dc input power is:

$$P_{dc} = 16 \times 63.7 \text{ mA} = 1.02 \text{ W}$$

Please notice that this is half of what it was when the amplifier was fully driven. The efficiency is also half of the fully driven value:

$$\text{Efficiency} = \frac{P_{ac}}{P_{dc}} \times 100\%$$
$$= \frac{0.4 \text{ W}}{1.02 \text{ W}} \times 100\% = 39.2\%$$

Efficiency decreases when the class B amplifier is not fully driven. However, this amplifier is more efficient than a class A amplifier driven to half of its maximum output swing.

Example 8-7

Can the efficiency of the amplifier shown in Fig. 8-10 be calculated for the condition of no input signal? With no input signal, the transistors are off and there is no current flow. With no current, the dc power is zero. The equation cannot be solved since division by zero is not defined:

$$\text{Efficiency} = \frac{P_{out}}{P_{in}} \times 100\% = \frac{0 \text{ W}}{0 \text{ W}} \times 100\%$$
$$= \textit{undefined}$$

Efficiency cannot be calculated in this case; however, we can reach a conclusion: The efficiency of a class B amplifier with no input signal does not equal zero as in the case of class A.

Class A power transistors require a *high wattage rating*. The reason is that the transistors are always biased on to half of the saturation current. For example, to build a 100-W class A amplifier, the transistor will need at least a 200-W rating. This is based on

$$\text{Efficiency} = \frac{P_{ac}}{P_{dc}} \times 100\%$$
$$= \frac{100 \text{ W}}{200 \text{ W}} \times 100\% = 50\%$$

Look at the above equation: 200 W goes into the transistor; 100 W comes out as signal power. The 100 W difference *heats* the transistor. What if the signal input is zero? The signal output is zero; *yet 200 W still goes into the transistor* and changes to heat.

The wattage rating needed for class B at a given power level is only *one-fifth* that needed for class A. To build a 100-W amplifier

About Electronics

Audio Power and Volume Power supplies for high-power audio amplifiers often use large filter capacitors to improve peak power capabilities. Some audio amplifiers use protection circuits to avoid damage at high volume levels and when output terminals are short-circuited.

requires a 200-W transistor in class A. In class B:

$$\frac{200}{5} = 40 \text{ W}$$

Two 20-W transistors operating in push-pull would provide 100 W output. Two 20-W transistors cost quite a bit less than one 200-W transistor. This is a marked advantage of class B push-pull over class A in high-power amplifiers.

The size of the heat sink is another factor. A transistor rating is based on some safe temperature. In high-power work, the transistor is mounted on a device which carries off the heat. A class B design will need only one-fifth the heat sink capacity for a given amount of power.

There is a very strong case for using class B in high-power work. However, there is still too much distortion for some applications. The push-pull circuit eliminates quite a bit of distortion, but some remains. The problem is called *crossover distortion.*

Crossover distortion

The base-emitter junction of a transistor behaves like a diode. It takes about 0.6 V to turn on the base-emitter junction in a silicon transistor. This means that the first ± 0.6 V of input signal in a class B push-pull amplifier *will not be amplified.* The amplifier has a *dead band* of about 1.2 V. The base-emitter junction is also very nonlinear near the turn-on point. Figure 8-13 shows the characteristic curve for a typical base-emitter junction. Note the curvature near the 0.6-V forward-bias region. As one transistor is turning off and the other is coming on in a push-pull design, this curvature distorts the output signal. The dead band and the nonlinearity make the class B push-pull circuit unacceptable for some applications.

Dead band

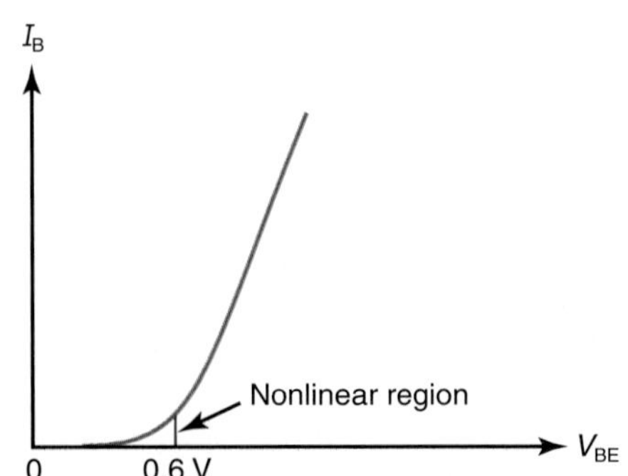

Fig. 8-13 Characteristic curve for a base-emitter junction.

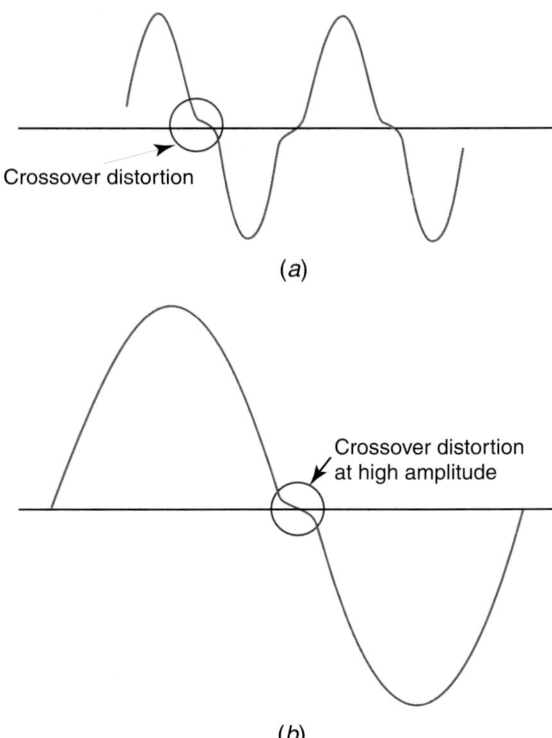

Fig. 8-14 Crossover distortion in the output signal.

The effect of crossover distortion on the output signal is shown in Fig. 8-14(*a*). It happens as the signal is *crossing over* from one transistor to the other. Crossover distortion is very noticeable when the input signal is small. In fact, if the input signal is 1 V_{p-p} or less, there won't be any output with silicon transistors. As shown in Fig. 8-14(*b*), the distortion is less noticeable for large signals. This can be a valuable clue when troubleshooting.

TEST

Determine whether each statement is true or false.

16. Refer to Fig. 8-9. Transistors Q_1 and Q_2 operate in parallel.
17. Refer to Fig. 8-10. Transistors Q_1 and Q_2 will never be on at the same time.
18. Crossover distortion is caused by the nonlinearity of the base-emitter junctions in the transistors.

Solve the following problems.

19. Refer to Fig. 8-10. Transformer T_2 has a turns ratio of 20:1. What is the load seen by the collector of Q_1? Q_2?

20. A class A power amplifier is designed to deliver 5 W of power. What is dissipated in the transistor at zero signal level?
21. A class B push-pull amplifier is designed to deliver 10 W of power. What is the most power that must be dissipated by each transistor?
22. Refer to Fig. 8-12. Assume the amplifier is being driven to only half its maximum swing. Calculate the rms power output.
23. Refer to Fig. 8-12. Assume the amplifier is driven to half its maximum swing. Calculate the average power input.
24. Refer to Fig. 8-12. Assume the amplifier is driven to half its maximum swing. Calculate the efficiency of the amplifier.

8-4 Class AB Power Amplifiers

The solution to the crossover distortion problem is to provide *some forward bias* for the base-emitter junctions. The forward bias will prevent the base-emitter voltage V_{BE} from ever reaching the nonlinear part of its curve. This is shown in Fig. 8-15. The forward bias is small and results in a *class AB* amplifier. It has characteristics between class A and class B.

The operating point for class AB is shown in Fig. 8-16. Note that class AB operates *near cutoff.*

Figure 8-17 is a class AB push-pull power amplifier. Resistors R_1 and R_2 form a voltage divider to forward-bias the base-emitter junctions. The bias current flows through both halves of the secondary of T_1. Capacitor C_1 grounds the center tap for ac signals. Without it, signal current will flow in R_1 and R_2 and a large part of the signal energy will be wasted (dissipated as heat).

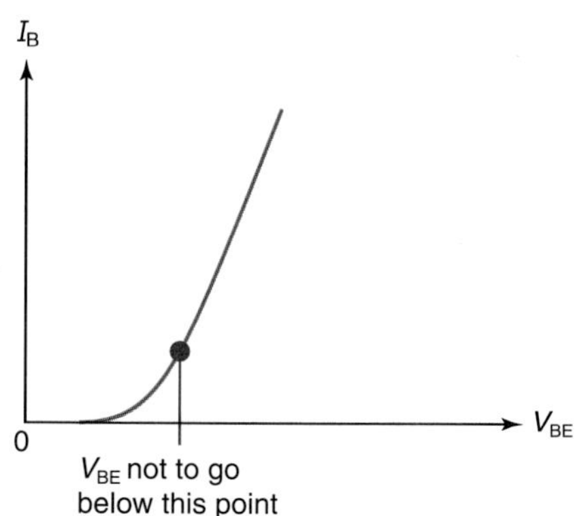

Fig. 8-15 Minimum value of V_{BE} to prevent crossover distortion.

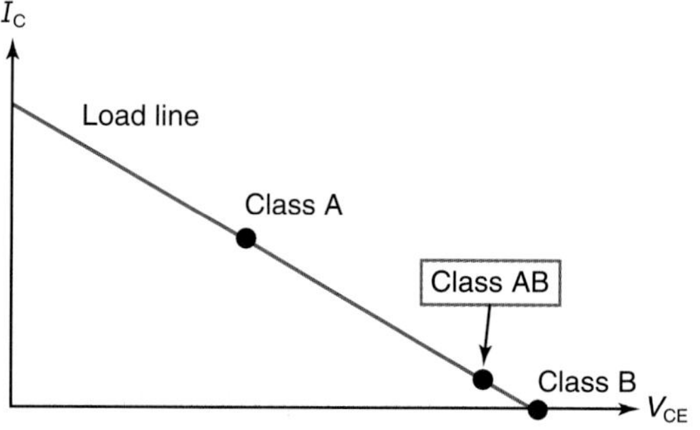

Fig. 8-16 Class AB operating point.

A class AB amplifier does not have as much efficiency as a class B amplifier. Its efficiency is better than that of a class A design. It is a compromise class that provides minimum distortion and reasonable efficiency. It is the most popular class for high-power audio work. Amplifiers such as the one shown in Fig. 8-17 are popular in portable radios and tape recorders.

Material safety data sheets are required in the workplace when hazardous materials are present. Read them.

Class AB

Example 8-8

Find a value for R_1 in Fig. 8-17 assuming class AB bias, a supply of 12 V, and a value for R_2 of 1 kΩ. Use the voltage divider equation and assume that a 0.6-V drop is desired for class AB:

$$0.6\text{ V} = \frac{1\text{ k}\Omega}{1\text{ k}\Omega + R_1} \times 12\text{ V}$$

Eliminate the fraction by multiplying both sides by the denominator:

$$0.6\text{ V}(1\text{ k}\Omega + R_1) = 1\text{ k}\Omega \times 12\text{ V}$$

Divide both sides by 12 V:

$$0.05(1\text{ k}\Omega + R_1) = 1\text{ k}\Omega$$

Multiply:

$$50\ \Omega + 0.05\ R_1 = 1\text{ k}\Omega$$

Subtract 50 Ω from both sides:

$$0.05\ R_1 = 950\ \Omega$$
$$R_1 = 19\text{ k}\Omega$$

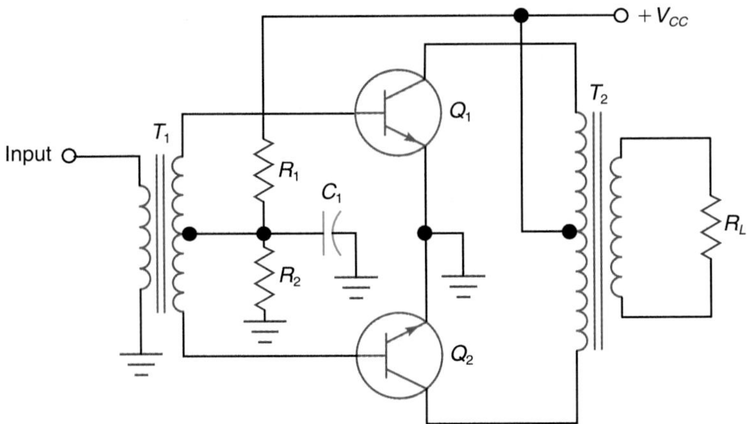

Fig. 8-17 Class AB push-pull power amplifier.

Now that the distortion problem is solved for push-pull amplifiers, it is time to look at the transformers. For high-power and high-quality work, the transformers are too expensive. They can be eliminated.

Driver transformers can be eliminated by using a combination of transistor polarities. A positive-going signal applied to the base of an NPN transistor tends to turn it on. A positive-going signal applied to the base of a PNP transistor tends to turn it off. This means that push-pull operation can be obtained without a center-tapped transformer.

Complementary symmetry

Output transformers can be eliminated by using a different amplifier configuration. The emitter-follower (common-collector) amplifier is noted for its low output impedance. This allows good matching to low-impedance loads such as loudspeakers.

Figure 8-18 shows an amplifier design that eliminates the transformers. Transistor Q_1 is an NPN transistor, and Q_2 is a PNP transistor. Push-pull operation is realized without a center-tapped driver transformer. Notice that the load is capacitively coupled to the emitter leads of the transistors. The transistors are operating as *emitter followers.*

The circuit of Fig. 8-18 is known as a *complementary symmetry amplifier.* The transistors are *electrical complements.* One is NPN, and the other is PNP. The curves in Fig. 8-19 show the symmetrical characteristics of NPN and PNP transistors. Good matching of characteristics is important in the complementary symmetry amplifier. For this reason, transistor manufacturers offer NPN-PNP pairs with good symmetry.

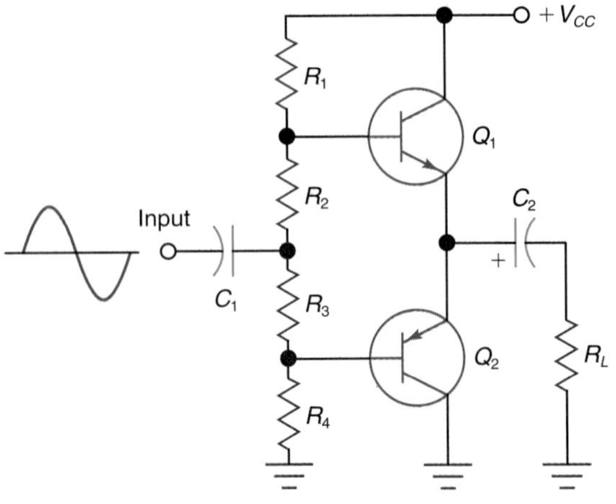

Fig. 8-18 A complementary symmetry amplifier.

Example 8-9

Select values for R_2 and R_3 for Fig. 8-18 assuming a 12-V supply and that R_1 and R_4 are each 10 kΩ. For class AB, we can assume a 0.6-V drop for *each* transistor, or a 1.2-V drop for *both.* Set up the divider formula to find the total value of R_2 and R_3 (R_T) first:

$$1.2 \text{ V} = \frac{R_T}{20 \text{ k}\Omega + R_T} \times 12 \text{ V}$$

Solve as before (the algebra is similar to example 8-8) to find that $R_T = 2.22$ kΩ. Each resistor should be half of that amount, or 1.11 kΩ.

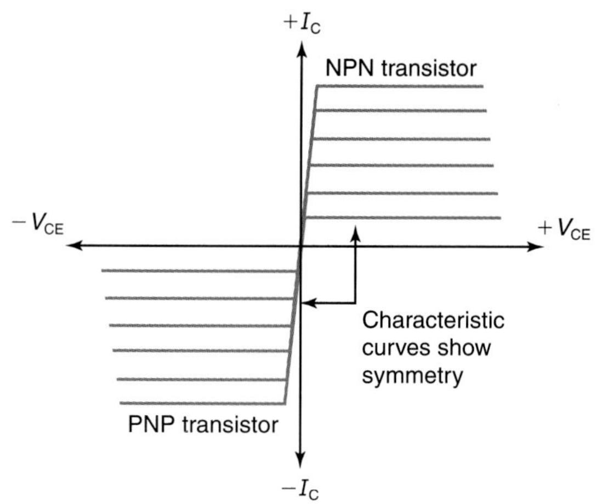

Fig. 8-19 NPN-PNP symmetry.

Figure 8-20 shows the output signal in a complementary symmetry amplifier when the input signal goes positive. Transistor Q_1, the NPN transistor, is turned on. Transistor Q_2, the PNP transistor, is turned off. Current flows through the load, through C_2, and through Q_1 into the power supply. This current charges C_2 as shown. Notice that there is *no phase inversion* in the amplifier. This is to be expected in an emitter follower.

When the input signal goes negative, the signal flow is as shown in Fig. 8-21 on page 210. Now Q_1 is off and Q_2 is on. This causes C_2 to discharge as shown. Again, the output is in phase with the input. Capacitor C_2 is usually a large capacitor (1000 μF or so). This is necessary for good low-frequency response with low values of R_L.

Another possibility is shown in Fig. 8-22 on page 210. This is known as a *quasi-complementary symmetry amplifier.* The output transistors Q_3 and Q_4 are not complementary. They are both NPN types. The *driver transistors* Q_1 and Q_2 are complementary. A positive-going input signal will turn on Q_1, the NPN driver, and it will turn off Q_2, the PNP driver. This results in a push-pull action since the drivers supply the base current for the output transistors. Again, no center-tapped transformer is necessary in the output.

Look for related information on the Intel website.

Notice the diodes used in the bias network of Fig. 8-22. These diodes provide *temperature compensation.* Transistors tend to conduct more current as temperature goes up. This is undesirable. The drop across a conducting diode *decreases* with an increase in temperature. If the diode drop is part of the bias voltage for the amplifier, compensation results. The decreasing voltage across the diode will lower the amplifier current. Thus, the operating point is more stable with this arrangement.

Temperature compensation

Integrated circuits (ICs) are available for power applications. Figure 8-23 on page 211 shows an IC power amplifier manufactured by National Semiconductor. It is designed for stereo audio and produces up to 10 W of output per channel. It provides a voltage gain of 316 (50 dB) when operated with negative feedback and a voltage gain of 31,620 (90 dB) when operated with no feedback. Its input impedance is 200 kΩ.

Figure 8-23(*a*) shows the LM2005 is a typical stereo application. A portion of the output signal from each channel is fed back to the minus inputs (pins 2 and 4). This is negative feedback. It reduces gain, improves frequency re-

Quasi-complementary symmetry

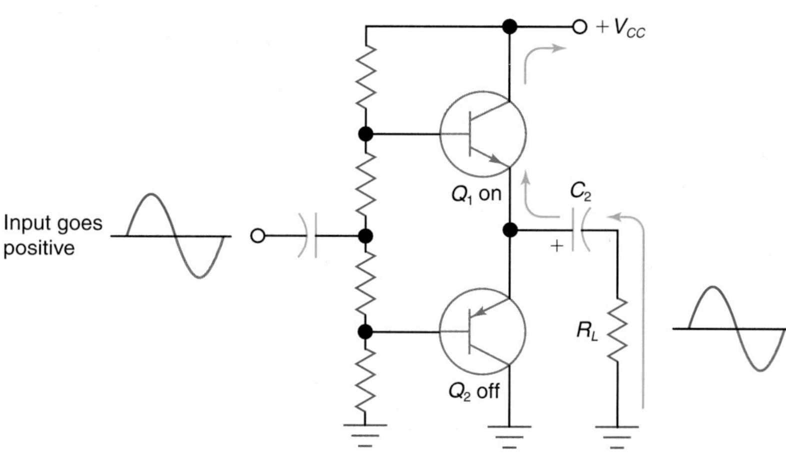

Fig. 8-20 A positive-going signal in a complementary symmetry amplifier.

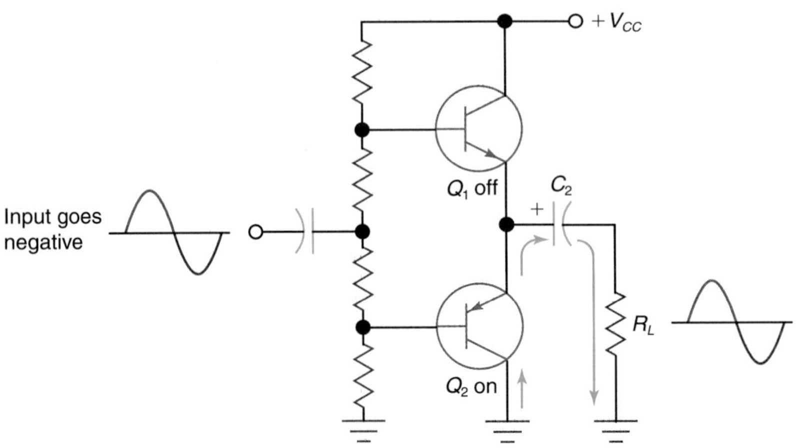

Fig. 8-21 A negative-going signal in a complementary symmetry amplifier.

sponse, and reduces noise and distortion. The output signal is coupled to the loudspeakers by large (2200-μF) electrolytic capacitors.

IC power amplifier

Figure 8-23(*b*) shows the package for the LM2005 *IC power amplifier*. The metal tab is connected to ground and must be mounted to a heat sink for most applications.

TEST

Solve the following problems.

25. Is the efficiency of class AB better than that of class A but not as good as that of class B?
26. Refer to Fig. 8-17. Assume that C_1 shorts. In what class will the amplifier operate?
27. Refer to Fig. 8-17. Assume that Q_1 and Q_2 are running very hot. Could C_1 be shorted? Why or why not?
28. Refer to Fig. 8-17. Assume that Q_1 and Q_2 are running very hot. Could R_2 be open? Why or why not?
29. Refer to Fig. 8-18. An input signal drives C_1 in a positive direction. In what direction will the top of R_L be driven?
30. Refer to Fig. 8-18. An input signal drives C_1 in a positive direction. Which transistor is turning off?
31. Refer to Fig. 8-18. Voltage $V_{CC} = 20$ V. With no input signal, what should the voltage be at the emitter of Q_1? At the base of Q_1? At the base of Q_2? (Hint: The transistors are silicon.)

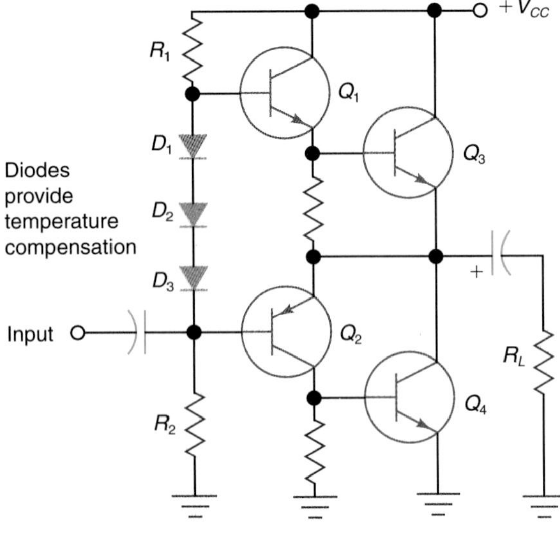

Fig. 8-22 A quasi-complementary symmetry amplifier.

8-5 Class C Power Amplifiers

Class C

Class C amplifiers are biased *beyond cutoff.* Figure 8-24 is a class C amplifier with a negative supply voltage V_{BB} applied to the base circuit. This negative voltage reverse-biases the base-emitter junction of the transistor. The transistor will not conduct until the input signal overcomes this reverse bias. This happens for only a small part of the input cycle. The transistor conducts for only a small part (90° or less) of the input waveform.

As shown in Fig. 8-24, the collector-current waveform is not a whole sine wave. It is not even half a sine wave. This extreme distortion means the class C amplifier *cannot* be used for audio work. Class C amplifiers are used at *radio frequencies.*

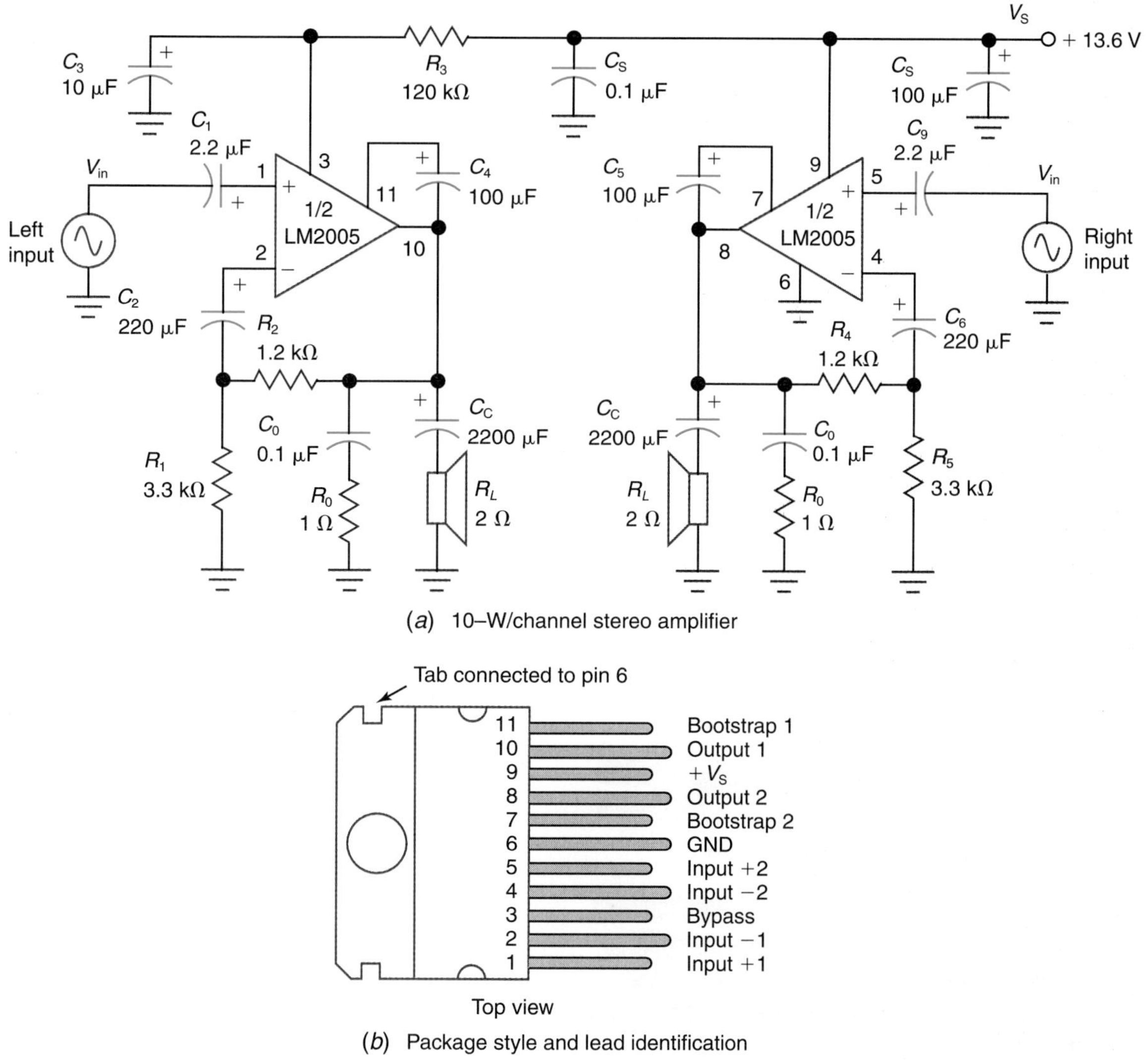

(a) 10–W/channel stereo amplifier

Tab connected to pin 6

Pin	Function
11	Bootstrap 1
10	Output 1
9	$+V_S$
8	Output 2
7	Bootstrap 2
6	GND
5	Input +2
4	Input −2
3	Bypass
2	Input −1
1	Input +1

Top view

(b) Package style and lead identification

Fig. 8-23 National LM2005 integrated power amplifier.

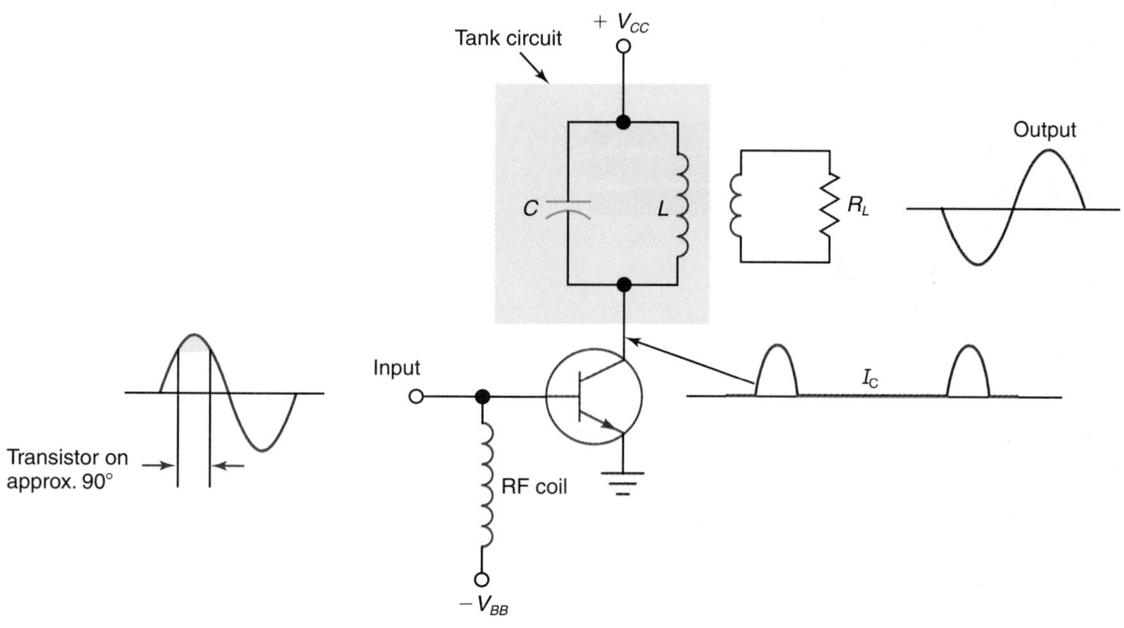

Fig. 8-24 A class C amplifier.

Tank circuit

Figure 8-24 shows the *tank circuit* in the collector circuit of the class C amplifier. This tank circuit restores the sine-wave input signal. Note that a sine wave is shown across R_L. Tank circuits can restore sine wave signals but not rectangular waves or complex audio waves.

Tank circuit action is explained in Fig. 8-25. The collector-current pulse charges the capacitor [Fig. 8-25(*a*)]. After the pulse charges, the capacitor discharges through the inductor [Fig. 8-25(*b*)]. Energy is stored in the field around the inductor. When the capacitor discharges to zero, the field collapses and keeps the current flowing [Fig. 8-25(*c*)]. This charges the capacitor again, but note that the polarity is opposite. After the field has collapsed, the capacitor again begins discharging through the inductor [Fig. 8-25(*d*)]. Note that current is now flowing in the opposite direction and the inductor field is expanding. Finally, the inductor field begins to collapse, and the capacitor is charged again to its original polarity [Fig. 8-25(*e*)].

Damped sine wave

Tank circuit action results from a capacitor discharging into an inductor which later discharges into the capacitor and so on. Both the capacitor and the inductor are energy storage devices. As the energy transfers from one to the other, a sine wave is produced. Circuit loss (resistance) will cause the sine wave to decrease gradually. This is shown in Fig. 8-26(*a*); the wave is called a *damped sine wave.* By pulsing the tank circuit every cycle, the sine wave amplitude can be kept constant. This is shown in Fig. 8-26(*b*). In a class C amplifier, the tank circuit is recharged by a collector-current pulse every cycle. This makes the sine wave output constant in amplitude.

Resonance

The values of inductance and capacitance are important in a class C amplifier tank circuit. They must *resonate* at the frequency of the input signal. The equation for resonance is

$$f_r = \frac{1}{2\pi\sqrt{LC}}$$

What is the resonant frequency of a tank circuit that has 100 pF of capacitance and 1 μH of inductance? Substituting the values into the equation, we get

$$f_r = \frac{1}{6.28\sqrt{1 \times 10^{-6} \times 100 \times 10^{-12}}}$$
$$= 15.9 \times 10^6 \text{ Hz}$$

The resonant frequency is 15.9 MHz.

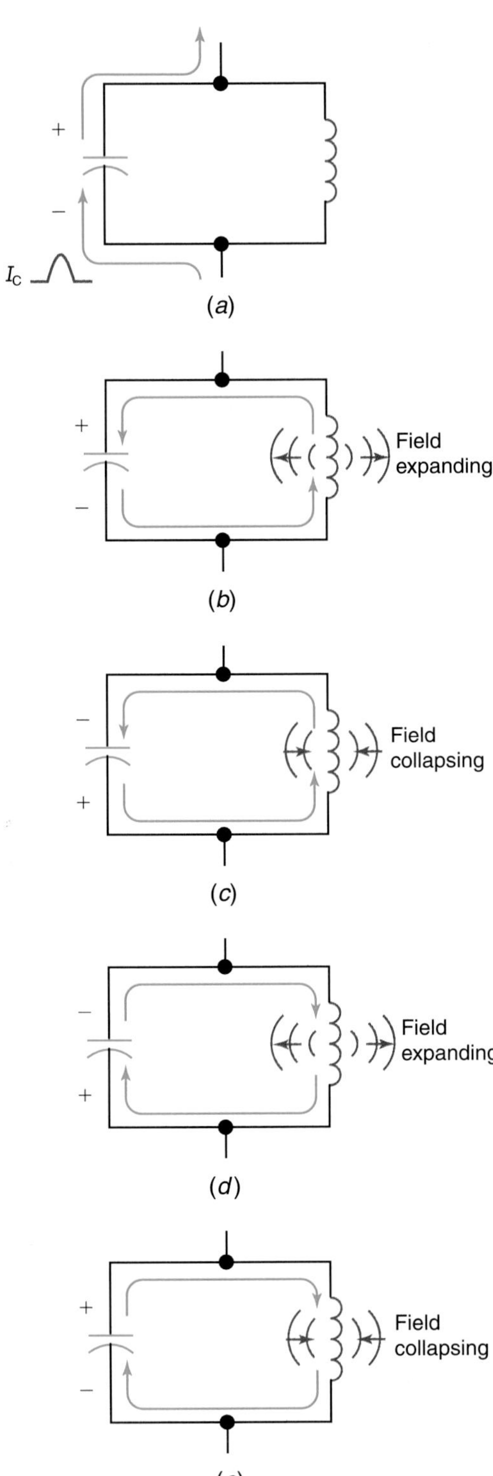

Fig. 8-25 Tank circuit action.

Frequency doubler

In some cases, the tank circuit is tuned to resonate at two or three times the frequency of the input signal. This produces an output signal two or three times the frequency of the input signal. Such circuits are called *frequency doublers* or *triplers.* They are commonly used where high-frequency signals are needed. For

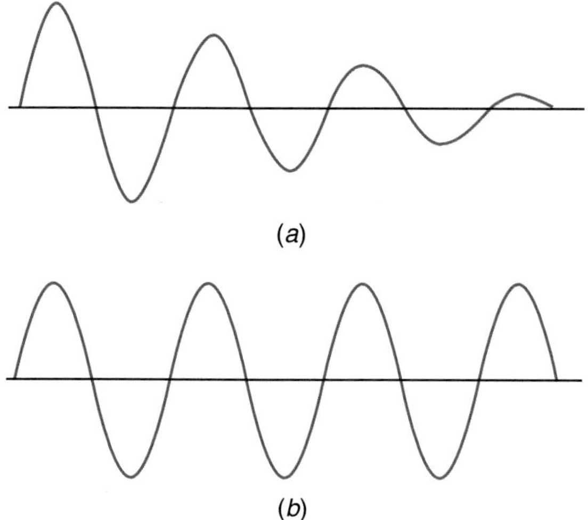

Fig. 8-26 Tank circuit waveforms.

example, suppose that a 150-MHz two-way transmitter is being designed. It may be easier to initially develop a lower frequency. The lower frequency can be multiplied up to the working frequency. Figure 8-27 shows the block diagram for such a transmitter.

The class C amplifier is the *most efficient* of all the analog amplifier classes. Its high efficiency is shown by the waveforms of Fig. 8-28. The top waveform is the input signal. Only the positive peak of the input forward-biases the base-emitter junction. This occurs at 0.6 V in a silicon transistor. Most of the input signal falls below this value because of the negative bias (V_{BE}). The middle waveform in Fig. 8-28 is the collector current I_C. The collector current is in the form of narrow pulses. The bottom waveform is the output signal. It is sinusoidal because of tank-circuit action. Note that the collector-current pulses occur when the output waveform is near zero. This means that little power will be dissipated in the transistor:

$$P_C = V_{CE} \times I_C = 0 \times I_C = 0 \text{ W}$$

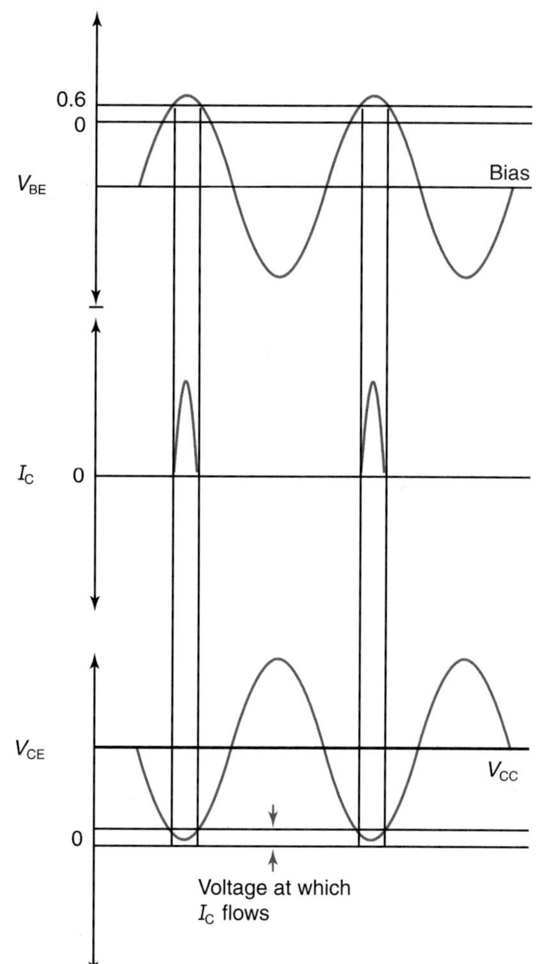

Fig. 8-28 Class C amplifier waveforms.

If no power is dissipated in the transistor, it must all become signal power. This leads to the conclusion that the class C amplifier is 100 percent efficient. Actually, there is power dissipated in the transistor. Voltage V_{CE} is not zero. The tank circuit will also cause some loss. Practical class C power amplifiers can achieve efficiencies as high as 85 percent. Class C amplifiers are very popular at radio frequen-

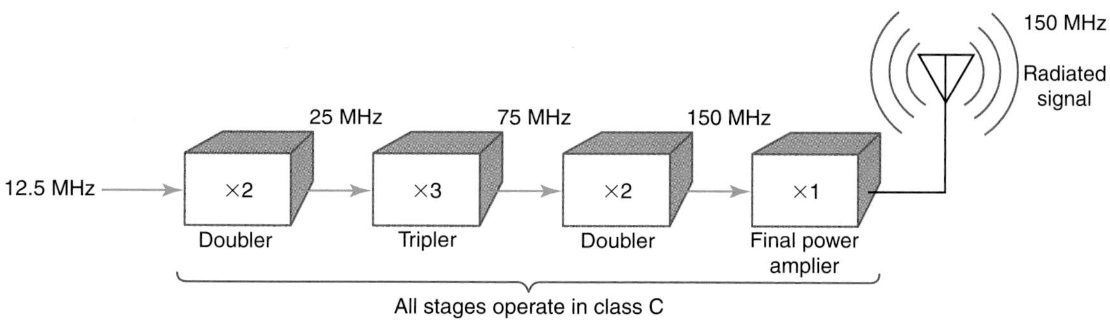

Fig. 8-27 Block diagram for a high-frequency transmitter.

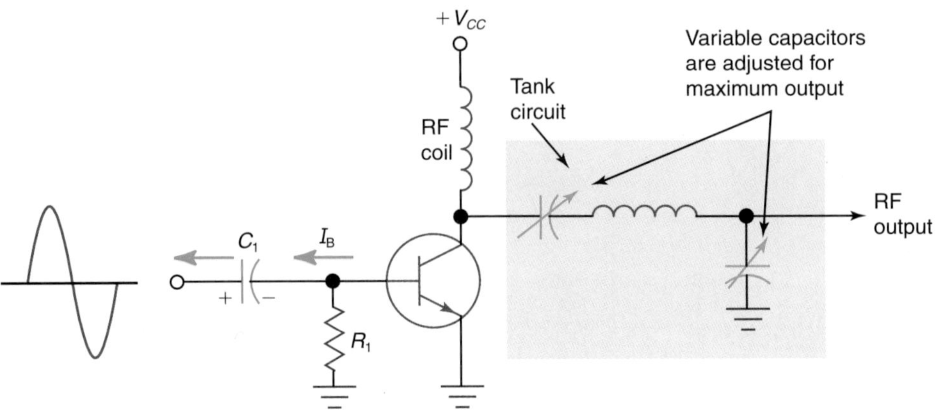

Fig. 8-29 Class C amplifier using signal bias.

cies where tank circuits can restore the sine-wave signal.

Practical class C power amplifiers seldom use a fixed bias supply for the base circuit. A better way to do it is to use *signal bias*. This is shown in Fig. 8-29. As the input signal goes positive, it forward-biases the base-emitter junction. Base current I_B flows as shown. The current charges C_1. Resistor R_1 discharges C_1 between positive peaks of the input signal. Resistor R_1 cannot completely discharge C_1, and the remaining voltage across C_1 acts as a bias supply. Capacitor C_1's polarity reverse-biases the base-emitter junction.

Signal bias

***L* network**

One of the advantages of signal bias is that it is self-adjusting according to the level of the input signal. Class C amplifiers are designed for a small conduction angle to make the efficiency high. If an amplifier uses fixed bias, the conduction angle will increase if the amplitude of the input signal increases. Figure 8-30 shows why. Two conduction angles can be seen for the fixed bias of $-V_{BE}$ shown on the graph. The angle is approximately 90° for the small signal and approximately 170° for the large signal. A 170° conduction angle is too large, and the amplifier efficiency would suffer. The amplifier might also overheat since the average current flow would be much greater. Signal bias overcomes this problem because the conduction angle tends to remain constant. For example, if the input signal in Fig. 8-29 becomes larger, the average charge on C_1 will increase. This will increase the reverse bias $-V_{BE}$. More reverse bias means a smaller conduction angle. The signal bias circuit automatically adjusts to changes in the amplitude of the input signal and tends to keep the conduction angle constant.

Crude and/or rude behavior is never acceptable on the job. It should be considered hazardous to one's professional future.

Figure 8-29 also shows a different type of tank circuit. It is known as an *L network*. It matches the impedance of the transistor to the load. Radio-frequency power transistors often have an output impedance of about 2 Ω. The standard load impedance in RF work is 50 Ω. Thus, the *L* network is necessary to match the transistor to 50 Ω.

Figure 8-31 shows a high-power RF amplifier using VFETs or power MOSFETs. The amplifier is a push-pull design and develops over 1 kW of output power over the range of 10 to 90 MHz. Its power gain ranges from 11 to 14 dB over this frequency range.

The circuit of Fig. 8-31 uses negative feedback to achieve such a wide frequency range.

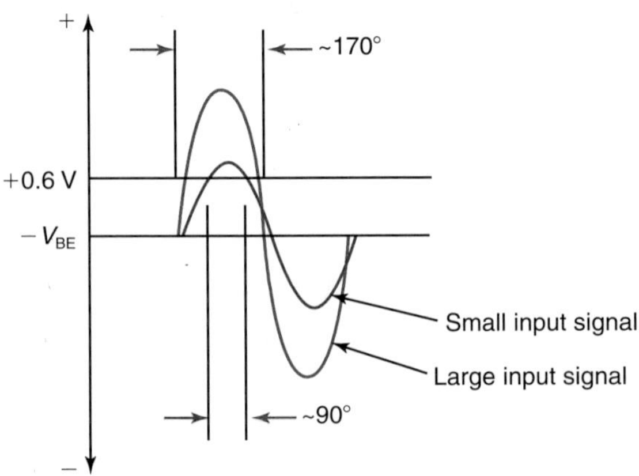

Fig. 8-30 Conduction angle changes with signal level.

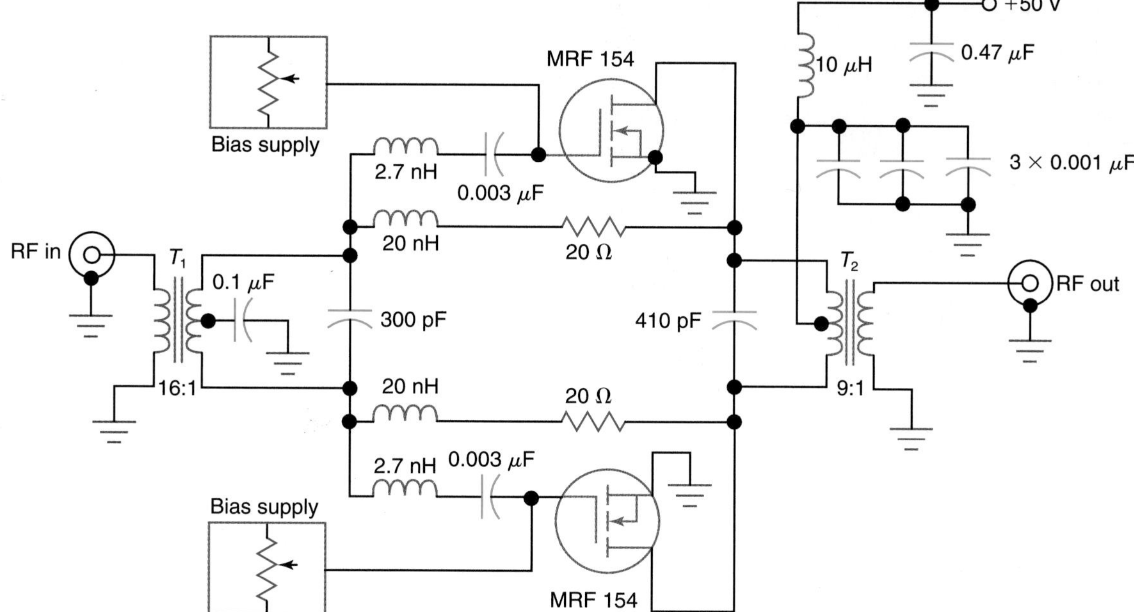

Fig. 8-31 Motorola 1-kW RF power amplifier.

The 20-nH inductors and 20-Ω resistors feed a part of the drain signal back into the gate circuit of each transistor. The transformers T_1 and T_2 are special wideband ferrite core types. These also contribute to the wide frequency range of the amplifier.

TEST

Solve the following problems.

32. Refer to Fig. 8-24. The transistor is silicon, and $-V_{BB} = 6$ V. How positive will the input signal have to swing to turn on the transistor?
33. Refer to Fig. 8-24. Assume the input signal is a square wave. What signal can be expected across R_L, assuming a high-Q tank circuit?
34. Is class C more efficient than class B?
35. Does class C have the smallest conduction angle?
36. The input frequency to an RF tripler stage is 10 MHz. What is the output frequency?
37. A tank circuit uses a 6.8-μH coil and a 47-pF capacitor. What is the resonant frequency of the tank?
38. Refer to Fig. 8-29. Assume that the amplifier is being driven by a signal and the voltage at the base of the transistor is negative. What should happen to the base voltage if the drive signal increases?

8-6 Switch-Mode Amplifiers

Class D

A switch-mode amplifier may also be called a *digital amplifier* or a *class D amplifier*. A digital power amplifier may be a better choice than an analog type. Digital amplifiers use the output transistors as switches. A switch has only two conditions: open or closed. Since the output devices are either on or off, these amplifiers are very efficient.

In a linear power amplifier, the transistors can produce a lot of heat. They may drop substantial voltage while conducting a large current. This heats the transistors and detracts from the efficiency of the amplifier. An ideal amplifier would convert all the energy it takes from its power supply into a useful output signal and not change any of it into heat. Digital amplifiers approach the ideal because of their on-off nature. Suppose an ideal digital amplifier supplies 50 V at 10 A. If we investigate the collector dissipation in the output devices:

$$P_{C(\text{off})} = V_{CE} \times I_C = 50\text{ V} \times 0\text{ A} = 0\text{ W}$$
$$P_{C(\text{sat})} = 0\text{ V} \times 10\text{ A} = 0\text{ W}$$

There is never any transistor dissipation in the ideal digital device. Of course, a real digital amplifier does show some dissipation. For one thing, V_{CE} is never zero even though a transistor may be turned on very hard. But, the dissipation is very small when compared to linear designs. Digital amplifiers are more efficient and smaller, run cooler, and tend to be more reliable than equivalent linear designs.

Switch-mode amplifiers often use pulse-width modulation (PWM). Look at Fig. 8-32. The rectangular waveform shows a varying pulse width over time. Note that the average value of the PWM wave is a sine wave in this case. It is possible to approximate any waveform by controlling pulse width. If the basic digital waveform is high enough in frequency, its average value can closely approach some desired analog waveform.

Switch-mode power amplifiers are often used with inductive loads, such as motors. In Fig. 8-33(*b*), the switch has been closed and the inductor is charging. Current flow increases linearly with time. The actual rate of current rise is proportional to the voltage and inversely proportional to inductance. For example, in the case of a 100-V supply and a 100-mH inductor:

$$\frac{\Delta I}{\Delta T} = \frac{V}{L} = \frac{100}{0.1} = 1000 \text{ A/s}$$

One thousand amperes is quite a current! However, if the switch is closed for only 1 ms, the current flow will reach 1 A. In other words, in a PWM amplifier the switch (or transistor) is on for a relatively short period of time. Another reason why the devices are only on briefly is core saturation. If the current flow becomes too high, the core of the inductor (or motor) will not be able to conduct any additional magnetic flux. This phenomenon is called *core saturation* and must be avoided. If the saturation point is reached, the inductance decreases drastically and the rate of current rise increases dramatically.

Freewheeling diode

Core saturation

Figure 8-33(*c*) shows what happens when the switch opens. The energy that has been stored in the inductor must be released. The field starts to collapse and maintains the inductor current in the same direction in which it was flowing when it was charging. That is the purpose of the diode. It becomes forward-biased when the switch opens and allows the current to decay back to zero, as shown in Fig. 8-33(*d*). Diodes used in this way are called *freewheeling diodes*. Without the diode, an extremely high voltage would be generated at the moment when the switch opens. This would cause arcing or device damage in the case of a solid-state switch. Power field-effect transistors may not need an external freewheeling diode since they have an integral body diode which serves the same purpose.

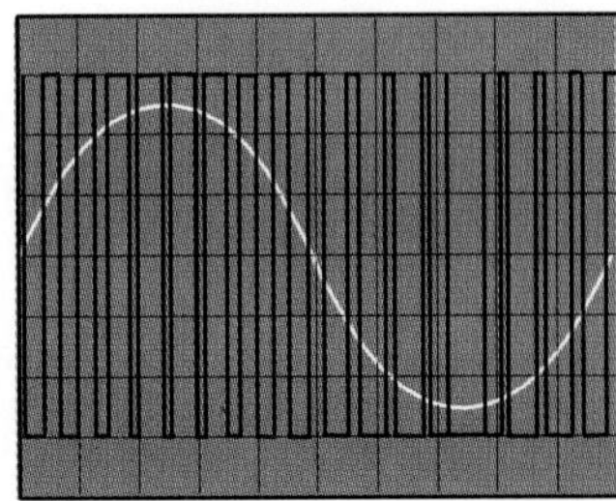

Fig. 8-32 A PWM waveform and its average value.

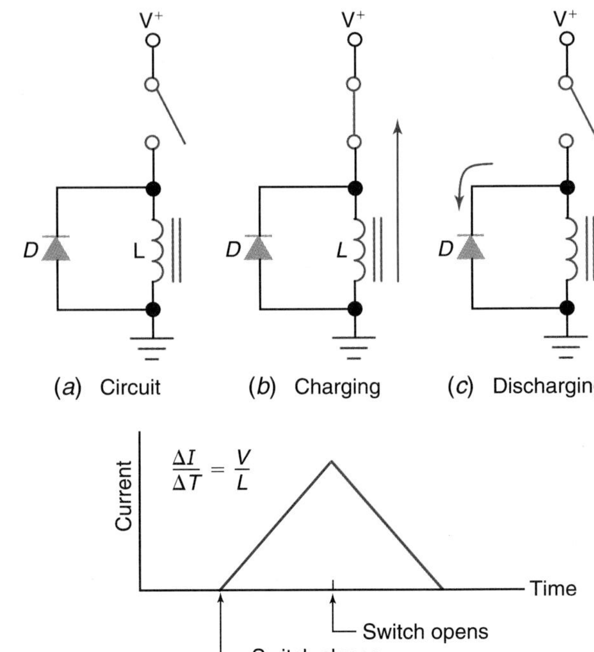

Fig. 8-33 Inductor charging and discharging.

Figure 8-34 shows a more practical arrangement. It is capable of providing load current in both directions. When Q_1 is on, the inductor current is to the positive supply. When Q_2 is on, the inductor current flows from the negative supply. Q_1 and Q_2 must never be on at the same time since this would provide a short-circuit path from $V-$ to $V+$.

Figure 8-34 also shows some possible waveforms. Starting at the top, Q_1 is on for a period of time. The inductor current increases from zero during this interval. When Q_1 switches off, D_2 is forward-biased and current continues to flow until the inductor is discharged. Then

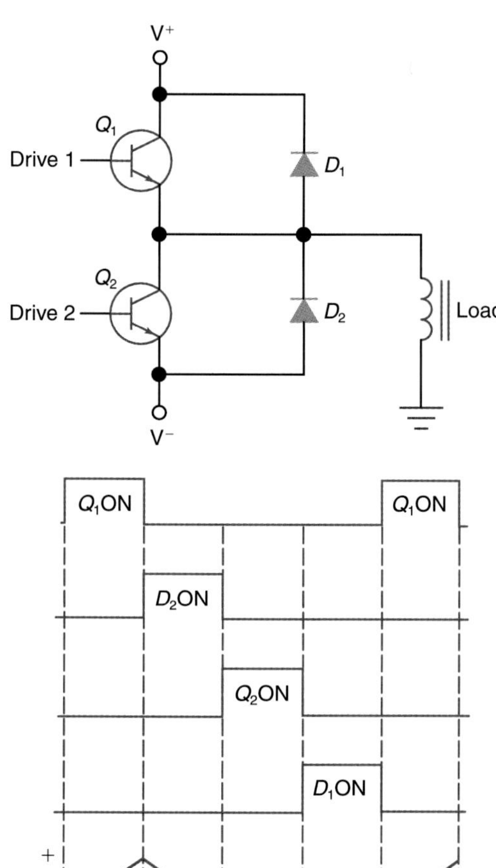

Fig. 8-34 Pulse-width modulation circuit and waveforms.

Q_2 is switched on and the load current again begins to increase, but now in an opposite direction. The negative supply is now providing the current. When Q_2 switches off, D_1 comes on and discharges the inductor. The load waveform shows that the average current is zero. Stated another way, there is no dc component in the load current.

If the positive and negative supplies of Fig. 8-34 have output filter capacitors, they are charged when the diodes are conducting. Or, if the supply uses rechargeable batteries, the batteries are charged by the diodes. This action is called *regeneration*. The energy stored in the inductors is returned and not wasted as heat. Electric vehicles might also use *regenerative braking*. This takes place when drive motors are turned into generators to slow down the vehicle.

Regeneration

TEST

Answer the following questions.

39. A digital amplifier applies 50-V pulses to a 150-mH load. Find the rate of current rise in the load.
40. When a digital amplifier is used with an inductive load, the output must not be on long enough to cause magnetic core ________.
41. The freewheeling components in Fig. 8-33 are ________.
42. The efficiency of digital amplifiers is significantly ________ than the efficiency of linear amplifiers.
43. In Fig. 8-34, transistors Q_1 and Q_2 should not be turned on at the ________.
44. When PWM is used to produce a sinusoidal load current, the digital switching frequency should be significantly ________ than the sinusoidal frequency.

Chapter 8 Review

Summary

1. All amplifiers are technically power amplifiers. Only those that handle large signals are called power amplifiers.
2. The power amplifier is usually the last stage in the signal processing system.
3. Power amplifiers should be efficient. Efficiency is a comparison of the signal power output to the dc power input.
4. Poor efficiency in a power amplifier means the power supply will have to be larger and more expensive. It also means that the amplifier will convert too much electrical energy to heat.
5. Class A amplifiers operate at the center of the load line. They have low distortion and a conduction angle of 360°.
6. Class B operates at cutoff. The conduction angle is 180°.
7. Class B amplifiers do not present a fixed drain on the power supply. The drain is zero with no input signal.
8. Class B is more efficient than class A.
9. Bias controls the operating point and the class of operation in any amplifier.
10. The maximum theoretical efficiency for class A operation is 25 percent. With transformer coupling, it is 50 percent.
11. In a transformer-coupled amplifier, the impedance ratio is equal to the square of the turns ratio.
12. The fixed drain on the power supply is a major drawback with class A circuits. Efficiency is very poor when signals are small.
13. A single class B transistor will amplify half the input signal.
14. Two class B transistors can be operated in push-pull.
15. The maximum theoretical efficiency for class B is 78.5 percent.
16. A class B amplifier draws less current from the supply for smaller signals.
17. For a given output power, class B transistors will require only one-fifth of the power rating needed for class A.
18. The biggest drawback to class B push-pull is crossover distortion.
19. Crossover distortion can be eliminated by providing some forward bias for the base-emitter junctions of the transistors.
20. Class AB amplifiers are forward-biased slightly above cutoff.
21. Class AB operation is the most popular for high-power audio work.
22. Push-pull operation can be obtained without center-tapped transformers by using a PNP-NPN pair.
23. An amplifier that uses a PNP-NPN pair for push-pull operation is called a complementary symmetry amplifier.
24. Complementary pairs have symmetrical characteristic curves.
25. Diodes may be used to stabilize the operating point in power amplifiers.
26. Integrated-circuit power amplifiers are available for some applications.
27. Class C amplifiers are biased beyond cutoff.
28. The conduction angle for class C is around 90°.
29. Class C amplifiers have too much distortion for audio work. They are useful at radio frequencies.
30. A tank circuit can be used to restore a sine-wave signal in a class C amplifier.
31. The tank circuit should resonate at the signal frequency. In a frequency multiplier, the tank resonates at some multiple of the signal input frequency.
32. The class C amplifier has a maximum theoretical efficiency of 100 percent. In practice, it can reach about 85 percent.
33. Switch-mode power amplifiers may also be called class D or digital power amplifiers.
34. Switch-mode amplifiers often use pulse-width modulation and are noted for their high efficiency.

Chapter Review Questions

Answer the following questions.

8-1. Refer to Fig. 8-1. In which of the three stages is efficiency the most important?

8-2. An amplifier delivers 60 W of signal power. Its power supply develops 28 V, and the current drain is 4 A. What is the efficiency of the amplifier?

8-3. An amplifier has an efficiency of 45 percent. It is rated at 5 W of output. How much current will it draw from a 12-V battery when delivering its rated output?

8-4. Which class of amplifier produces the least distortion?

8-5. What is the conduction angle of a class B amplifier?

8-6. The operating point for an amplifier is at the center of the load line. What class is the amplifier?

8-7. Refer to Fig. 8-7. What is the maximum theoretical efficiency for this circuit? At what signal level is this efficiency achieved?

8-8. What will happen to the efficiency of the amplifier in Fig. 8-7 as the signal level decreases?

8-9. Refer to Fig. 8-7. What turns ratio will be required to transform the 80-Ω load to a collector load of 1.28 kΩ?

8-10. A class A transformer-coupled amplifier operates from a 9-V supply. What is the maximum peak-to-peak voltage swing at the collector?

8-11. Refer to Fig. 8-9. What would have to be done to V_{CC} so that the circuit could use PNP transistors?

8-12. Refer to Fig. 8-9. With zero signal level, how much current will be taken from the 16-V supply?

8-13. Refer to Fig. 8-9. What is the phase of the signal at the base of Q_1 as compared to the base of Q_2? What component causes this?

8-14. Refer to Fig. 8-9. Assume the peak-to-peak sine wave swing across the collectors is 24 V. The transformer is 100 percent efficient. Calculate

a. $V_{\text{p-p}}$ across the load (do not forget to use half the turns ratio)

b. V_{rms} across the load

c. P_L (load power)

8-15. Calculate the minimum wattage rating for each transistor in a push-pull class B amplifier designed for 100-W output.

8-16. Calculate the minimum wattage rating for a class A power transistor that is transformer-coupled and rated at 25-W output.

8-17. At what signal level is crossover distortion most noticeable?

8-18. Refer to Fig. 8-17. Which two components set the forward bias on the base-emitter junctions of Q_1 and Q_2?

8-19. Refer to Fig. 8-17. There is no input signal. Will the amplifier take any current from the power supply?

8-20. Refer to Fig. 8-17. What will happen to the current taken from the power supply as the signal level increases?

8-21. Refer to Fig. 8-18. What is the configuration of Q_1?

8-22. Refer to Fig. 8-18. What is the configuration of Q_2?

8-23. Refer to Fig. 8-21. When the input signal goes negative, what supplies the energy to the load?

8-24. The major reason for using class AB in a push-pull amplifier is to eliminate distortion. What name is given to this particular type of distortion?

8-25. Refer to Fig. 8-22. Assume that a signal drives the input positive. What happens to the current flow in Q_1 and Q_3?

8-26. Refer to Fig. 8-22. Assume that a signal drives the input positive. What happens at the top of R_L?

8-27. Refer to Fig. 8-22. Assume that Q_2 is turned on harder (conducts more). What should happen to Q_4?

8-28. Refer to Fig. 8-22. Which two transistors operate in complementary symmetry?

8-29. Which nondigital amplifier class has the best efficiency?

8-30. Refer to Fig. 8-24. What allows the output signal across the load resistor to be a sine wave?

8-31. Refer to Fig. 8-24. What makes the conduction angle of the amplifier so small?

8-32. Refer to Fig. 8-29. What does the charge on C_1 accomplish?

8-33. Refer to Fig. 8-29. What two things does the tank circuit accomplish?

8-34. Refer to Fig. 8-29. What will happen to the reverse bias at the base of the transistor if the drive level increases?

Critical Thinking Questions

8-1. Why can't the theoretical efficiency of an amplifier exceed 100 percent?

8-2. Can you identify any problem which could occur when the power transistors in a push-pull amplifier are poorly matched?

8-3. Can you think of any way to alleviate the problem identified in question 2?

8-4. Why is the failure rate for power amplifiers greater than that of small-signal amplifiers?

8-5. Amplifiers capable of power output levels in excess of a kilowatt are often based on vacuum tube technology. Why?

8-6. The diodes used to thermally compensate transistor power amplifiers are sometimes physically mounted to the same heat sink that is used to cool the transistors. Why?

8-7. Suppose you are working on an RF power amplifier similar to the one shown in Fig. 8-29. The amplifier repeatedly blows the V_{CC} fuse. What could be wrong?

Answers to Tests

1. F
2. T
3. F
4. F
5. T
6. T
7. 40 percent
8. 167 W
9. 0.768 W
10. 0.6 W
11. 720 Ω
12. 7.5 V peak-to-peak; 88 mW
13. F
14. T
15. F
16. F
17. T
18. T
19. 800 Ω; 800 Ω
20. 10 W
21. 2 W
22. 0.4 W
23. 1.02 W
24. 39.2 percent (Note: This is half the efficiency achieved for driving the amplifier to its maximum swing.)
25. Yes
26. class B
27. No, because this would remove forward bias and tend to make them run cooler.
28. Yes, because this would increase forward bias.
29. positive
30 Q_2
31. 10, 10.7, and 9.3 V
32. 6.6 to 6.7 V
33. sine wave
34. Yes
35. Yes
36. 30 MHz
37. 8.9 MHz
38. It should increase (go more negative).
39. 333 A/s
40. saturation
41. diodes
42. better
43. same time
44. greater

Chapter 9

Operational Amplifiers

Chapter Objectives

This chapter will help you to:

1. *Predict* the phase relationships in differential amplifiers.
2. *Determine* the CMRR for differential amplifiers.
3. *Calculate* the power bandwidth for operational amplifiers.
4. *Find* voltage gain for operational amplifiers.
5. *Determine* the small-signal bandwidth for operational amplifiers.
6. *Identify* various applications for operational amplifiers.

Thanks to integrated-circuit technology, differential and operational amplifiers are inexpensive, offer excellent performance, and are easy to apply. This chapter deals with the theory and characteristics of these amplifiers. It also covers some of their many applications.

9-1 The Differential Amplifier

An amplifier can be designed to respond to the *difference* between two input signals. Such an amplifier has *two inputs* and is called a difference, or *differential,* amplifier. Figure 9-1 on page 222 shows the basic arrangement. The $-V_{EE}$ supply provides forward bias for the base-emitter junctions and $+V_{CC}$ reverse-biases the collectors. Such supplies are called *dual supplies* or *bipolar supplies.* Two batteries can be used to form a bipolar supply, as in Fig. 9-2 on page 222. Figure 9-3 on page 222 shows a bipolar rectifier circuit.

A differential amplifier can be driven at *one* of its inputs. This is shown in Fig. 9-4 on page 223. An output signal will appear at *both* collectors. Assume that the input drives the base of Q_1 in a positive direction. The conduction in Q_1 will increase since it is an NPN device. More voltage will drop across Q_1's load resistor because of the increase in current. This will cause the collector of Q_1 to go less positive. Thus, an inverted output is available at the collector of Q_1.

What causes the signal at the collector of Q_2? As Q_1 is turned on harder by the positive-going input, the current through R_E will increase. This makes the drop across R_E increase. The emitters of both transistors will go in a positive direction. With Q_2's emitter going in a positive direction, it responds by conducting less current. Less voltage drops across Q_2's load resistor, and its collector goes in a positive direction.

Difference amplifier

Later, we are going to learn that the total emitter current in a differential amplifier should be constant. Let's use some numbers to clarify why both outputs are active in Fig. 9-4 *without resorting to changing the current in* R_E. Assume that the quiescent emitter voltage is -0.7 V. For simplicity, we can also assume that V_B for both transistors is zero. This is reasonable because the base current is very small; so the voltage drop across the base resistors will be close to zero. Putting the two assumptions together suggests that V_{BE} is $+0.7$ V for both transistors when there is no input signal. Now, a positive-going input signal is applied to Q_1 and increases its base voltage to $+0.05$ V.

Bipolar supply

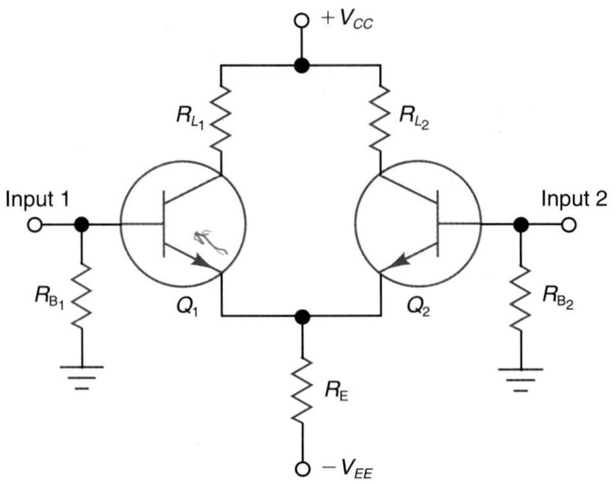

Fig. 9-1 A differential amplifier.

The emitter of Q_1 responds by going in a positive direction, say to -0.67 V. Since the emitters are wired together, V_{BE} for Q_2 is now $^{+}0.67$ V. Q_2 now conducts less current because it has less forward bias. So, the positive-going signal applied to Q_1 has increased its V_{BE} to 0.72 and has decreased the V_{BE} for Q_2 to 0.67. If the increase in current in Q_1 is equal to the decrease in current in Q_2, then the current in R_E does not change.

Which explanation is correct? Does the current in R_E change or not? The current in R_E does change somewhat in Fig. 9-4, and the best explanation is a combination of the two presented before. However, when R_E is replaced with a constant current source, the output from Q_2 must be explained with changes in both V_{BE} drops.

Single-ended output

Differential output

Common-mode signal

As can be seen in Fig. 9-4, both *inverted* (out-of-phase) and *noninverted* (in-phase) outputs are available. These appear from ground to either collector terminal and may be called *single-ended outputs.* A *differential output* is also available. This output is taken from the collector of Q_1 to the collector of Q_2. The differential output has twice the swing of either single-ended output. If, for example, Q_1's collector has gone 2 V negative and Q_2's collector has gone 2 V positive, the difference is $(+2) - (-2) = 4$ V.

The amplifier can also be *driven differentially,* as shown in Fig. 9-5. The advantage to this connection is that hum and noise can be significantly reduced. Power-line hum is a common problem in electronics, especially where high-gain amplifiers are involved. The 60-Hz power circuits radiate signals which are picked up by sensitive electronic circuits. If the hum is common to both inputs (same phase), it will not affect the differential output.

Figure 9-6 shows how hum can affect a desired signal. The result is a noisy signal of poor quality. The hum and noise can be stronger than the desired signal.

Refer to Fig. 9-7 on page 224. A noisy differential signal is shown. Note that the phase of the hum signal is common. The hum goes positive to both inputs at the same time. Later, both inputs see a negative-going hum signal. This is called a *common-mode signal.* Common-mode signals can be greatly attenuated (made smaller) in differential amplifiers.

Refer to Fig. 9-5. A common-mode signal will drive both inputs in the same direction.

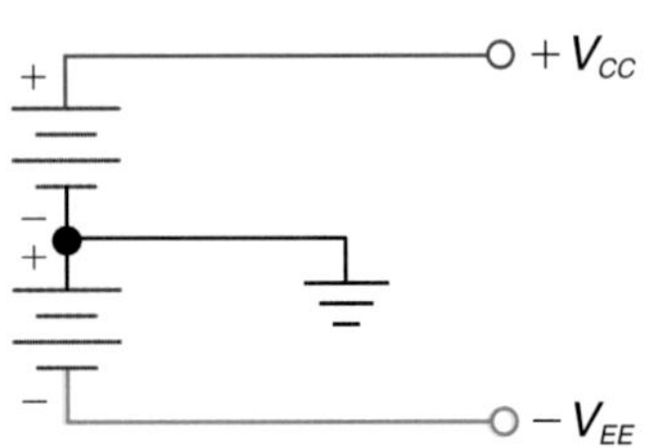

Fig. 9-2 A battery dual supply.

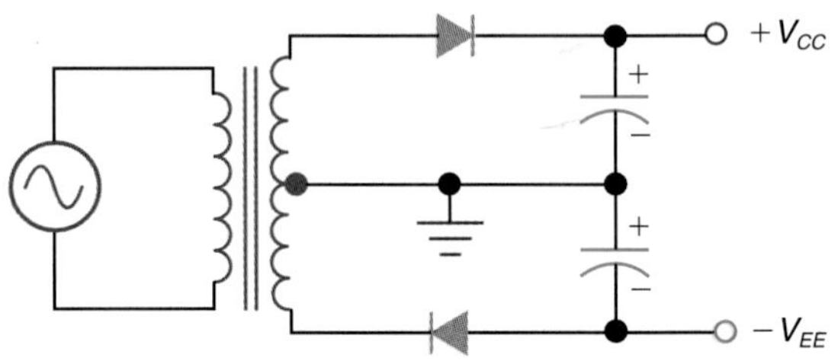

Fig. 9-3 A rectifier dual supply.

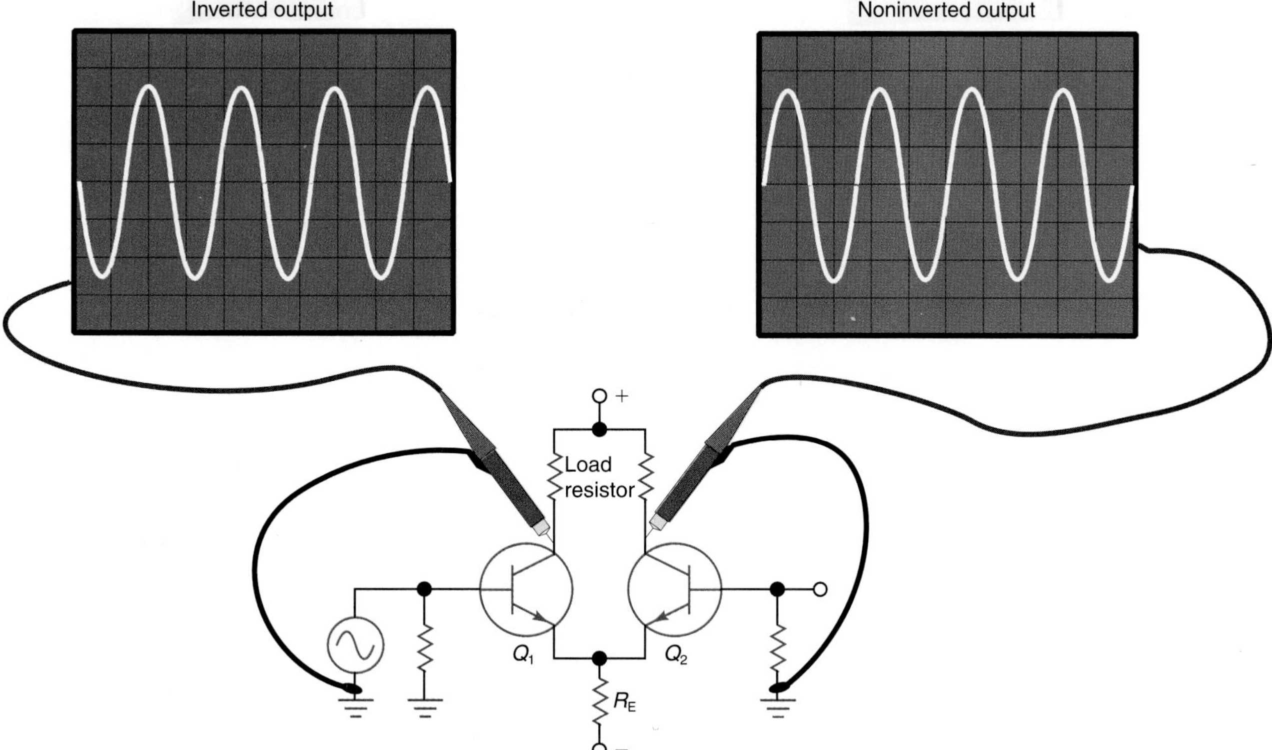

Fig. 9-4 Driving a differential amplifier at one input.

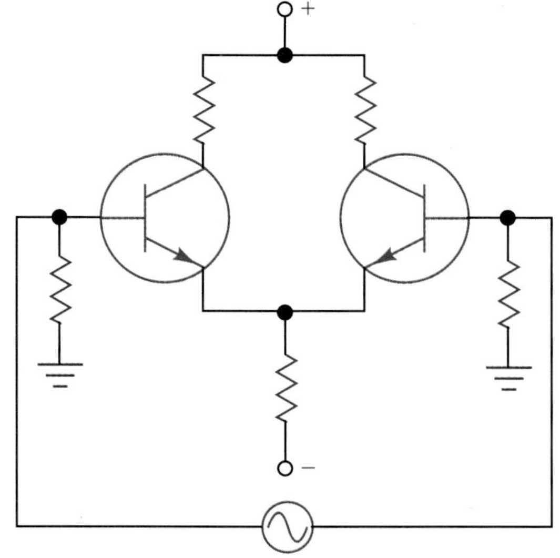

Fig. 9-5 Driving the amplifier differentially.

For example, both inputs can go positive, and this will tend to increase the conduction in both transistors. This can occur to some extent when the amplifier is biased with an emitter resistor. So there will be some decrease in collector voltage at both transistors. But the change in collector voltage will be quite a bit less than what is produced by an equivalent differential input. The amplifier shows considerably less gain for common-mode signals than it does for differential signals. Figure 9-7 shows that the output does not contain the common-mode signal.

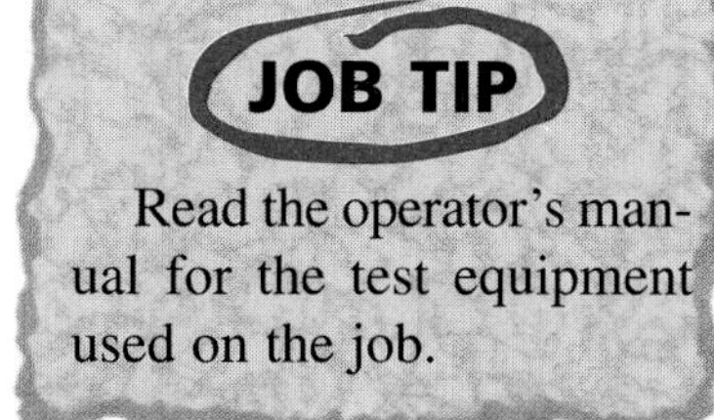

If the emitter resistor in Fig. 9-5 is replaced with a constant-current source, the transistor

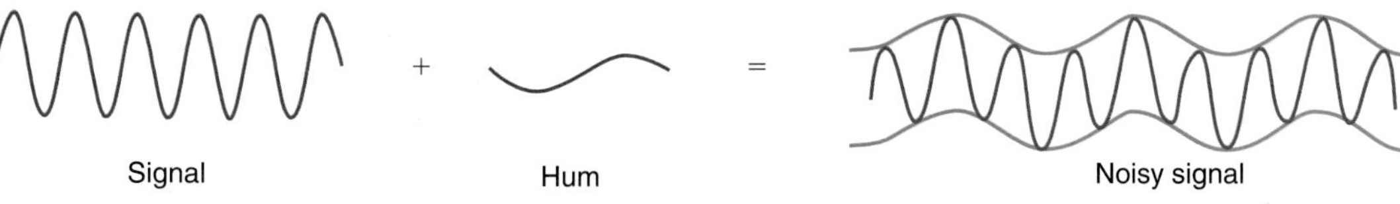

Fig. 9-6 Hum voltage can add to a signal.

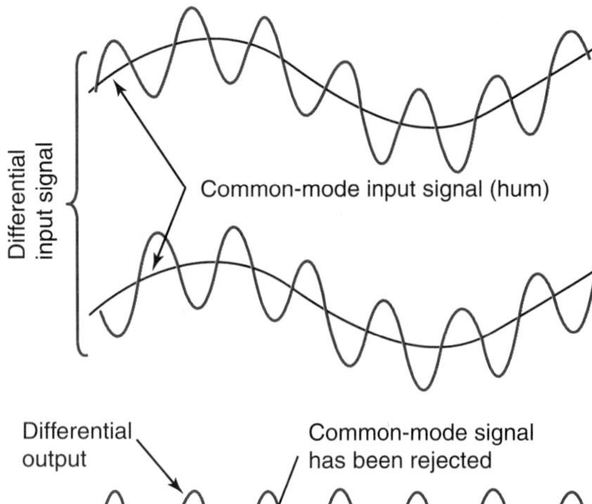

Fig. 9-7 The common-mode hum can be rejected.

currents will remain the same with a common-mode input signal. If the currents do not change, there will be no change in voltage at either collector. Ideally, there is zero signal output for common-mode input signals. The ideal is never quite achieved because this would require perfectly matched transistors and resistors. Also, there is no such thing as an ideal current source, and some slight change in currents does occur with a common-mode input.

In practice, a differential amplifier will not have perfect balance. For example, one transistor may show a little more gain than the other. This means that some common-mode signal will appear at the output. The ability to reject the common-mode signal is given by the *common-mode rejection ratio* (CMRR):

CMRR

$$\text{CMRR} = \frac{A_{V(\text{dif})}}{A_{V(\text{com})}}$$

where $A_{V(\text{dif})}$ = voltage gain of amplifier for differential signals

$A_{V(\text{com})}$ = voltage gain of amplifier for common-mode signals

About Electronics

Plastic Op-Amp Packages Op amps housed in plastic packages are not acceptable for some applications in areas such as aerospace, military, and medical.

Assume that a common-mode input signal is 1 V and that it produces a 0.05-V output signal. The common-mode voltage gain is

$$A_{V(\text{com})} = \frac{\text{signal out}}{\text{signal in}} = \frac{0.05\text{ V}}{1\text{ V}} = 0.05$$

Also assume that a differential input signal is 0.1 V and produces an output of 10 V. The differential voltage gain is

$$A_{V(\text{dif})} = \frac{\text{signal out}}{\text{signal in}} = \frac{10\text{ V}}{0.1\text{ V}} = 100$$

The common-mode rejection ratio of the amplifier is

$$\text{CMRR} = \frac{100}{0.05} = 2000$$

The amplifier shows 2000 times as much gain for differential signals as it does for common-mode signals. The CMRR is usually specified in decibels:

$$\text{CMRR}_{(\text{dB})} = 20 \times \log 2000 = 66\text{ dB}$$

Example 9-1

An amplifier has a differential gain of 40 dB and a common mode gain of −26 dB. What is the CMRR for this amplifier? When the differential and common-mode gains are expressed in decibels, the CMRR is found by subtracting:

$$\text{CMRR} = 40\text{ dB} - (-26\text{ dB}) = 66\text{ dB}$$

Some differential amplifiers have common-mode rejection ratios over 100 dB. They are very effective in rejecting common-mode signals.

TEST

Solve the following problems.

1. Refer to Fig. 9-1. What name is given to the energy source marked $+V_{CC}$ and $-V_{EE}$?
2. Refer to Fig. 9-1. Assume that input 1 and input 2 are driven 1 V positive. If the amplifier has perfect balance, what voltage difference should appear across the two collectors?
3. Refer to Fig. 9-1. Assume that a signal appears at input 1 and drives it positive.

In what direction will the collector of Q_1 be driven? The collector of Q_2?

4. When a signal drives input 2 in Fig. 9-1, why is there a collector voltage change at Q_1?
5. Assume that in Fig. 9-1 a signal drives input 1 and produces an output at the collector of Q_1. This output measures 2 V peak-to-peak with respect to ground? What signal amplitude should appear across the two collectors?
6. Refer to Fig. 9-2. Both batteries are 12 V. What is V_{CC} with respect to ground? What is V_{EE} with respect to ground? What is V_{CC} with respect to V_{EE}?
7. Refer to Fig. 9-4. What reference point is used to establish the inverted output and the noninverted output?
8. Refer to Fig. 9-5. Assume the signal source supplies 120 mV. The differential output signal is 12 V. Calculate the differential voltage gain of the amplifier.
9. Refer to Fig. 9-5. The differential voltage gain is 80. A common-mode hum voltage of 80 mV is applied to both inputs. The differential hum output is 8 mV. Calculate the CMRR for the amplifier.

9-2 Differential Amplifier Analysis

The properties of differential amplifiers can be demonstrated by working through the dc and ac conditions of a typical circuit. Figure 9-8 shows a circuit with all the values necessary to determine the dc and ac conditions.

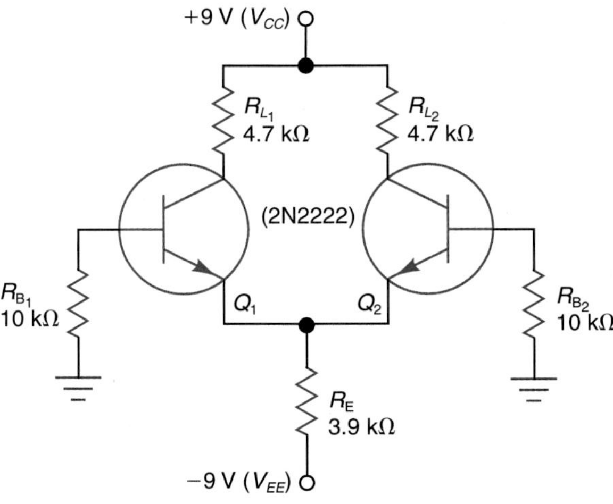

Fig. 9-8 A differential-amplifier circuit.

Analyzing circuits like Fig. 9-8 is made easier by making a few assumptions. One assumption is that the base leads of the transistors are at ground potential. This is reasonable since the base currents are very small, which makes the drops across R_{B_1} and R_{B_2} close to 0 V. The next assumption is that the transistors are turned on. If the bases are at 0 V, then the emitters must be at −0.7 V. This condition is required to forward-bias the base-emitter junctions and turn on the transistors. Saying that the emitter is −0.7 V with respect to the base is the same as saying that the base is +0.7 V with respect to the emitter. This satisfies the bias requirements for NPN transistors.

Now that we have made our basic assumptions, we can begin the dc analysis. Knowing the voltage at both ends of R_E allows us to find its drop:

$$V_{R_E} = -9\text{ V} - (-0.7\text{ V}) = -8.3\text{ V}$$

Knowing that there is a drop of 8.3 V (we now discard its sign) allows us to solve for the current flow in the emitter resistor:

$$I_{R_E} = \frac{V_{R_E}}{R_E} = \frac{8.3\text{ V}}{3.9\text{ k}\Omega} = 2.13\text{ mA}$$

Assuming balance, each transistor will support half of this current. The emitter current, for each transistor, is

$$I_E = \frac{2.13\text{ mA}}{2} = 1.06\text{ mA}$$

As usual, we assume the collector currents to be equal to the emitter currents. The drop across each load resistor is

$$V_{R_L} = 1.06\text{ mA} \times 4.7\text{ k}\Omega = 4.98\text{ V}$$

V_{CE} is found by Kirchhoff's voltage law:

$$\begin{aligned} V_{CE} &= V_{CC} - V_{R_L} - V_E \\ &= 9 - 4.98 - (-0.7) = 4.72\text{ V} \end{aligned}$$

The above dc analysis shows that the dc conditions of the differential amplifier of Fig. 9-8 are good for linear operation. Note that the collector-to-emitter voltage is about half of the collector supply. Before leaving the dc analysis, we will make two more calculations. We can estimate the base current by guessing that β is 200. This is reasonable for 2N2222 transistors. The base current is

$$I_B = \frac{I_C}{\beta} = \frac{1.06\text{ mA}}{200} = 5.3\ \mu\text{A}$$

This current flows in each of the 10-kΩ base resistors. The voltage drop across each resistor is

$$V_{RB} = 5.3\ \mu\text{A} \times 10\ \text{k}\Omega = 53\ \text{mV}$$

Each base is 53 mV negative with respect to ground. Remember that base current flows out of an NPN transistor. This direction of flow makes the bases in Fig. 9-8 slightly negative with respect to ground. The 53 mV value is very small, so the initial assumption was valid.

We are now prepared to do an *ac analysis* of the circuit. The first step is to find the ac resistance of the emitters:

$$r_E = \frac{50}{I_E} = \frac{50}{1.06} = 47\ \Omega$$

You may recall that ac emitter resistance can be estimated by using a 25- or a 50-mV drop. The higher estimate is more accurate for circuits like Fig. 9-8.

Knowing r_E allows us to find the voltage gain for the differential amplifier. There are actually two gains to find: (1) the differential voltage gain and (2) the common-mode voltage gain. Figure 9-9 shows an ac equivalent circuit that is appropriate when the amplifier is driven at one input. Notice that r_E is shown in both emitter circuits. The differential voltage gain (A_D) is equal to the collector load resistance divided by 2 times r_E.

In Fig. 9-9, very little signal current flows in R_E, so it does not appear in the voltage gain equation. Q_1 is driven at its base by the signal source. Its emitter signal current must flow through its 47 Ω of ac emitter resistance. This emitter signal also drives the emitter of Q_2 and must overcome its 47 Ω of ac emitter resistance. Q_2 acts as a common-base amplifier in this circuit and it is driven at its emitter by the emitter of Q_1. This is why the denominator of the gain equation contains $2 \times r_E$ (the two 47-Ω resistances are acting in series for signal current). R_E is much larger than the ac emitter resistances and its effect is small enough to be ignored.

The base resistors in Fig. 9-9 can also affect the differential voltage gain. When these resistors are small, they can be ignored. If the resistors are large, they will reduce the gain. The reason this happens is that the signal current flows in the base-emitter circuit, and so the base resistor offers additional opposition. However, the effect of the base resistor is reduced by the current gain of the transistor. When viewed from the emitter, the base resistor appears smaller to the ac signal current. So, if the base resistors in Fig. 9-9 are fairly large, say 10 kΩ, the ac base resistance is found by:

$$r_B = \frac{R_B}{\beta} = \frac{10\ \text{k}\Omega}{200} = 50\ \Omega$$

The ac base resistance decreases the differential gain:

$$A_{V(\text{dif})} = \frac{R_L}{(2 \times r_E) + r_B} = \frac{4.7\ \text{k}\Omega}{(2 \times 47\ \Omega) + 50\ \Omega} = 32.6$$

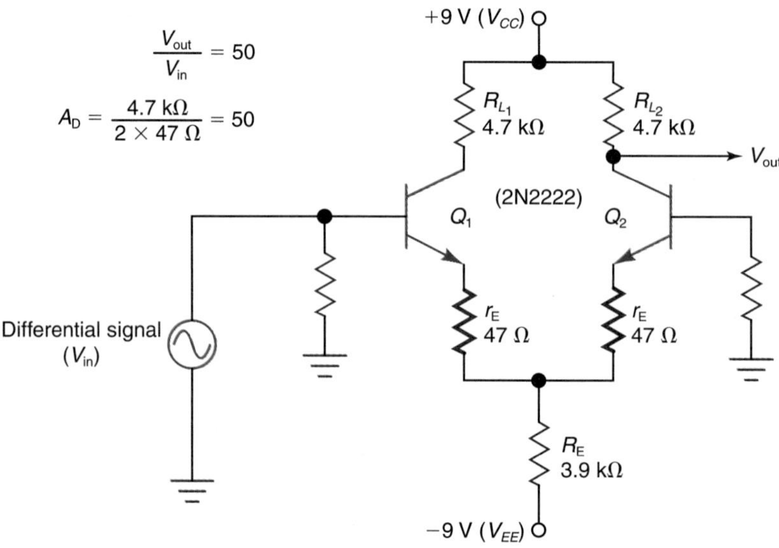

Fig. 9-9 AC equivalent circuit for differential signal gain.

The base resistance is used only once in the gain equation (not multiplied by 2) because the signal source is applied directly to one base. In Fig. 9-9, only the base resistor on the right affects the signal current.

Considering the ac base resistance can provide a more accurate estimate of *differential gain*. However, it may not be necessary. Since we used the conservative 50-mV value for estimating r_E, the gain will probably be closer to 50 for Fig. 9-9. Designers often use a very conservative approach to ensure that the actual circuit gain will be at least as high as their calculated value. Too much gain is an easier problem to solve than not enough gain.

A differential gain of 50 is very respectable. As we will see, the common-mode gain is much less. Figure 9-10 shows the ac equivalent circuit for *common-mode gain*. Here, the 47-Ω emitter resistances of the transistors are eliminated. They are so small compared to 7.8 kΩ that they can be ignored. R_E is physically a 3.9-kΩ resistor. However, it appears to be twice that value because it supports both transistor currents. A common-mode signal drives both bases in the same direction. So, considering the case when the signal is going positive, both transistors are turned on harder. R_E must support twice the increase in current that it would if it were serving just a single transistor. The output signal is taken from Q_2 in Fig. 9-10. As far as Q_2 is concerned, its emitter is loaded by 7.8 kΩ. This large resistance makes the common-mode gain less than 1.

Our ac analysis of the differential amplifier has shown a differential gain of 50 and a common-mode gain of 0.603. The ratio is

Differential gain

$$\frac{50}{0.603} = 82.9$$

We can expect this differential amplifier to produce almost 83 times more gain for a differential signal than for a common-mode signal. This will go a long way toward eliminating noise in many applications. The decibel CMRR is

$$\text{CMRR} = 20 \times \log 82.9 = 38.4 \text{ dB}$$

Common-mode gain

A CMRR of 38.4 dB is good for such a simple circuit. However, some applications are very demanding and even better performance may be required. A look at the equation for common-mode gain will provide insight into what can be done to improve performance.

$$A_{V(\text{com})} = \frac{R_L}{2 \times R_E}$$

This equation shows that if R_E is made very large, then the common-mode voltage gain is very small. Using a high voltage for the emitter supply (V_{EE}) is one way to accomplish this.

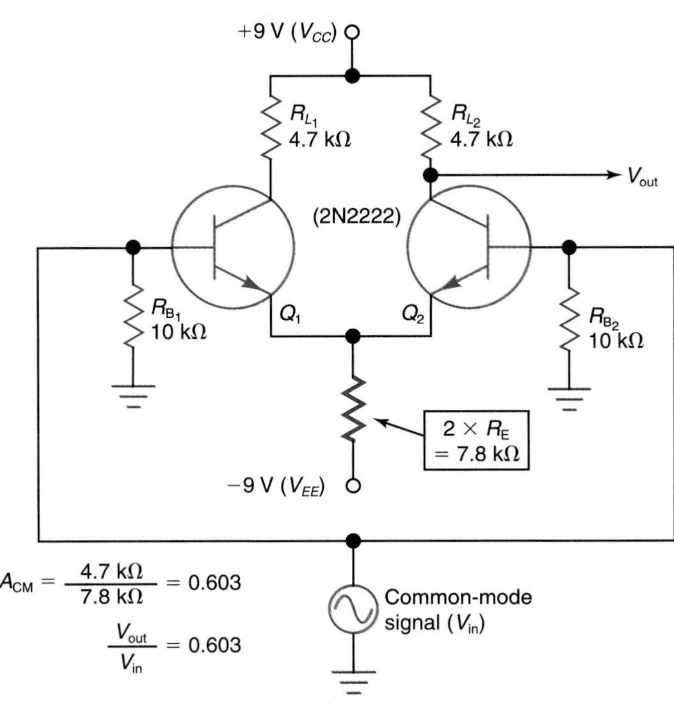

Fig. 9-10 AC equivalent circuit for common-mode signal gain.

For example, if V_{EE} were 90 V, R_E would be much larger. Assuming that we still need about 2 mA of total emitter current:

$$R_E = \frac{90 \text{ V}}{2 \text{ mA}} = 45 \text{ k}\Omega$$

Using this resistor in the differential amplifier would decrease the common-mode gain to 0.052. Unfortunately, a 90-V power supply is not desirable in most applications.

Example 9-2

What is the dB CMRR for Fig. 9-10 if R_E is 45 kΩ? The differential gain is not affected by the emitter resistor. The differential gain remains at 50 and:

$$\text{CMRR} = 20 \times \log \frac{50}{0.0522} = 59.6 \text{ dB}$$

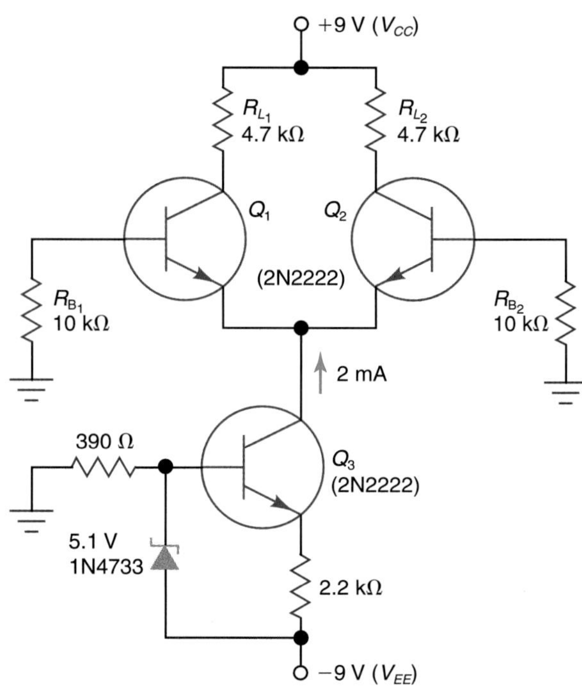

Fig. 9-11 A differential amplifier with current source biasing.

Current source

Figure 9-11 shows a better solution. R_E has been replaced with a *current source* consisting of two resistors, a zener diode, and transistor Q_3. The zener diode is biased by the −9-V supply. The 390-Ω resistor limits the zener current. The zener cathode is 5.1 V positive with respect to its anode. This drop forward-biases the base-emitter circuit of Q_3. If we subtract for V_{BE}, we can determine the current flow in the 2.2-kΩ resistor:

$$I = \frac{5.1 \text{ V} - 0.7 \text{ V}}{2200 \text{ }\Omega} = 2 \text{ mA}$$

The emitter current of Q_3 is 2 mA. We can make the usual assumption that the collector current is equal to the emitter current. Thus, the collector of Q_3 in Fig. 9-11 supplies 2 mA to the emitters of the differential amplifier.

A current source such as the one shown in Fig. 9-11 has a very high ac resistance. This resistance is a function of Q_3's ac collector and ac emitter resistances and the 2.2-kΩ emitter resistor.

The ac collector resistance of small transistors typically ranges from 50 to 200 kΩ. As Fig. 9-12 shows, the collector curve is relatively flat. The collector current changes only a small amount over the 20-V range of the graph. The ac collector resistance can be found from the graph by using Ohm's law. The graph shows that the collector current change is 0.2 mA for a collector-to-emitter change of 20 V:

$$r_C = \frac{\Delta V_{CE}}{\Delta I_C} = \frac{20 \text{ V}}{0.2 \text{ mA}} = 100 \text{ k}\Omega$$

The ac emitter resistance is estimated with the familiar:

$$r_E = \frac{50 \text{ mV}}{I_E} = \frac{50 \text{ mV}}{2 \text{ mA}} = 25 \text{ }\Omega$$

The following formula can be used to estimate the ac resistance of a constant current source such as the one shown in Fig. 9-11:

$$\begin{aligned} r_{EE} &= r_C \times \left(1 + \frac{R_E}{r_E}\right) \\ &= 100 \text{ k}\Omega \times \left(1 + \frac{2.2 \text{ k}\Omega}{25}\right) \\ &= 8.9 \text{ M}\Omega \end{aligned}$$

This rather high value of ac resistance makes the common-mode gain of the amplifier in Fig. 9-11 very small:

$$\begin{aligned} A_{V(\text{com})} &= \frac{R_L}{2 \times r_{EE}} \\ &= \frac{4.7 \text{ k}\Omega}{2 \times 8.9 \text{ M}\Omega} \\ &= 0.264 \times 10^{-3} \end{aligned}$$

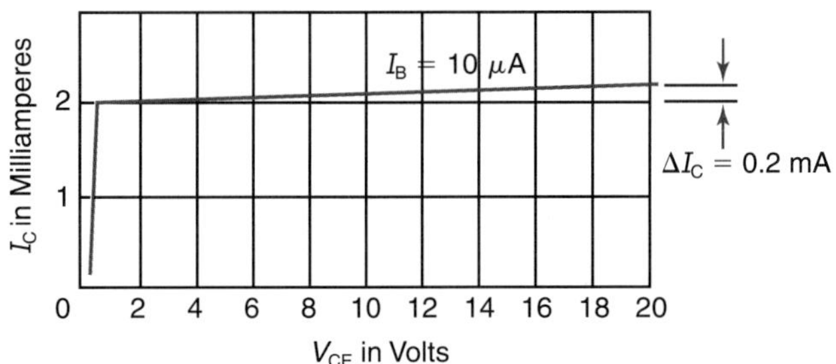

Fig. 9-12 Typical collector curve for a 2N2222 transistor.

The current source biases the amplifier of Fig. 9-11 at about the same level of current as the circuit of Fig. 9-8. Therefore, the differential gain will be about the same (50). The CMRR for Fig. 9-11 is quite large:

$$\text{CMRR} = 20 \times \log \frac{50}{0.264 \times 10^{-3}} = 106 \text{ dB}$$

In practice, it is difficult to achieve such a high CMRR. However, the circuit of Fig. 9-11 is substantially better than the circuit of Fig. 9-8. When the CMRR must be optimized, matched components and laser-trimmed components can be used. Integrated circuit (IC) amplifiers with differential inputs usually have good CMRRs because the transistors and resistors tend to be well matched and they track thermally (change temperature by the same amount).

TEST

Solve the following problem.

10. Use Fig. 9-8 as a guide with the following changes and assumptions: the transistors are identical and $\beta = 200$, $R_{L_1} = R_{L_2} = 10$ kΩ, $R_E = 8.9$ kΩ, $R_{B_1} = R_{B_1} = 100$ kΩ. Use 50 mV to estimate r_E. Find: I_{R_E}, $I_{E(Q_1)}$, $I_{E(Q_2)}$, $V_{CE(Q_1)}$, $V_{CE(Q_2)}$, $V_{B(Q_1)}$, $V_{B(Q_2)}$, $A_{V(\text{dif})}$, $A_{V(\text{com})}$, and CMRR.

9-3 Operational Amplifiers

Operational amplifiers (op amps) use differential input stages. They have characteristics that make them very useful in electronic circuits. Some of their characteristics are as follows:

1. *Common-mode rejection:* This gives them the ability to reduce hum and noise.
2. *High input impedance:* They will not "load down" a high-impedance signal source.
3. *High gain:* They have "gain to burn" which is usually reduced by using negative feedback.
4. *Low output impedance:* They are able to deliver a signal to a low-impedance load.

No single-stage amplifier circuit can rate high in all the above characteristics. An operational amplifier is actually a combination of several amplifier stages. Refer to Fig. 9-13 on page 230. The first section of this multistage circuit is a differential amplifier. Differential amplifiers have common-mode rejection and a high input impedance. Some operational amplifiers may use field-effect transistors in this first section for an even higher input impedance. Operational amplifiers that combine *bi*polar devices with *FET* devices are called *BIFET* op amps.

Did You Know?

Op-Amps/EMI
There are devices called programmable op amps. Some op amps have very high gain. EMI (electromagnetic interference) tends to be more troublesome in high-gain circuits than in other circuits.

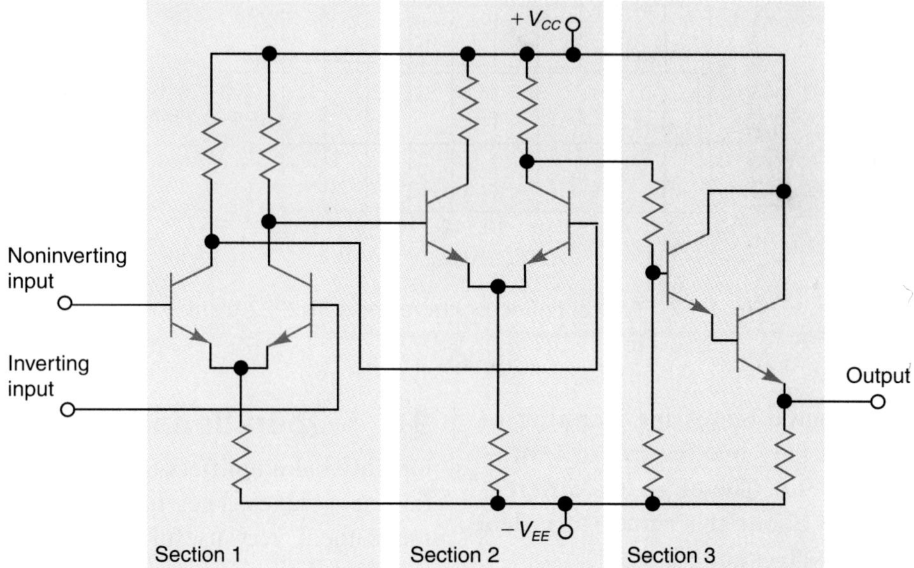

Fig. 9-13 The major sections of an operational amplifier.

Noninverting input

Inverting input

Single-ended output

The second section of Fig. 9-13 is another differential amplifier. This allows the differential output of the first section to be used. This provides the best common-mode rejection and differential voltage gain.

The third section of Fig. 9-13 is a common-collector, or emitter-follower, stage. This configuration is known for its low output impedance. Notice that the output is a single terminal. No differential output is possible. This is usually referred to as a *single-ended output.* Most electronic applications require only a single-ended output.

A single-ended terminal can show only one phase with respect to ground. This is why Fig. 9-13 shows one input as *noninverting* and the other as *inverting.* The noninverting input will be in phase with the output terminal. The inverting input will be 180° out of phase with the output terminal.

Figure 9-14 shows the amplifier in a simplified way. Notice the triangle. Electronic di-

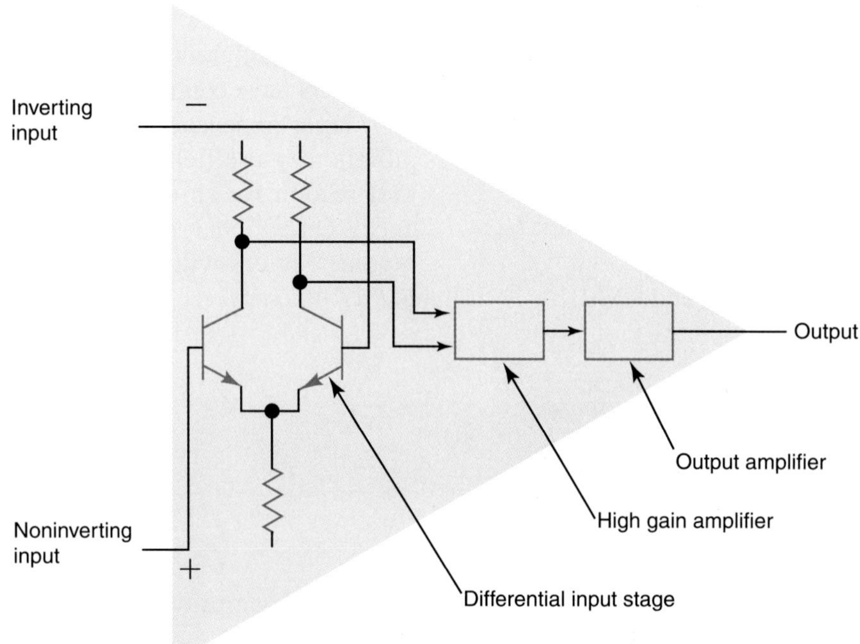

Fig. 9-14 A simplified way of showing an operational amplifier.

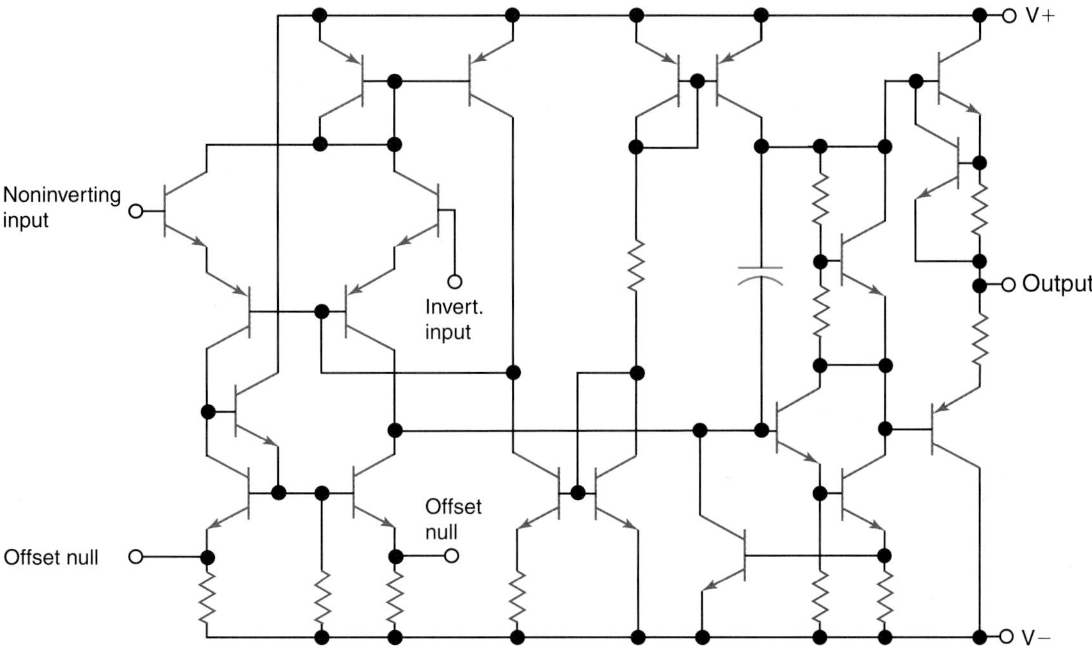

Fig. 9-15 A schematic of an operational amplifier.

agrams often use triangles to represent amplifiers. Also notice that the inverting input is marked with a minus (−) sign and that the noninverting input is marked with a plus (+) sign. This is standard practice.

Figure 9-15 shows the schematic diagram of a common integrated-circuit op amp. This device has a noninverting input, an inverting input, and a single-ended output. It also has two terminals marked "offset null." These terminals can be used in those applications where it is necessary to correct for dc offset error. It is not possible to manufacture amplifiers with perfectly matched transistors and resistors. The mismatch creates a dc offset error in the output. With no differential dc input, the dc output of an op amp should ideally be zero volts with respect to ground. Any deviation from this is known as dc offset error. Figure 9-16 shows a typical application for nulling (eliminating) the offset.

The potentiometer in Fig. 9-16 is adjusted so that the output terminal is at dc ground potential with no differential dc input voltage. This potentiometer has a limited range. The null circuit is designed to overcome an internal offset in the millivolt range. It is not designed to null the output when there is a large dc differential input applied to the op amp by external circuit conditions. In many applications, a small offset does not present a problem. The offset null terminals are not connected in such applications.

Most op amps are integrated. A technician cannot see what is inside an integrated circuit or make any internal measurements. Therefore, it is seldom necessary to show the schematic details of the internal circuitry. Figure 9-17 on page 232 is the standard way of showing an operational amplifier on a schematic diagram. The supply terminals and offset null terminals may also be shown on some diagrams.

Offset null

Quite a variety of integrated circuit operational amplifiers are available. Some use bipolar transistors, and some use field-effect transistors in combination with bipolars. Special op amps are available with enhanced characteristics in areas such as input impedance and high-frequency performance. It is not possible

\+ Supply terminal
Inverting input terminal
Noninverting input terminal
Output terminal
− Supply terminal
Offset null terminals
10 kΩ
Offset adjust potentiometer

Fig. 9-16 Using the offset null terminals.

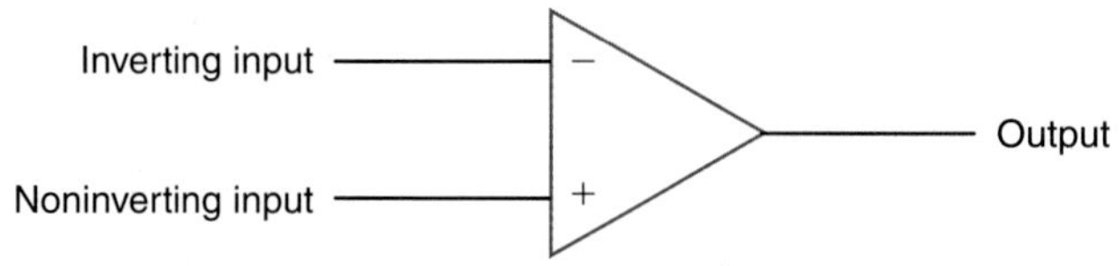

Fig. 9-17 The standard way of showing an operational amplifier.

to list all their characteristics here. The following list represents some general characteristics for a typical op amp like the LM741C.

- Voltage gain: 200,000 (106 dB)
- Output impedance: 75 Ω
- Input impedance: 2 MΩ
- CMRR: 90 dB
- Offset adjustment range: ±15 mV
- Output voltage swing: ±13 V
- Small-signal bandwidth: 1 MHz
- Slew rate: 0.5 V/μs

Slew rate

The last characteristic in the list is *slew rate.* This is the maximum rate of change for the output voltage of an op amp. Fig. 9-18 shows what happens when the input voltage changes suddenly. The output cannot produce an instantaneous voltage change. It *slews* (changes) a certain number of volts in a given period of time. The unit of time used to rate op amps is the microsecond ($1\ \mu s = 1 \times 10^{-6}$ s). Some op amps have *slew rates* as low as 0.04 V/μs, and others are rated at 70 V/μs.

Distortion

Power bandwidth

Slew rate is an important consideration for high-frequency operation. High frequency means rapid change. An op amp may not be able to slew fast enough to reproduce its input signal. Fig. 9-19 shows an example of slew rate *distortion.* Note that the input signal is sinusoidal and that the output signal is triangular. The output signal from a linear amplifier is supposed to have the same shape as the input signal. Any deviation is called distortion.

In addition to causing distortion, slew rate may prevent an op amp from producing its full output voltage swing. Large output signals are more likely to be limited than small signals. So, the factors are signal frequency, output swing, and the slew rate specification of the op amp. The following equation predicts the maximum frequency of operation for sinusoidal input signals:

$$f_{max} = \frac{SR}{6.28 \times V_p}$$

where SR is the slew rate, V/μs, and V_p is the peak output, V.

A general-purpose op amp like the LM741C can produce a maximum peak output swing of 13 V when powered by a 15-V supply. Let's see what the maximum sine-wave frequency is if its slew rate is 0.5 V/μs:

$$f_{max} = \frac{0.5\ V/\mu s}{6.28 \times 13\ V} = \frac{1}{6.28 \times 13\ V} \times \frac{0.5\ V}{1 \times 10^{-6}\ s} = 6.12\ kHz$$

The voltage units cancel and the expression reduces to the reciprocal of time, which is equal to frequency; 6.12 kHz may be called the *power bandwidth* of the op amp. Two things will happen if a sinusoidal input signal is significantly greater than 6.12 kHz and is large enough to drive the output to 13 V peak: (1) the output signal will show distortion (as in Fig. 9-19),

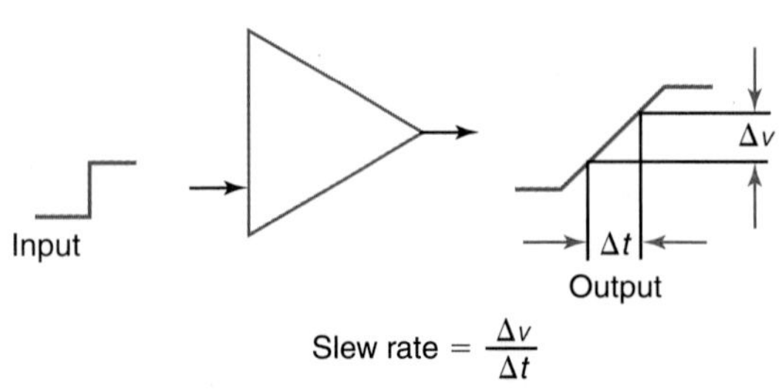

Fig. 9-18 Op-amp response to a sudden input change.

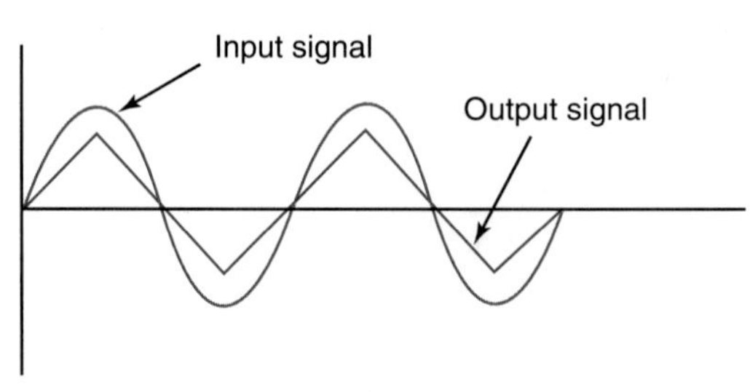

Fig. 9-19 Slew-rate distortion.

and (2) the peak output swing will be less than 13 V.

Example 9-3

Calculate the power bandwidth of a high-speed op amp with a slew rate of 70 V/μs when the output signal is 20 V_{p-p}. Apply the formula:

$$f_{max} = \frac{70\ \text{V}/\mu\text{s}}{6.28 \times 10\ \text{V}}$$

$$= \frac{1}{6.28 \times 10\ \text{V}} \times \frac{70\ \text{V}}{1 \times 10^{-6}\ \text{s}}$$

$$= 1.11\ \text{MHz}$$

A typical op amp like the LM741C has a *small-signal* bandwidth of 1 MHz. Large, high-frequency signals will be slew-rate-limited. The power bandwidth of an operational amplifier is less than its small-signal bandwidth. Table 9-1 lists several types of op amps along with a few of their specifications.

TEST

Solve the following problems.

11. Refer to Fig. 9-13. Which section of the amplifier (1, 2, or 3) operates as an emitter follower to produce a low output impedance?
12. Refer to Fig. 9-13. A signal is applied to the inverting input terminal. What is the phase of the signal that appears at the output terminal as compared to the input signal?
13. Refer to Fig. 9-13. Is the output of the amplifier differential or single-ended?
14. What is the name of the terminals used to null the effect of internal dc imbalance in an op amp?
15. What are two possible output effects if an input signal exceeds the power bandwidth of an op amp?
16. Refer to Table 9-1. What is the power bandwidth of the TL070 op amp, assuming a peak output of 10 V? How does this compare to the small-signal bandwidth of the device?

9-4 Setting Op-Amp Gain

A general-purpose op amp has an *open-loop* voltage gain of 200,000. Open loop means *without feedback.* Op amps are usually operated *closed-loop.* The output, or a part of it, is fed back to the inverting (−) input. This is *negative* feedback which reduces the gain and increases the bandwidth of the amplifier.

Open loop

Closed loop

Figure 9-20 on page 234 shows a closed-loop operational amplifier circuit. The output is fed back to the inverting input. The input signal drives the noninverting (+) input. The circuit is easy to analyze if we make an assumption: There is no difference in voltage across the op-amp inputs. What is the basis for this assumption? Considering a typical gain of 200,000, the assumption is reasonable. For example, if the output is at its maximum positive value, say 10 V, the differential input is only

$$V_{in(dif)} = \frac{V_{out}}{A_V} = \frac{10\ \text{V}}{2 \times 10^5} = 50\ \mu\text{V}$$

Fifty microvolts is close to zero so the assumption is valid. This is a *key point* for understanding op-amp circuits. The differential gain is *so large* that the differential input voltage can be assumed to be zero when making many practical calculations.

Table 9-1 A Sample of Op-Amp Specifications

Device	Description	A_V, dB	Z_{in}, Ω	CMRR, dB	Bandwidth, MHz	Slew Rate, V/μs
TL070	BIFET, low noise	106	10^{12}	86	3	13
TL080	BIFET, low power	106	10^{12}	86	3	13
TLC277	CMOS	92	10^{12}	88	2.3	4.5
LM308	High performance	110	40×10^6	100	1	0.3
LM318	High performance	106	3×10^6	100	15	70
LM741C	General purpose	106	2×10^6	90	1	0.5
TLC27L7	CMOS, low bias	114	10^{12}	88	0.1	0.04

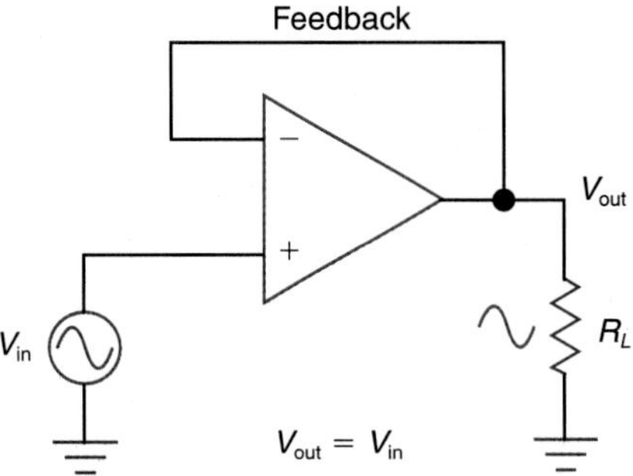

Fig. 9-20 An op amp with negative feedback.

Now let's apply the assumption to the circuit shown in Fig. 9-20. The feedback will eliminate any voltage difference across the input terminals. If the input signal swings 1 V positive, the output terminal will do exactly the same. Since the output is fed back to the inverting input, both inputs will be at +1 V and the differential input will be 0. If the input signal swings 5 V negative, the output terminal will do exactly the same. Again, the differential input is 0 because of the feedback. It should be clear that in Fig. 9-20 the output *follows* the input signal. In fact, this circuit is called a *voltage follower.* $V_{out} = V_{in}$, so the circuit has a voltage gain of 1.

Voltage follower

At first glance, an amplifier with a gain of 1 may seem to be no better than a piece of wire! However, such an amplifier can be useful if it has a high input impedance and a low output impedance. The input impedance of the voltage follower in Fig. 9-20 is equal to the input impedance of the op amp itself. If it is a general-purpose type, such as a 741, the input impedance is 2 MΩ. The output impedance of a voltage follower is approximately equal to the basic output impedance of the op amp divided by its open-loop gain. Because the open-loop gain is so high, the output impedance is 0 Ω for practical purposes:

$$Z_{out(CL)} \approx \frac{75\ \Omega}{200 \times 10^3} = 0.375\ \text{m}\Omega$$

Buffer

An amplifier that has an input impedance of 2 MΩ and an output impedance near 0 Ω makes an excellent *buffer.* Buffer amplifiers are used to isolate signal sources from any loading effects. They are also useful when working with signal sources that have rather high internal impedances.

Figure 9-21 shows an op-amp circuit that has a voltage gain greater than 1. The actual value of the gain is easy to determine. R_1 and the feedback resistor R_F form a voltage divider for the output voltage. The divided output voltage must be equal to the input voltage to satisfy the assumption that the differential input voltage is 0:

$$V_{in} = V_{out} \times \frac{R_1}{R_1 + R_F}$$

Dividing both sides by V_{out} and inverting gives

$$A_V = \frac{V_{out}}{V_{in}}$$

$$= \frac{R_1 + R_F}{R_1} = 1 + \frac{R_F}{R_1}$$

Let's apply this gain equation to Fig. 9-21:

$$A_V = 1 + \frac{R_F}{R_1}$$

$$= 1 + \frac{100\ \text{k}\Omega}{10\ \text{k}\Omega} = 11$$

Noninverting amplifier

The circuit of Fig. 9-21 is a *noninverting amplifier.* The input signal is applied to the + input of the op amp. An ac output signal will

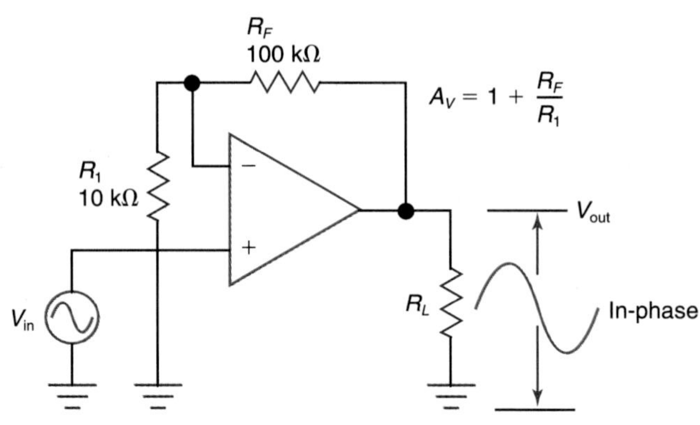

Fig. 9-21 A noninverting circuit with gain.

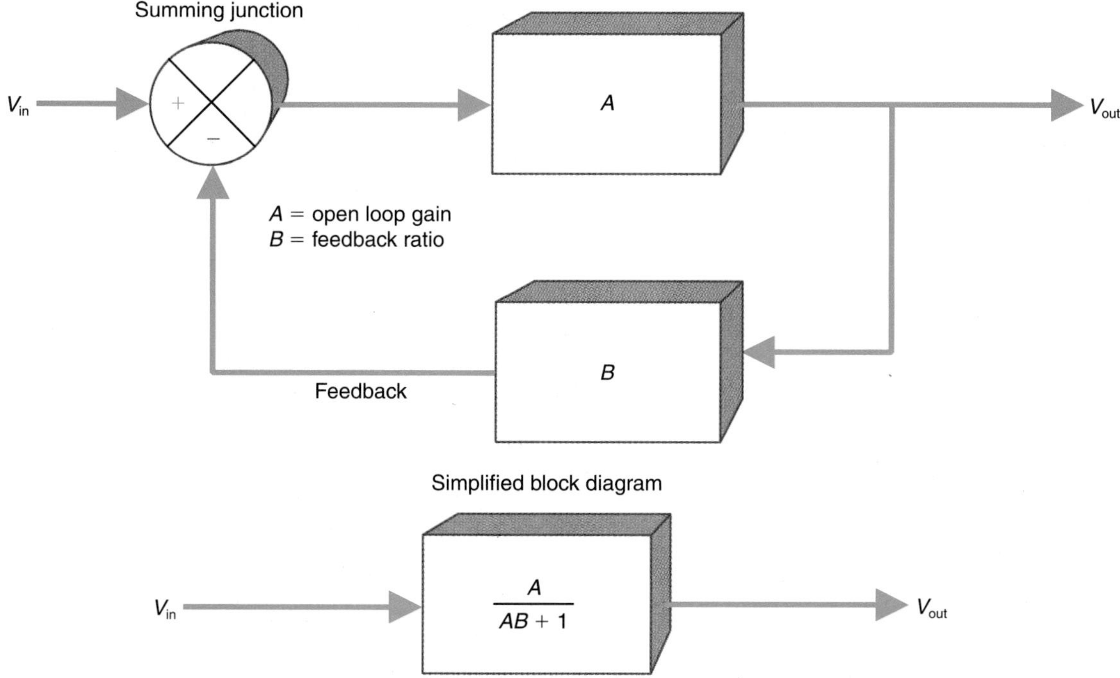

Fig. 9-22 Another model of negative feedback.

be in phase with the input signal. A dc input signal will create a dc output signal of the same polarity. For example, if the input is −1 V, the output will be −11 V (−1 V × 11 = −11 V).

Figure 9-22 shows another model for amplifiers with negative feedback. This model was presented in Chap. 7.

Example 9-4

Calculate the gain for Fig. 9-21 using the other model for closed-loop gain. The open-loop gain of the typical op amp is very high ($A = 200{,}000$). The feedback ratio (B) in Fig. 9-21 is determined by R_F and R_1, which form a voltage divider:

$$B = \frac{R_1}{R_F + R_1} = \frac{1\text{ k}\Omega}{10\text{ k}\Omega + 1\text{ k}\Omega} = 0.091$$

Applying the equation from the other model:

$$A_{CL} = \frac{A}{AB + 1} = \frac{200{,}000}{(200{,}000)(0.091) + 1} = 11$$

This is the same closed-loop gain determined before.

Example 9-5

Determine the output signal (amplitude and phase) for Fig. 9-21 if R_1 is changed to 22 kΩ and V_{in} is 100 $\text{mV}_{\text{p-p}}$. First, determine the gain for the amplifier:

$$A_V = 1 + \frac{R_F}{R_1} = 1 + \frac{100\text{ k}\Omega}{22\text{ k}\Omega} = 5.55$$

The output signal will be in phase with the input, and the amplitude will be:

$$\begin{aligned} V_{out} &= V_{in} \times A_V \\ &= 100\text{ mV}_{\text{p-p}} \times 5.55 \\ &= 555\text{ mV}_{\text{p-p}} \end{aligned}$$

Inverting amplifier

Figure 9-23(*a*) on page 236 shows an *inverting amplifier.* Here, the signal is fed to the − input of the op amp. The output signal will be 180° out of phase with the input signal.

Virtual ground

The gain equation is a little different for the inverting circuit. As Fig. 9-23(*a*) shows, the noninverting input is at ground potential. Therefore the inverting input is also at ground potential because we again can assume that there is no difference across the inputs. The inverting input is known as a *virtual ground.* With the right end of R_1 effectively grounded (it is connected to the virtual ground), any in-

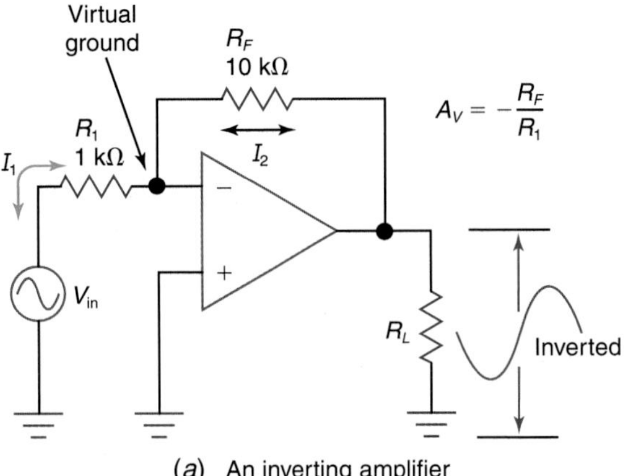

(*a*) An inverting amplifier

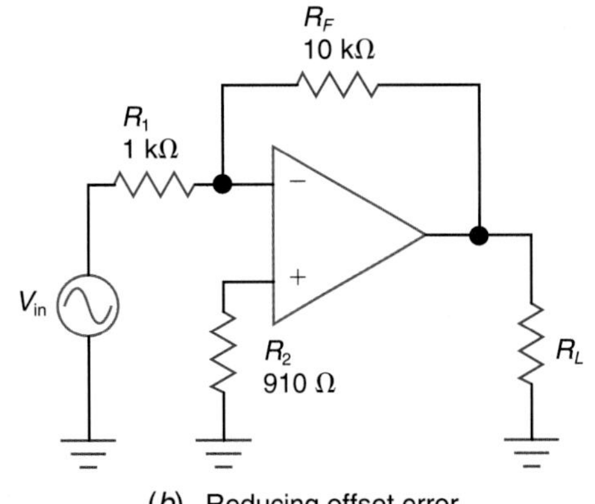

(*b*) Reducing offset error

Fig. 9-23 Inverting amplifier circuits.

put signal will cause a current to flow in R_1. According to Ohm's law:

$$I_1 = \frac{V_{in}}{R_1}$$

And any output signal will cause a current to flow in R_F:

$$I_2 = \frac{-V_{out}}{R_F}$$

V_{out} is negative above because the amplifier *inverts*. The current into or out of the − terminal of the op amp is so small that it is effectively 0. Thus, $I_2 = I_1$, and by substitution:

$$\frac{-V_{out}}{R_F} = \frac{V_{in}}{R_1}$$

Rearranging gives

$$A_V = \frac{V_{out}}{V_{in}} = -\frac{R_F}{R_1}$$

Applying the inverting gain equation to Fig. 9-23(*a*) gives

$$A_V = -\frac{R_F}{R_1}$$

$$= -\frac{10\ \text{k}\Omega}{1\ \text{k}\Omega} = -10$$

A gain of −10 means that an ac output signal will be 10 times the amplitude of the input signal, but opposite in phase. If the input signal is dc, then the output will also be dc but with opposite polarity. For example, if the input signal is −1 V, the output will be +10 V(−1 V × −10 = +10 V).

Figure 9-23(*b*) shows an inverting amplifier with an additional resistor. R_2 is included to reduce any offset error caused by amplifier bias current. The value of this resistor should be equal to the parallel equivalent of the resistors connected to the inverting input. From the standard product-over-sum equation:

$$R_2 = \frac{R_1 \times R_F}{R_1 + R_F}$$

$$= \frac{1\ \text{k}\Omega \times 10\ \text{k}\Omega}{1\ \text{k}\Omega + 10\ \text{k}\Omega} = 909\ \Omega$$

The closest standard value is 910 Ω. The amplifier bias currents will find the same effective resistance at both inputs. This will equalize the resulting dc voltage drops and eliminate any dc difference between the inputs caused by bias currents. One manufacturer of the 741 op amp lists the typical input bias current at 80 nA and the maximum value at 500 nA at room temperature.

The addition of R_2 in Fig. 9-23(*b*) does *not* substantially affect the signal voltage gain or the virtual ground. The current flowing in R_2 is so small that the drop across it is effectively 0. For example, if we use 80 nA and 910 Ω:

$$V = 80 \times 10^{-9}\ \text{A} \times 910\ \Omega$$

$$= 72.8\ \mu\text{V}$$

Therefore, the noninverting input is still effectively at ground potential and the inverting input is still a virtual ground.

AC-coupled

Figure 9-24 shows an *ac-coupled* noninverting amplifier. This situation mandates the use of R_2 to provide a dc path for the input bias current. To minimize any offset effect, R_2 is again chosen to be equal to the parallel equivalent of the resistors connected to the other op-amp input. R_2 in Fig. 9-24 sets the input impedance of the amplifier. Thus, the signal source sees a load of 9.1 kΩ. The op-amp input resistance is in the megohm range, so its effect can be ignored.

Input impedance

In an inverting amplifier, the − input of the op amp is a virtual ground. Therefore, the *input impedance* of this type of amplifier is equal to the resistor connected between the signal source and the inverting input. The signal source in Fig. 9-23 sees a load of 1 kΩ.

Example 9-6

Determine the output signal (amplitude and phase) for Fig. 9-23 if the signal source has an internal resistance of 600 Ω and V_{in} is 100 mV_{p-p} open circuit. *Open circuit* means that the signal source is not loaded. By inspection, it can be seen that the amplifier has an input resistance of 1 kΩ. The loading effect of the input must be taken into account. Using the voltage divider formula:

$$V_{in(closed\ circuit)} = V_{in(open\ circuit)} \times \frac{R_{amp}}{R_{amp} + R_{source}}$$
$$= 100\ mV_{p-p} \times \frac{1\ k\Omega}{1\ k\Omega + 600\ \Omega}$$
$$= 62.5\ mV_{p-p}$$

The amplifier has a gain of −10, so the output signal is 180° out of phase with the input and has an amplitude of:

$$V_{out} = 62.5\ mV_{p-p} \times 10 = 625\ mV_{p-p}$$

The negative gain has been accounted for by expressing the phase relationship as 180°.

All op-amp circuits have limits. Two of these limits are set by the *rail voltages.* A rail is simply another name for the power supply in an op-amp circuit. If a circuit is powered by ±12 V, the positive rail is +12 V and the negative rail is −12 V. The rail voltages cannot be exceeded by the output. In fact, the output voltage is usually limited to at least 1 V less than the rail. The most output that can be expected from an op amp powered by ±12V is about ±11V.

Rail voltage

Suppose you are asked to calculate the output voltage for an inverting amplifier with a gain of −50 and an input signal of 500 mV dc. The supply is specified at ±15 V.

$$V_{out} = V_{in} \times A_V$$
$$= 500\ mV \times -50$$
$$= -25\ V$$

This output is *not* possible. The amplifier will *saturate* within about 1 V of the negative rail. The output will be about −14 V dc.

Saturate

As another example, find the peak-to-peak output voltage for an op amp with a gain of 100. Assume a ±9 V supply and an ac input signal of 250 mV peak-to-peak.

$$V_{out} = V_{in} \times A_V$$
$$= 100 \times 250\ mV\ \text{peak-to-peak}$$
$$= 25\ V\ \text{peak-to-peak}$$

This output *cannot* be achieved. The maximum output swing will be from about −8 V to about +8 V, which is 16 V peak-to-peak. The output signal will be *clipped* in cases like this.

Clipped

The graph of Fig. 9-25 shows gain versus frequency for a typical integrated-circuit op

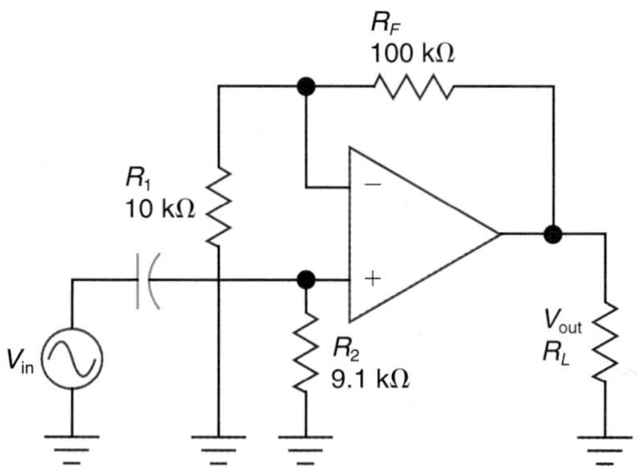

Fig. 9-24 An ac-coupled noninverting amplifier.

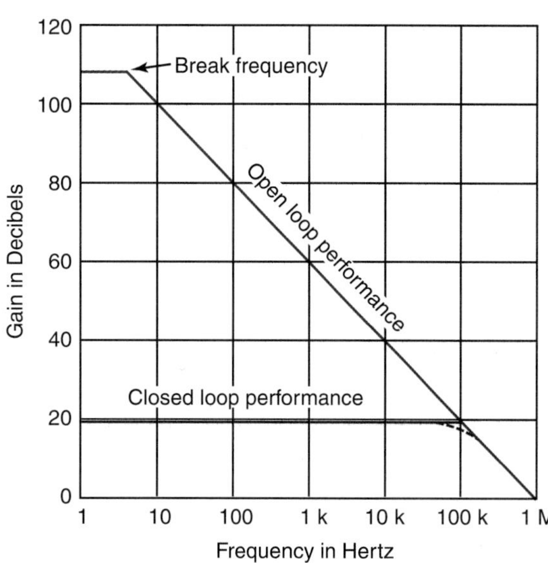

Fig. 9-25 Typical op-amp Bode plot.

Bode plot

Break frequency

f_b

20 dB per decade

amp. Graphs of this type are known as *Bode plots.* They show how gain decreases as frequency increases. Notice in Fig. 9-25 that the open-loop performance curve shows a *break frequency* at about 7 Hz. This frequency is designated as f_b. The gain will *decrease at a uniform rate* as frequency is *increased* beyond the break frequency. Most op amps show a gain decrease of *20 dB per decade* above f_b.

Check the open-loop gain in Fig. 9-25 at 10 Hz and note that it is 100 dB. A *decade* increase in frequency means an increase of *10 times.* Now check the gain at 100 Hz and verify that it drops to 80 dB. The loss in gain is 100 dB − 80 dB = 20 dB. Beyond the f_b, gain drops at 20 dB per decade.

Bode plots are approximate. Figure 9-26 shows that the actual performance of an amplifier is 3 dB less at f_b. This is the point of worst error, and Bode plots are accurate for frequencies significantly higher or lower than f_b. To find the true gain at f_b, subtract 3 dB.

JOB TIP

Periodic instrument calibration is required in many companies. Be aware of items such as calibration labels.

The open-loop gain shown in Fig. 9-25 indicates a break frequency lower than 10 Hz. This is a Bode plot, so we know that the gain is already 3 dB less at this point. The gain of the general-purpose op amp begins to decrease around 5 Hz. Obviously, it is not a wideband amplifier when operated open-loop. Op amps are usually operated closed-loop, and the *negative feedback increases the bandwidth* of the op amp. For example, the gain can be decreased to 20 dB. Now, the bandwidth of the amplifier increases to 100 kHz. This closed-loop performance is also shown in Fig. 9-25.

Negative feedback increases bandwidth

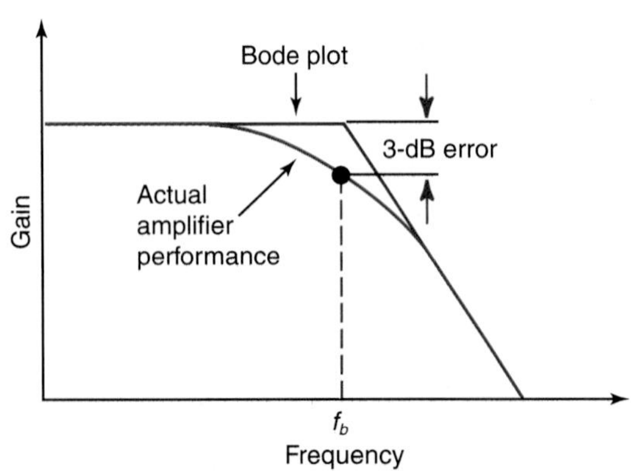

Fig. 9-26 Bode plot error is greatest at the break frequency.

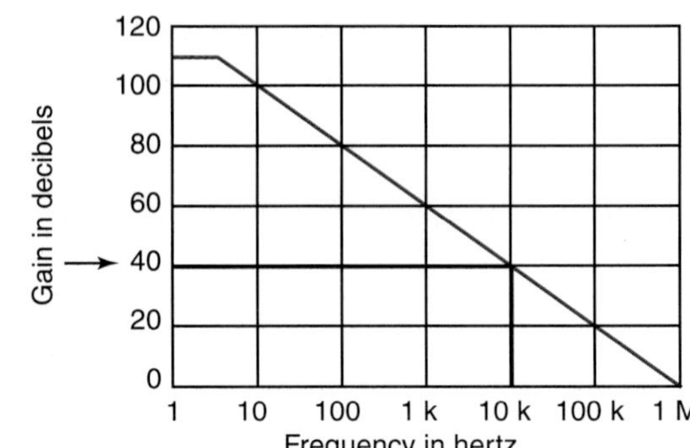

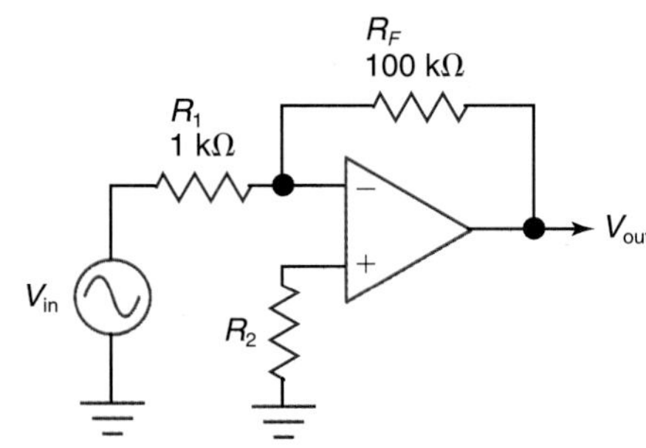

Fig. 9-27 Finding the bandwidth for a closed-loop amplifier.

Bode plots make it easy to predict the bandwidth for an op amp that is operating with negative feedback. Figure 9-27 shows an example. The first step is to find the closed-loop voltage gain. The appropriate equation is

$$A_V = -\frac{R_F}{R_1} = -\frac{100\ \text{k}\Omega}{1\ \text{k}\Omega} = -100$$

The negative gain indicates that the amplifier inverts. The negative sign is *eliminated* to find the dB gain:

$$A_V = 20 \times \log 100 = 40\ \text{dB}$$

The dB gain is located on the vertical axis of the Bode plot. Projecting to the right produces an intersection with the open-loop plot at 10 kHz. This is f_b (the break frequency), and the bandwidth of the amplifier is 10 kHz. Above f_b, the gain drops at 20 dB per decade. So the gain will be 40 dB − 20 dB = 20 dB at 100 kHz. The gain at f_b is down 3 dB and 40 dB − 3 dB = 37 dB at 10 kHz.

Earlier in this chapter, it was determined that the *power bandwidth* of an op amp is established by its slew rate and output amplitude. Here we find that another bandwidth is determined by the Bode plot of an op amp. To avoid confusion, this is called the *small-signal band-*

width. The small-signal bandwidth can be determined from the Bode plot or from the *gain-bandwidth product,* which is called f_{unity}. In Fig. 9-27, f_{unity} is 1 MHz. This is the frequency at which the gain of the amplifier is unity. A gain of unity means that the gain is 1, which corresponds to 0 dB. If you know f_{unity} for an op amp, you can determine the small-signal bandwidth without resorting to a Bode plot. The break frequency can be found by dividing f_{unity} by the ratio gain:

$$f_b = \frac{f_{unity}}{A_V}$$

Example 9-7

Find the small-signal bandwidth for an op amp with a gain-bandwidth product of 1 MHz if the closed-loop voltage gain is 60 dB. The first step is to convert 60 dB to the ratio gain (A_V):

$$60 \text{ dB} = 20 \times \log A_V$$

Divide both sides of the equation by 20:

$$3 = \log A_V$$

Take the inverse log of both sides:

$$A_V = 1000$$

Find the break frequency:

$$f_b = \frac{1 \text{ MHz}}{1000} = 1 \text{ kHz}$$

The small-signal bandwidth of the amplifier is 1 kHz. Please refer to Fig. 9-27 and verify that this agrees with the Bode plot for a gain of 60 dB.

TEST

Solve the following problems.

17. Refer to Fig. 9-20. Is the amplifier operating in open loop or closed loop?
18. Refer to Fig. 9-23. Assume that $R_1 =$ 470 Ω and $R_F = 47$ kΩ. What is the low-frequency voltage gain of the amplifier? What is the input impedance of the amplifier?
19. Refer to Fig. 9-23. It is desired to use this circuit for an amplifier that has an input impedance of 3300 Ω and a voltage gain of −10. What value would you choose for R_1? For R_F?
20. Refer to Fig. 9-24. $R_1 = 47$ kΩ, $R_2 = 22$ kΩ, $R_F = 47$ kΩ. Calculate the voltage gain of the amplifier. What is the input impedance of the amplifier?

Questions 21 to 27 use Fig. 9-27 as a guide.

21. The desired amplifier characteristics are a voltage gain of 80 dB and an input impedance of 100 Ω. Select a value for R_1.
22. Select a value for R_F.
23. Select a value for R_2 that will minimize dc offset error.
24. What is the small-signal bandwidth of the amplifier?
25. What is the gain of the amplifier at f_b?
26. What is the gain of the amplifier at 10 Hz?
27. What is the gain of the amplifier at 1 kHz?

9-5 Frequency Effects in Op Amps

We have learned that the open-loop gain of general-purpose operational amplifiers starts decreasing at a rate of 20 dB per decade at some relatively low frequency. This is caused by an internal *RC lag network* in the op amp. If you refer back to Fig. 9-15, you will find a single capacitor in the diagram. This capacitor forms one part of the lag network that determines the break frequency f_b. This capacitor is also one of the major factors that limits the slew rate of the amplifier.

Figure 9-28 on page 240 summarizes *RC lag networks.* The *RC* circuit is shown in Fig. 9-28(*a*). It consists of a series resistor and a capacitor connected to ground. A lag network does two things: (1) it causes the output voltage to drop with increasing frequency and (2) it causes the output voltage to *lag behind* the input voltage. Figure 9-28(*b*) shows the vector diagram for an *RC* lag network that is operating at its break frequency f_b. The resistance R and the capacitive reactance X_C are equal in this case, and the phase angle of the circuit is −45°. Figure 9-28(*c*) shows two Bode plots for the *RC* lag network. The one at the top is about the same as those shown in the last section. The change in amplitude from the break frequency to a frequency 10 times higher ($10 f_b$)

***RC* lag network**

Miller effect

$f_b = \frac{1}{2\pi RC}$

at f_b: $V_{out} = 0.707 \times V_{in} = -3$ dB, $\angle -45°$

(*a*) *RC* lag network

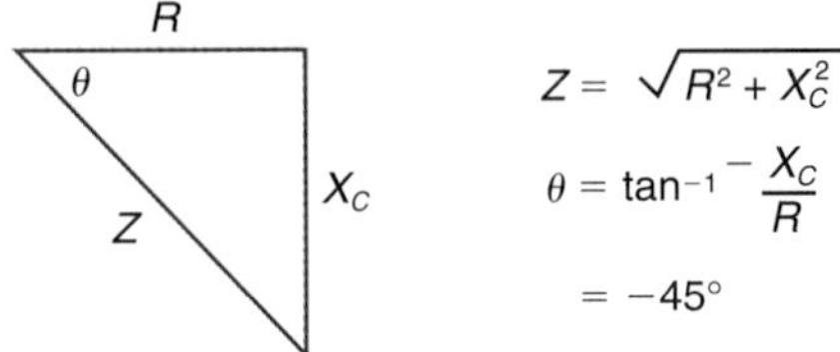

(*b*) Vector diagram for lag network at f_b

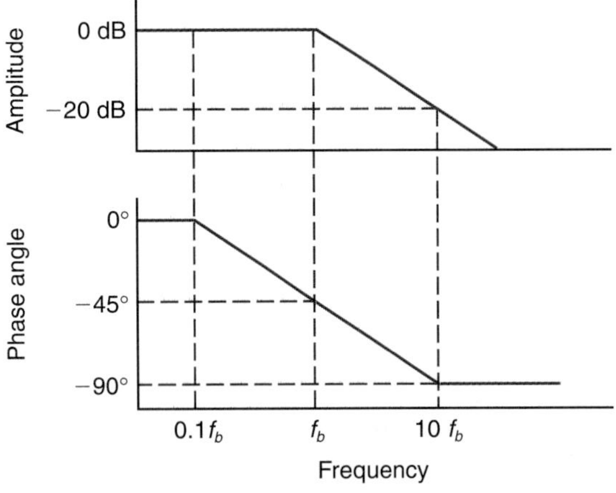

(*c*) Bode plots for lag network

Fig. 9-28 *RC* lag network.

is −20 dB. The bottom Bode plot shows the *phase angle response* for the network. You can see that the angle is −45° at f_b. It also shows that the angle is 0° for frequencies equal to or less than 0.1 f_b and that the angle is −90° for frequencies equal to or greater than 10 f_b. As stated before, Bode plots are approximate. The maximum error points are at 0.1 f_b where the angle is actually −6° and at 10 f_b where the angle is actually −84°.

Interelectrode capacitance

RC lag networks are *inherent in all amplifiers.* Transistors have *interelectrode capacitances* that form lag networks with certain resistances within the amplifier. Figure 9-29 shows how capacitive loading affects the input circuit of an NPN transistor amplifier. As shown in Fig. 9-29(*a*), there is a capacitor from base to collector (C_{BC}) and a capacitor from base to emitter (C_{BE}). All devices have interelectrode capacitance.

Because of voltage gain, the collector-base capacitor appears to be larger in the input circuit. This is known as the *Miller effect* and is shown in Fig. 9-29(*b*). Assuming a voltage gain of 100 (40 dB) from base to collector, the 5-pF interelectrode capacitance appears to be approximately 100 times greater in the base circuit. Thus, the total capacitance is 500 pF + 200 pF = 700 pF in the equivalent input circuit. If we assume the equivalent input resistance in Fig. 9-29(*b*) is 200 Ω, we can evaluate the circuit as a lag network and find its break frequency:

$$f_b = \frac{1}{2\pi RC} = \frac{1}{6.28 \times 200\ \Omega \times 700\ \text{pF}} = 1.14\ \text{MHz}$$

Knowing f_b allows us to predict the frequency response of the amplifier. If we use the last example, we know that the gain will be 100 (40 dB) for frequencies below 1 MHz. We also know that it will be 37 dB at 1.14 MHz and that it will be 20 dB at 11.4 MHz and 0 dB at 111 MHz. However, we have considered only the amplifier input circuit. The actual break frequency could be lower, depending on the output circuit.

Since it may have been some time since you worked with ac circuits, let's check the numbers another way. We will use the data from the last example: 700 pF, 200 Ω, and 1.14 MHz. Find the capacitive reactance:

$$X_C = \frac{1}{2\pi fC} = \frac{1}{6.28 \times 1.14\ \text{MHz} \times 700\ \text{pF}} = 200\ \Omega$$

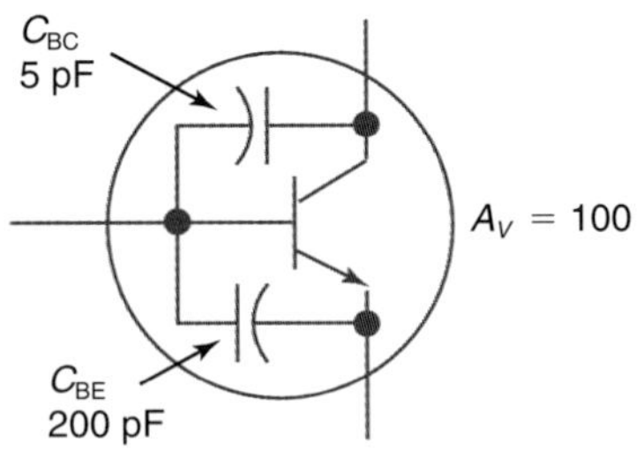

(*a*) Transistor interelectrode capacitances

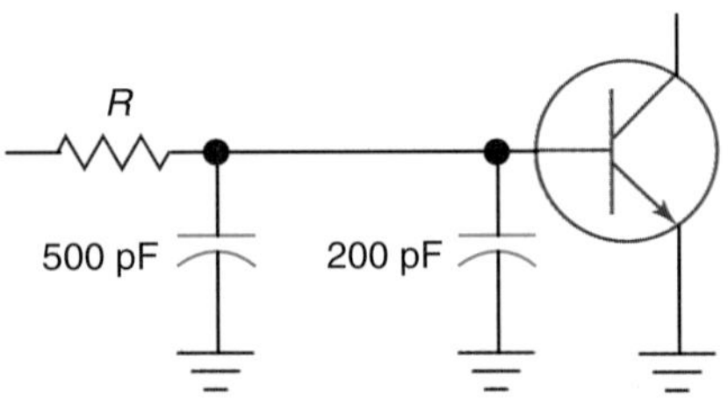

(*b*) Miller equivalent input circuit

Fig. 9-29 Capacitive loading in a transistor amplifier.

Find the impedance:

$$Z = \sqrt{R^2 + X^2} = \sqrt{200^2 + 200^2} = 283\ \Omega$$

Now, if you will refer back to Fig. 9-28(*a*), you can see that the capacitor and resistor form a voltage divider. We can use the voltage divider equation along with the impedance and capacitive reactance:

$$V_{\text{out}} = \frac{X_C}{Z} \times V_{\text{in}} = \frac{200\ \Omega}{283\ \Omega} \times V_{\text{in}}$$
$$= 0.707 \times V_{\text{in}}$$

This demonstrates that the output voltage is 0.707 or −3 dB at f_b. The phase angle can be determined with:

$$\phi = \tan^{-1}\frac{-X_C}{R} = \tan^{-1}\frac{-200\ \Omega}{200\ \Omega} = -45^\circ$$

The vector diagram of Fig. 9-28(*b*) shows that X_C is negative and that the phase angle is also negative (it lags).

As the schematic of the op amp shows (Fig. 9-15), there are quite a few transistors. Each one of them has interelectrode capacitances. *There are many lag networks in any operational amplifier.* With many lag networks there are going to be several break points. Figure 9-30 shows a Bode plot with several break frequencies. The gain will drop at 20 dB per decade between f_{b_1} and f_{b_2}. It will drop at 40 dB per decade between f_{b_2} and f_{b_3}. It will drop at 60 dB per decade for frequencies beyond f_{b_3}. The effect of *multiple lag networks accumulates.*

The phase angle with multiple lag networks also accumulates. It can be −100°, −150°, or −180°. This is a problem when an amplifier uses negative feedback. If the inherent lags add up to −180°, the amplifier can become *unstable.* Figure 9-31 shows this situation. The op amp uses a connection from its output to its

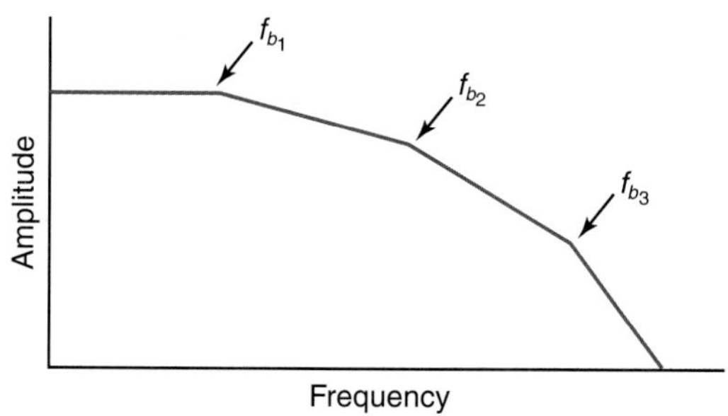

Fig. 9-30 Bode plot with several break points.

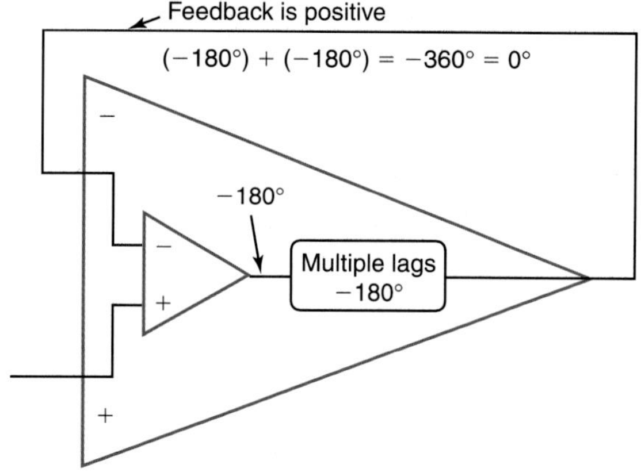

Fig. 9-31 How negative feedback can become positive.

inverting input. This normally provides negative feedback. However, if internal lags accumulate to −180°, the overall feedback goes to 0°. A phase angle of 0° is in-phase or *positive feedback.*

Positive feedback

This is what can happen with positive feedback: An input signal drives the amplifier. If the amplifier has gain, a larger signal appears at its output. This larger signal is returned to the − (minus) input with a phase that *reinforces* the input signal. The input and the feedback phase add for an even greater effective input. The output responds by increasing even more. This increases the input even more. The amplifier is no longer controlled by the input signal but by its own output. It is unstable and *useless* as an amplifier.

Multiple lag network

Instability

Internally compensated

Instability is not acceptable in any amplifier. One solution is that most op amps are *internally compensated.* They have a *dominant lag network* that begins rolling off the gain at a low frequency. By the time that the other lag networks (due to transistor capacitances) start to take effect, the gain has dropped below 0 dB. With the gain less than 0 dB, the amplifier cannot become unstable regardless of the actual feedback phase. Now you know why the open-loop Bode plot for the general-purpose op amp has such a low value of f_b.

External compensation

Unfortunately, the internal frequency compensation limits high-frequency gain and slew rate. For this reason, some op amps are available with *external frequency compensation.* The designer must compensate the amplifier in such a way that it is always stable. This is more involved and requires more components.

Figure 9-32 shows an example of an externally compensated op amp.

Another possibility is to use a high-performance op amp. This type is more costly, but it has a better slew rate and wider open-loop bandwidth than the general-purpose device. Figure 9-33 shows the Bode plot for one of these amplifiers. Notice that the open-loop gain does not reach 0 dB until a frequency of 10 MHz is reached. The small-signal bandwidth of this device is 10 times that of a general-purpose operational amplifier.

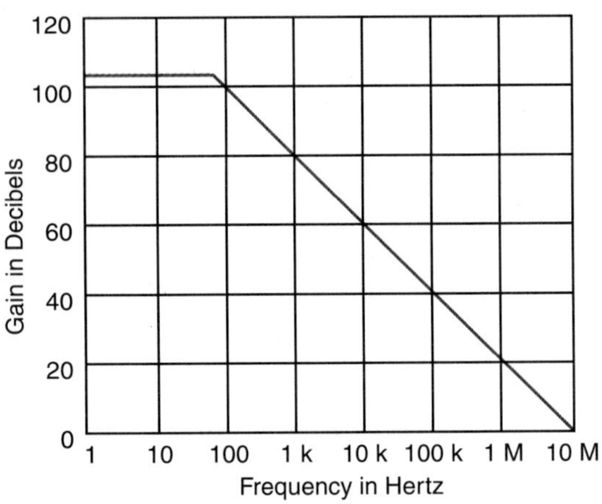

Fig. 9-33 Bode plot for a high-performance op amp.

Example 9-8

The Bode plot for a high-performance op amp illustrated in Fig. 9-33 indicates that f_{unity} is 10 MHz. Find the small-signal bandwidth for this amplifier when it operates with a closed-loop gain of 40 dB. Using Fig. 9-33 for a graphical solution is straightforward. Project across from 40 dB and then down to verify that the bandwidth is 100 kHz. The other method is to find the ratio gain and divide into f_{unity}.

$$40 \text{ dB} = 20 \times \log A_V$$
$$2 = \log A_V$$
$$A_V = 100$$
$$f_b = \frac{10 \text{ MHz}}{100} = 100 \text{ kHz}$$

TEST

Solve the following problems.

28. What does a lag network do to the amplitude of an ac signal as it increases in frequency?
29. What does a lag network do to the phase of an ac signal as it increases in frequency?
30. The compensation capacitor in a general-purpose op amp is 30 pF. However, because of the Miller effect it is effectively 300 times larger. If the effective resistance in series with this capacitance is 2.53 MΩ, find f_b for the op amp.
31. Why does the lag network of question 30 dominate the op amp?
32. If an op amp is designed for external compensation, why can it become unstable if the circuit is not designed correctly?

9-6 Op-Amp Applications

Figure 9-34 shows an operational amplifier used in the *summing mode.* Two input signals V_1 and V_2 are applied to the inverting input. The output will be the *inverted sum* of the two input signals. Summing amplifiers can be used

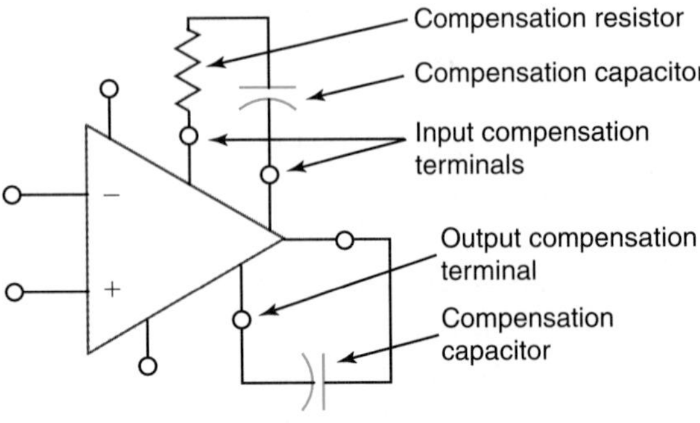

Fig. 9-32 An externally compensated op amp.

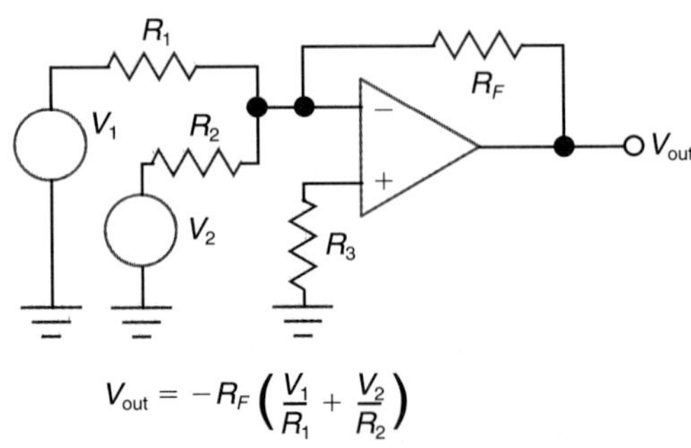

Fig. 9-34 An operational summing amplifier.

to add ac or dc signals. The output signal is given by:

$$V_{out} = -R_F\left(\frac{V_1}{R_1} + \frac{V_2}{R_2}\right)$$

Suppose, in Fig. 9-34, that all the resistors are 10 kΩ. Assume also that V_1 is 2 V and V_2 is 4 V. The output will be

$$\begin{aligned} V_{out} &= -10\text{ k}\Omega\left(\frac{2\text{ V}}{10\text{ k}\Omega} + \frac{4\text{ V}}{10\text{ k}\Omega}\right) \\ &= -\left(\frac{2\text{ V} \times 10\text{ k}\Omega}{10\text{ k}\Omega} + \frac{4\text{ V} \times 10\text{ k}\Omega}{10\text{ k}\Omega}\right) \\ &= -(2\text{ V} + 4\text{ V}) = -6\text{ V} \end{aligned}$$

The output voltage is negative because the two inputs are summed at the inverting input.

The circuit of Fig. 9-34 can be changed to scale the inputs. For example, R_1 could be changed to 5 kΩ. The output voltage will now be

$$\begin{aligned} V_{out} &= -10\text{ k}\Omega\left(\frac{2\text{ V}}{5\text{ k}\Omega} + \frac{4\text{ V}}{10\text{ k}\Omega}\right) \\ &= -(4\text{V} + 4\text{ V}) = -8\text{ V} \end{aligned}$$

The amplifier has scaled V_1 to 2 times its value and then added it to V_2.

Figure 9-34 could be expanded for more than two inputs. A third, fourth, and even a tenth input can be summed at the inverting input. Scaling of some or all the inputs is possible by selecting the input resistors and the feedback resistor.

Op-amp *summing amplifiers* are also called *mixers.* An audio mixer could be used to add the outputs of four microphones during a recording session. One of the advantages of inverting op-amp mixers is that there is no interaction between inputs. The inverting input is a *virtual ground.* This prevents one input signal from appearing at the other inputs. Figure 9-35

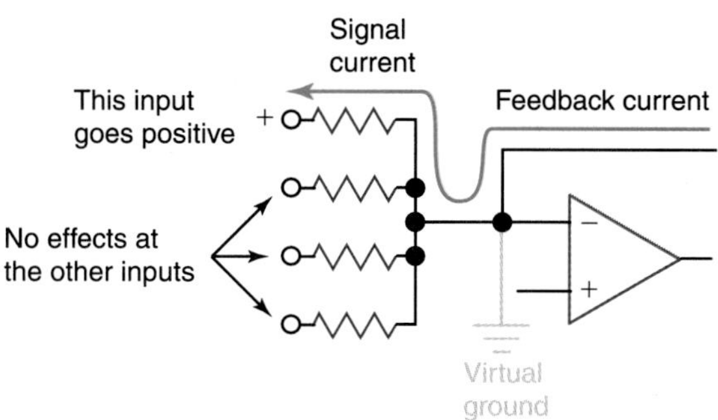

Fig. 9-35 The virtual ground isolates the inputs from one another.

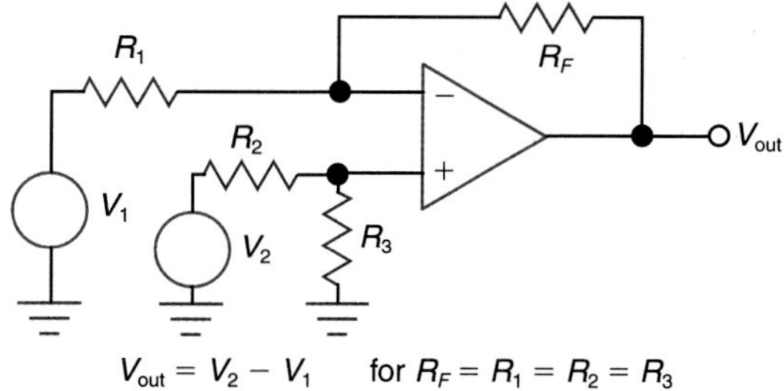

Fig. 9-36 An operational subtracting amplifier.

shows that the virtual ground isolates the inputs.

Op amps can be used in a *subtracting mode.* Figure 9-36 shows a circuit that can provide the difference between two inputs. With all resistors equal, the output is the nonscaled difference of the two inputs. If V_1 = 2 V and V_2 = 5 V, then

Subtracting amplifier

$$V_{out} = V_2 - V_1 = 5\text{ V} - 2\text{ V} = 3\text{ V}$$

It is possible to have a negative output if the voltage to the inverting input is greater than the voltage to the noninverting input. If V_1 = 6 V and V_2 = 5 V,

$$V_{out} = 5\text{ V} - 6\text{ V} = -1\text{ V}$$

Figure 9-36 can be modified to scale the inputs. Changing R_1 or R_2 would accomplish this.

Operational amplifiers can be used to *filter* signals. A filter is a circuit or device that allows some frequencies to pass through and attenuates other frequencies. Filters that use only resistors, capacitors, and inductors are called *passive filters.* Filter performance can often be improved by adding active devices such as transistors or op amps. Filters that use active devices are called *active filters.* Low-cost integrated op amps have helped to make active filters popular.

Summing amplifier

Mixers

Passive filter

Active filter

Figure 9-37(*a*) on page 244 shows an active low-pass filter circuit. The graph in Fig. 9-37(*b*) shows that the gain is at its maximum at low frequencies. As frequency increases, the gain starts to drop. The filter rejects, or attenuates, the high frequencies and passes the low frequencies.

Active filters using op amps have an advantage at lower frequencies. A passive *low-pass filter* designed with a cutoff point around 800 Hz requires a relatively high value inductor. Such an inductor tends to be physically large

Low-pass filter

and expensive. Inductance can be *simulated* by using feedback of the correct phase. You may recall that inductors cause current to lag the applied voltage by 90°. The feedback through the capacitor in Fig. 9-37(*a*) produces this same phase effect at the cutoff frequency. The active filter provides a response very similar to an inductor-capacitor low-pass filter.

Cascading filters

It is possible to *cascade* active filters for a *sharper cutoff.* Figure 9-38(*a*) shows two active low-pass filters in cascade (one after the other). The improved cutoff can be seen in the graph [Fig. 9-38(*b*)]. The active filter seems to be more complex than an equivalent passive inductor-capacitor filter. It is, but at low frequencies it will be smaller and cost less to manufacture.

High-pass filter

Figure 9-39(*a*) shows an active *high-pass filter.* The graph [Fig. 9-39(*b*)] indicates attenuation of low frequencies. This action is opposite to that of the low-pass circuit. Notice that the resistors and capacitors have exchanged positions. You can compare Fig. 9-39 with Fig. 9-37 to verify this.

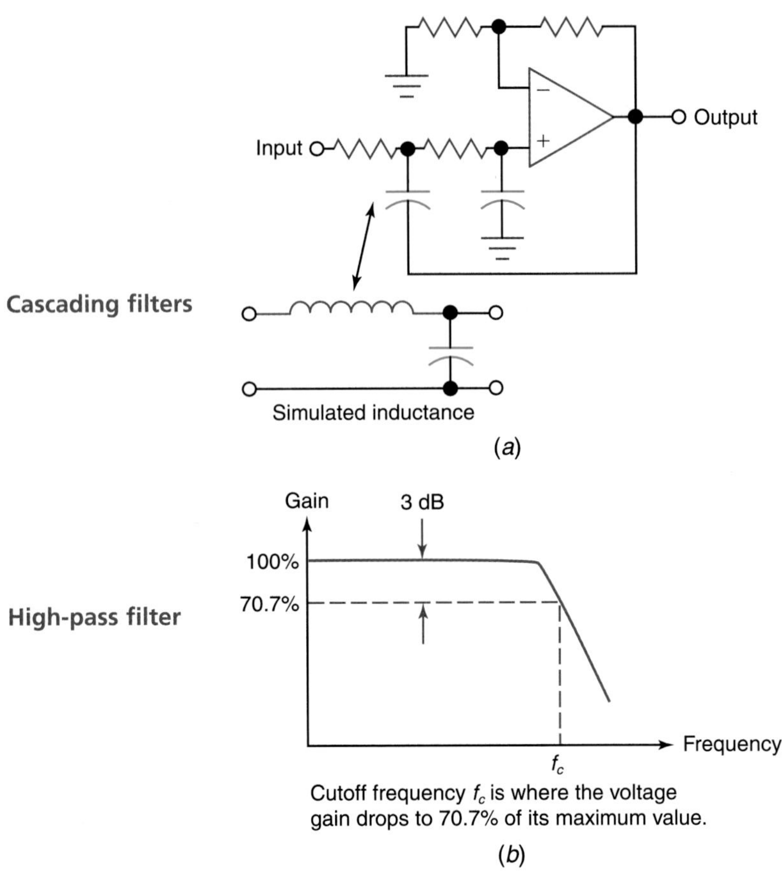

Fig. 9-37 An active low-pass filter.

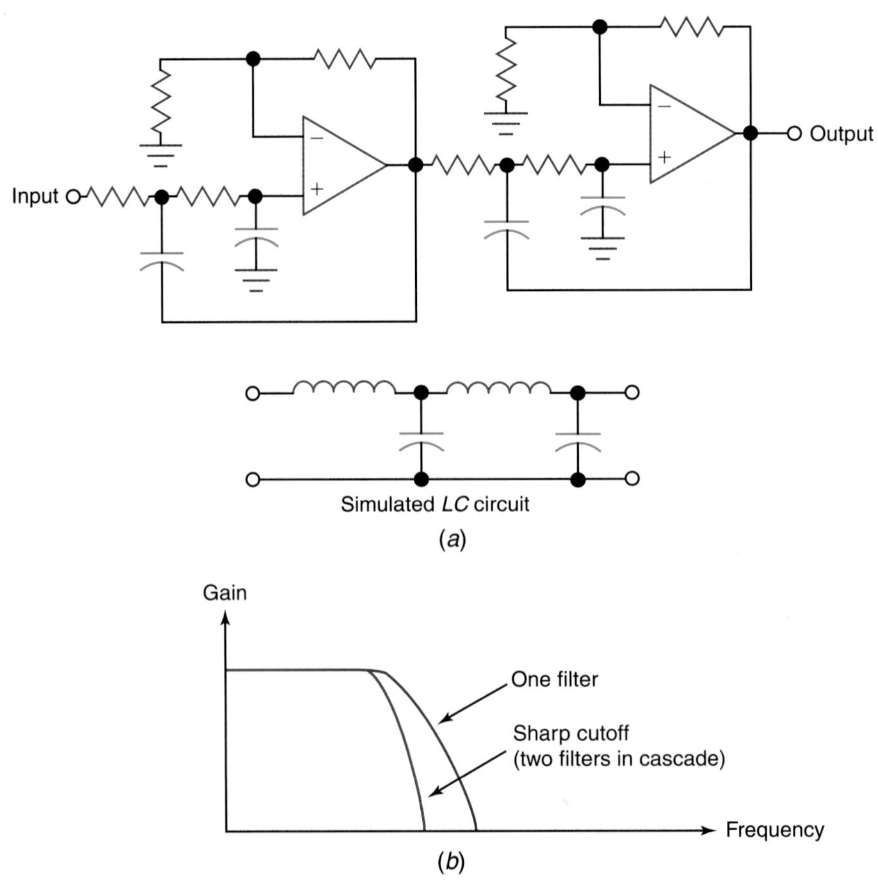

Fig. 9-38 Cascade filters give a sharper cutoff.

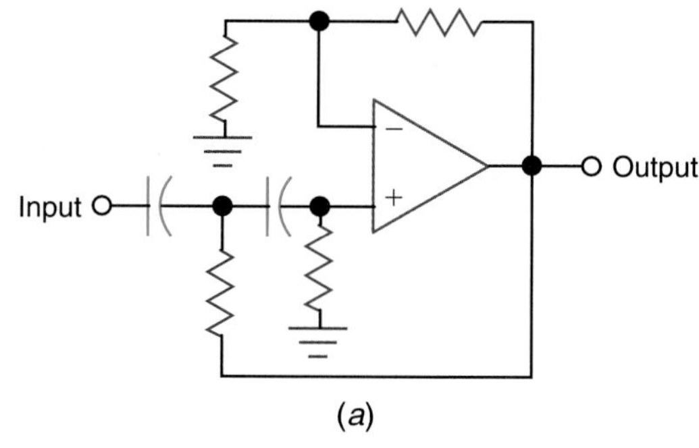

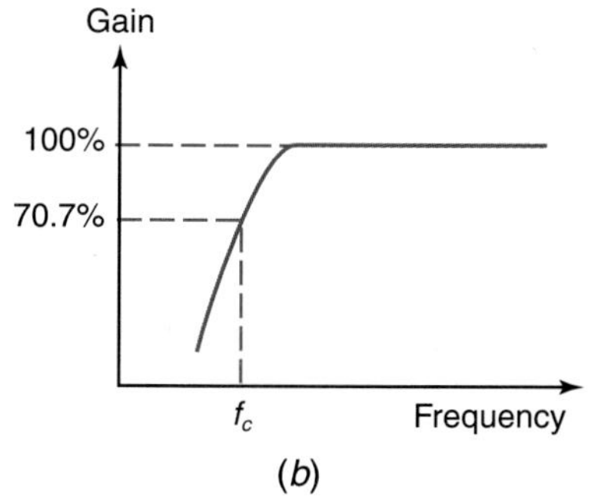

Fig. 9-39 An active high-pass filter.

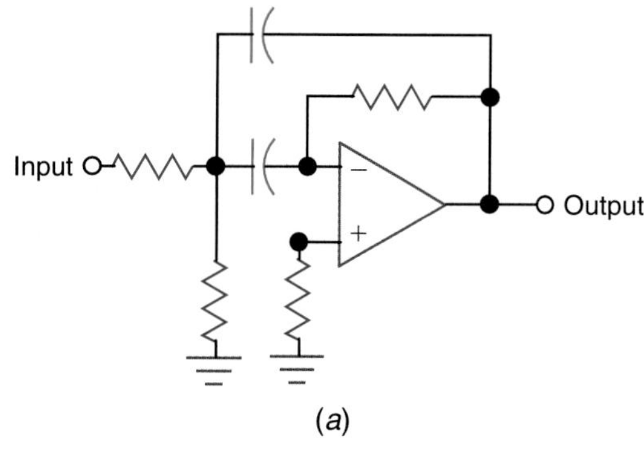

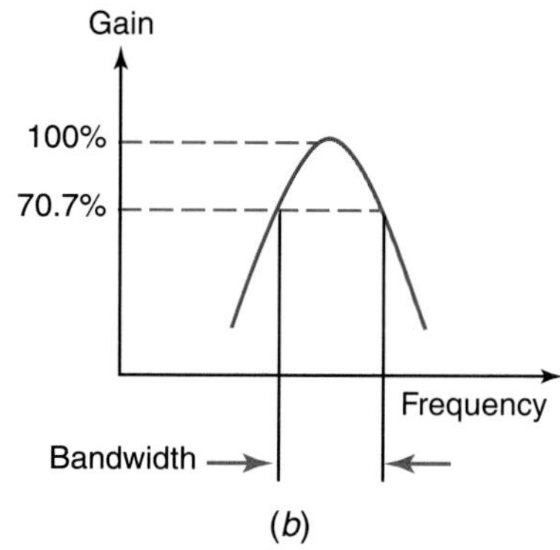

Fig. 9-40 An active band-pass filter.

Band-pass filter

An active *band-pass filter* is shown in Fig. 9-40(*a*). The graph [Fig. 9-40(*b*)] shows maximum gain for one frequency. Higher or lower frequencies receive less gain. Filters of this type are useful for separating signals. For example, different frequencies can be sent for control purposes. Touch-tone dialing is an example. Band-pass filters can be used to decode the various tones. The filter outputs will determine which tones were sent.

Band-stop filter

Figure 9-41 shows a 60-Hz *band-stop filter.* It may also be called a *notch filter,* or a trap. It produces maximum attenuation (or minimum gain) at a single frequency; 60 Hz in this case. Frequencies significantly above or below 60 Hz will pass through the notch filter with no attenuation. Band-stop filters are useful when a signal at one particular frequency is causing problems. For example, the 60-Hz notch filter could be used to eliminate power-line hum.

Operational amplifiers are sometimes used as *comparators.* Also, special comparator ICs are available. These are introduced in the last section of this chapter. A comparator operates open loop. This makes the gain very high, and the output is normally saturated in either a high or a low state. The output of a comparator is therefore a *digital signal* (has only two states). Comparators are therefore nonlinear circuits.

Comparators are used to provide an indication of the relative state of two inputs. If the positive (noninverting) input is more positive than the negative (inverting) input, the comparator output will be at positive saturation. If the positive input is less positive than the negative input, the output will be at negative saturation. Generally, a fixed reference voltage is applied to one input. The output will then be an indication of the relative magnitude of any signal applied to the other input. Comparators are often used to determine whether a signal is above or below the reference level. Several comparator applications in this category are presented later in this section.

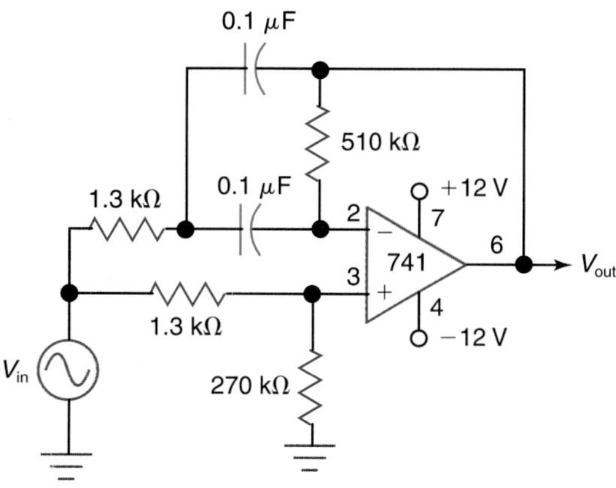

Fig. 9-41 A 60-Hz active band-stop filter.

When the reference voltage is zero, a comparator may be called a *zero-crossing detector.* A zero-crossing detector can be used to convert a sine wave into a square wave. Two comparators can be used in a "window" circuit that is used to determine whether a signal is between two prescribed limits as shown in the last section of this chapter.

It is often desired that the output of a comparator change states as quickly as possible. Another requirement is that the comparator output be compatible with logic inputs. A special *strobe* input may be needed in some applications so that the comparator output is active only at selected times. Special comparator ICs offer enhanced performance and additional features over op amps and are used in favor of op amps in some applications. These are most commonly operated from a single supply voltage.

Integrator

Voltage-to-frequency converter

Another way the operational amplifier may be used is in *integrator circuits.* Integration is a mathematical operation. It is a process of continuous addition. Integrators were used in analog computers. As we will see, there are other uses for integrators.

An op-amp integrator is shown in Fig. 9-42. Notice the capacitor in the feedback circuit. Suppose a positive-going signal is applied to the input. The output must go negative because the inverting input is used. The feedback keeps the inverting input at virtual ground. The current through resistor *R* is supplied by charging the feedback capacitor as shown.

If the input signal in Fig. 9-42 is at some constant positive value, the feedback current will also be constant. We can assume that the capacitor is being charged by a constant current. When a capacitor is charged by a constant current, the voltage across the capacitor increases in a linear fashion. Figure 9-42 shows that the output of the integrator is ramping negative, and that the ramp is linear.

Now look at Fig. 9-43 on page 247. This circuit is a *voltage-to-frequency converter.* It is a very useful circuit. It uses an op-amp integrator to convert positive voltages to a fre-

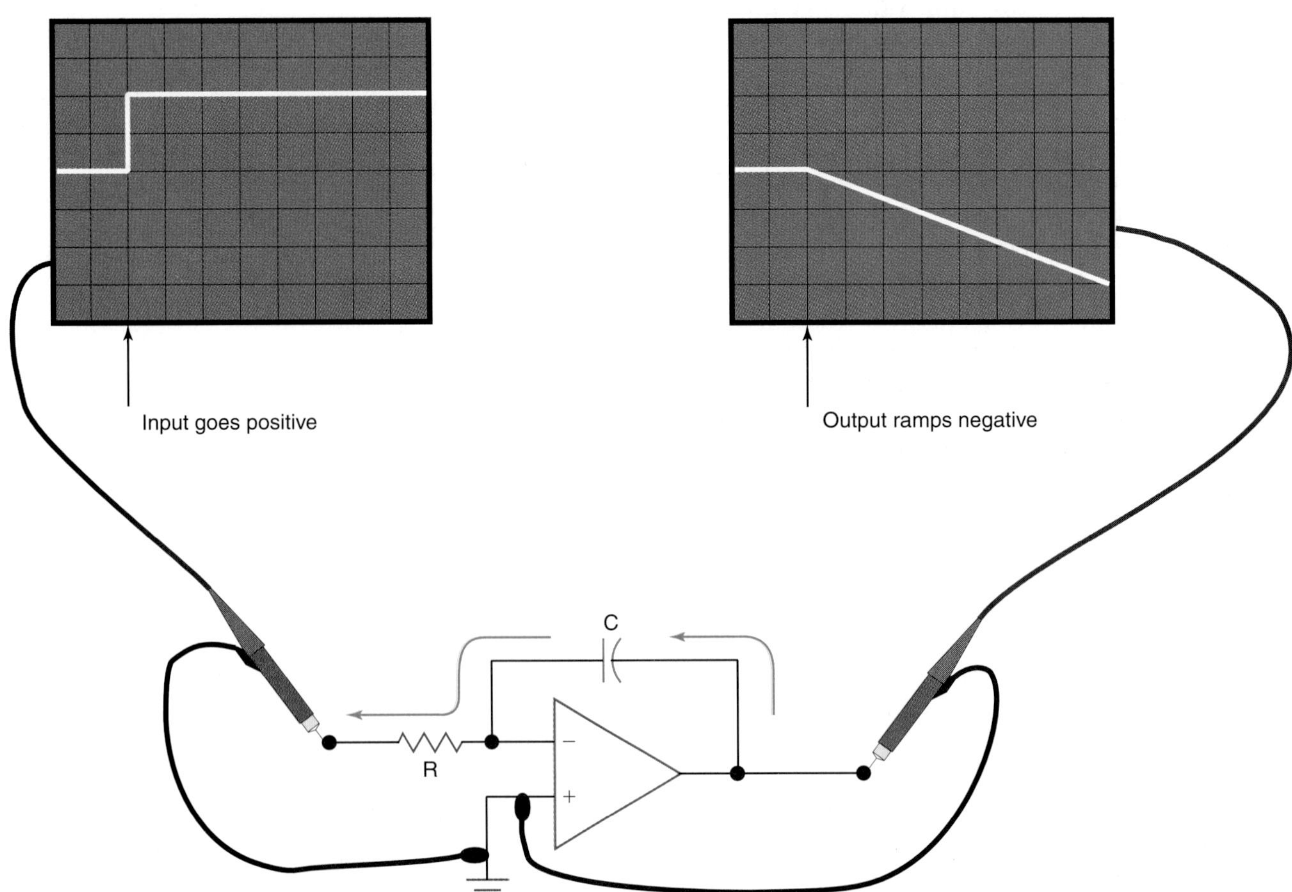

Fig. 9-42 An op-amp integrator.

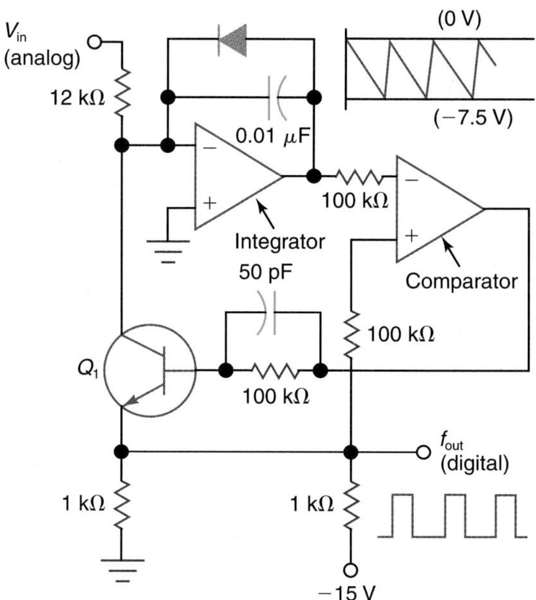

Fig. 9-43 A voltage-to-frequency converter.

quency. If the frequency is sent to a digital counter, a *digital voltmeter* is the result. If the voltage V_{in} represents a temperature, a digital thermometer is the result. Voltage-to-frequency converters form the basis for many of the digital instruments now in use.

What happens when a dc voltage is applied in the circuit of Fig. 9-43? If the voltage is positive, we know that the integrator will ramp in a negative direction. Note that the output of the integrator goes to a second op amp used as a *comparator.* It compares two inputs. One comparator input is a fixed −7.5 V which comes from the voltage divider formed by the two 1-kΩ resistors.

The integrator output in Fig. 9-43 will continue to ramp negative until its level exceeds −7.5 V. At this time, the comparator sees a greater negative voltage at its inverting input. This will cause the output of the comparator to go positive. This positive-going output then turns on Q_1. Since the emitter of Q_1 is negative, the input of the integrator is now quickly driven in a negative direction. This makes the integrator output go in a positive direction. Finally, the comparator again sees a greater negative voltage at its noninverting input. The comparator output goes negative, which switches off Q_1.

The waveforms of Fig. 9-43 explain the voltage-to-frequency conversion process. With a constant positive dc voltage applied to the input, a series of negative ramps appears at the integrator output. When each ramp exceeds −7.5 V, Q_1 is switched on. The current through the transistor causes a voltage pulse across the emitter resistors. The transistor is on for a very short time. The output is a series of narrow pulses.

Analog-to-digital converter

Figure 9-43 is one type of *analog-to-digital converter.* It converts a positive analog dc input voltage to a rectangular (digital) output. Ideally, circuits like this should show a linear relationship between the analog input and the digital output. For example, if the dc input voltage is exactly doubled, the output frequency should double. This means the output frequency is a linear function of the input voltage.

Why does the frequency output double when the input voltage doubles in Fig. 9-43? The input voltage causes a current to flow in the 12-kΩ resistor. If the input voltage increases, so will the input current. The − (minus) input of the op amp is a virtual ground and this current is supplied by charging the 0.01-μF integrator capacitor. We can assume that the charging current is now twice what it was (because the analog input voltage *doubled*). This means that the voltage across the capacitor will increase twice as fast. It will take only half the time to reach −7.5 V and switch the comparator. This doubles the output frequency. Figure 9-44 shows the graph of output frequency versus input voltage for the voltage-

Interviewers will file a negative report about applicants whom they feel might ignore safety regulations and company policies.

Comparator

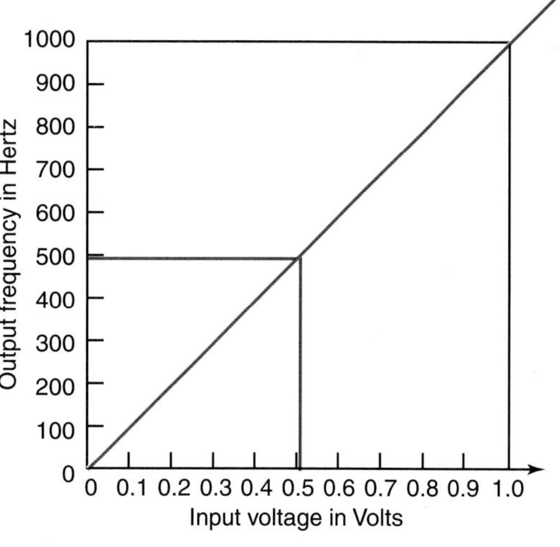

Fig. 9-44 Performance of the voltage-to-frequency converter.

to-frequency converter. Note the straight-line (linear) relationship.

Example 9-9

For an op-amp integrator, the rate at which its output voltage can change is proportional to its input voltage and to $1/RC$. Use this information to find the output frequency for the circuit of Fig. 9-43 when the input is +1 V. The output slope is negative-going since this is an inverting integrator:

$$\text{V/s} = \text{slope} = -V_{\text{in}} \times \frac{1}{RC} = -1\text{ V} \times \frac{1}{12\text{ k}\Omega \times 0.01\ \mu\text{F}} = -8330\text{ V/s}$$

Since the integrator output ramps from about 0 V to −7.5 volts, the output frequency can be found with:

$$f_{\text{out}} = \frac{-8330\text{ V/s}}{-7.5\text{ V}} = 1.11\text{ kHz}$$

This agrees reasonably well with the graph of Fig. 9-44.

Figure 9-45 shows another application for an op amp integrator followed by an op amp comparator. The circuit is called a *light integrator* because it is used to *sum* the amount of light received by a sensor in order to achieve some desired total exposure. Light integrators have applications in areas such as photography, where exposures are critical. A simple timer could be used to control exposure, but there are problems with this approach if the light intensity varies. For example, the intensity of a light source may change with the power-supply voltage and the temperature and age of the lamp. Another problem is that a filter may be used for some exposures. A filter decreases light intensity. With all of these possible variations in light, a simple timer may not offer adequate control.

The circuit in Fig. 9-45 uses a *light-dependent resistor* (LDR) to measure light intensity. Its resistance drops as brightness increases. The LDR and the 100-Ω resistor form a voltage divider for the −12 V power supply. The divided negative voltage is applied to the input of the integrator. The output of the integrator ramps positive at a rate that is directly proportional to light intensity. A second op amp is used as a *comparator* in Fig. 9-45. Its inverting input is biased at +6 V by the voltage divider formed by the two 1-kΩ resistors. As the integrator is ramping positive, the output of the comparator will be negative until the +6-V reference threshold is crossed. Notice in Fig. 9-45 that the comparator output is applied to the base of the PNP relay amplifier. As long as the comparator output is negative, the transistor is on, and the relay contacts are closed. The closed relay contacts keep the lamp energized

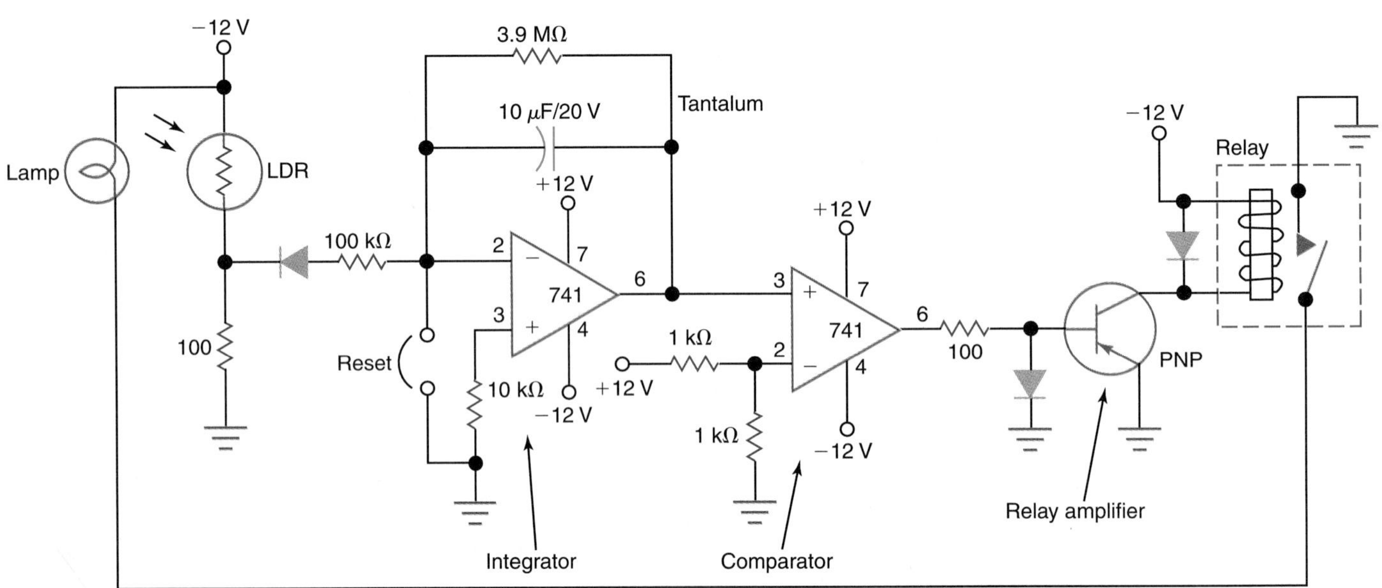

Fig. 9-45 Light-integrating circuit.

and the exposure continues. However, when the integrator output crosses the +6-V reference threshold, the comparator will suddenly switch to a positive output (its inverting input is now negative with respect to its noninverting input), and the relay will open. The light will now remain off until the reset button is pressed, which discharges the integrator capacitor and begins another exposure cycle.

The circuit of Fig. 9-45 can produce very accurate exposures. Changes in light intensity are compensated for by the amount of time that the relay remains closed. For example, if the light source was momentarily interrupted, an accurate exposure would still result. The integrator would stop ramping at the time of the interruption. It would hold its output voltage level until light once again reached the LDR.

The diode in the input circuit of the integrator in Fig. 9-45 prevents the integrator from being discharged if the light source is fluctuating. The diode in the base circuit of the transistor protects the transistor when the comparator output is positive. The diode will come on and prevent the base voltage from exceeding approximately +0.7 V. The diode across the relay coil prevents the inductive "kick" from damaging the transistor when it turns off.

There are a few op-amp circuits that use *positive feedback*. For example, Fig. 9-46 shows a signal-conditioning circuit known as a *Schmitt trigger*. This circuit is similar to a comparator, but the positive feedback gives it *two threshold points*. Assume that the op amp is powered by ±20 V and can swing about ±18 V at its output. Resistors R_1 and R_2 divide the output and establish the voltage that is applied to the noninverting input of the op amp.

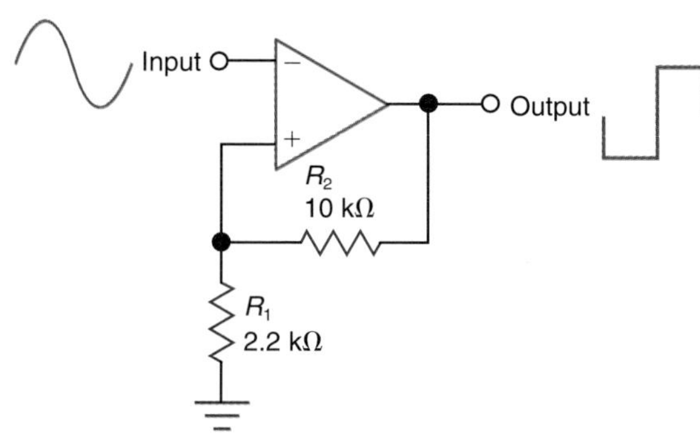

Fig. 9-46 Using an op amp as a Schmitt trigger.

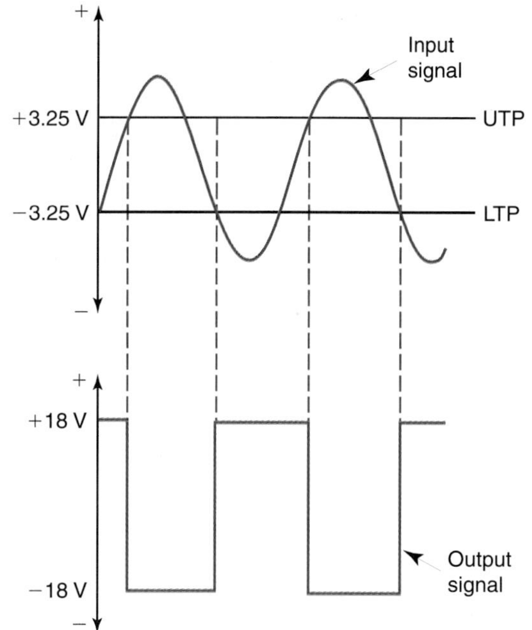

Fig. 9-47 Schmitt trigger operation.

When the output of Fig. 9-46 is maximum positive, the voltage divider will produce the upper threshold point (UTP):

$$\text{UTP} = V_{\text{max}}\left(\frac{R_1}{R_1 + R_2}\right)$$
$$= +18\text{ V}\left(\frac{2.2\text{ k}\Omega}{2.2\text{ k}\Omega + 10\text{ k}\Omega}\right)$$
$$= +3.25\text{ V}$$

When the output of Fig. 9-46 is maximum negative (V_{min}), the voltage divider will produce the lower threshold point (LTP):

$$\text{LTP} = V_{\text{min}}\left(\frac{R_1}{R_1 + R_2}\right)$$
$$= -18\text{ V}\left(\frac{2.2\text{ k}\Omega}{2.2\text{ k}\Omega + 10\text{ k}\Omega}\right)$$
$$= -3.25\text{ V}$$

Positive feedback

Schmitt trigger

Threshold points

Figure 9-47 shows the Schmitt trigger in operation with an input signal that exceeds the upper and lower threshold points. As the input signal is going positive, it eventually crosses the upper threshold point of +3.25 V. The inverting input of the op amp is now more positive than the noninverting input; therefore, the output rapidly switches to −18 V. Later, the input signal starts going negative and eventually crosses the lower threshold point of −3.25 V. At this time the Schmitt trigger output goes positive to +18 V, which reestablishes the UTP. The difference between the two

Hysteresis

threshold points is called *hysteresis* and in our example is

$$\begin{aligned}\text{Hysteresis} &= \text{UTP} - \text{LTP}\\ &= +3.25 - (-3.25)\\ &= 6.5\ \text{V}\end{aligned}$$

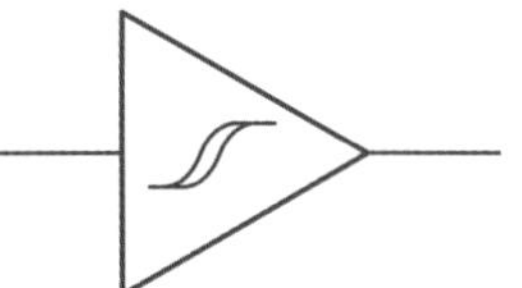

Fig. 9-48 Schmitt trigger symbol.

Example 9-10

Calculate the hysteresis voltage for Fig. 9-46 if the op amp is powered by a bipolar 9-V supply. We will assume that the output will swing ±8 V. The trip points are:

$$\text{UTP} = +8\ \text{V} \times \frac{2.2\ \text{k}\Omega}{2.2\ \text{k}\Omega + 10\ \text{k}\Omega} = 1.44\ \text{V}$$

$$\text{LTP} = -8\ \text{V} \times \frac{2.2\ \text{k}\Omega}{2.2\ \text{k}\Omega + 10\ \text{k}\Omega} = -1.44\ \text{V}$$

The hysteresis is the difference between the two trip points:

$$\text{Hysteresis} = 1.44\ \text{V} - (-1.44\ \text{V}) = 2.88\ \text{V}$$

Figure 9-48 shows a schematic symbol for a Schmitt trigger. You may recognize the *hysteresis loop* inside the general amplifier symbol.

Signal conditioning

Hysteresis is valuable when *conditioning* noisy *signals* for use in a digital circuit or system. Figure 9-49 shows why. It shows how a Schmitt trigger output can differ from the output of a comparator. The Schmitt trigger has hysteresis. The noise on the signal does not cause false triggering, and the output is at the *same frequency* as the input. However, the comparator output is at a higher frequency than the input signal because of false triggering on signal noise. Note that the comparator has a *single threshold point* (no hysteresis). The noise on the signal causes extra crossings back and forth through the threshold point (TP), and extra pulses appear in the output.

Single supply

Op amps normally require a bipolar power supply. However, for some applications they can be powered by a *single supply*. Figure 9-50 shows a typical circuit. Two 10-kΩ resistors divide the +15-V supply to +7.5 V which is applied to the noninverting inputs of the op amps. Terminal 4 of each amplifier, which is normally connected to the negative supply, is grounded. With no input signal, both amplifier outputs will be at +7.5 V. With an input signal, the outputs can swing from approximately +14 V to +1 V. The 4.7-μF capacitor bypasses any power-supply noise to ground.

Single supply circuits are often used in ac amplifiers. As Fig. 9-50 shows, the signal source is capacitively coupled. Since the noninverting inputs are at +7.5 V, the inverting inputs are also at +7.5 V. The input coupling ca-

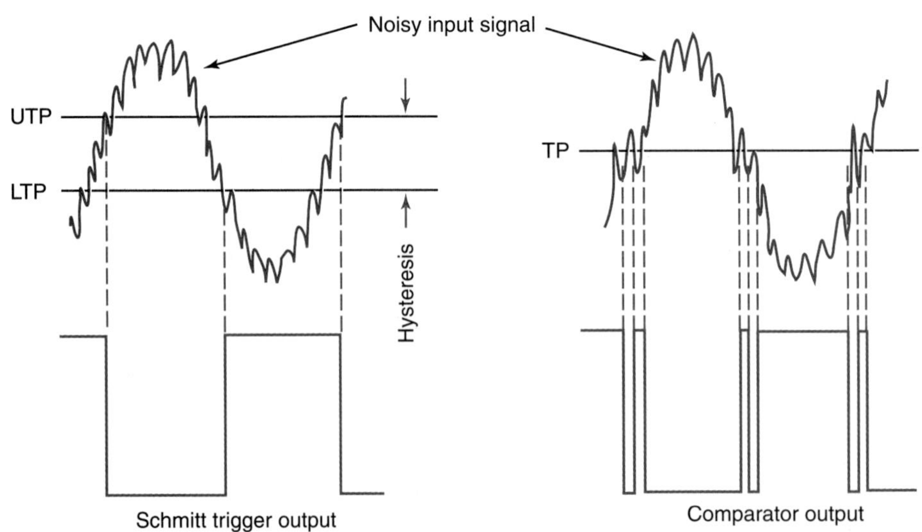

Fig. 9-49 A comparison of Schmitt trigger output with comparator output when the input is noisy.

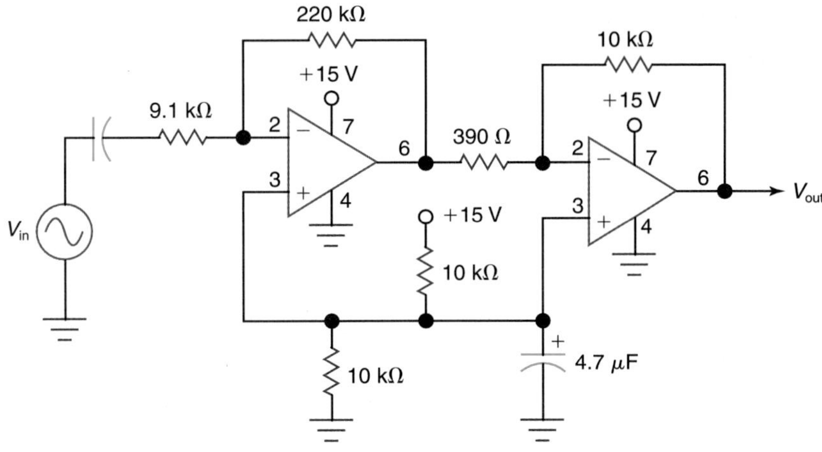

Fig. 9-50 Powering op amps from a single supply.

pacitor prevents the signal source from changing this dc voltage.

TEST

Solve the following problems.

33. Refer to Fig. 9-34. All resistors are the same value. If $V_1 = +1$ V and $V_2 = +2$ V, what will the output voltage be (value and polarity)?
34. Refer to Fig. 9-34. All resistors are the same value. If $V_1 = -2$ V and $V_2 = -3$ V, what will the output voltage be?
35. Refer to Fig. 9-34. All resistors are the same value. If $V_1 = +2$ V and $V_2 = -3$ V, what will the output voltage be?
36. Refer to Fig. 9-34. $R_F = 20$ kΩ, $R_1 = 10$ kΩ, and $R_2 = 5$ kΩ. If $V_1 = 2$ V and $V_2 = 1$ V, what will the output voltage be?
37. Refer to Fig. 9-35. What circuit feature prevents a signal at one of the inputs from appearing at the other inputs?
38. Refer to Fig. 9-36. All the resistors are the same value. If $V_1 = 3$ V and $V_2 = 5$ V, what will the output voltage be?
39. Refer to Fig. 9-36. All resistors are the same value. If $V_1 = 5$ V and $V_2 = 5$ V, what will the output voltage be?
40. Refer to Fig. 9-36. All resistors are the same value. If $V_1 = -2$ V and $V_2 = 1$ V, what will the output voltage be?
41. Refer to Fig. 9-37. This circuit is checked with a variable-frequency signal generator and an oscilloscope. The following data are collected:

    $V_{out} = 10$ V peak-to-peak at 100 Hz
    $V_{out} = 10$ V peak-to-peak at 1 kHz
    $V_{out} = 7$ V peak-to-peak at 10 kHz
    $V_{out} = 1$ V peak-to-peak at 20 kHz

    What is the cutoff frequency (f_c) of the filter?
42. Refer to Fig. 9-39. What can you expect to happen to the circuit gain as the signal frequency drops below f_c?
43. Refer to Fig. 9-40. Assume the gain to be maximum at 2500 Hz. Also assume that the gain drops 3 dB at 2800 Hz and at 2200 Hz. What is the filter bandwidth?
44. Suppose the input to the integrator shown in Fig. 9-42 goes negative. What will the output do?
45. Refer to Fig. 9-43. Assume the converter is perfectly linear. If $f_{out} = 300$ Hz when $V_{in} = 0.3$ V, what should f_{out} be at $V_{in} = 0.6$ V?
46. Refer to Fig. 9-45. Assume linear operation. If the relay remains energized for 2 s, how long will it remain energized after the circuit is reset if the light intensity falls to one-half its original level?
47. Refer to Fig. 9-46. Assume that the output of the op amp can swing ±13 V and that R_1 is changed to a 1-kΩ resistor. What is the value of UTP, LTP, and the hysteresis?

9-7 Comparators

As already discussed in this chapter, op amps can be used as comparators. However, this doesn't work in some cases. For example, when a *TTL-compatible* output is needed, an op amp used as a comparator may not meet circuit requirements. *TTL* stands for "transistor-transis- **TTL-compatible** tor logic," and it represents a widely applied digital family of parts that are used in many products and systems. More and more *hybrid* electronics applications are emerging that are a combination of both analog and digital electronics circuits and devices. A comparator is often the place where analog meets digital.

Hybrid

Comparator IC

Special *comparator ICs* are available. They are better for joining the analog and digital worlds. They provide a logic state output that indicates the relative state of two analog voltages, one of which is often a fixed reference. Comparators can signal when a voltage exceeds a reference, when a voltage is less than a reference, or when a voltage is within a specified range. Comparator ICs must change output states rapidly. They are optimized for high gain, wide bandwidth, and a fast slew rate. The *switching time* of a digital signal usually must be very fast. As Fig. 9-51 shows, the critical voltages for TTL circuits are 0.8 and 2 V. Any transition between these two points must be fast. The white oscilloscope trace is marginal, whereas the red trace is well within the required switching time. The LM311 is a popular IC comparator.

The signal source driving a comparator can also be critical. If the signal source has a slow switching time, a Schmitt trigger configuration may be necessary. This configuration uses positive feedback and was discussed in the preceding section of this chapter. Another factor is the impedance of the signal source that drives a comparator. In the case of high-impedance sources, extra output pulses called *glitches* are possible. Again, a Schmitt trigger is a possible solution.

Some IC comparators can operate from single or dual supplies from 5 to 30 V (or ±15 V). Others require dual supplies like op amps. Some, like the popular LM311, have an uncommitted output transistor with both the collector and the emitter terminals available. This makes the output of this device very flexible. It can be used to drive many kinds of logic circuits. Figure 9-52 shows the pinout of the

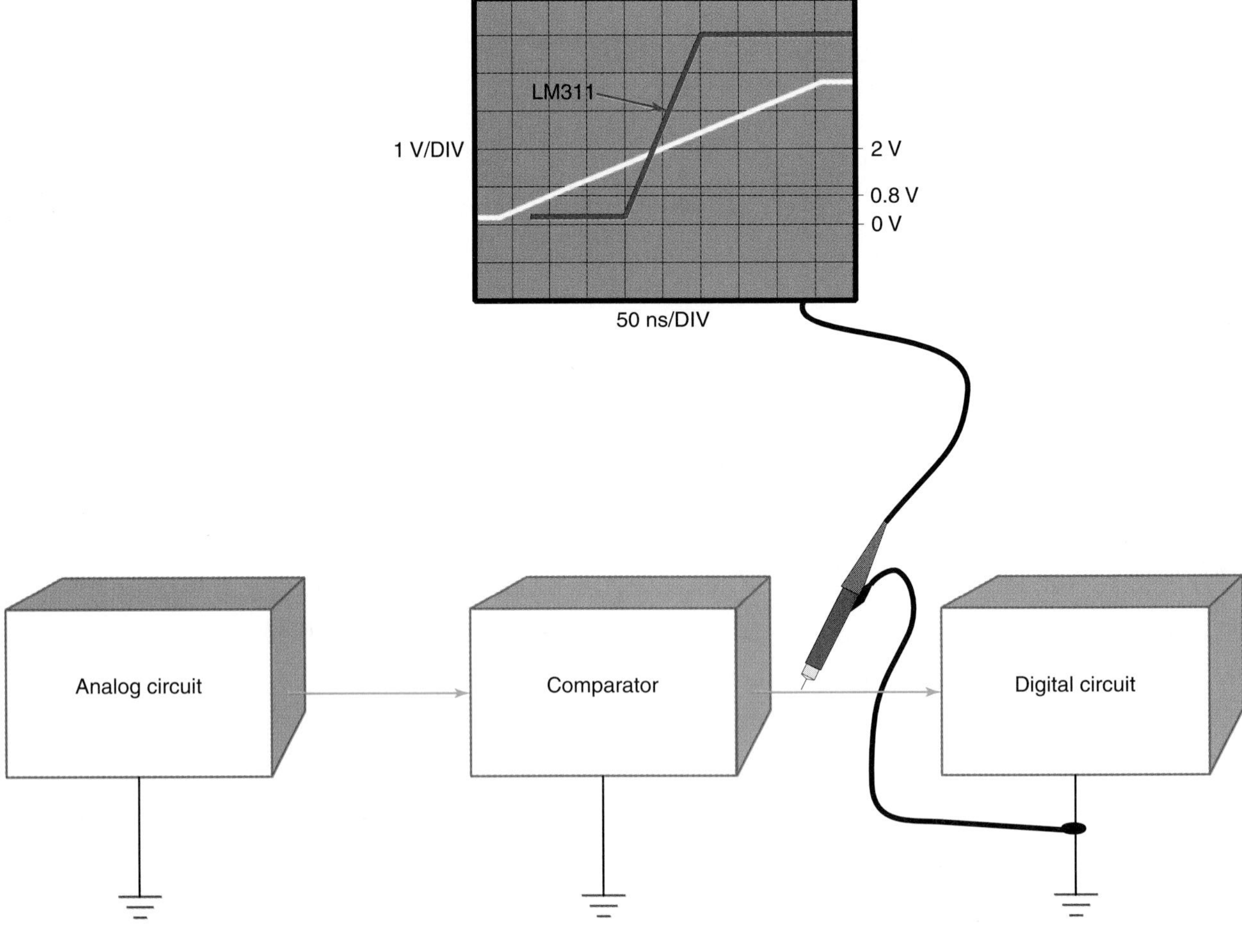

Fig. 9-51 Digital circuits may require switching times of less than 150 ns.

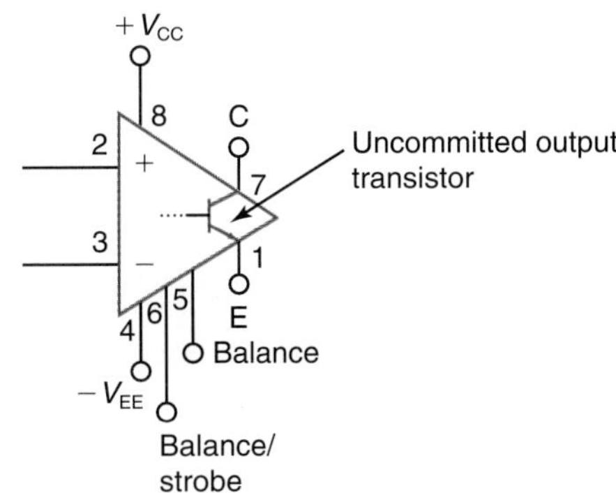

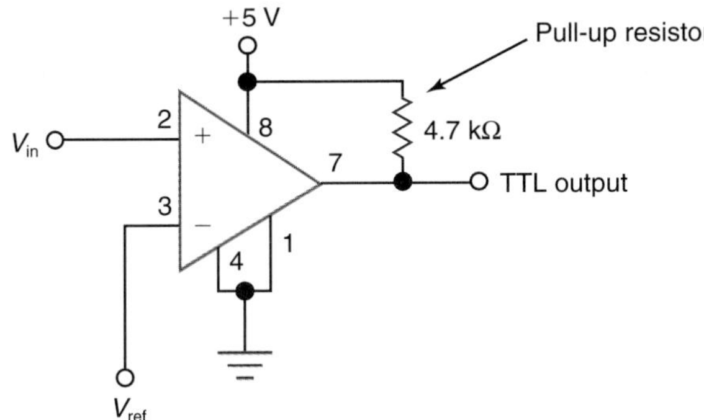

Fig. 9-52 LM311 or TLC311 comparator.

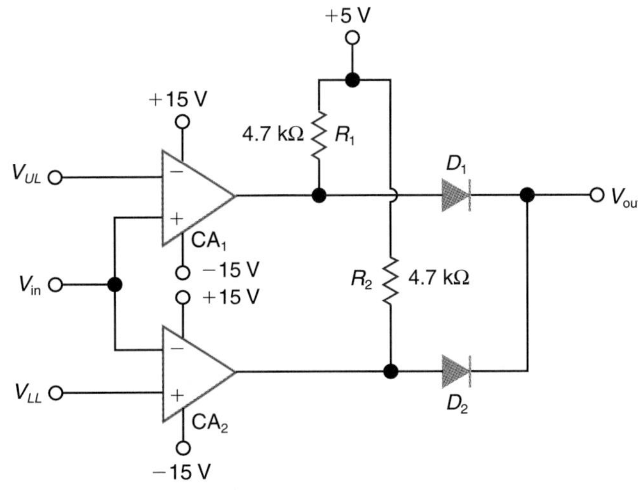

Fig. 9-53 Window comparator.

LM311. The balance inputs work pretty much the same as the offset null terminals on some op amps. The strobe input is for those cases when the output is to be active only during a specified time called the *strobe interval.* Notice in Fig. 9-52 that a *pull-up resistor* may be required because of the uncommitted transistor at the output of this device.

Figure 9-53 shows a window comparator. This circuit determines whether a signal voltage is within a defined *window.* Such circuits are useful in test or production situations to select parts or assemblies that are within a specific set of limits. They are also useful in automatic equipment such as battery chargers. The diodes in Fig. 9-53 logically combine the two comparator outputs. When V_{in} is between the upper limit (V_{UL}) and the lower limit (V_{LL}), then V_{out} is zero (logic LOW). If V_{in} is above V_{UL}, or if V_{in} is below V_{LL}, the output is near +5 V (logic HIGH).

Strobe interval

Window

TEST

Answer the following questions.

48. Inspect Fig. 9-51 and determine how long it takes the output of an LM311 to change from 0.8 to 2 V.
49. Refer to Fig. 9-52. If pin 1 is grounded and nothing is connected to pin 7, what will the output condition at pin 7 be, regardless of the input conditions?
50. Refer to Fig. 9-52. If pin 7 is connected to V_{CC}, where would an output load be connected? What output amplifier configuration does this represent?
51. Refer to Fig. 9-53. What are R_1 and R_2 called?
52. Refer to Fig. 9-53. Assume $V_{UL} = 12.9$ V and $V_{LL} = 11.9$ V. What is V_{out} when $V_{in} = 12.6$ V?
53. Refer to Fig. 9-53. Assume $V_{UL} = 12.9$ V and $V_{LL} = 11.9$ V. What is V_{out} when $V_{in} = 13.0$ V?
54. Refer to Fig. 9-53. Assume $V_{UL} = 12.9$ V and $V_{LL} = 11.9$ V. Which diode(s) is(are) forward-biased when $V_{in} = 13.0$ V?

Chapter 9 Review

Summary

1. A differential amplifier responds to the difference between two input signals.
2. A dual (or bipolar) supply develops both positive and negative voltages with respect to ground.
3. A differential amplifier can be driven at one of its inputs.
4. It is possible to use a differential amplifier as an inverting or as a noninverting amplifier.
5. A differential amplifier rejects common-mode signals.
6. The common-mode rejection ratio is the ratio of differential gain to common-mode gain.
7. A differential amplifier can show high CMRR for a single-ended output if the resistance of the emitter supply is very high.
8. Current sources have a very high output impedance.
9. Most op amps have a single-ended output (one output terminal).
10. Op amps have two inputs. One is the inverting input, and the other is the noninverting input. The inverting input is marked with a minus (−) sign, and the noninverting input is marked with a plus (+) sign.
11. An op amp's offset null terminals can be used to reduce dc error in the output. With no dc differential input, the output terminal is adjusted to 0 V with respect to ground.
12. Slew rate can limit the amplitude of an op-amp output and cause waveform distortion.
13. General-purpose op amps work best for dc and low ac frequencies.
14. The open-loop (no-feedback) gain of op amps is very high at 0 Hz (dc frequency). It drops off rapidly as frequency increases.
15. Op amps are operated closed-loop (with feedback).
16. Negative feedback decreases the voltage gain and increases the bandwidth of the amplifier.
17. Op-amp gain is set by the ratio of feedback resistance to input resistance.
18. Negative feedback makes the impedance of the inverting input very low. The terminal is called a virtual ground.
19. The impedance of the noninverting input is very high.
20. The Bode plot for a standard op amp shows the gain decreasing at 20 dB per decade beyond the break frequency.
21. The actual gain at the break frequency is 3 dB lower than shown on the Bode plot.
22. The high-frequency performance of an op amp is limited by both its Bode plot and its slew rate.
23. An *RC* lag network causes amplitude to drop at 20 dB per decade beyond the break frequency.
24. An *RC* lag network causes the output to phase-lag the input by 45° at the break frequency and as much as 90° for higher frequencies.
25. Because of device interelectrode capacitance, *RC* lag networks are inherent in any amplifier.
26. Because of the inherent lag networks, the total phase error will be −180° at some frequency. This will cause instability in an amplifier using negative feedback unless the gain is less than 1.
27. Most op amps are internally compensated to prevent instability.
28. Some op amps use external compensation to allow circuit designers to achieve better high-frequency gain and better slew rate.
29. Internally compensated op amps are easier to use and are more popular.
30. Op amps can be used as summing amplifiers.
31. By adjusting input resistors, a summing amplifier can scale some, or all, of the inputs.
32. Summing amplifiers may be called mixers. A mixer can sum several audio inputs.
33. Op amps can be used as subtracting amplifiers. The signal at the inverting input is subtracted from the signal at the noninverting input.
34. Op amps are used in active filter circuits. One of their advantages is that they eliminate the need for inductors.
35. Active filters can be cascaded (connected in series) for sharper cutoff.
36. Op-amp integrators use capacitive feedback. The output of an integrator produces a linear ramp in response to a dc input signal.

37. A comparator is a circuit that looks at two input signals and switches its output according to which of the inputs is greater.
38. An op-amp integrator and an op-amp comparator can be combined to form a voltage-to-frequency converter. This is one way to achieve analog-to-digital conversion.
39. A Schmitt trigger is a signal-conditioning circuit with two threshold points.
40. In a Schmitt trigger, the difference between the two threshold points is called hysteresis.
41. Hysteresis can prevent noise from false-triggering a circuit.
42. Op amps can be powered from a single supply voltage by using a voltage divider to bias the inputs at half the supply voltage.

Chapter Review Questions

Answer the following questions.

9-1. What name is given to an amplifier that responds to the difference between two input signals?

9-2. A bipolar power supply provides how many polarities with respect to ground?

9-3. Refer to Fig. 9-4. Assume that the input signal drives the base of Q_1 in a positive direction. What effect does this have on the emitter of Q_2? On the collector of Q_2?

9-4. Refer to Fig. 9-5. Assume that both wires that connect the signal source to the amplifier pick up a hum voltage. Why can the amplifier greatly reduce this hum voltage?

9-5. Refer to Fig. 9-5. If the output signal is taken across the two collectors, what is the output called? Will this output connection reduce common-mode signals?

9-6. Refer to Fig. 9-5. If the single-ended output signal is 2.3 V peak-to-peak, what will the differential output be?

9-7. The differential input of an amplifier is 150 mV, and the output is 9 V. What is the differential gain of the amplifier?

9-8. In using the same amplifier as in question 9-7, it is noted that a 2-V common-mode signal is reduced to 50 mV in the output. What is the CMRR?

9-9. Refer to Fig. 9-11. Will this circuit show good common-mode rejection at either one of its single-ended outputs?

9-10. Refer to Fig. 9-13. Does this operational amplifier provide a single-ended or differential output?

9-11. What geometric shape is often used on schematic diagrams to represent an amplifier?

9-12. What polarity sign will be used to mark the noninverting input of an operational amplifier?

9-13. Which op-amp terminals can be used to correct for slight dc internal imbalances?

9-14. An op amp has a slew rate of 5 V/μs. What is the power bandwidth of the op amp for a 16-V peak-to-peak output swing? (Hint: Don't forget to use the peak value in your calculation.)

9-15. What is the gain of an op amp called when there is no feedback?

9-16. What does negative feedback do to the open-loop gain of an op amp?

9-17. What does negative feedback do to the bandwidth of an op amp?

9-18. Refer to Fig. 9-21. To what value will R_F have to be changed in order to produce a voltage gain of 33?

9-19. Refer to Fig. 9-23. Change R_1 to 470 Ω. What is the voltage gain? What is the input impedance?

9-20. Refer to Fig. 9-23. What component sets the input impedance of this amplifier?

9-21. Refer to Fig. 9-23(*b*). What can happen to the op amp if R_2 is very different in value compared to the parallel equivalent of R_1 and R_F?

9-22. Refer to Fig. 9-23. Resistors R_1 and R_2 are 2200 Ω, and R_F is 220 kΩ. What is the voltage gain of the amplifier? What is the input impedance?

9-23. Refer to Fig. 9-25. Where does the maximum error occur in a Bode plot? What is the magnitude of this error?

9-24. Refer to Fig. 9-25. The gain of the op amp is to be set at 80 dB by using negative feedback. Where will the break frequency be?

9-25. Refer to Fig. 9-25. Is it possible to use this op amp to obtain a 30-dB gain at 100 Hz?

9-26. Refer to Fig. 9-25. Is it possible to use this op amp to obtain a 70-dB gain at 1 kHz?

9-27. Refer to Fig. 9-34. Assume that R_1 and R_2 are 4.7 kΩ. What impedance does source V_1 see? Source V_2?

9-28. Refer to Fig. 9-34. Resistors R_1 and R_2 are both 10 kΩ, and R_F is 68 kΩ. If $V_1 = 0.3$ V and $V_2 = 0.5$ V, what will V_{out} be?

9-29. Refer to Fig. 9-36. All the resistors are 1 kΩ. If $V_1 = 2$ V and $V_2 = 2$ V, what will the output voltage be?

9-30. The cutoff frequency of a filter can be defined as the frequency at which the output drops to 70.7 percent of its maximum value. What does this represent in decibels?

9-31. Find the break frequency for an *RC* lag network with 22 kΩ of resistance and 0.1 μF of capacitance. What is the phase angle of the output at f_b?

9-32. What is the phase angle of the output from a lag network operating at 10 times its break frequency?

9-33. What can happen in a negative-feedback amplifier if the internal lags accumulate to −180°?

9-34. Refer to Fig. 9-43. What component is used to discharge the integrator?

9-35. Refer to Fig. 9-43. Which op amp is used to turn on Q_1?

9-36. Refer to Fig. 9-44. Is the relationship between input voltage and output frequency linear?

9-37. Refer to Fig. 9-46. What happens to the hysteresis as R_1 is made larger? Smaller?

9-38. Refer to Fig. 9-49. How does the output frequency from the Schmitt trigger compare to the input frequency?

9-39. Refer to Fig. 9-49. How does the output frequency from the comparator compare to the input frequency?

9-40. Refer to Fig. 9-50. What is the load on the signal source and what is the overall gain of the two stages?

Critical Thinking Questions

9-1. Why is CMRR a critical specification for some medical electronic equipment?

9-2. What advantage could be offered by cascading differential amplifiers?

9-3. An amplifier uses three op-amp stages in cascade. The break frequency of each individual stage is 10 kHz. Why is the small-signal bandwidth of the cascade circuit less than 10 kHz?

9-4. People who work around radiation sources may be required to wear a film badge. The purpose of the badge is to accumulate a measurement of their total dose of radiation exposure. Can you think of an electronic replacement for the film badge?

9-5. What would be the advantages of the electronic gadget described in question 9-4?

9-6. What would be the disadvantage of the electronic gadget described in question 9-4?

9-7. Would the output of a Schmitt trigger ever show any noise? Why?

Answers to Tests

1. dual or bipolar supply
2. 0 V
3. negative; positive
4. because both transistor currents change
5. 4 V peak-to-peak
6. +12 V; −12 V; +24 V
7. ground
8. 100
9. 800 (58 dB)
10. $I_{R_E} = 0.933$ mA
    $I_{E(Q_1)} = 0.466$ mA
    $I_{E(Q_2)} = 0.466$ mA
    $V_{CE(Q_1)} = 5.04$ V
    $V_{CE(Q_2)} = 5.04$ V
    $V_{B(Q_1)} = -0.233$ V
    $V_{B(Q_2)} = -0.233$ V
    $A_{V(dif)} = 46.6$
    $A_{V(com)} = 0.562$
    CMRR = 82.9 (38.4 dB)
11. section 3
12. 180° out of phase
13. single-ended
14. offset null
15. amplitude reduction and waveform distortion
16. 207 kHz; the small-signal bandwidth is greater (3 MHz)
17. closed loop
18. −100; 470 Ω
19. 3300 Ω; 33 kΩ
20. 2; 22 kΩ
21. 100 Ω
22. 1 MΩ
23. 100 Ω
24. 100 Hz
25. 77 dB
26. 80 dB
27. 60 dB
28. decreases it
29. shifts it more negative
30. 6.99 Hz
31. It occurs at a very low frequency so gain rolls off to less than 1 before the inherent lag net-

works can cause a problem.

32. The feedback can become positive at a frequency where the gain is greater than one.
33. −3 V
34. +5 V
35. +1 V
36. −8 V
37. the virtual ground
38. +2 V
39. 0 V
40. +3 V
41. 10 kHz
42. It will decrease.
43. 600 Hz
44. It will ramp positive.
45. 600 Hz
46. 4 s
47. +1.18 V; −1.18 V; 2.36 V
48. 25 ns
49. zero volts (LOW)
50. from pin 1 to ground; emitter follower
51. pull-up resistors
52. zero volts (LOW)
53. ≈ 5 V (HIGH)
54. D_1

Chapter 10

Troubleshooting

Chapter Objectives

This chapter will help you to:

1. *Identify* symptoms in malfunctioning equipment and systems.
2. *Perform* preliminary checks and eliminate obvious problems.
3. *Localize* circuit defects by using signal injection and signal tracing.
4. *Find* component-level defects by using voltage analysis.
5. *Troubleshoot* intermittent systems.
6. *Troubleshoot* operational-amplifier (op-amp) circuits.

Electronic components sometimes fail, and part of the troubleshooting process involves identifying the failed components. This is accomplished by using logic based on an understanding of circuits and by using test equipment. Today, this part of the troubleshooting process must be based on a system viewpoint and should be the next step after some very important preliminary checks have been made.

10-1 Preliminary Checks

GOAL

When troubleshooting, remember the word *GOAL.* Good troubleshooting is a matter of:

1. Observing the symptoms
2. Analyzing the possible causes
3. Limiting the possibilities

System view

The letter *L* is last in GOAL. Keep that in mind. Don't be too quick to *Limit* the possibilities. Troubleshooting requires a *system view.* There is more than just a chance that the real cause of a problem is not where the symptoms appear. Medical doctors, who must find and deal with root cause, know this. In the human body, *referred pain* can occur. That is, a pain in one location may have its origin in an entirely different location.

Look at Fig. 10-1. It shows a piece of equipment that requires troubleshooting because it is not working properly. A good technician knows to use a system view. There are all sorts of things that can go wrong with systems:

- Faulty power sources (including dead batteries)
- Bad connectors and loose connectors
- Open cables and cables connected incorrectly
- Input signals missing
- Incorrectly set controls
- Component failures
- Network problems
- Software problems

The last two items are becoming more common than in the past. Equipment is often in "communication" with other pieces of equipment, and software runs behind the scenes and affects how things work. It is very important to take a system view and to make some preliminary checks before taking a piece of equipment apart. Overlooking software problems can cause a great loss of time and money.

Not all equipment is networked to other equipment, but it still pays to use a system view. Troubleshooters who have this mindset know how important preliminary checking is. It is amazing but true that many technicians have found that the cause of a reported "problem" was that a piece of equipment was not

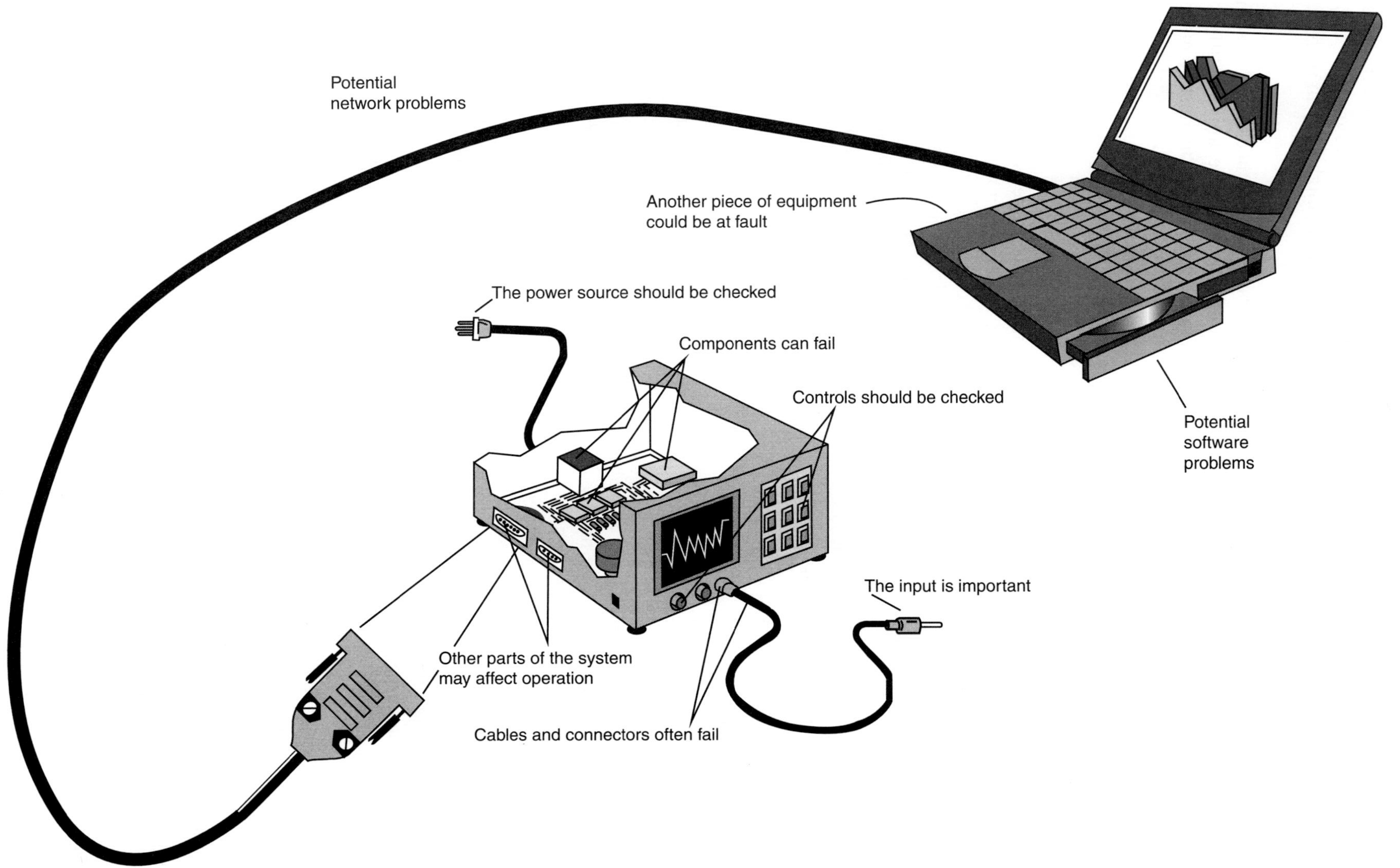

Fig. 10-1 Troubleshooting often requires a system view.

plugged in. Don't assume anything when troubleshooting. Don't forget to look at back panels, indicator lights, ready-access fuse holders, switch and selector settings, plugs and cables, and so on. Check everything, and always begin with the most obvious items. This takes a little time, but it takes more time to reassemble something that should not have been taken apart in the first place.

Ten percent rule

Experience is extremely valuable. Many technical workers believe in what is called the *10 percent rule.* This rule says that 10 percent of the possibilities cause 90 percent of the problems. When you know the items that are likely to fail, you can check them early in the troubleshooting process. Obviously, beginners do not often know what belongs in the 10 percent category. Don't be reluctant to ask questions. Ask coworkers and supervisors direct questions such as: "What could be wrong with a model 360L that it won't pass diagnostic check number 4?" Ask customers indirect questions such as: "Has anything like this happened before?" Another good question is: "Was anything changed or did anything odd happen before it failed?"

Service literature

Find the technical reference material and use it. Although this seems rather obvious, it is a fact that egos, laziness, and other human weaknesses waste an awful lot of time and money. Learn to use the table of contents to scan manuals for relevant sections. Learn to use the index, if there is one, and don't forget the appendix material because sometimes the best troubleshooting information can be found there.

Companies whose business has no direct relationship to electronics might hire electronic technicians.

Visual inspection

Ignorance is deadly in the troubleshooting business. The old saying "What you don't know can't hurt you" is totally wrong. What you don't know, that you *should know,* will cause you endless grief. When technicians are troubleshooting they *should know:*

- All about relevant safety issues
- All about relevant regulatory issues (environmental impact laws, codes, etc.)
- What is normal behavior
- About various modes of operation (automatic modes, programming modes, etc.)
- What the various parts of a system do
- What the controls are supposed to do
- What inputs and outputs are for and how they should be connected
- Whether a device can be reasonably tested when it is removed from a system
- What role software might play in performance

When the preliminary checks and system tests are completed and the unit is still not working, an internal inspection must be made. Do not attempt to remove the unit from its cabinet until you have disconnected it from the ac line. Be wary of charged filter capacitors. Use a voltmeter and *verify* that they are discharged.

Follow the manufacturer's procedures when taking apart equipment. Often *service literature* will show exactly how to do it. Many technicians overlook this and just start removing parts and fasteners. This may cause internal assemblies to fall apart. Damage and long delays in reassembly can result. It saves time in the long run to work carefully and use the service literature.

Use the proper tools. The wrong wrench or screwdriver can slip and damage fasteners or other parts. A scratched front panel is not nice to look at. It may take weeks to get a new one and several hours to replace. Or, it may not be possible to obtain a new one. The old saying "haste makes waste" fits perfectly in electronic repair.

Sort and save all fasteners and other parts. There is nothing more disturbing to a customer or a supervisor than to find an expensive piece of equipment with missing screws, shields, and other parts. The manufacturer includes all those pieces for a very good reason: they are necessary for proper and safe operation of the equipment.

The next part of the preliminary check is a *visual inspection* of the interior of the equipment. Look for the following:

1. Burned and discolored components
2. Broken wires and components
3. Cracked or burned circuit boards
4. Foreign objects (paper clips, etc.)
5. Bent transistor leads that may be touching (this includes other noninsulated leads as well)
6. Parts falling out of sockets or only partly seated
7. Loose or partly seated connectors

8. Leaking components (especially electrolytic capacitors and batteries)

Obvious damage can be repaired at this point. However, do not energize the unit immediately. For example, suppose a resistor is burned black. In many cases, the new one will quickly do the same. Inspect the schematic to see what the resistor does in the circuit. Try to determine what kinds of problems could have caused the overload.

Refer to Fig. 10-2. Suppose a visual inspection revealed that R_1 was badly burned. What kinds of other problems are likely? There are several:

1. Capacitor C_1 could be shorted.
2. Zener diode D_1 could be shorted.
3. There is a short somewhere in the regulated output circuit.
4. The unregulated input voltage is too high (this is not too likely, but it is a possibility).

Electrical check

When the preliminary visual inspection is complete, a preliminary *electrical check* should be made. Be sure to use an isolation transformer in those cases that require it. Remember that most variable transformers are autotransformers, and they do not provide line isolation. In some cases, it is possible to safely make *floating measurements*. However, remember that floating measurements require test equipment designed for this purpose. Line isolation and floating measurements were covered in Chap. 4 (see Figs. 4-25 and 4-26).

Overheating

The first part of the preliminary electrical check involves any signs of *overheating*. Your nose may give important information. Hot electronic components often give off a distinct odor. Many technicians use a finger to test for excessive heat. This must be done with *extreme caution*. It must never be done if there is any chance of high voltage being present. Electronic temperature probes are available for accurately measuring heat. Noncontact temperature probes make it possible to measure the temperature of a general area or component by aiming and squeezing a trigger.

Verify voltage

The next part of the electrical check is to *verify* the power-supply *voltages*. Power-supply problems can produce an entire range of symptoms. This is why it can save a lot of time to check here first. Consult the manufacturer's specifications. The proper voltages are usually indicated on the schematic diagram. Some error is usually allowed. A 20 percent variation is not unusual in many circuits. Of course, if a precision voltage regulator is in use, this much error is not acceptable. Be sure to check all the supply voltages. Remember, it only takes one incorrect voltage to keep a system from working.

Static discharge

When making preliminary checks, a technician must be careful to avoid *static discharge* into a sensitive circuit or component. Later, if parts are being replaced, precautions must again be taken to avoid damage by static discharge. A technician's body can become charged. If it discharges into electronic devices, the devices may be damaged or destroyed. Components that are not connected to anything are especially vulnerable (Fig. 10-3 on page 262). That's why it's best to leave components in their protective containers until they are actually going to be used.

Career and other information can be found at the website for the Consumer Electronics Manufacturer's Association (CEMA).

Table 10-1 on page 262 lists some of the voltages that can be generated by ordinary movements and actions. Notice that very high voltages are possible, especially when the relative humidity is low (10 to 20 percent).

Wrist strap

Some metal-oxide semiconductor (MOS) and integrated devices may be damaged with static charges as low as several hundred volts. People who handle sensitive electronic components may be required to wear a conductive wrist strap. The *wrist strap* is connected to an equipment ground or work-surface ground though a flexible cord that has an integral 1-MΩ resistor (Fig. 10-4 on page 263). *Never ground any part of your body* when working on electronic equipment unless it is with an approved device and in an approved environment. Such an environment often includes special conductive mats for the floor and for work surfaces. The following guidelines are recommended for avoiding damage by static discharge:

1. When possible, maintain the relative humidity in the work environment at 50 percent.

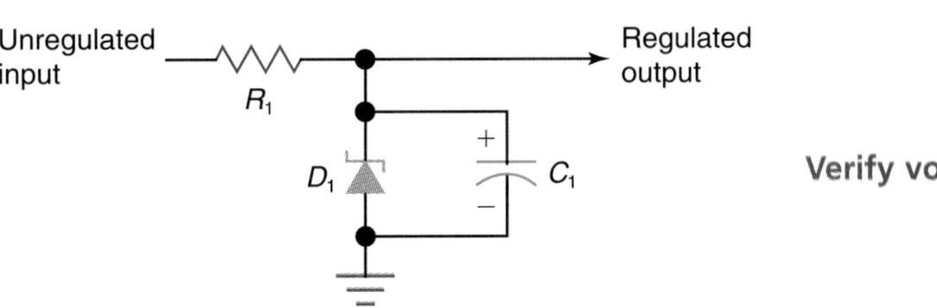

Fig. 10-2 Zener shunt regulator.

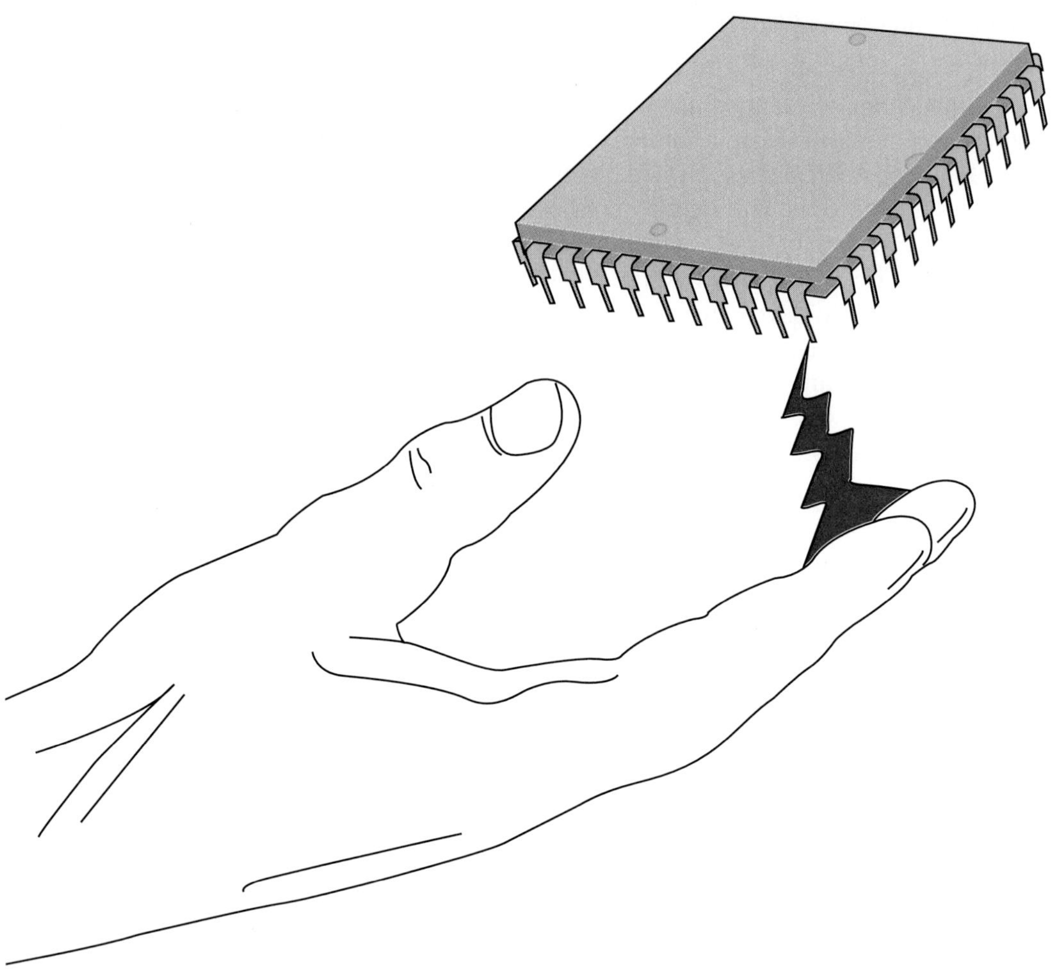

Fig. 10-3 Unconnected solid-state components are especially vulnerable to electrostatic discharge.

2. Assume that all components are sensitive to damage by electrostatic discharge.
3. When possible, make sure the equipment is off before touching sensitive parts. When instruments must be connected to live circuits, always connect the ground lead of the instrument first.
4. Do not handle sensitive parts any more than is necessary. They must be kept in their protective carriers or remain in special conductive foam until it is time to use them.
5. Touch the conductive package or foam to ground before removing the part.
6. Touch a grounded part of the chassis before removing or inserting a part.
7. Immediately place removed parts into a protective carrier or conductive foam.
8. Use only antistatic spray materials.
9. Use as little motion as possible when working with sensitive circuitry and parts (Table 10-1).

Table 10-1 Electrostatic Charges Generated by Technical Personnel

	Generated Charge in Volts	
Action	**High Humidity**	**Low Humidity**
Walking across carpet	1500	35,000
Removing item from plastic bag	1200	20,000
Sliding off or onto plastic chair	1500	18,000
Walking across vinyl floor	250	12,000
Sliding sleeve across laminated bench	100	6000

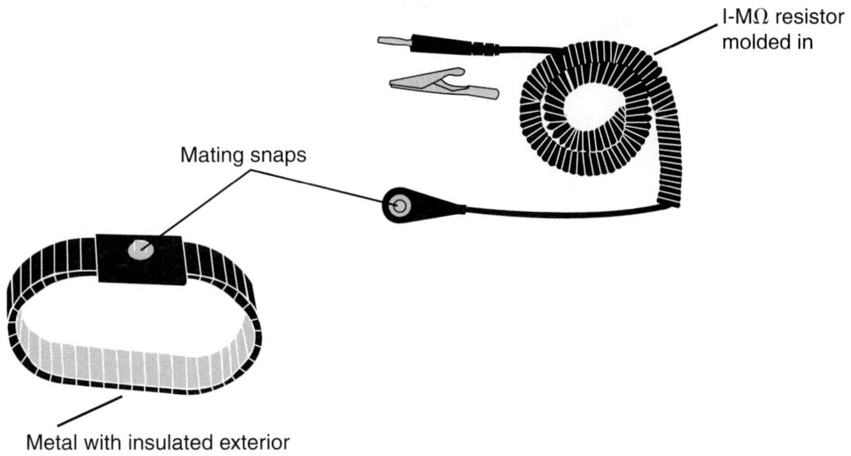

Fig. 10-4 Wrist strap for electrostatic discharge control.

TEST

Choose the letter that best completes each statement.

1. The first step in troubleshooting is to take:
   a. Resistance readings
   b. Voltage readings
   c. A look at the overall system
   d. The covers off all defective units
2. In some cases, troubleshooting electronic parts inside one piece of equipment can be a waste of time because:
   a. The problem might be in another part of the system
   b. The problem might be with the system software
   c. An external connector might be loose
   d. All of the above
3. A system point of view will help:
   a. Prevent a technician from working on things that are not broken
   b. Prevent time from being wasted
   c. Guide an observant technician to the correct conclusion
   d. All of the above
4. Refer to Fig. 10-2. A visual check shows that C_1 is bulging. This may be a sign of excessive voltage. This could have been caused by
   a. A short in D_1
   b. An open in R_1
   c. The output being shorted to ground
   d. An open in D_1
5. After a piece of equipment has been removed from its cabinet and inspected visually, the next step should be
   a. To check supply voltages
   b. To check the transistors
   c. To check the integrated circuits
   d. To check the electrolytic capacitors
6. Which components are most likely to be damaged by static discharge?
   a. Resistors
   b. CMOS devices
   c. Capacitors
   d. Printed circuit boards

10-2 No Output

No input

There are several causes for no output in an amplifier. Perhaps the most obvious is *no input.* It is worth the effort to check this early. You may find that a wire or a connector has been pulled loose. With no input signal, there can be no output signal.

The output device may be defective. For example, in an audio amplifier, the output is sent to a loudspeaker or perhaps headphones. These devices can fail and are easy to check. An ordinary flashlight cell can be used to make the test. One cell with two test leads will allow an easy way to temporarily energize a speaker. A

Sensors/CETs
A malfunctioning sensor can cause increased emission from an automobile and yet produce no noticeable change in performance. *CET* stands for certified electronic technician.

AC coupling

good speaker will make a clicking sound when the test leads touch the speaker terminals. Analog ohmmeters, on the $R \times 1$ range, will make the same click when connected across a speaker. Either technique tells you the speaker is capable of changing electricity into sound. This test is a simple one and cannot be used to check the quality of a speaker.

Signal chain

If there is nothing wrong with the output device, the power supply, or the input signal, then there is a break in the *signal chain.* Figure 10-5 illustrates this. The signal must travel the chain, stage by stage, to reach the load. A break at any point in the chain will usually cause the no-output symptom.

Signal injection

A four-stage amplifier contains many parts. There are many measurements that can be taken. Therefore, the efficient way to troubleshoot is to *isolate* the problem to one stage. One way to do this is to use *signal injection.* Figure 10-6 shows what needs to be done. A signal generator is used to provide a test signal. The test signal is injected at the input to the last stage. If an output signal appears, then the last stage is good. The test signal is then moved to the input of the next-to-last stage. When the signal is injected to the input of the broken stage, no output will be noticed. This eliminates the other stages, and you can zero in on the defective circuit.

Click test

Signal injection must be done carefully. One danger is the possibility of overdriving an amplifier and damaging something. More than one technician has ruined a loudspeaker by feeding too large a signal into an audio amplifier. A high-power amplifier must be treated with respect!

Signal tracing

Another danger in signal injection is improper connection. A schematic diagram is a must. The common ground is generally used for connecting the ground lead from the generator. Assuming that the chassis is common will not always work out. If the common connection is made in error, a large hum voltage may be injected into the system. Damage can result.

Many amplifiers can be tested with *ac* signals. The signal should be capacitively *coupled* to avoid upsetting the bias on a transistor or integrated circuit. If the generator is dc-coupled, a capacitor must be used in series with the hot lead. This capacitor will block the dc component yet allow the ac signal to be injected. A 0.1-μF capacitor is usually good for audio work, and a 0.001-μF capacitor can be used for radio frequencies (be sure that the voltage rating of the capacitor is adequate).

The test frequency varies, depending on the amplifier being tested. A frequency of 400 to 1000 Hz is often used for audio work. A radio-frequency amplifier should be tested at its design frequency. This is especially important in band-pass amplifiers. Some are so narrow that an error of a few kilohertz will cause the signal to be blocked. It may be necessary to vary the generator frequency slowly while watching for output.

Signal injection can be performed without a signal generator in many amplifiers. A resistor can be used to inject a *click* into the signal chain. The click is really a *signal pulse* caused by suddenly changing the bias on a transistor. A resistor is connected momentarily from the collector lead to the base lead as in Fig. 10-7. This will cause a sudden increase in transistor current. The collector voltage will drop suddenly, and a pulse travels down the chain and reaches the output device. When the stage is reached where the click cannot reach the output, the problem has been isolated. As with other types of signal injection, start at the last stage and work toward the first stage.

The *click test* must be used *carefully.* Use only a resistor of several thousand ohms. Never use a screwdriver or a jumper wire. This can cause severe damage to the equipment. *Never* use a click test in high-voltage/high-power equipment. It is not safe for you or for the equipment. Always be careful when probing in live circuits. If you slip and short two leads, damage often results.

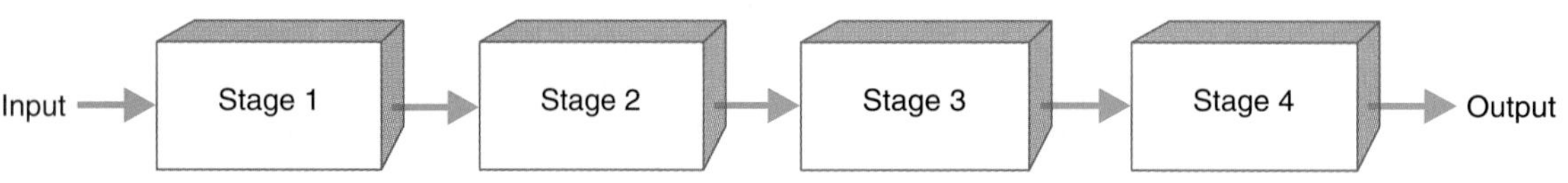

Fig. 10-5 The signal chain.

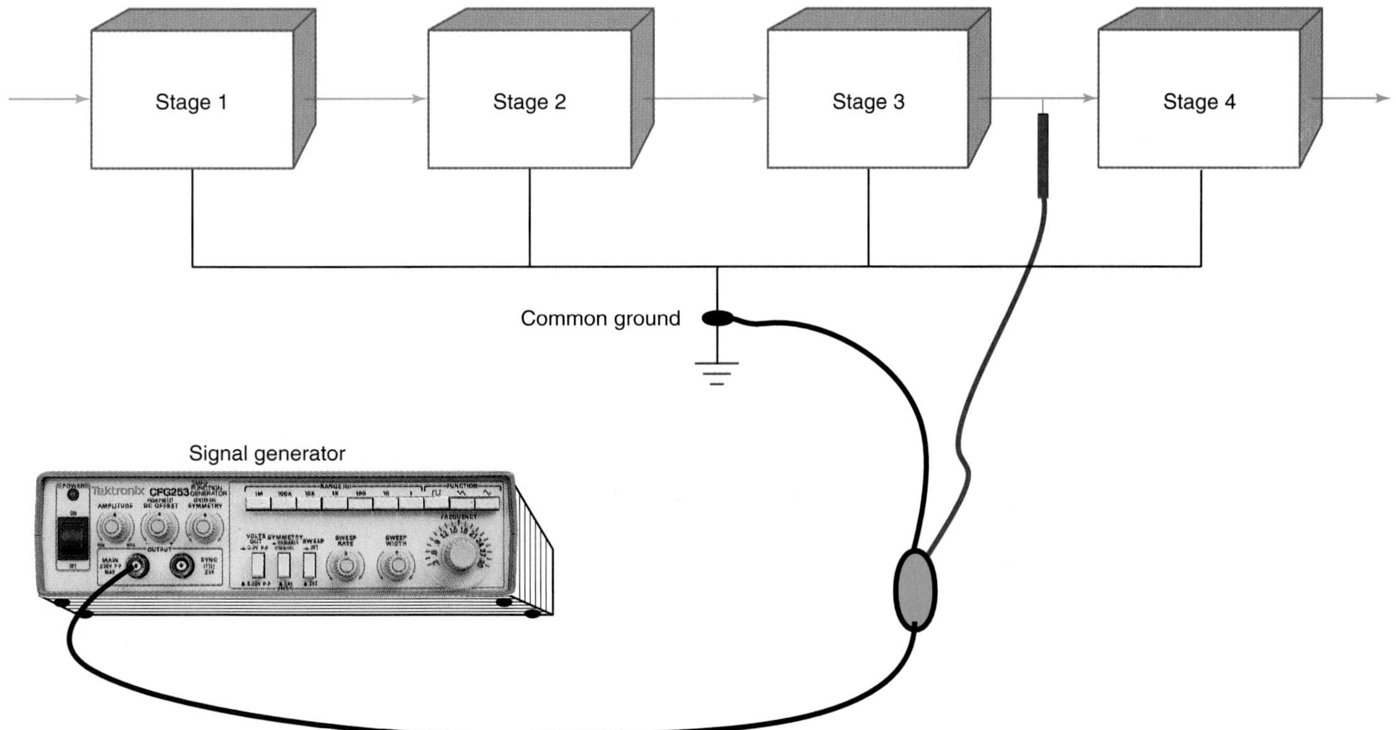

Fig. 10-6 Using signal injection.

Signal tracing is another way to isolate the defective stage. This technique may use a meter, an oscilloscope, a signal tracer, or some related instrument. Signal tracing starts at the input to the first stage of the amplifier chain. Then the tracing instrument is moved to the input of the second stage, and so on. Suppose that a signal is found at the input to the third stage but not at the input to the fourth stage. This would mean that the signal is being lost in the third stage. The third stage is probably defective.

The important thing to remember in signal tracing is the gain and frequency response of the instrument being used. For example, do not expect to see a low-level audio signal on an ordinary ac voltmeter. Also, do not expect to see a low-level RF signal on an oscilloscope. Even if the signal is in the frequency range of the oscilloscope, the signal must be in the millivolt range to be detectable. Some radio signals are in the microvolt range. Not knowing the limitations of your test equipment will cause you to reach false conclusions!

Once the fault has been localized to a particular stage, it is time to determine which part

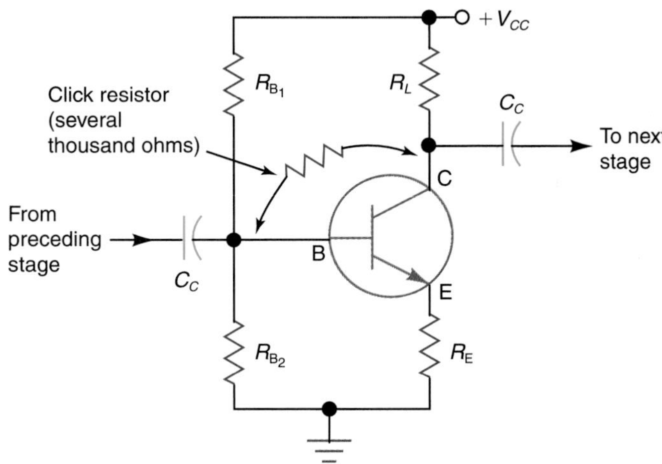

Fig. 10-7 The click test.

has failed. Of course, it is possible that more than one part is defective. More often than not, one component will be found defective.

Voltage analysis

Most technicians use *voltage analysis* and their knowledge of circuits. Study Fig. 10-8. Suppose the collector of Q_2 measures 20 V. The manufacturer's schematic shows that the collector of Q_2 should be 12 V with respect to ground. What could cause this large error? It is likely that Q_2 is in *cutoff.* A 20-V reading at the collector tells us that the voltage is almost the same on both ends of R_6. Ohm's law tells us that a low voltage drop means little current flow. Transistor Q_2 must be cut off.

Now, what are some possible causes for Q_2 to be cut off? First, the transistor could be defective. Second, R_7 could be open. Resistor R_7 supplies the base current for Q_2. If it opens, no base current will flow. This cuts off the transistor. This can be checked by measuring the base voltage of Q_2. With R_7 open, the base voltage will be zero. Third, R_9 could be open. If it opens, there will be no emitter current. This cuts off the transistor. A voltage check at the collector of Q_2 will show a little less than 21 V. The actual voltage will be determined by the divider formed by R_6, R_7, and R_8. Fourth, R_8 could be shorted. This seldom happens, but a troubleshooter soon learns that all things are possible. With R_8 shorted, no base current can flow and the transistor is cut off. The base voltage will measure zero.

Let us try another symptom. Suppose the collector voltage at Q_1 measures 0 V. A check on the manufacturer's service notes shows that it is supposed to be 11 V. What could be wrong? First, C_1 could be shorted. The combination of R_1 and C_1 acts as a low-pass filter to prevent any hum or other unwanted ac signal from reaching Q_1. If C_1 shorts, R_1 will drop the entire 21-V supply. This can be checked by measuring the voltage at the junction of R_2 and C_1. With C_1 shorted, it will be 0 V. Second, R_2 could be open. This can also be checked by measuring the voltage at the junction of R_2 and C_1. A 21-V reading here indicates R_2 must be open. Could Q_1 be shorted? The answer is no. Resistor R_5 would drop at least a small voltage, and the collector would be above 0 V.

Sometimes it helps to ask yourself what might happen to the circuit given a specific component failure. This thoughtful question-and-answer game is used by most technicians. Again, refer to Fig. 10-8. What would happen if C_4 shorts? This short circuit would apply the dc collector potential of Q_1 to the base of Q_2. Chances are that this would greatly increase the base voltage and drive Q_2 into saturation. The collector voltage at Q_2 will drop to some low value.

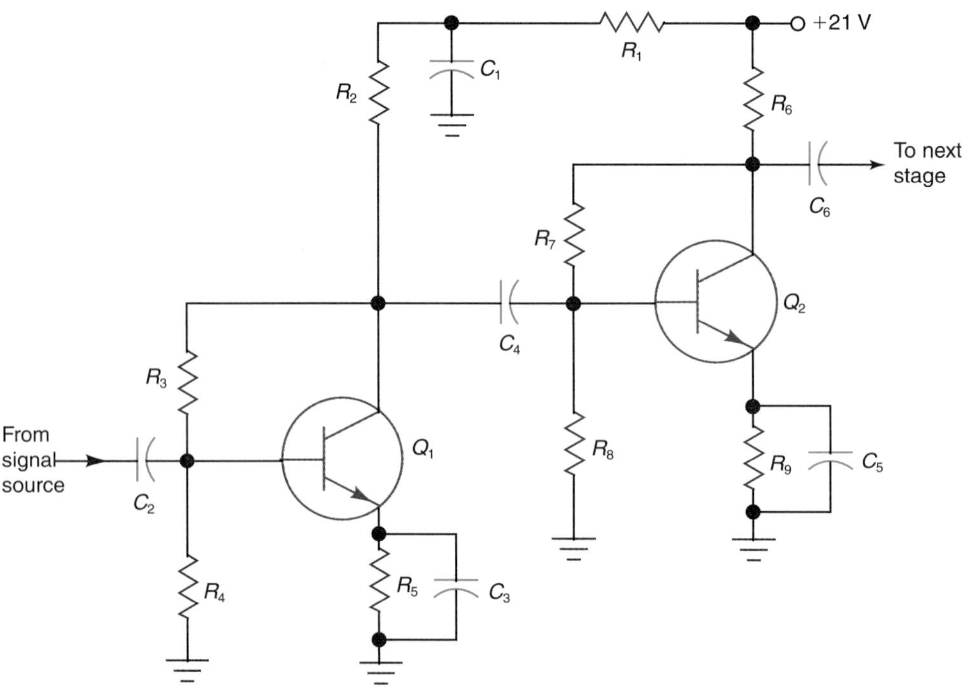

Fig. 10-8 Troubleshooting with voltage analysis.

What if C_2 in Fig. 10-8 shorts? Transistor Q_1 could be driven to cutoff or to saturation. If the signal source has a ground or negative dc potential, the transistor will be cut off. If the signal source has a positive dc potential, the transistor will be driven toward saturation.

The advantage of voltage analysis is that it is easy to make the measurements. Often, the expected voltages are indicated on the schematic diagram. A small error is usually not a sign of trouble. Many schematics will indicate that all voltages are to be within a ±10 percent range.

Current analysis

Current analysis is not easy. Circuits must be broken to measure current. Sometimes, a technician can find a known resistance in the circuit where current is to be measured. A voltage reading can be converted to current by Ohm's law. However, if the resistance value is wrong, the calculated current will also be wrong.

Resistance analysis

Resistance analysis can also be used to isolate defective components. This can be tricky, however. Multiple paths may produce confusing readings. Refer to Fig. 10-9. An ohmmeter check is being made to verify the value of the 3.3-kΩ resistor. The reading is incorrect because the internal supply in the ohmmeter forward-biases the base-emitter junction. Current flows in both the resistor and the transistor. This makes the reading *lower* than 3.3 kΩ. The ohmmeter polarity can be reversed. This will reverse-bias the transistor and allow the 3.3-kΩ resistor to be measured by itself. Most DMMs, when used as ohmmeters, will not turn on semiconductor junctions.

Even if a junction is not turned on by the ohmmeter, in many cases it is still impossible to obtain useful resistance readings. There will be other components in the circuit to draw current from the ohmmeter. Any time that you are using resistance analysis, remember that a *low reading* could be caused by multiple paths.

It is usually poor practice to unsolder parts for resistance analysis unless you are reasonably sure the part is defective. Unsoldering can cause damage to circuit boards and to the parts. It is also time-consuming.

As mentioned before, most technicians use voltage analysis to locate defective parts. This is valid and effective since most circuit faults will change at least one dc voltage. However, there is the possibility of an ac fault that breaks the signal chain without changing any of the dc readings. Some ac faults are

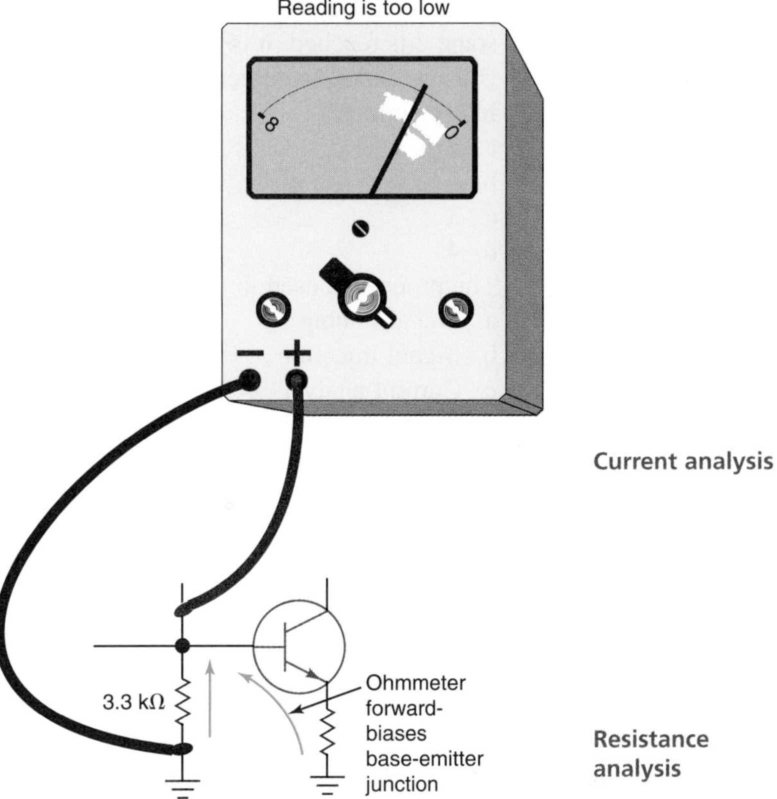

Fig. 10-9 An ohmmeter may forward-bias a PN junction.

1. An open coupling capacitor
2. A defective coupling coil or transformer
3. A break in a printed circuit board
4. A dirty or bent connector (plug-in modules often suffer this fault)
5. An open switch or control such as a relay

To find this type of fault, signal tracing or signal injection can be used. You will find different conditions at either end of the break in the chain. Some technicians use a coupling capacitor to bypass the signal around the suspected part. The value of the capacitor can usually be 0.1 μF for audio work and 0.001 μF for radio circuits. *Do not* use this approach in high-voltage circuits. *Never* use a jumper wire. Severe circuit damage may result from jumping the wrong two points.

TEST

Choose the letter that best answers each question.

7. Refer to Fig. 10-5. A signal generator is applied to the input of stage 4, then stage

3, and then stage 2. When the input of stage 2 is reached, it is noticed that there is no output. The defective stage is most likely number
a. 1
b. 2
c. 3
d. 4
8. The procedure used in question 7 is called
a. Signal tracing
b. Signal injection
c. Current analysis
d. Voltage analysis
9. Refer to Fig. 10-5. A signal generator is first applied to the input of stage 1. It is noticed that nothing reaches the output. The difficulty is in
a. Stage 1
b. Stage 2
c. Stage 3
d. Any of the stages
10. A good test frequency for audio troubleshooting is
a. 2 Hz
b. 1 kHz
c. 455 kHz
d. 10 MHz
11. It is a good idea to use a coupling capacitor in signal injection to
a. Improve the frequency response
b. Block the ac signal
c. Provide an impedance match
d. Prevent any dc shift or loading effect
12. Refer to Fig. 10-7. When the click resistor is added, the collector voltage should
a. Not change
b. Go in a positive direction
c. Go in a negative direction
d. Change for a moment and then settle back to normal
13. Refer to Fig. 10-7. When the click resistor is added, the emitter voltage should
a. Not change
b. Go in a positive direction
c. Go in a negative direction
d. Change for a moment and then settle back to normal
14. Refer to Fig. 10-8. Assume that C_4 is open. It is most likely that
a. The collector voltage of Q_1 will read high
b. The collector voltage of Q_1 will read low
c. The base of Q_2 will be 0 V
d. All the dc voltages will be correct
15. Refer to Fig. 10-8. Resistor R_7 is open. It is most likely that the
a. Collector voltage of Q_2 will be high (near 21 V)
b. Collector voltage of Q_2 will be low (near 1 V)
c. Case of Q_2 will run hot
d. Transistor will go into saturation
16. Refer to Fig. 10-8. Suppose it is necessary to know the collector current of Q_1. The easiest technique is to
a. Break the circuit and measure it
b. Measure the voltage drop across R_2 and use Ohm's law
c. Measure the collector voltage
d. Measure the emitter voltage
17. Refer to Fig. 10-8. Transistor Q_1 is defective. This will not affect the dc voltages at Q_2 because
a. C_1 isolates the two stages from the power supply
b. The two stages are not dc-coupled
c. Both transistors are NPN devices
d. All of the above
18. Refer to Fig. 10-8. It is desired to check the value of R_8. The power is turned off; the negative lead of the ohmmeter is applied at the top of R_8, and the positive lead is applied at the bottom. This will prevent the ohmmeter from forward-biasing the base-emitter junction. The ohmmeter reading is still going to be less than the actual value of R_8 because
a. Capacitor C_4 is in the circuit
b. Transistor Q_1 is in the circuit
c. Resistors R_7 and R_6 and the power supply provide a path to ground
d. None of the above

10-3 Reduced Output

Low output from an amplifier tells us there is lack of gain in the system. In an audio amplifier, for example, normal volume cannot be reached at the maximum setting of the volume control. Do not troubleshoot for low output until you have made the preliminary checks described in Sec. 10-1.

Low output from an amplifier can be caused by *low input* to the amplifier. The signal source is weak for some reason. A microphone may deteriorate with time and rough treatment. The same thing can happen with some sensors. To check, try a new signal source or substitute a signal generator.

Another possible cause for low output is reduced performance in the output device. A loudspeaker defect or a poor connection may prevent normal loudness. In a video system, there may be a difficulty in the cathode-ray tube (picture tube) which causes poor contrast. This can be checked by substituting for the output device or replacing the device with a known load and measuring the output.

In Fig. 10-10 a loudspeaker has been replaced with an 8-Ω resistor in order to measure the output power of an audio amplifier. This resistor must match the output impedance requirement of the amplifier. It must also be rated to safely dissipate the output power of the amplifier. The signal generator is usually adjusted for a sinusoidal output of 1 kHz. The signal level is set carefully so as not to overdrive the amplifier being tested.

Suppose you want to check an audio amplifier for rated output power with an oscilloscope and signal generator. The specifications for the amplifier rate it at 100 W of continuous sine-wave power output. How could you be sure the amplifier meets its specification and does not suffer from low output? The power formula shows the relationship between output voltage and the output resistance:

$$P = \frac{V^2}{R}$$

In this case, *P* is known from the specifications and *R* is the substitute resistor. What you must determine is the expected output voltage:

$$V^2 = PR \qquad \text{or} \qquad V = \sqrt{PR}$$

From the known data:

$$V = \sqrt{100\ \text{W} \times 8\ \Omega} = 28.28\ \text{V}$$

The 100-W amplifier can be expected to develop 28.28 V across the 8-Ω load resistor. The oscilloscope measures peak-to-peak. Thus, it would be a good idea to convert 28.28 V to its peak-to-peak value:

$$V_{\text{p-p}} = V_{\text{rms}} \times 1.414 \times 2 = 80\ \text{V}$$

To test the 100-W amplifier, the gain control would be advanced until the oscilloscope showed an output sine wave of 80 V peak-to-peak. There should be no sign of clipping on the peaks of the waveform. If the amplifier passes this test, you know it is within specifications.

Figure 10-11 on page 270 shows another method of testing amplifiers that is often used in the radio communications industry. Normally, the two-way radio delivers its RF output power to an antenna. For test purposes the antenna has been replaced by a *dummy load.* This load is a noninductive resistor of usually 50 Ω. This provides a way of testing without producing interference and ensures the proper load for the transmitter. The RF wattmeter indicates the output power.

Dummy load

If the input signal and output device are both normal, the problem must be in the amplifier itself. One or more stages are giving less than normal gain. You can expect the problem to be limited to one stage in most cases. It is more difficult to isolate a low-gain stage than it is to find a total break in the signal chain. Signal tracing and signal injection can both give misleading results.

JOB TIP

Many organizations look for employees who will have a customer focus.

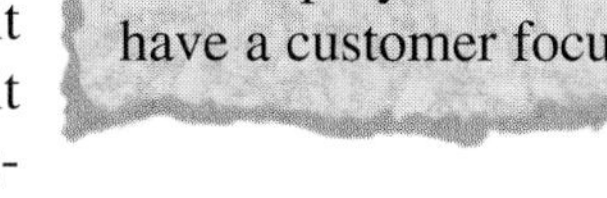

Suppose you are troubleshooting the three-stage amplifier shown in Fig. 10-12 on page 271. Your oscilloscope shows that the input to stage 1 is 0.1 V and the output is 1.5 V. A quick calculation gives a gain of 15:

$$A_V = \frac{1.5\ \text{V}}{0.1\ \text{V}} = 15$$

Fig. 10-10 Replacing the speaker with a resistor.

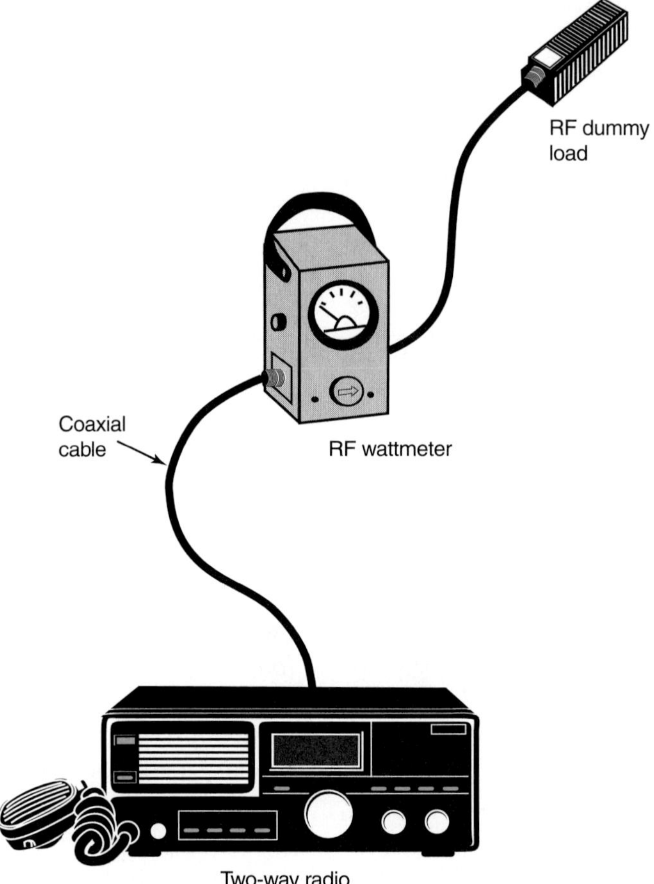

Fig. 10-11 Measuring transmitter power output.

This seems acceptable, so you move the probe to the output of stage 2. The voltage here also measures 1.5 V. This seems strange. Stage 2 is not giving any voltage gain. However, a close inspection of the schematic shows that stage 2 is an *emitter follower.* You should remember that emitter followers do not produce any voltage gain. Perhaps the problem is really in stage 1. The normal gain for this stage could be 150 rather than 15. Knowledge of the circuit is required for troubleshooting the low-output symptom.

Emitter follower

About Electronics

Troubleshooting Tips

Some equipment can be diagnosed via a modem. Some equipment runs a series of self-diagnostic checks when powered on. Block diagrams can be more important than schematic diagrams for troubleshooting some systems.

Some beginners believe that a dead circuit is going to be more difficult to fix than a weak one. Just the opposite is generally true. It is usually easier to find a broken link in the signal chain because the symptoms are more definite.

With experience, the low-output problem is usually not too difficult because the technician knows approximately what to expect from each stage. In addition to experience, service notes and schematics can be a tremendous help. They often include pictures of the expected waveforms at various circuit points. If the oscilloscope shows a low output at one stage, then the input can be checked. If the input signal is normal, it is safe to assume that the low-gain stage has been found.

Sometimes the needed information is found in the equipment itself. A good example is a stereo amplifier (Fig. 10-13) on page 272. Assume that the left channel is weak. By checking back and forth between the right channel and the left channel, the low-gain stage can be isolated. Remember when you are signal-tracing to work toward the output.

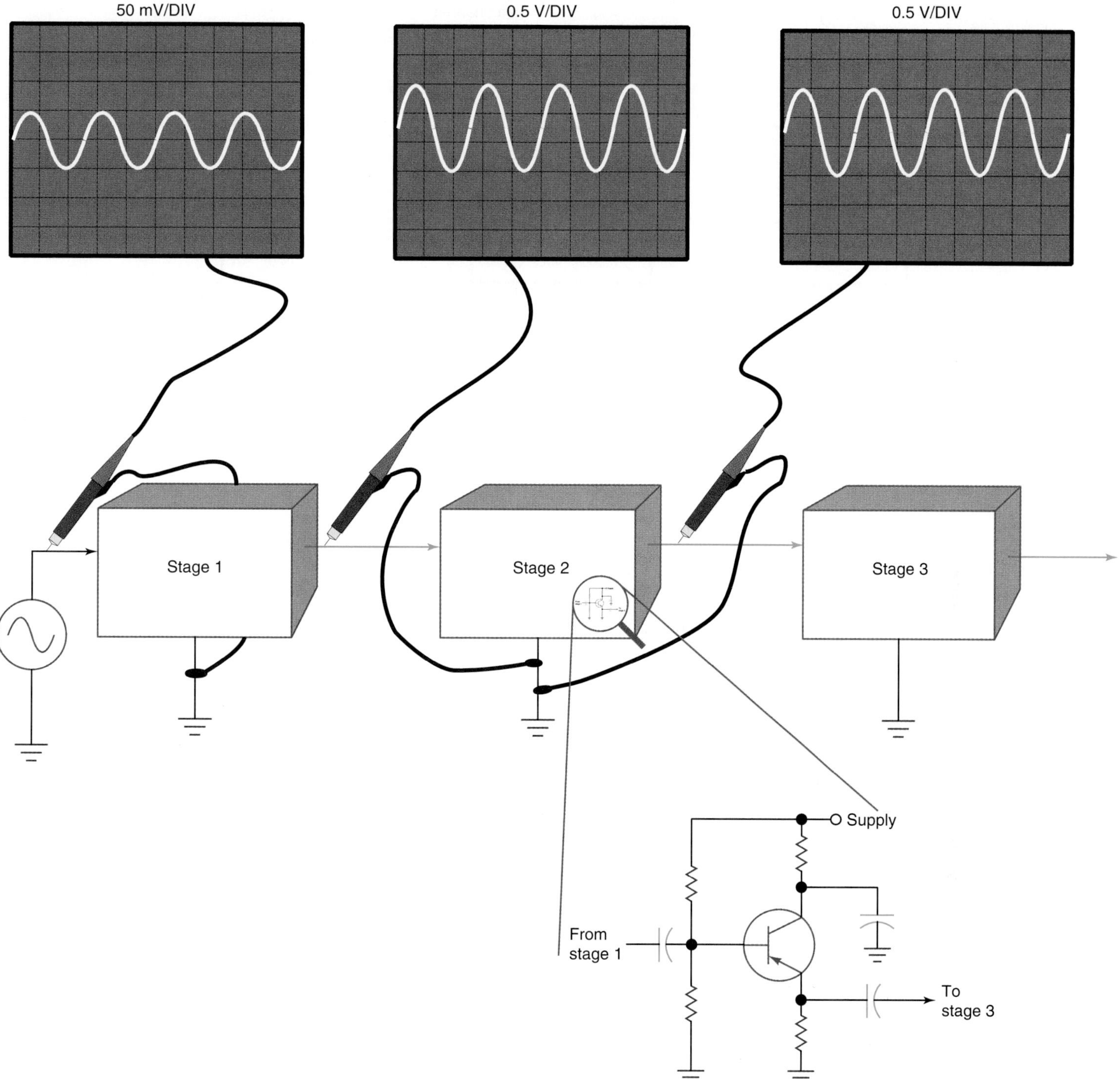

Fig. 10-12 Troubleshooting a three-stage amplifier.

Once the weak stage is located, the fault can often be isolated to a single component. Some possible causes for low gain are

1. Low supply voltage
2. Open bypass capacitor
3. Partially open coupling capacitor
4. Improper transistor bias
5. Defective transistor
6. Defective coupling transformer
7. Misaligned or defective tuned circuit

Supply voltages are supposed to be verified earlier in the troubleshooting process. However, it is still possible that a stage will not receive the proper voltage. Note the resistor and capacitor in the supply line in Fig. 10-14 on page 272. This *RC decoupling network* may be defec-

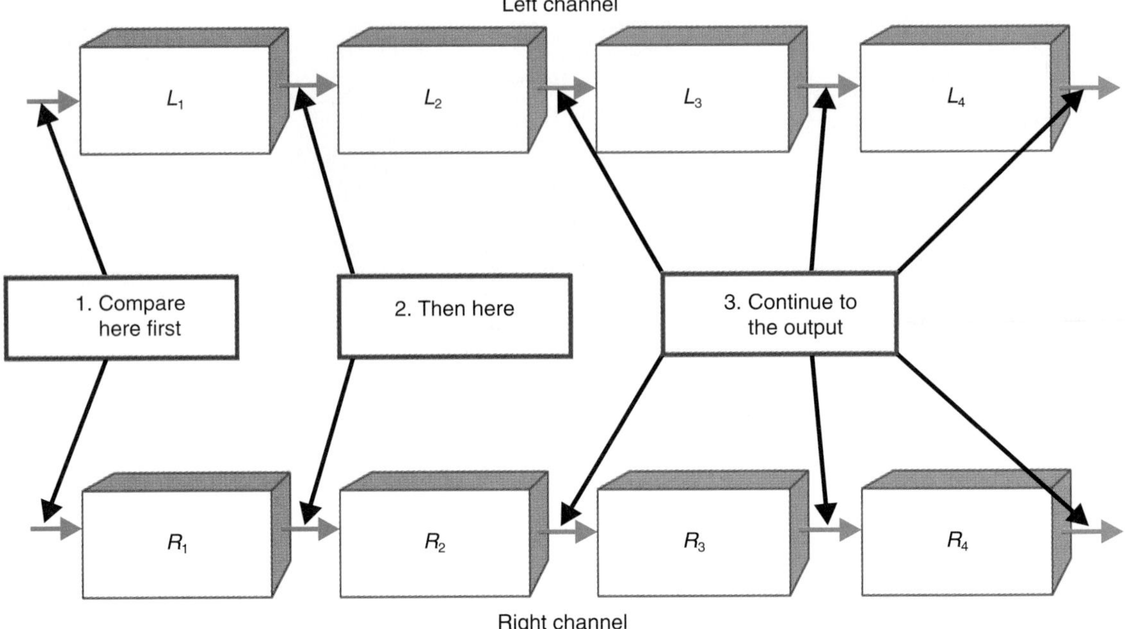

Fig. 10-13 A stereo amplifier.

tive. R_3 may have increased in value or C_2 may be leaky. These defects can significantly lower the collector voltage which can decrease gain.

Figure 10-14 also shows that the emitter bypass capacitor may be open. This can lower the voltage gain from over 100 to less than 4. The coupling capacitors may have lost capacity, causing loss of signal. The dc voltage checks at the transistor terminals shown in Fig. 10-14 will determine whether the bias is correct but they will not determine any open capacitors.

AGC circuit

In the dual-gate MOSFET RF amplifier of Fig. 10-15, the input signal is applied to gate 1 of the transistor. Gate 2 is connected to the supply through a resistor and to a separate *AGC circuit.* The letters *AGC* stand for automatic gain control. An AGC fault will often reduce the gain of an amplifier. The gain reduction can be more than 20 dB. Thus, if an amplifier is controlled by AGC, this control voltage must be measured to determine if it is normal.

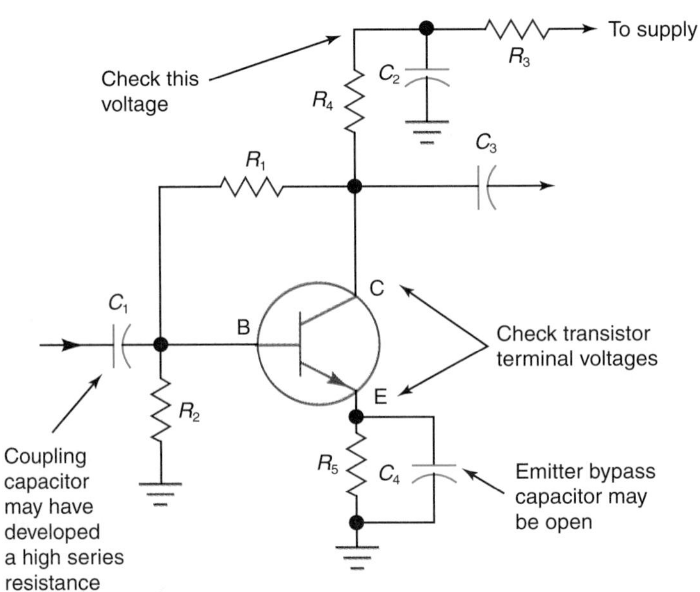

Fig. 10-14 Checking for the cause of low gain.

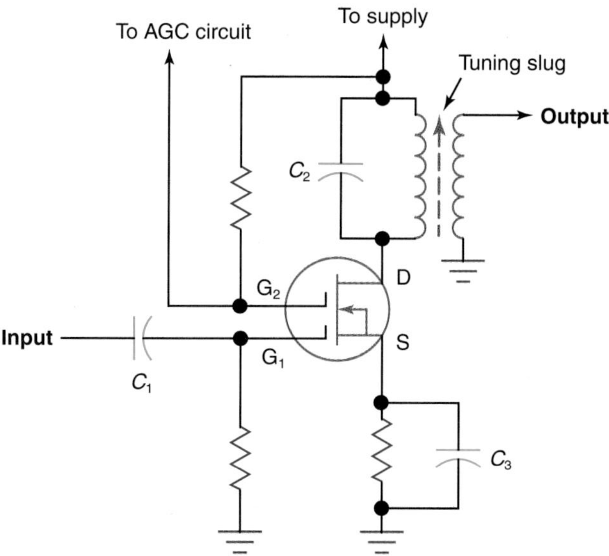

Fig. 10-15 A MOSFET RF amplifier.

Figure 10-15 shows that the drain load is a tuned circuit. This circuit is adjusted for the correct resonant frequency by moving a tuning slug in the transformer. The possibility exists that someone turned the slug. This can produce a severe loss of gain at the operating frequency of the amplifier. In such cases, refer to the service notes for the proper adjustment procedure.

Troubleshooting for loss of gain in amplifiers can be difficult. Many things can give this symptom. Voltage analysis will locate some of them. Others must be found by substitution. For example, a good capacitor can be temporarily placed in parallel with one that is suspected of being open. If gain is restored, the technician's suspicion that the original capacitor was defective is correct.

TEST

Choose the letter that best answers each question.

19. Refer to Fig. 10-10. The amplifier is rated at 35 W power output. Assuming a sine-wave test signal, the oscilloscope should show at least
    a. 17.9 V peak-to-peak before clipping
    b. 47.2 V peak-to-peak before clipping
    c. 75.8 V peak-to-peak before clipping
    d. 99.6 V peak-to-peak before clipping
20. The normal voltage gain for an emitter-follower amplifier is
    a. 250
    b. 150
    c. 50
    d. Less than 1
21. Refer to Fig. 10-14. The power supply has been checked, and it is normal; yet, the collector of the transistor is quite low in voltage. This could be caused by
    a. An open in R_1
    b. A leaky C_2
    c. An open in C_3
    d. An open in C_4
22. Refer to Fig. 10-14. The stage is supposed to have a gain of 50, but a test shows that it is much less. It is least likely that the cause is
    a. A defective transistor
    b. A short in C_4
    c. An open in C_4
    d. An open in C_1
23. Refer to Fig. 10-15. The stage is suffering from low gain. This could be caused by
    a. An incorrect AGC voltage
    b. A misadjusted tuning slug
    c. A short in C_2
    d. Any of the above

10-4 Distortion and Noise

Distortion and noise in an amplifier mean that the output signal contains different information from the input signal. A linear amplifier is not supposed to change the quality of the signal.

The amplifier is used to increase the amplitude of the signal.

Noise can produce a variety of symptoms. Some noise problems that may be found in an audio amplifier are

1. Constant frying or hissing noise
2. Popping or scratching sound
3. Hum
4. Motorboating (a "putt-putt" sound)

Noise problems can often be traced to the power supply. In troubleshooting for this symptom, it is a very good idea to use an oscilloscope to check the various supply lines in the equipment. The checks detailed in Sec. 10-1 rely on meter readings to check the supply. An oscilloscope will show things that a meter cannot. For example, Fig. 10-16 shows a power-supply waveform with *excess* ac *ripple*. The average dc value of the waveform is correct. This means that the meter reading will be acceptable.

Excess ripple

It is possible to obtain some idea of the ac content in a power-supply line without an oscilloscope. Many volt-ohm-milliammeters have a separate jack or function marked "output." This places a dc blocking capacitor in series with the test leads and allows a measurement of just the ac content on the supply line. Other VOMs have a selector switch. Figure 10-17 shows both possibilities. Most DMMs block dc when an ac range is selected and can also be used to make this test. The oscilloscope is preferred since it gives more information.

Shielded cable

Hum

The most common noise problem is *hum*. Hum refers to the introduction of a 60-Hz signal into the amplifier. It can also refer to 120-Hz interference. If the hum is coming from the power supply, it will be 60 Hz for half-wave supplies and 120 Hz for full-wave supplies. Hum can also get into the amplifier because of a broken ground connection. High-gain amplifiers often use shielded cables in areas where the ac line frequency can induce signals into the circuitry. *Shielded cable* is used in a stereo amplifier system for connections between components (Fig. 10-18). Be sure to check all shielded cables when hum is a problem. The braid may be open, which allows hum to get into the amplifier. Check connectors since the break in the ground shield is often found there.

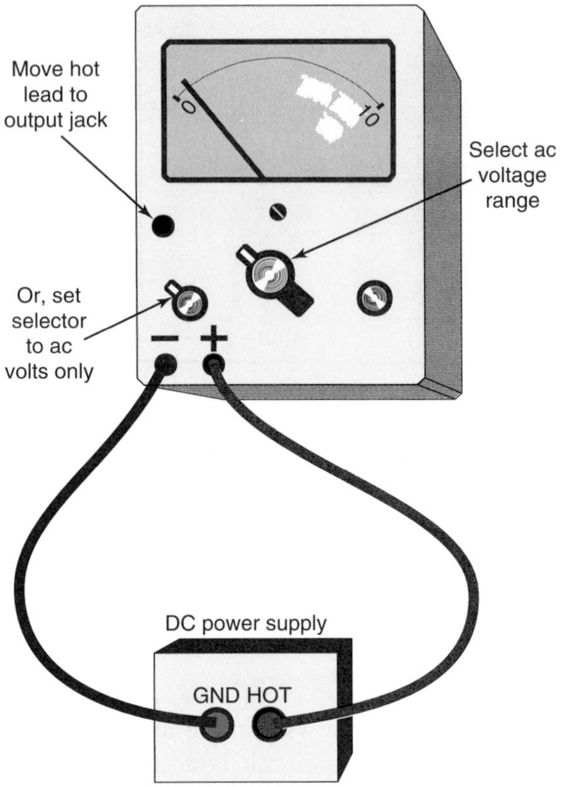

Some high-gain amplifiers use metal shields around circuits to keep hum and noise out. These shields must be in the proper position and fastened securely.

Another cause for hum is poor grounding of circuit boards. In some equipment, the fasteners do double duty. They mechanically hold the board and provide an electrical contact to the chassis. Check the fasteners to make sure they are secure.

Sometimes amplifier noise can be limited to general sections of the circuit by checking the effect of the various controls. Figure 10-19 is

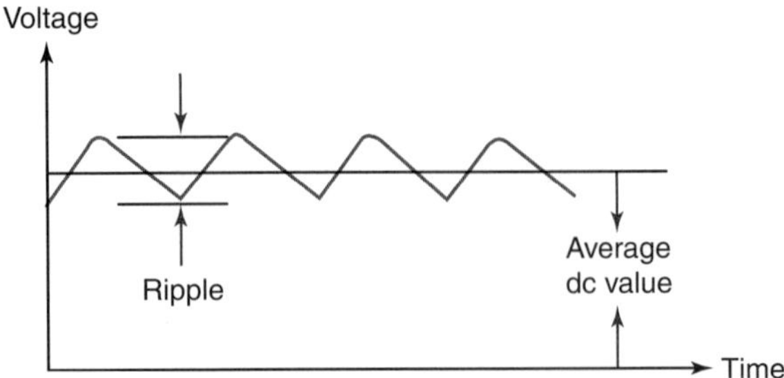

Fig. 10-16 Excessive ripple.

Move hot lead to output jack

Select ac voltage range

Or, set selector to ac volts only

DC power supply

GND HOT

Fig. 10-17 Blocking the dc component of a supply voltage.

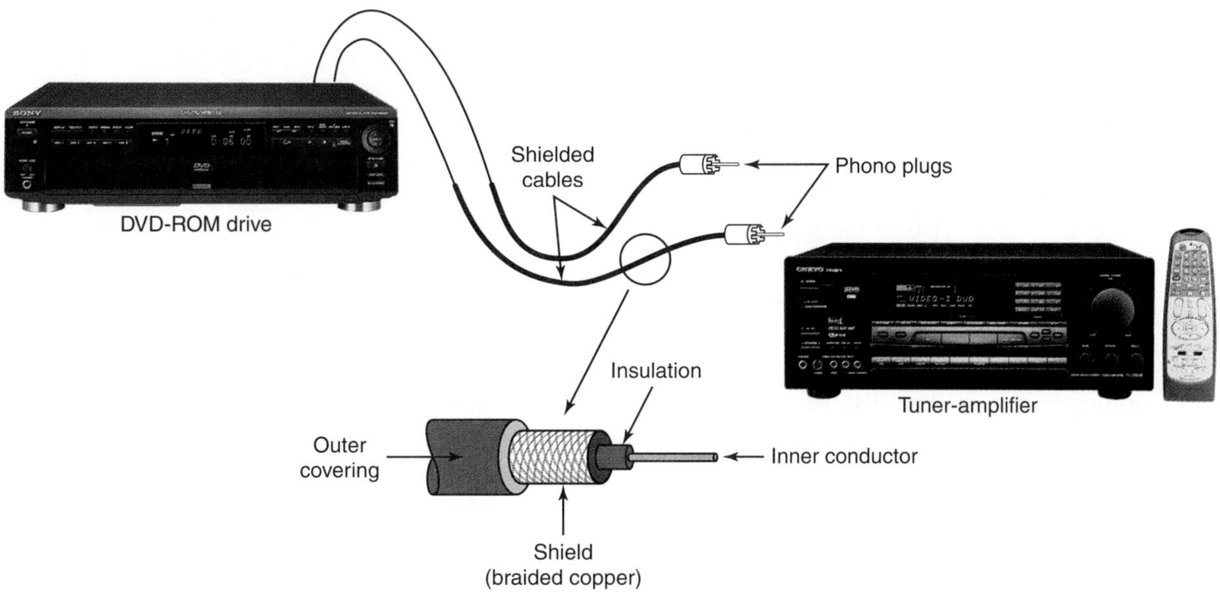

Fig. 10-18 Signal circuits often use shielded cables.

a block diagram of a four-stage amplifier. The gain control is located between stage 2 and stage 3. It is a good idea to operate this control to see whether it affects the noise reaching the output. If it does, then the noise is most likely originating in stage 1, stage 2, or in the signal source. Of course, if the control has no effect on the noise level, then it is probably originating in stage 3 or stage 4.

Another good reason for checking the controls is that they are often the source of the noise. Scratchy noises and popping sounds can often be traced to variable resistors. Special cleaner sprays are available for reducing or eliminating noise in controls. However, the noise often returns. The best approach is to replace noisy controls.

A constant frying or hissing noise usually indicates a defective transistor or integrated circuit. Signal tracing is effective in finding out where the noise is originating. Resistors can also become somewhat noisy. The problem is generally limited to early stages in the signal chain. Because of the high gain, it does not take a large noise signal to cause problems at the output.

It is worth mentioning that the noise may be coming from the signal source itself. It may be necessary to substitute another source or disconnect the signal. If the noise disappears, the problem has been found.

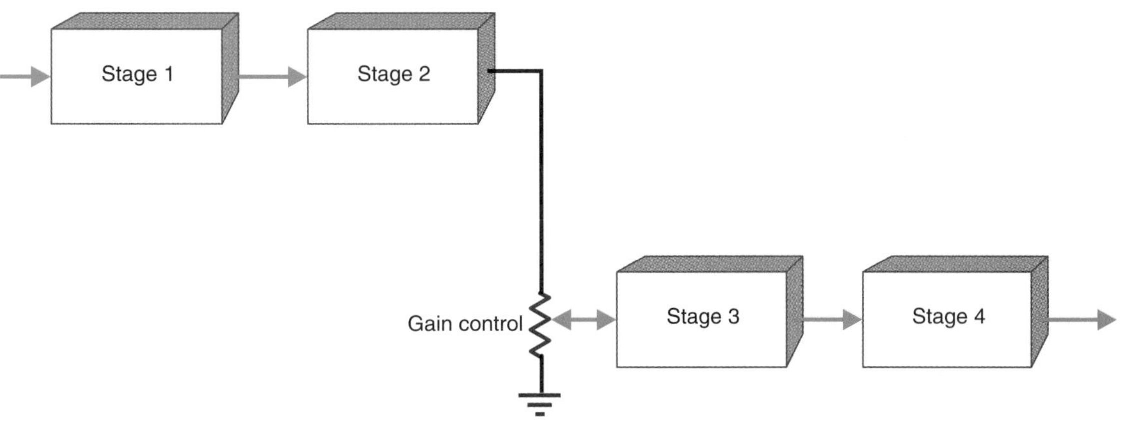

Fig. 10-19 A four-stage amplifier.

Motorboating

Motorboating is a problem that usually indicates an open filter capacitor, an open bypass capacitor, or a defect in the feedback circuit of the amplifier. An amplifier can become unstable and turn into an *oscillator* (make its own signal) under certain conditions. This topic is covered in detail in Sec. 11-6 of the next chapter.

Bias error

Triangular waveform

Amplifier distortion may be caused by *bias error* in one of the stages. Remember that bias sets the operating point for an amplifier. Incorrect bias can shift the operating point to a nonlinear region, and distortion will result. Of course, a transistor or integrated circuit can be defective and produce severe distortion.

It may help to determine whether the distortion is present at all times or just on some signals. A large-signal distortion may indicate a defect in the power (large-signal) stage. Similarly, a distortion that is more noticeable at low signal levels may indicate a bias problem in a push-pull power stage. You may wish to review crossover distortion in Chap. 8.

Another way to isolate the stage causing distortion is to feed a test signal into the amplifier and "walk" through the circuit with an oscilloscope. Many technicians prefer using a *triangular waveform* (Fig. 10-20) for making this test. The sharp peaks of the triangle make it very easy to spot any clipping or compression. The straight lines make it very easy to see any crossover or other types of distortion. In using such a test, it is a good idea to try different signal levels. Some problems show up at low levels and some at high levels. For example: (1) Crossover distortion in a push-pull amplifier is most noticeable at low levels and (2) operating point error in an early class A stage is most noticeable at high levels.

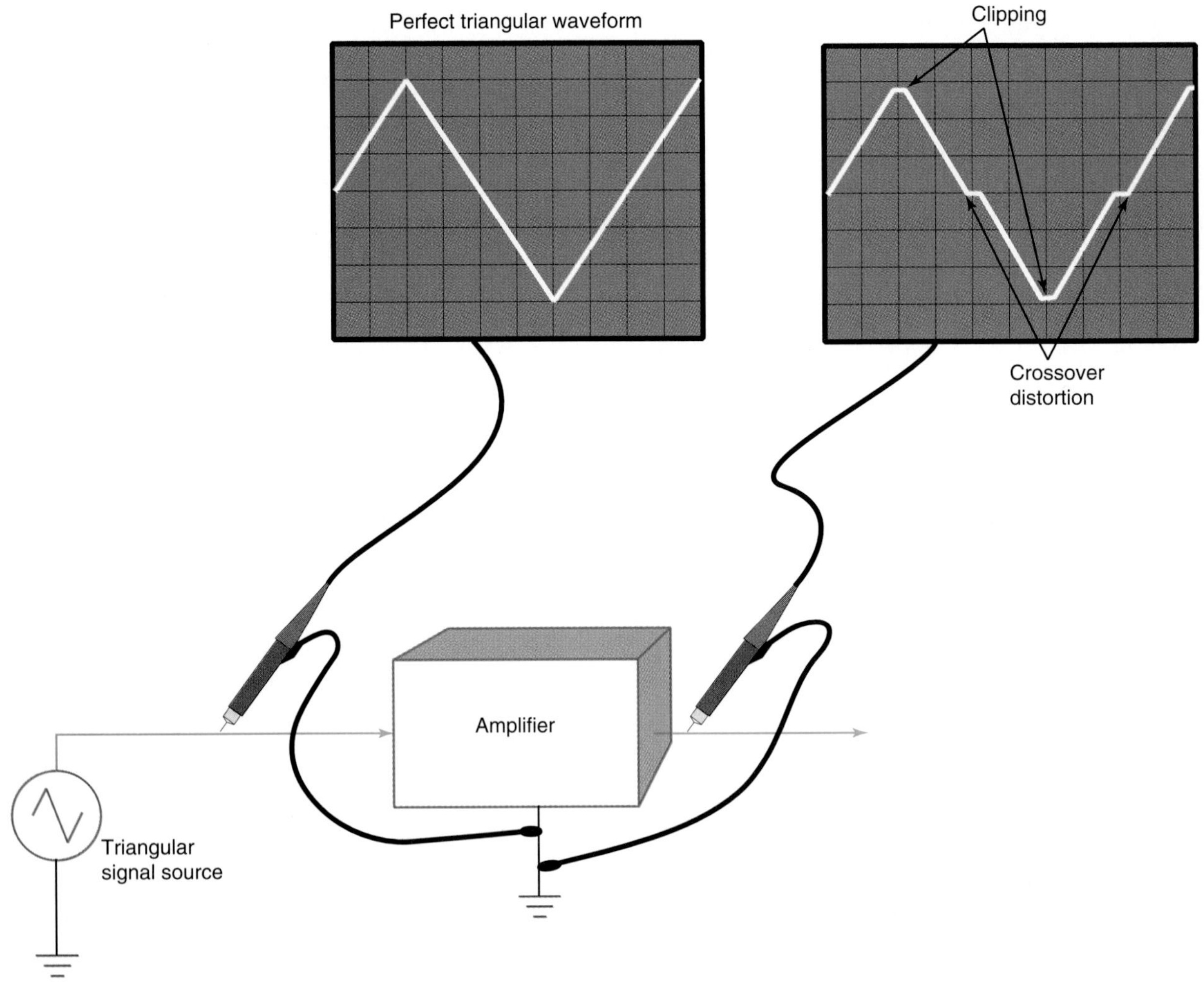

Fig. 10-20 Triangular waveform used for distortion analysis.

TEST

Choose the letter that best answers each question.

24. A stereo amplifier has severe hum in the output only when the selector is switched to TAPE. Which of the following is least likely to be wrong?
    a. A defective tape jack
    b. A defective shielded cable to the tape player.
    c. A bad filter capacitor in the power supply
    d. An open ground in the tape player
25. An audio amplifier has severe hum all the time. Which of the following is most likely at fault?
    a. The volume control
    b. The power cord
    c. The filter in the power supply
    d. The output transistor
26. Refer to Fig. 10-19. The amplifier has a loud, hissing sound only when the volume control is turned up. Which of the following conclusions is the best?
    a. The problem is in stage 3
    b. The problem is in stage 4
    c. The power supply is defective
    d. The problem is in stage 1 or 2
27. An audio amplifier has bad distortion when played at low volume. At high volume, it is noticed that the distortion is only slight. Which of the following is most likely to be the cause?
    a. A bias error in the push-pull output stage
    b. A defective volume control
    c. A defective tone control
    d. The power supply voltage is too low
28. An amplifier makes a putt-putt sound at high volume levels (motorboating). Which of the following is most likely to be the cause?
    a. Crossover distortion
    b. A defective transistor
    c. An open filter or bypass capacitor
    d. A defective speaker
29. An amplifier has bad distortion when played at high volume. Which of the following is most likely to be the cause?
    a. A cracked circuit board
    b. A bias error in one of the amplifiers
    c. A defective volume control
    d. A defective tone control

10-5 Intermittents

An electronic device is *intermittent* when it will work only some of the time. It may become defective after being on for a few minutes. It may come and go with vibration. The source of these kinds of problems can be very difficult to locate. Technicians generally agree that intermittents are the most difficult to troubleshoot.

There are two basic ways to find the cause of an intermittent problem. One way is to run the equipment until the problem appears and then use ordinary troubleshooting practice to isolate it. The second way is to use various procedures to force the problem to show up. Some of these are

1. Heat various parts of the circuit.
2. Cool various parts of the circuit.
3. Change the supply voltage.
4. Vibrate various parts of the circuit.

The actual technique used will depend on the symptoms and how much time is available to service the equipment. Some intermittents will not show up in a week of continuous operation. In such a case, it is probably best to try to make the problem occur.

Thermal intermittents

Many *intermittents* are *thermal.* That is, they appear at one temperature extreme or another. If the problem shows up only at a high temperature, it may be very difficult to find with the cabinet removed. With the cabinet removed the circuits usually run much cooler, and a thermal intermittent will not show. In such a case, it may be necessary to use a little heat to find the problem.

Figure 10-21 on page 278 shows some of the ways that electronic equipment and components can be safely heated to check for thermal intermittents. The bench lamp is useful for heating many components at one time. By placing a 100-W lamp near the equipment, the circuits will become quite warm after a few minutes. Be careful not to overheat the circuits. Certain plastic materials can be easily damaged. A vacuum desoldering tool makes a good heat source for small areas. Squeezing the bulb will direct a stream of hot air where needed. Be careful not to spray solder onto the circuit.

During the interview, avoid using too many short answers but do not elaborate in every case. Also, ask the interviewer whether it is permissible to call a day or two later to ask for an assessment.

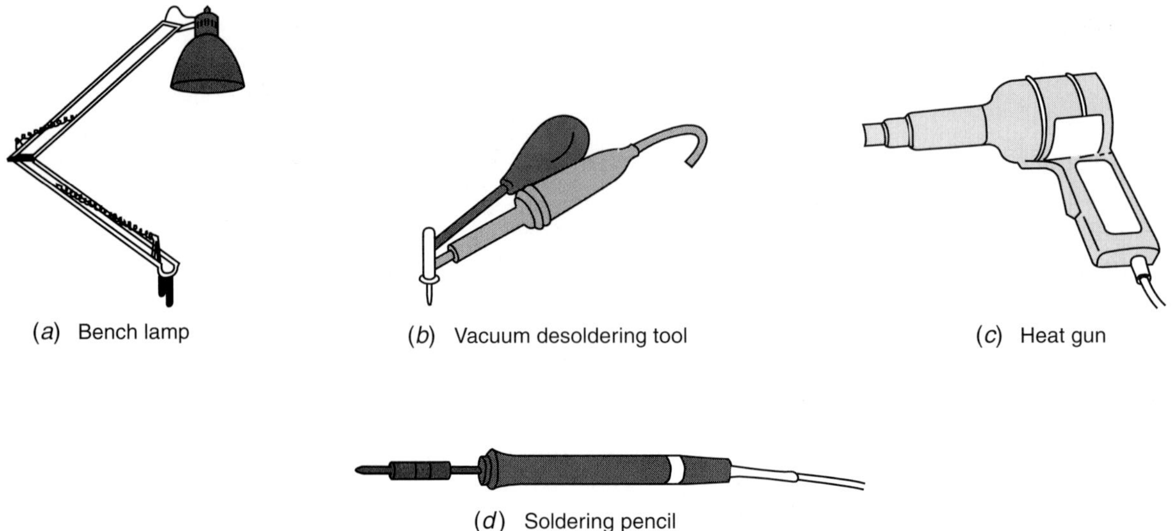

Fig. 10-21 Methods for heating components and circuits.

A heat gun is useful for heating larger components and several parts at one time. Be careful because some heat guns can damage circuit boards and parts. Finally, an ordinary soldering pencil may be used by touching the tip to a component lead or to a metal case.

Spray coolers are available for tracing thermal intermittents. A spray tube is included to control the application closely (Fig. 10-22). This makes it easy to confine the spray to one component at a time. Be very careful not to use just any spray coolant. Some types can generate static charges in the thousands of volts when they are used and others may damage the environment. Sensitive devices can be damaged by static discharges, as discussed earlier.

Some intermittents are voltage-sensitive. The ac line voltage is nominally rated at 117 V. However, it can and does fluctuate. It may go below 105 V, and it may go above 130 V. Most electronic equipment is designed to work over this range. In some cases, a circuit or a component can become critical and voltage-sensitive. This type of situation may show up as an intermittent. Figure 10-23 shows one test arrangement. The equipment is connected to a

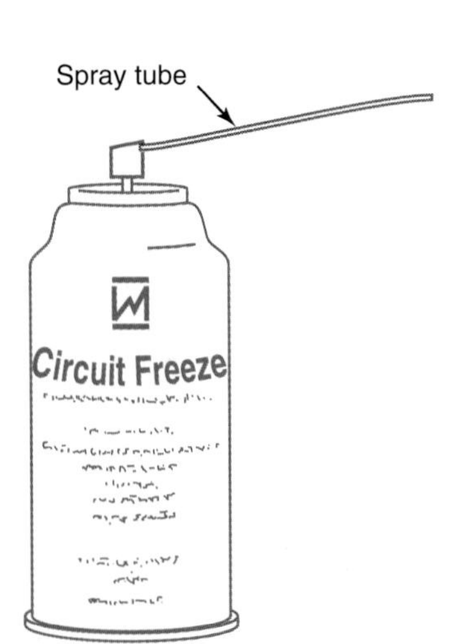

Fig. 10-22 Spray cooler with tube for localizing spray.

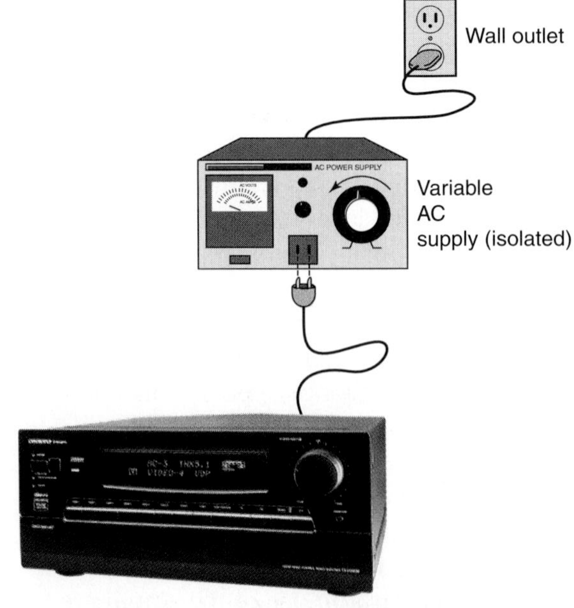

Fig. 10-23 Checking for a voltage-sensitive intermittent.

variable ac supply. This forces a voltage-sensitive problem to appear.

Intermittents are often sensitive to *vibration*. This may be caused by a bad solder joint, a bad connector, or a defective component. The only way to trace this kind of a problem is to use vibration or physical pressure. Careful tapping with an insulated tool may allow you to isolate the defect. In addition to tapping, try flexing the circuit boards and the connectors. These tests are made with the power on, *so be very careful*.

You may find it impossible to localize the intermittent to a single point in the circuit. Turn off the power and use some fresh solder to reflow every joint in the suspected area. Joints can fail electrically and yet appear to be good. Resoldering is the only way to be sure.

Do not overlook sockets. Try plugging and unplugging several times to clean the sliding contacts. The power *must be off. Never* plug or unplug connectors, devices, or circuit boards with the power on. Severe damage may result.

Circuit board connectors may require cleaning. An ordinary pencil eraser does a good job on the board contacts. Do not use an ink eraser as it is too abrasive. Use just enough pressure to brighten the contacts. Clear away any debris left by the eraser before reconnecting the board.

Another interesting type of intermittent is possible with one of the newer protection devices. Some audio speaker systems use series-connected, self-resetting fuses such as the PolySwitch made by the Raychem Corporation. These devices provide soft switching into a high-resistance tripped state and then automatically reset to a low-resistance state when power is removed. They are made from conductive polymers that expand with heat when there is excess current flow. The expansion causes a separation of conductive paths and a dramatic increase in resistance that protects the speaker. These devices may also be wired in parallel with positive temperature coefficient devices such as light bulbs. When this is done and the PolySwitch device trips, the speaker current now flows through the bulb, which heats and increases in resistance. So, the symptom in this case is a decrease in speaker volume. Without the shunt light bulb, the symptom is a total loss of volume. In either case, the volume is restored when the amplifier is turned off (or down) and the fuse is allowed to cool.

Intermittents can be tough to work on, but they are not impossible. Use every clue and test possible to localize the problem. It is far easier to check a few things than to check every joint, contact, and component in the system.

Vibration

TEST

Choose the letter that best completes each statement.

30. A circuit works intermittently as the chassis is tapped with a screwdriver. The problem may be
    a. Thermal
    b. An open filter capacitor
    c. A cold solder joint
    d. Low supply voltage
31. A problem appears as one section of a large circuit board is flexed. The proper procedure is to
    a. Replace the components in that part of the board
    b. Reflow the solder joints in that part of the board
    c. Heat that part of the board
    d. Cool that part of the board
32. A circuit always works normally when first turned on but then fails after about 20 minutes of operation. Out of its cabinet, it works indefinitely. The problem is
    a. Thermal
    b. An open ground
    c. High line voltage
    d. The on-off switch
33. The correct procedure to isolate the defect in question 32 is to
    a. Run the circuit at reduced line voltage
    b. Replace the electrolytic capacitors, one at a time
    c. Remove the cabinet and heat various parts
    d. Resolder the entire circuit

10-6 Operational Amplifiers

The techniques used when troubleshooting circuits with op amps are similar to those already presented in this chapter. As always, check the obvious things first. Op amps often use a bipolar supply. You should verify *both* supply voltages early in the troubleshooting process.

In general, component failures follow a pattern. The following list is presented as a rough

Failure rate

guide to *average failure rates*. Many devices, such as resistors, have very low failure rates and are not listed. Also, parts such as cells and batteries are not listed since they are expected to be replaced on a regular basis. The items listed first have the highest failure rates:

1. High-power devices and devices subject to transients
2. Incandescent lamps
3. Complex devices such as integrated circuits
4. Mechanical devices such as connectors, switches, and relays
5. Electrolytic capacitors

Latch-up

High-power devices run hot. If equipment is not powered all the time, a large number of hot-cold cycles can accumulate. The repeated expansion and contraction tends to weaken the internal connections of electronic devices. Most op amps are small signal devices. However, a few do dissipate enough power to run hot and some may be located where they are heated by other devices. Remember that *heat* is one of the leading causes of failures in electronic systems.

Transient

A *transient* is a brief and abnormally high voltage. Transients are tough on solid-state devices because they can break down PN junctions and degrade or destroy them. Op amps are sometimes connected to sensors through long runs of wire. These wires can act as antennas and pick up transients caused by lightning or by surges in other wires in the near vicinity. Watch for repeated failures in such cases. They can indicate the need for a different wiring arrangement or the addition of transient protection devices.

Op amps are almost always integrated circuits. This makes them complex and increases their failure rate over simple devices. So if you are working on a defective circuit that contains some transistors and some integrated circuits, it is more likely that an IC is bad than a transistor (unless the transistors are high-power ones or subject to transients). Please remember that the above list is a *rough* guideline. Experienced technicians know that anything can go wrong and expect that it will sooner or later. The idea is that their experience makes them more efficient because it tells them where to look first.

Companies place a high value on efficient troubleshooters. However, attempting to repair electronic systems without knowledge and/or the proper test equipment can be risky.

Please don't form the opinion that integrated circuits are not reliable. They are actually *more reliable* than equivalent circuitry using separate parts. The equivalent circuit for an ordinary op amp might contain 40 parts. A circuit with 40 parts is usually not as reliable as a single IC because it has so many more places for failures to occur. Just remember that a complicated device is more likely to fail than *one* simple device.

Occasionally, an op amp can *latch up*. This is not a failure but can be a troubleshooting problem. When an op amp latches, its output gets stuck at its maximum value (either positive or negative). The only way to get it back to normal is to power down and then power up again. Normal operation will be restored if the initial cause (such as an abnormally large input signal) has been removed. Op amps have a maximum common-mode range which is the maximum signal that can be applied to both inputs without saturating or cutting off the amplifier. Latch-up usually occurs in voltage follower stages where saturation has occurred.

When a nonfeedback amplifier is driven into saturation, it will resume normal operation when the abnormal input signal is removed. This is always true unless the abnormal input was large enough to damage the amplifier. When feedback is used, the situation can be different. A saturated stage no longer acts as an amplifier. It acts as a resistor and passes some part of the signal through to the next stage. If the saturated stage was an inverting amplifier, *the inversion is lost.* When this happens in an amplifier with intended negative feedback, the feedback goes positive and the amplifier may then keep itself in saturation. Op amps are candidates for latch-up since they are usually operated with negative feedback.

Latch-up in operational amplifiers is not as probable as it once was. The designers of linear ICs have made changes that make it less likely to occur. Try powering down and back on again if you suspect latch-up. If the symptoms seem to indicate it, then you will have to investigate the input signal to determine if it is exceeding the common mode range of the device. Also make sure that the power supply is normal and that the positive and negative voltages are applied at the same time when the circuit is turned on.

The output of an op amp can usually go within a volt or so of the power supply values. If it is powered by ±12 V, then the output could approach +11 V or −11 V. There is usually a problem when the output of an op amp is at or near one of its extremes. It could be latch-up but it is more likely to be a dc error at its input or in a prior stage. The dc gain in some circuitry is high. A moderate error in an early stage can drive an output stage to one of its extremes. Use a meter and check the dc voltages. When stages are dc-coupled, check earlier stages as well. Do this even though the problem may appear on an ac signal. Always remember that an amplifier near saturation or cutoff cannot provide a normal linear output swing for any signal.

Don't forget to check all of the relevant dc levels. Op amps can sum and subtract several different dc signals. All it takes is one of them to be missing or in error to throw off the dc balance of the entire circuit. Refer to Fig. 10-24. It shows a circuit with two stages. The signal source has a +1 V dc offset that must be eliminated. This is accomplished by the first stage, where a −5-V reference is summed with the signal source. Assuming that the 10-kΩ trimmer is adjusted for 5 kΩ, we can calculate the dc output of the first stage:

$$V_{\text{out}} = -100\text{ k}\Omega\left(\frac{+1\text{ V}}{1\text{ k}\Omega} + \frac{-5\text{ V}}{5\text{ k}\Omega}\right)$$
$$= 0\text{ V}$$

This calculation demonstrates that the amplifier is properly designed and adjusted to eliminate the dc offset of the signal source.

What will happen in Fig. 10-24 if the 10-kΩ trimmer resistor develops an open? This will effectively remove the second term from inside the parentheses above, and the output now calculates to

$$V_{\text{out}} = -100\text{ k}\Omega\left(\frac{+1\text{ V}}{1\text{ k}\Omega}\right)$$
$$= -100\text{ V}$$

Obviously, the amplifier cannot achieve this output. It will saturate at approximately −14 V. The second stage is a voltage follower and its output will also be near −14 V.

When you are troubleshooting comparators, observe the symptoms carefully. If there is no output at all, make sure that the power supply is OK and that there is a normal signal connected to the comparator input. Other possibilities for no output include a shorted output cable, a bad comparator IC, and an open pull-up resistor.

If the symptom is noise in the output or extra output pulses, the input signal should be checked. Many comparators do not work well with high-impedance signal sources, so verify the source and verify a good connection. Finally, a comparator might need positive feedback to prevent a noisy output. Investigate whether there is a resistor that feeds back from the output to the + input and make sure that it is not open.

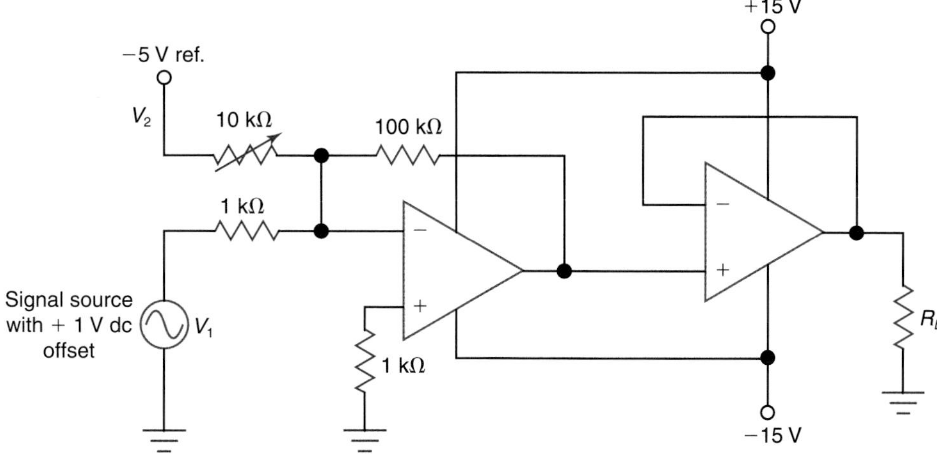

Fig. 10-24 Op-amp troubleshooting example.

TEST

Choose the letter that best completes each statement.

34. Of the following components, the *least* likely to fail is
    a. A high-power output transistor
    b. An integrated circuit
    c. A device that runs hot
    d. A small-signal transistor
35. Latch-up in a negative feedback amplifier is caused by
    a. An open bypass capacitor
    b. An open coupling capacitor
    c. A stage driven into saturation
    d. An open in the feedback network
36. Refer to Fig. 10-24. The purpose of the −5 V reference and 10-kΩ potentiometer is
    a. To set the gain of the first stage
    b. To null the dc offset in the source
    c. To power the first op amp
    d. Adjust the frequency response of the amplifiers
37. Refer to Fig. 10-24. The connection to the signal source is defective (open). The dc across R_L will be about
    a. +14 V
    b. −14 V
    c. 0 V
    d. ±7 V

Chapter 10 Review

Summary

1. When troubleshooting, always use a system point of view. Don't ignore other equipment and/or software that could be affecting performance, and always check the obvious.
2. If the unit is ac-operated, disconnect it from the line before taking it apart.
3. Use service literature and the proper tools.
4. Sort and save all fasteners, knobs, and other small parts.
5. Make a thorough visual inspection of the interior of the equipment.
6. Try to determine why a component failed before turning on the power.
7. Check for overheating.
8. Verify all power-supply voltages.
9. Lack of amplifier output may not be in the amplifier itself. There could be a defective output device or no input signal.
10. A multistage amplifier can be viewed as a signal chain.
11. Signal injection begins at the load end of the chain.
12. Signal tracing begins at the input end of the chain.
13. Voltage analysis is generally used to limit the possibilities to one defective component.
14. Some circuit defects cannot be found by dc voltage analysis. These defects are usually the result of an open device or coupling component.
15. Low output from an amplifier may be due to low input.
16. A dummy load resistor is often substituted for the output device when amplifier performance is measured.
17. Both signal tracing and signal injection can give misleading results when troubleshooting for the low-gain stage.
18. Voltage analysis will lead to some causes of low gain.
19. A capacitor suspected of being open can be checked by bridging it with a new one.
20. A linear amplifier is not supposed to change anything but the amplitude of the signal.
21. Noise may be originating in the power supply.
22. Hum refers to a 60-Hz or a 120-Hz signal in the output.
23. Hum may be caused by a defective power supply, an open shield, or a poor ground.
24. Operate all controls to see if the noise occurs before or after the control.
25. Motorboating noise means the amplifier is oscillating.
26. Distortion can be caused by bias error, defective transistors, or an input signal that is too large.
27. Thermal intermittents may show up after the equipment is turned on for some time.
28. Use heat or cold to localize thermal problems.
29. Vibration intermittents can be isolated by careful tapping with an insulated tool.
30. Failure rates are directly related to device temperature and complexity.
31. Transients can and often do damage solid-state devices.
32. An amplifier with negative feedback may latch up if it is driven into saturation.

Chapter Review Questions

Choose the letter that best answers each question.

10-1. When troubleshooting, which of the following questions is not part of a preliminary check?
   a. Are all cables plugged in?
   b. Are all controls set properly?
   c. Are all transistors good?
   d. Is the power supply on?

10-2. What is the quickest way to check a speaker for operation (not for quality)?
   a. A click test using a dry cell
   b. Substitution with a good speaker
   c. Connecting an ammeter in series with the speaker
   d. Connecting an oscilloscope across the speaker

10-3. Refer to Fig. 10-2. The regulated output is zero. The unregulated input is normal. Which of the following could be the cause of the problem?
   a. C_1 is open
   b. D_1 is open
   c. R_1 is open
   d. R_1 is shorted

10-4. Refer to Fig. 10-2. The regulated output is low. The unregulated input is normal. Which of the following could be the cause of the problem?
   a. D_1 is open
   b. D_1 is shorted
   c. C_1 is open
   d. Excessive current at the output

10-5. Refer to Fig. 10-4. Where does the plug end of the cord connect?
   a. To $+V_{CC}$
   b. To $-V_{EE}$
   c. To an approved ground
   d. To any ground

10-6. Refer to Fig. 10-5. The output is zero. Where should the signal be injected first?
   a. At the input of stage 1
   b. At the input of stage 2
   c. At the input of stage 3
   d. At the input of stage 4

10-7. Refer to Fig. 10-5. The amplifier is dead. A known good signal has been connected to the input of stage 1. Signal tracing should begin at
   a. The output of stage 1
   b. The output of stage 2
   c. The output of stage 3
   d. The output of stage 4

10-8. Refer to Fig. 10-8. The collector of Q_1 measures almost 21 V, and it should be 12 V. Which of the following is most likely to be wrong?
   a. Q_1 is open
   b. C_3 is shorted
   c. R_4 is open
   d. Q_2 is shorted

10-9. Refer to Fig. 10-8. The collector of Q_1 measures 2 V, and it is supposed to be 12 V. Which of the following could be the problem?
   a. Q_2 is shorted
   b. C_1 is open
   c. R_2 is open
   d. C_4 is shorted

10-10. Refer to Fig. 10-8. Resistor R_1 is open. Which of the following statements is correct?
   a. The collector of Q_1 will be at 0 V
   b. Q_2 will run hot
   c. Q_2 will go into cutoff
   d. Q_1 will run hot

10-11. Refer to Fig. 10-8. Resistor R_9 is open. Which of the following statements is correct?
   a. The collector of Q_2 will be at 0 V
   b. Q_2 will go into saturation
   c. Q_2 will go into cutoff
   d. Q_2 will run hot

10-12. Refer to Fig. 10-14. Resistor R_1 is open. Which of the following is correct?
   a. The collector voltage will be very low
   b. The collector voltage will be very high
   c. The transistor will be in saturation
   d. The emitter voltage will be very high

10-13. Refer to Fig. 10-15. Capacitor C_3 is open. Which of the following is correct?
   a. The dc voltages will all be wrong
   b. The transistor will run hot
   c. Extreme distortion will result
   d. The gain will be low

10-14. Refer to Fig. 10-19. A scratching sound is heard as the gain control is rotated. Where is the problem likely to be?
   a. Stage 1 or 2
   b. The volume control
   c. Stage 3 or 4
   d. The speaker

10-15. Refer to Fig. 10-19. There is severe hum in the output, but turning down the volume control makes it stop completely. Where is the problem likely to be?
a. Third stage or fourth stage
b. Power-supply filter
c. The volume control
d. Input cable (broken ground)

10-16. An amplifier is capacitively coupled. What is the best way to find an open coupling capacitor?
a. Look for transistors in cutoff
b. Look for transistors in saturation
c. Look for dc bias errors on the bases
d. Look for a break in the signal chain

10-17. An amplifier has a push-pull output stage. Bad distortion is noted at high volume levels only. The problem could be which of the following?
a. Bias error in an earlier stage
b. A shorted output transistor
c. Crossover distortion
d. A defective volume control

10-18. What is probably the slowest way to find an intermittent problem?
a. Try to make it show by using vibration
b. Use heat
c. Use cold
d. Wait until it shows up by itself

10-19. An automobile radio works normally except while traveling over a bumpy road. What is likely to be the cause of the problem?
a. Thermal
b. An open capacitor
c. A low battery
d. A loose antenna connection

10-20. When working on electronic equipment, a grounded wrist strap may be used to prevent
a. Ground loops
b. Electrostatic discharge
c. Thermal damage
d. Loading effect

10-21. Refer to Fig. 10-24. Suppose the −5-V reference supply fails and goes to 0 V. The voltage across R_L will be
a. 0 V
b. +14 V
c. −14 V
d. +30 V

Critical Thinking Questions

10-1. You are visiting a friend and notice that the sound coming from the left speaker of her stereo is distorted. You have no test equipment with you. Is there anything that you can do to help her find out what is wrong?

10-2. Your automobile often won't start on Monday mornings and never fails to start at other times. Is this some sort of weird coincidence?

10-3. Can you think of any reason why stereo amplifiers sometimes fail during parties?

10-4. Technicians often put batteries in portable equipment before performing other tests even though the customer has stated that the batteries are new. Do technicians think their customers are crazy?

10-5. Where do technicians look when they are working on equipment that failed during a lightning storm?

10-6. Can you think of an op-amp application in which it is normal for the output to be saturated?

10-7. Component-level repair is not a widespread practice today. Is it still worthwhile to learn how electronic circuits operate?

Answers to Tests

1. *c*
2. *d*
3. *d*
4. *d*
5. *a*
6. *b*
7. *b*
8. *b*
9. *d*
10. *b*
11. *d*
12. *c*
13. *b*
14. *d*
15. *a*
16. *b*
17. *b*
18. *c*
19. *b*
20. *d*

21. *b*
22. *b*
23. *d*
24. *c*
25. *c*
26. *d*
27. *a*
28. *c*
29. *b*
30. *c*
31. *b*
32. *a*
33. *c*
34. *d*
35. *c*
36. *b*
37. *a*

Oscillators

Chapter Objectives

This chapter will help you to:

1. *Identify* oscillator circuits.
2. *Apply* the concepts of gain and feedback to oscillators.
3. *Predict* the frequency of operation for oscillators.
4. *List* causes of undesired oscillations.
5. *Identify* techniques used to prevent undesired oscillation.
6. *Troubleshoot* oscillators.

An amplifier needs an ac input signal to produce an ac output signal but an oscillator doesn't. An oscillator is a circuit that creates an ac signal. Oscillators can be designed to produce many kinds of waveforms such as sine, rectangular, triangular, or sawtooth. The range of frequencies that oscillators can generate is from less than 1 Hz to well over 10 gigahertz (10 GHz $= 1 \times 10^{10}$ Hz). Depending on the waveform and frequency requirements, oscillators are designed in different ways. This chapter covers some of the most popular circuits, and it also discusses undesired oscillations.

11-1 Oscillator Characteristics

An oscillator is a circuit that converts dc to ac as shown in Fig. 11-1(*a*) on page 288. The only input to the oscillator is a dc power supply and the output is ac. Most oscillators are amplifiers with *feedback* as shown in Fig. 11-1(*b*). If the feedback is *positive,* the amplifier may oscillate (produce alternating current).

Many amplifiers will oscillate if the conditions are correct. For example, you probably know what happens when someone adjusts the volume control too high on a public address system. The squeals and howls that are heard are oscillations. The feedback in this case is the sound (acoustical) waves from the loudspeakers that enter the microphone (Fig. 11-2) on page 288. Although acoustical feedback can produce oscillations, almost all practical oscillators use electrical feedback. The feedback circuit uses components such as resistors, capacitors, coils, or transformers to connect the input of the amplifier to the output of the amplifier.

Feedback alone will not guarantee oscillations. Look at Fig. 11-2 again. You probably know that turning down the volume control will stop the public address system from oscillating. The feedback is still present. But now there is not enough to gain to overcome the loss in the feedback path. This is one of the two basic criteria that must be met if an amplifier is to oscillate: The amplifier *gain* must be *greater* than the loss in the feedback path. The other criterion is that the signal fed back to the input of the amplifier must be *in phase.* In-phase feedback is also called *positive feedback,* or regenerative feedback. When the amplifier input and output are normally out of phase (such as in a common-emitter amplifier), the feedback circuit will have to produce a phase reversal. Figure 11-3 on page 289 summarizes what the requirements are. The total phase shift is 180° + 180° = 360°. 360° is the

Feedback

Gain

In phase

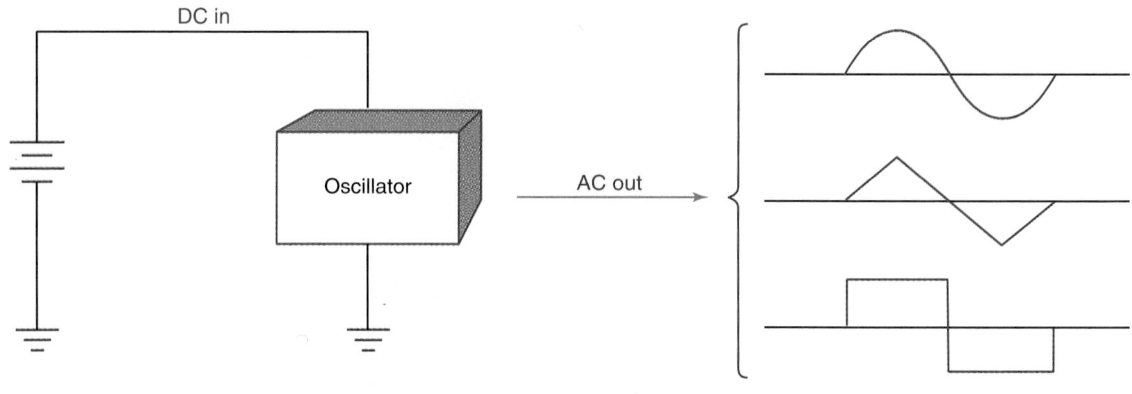

(*a*) Oscillators convert direct to alternating current

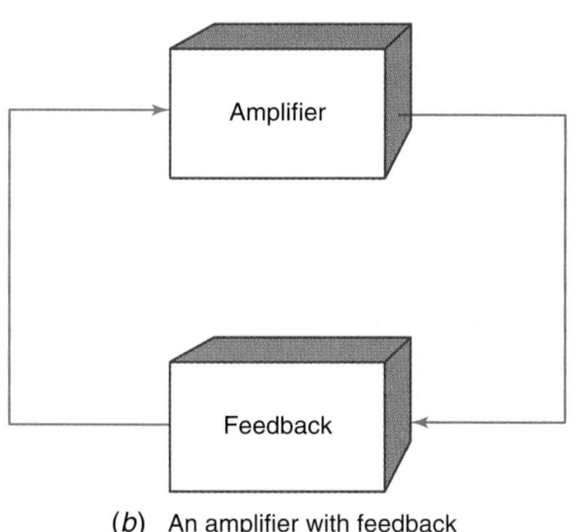

(*b*) An amplifier with feedback

Fig. 11-1 Oscillator basics.

same phase as 0°. The feedback circuit provides the needed phase shift at the desired frequency of oscillation (f_o).

Oscillators are widely applied, for example:

1. Many digital devices such as computers, calculators, and watches use oscillators to generate rectangular waveforms that time and coordinate the various logic circuits.
2. Signal generators use oscillators to produce the frequencies and waveforms required for testing, calibrating, or troubleshooting other electronic systems.
3. Touch-tone telephones, musical instruments, and remote control transmitters can use them to produce the various frequencies needed.
4. Radio and television transmitters use oscillators to develop the basic signals sent to the receivers.

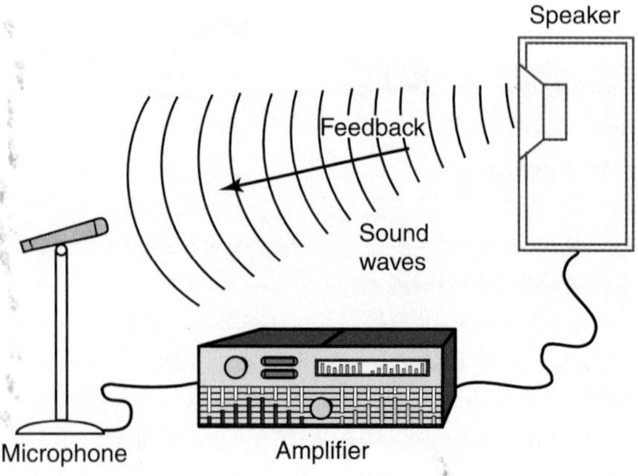

Fig. 11-2 Feedback can make an amplifier unintentionally oscillate.

Stability

The various oscillator applications have different requirements. In addition to frequency and waveform, there is the question of *stability.* A stable oscillator will produce a signal of constant amplitude and frequency. Another requirement for some oscillators is the capability to produce a range of frequencies. Variable-frequency oscil-

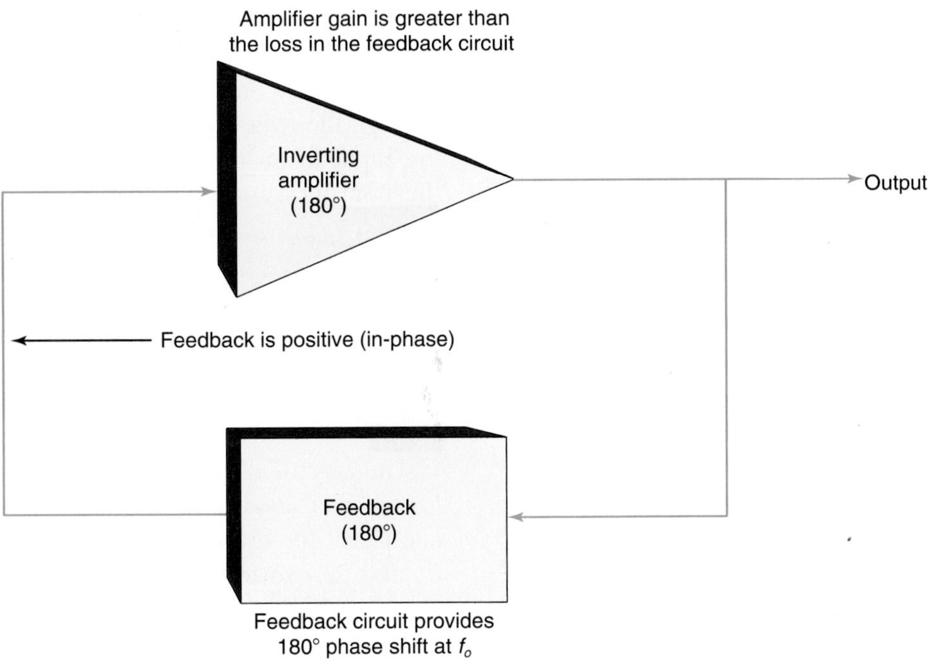

Fig. 11-3 What's required to make an amplifier oscillate.

lators (VFOs) meet this need and so do voltage-controlled oscillators (VCOs).

TEST

Choose the letter that best answers each question.

1. What conditions are required for an amplifier to oscillate?
   a. There must be feedback.
   b. The feedback must be in phase.
   c. The gain must be greater than the loss.
   d. All of the above are required.
2. Which of the following statements is not true?
   a. An oscillator is a circuit that converts dc to ac.
   b. An oscillator is an amplifier that supplies its own input signal.
   c. All oscillators generate sine waves.
   d. In-phase feedback is called positive feedback.
3. Refer to Fig. 11-2. The system oscillates. Which of the following is most likely to correct the problem?
   a. Increase the gain of the amplifier.
   b. Use a more sensitive microphone.
   c. Move the microphone closer to the speaker.
   d. Decrease the acoustical feedback by adding sound absorbing materials to the room.

11-2 *RC* Circuits

VFO

VCO

It is possible to control the frequency of an oscillator by using resistive-capacitive components. One *RC* circuit that can be used for frequency control in oscillators is shown in Fig. 11-4. This circuit is called a *lead-lag network* and shows maximum output and zero phase shift at one frequency. This frequency is called the resonant frequency f_r. It can be found with this equation:

Lead-lag network

$$f_r = \frac{1}{2\pi RC}$$

f_r

The series and shunt values of *R* and *C* in Fig. 11-4 are equal.

Example 11-1

In Fig. 11-4 both resistors are 10 kΩ and both capacitors are 0.01 μF. Determine the resonant frequency of the lead-lag network. Use the equation:

$$f_r = \frac{1}{2\pi RC}$$

$$= \frac{1}{6.28 \times 10 \times 10^3 \times 0.01 \times 10^{-6}}$$

$$= 1.59 \text{ kHz}$$

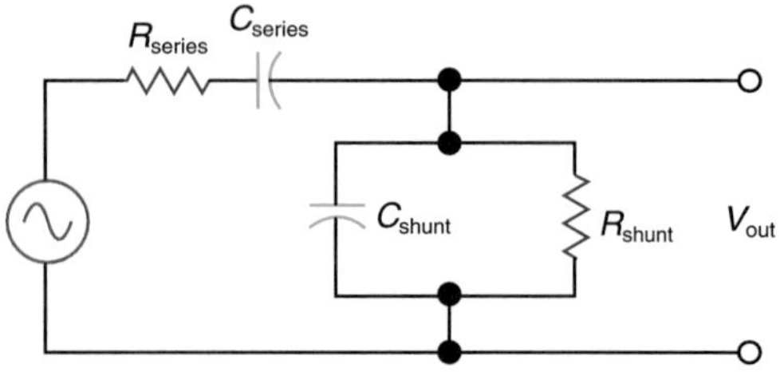

Fig. 11-4 An *RC* lead-lag network.

Wien bridge oscillator

0° at f_r

Figure 11-5 illustrates a computer-generated amplitude and phase response for the 1.59-kHz lead-lag network. The amplitude plot rises as the frequency increases from 100 Hz until the resonant frequency is reached. The amplitude plot drops for frequencies above resonance. The phase angle plot leads for frequencies below resonance and lags for frequencies above resonance. Note that the phase response is 0° at resonance.

The lead-lag network, at resonance, shows an output voltage that is one-third the input voltage:

$$V_{\text{dB}} = 20 \times \log \frac{V_{\text{out}}}{V_{\text{in}}} = 20 \times \log \frac{1}{3}$$
$$= -9.54 \text{ dB}$$

This result agrees with the amplitude plot of Fig. 11-5. If an oscillator uses a lead-lag network in its feedback circuit, then its amplifier section will require a voltage gain greater than three (9.54 dB).

Figure 11-6 shows how a lead-lag network can be used to control the frequency of an oscillator. Note that the feedback is applied through the lead-lag network to the noninverting input of an op amp. Feedback applied to the noninverting input is positive feedback. However, *only one frequency* will arrive at the noninverting input exactly in phase. That frequency is the *resonant frequency* f_r of the network. All other frequencies will lead or lag. This means that the oscillator will operate at a single frequency and the output will be sinusoidal.

The circuit of Fig. 11-6 is called a *Wien bridge oscillator.* The lead-lag network forms one leg of the bridge, and the resistors marked R' and $2R'$ form the other. The operational-amplifier inputs are connected across the legs of the bridge. Resistor R' is a device with a large positive temperature coefficient such as a tungsten filament lamp. The purpose of the R' leg of the bridge is to adjust the gain of the amplifier so that it is just greater than the loss in the lead-lag network. If the gain is too small, the circuit will not oscillate. If the gain is too large, the output waveform will be clipped.

At the moment the circuit in Fig. 11-6 is first turned on, R' will be cold and relatively low in resistance. The circuit will begin oscillating due to the positive feedback through the lead-lag network. The resulting signal across R' will heat it and its resistance will increase. Resistors R' and $2R'$ form a voltage divider. As

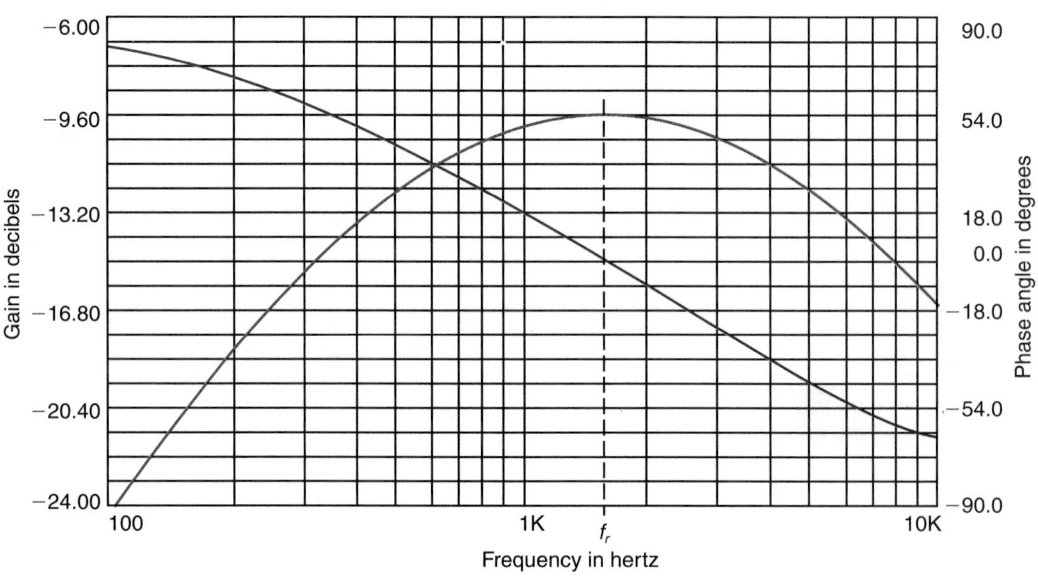

Fig. 11-5 Computer-generated amplitude and phase response of a lead-lag network.

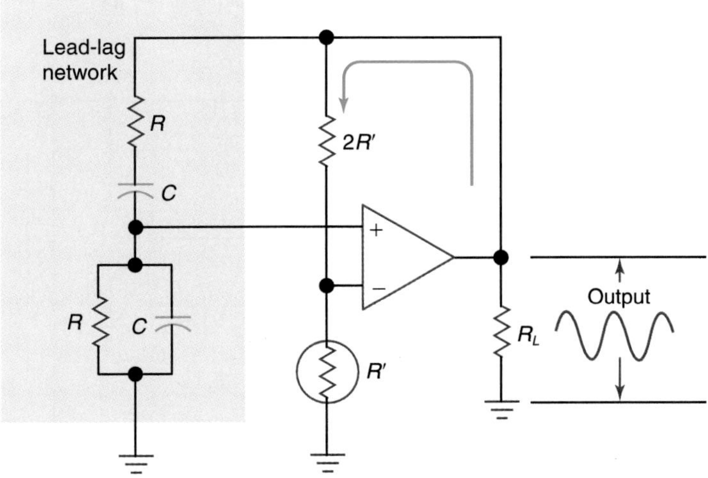

Fig. 11-6 A Wien bridge oscillator.

R' increases, the voltage applied to the inverting input of the operational amplifier will increase. As we learned earlier, negative feedback decreases the gain of an op amp. If the circuit is properly designed, the gain will decrease to a value that prevents clipping but which is large enough to sustain oscillation.

The Wien bridge circuit satisfies the basic demands of all oscillator circuits: (1) The gain is adequate to overcome the loss in the feedback circuit, and (2) the feedback is in phase. The gain of the circuit is high at the moment of power on. This ensures rapid starting of the oscillator. After that, the gain decreases because of the heating of R'. This eliminates amplifier clipping. Wien bridge oscillators are noted for their low-distortion sinusoidal output.

It is possible to make a *variable-frequency* Wien bridge oscillator (Fig. 11-7). The variable capacitors shown are *ganged.* One control shaft operates both capacitors. What is the frequency range of this circuit? It will be necessary to use the resonant frequency formula twice to answer this question:

$$f_{r(\text{HI})} = \frac{1}{6.28 \times 47 \times 10^3 \times 100 \times 10^{-12}}$$
$$= 33{,}880 \text{ Hz}$$

$$f_{r(\text{LO})} = \frac{1}{6.28 \times 47 \times 10^3 \times 500 \times 10^{-12}}$$
$$= 6776 \text{ Hz}$$

The range is from 6776 to 33,880 Hz. A 5:1 capacitor range produces a 5:1 frequency range:

$$6776 \text{ Hz} \times 5 = 33{,}880 \text{ Hz}$$

There is another way to use *RC* networks to control the frequency of an oscillator. They can be used to produce a 180° phase shift at the desired frequency. This is useful when the common-emitter amplifier configuration is used. Figure 11-8 on page 292 shows the circuit for a *phase-shift oscillator.* The signal at the collector is 180° out of phase with the signal at the base. By including a network which gives an additional 180° phase shift, the base receives in-phase feedback. This is because 180° + 180° = 360°, and 360° is the same as 0° in circular measurement.

Phase-shift oscillator

In Fig. 11-8 the phase-shift network is divided into three separate sections. Each section is designed to produce a 60° phase shift, and the total phase shift will

Expect to be rejected several times before you receive an offer. Becoming discouraged will work against you.

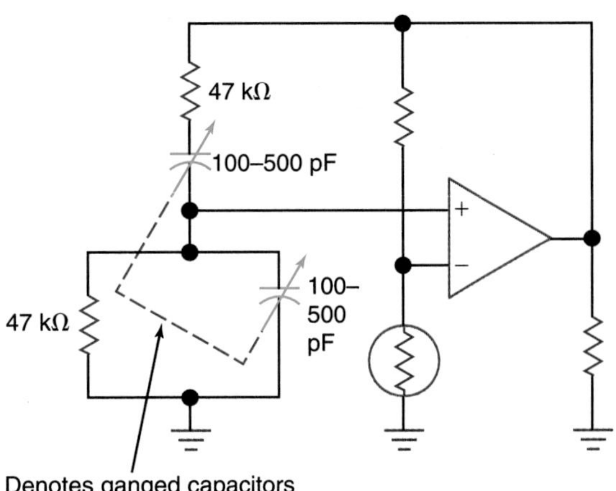

Fig. 11-7 A variable-frequency oscillator.

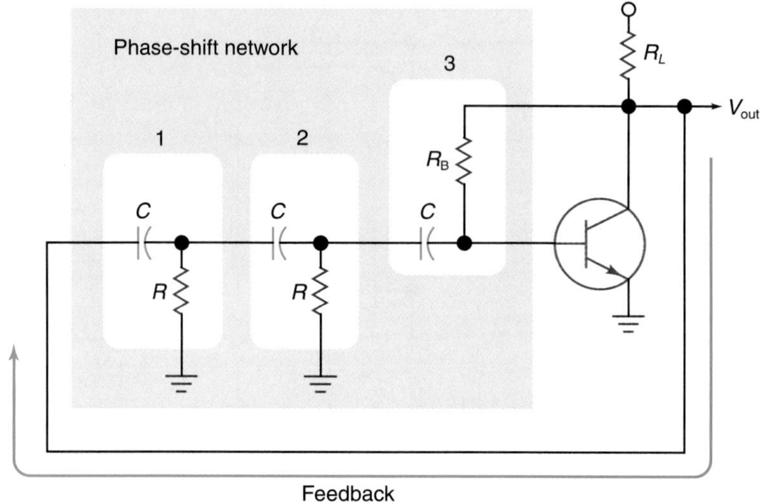

Fig. 11-8 A phase-shift oscillator.

be 3 × 60°, or 180°. The frequency of oscillations can be found with

$$f = \frac{1}{15.39RC}$$

Example 11-2

The phase-shift components in Fig. 11-8 are 0.1-μF capacitors and 18-kΩ resistors. At what frequency will the network produce a phase shift of 180°? Use the equation:

$$f = \frac{1}{15.39RC}$$

$$= \frac{1}{15.39 \times 18 \times 10^3 \times 0.1 \times 10^{-6}}$$

$$= 36.1 \text{ Hz}$$

Figure 11-9 shows the schematic of a phase-shift oscillator circuit with the component values given. Each of the three phase-shift networks has been designed to produce a 60° response at the desired output frequency. Notice, however, that the value of R_B is 100 times higher than the values of the other two resistors in the network. This may seem to be an error since all three networks should be the same. Actually, R_B does appear to be much lower in value as far as the ac signal is concerned. This is because it is connected to the collector of the transistor. There is an ac signal present at the collector when the oscillator is running which is 180° out of phase with the base signal. This makes the voltage difference across R_B much higher than would be produced by the base signal alone. Thus, more signal current flows through R_B. Resistor R_B produces an ac loading effect at the base that

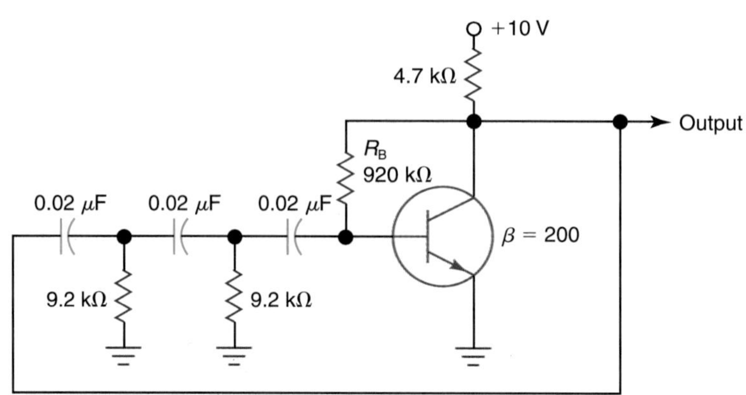

Fig. 11-9 A phase-shift oscillator with component values.

is set by the voltage gain of the amplifier and the value of R_B:

$$r_B = \frac{R_B}{A_V}$$

This equation tells us that the actual ac loading effect r_B is equal to R_B divided by the voltage gain of the amplifier. If we assume that the gain of the amplifier is 100, then

$$r_B = \frac{920 \text{ k}\Omega}{100} = 9.2 \text{ k}\Omega$$

We may conclude that all three phase-shift networks are the same. The frequency of oscillation for Fig. 11-9 will be

$$f = \frac{1}{15.39RC}$$
$$= \frac{1}{15.39 \times 9.2 \times 10^3 \times 0.02 \times 10^{-6}}$$
$$= 353 \text{ Hz}$$

Figure 11-10 shows a computer-generated amplitude and phase plot for the *RC* phase-shift part of Fig. 11-9. Networks of this type produce an output voltage equal to 1⁄29 of the input voltage at that frequency where the phase shift is 180°. This represents a feedback network gain of

$$V_{dB} = 20 \times \log \frac{V_{out}}{V_{in}} = 10 \times \log \frac{1}{29}$$
$$= -29.2 \text{ dB}$$

The common-emitter amplifier in Fig. 11-9 must have a gain greater than 29.2 dB in order for the circuit to oscillate.

The circuit of Fig. 11-9 will not oscillate at exactly 353 Hz. The formula deals with only the values of the *RC* network. It ignores some other effects caused by the transistor. The formula is adequate for practical work.

Twin-T oscillator

Figure 11-11 on page 294 shows another type of *RC* oscillator that is based on the *twin-T network.* These networks act as notch filters and provide minimum output amplitude and a phase lag of 180° at their resonant frequency. The resonant frequency of a twin-T network can be found by examining the circuit to determine which components are in series with the signal flow. Then use the values of the series components in this equation:

$$f_r = \frac{1}{2\pi RC}$$

Example 11-3

Find the resonant frequency for the twin-T network in Fig. 11-11. The series components in the network are the 10-kΩ resistors and the 0.033-μF capacitors. Use the equation:

$$f_r = \frac{1}{2\pi RC}$$
$$= \frac{1}{6.28 \times 10 \times 10^3 \times 0.033 \times 10^{-6}}$$
$$= 483 \text{ Hz}$$

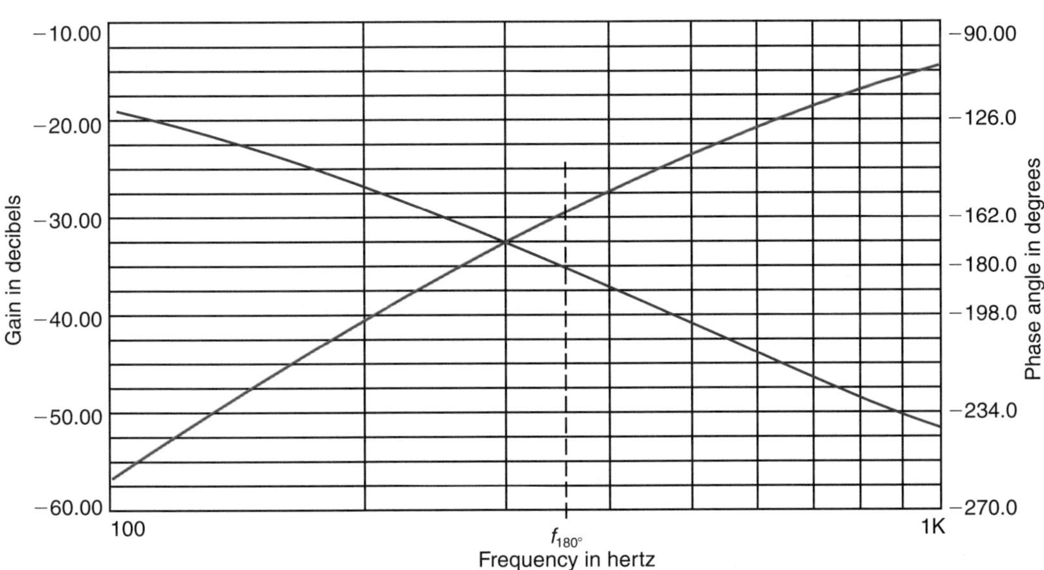

Fig. 11-10 Computer-generated response for a phase-shift network.

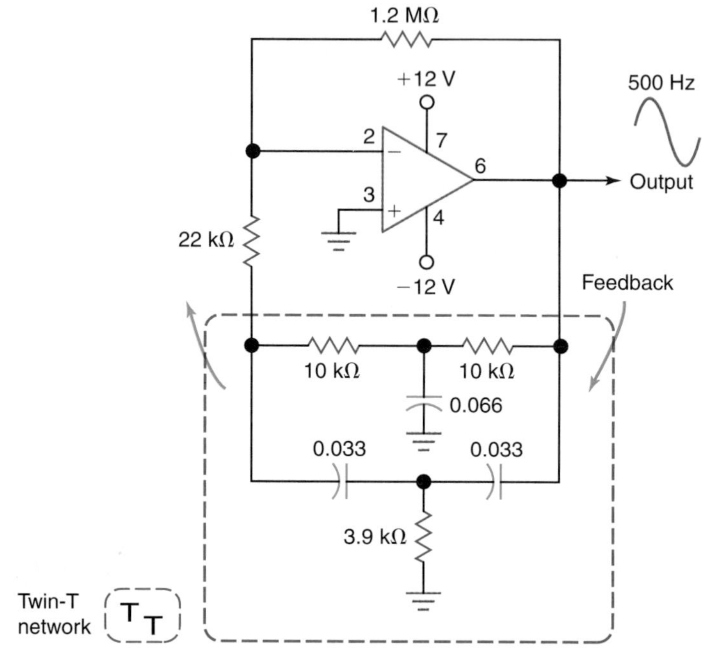

Fig. 11-11 Op-amp twin-T oscillator.

The twin-T network provides feedback from the output to the inverting input of the op amp. That feedback becomes positive at f_r because the twin-T network shows a 180° phase shift at this particular frequency. Notice that the 0.066-μF network capacitor is twice the value of each series capacitor. This is standard in a twin-T network. However, the 3.9-kΩ resistor is not standard. It is normally equal to one-half the value of each series resistor, or 5 kΩ in this case. A perfectly balanced twin-T network would provide no feedback at f_r. The intentional error allows enough positive feedback to reach pin 2 of the op amp, and a sine-wave signal of approximately 500 Hz appears at the output.

Figure 11-12 shows a computer-generated response for the unbalanced twin-T network of Fig. 11-11. Because of the unbalancing, the actual resonant frequency is about 520 Hz and the amplitude is −31 dB at that point. The op amp must provide a voltage gain greater than 35.5 (31 dB) in order for the circuit to oscillate.

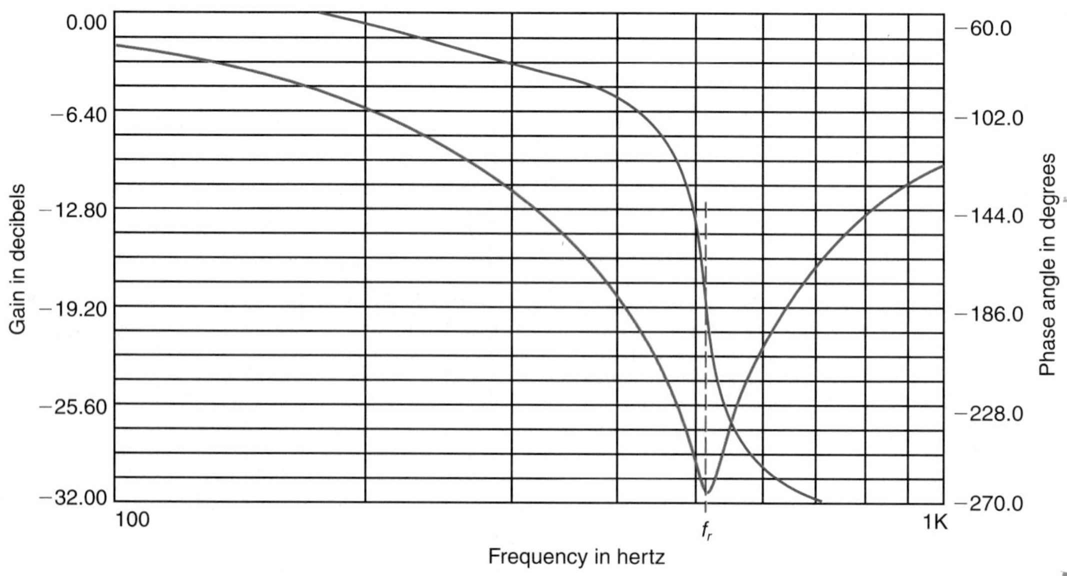

Fig. 11-12 Computer-generated response for the unbalanced twin-T network.

Did You Know?

Clock Signals and the NIST
The clock signal in many digital systems comes from a crystal-controlled oscillator. Standard time signals are produced by atomic clocks maintained at the National Institute of Standards and Technology (NIST). Many organizations use calibration standards traceable to NIST.

TEST

Choose the letter that best answers each question.

4. Refer to Fig. 11-4, where $R = 4700\ \Omega$ and $C = 0.02\ \mu F$. At what frequency will V_{out} be in phase with the signal source?
   a. 486 Hz
   b. 1693 Hz
   c. 3386 Hz
   d. 9834 Hz
5. Refer to Fig. 11-6, where $R = 6800\ \Omega$ and $C = 0.002\ \mu F$. What will the frequency of the output signal be?
   a. 11.70 kHz
   b. 46.79 kHz
   c. 78.90 kHz
   d. 98.94 kHz
6. Refer to Fig. 11-6. What is the phase relationship of the output signal and the signal at the noninverting (+) input of the amplifier?
   a. 180°
   b. 0°
   c. 90°
   d. 270°
7. Refer to Fig. 11-8. What is the configuration of the transistor amplifier?
   a. Common emitter
   b. Common collector
   c. Common base
   d. Emitter follower
8. Refer to Fig. 11-9 and assume that $R_B = 820\ k\Omega$ and that the voltage gain of the circuit is 120. What is the actual loading effect of R_B as far as the phase-shift network is concerned?
   a. 1 MΩ
   b. 500 kΩ
   c. 6833 Ω
   d. 384 Ω
9. Refer to Fig. 11-9. The capacitors are all changed to 0.05 μF. What is the frequency of oscillation?
   a. 60 Hz
   b. 141 Hz
   c. 1.84 kHz
   d. 0.95 MHz
10. Refer to Fig. 11-9. What is the phase relationship of the signal arriving at the base compared to the output signal?
    a. 0°
    b. 90°
    c. 180°
    d. 270°
11. What do phase-shift oscillators, twin-T oscillators, and Wien bridge oscillators have in common?
    a. They use *RC* frequency control.
    b. They have a sinusoidal output.
    c. They use amplifier gain to overcome feedback loss.
    d. All of the above.

11-3 *LC* Circuits

The *RC* oscillators are limited to frequencies below 1 MHz. Higher frequencies require a different approach to oscillator construction. Inductive-capacitive (*LC*) circuits can be used to design oscillators that operate at hundreds of megahertz (MHz). These *LC* networks are often called *tank circuits,* or flywheel circuits. **Tank circuit**

Figure 11-13 on page 296 shows how a tank circuit can be used to develop sinusoidal oscillations. Figure 11-13(*a*) assumes that the capacitor is charged. As the capacitor discharges through the inductor, a field expands about the turns of the inductor. After the capacitor has been discharged, the field collapses and current continues to flow. This is shown in Fig.

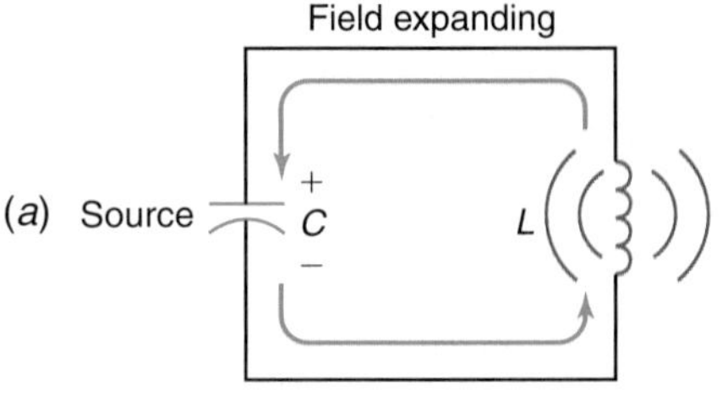

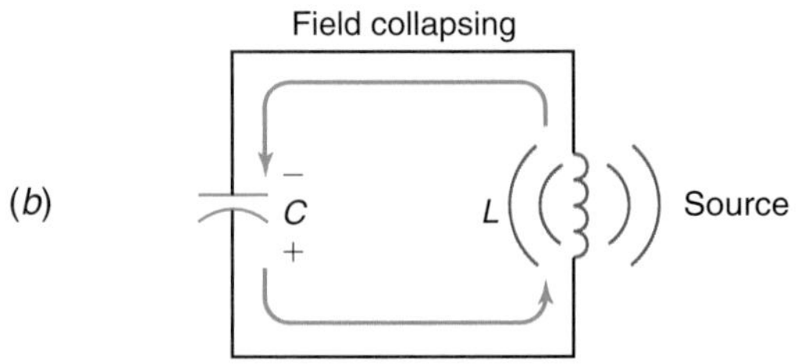

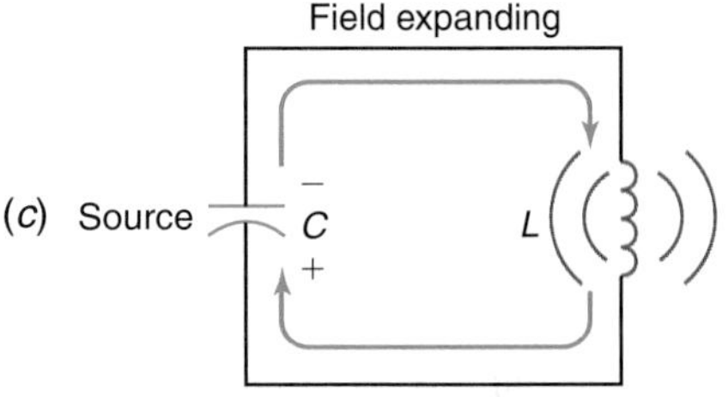

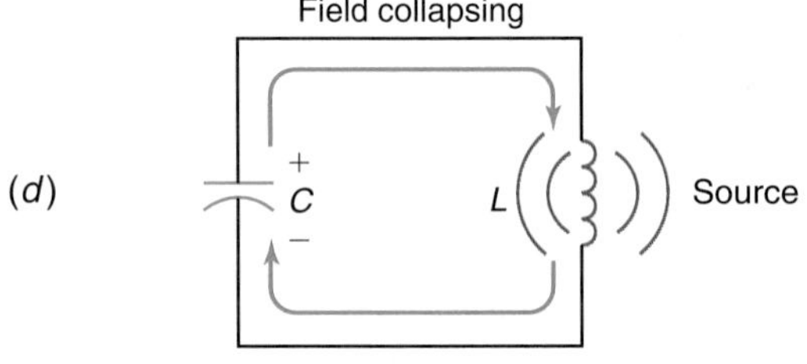

Fig. 11-13 Tank circuit action.

11-13(*b*). Note that the capacitor is now being charged in the opposite polarity. After the field collapses, the capacitor again acts as the source. Now, the current is flowing in the opposite direction. Figure 11-13(*c*) shows the second capacitor discharge. Finally, Fig. 11-13(*d*) shows the inductor acting as the source and charging the capacitor back to the original polarity shown in Fig. 11-13(*a*). The cycle will repeat over and over.

Inductors and capacitors are both energy storage devices. In a tank circuit, they exchange energy back and forth at a rate fixed by the values of inductance and capacitance. The frequency of oscillations is given by

$$f_r = \frac{1}{2\pi\sqrt{LC}}$$

You should recognize this formula. It is the resonance equation for an inductor and a capacitor. It is based on the resonant frequency, where the inductive reactance and the capacitive reactance are equal. An energized *LC* tank circuit will oscillate at its resonant frequency.

Example 11-4

What is the resonant frequency of the tank circuit in Fig. 11-13 if the coil is 1 μH and the capacitor is 180 pF? Apply the equation:

$$f_r = \frac{1}{6.28\times \sqrt{1 \times 10^{-6} \times 180 \times 10^{-12}}}$$
$$= 11.9 \text{ MHz}$$

Real tank circuits have resistance in addition to inductance and capacitance. This resistance will cause the tank circuit oscillations to decay with time. To build a practical *LC* oscillator, an amplifier must be added. The gain of the amplifier will overcome resistive losses, and a sine wave of constant amplitude can be generated.

Hartley oscillator

Tapped inductor

One way to combine an amplifier with an *LC* tank circuit is shown in Fig. 11-14. The circuit is called a *Hartley oscillator.* Note that the *inductor* is *tapped.* The tap position is important since the ratio of L_A to L_B determines the *feedback ratio* for the circuit. In practice, the feedback ratio is selected for reliable operation. This ensures that the oscillator will start every time the power is turned on. Too much

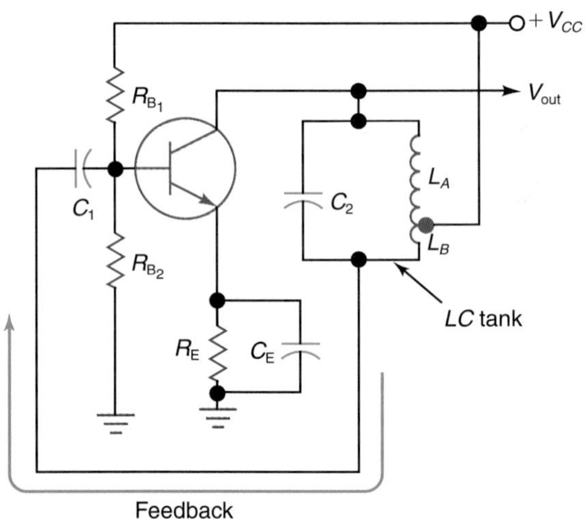

Fig. 11-14 The Hartley oscillator.

feedback will cause clipping and distort the output waveform.

The transistor amplifier of Fig. 11-14 is in the common-emitter configuration. This means that a 180° phase shift will be required somewhere in the feedback path. The tank circuit provides this phase shift. Note that the coil is tapped and that the tap connects to $+V_{CC}$. The tap is at ac ground, and there is a phase reversal across the tank. Thus, the collector signal arrives in phase at the base. Knowing the total inductance and the capacitance of the tank circuit will allow a solution for the resonant frequency. For example, if the total inductance $L_A + L_B$ is 20 μH and the capacitance C_2 is 400 pF, then

$$f_r = \frac{1}{6.28 \times \sqrt{20 \times 10^{-6} \times 400 \times 10^{-12}}}$$
$$= 1.78 \text{ MHz}$$

Another way to control the feedback of an *LC* oscillator is to *tap* the *capacitive* leg of the tank circuit. When this is done, the circuit is called a *Colpitts oscillator* (Fig. 11-15). Capacitor C_1 grounds the base of the transistor for ac signals and the transistor is operating as a common-base amplifier. You may recall that the input (the emitter) and the output (the collector) are in phase for this amplifier configuration. The feedback is in phase for the common-base configuration (shown in Fig. 11-15).

Tapped capacitor

Colpitts oscillator

Capacitors C_2 and C_3 in Fig. 11-15 act in series as far as the tank circuit is concerned. Assume that $C_2 = 1000$ pF and $C_3 = 100$ pF. Let us use the series capacitor formula to determine the effect of the series connection:

$$C_T = \frac{C_2 \times C_3}{C_2 + C_3} = \frac{1000 \text{ pF} \times 100 \text{ pF}}{1000 \text{ pF} + 100 \text{ pF}}$$
$$= 90.91 \text{ pF}$$

This means that 90.91 pF, along with the value of *L*, would be used to predict the frequency of oscillation. If $L = 1\ \mu$H, the circuit will oscillate at

$$f_r = \frac{1}{6.28 \times \sqrt{1 \times 10^{-6} \times 90.9 \times 10^{-12}}}$$
$$= 16.7 \text{ MHz}$$

Figure 11-16 on page 298 shows a *VFO* followed by a *buffer amplifier.* Both stages are operating in the common-drain configuration and use insulated-gate field-effect transistors. This circuit represents a design that can be used when maximum frequency stability is needed.

VFO

Buffer amplifier

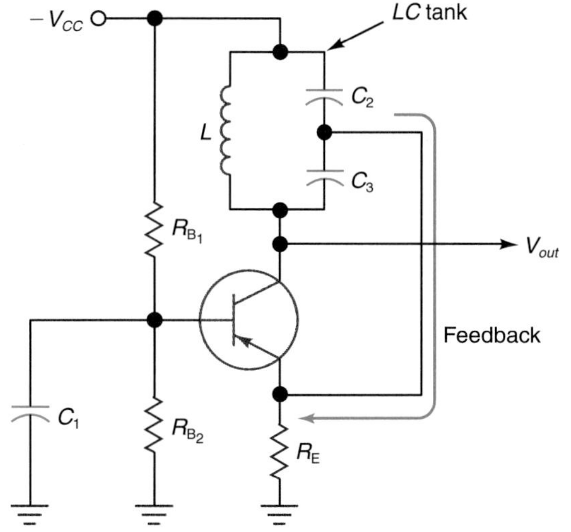

Fig. 11-15 The Colpitts oscillator.

Transistor Q_1 in Fig. 11-16 provides the needed gain to sustain the oscillations. Transistor Q_2 serves as a buffer amplifier. This protects the oscillator circuit from loading effects. Changing the load on an oscillator tends to change both the amplitude and the frequency of the output. For best stability, the oscillator circuit should be isolated from the stages that follow. Transistor Q_2 has a very high input impedance and a low output impedance. This allows the buffer amplifier to isolate the oscillator from any loading effects.

JOB TIP

The best interview policy is to be open and honest. Never exaggerate, and do not sell yourself short.

The tank circuit of Fig. 11-16 is made up of *L*, C_1, C_2, and C_3. This arrangement is known as a series tuned Colpitts, or *Clapp,* circuit. It is one of the most stable of all *LC* oscillators. Assume that C_1 varies from 10 to 100 pF and that C_2 and C_3 are both 1000 pF. We will use the series capacitor formula to determine the capacitive range of the tank circuit. When $C_1 = 10$ pF,

Clapp oscillator

$$C_T = \frac{1}{1/C_1 + 1/C_2 + 1/C_3}$$
$$= \frac{1}{1/10 \text{ pF} + 1/1000 \text{ pF} + 1/1000 \text{ pF}}$$
$$= 9.8 \text{ pF}$$

When $C_1 = 100$ pF,

$$C_T = \frac{1}{1/100 \text{ pF} + 1/1000 \text{ pF} + 1/1000 \text{ pF}}$$
$$= 83.3 \text{ pF}$$

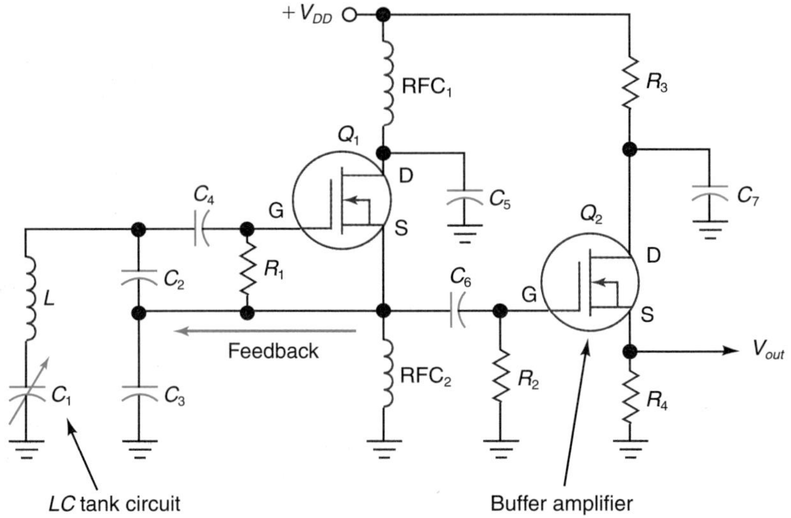

Fig. 11-16 A highly stable oscillator design.

The calculations show that the effective value C_T of the capacitors is determined mainly by C_1. The stray and shunt capacities of Fig. 11-16 appear in parallel with C_2 and C_3. These stray and shunt capacities can change and cause frequency drift in *LC* oscillator circuits. The Clapp design minimizes these effects because the series-tuning capacitor has the major effect on the tank circuit.

Varicap diode

Variable-frequency oscillators can be tuned by variable capacitors. However, variable capacitors are expensive and tend to be large. Many designs now replace the variable capacitor with a *varicap diode.* These diodes were covered in Sec. 3-4 of Chap. 3. As an example, variable capacitor C_1 in Fig. 11-16 could be replaced with a varicap diode and a bias circuit. Varying the bias voltage would tune the oscillator to various frequencies. Such a circuit would be called a voltage-controlled oscillator (*VCO*).

VCO

TEST

Choose the letter that best answers each question.

12. Refer to Fig. 11-14. What is the configuration of the amplifier?
    a. Common emitter
    b. Common base
    c. Common collector
    d. Emitter follower
13. Refer to Fig. 11-14. Where is the feedback signal shifted 180°?
    a. Across C_1
    b. Across R_{B_2}
    c. Across R_E
    d. Across the tank circuit
14. Refer to Fig. 11-14, where $C_2 = 120$ pF and $L_A + L_B = 1.8$ μH. Calculate the frequency of the output signal.
    a. 484 kHz
    b. 1.85 MHz
    c. 5.58 MHz
    d. 10.8 MHz
15. Refer to Fig. 11-14. What is the waveform of V_{out}?
    a. Sawtooth wave
    b. Sine wave
    c. Square wave
    d. Triangle wave
16. Refer to Fig. 11-15. The amplifier is in what configuration?
    a. Common emitter
    b. Common base
    c. Common collector
    d. Emitter follower
17. Refer to Fig. 11-15, where $C_2 = 330$ pF, $C_3 = 47$ pF, and $L = 0.8$ μH. What is the frequency of oscillation?
    a. 1.85 MHz
    b. 9.44 MHz
    c. 23.1 MHz
    d. 27.7 MHz
18. Refer to Fig. 11-16. Transistor Q_1 is operating as what type of oscillator?
    a. Clapp oscillator
    b. Hartley oscillator
    c. Phase-shift oscillator
    d. Buffer oscillator

19. Refer to Fig. 11-16. What is the major function of Q_2?
    a. It provides voltage gain.
    b. It provides the feedback signal.
    c. It isolates the oscillator from loading effects.
    d. It provides a phase shift.

11-4 Crystal Circuits

Another way to control the frequency of an oscillator is to use a *quartz crystal.* Quartz is a *piezoelectric material.* Such materials can change electric energy into mechanical energy. They can also change mechanical energy into electric energy. A quartz crystal will tend to vibrate at its resonant frequency. The resonant frequency is determined by the physical characteristics of the crystal. Crystal thickness is the major determining factor for the resonant point.

Figure 11-17(*a*) shows the construction of a quartz crystal. The quartz disk is usually very thin, especially for high-frequency operation. A metal electrode is fused to each side of the disk. When an ac signal is applied across the electrodes, the crystal vibrates. The vibrations will be strongest at the resonant frequency of the crystal. When a crystal is vibrating at this frequency, a large voltage appears across the electrodes. The schematic symbol for a crystal is shown in Fig. 11-17(*b*).

Crystals can become the frequency-determining components in high-frequency oscillator circuits. They can replace *LC* tank circuits. Crystals have the advantage of producing very *stable* output frequencies. A crystal oscillator can have a stability better than 1 part in 10^6 per day. This is equal to an accuracy of 0.0001 percent. A crystal oscillator can be placed in a temperature-controlled oven to provide a stability better than 1 part in 10^8 per day.

Stability

An *LC* oscillator circuit is subject to frequency variations. Some things that can cause a change in oscillator output frequency are

Piezoelectric

1. Temperature
2. Supply voltage
3. Mechanical stress and vibration
4. Component drift
5. Movement of metal parts near the oscillator circuit

Crystal-controlled circuits can greatly reduce all these effects.

A quartz crystal can be represented by an equivalent circuit (Fig. 11-18) on page 300. The *L* and *C* values of the quartz equivalent circuit represent the resonant action of the crystal and determine what is known as the *series resonance* of the crystal. The electrode capacitance causes the crystal to also show a *parallel resonant* point. Since the capacitors act in series, the net capacitance is a little lower for parallel resonance. This makes the parallel res-

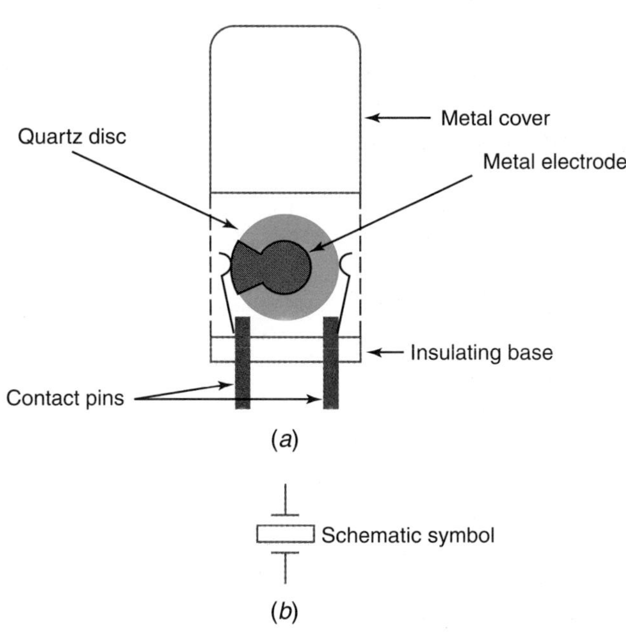

Fig. 11-17 A crystal used for frequency control.

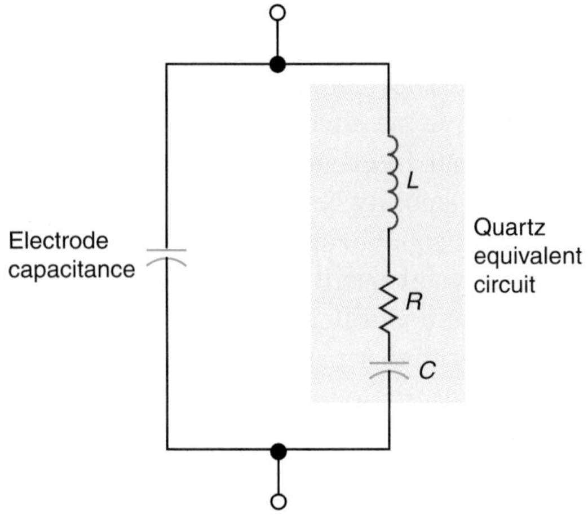

Fig. 11-18 Quartz-crystal equivalent oscillator.

onant frequency slightly higher than the series resonant frequency.

The equivalent circuit of a crystal predicts that oscillations can occur in two modes: parallel and series. In practice, the parallel mode is from 2 to 15 kHz higher. Oscillator circuits may be designed to use either mode. When a crystal is replaced, it is very important to obtain the correct type. For example, if a series-mode crystal is substituted in a parallel-mode circuit, the oscillator will run high in frequency.

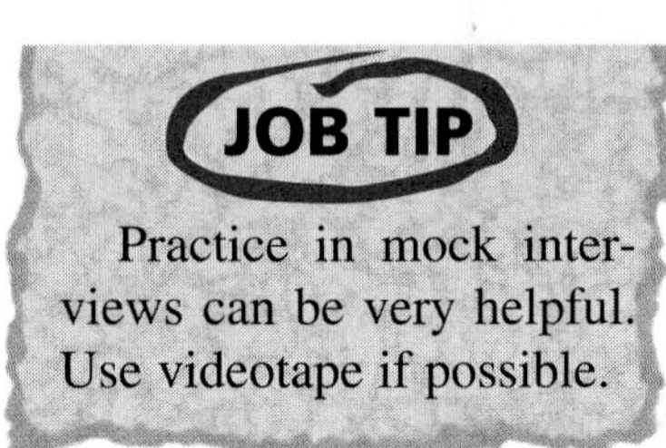

Refer again to Fig. 11-18. Note that the quartz equivalent circuit also contains resistance *R*. This represents losses in the quartz. Most crystals have small losses. In fact, the losses are small enough to give crystals a very high *Q*. Circuit *Q* is very important in an oscillator circuit. *High Q* gives frequency stability. Crystal *Q*s can be in excess of 3000. By comparison, *LC* tank circuit Qs seldom exceed 200. This is why a crystal oscillator is so much more stable than an *LC* oscillator.

High Q

Crystal oscillator

Figure 11-19 shows the schematic diagram of a *crystal oscillator*. The amplifier configuration is common emitter. This means that the feedback path must provide a 180° phase shift for oscillations to occur. This phase shift is produced by capacitors C_1 and C_2. Capacitors C_1 and C_2 form a voltage divider to control the amount of feedback. Excess feedback causes distortion and drift. Too little feedback causes unreliable operation: For example, the circuit may not start every time it is turned on. Capacitor C_3 is a *trimmer capacitor*. It is used to precisely set the frequency of oscillation. The remaining components in Fig. 11-19 are standard for the common-emitter configuration.

Overtone crystal

Harmonics

Very high frequency crystals present problems. The thickness of the quartz must decrease as frequency goes up. Above 15 MHz, the quartz becomes so thin that it is too fragile. Higher frequencies require the use of *overtone crystals,* which operate at *harmonics* of the fundamental frequency. Harmonics are integer multiples of a frequency. For example, the second harmonic of 10 MHz is 20 MHz, the third harmonic is 30 MHz, and so on. The use of harmonics can extend the range of crystal oscillators to around 150 MHz.

Oscillator circuits designed to use overtone crystals must include an *LC* tuned circuit. This *LC* tuned circuit must be tuned to the correct harmonic. This ensures that the crystal will vibrate in the proper mode. Otherwise, it would tend to oscillate at a lower frequency.

Overtone oscillator

Figure 11-20 is an *overtone oscillator* circuit. Capacitor C_1 grounds the base for ac signals. The transistor is in the common-base configuration, so no phase reversal is required in the feedback circuit. Capacitors C_3 and C_4 form a divider to set the amount of feedback from collector to emitter. Crystal X_1 is in the feedback path. It is operating in the series mode. No phase reversal occurs across a series resonant circuit. All overtone crystals operate in the series mode.

Inductor L_1 of Fig. 11-20 is part of the tuned circuit used to select the proper overtone. It resonates with C_3, C_4, and C_5 to form a tank circuit. Coil L_1 is adjusted to the correct overtone frequency. Capacitor C_2 is a trimmer capacitor used to set the crystal frequency. In

Fig. 11-19 A crystal-controlled oscillator.

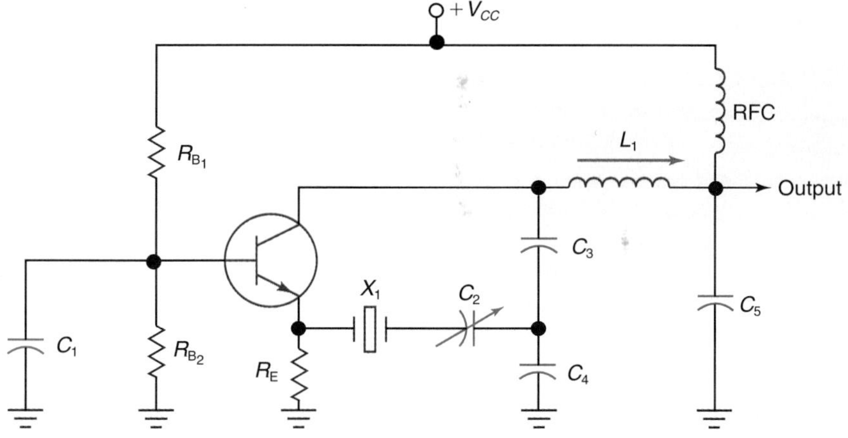

Fig. 11-20 An overtone crystal oscillator.

practice, L_1 is adjusted first until the oscillator starts and works reliably. Then C_2 is adjusted for the exact frequency required.

Crystals increase the cost of oscillator circuits. This can become quite a problem in equipment such as a multichannel transmitter. A separate crystal will be required for every channel. The cost soon reaches the point where another solution must be found. This solution is a *frequency synthesizer.* There are combination digital and analog circuits that can synthesize many frequencies from one or more crystals.

Frequency synthesizer

TEST

Choose the letter that best completes each statement.

20. The quartz crystals used in oscillators show a
    a. Piezoelectric effect
    b. Semiconductor effect
    c. Diode action
    d. Transistor action
21. An oscillator that uses crystal control should be
    a. Frequency-stable
    b. Useful only at low frequencies
    c. A VFO
    d. None of the above
22. A 6-MHz crystal oscillator has a stability of 1 part in 10^6. The largest frequency error expected of this circuit is
    a. 0.06 Hz
    b. 0.6 Hz
    c. 6 Hz
    d. 60 Hz
23. A series-mode crystal is marked 10.000 MHz. It is used in a circuit that operates the crystal in its parallel mode. The circuit can be expected to
    a. Run below 10 MHz
    b. Run at 10 MHz
    c. Run above 10 MHz
    d. Not oscillate
24. Refer to Fig. 11-19. The phase relationship of the signal at the collector of Q_1 compared to the base signal is
    a. 0°
    b. 90°
    c. 180°
    d. 360°
25. Refer to Fig. 11-19. The required phase shift is produced by
    a. R_L
    b. C_1 and C_2
    c. R_{B1}
    d. C_3
26. Refer to Fig. 11-20. The configuration of the amplifier is
    a. Common emitter
    b. Common collector
    c. Common base
    d. Emitter follower
27. Refer to Fig. 11-20. The function of C_2 is to
    a. Act as an emitter bypass
    b. Adjust the tank circuit to the crystal harmonic
    c. Produce the required phase shift
    d. Set the exact frequency of oscillation

11-5 Relaxation Oscillators

All the oscillator circuits discussed so far produce a sinusoidal output. There is another major class of oscillators that do not produce sine waves. They are known as *relaxation oscillators.* The outputs for these circuits are sawtooth or rectangular waveforms.

Relaxation oscillator

UJT

Negative resistance

Figure 11-21 shows one type of relaxation oscillator. A unijunction transistor is the key component. You may recall that a *UJT* is a *negative-resistance* device. Specifically, the resistance from the emitter terminal to the B_1 terminal will drop when the transistor fires.

Pulse waveform

When the power is applied to the oscillator circuit of Fig. 11-21, the capacitor begins charging through R_1. As the capacitor voltage increases, the emitter voltage of the transistor also increases. Eventually, the emitter voltage will reach the firing point, and the emitter diode begins to conduct. At this time, the resistance of the UJT suddenly decreases. This quickly discharges the capacitor. With the capacitor discharged, the UJT switches back to its high resistance state, and the capacitor begins charging again. This cycle will repeat over and over.

Sawtooth waveform

Intrinsic standoff ratio

Figure 11-22 shows two of the waveforms that can be expected from the UJT oscillator. Notice that a *sawtooth wave* appears at the emitter of the transistor. This wave shows the gradual increase of capacitor voltage. When the firing voltage V_P is reached, the capacitor is rapidly discharged. The discharge current flows through the base 1 resistor, causing a voltage drop. Therefore the base 1 waveform shows narrow voltage pulses that correspond to the falling edge of the sawtooth. The pulses are narrow since the capacitor is discharged rapidly.

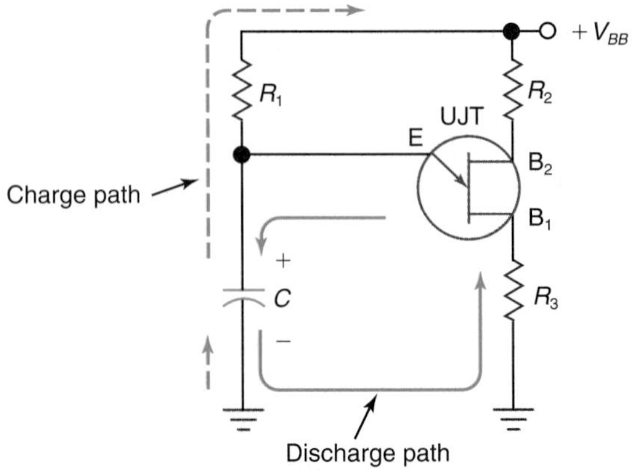

Fig. 11-21 A UJT relaxation oscillator.

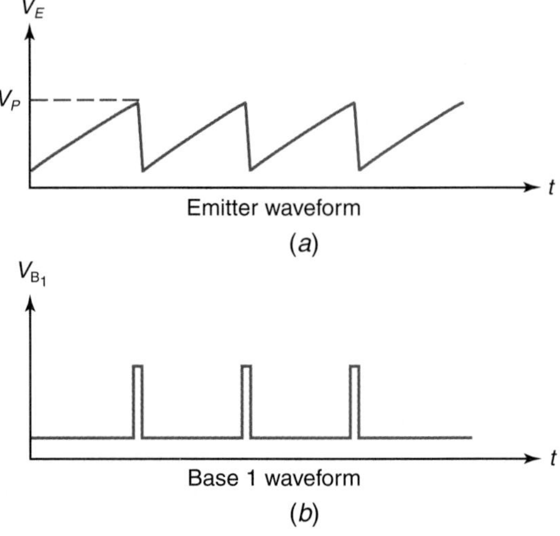

Fig. 11-22 Unijunction transistor oscillator waveforms.

In practice, either the sawtooth or the *pulse* or both waveforms may become the output of the circuit. Such circuits are very useful in timing and control applications. The frequency of oscillations may be roughly predicted by

$$f = \frac{1}{RC}$$

Assume that $R_1 = 10{,}000\ \Omega$ and $C = 10\ \mu\text{F}$ in Fig. 11-21. The approximate frequency of oscillation is given by

$$f = \frac{1}{10 \times 10^3 \times 10 \times 10^{-6}} = 10\ \text{Hz}$$

The UJT itself will also have an effect on the frequency of oscillation. The most important UJT parameter is the *intrinsic standoff ratio.* This ratio is a measure of how the transistor will internally divide the supply voltage and thus bias the emitter terminal. The ratio will set the firing voltage V_P. Standard UJTs have intrinsic standoff ratios from around 0.4 to 0.85. If the intrinsic standoff ratio is near 0.63, then the formula given previously will be accurate. This is because the capacitor in an *RC* network will charge to 63 percent of the supply voltage in the first time constant. The time constant T is given by

$$T = RC$$

PUT

The variation in intrinsic standoff ratio can be overcome with a device called a *programmable unijunction transistor* (*PUT*). In this device, the ratio is programmed with external resistors. This makes it possible to build a circuit

Example 11-5

Assume that the intrinsic standoff ratio for the UJT in Fig. 11-21 is close to 0.63. If R_1 is 10 kΩ, pick a value for the capacitor so that the frequency of oscillation will be equal to 100 Hz. Begin with the equation:

$$f = \frac{1}{RC}$$

Rearrange the equation to solve for C:

$$C = \frac{1}{Rf} = \frac{1}{10 \times 10^3 \times 100} = 1\ \mu\text{F}$$

and have it oscillate very near the design frequency. Figure 11-23 shows a PUT oscillator. Resistor R_1 and capacitor C will set the frequency of oscillation along with R_3 and R_4, which determine the intrinsic standoff ratio.

Figure 11-24 shows another type of relaxation oscillator, the *astable multivibrator.* The circuit has no stable states. The circuit voltages switch constantly as it oscillates. This is in contrast to the *monostable* version, which has one stable state, and the *bistable* circuit, with two stable states. The monostable and bistable circuits will not be discussed since this chapter is limited to oscillators.

Astable multivibrators are also called *free-running flip-flops.* This name is more descriptive of how the circuit behaves. Notice in Fig. 11-24 that two transistors are used. If Q_1 is on (conducting), Q_2 will be off. After a period of time, the circuit flips and Q_1 goes off while Q_2 comes on. After a second period, the circuit flops, turning on Q_1 again and turning off Q_2.

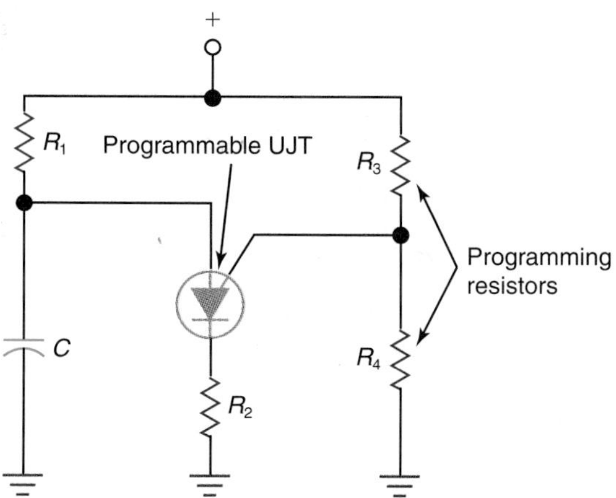

Fig. 11-23 Using a programmable UJT.

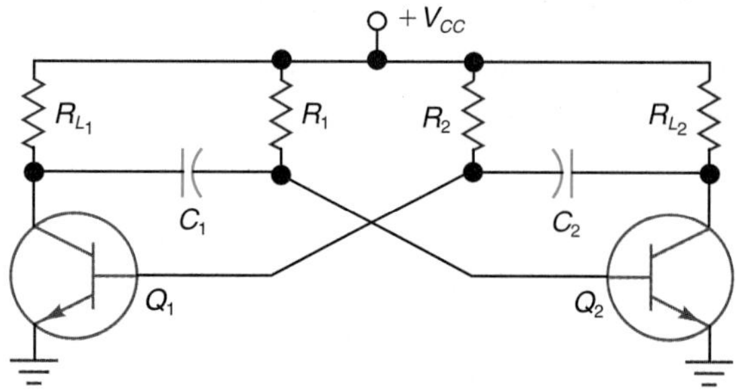

Fig. 11-24 The astable multivibrator.

The flip-flop action continues as long as the power is applied.

Study the waveforms shown in Fig. 11-25. They are for transistor Q_1 of Fig. 11-24. Transistor Q_2's waveforms will look the same, but they will be inverted. Suppose that Q_2 has just turned on, making its collector less positive. This means the collector of Q_2 is going in a negative direction. This negative signal is coupled by C_2 to the base of Q_1. This turns off Q_1. Capacitor C_2 will hold off Q_1 until R_2 can allow the capacitor to charge sufficiently positive to allow Q_1 to come on. The circuit works on RC time constants. Transistor Q_1 is being held in the off state by the time constant of R_2 and C_2.

Astable multivibrator

As Q_1 is turning on, its collector will be going less positive. This negative-going signal is coupled by C_1 to the base of Q_2, and Q_2 is turned off. It will stay off for a period determined by the *RC time constant* of R_1 and C_1.

Flip-flop

***RC* time constant**

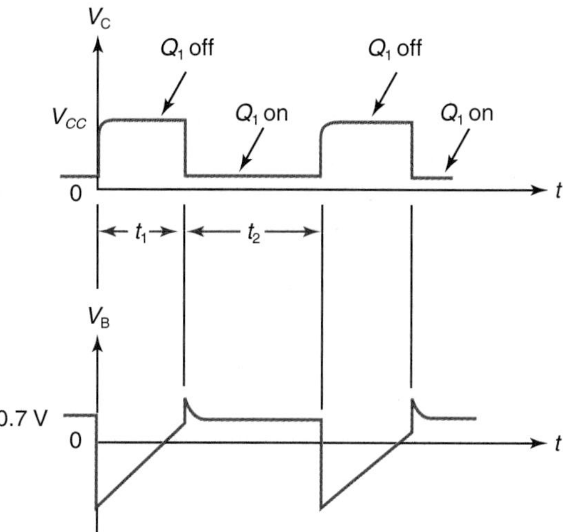

Fig. 11-25 Multivibrator waveforms.

Again, refer to Fig. 11-25. One rectangular cycle will be produced during one period. The period has two parts; thus it is equal to

$$T = t_1 + t_2$$

It takes 0.69 time constants for the *RC* network to reach the base turn-on voltage. This gives us a way to estimate the time that each transistor will be held in the off state:

$$t = 0.69RC$$

Assume that R_1 and R_2 are both 47-kΩ resistors and C_1 and C_2 are both 0.05-μF capacitors. Each transistor should be held off for

$$\begin{aligned} t &= 0.69 \times 47 \times 10^3 \times 0.05 \times 10^{-6} \\ &= 1.62 \times 10^{-3}\ \text{s} \end{aligned}$$

The period will be twice this value:

$$T = 2 \times 1.62\ \text{ms} = 3.24\ \text{ms}$$

It will take 3.24 ms for the oscillator to produce one cycle. Now that the period is known, it will be easy to calculate the frequency of oscillation:

$$f = \frac{1}{T} = \frac{1}{3.24 \times 10^{-3}} = 309\ \text{Hz}$$

Square wave

Rectangular wave

With $R_1 = R_2$ and $C_1 = C_2$, the oscillator can be expected to produce a *square waveform.* A square wave is a special case of a *rectangular wave* in which each alternation consumes the same time interval. Connecting an oscilloscope to either collector will show the positive-going part of the signal equal in time to the negative-going part.

What happens when the timing components are not equal? Assume in Fig. 11-24 that R_1 and R_2 are 10 kΩ, $C_1 = 0.01\ \mu$F, and $C_2 = 0.1\ \mu$F. What waveform can be expected at the collector of Q_2? Computing both time constants will answer the questions:

$$\begin{aligned} t_1 &= 0.69 \times 10 \times 10^3 \times 0.1 \times 10^{-6} \\ &= 0.69 \times 10^{-3}\ \text{s} \\ t_2 &= 0.69 \times 10 \times 10^3 \times 0.01 \times 10^{-6} \\ &= 0.069 \times 10^{-3}\ \text{s} \end{aligned}$$

Transistor Q_1 will be held in the off mode 10 times longer than Q_2. Figure 11-26 shows the expected collector waveform for Q_1. Such a circuit is *nonsymmetrical* and the output waveform is considered rectangular but is not a square wave.

What is the frequency of the rectangular waveform of Fig. 11-26? First, the period must be determined:

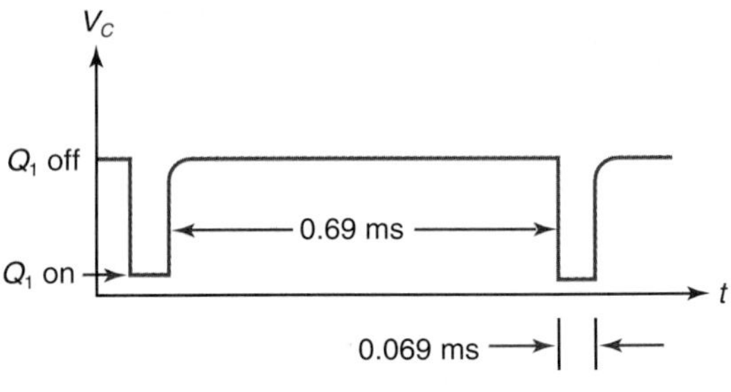

Fig. 11-26 Waveform for a nonsymmetrical multivibrator.

$$\begin{aligned} T &= 0.69 \times 10^{-3}\ \text{s} + 0.069 \times 10^{-3}\ \text{s} \\ &= 0.759 \times 10^{-3}\ \text{s} \end{aligned}$$

The frequency will be given by

$$f = \frac{1}{0.759 \times 10^{-3}} = 1318\ \text{Hz}$$

TEST

Choose the letter that best answers each question.

28. Refer to Fig. 11-21. What waveform should appear across the capacitor?
    a. Sawtooth
    b. Pulse
    c. Sinusoid
    d. Square
29. Refer to Fig. 11-21. What waveform should appear across R_3?
    a. Sawtooth
    b. Pulse
    c. Sinusoid
    d. Square
30. Refer to Fig. 11-21, where $R_1 = 10{,}000\ \Omega$ and $C = 0.5\ \mu$F. What is the approximate frequency of operation?
    a. 200 Hz
    b. 1000 Hz
    c. 2000 Hz
    d. 20 kHz
31. Refer to Fig. 11-23. What is the purpose of resistors R_3 and R_4?
    a. To set the desired intrinsic standoff ratio
    b. To set the exact frequency of oscillation
    c. Both of the above
    d. None of the above
32. Refer to Fig. 11-24. What waveform can be expected at the collector of Q_1?
    a. Sawtooth
    b. Triangular

c. Sinusoid
d. Rectangular

33. Refer to Fig. 11-24. What waveform can be expected at the collector of Q_2?
a. Sawtooth
b. Triangular
c. Sinusoid
d. Rectangular

34. Refer to Fig. 11-24. What is the phase relationship of the signal at the collector of Q_2 to the signal at the collector of Q_1?
a. 0°
b. 90°
c. 180°
d. 360°

35. Refer to Fig. 11-24 and assume that $C_1 = C_2 = 0.5\ \mu\text{F}$ and $R_1 = R_2 = 22\ \text{k}\Omega$. What is the frequency of oscillation?
a. 16 Hz
b. 33 Hz
c. 66 Hz
d. 99 Hz

11-6 Undesired Oscillations

It was mentioned earlier that a public address system can oscillate if the gain is too high. Such oscillations are undesired. Now that you have studied oscillators, it will be easier to understand how amplifiers can oscillate and what can be done to prevent it.

Negative feedback is often used in amplifiers to decrease distortion and improve frequency response. A three-stage amplifier is shown in simplified form in Fig. 11-27. Each stage uses the common-emitter configuration, and each will produce a 180° phase shift. This makes the feedback from stage 3 to stage 1 negative. Positive feedback is required for oscillation; therefore, the amplifier should be stable. But, at very high or very low frequencies, the feedback can become positive. Transistor interelectrode capacitances form *lag* networks that can cause a phase error at high frequencies. Coupling capacitors form *lead* networks that cause phase errors at low frequencies. These effects accumulate in multistage amplifiers. The overall phase error will reach −180° at some high frequency and it will reach +180° at some low frequency if the amplifier uses capacitive coupling.

A system such as the one shown in Fig. 11-27 can become an oscillator at a frequency

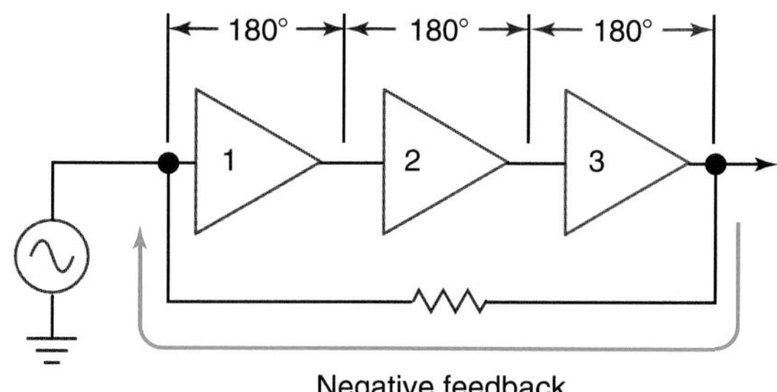

Fig. 11-27 A three-stage amplifier with negative feedback.

where the internal phase errors sum to ±180°. If amplifier gain is high enough at that frequency, the amplifier will oscillate. Such an amplifier is unstable and useless. *Frequency compensation* can be used to make such an amplifier stable. A compensated amplifier has one or more networks added which decrease gain at the frequency extremes. Thus, by the time the frequency is reached where the phase errors total ±180°, the gain is too low for oscillations to occur. A good example of this technique is modern operational amplifiers. They are internally compensated for gain reductions of 20 dB per decade. At the higher frequencies where the phase errors total −180°, the gain is too low for oscillation to occur. This has already been discussed in Sec. 9-5 of Chap. 9.

Frequency compensation

Another way that amplifiers can become unstable is when feedback paths occur which do not show on the schematic diagram. For example, a good power supply is expected to have a very low internal impedance. This will make it very difficult for ac signals to appear across it. However, a power supply might have a high impedance. This can be caused by a defective filter capacitor. An old battery power supply may develop a high internal impedance because it is drying out. The impedance of the power supply can provide a common load where signals are developed.

Phase lag and lead

In the simplified three-stage amplifier of Fig. 11-28 on page 306. Z_P represents the internal impedance of the power supply. Suppose that stage 3 is drawing varying amounts of current because it is amplifying an ac signal. The varying current will produce a signal across Z_P. This signal will obviously affect stage 1 and stage 2. It is a form of unwanted feedback, and it may cause the circuit to oscillate.

Unwanted feedback

Figure 11-29 on page 306 shows a solution for the *unwanted feedback* problem. An *RC*

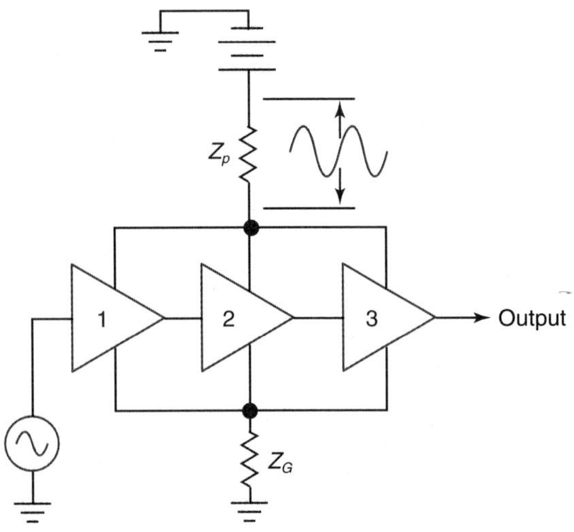

Fig. 11-28 The effect of supply and ground impedances.

network has been added in the power-supply lines to each amplifier. These networks act as low-pass filters. The capacitors are chosen to have a low reactance at the signal frequency. They are called *bypass capacitors,* and they effectively short any ac signal appearing on the supply lines to ground. In some cases the resistors are eliminated and only bypass capacitors are used to filter the supply lines.

Bypass capacitor

Neutralization

Ground impedances can also produce feedback paths that do not appear on schematics. Heavy currents flowing through printed circuit foils or the metal chassis can cause voltage drops. The voltage drop from one amplifier may be fed back to another amplifier. Refer again to Fig. 11-28. The impedance of the ground path is Z_G. As before, signal currents from stage 3 could produce a voltage across Z_G that will be fed back to the other stages. Ground currents cannot be eliminated, but proper layout can prevent them from producing feedback. The idea is to prevent later stages from sharing ground paths with earlier stages.

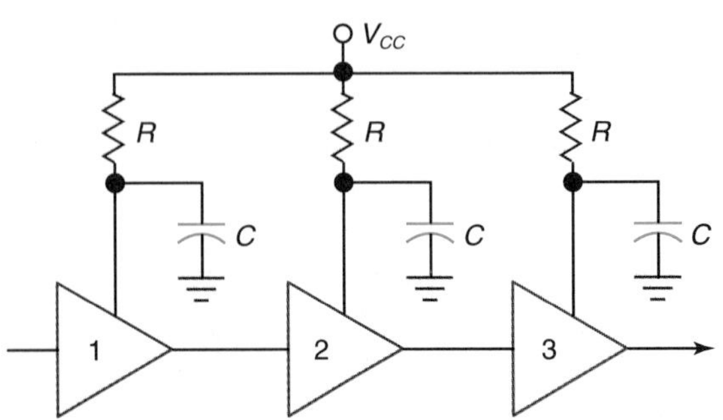

Fig. 11-29 Preventing supply feedback.

Shielding

High-frequency amplifiers such as those used in radio receivers and transmitters are often prone to oscillation. These circuits can be coupled by stray capacitive and magnetic paths. When such circuits can "see" each other in the electrical sense, oscillations are likely to occur. These circuits must be *shielded.* Metal partitions and covers are used to keep the circuits isolated and prevent feedback.

Another feedback path often found in high-frequency amplifiers lies within the transistor itself. This path can also produce oscillations and make the amplifier useless. In Fig. 11-30, C_{bc} represents the capacitance from the collector to the base of the transistor in a tuned high-frequency amplifier. This capacitance will feed some signal back. The feedback can become positive at a frequency where enough internal phase shift is produced.

Nothing can be done to eliminate the feedback inside a transistor. However, it is possible to create a second path external to the transistor. If the phase of the external feedback is correct, it can cancel the internal feedback. This is called *neutralization.* Figure 11-30 shows how a capacitor can be used to cancel the feedback of C_{bc}. Capacitor C_N feeds back from the collector circuit to the base of the transistor. The phase of the signal fed back by C_N is opposite to the phase fed back by C_{bc}. This stabilizes the amplifier. Notice that the required phase reversal is produced across the tuned circuit. Another possibility is to use a separate neutralization winding that is coupled to the tuned circuit.

Figure 11-31 is an actual radio-frequency amplifier used in a frequency modulation (FM) tuner. You will note that several of the tech-

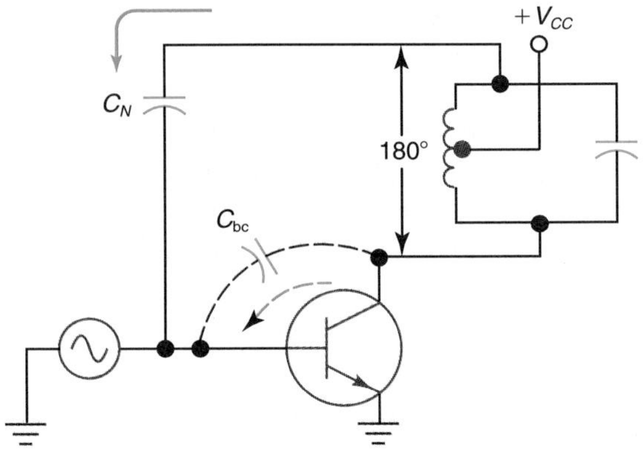

Fig. 11-30 Feedback inside and outside the transistor.

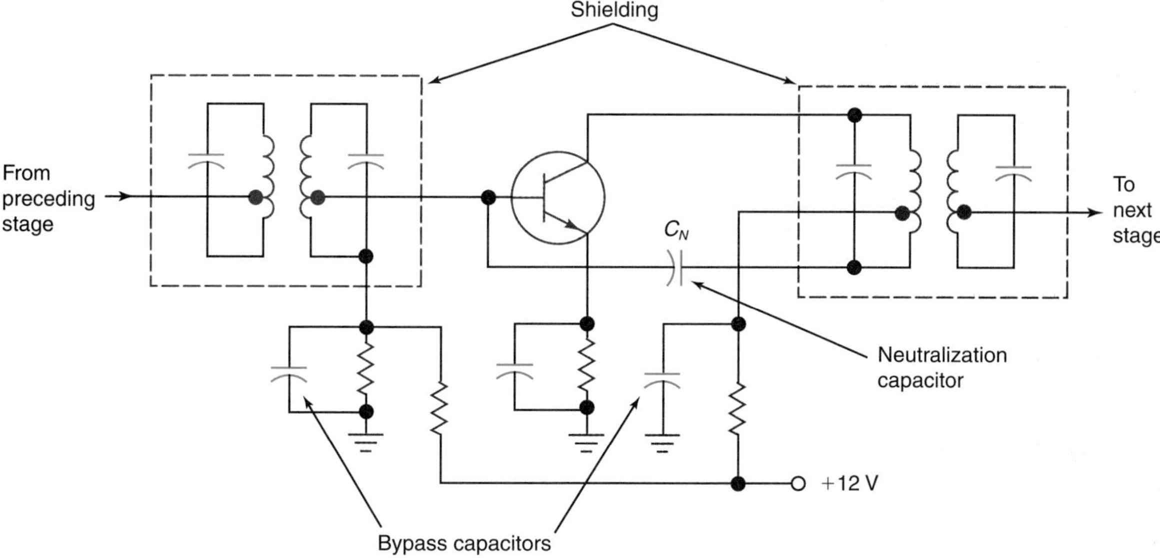

Fig. 11-31 A stabilized RF amplifier.

niques discussed in this section have been employed to stabilize the amplifier.

TEST

Choose the letter that best answers each question.

36. Examine Fig. 11-27. Assume at some frequency extreme that the actual phase shift in each stage is 240°. What happens to the feedback at that frequency?
    a. It does not exist.
    b. It becomes positive.
    c. It decreases the gain for that frequency.
    d. None of the above.
37. Refer to question 36. Assume that the amplifier has more gain at that frequency than it has loss in the feedback path. What happens to the amplifier?
    a. It burns out.
    b. It short-circuits the signal source.
    c. It becomes unstable (oscillates).
    d. It can no longer deliver an output signal.
38. Why are most operational amplifiers internally compensated for gain reductions of 20 dB per decade?
    a. To prevent them from becoming unstable
    b. To prevent any phase error at any frequency
    c. To prevent signal distortion
    d. To increase their gain at high frequencies
39. Refer to Fig. 11-28, where $Z_P = 10\ \Omega$ and stage 3 is taking a current from the supply that fluctuates 50 mA peak-to-peak. What signal voltage is developed across Z_P?
    a. 100 mV peak-to-peak
    b. 1 V peak-to-peak
    c. 10 V peak-to-peak
    d. None of the above
40. Refer to Fig. 11-28. Stage 3 draws current from the power supply, and a signal is produced across Z_P. This signal
    a. Is delivered to the output
    b. Is canceled in stage 1
    c. Is dissipated in Z_G
    d. Becomes feedback to stage 1 and stage 2
41. Refer to Fig. 11-29. The *RC* networks shown are often called decoupling networks. This is because they
    a. Prevent unwanted signal coupling
    b. Bypass any dc to ground
    c. Act as high-pass filters
    d. Disconnect each stage from V_{CC}
42. Refer to Fig. 11-30. The function of C_N is to
    a. Bypass the base of the transistor
    b. Filter V_{CC}
    c. Tune the tank circuit
    d. Cancel the effect of C_{bc}
43. Refer to Fig. 11-31. How many techniques are shown for ensuring the stability of the amplifier?
    a. One
    b. Two
    c. Three
    d. Four

11-7 Oscillator Troubleshooting

Oscillator troubleshooting uses the same skills needed for amplifier troubleshooting. Since most oscillators are amplifiers with positive feedback added, many of the faults are the same. When troubleshooting an electronic circuit, remember the word "GOAL." Good troubleshooting involves:

GOAL

1. Observing the symptoms
2. Analyzing the possible causes
3. Limiting the possibilities

It is possible to observe the following symptoms when troubleshooting oscillators:

1. No output
2. Reduced amplitude
3. Unstable frequency
4. Frequency error

It is also possible that two symptoms may be observed at the same time. For example, an oscillator circuit may show reduced amplitude and frequency error.

Certain instruments are very useful for proper symptom identification. A digital frequency counter is very valuable when troubleshooting for frequency error. An oscilloscope is also a good instrument for oscillator troubleshooting. As always, a voltmeter is needed for power-supply and bias-voltage checks. When using instruments in and around oscillator circuits, always remember this: Oscillators can be subject to *loading effects.* More than one technician has been misled because connecting test equipment pulled the oscillator off frequency or reduced the amplitude. In some cases an instrument may load an oscillator to the point at which it will stop working altogether.

Loading effects can be reduced by using high-impedance instruments. It is also possible to reduce loading effects by taking readings at the proper point. If an oscillator is followed by a buffer stage, frequency and waveform readings should be taken at the output of the buffer. The buffer will minimize the loading effect of the test equipment.

Do not forget to check the effect of any and all controls when troubleshooting. If the circuit is a VFO, it is a good idea to tune it over its entire range. You may find that the trouble appears and disappears as the oscillator is tuned. Variable capacitors can short over a portion of their range. If the circuit is a VCO, it may be necessary to override the tuning voltage with an external power supply to verify proper operation and frequency range. Use a current-limiting resistor of around 100 kΩ to avoid loading effects and circuit damage when running this type of test.

The power supply can have several effects on oscillator performance. Frequency and amplitude are both sensitive to the power-supply voltage. It is worth knowing if the power supply is correct and if it is stable. Power-supply checks should be made early in the troubleshooting process. They are easy to make and can save a lot of time.

It is important to review the theory of the circuit when troubleshooting. This will help you analyze possible causes. Determine what controls the operating frequency. Is it a lead-lag network, an *RC* network, a tank circuit, or a crystal? Is there a varicap diode in the frequency-determining network? Remember that loading effects can pull an oscillator off frequency. The problem could be in the next stage which is fed by the oscillator circuit.

Unstable oscillators can be quite a challenge. Technicians often resort to tapping components and circuit boards with an insulated tool to localize the difficulty. If this fails, they may use heat or cold to isolate a sensitive component. Desoldering pencils make excellent heat sources. A squeeze on the bulb will direct a stream of hot air just where it is needed. Chemical "cool sprays" are available for selective cooling of components.

Table 11-1 is a summary of causes and effects to help you troubleshoot oscillators.

TEST

Choose the letter that best answers each question.

44. What can loading do to an oscillator?
    a. Cause a frequency error
    b. Reduce the amplitude of the output
    c. Kill the oscillations completely
    d. All of the above
45. An astable multivibrator is a little off frequency. Which of the following is least likely to be the cause?
    a. The power-supply voltage is wrong.
    b. A resistor has changed value.

Table 11-1 Troubleshooting Oscillators

Problem	Possible Cause
No output	Power-supply voltage. Defective transistor. Shorted component (check tuning capacitor in VFO). Open component. Severe load (check buffer amplifier). Defective crystal. Defective joint (check printed circuit board).
Reduced amplitude	Power-supply voltage low. Transistor bias (check resistors). Circuit loaded down (check buffer amplifier). Defective transistor.
Frequency unstable	Power-supply voltage changes. Defective connection (vibration test). Temperature sensitive (check with heat and/or cold spray). Tank circuit fault. Defect in *RC* network. Defective crystal. Load change (check buffer amplifier). Defective transistor.
Frequency error	Wrong power-supply voltage. Loading error (check buffer amplifier). Tank circuit fault (check trimmers and/or variable inductors). Defect in *RC* net work. Defective crystal. Transistor bias (check resistors).

c. A capacitor has changed value.
d. The transistors are defective.

46. A technician notes that a tool or a finger brought near a high-frequency oscillator tank circuit causes the output frequency to change.
    a. This is to be expected.
    b. It is a sign that the power supply is unstable.
    c. The tank circuit is defective.
    d. There is a bad transistor in the circuit.
47. A technician replaces the UJT in a relaxation oscillator. The circuit works, but the frequency is off a little. What is wrong?
    a. The new transistor is defective.
    b. The intrinsic standoff ratio is different.
    c. The resistors are burned out.
    d. The circuit is wired incorrectly.

11-8 Direct Digital Synthesis

There was a time when crystal-controlled oscillators were the best choice when accurate and stable high-frequency signals were needed. They are still a good choice when a small number of frequencies are required. When a large number of frequencies are needed, a phase-locked loop (PLL) frequency synthesizer or a direct digital synthesizer (*DDS*) is probably a better choice. Phase-locked loop synthesis is covered in Chap. 13.

A direct digital synthesizer can be used in place of a crystal-controlled oscillator. The basic advantage of direct digital synthesis is frequency agility. A DDS can be programmed to produce a large number of high-resolution frequencies. Sometimes, DDS systems are called *numerically controlled oscillators.* A DDS generates an output waveform that is a function of a clock signal and a tuning word that is in the form of a binary number. Figure 11-32 on page 310 shows the major parts of a DDS. A frequency-tuning word sets the value for the phase increment. On each clock pulse, the phase accumulator jumps to a new location in the sine lookup table. Each sine value is then sent from the lookup table to the digital-to-analog (D/A) converter, which produces a voltage that corresponds to a sine wave at some particular phase value. Note that the D/A output shown in Fig. 11-32 is an approximation of a sine wave. After the high-frequency components are filtered out, the output is close to being sinusoidal.

Figure 11-33 shows how the output frequency of a DDS is controlled. The clock frequency is fixed. What changes is the phase increment value. In the case of the top waveform, the phase increment is 30°. The phase increment is 45° for the bottom waveform. Notice that the smaller phase increment produces a lower output frequency. In Fig. 11-33, for the same number of clock pulses, the smaller phase increment produces exactly 1½ cycles of output, and the larger phase increment produces 2¼ cycles of output. **DDS**

The output frequency for a DDS is given by:

$$f_{\text{out}} = \frac{f_{\text{clock}} \cdot \Delta\phi}{2^N}$$

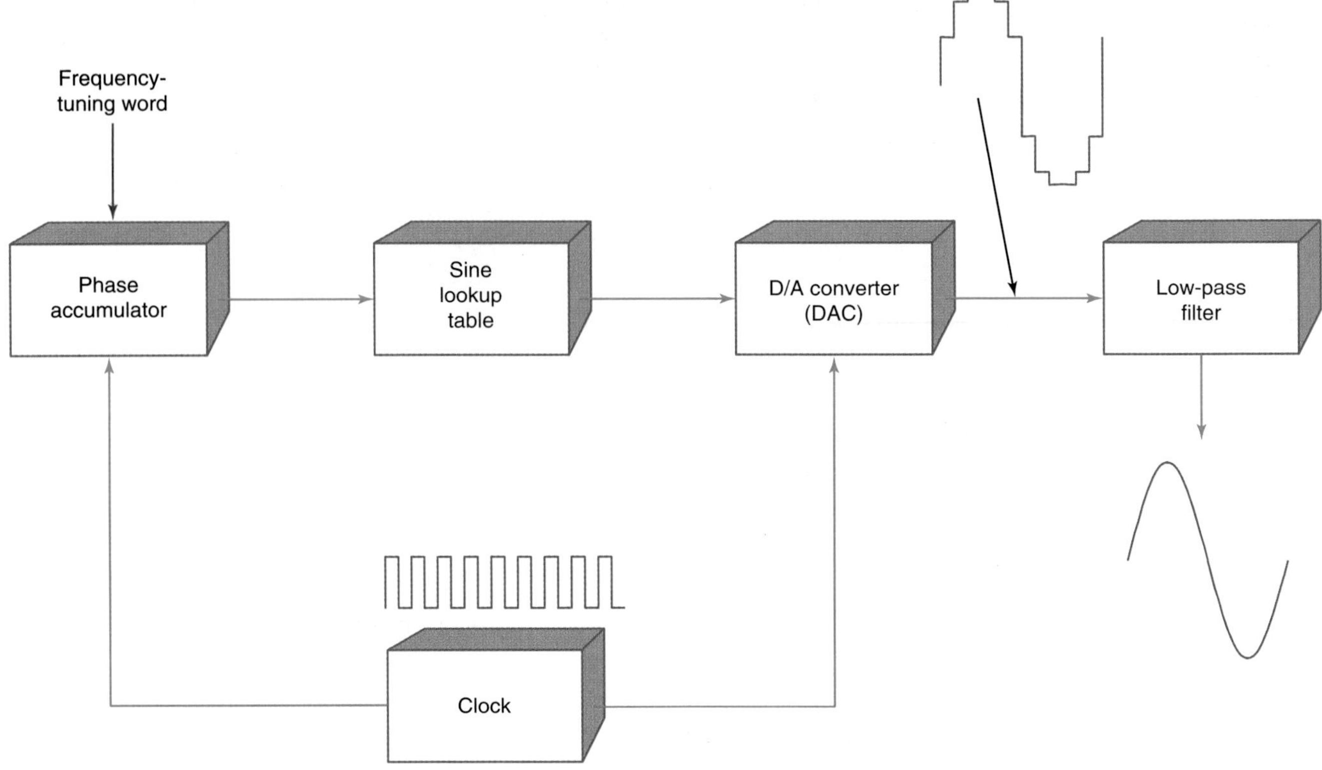

Fig. 11-32 DDS block diagram.

where f_{out} = the output frequency
f_{clock} = the clock frequency
$\Delta\phi$ = the phase increment value
N = the bit size of the phase accumulator

The word *bit* is a contraction for *b*inary dig*it*. Binary numbers have only two characters: 0 and 1. Some commercial DDS ICs have 32-bit phase accumulators and can operate at clock frequencies as high as 100 MHz or more. The

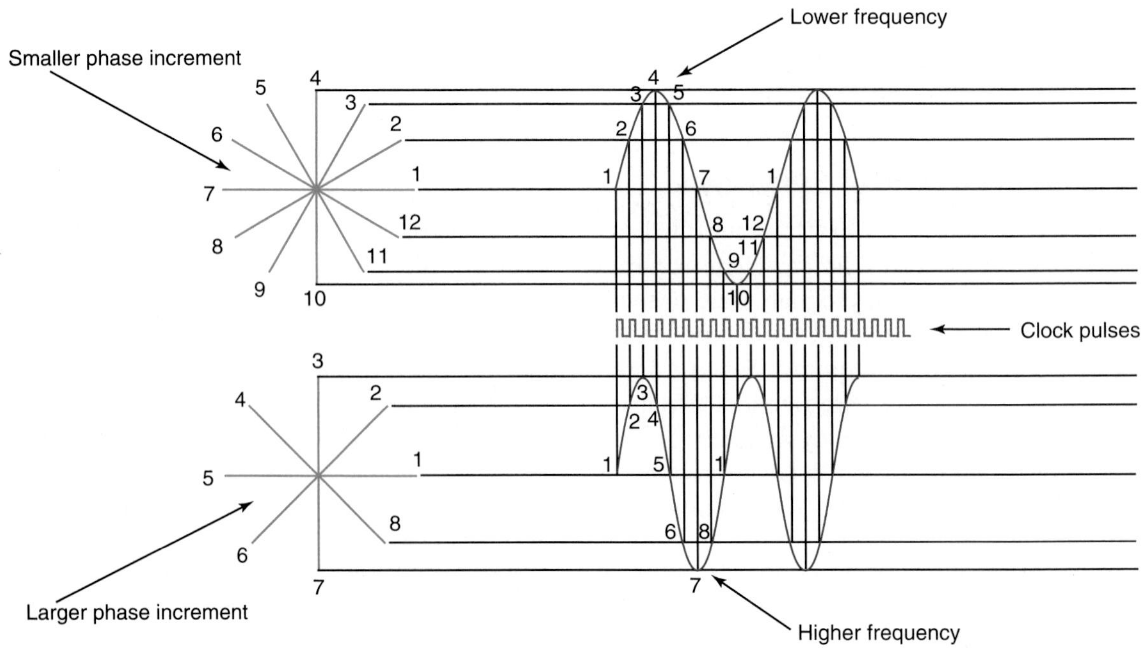

Fig. 11-33 How the phase increment value controls output frequency.

largest phase increment value would be equal to the size of the phase accumulator. In that case, the output frequency would be equal to the clock frequency. In practice, this is never done, and the phase increment value is always some integer value that is smaller than 2^N.

Example 11-6

Determine the output frequency for a DDS chip with a 32-bit phase accumulator and a clock frequency of 30 MHz. Assume that the frequency-tuning word programs the accumulator for a phase increment of 2^{30}.

$$f_{\text{out}} = \frac{f_{\text{clock}} \cdot \Delta\phi}{2^N} = \frac{30 \text{ MHz} \cdot 2^{30}}{2^{32}}$$
$$= 7.5 \text{ MHz}$$

Values like 2^{30} can be handled with a calculator that has an x^y key. Enter 2, then press the x^y key and then enter 30 and press the = key (display shows 1073741824). Also, you may notice in this example that exponents are subtracted when dividing, so the denominator becomes 2^2 (4), and 30 divided by 4 is 7.5.

Any frequency can be produced by programming the phase increment value with any integer value that is within the bit resolution of the phase accumulator. The frequency resolution of a DDS is the smallest possible frequency change:

$$f_{\text{resolution}} = \frac{f_{\text{clock}}}{2^N}$$

Example 11-7

Find the frequency resolution for a DDS with a 32-bit phase accumulator and a clock frequency of 100 MHz. Applying the equation:

$$f_{\text{resolution}} = \frac{f_{\text{clock}}}{2^N} = \frac{100 \times 10^6}{2^{32}}$$
$$= 0.0233 \text{ Hz}$$

This example is important because it demonstrates one of the most important features of DDS: *the ability to produce accurate, high-frequency signals in programmable millihertz steps.* Currently, DDS is the only technology available that can accomplish this.

As with so many other technologies that are gaining widespread use, DDS is now available using a single IC. Or, designs might use two or three ICs. In any case, the cost has decreased dramatically. Decreasing costs and increasing use go hand in hand when people are looking for technical solutions.

11-9 DDS Troubleshooting

As always, when troubleshooting, remember the word "GOAL." Exactly what are the symptoms that you observe? Some possibilities are:

- No output
- Reduced output amplitude
- Some frequencies are wrong
- All frequencies are wrong by a modest amount

After you observe, analyze, and then limit. Don't forget to use a system viewpoint. Look at Fig. 11-32. The frequency-tuning word is applied to the DDS chip via a parallel bus or a serial bus. In the case of a parallel bus, there might be as many as 22 frequency-control pins on the DDS IC. These pins will typically be controlled by a microprocessor. So, when some frequencies are wrong, it is possible that there is a bad connection or solder joint at one or more of the frequency control pins, or the DDS chip itself could be defective, or there could be a problem with the microprocessor (and that could be a software problem).

In the case of a serial bus, the DDS is programmed at one pin with a bit stream. A *bit stream* is a string of 0s and 1s that are sent to the serial pin one bit at a time. The bit stream is often supplied by a microprocessor. Here again, if the problem seems to be related to the frequency-control word, the source could be the DDS chip, the microprocessor, or the software.

Bit stream

In the case of no output, don't forget to check power supply voltages. Then, if they are OK, snoop around with an oscilloscope to verify the clock signal and the D/A output. In case these are both normal, the problem would be in the low-pass filter (Fig. 11-32). The low-pass filter is also worth checking in the case of reduced output amplitude. However, make sure that the output frequency is correct because, if it happens to be higher than normal, the low-pass filter will reduce the amplitude.

As Fig. 11-32 shows, the clock is applied to the phase accumulator and to the D/A converter. In some designs, the D/A converter is a separate IC. In any case, use an oscilloscope and/or a frequency counter to verify that the clock signals are present and that the clock frequency is correct. Since a relatively minor error in clock frequency will cause problems, the use of an accurate frequency counter is recommended.

TEST

Answer the following questions.

48. List three methods of producing high-stability, high-frequency signals.
49. What is another name for DDS?
50. What is used to smooth the output of the D/A converter in a DDS?
51. In any given DDS design, the clock frequency does not ________.
52. In any given DDS, the output frequency is changed by varying the ________.
53. What happens to the output frequency of a DDS when the phase increment value increases?
54. What is the output frequency for a DDS with a clock frequency of 5 MHz, a 32-bit phase accumulator, and a phase increment value of 85899346?
55. If the clock frequency in a DDS is out of tolerance, which output frequencies will suffer?

Chapter 11 Review

Summary

1. Oscillators convert direct current to alternating current.
2. Many oscillators are based on amplifiers with positive feedback.
3. The gain of the amplifier must be greater than the loss in the feedback circuit to produce oscillation.
4. The feedback must be in phase (positive) to produce oscillation.
5. It is possible to control the frequency of an oscillator by using the appropriate *RC* network.
6. The resonant frequency of a lead-lag network produces maximum output voltage and a 0° phase angle.
7. The Wien bridge oscillator uses a lead-lag network for frequency control.
8. It is possible to make the lead-lag network tunable by using variable capacitors or variable resistors.
9. Phase-shift oscillators use three *RC* networks, each giving a 60° phase angle.
10. An *LC* tank circuit can be used in very high frequency oscillator circuits.
11. A Hartley oscillator uses a tapped inductor in the tank circuit.
12. The Colpitts oscillator uses a tapped capacitive leg in the tank circuit.
13. A buffer amplifier will improve the frequency stability of an oscillator.
14. The series tuned Colpitts, or Clapp, circuit is noted for good frequency stability.
15. A varicap diode can be added to an oscillator circuit to provide a voltage-controlled oscillator.
16. A quartz crystal can be used to control the frequency of an oscillator.
17. Crystal oscillators are more stable in frequency than *LC* oscillators.
18. Crystals can operate in a series mode or a parallel mode.
19. The parallel frequency of a crystal is a little above the series frequency.
20. Crystals have a very high *Q*.
21. Relaxation oscillators produce nonsinusoidal outputs.
22. Relaxation oscillators can be based on negative-resistance devices such as the UJT.
23. Relaxation oscillator frequency can be predicted by *RC* time constants.
24. The intrinsic standoff ratio of a UJT will affect the frequency of oscillation.
25. The intrinsic standoff ratio of a programmable UJT can be set by the use of external resistors.
26. The astable multivibrator produces rectangular waves.
27. A nonsymmetrical multivibrator is produced by using different *RC* time constants for each base circuit.
28. Feedback amplifiers use frequency compensation to achieve stability.
29. Feedback signals can develop across the internal impedance of the power supply.
30. An *RC* network or a bypass capacitor is used to prevent feedback on power-supply lines.
31. High-frequency circuits often must be shielded to prevent feedback.
32. Oscillator symptoms include no output, reduced amplitude, instability, and frequency error.
33. Test instruments can load an oscillator circuit and cause errors.
34. Unstable circuits can be checked with vibration, heat, or cold.
35. A direct digital synthesizer produces a large number of precise frequencies.
36. Direct digital synthesizers are sometimes called numerically controlled oscillators.

Chapter Review Questions

Choose the letter that best answers each question.

11-1. An amplifier will oscillate if
a. There is feedback from output to input
b. The feedback is in phase (positive)
c. The gain is greater than the loss
d. All of the above are true

11-2. It is desired to build a common-emitter oscillator that operates at frequency f. The feedback circuit will be required to provide
a. 180° phase shift at f
b. 0° phase shift at f
c. 90° phase shift at f
d. Band-stop action for f

11-3. In Fig. 11-4, $R = 3300\ \Omega$ and $C = 0.1\ \mu$F. What is f_r?
a. 48 Hz
b. 120 Hz
c. 482 Hz
d. 914 Hz

11-4. Examine Fig. 11-4. Assume the signal source develops a frequency above f_r. What is the phase relationship of V_{out} to the source?
a. Positive (leading)
b. Negative (lagging)
c. In phase (0°)
d. None of the above

11-5. In Fig. 11-6, $R = 8200\ \Omega$ and $C = 0.05\ \mu$F. What is the frequency of oscillation?
a. 39 Hz
b. 60 Hz
c. 194 Hz
d. 388 Hz

11-6. Refer to Fig. 11-6. What is the function of R'?
a. It provides the required phase shift.
b. It prevents clipping and distortion.
c. It controls the frequency of oscillation.
d. None of the above.

11-7. In Fig. 11-9, assume that $R_B = 470$ kΩ and the voltage gain of the amplifier is 90. What is the actual loading effect of R_B to a signal arriving at the base?
a. 5222 Ω
b. 8333 Ω
c. 1 MΩ
d. Infinite

11-8. Refer to Fig. 11-9. Assume the phase-shift capacitors are changed to 0.1 μF. What is the frequency of oscillation?
a. 10 Hz
b. 40 Hz
c. 75 Hz
d. 71 Hz

11-9. Refer to Fig. 11-9. How many frequencies will produce exactly the phase response needed for the circuit to oscillate?
a. One
b. Two
c. Three
d. An infinite number

11-10. Refer to Fig. 11-14. What is the major effect of C_E?
a. It increases the frequency of oscillation.
b. It decreases the frequency of oscillation.
c. It makes the transistor operate common base.
d. It increases voltage gain.

11-11. Refer to Fig. 11-14. What would happen if C_2 were increased in capacity?
a. The frequency of oscillation would increase.
b. The frequency of oscillation would decrease.
c. The inductance of L_A and L_B would change.
d. Not possible to determine.

11-12. In Fig. 11-15, $L = 1.8\ \mu$H, $C_2 = 270$ pF, and $C_3 = 33$ pF. What is the frequency of oscillation?
a. 11 MHz
b. 22 MHz
c. 33 MHz
d. 41 MHz

11-13. Refer to Fig. 11-15. What is the purpose of C_1?
a. It bypasses power-supply noise to ground.
b. It determines the frequency of oscillation.
c. It provides an ac ground for the base.
d. It filters V_{out}.

11-14. Refer to Fig. 11-16. What is the configuration of Q_1?
a. Common square
b. Common gate
c. Common drain
d. Drain follower

11-15. Crystal-controlled oscillators, as compared to *LC*-controlled oscillators, are generally
a. Less expensive
b. Capable of a better output power
c. Superior for VFO designs
d. Superior for frequency stability

11-16. Why can the circuit of Fig. 11-19 not be used for overtone operation?
a. The common-emitter configuration is used.
b. Trimmer C_3 makes it impossible.

c. The feedback is wrong.
d. There is no *LC* circuit to select the overtone.

11-17. The *Q* of a crystal, as compared to the *Q* of an *LC* tuned circuit, will be
a. Much higher
b. About the same
c. Lower
d. Impossible to determine

11-18. In Fig. 11-21, $R_1 = 47\ \text{k}\Omega$ and $C = 10\ \mu\text{F}$. What is the approximate frequency of oscillation?
a. 0.21 Hz
b. 2.13 Hz
c. 200 Hz
d. 382 Hz

11-19. In Fig. 11-24, $R_1 = R_2 = 10{,}000\ \Omega$, $C_1 = 0.5\ \mu\text{F}$, and $C_2 = 0.02\ \mu\text{F}$. What is the frequency of oscillation?
a. 112 Hz
b. 279 Hz
c. 312 Hz
d. 989 Hz

11-20. In question 11-19, what will the rectangular output waveform show?
a. Symmetry
b. Nonsymmetry
c. Poor rise time
d. None of the above

11-21. Refer to Fig. 11-27. How may the stability of such a circuit be ensured?
a. Operate each stage at maximum gain.
b. Decrease losses in the feedback circuit.
c. Use more stages.
d. Compensate the circuit so the gain is low for those frequencies that give a critical phase error.

11-22. Refer to Fig. 11-28. How may signal coupling across Z_G be reduced?
a. By not allowing stages to share a ground path
b. By careful circuit layout
c. By using low-loss grounds
d. All of the above

11-23. What is the purpose of neutralization?
a. To ensure oscillations
b. To stabilize an amplifier
c. To decrease amplifier output
d. To prevent amplifier overload

Critical Thinking Questions

11-1. Are there any other ways to keep a PA system from oscillating aside from turning down the volume?

11-2. Can digital computer technology replace oscillators?

11-3. How could an oscillator be used as a metal detector?

11-4. Almost all timekeeping instruments use some form of an oscillator. Can you think of any that do not?

11-5. Quartz is not the only piezoelectric material. Does this fact suggest anything to you?

11-6. Can you name any electronic products that are oscillators but are called something else?

11-7. What is the most powerful electronic oscillator commonly found in homes and apartments?

Answers to Tests

1. *d*	11. *d*	21. *a*	31. *c*
2. *c*	12. *a*	22. *c*	32. *d*
3. *d*	13. *d*	23. *c*	33. *d*
4. *b*	14. *d*	24. *c*	34. *c*
5. *a*	15. *b*	25. *b*	35. *c*
6. *b*	16. *b*	26. *c*	36. *b*
7. *a*	17. *d*	27. *d*	37. *c*
8. *c*	18. *a*	28. *a*	38. *a*
9. *b*	19. *c*	29. *b*	39. *d*
10. *c*	20. *a*	30. *a*	40. *d*

41. *a*
42. *d*
43. *c*
44. *d*
45. *d*
46. *a*
47. *b*
48. crystal-controlled oscillator, PLL, DDS
49. numerically controlled oscillator
50. low-pass filter
51. change
52. frequency-tuning word (or phase increment value)
53. it increases
54. 0.1 MHz
55. all of them

Chapter 12 Radio Receivers

Chapter Objectives

This chapter will help you to:

1. *Define* modulation and demodulation.
2. *List* the characteristics of AM, SSB, and FM.
3. *Explain* the operation of basic radio receivers.
4. *Predict* the bandwidth of AM signals.
5. *Calculate* the oscillator frequency for superheterodyne receivers.
6. *Calculate* the image frequency for superheterodyne receivers.
7. *Troubleshoot* receivers.

Communications represents a large part of the electronics industry. This chapter introduces the basic ideas used in electronic communications. Once these basics are learned, it is easier to understand other applications such as television, two-way radio, telemetry, and digital data transmission. Modulation is the fundamental process of electronic communication. It allows voice, pictures, and other information to be transferred from one point to another. The modulation process is reversed at the receiver to recover the information. This chapter covers the basic theory and some of the circuits used in radio receivers and transmitters.

12-1 Modulation and Demodulation

Any high-frequency oscillator can be used to produce a *radio wave*. Figure 12-1 on page 318 shows an oscillator that feeds its output energy to an antenna. The antenna converts the high-frequency alternating current to a radio wave.

A radio wave travels through the atmosphere or space at the speed of light (3×10^8 meters per second). If a radio wave strikes another antenna, a high-frequency current will be induced that is a replica of the current flowing in the transmitting antenna. Thus, it is possible to transfer high-frequency electrical energy from one point to another without using wires. The energy in the receiving antenna is typically only a tiny fraction of the energy delivered to the transmitting antenna.

A radio wave can be used to carry information by a process called *modulation*. Figure 12-2 on page 318 shows a very simple type of modulation. A key switch is used to turn the antenna current (and thus the radio wave) on and off. This is the basic scheme of *radio telegraphy*. The key switch is opened and closed according to some pattern or *code*. For example, the Morse code can be used to represent numbers, letters, and punctuation. This basic modulation form is known as interrupted *continuous wave,* or CW. Continuous-wave modulation is very simple, but it has disadvantages. A code such as the Morse code is difficult to learn, transmission is slower than in voice communications, and CW cannot be used for music, pictures, and other kinds of information. Today, CW is used only by some amateur radio operators.

Radio telegraphy

Radio wave

Figure 12-3 on page 318 shows *amplitude modulation,* or *AM*. In this modulation system, the intelligence or information is used to *control the amplitude* of the RF signal. Amplitude modulation overcomes the disadvantages of CW

AM

Modulation

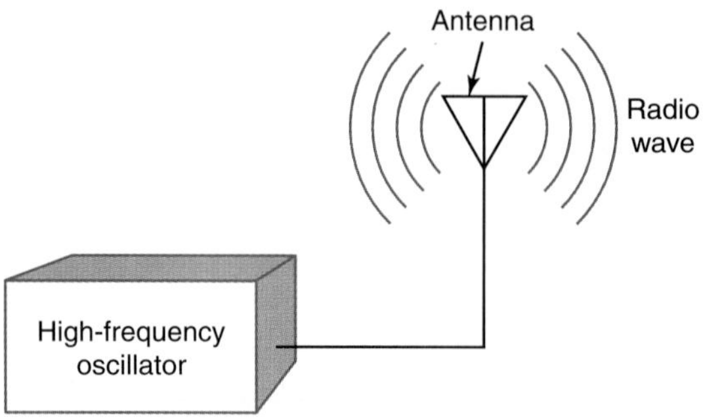

Fig. 12-1 A basic radio transmitter.

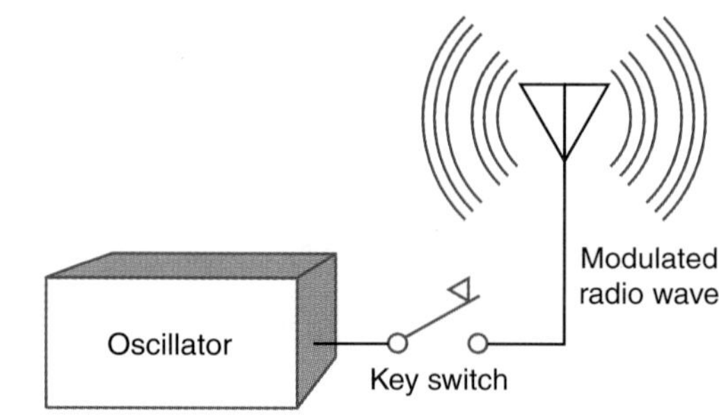

Fig. 12-2 A CW transmitter.

modulation. It can be used to transmit voice, music, data, or even picture information (video). The oscilloscope display in Fig. 12-3 shows that the RF signal amplitude varies in accordance with the audio frequency (AF) signal. The RF signal could just as well be amplitude-modulated by a video signal or digital (on-off) data.

Figure 12-4 shows a typical circuit for an amplitude modulator. The audio information is coupled by T_1 to the collector circuit of the transistor. The audio voltage induced across the secondary of T_1 can either aid V_{CC} or oppose V_{CC} depending on its phase at any given moment. This means that the collector supply

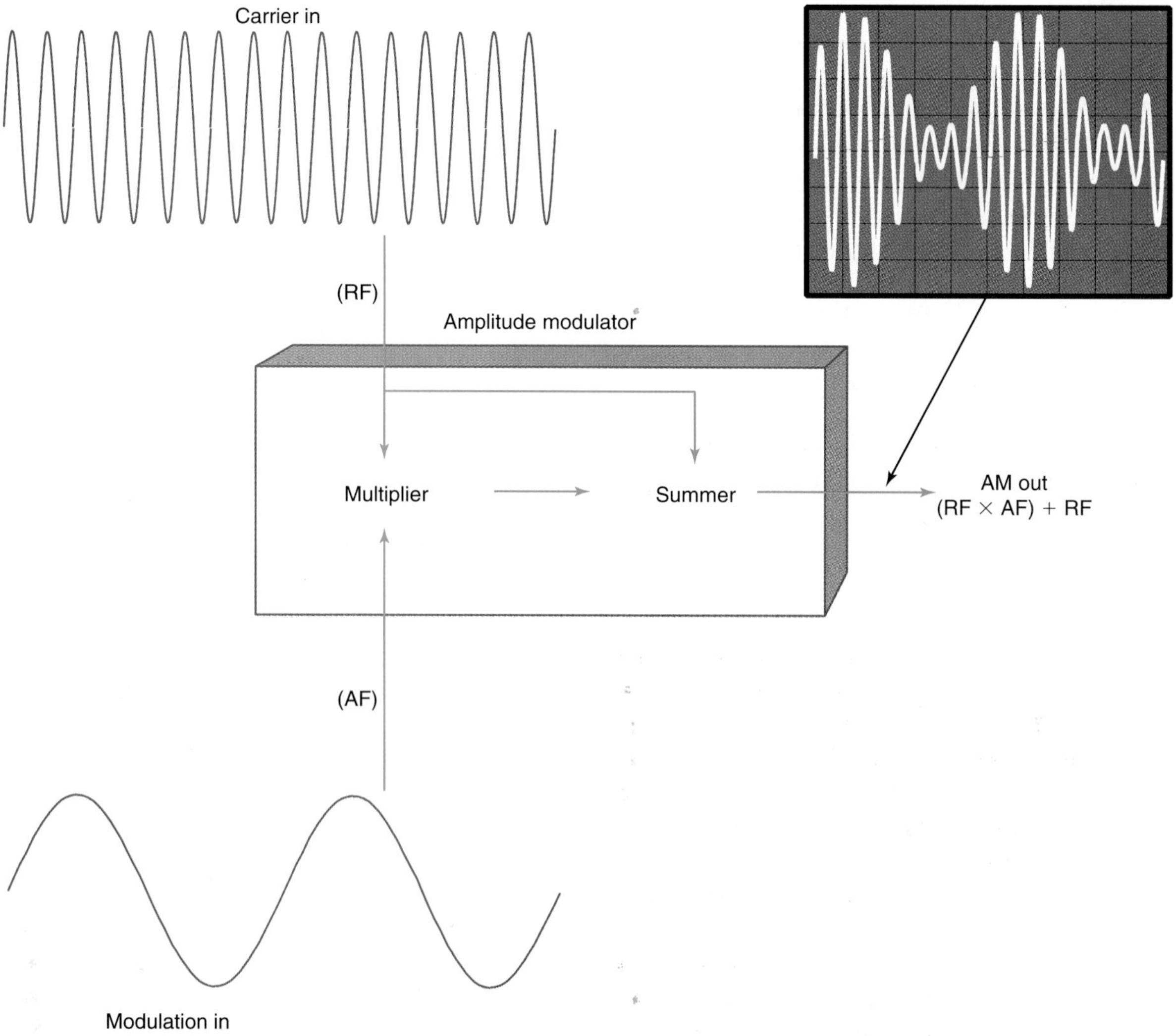

Fig. 12-3 Amplitude modulation.

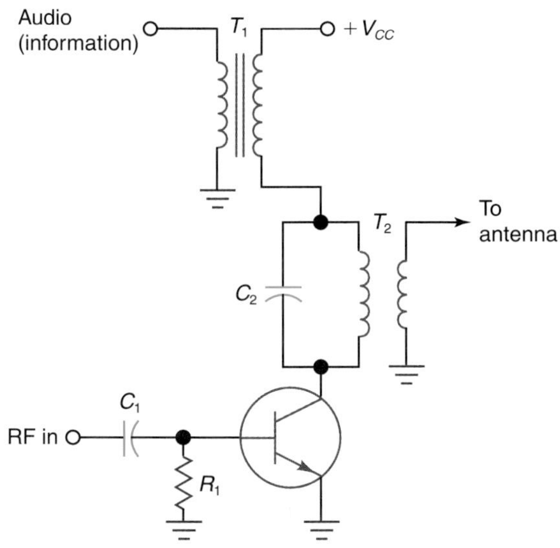

Fig. 12-4 An amplitude modulator.

for the transistor is not constant. It is varying with the audio input. This is how the amplitude control is achieved.

As an example, suppose that V_{CC} in Fig. 12-4 is 12 V and that the induced audio signal across the secondary of T_1 is 24 V peak-to-peak. When the audio peaks negative at the top of the secondary winding, 12 V will be added to V_{CC} and the transistor will see 24 V. When the audio peaks positive at the top of the secondary winding, 12 V will be subtracted from V_{CC} and the transistor will see 0 V.

Transformer T_2 and capacitor C_2 in Fig. 12-4 form a resonant *tank circuit.* The resonant frequency will match the RF input. Capacitor C_1 and resistor R_1 form the input circuit for the transistor. Reverse bias is developed by the base-emitter junction, and the amplifier operates in class C. Signal bias was covered in Chap. 8.

An amplitude-modulated signal consists of several frequencies. Suppose that the signal from a 500-kHz oscillator is modulated by a 3-kHz audio tone. Three frequencies will be present at the output of the modulator. The original RF oscillator signal, called the *carrier,* is shown at 500 kHz on the frequency axis in Fig. 12-5. Also note that an *upper sideband* (USB) appears at 503 kHz and a *lower sideband* (LSB) appears at 497 kHz. An AM signal consists of a carrier plus two sidebands.

Figure 12-5 is in the type of display shown on a *spectrum analyzer.* A spectrum analyzer uses a cathode-ray-tube display similar to an oscilloscope. The difference is that the spectrum analyzer draws a graph of amplitude versus *frequency.* An oscilloscope draws a graph of amplitude versus *time.* Spectrum analyzers display the *frequency domain* while oscilloscopes display the *time domain.* Figure 12-6 shows how an AM signal looks on an oscilloscope. In Fig. 12-6(a) the carrier frequency is relatively low, so the individual cycles can be seen. In practice, the carrier frequency is relatively high and the individual cycles cannot be seen [Fig. 12-6(b)]. Spectrum analyzers are generally more costly than oscilloscopes. They

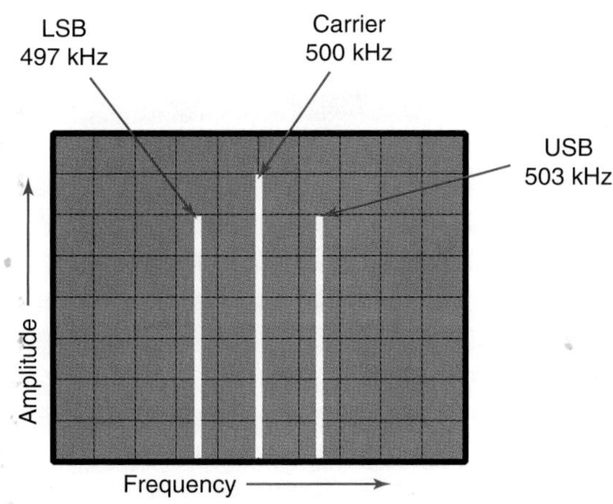

Fig. 12-5 AM on a spectrum analyzer.

Tank circuit

Carrier

Sidebands

Spectrum analyzer

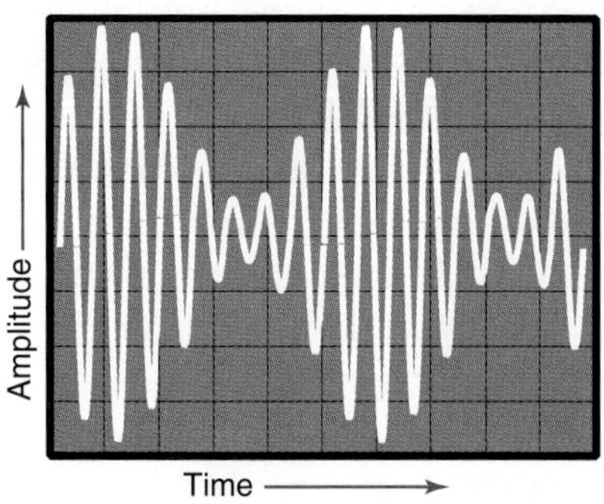

(a) When the carrier frequency is relatively low

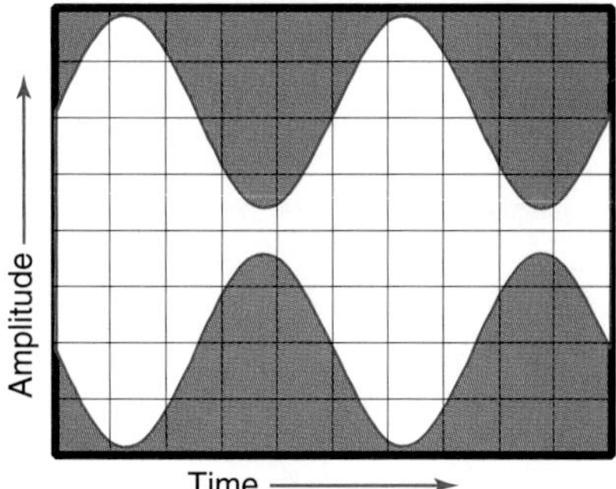

(b) When the carrier frequency is relatively high

Fig. 12-6 AM waveforms on an oscilloscope.

are very useful instruments for evaluating the frequency content of signals.

Bandwidth

Since AM signals have sidebands, they must also have *bandwidth.* An amplitude-modulated signal will occupy a given portion of the available spectrum of frequencies. The sidebands appear above and below the carrier according to the frequency of the modulating information. If someone whistles at 1 kHz into the microphone of an AM transmitter, an upper sideband will appear 1 kHz above the carrier frequency and a lower sideband will appear 1 kHz below the carrier frequency. The bandwidth of an AM signal is *twice* the modulating frequency. For example, frequencies up to about 3.5 kHz are required for speech. An AM voice transmitter will have a minimum bandwidth of 7 kHz (2 × 3.5 kHz).

Example 12-1

What is the bandwidth of the AM signal shown in Fig. 12-5? It can be found by subtracting:

$$\begin{aligned} \text{BW} &= f_{\text{high}} - f_{\text{low}} \\ &= 503 \text{ kHz} - 497 \text{ kHz} = 6 \text{ KHz} \end{aligned}$$

Bandwidth is important because it limits the number of stations that can use a range of frequencies without interference. For example, the standard AM broadcast band has its lowest channel assigned to a carrier frequency of 540 kHz and its highest channel assigned to a carrier frequency of 1600 kHz. The channels are spaced 10 kHz apart, and this allows:

$$\begin{aligned} \text{No. of channels} &= \frac{1600 \text{ kHz} - 540 \text{ kHz}}{10 \text{ kHz}} + 1 \\ &= 107 \text{ channels} \end{aligned}$$

However, each station may modulate with audio frequencies up to 15 kHz, so the total bandwidth required for one station is twice this frequency or 30 kHz. With 107 stations on the air, the total bandwidth required would be 107 × 30 kHz = 3210 kHz. This far exceeds the width of the AM broadcast band.

FCC

One solution would be to limit the maximum audio frequency to 5 kHz. This would allow the 107 stations to fit the AM band. This is not an acceptable solution since 5 kHz is not adequate for reproduction of quality music. A better solution is to assign channels on the basis of geographical area. The Federal Communications Commission (*FCC*) assigns carrier frequencies that are spaced at least three channels apart in any given geographical region. The three-channel spacing separates the carriers by 30 kHz and prevents the upper sideband from a lower channel from spilling into the lower sideband of the channel above it.

Demodulation

Detection

An AM radio receiver must recover the information from the modulated signal. This process reverses what happened in the modulator section of the transmitter. It is called *demodulation* or *detection.*

The most common AM detector is a diode (Fig. 12-7). The modulated signal is applied across the primary of T_1. Transformer T_1 is tuned by capacitor C_1 to the carrier frequency. The passband of the tuned circuit is wide enough to pass the carrier and both sidebands. The diode *detects* the signal and recovers the original information used to modulate the carrier at the transmitter. Capacitor C_2 is a low-pass filter. It removes the carrier and sideband frequencies since they are no longer needed. Resistor R_L serves as the load for the information signal.

Nonlinear

A diode makes a good detector because it is a *nonlinear* device. All nonlinear devices can

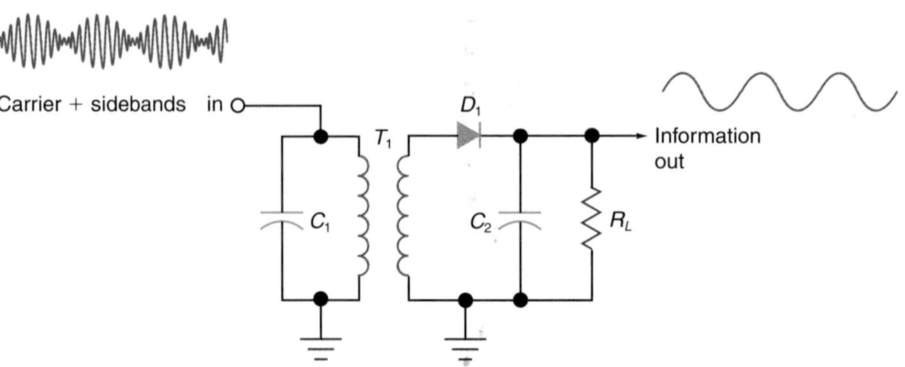

Fig. 12-7 An AM detector.

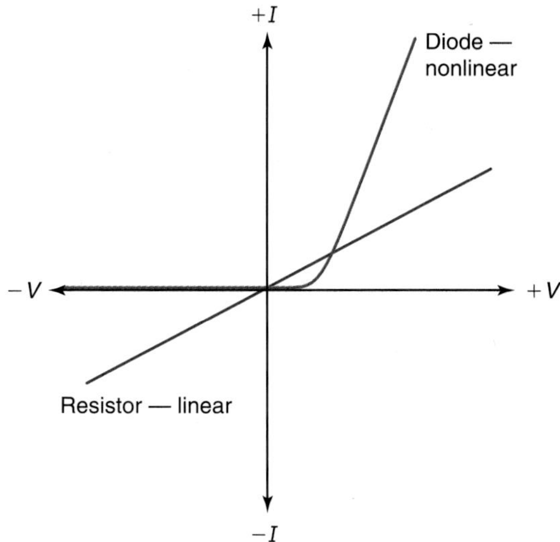

Fig. 12-8 Diodes are nonlinear devices.

be used to detect AM. Figure 12-8 is a volt-ampere characteristic curve of a solid-state diode. It shows that a diode will make a good detector and a resistor will not.

Nonlinear devices produce *sum and difference* frequencies. For example, if a 500-kHz signal and a 503-kHz signal both arrive at a nonlinear device, several new frequencies will be generated. One of these is the sum frequency at 1003 kHz. In detectors, the important one is the *difference* frequency, which will be at 3 kHz for our example. Refer again to Fig. 12-5. The spectrum display shows that modulating a 500-kHz signal with a 3-kHz signal produces an upper sideband at 503 kHz. When this signal is detected, the modulation process is reversed and the original 3-kHz signal is recovered.

Sum and difference frequencies

The lower sideband will also interact with the carrier. It, too, will produce a difference frequency of 3 kHz (500 kHz − 497 kHz = 3 kHz). The two 3-kHz difference signals add in phase in the detector. Thus, in an AM detector, both sidebands mix with the carrier and reproduce the original information frequencies.

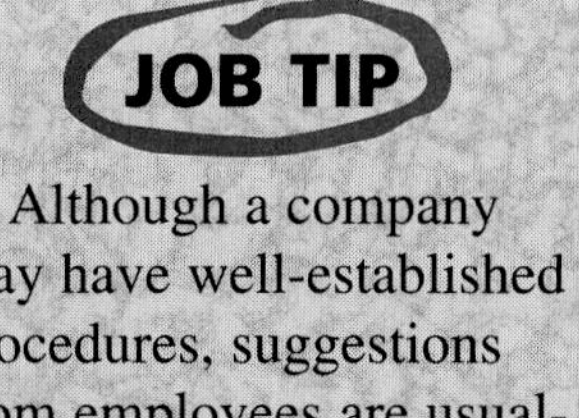

A transistor can also serve as an AM detector (Fig. 12-9). The circuit shown is a common-emitter amplifier. Transformer T_1 and capacitor C_1 form a resonant circuit to pass the modulated signal (carrier plus sidebands). Capacitor C_4 is added to give a low-pass filter action, since the high-frequency carrier and the sidebands are no longer needed after detection.

BJTs can demodulate signals because they are also nonlinear devices. The base-emitter junction is a diode. The transistor detector has the advantage of producing gain. This means that the circuit of Fig. 12-9 will produce more information amplitude than the simple diode detector of Fig. 12-7. Both circuits are useful for detecting AM signals.

The modulation-demodulation process is the basis of all electronic communication. It allows high-frequency carriers to be placed at different frequencies in the RF spectrum. By spacing the carriers, interference can be controlled. The use of different frequencies also allows different communication distances to be covered. Some frequencies lend themselves to short-range work and others are better for long-range communications.

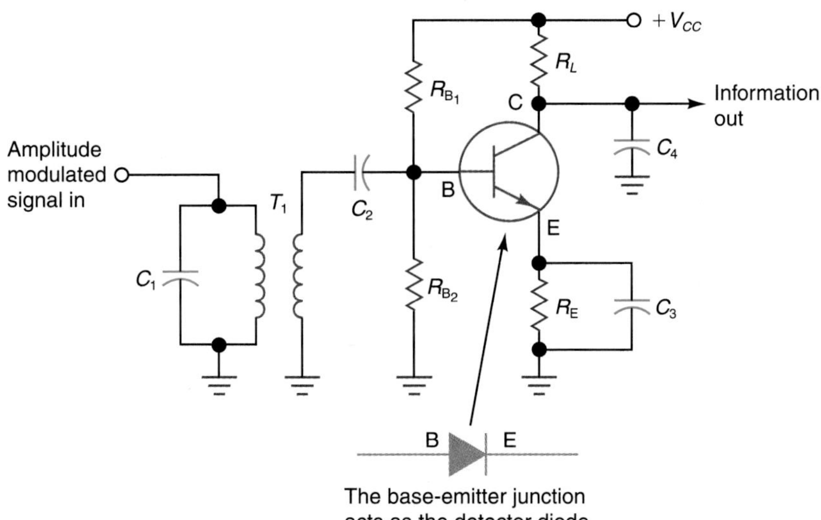

Fig. 12-9 A transistor detector.

TEST

Choose the letter that best completes each statement.

1. A circuit used to place information on a radio signal is called
   a. An oscillator
   b. A detector
   c. An antenna
   d. A modulator
2. A CW transmitter sends information by
   a. Varying the frequency of the audio signal
   b. Interrupting the radio signal
   c. Use of a microphone
   d. Use of a camera
3. Refer to Fig. 12-4. The voltage at the top of C_2 will
   a. Always be equal to V_{CC}
   b. Vary with the information signal
   c. Be controlled by the transistor
   d. Be a constant 0 V with respect to ground
4. Refer to Fig. 12-4. Capacitor C_2 will resonate the primary of T_2
   a. At the radio frequency
   b. At the audio frequency
   c. At all frequencies
   d. None of the above
5. A 2-MHz radio signal is amplitude-modulated by an 8-kHz sine wave. The frequency of the lower sideband is
   a. 2.004 MHz
   b. 2.000 MHz
   c. 1.996 MHz
   d. 1.992 MHz
6. A 1.2-MHz radio transmitter is to be amplitude-modulated by audio frequencies up to 9 kHz. The bandwidth required for the signal is
   a. 9 kHz
   b. 18 kHz
   c. 27 kHz
   d. 1.2 MHz
7. The electronic instrument used to show

About Electronics

Communications Technology

Digital communication systems might use a combination of phase and amplitude modulation. The GPS (global positioning satellite) system requires highly accurate time signals from an atomic clock. Using GPS, passenger cars can now be equipped with dashboard-mounted navigation systems such as the device pictured.

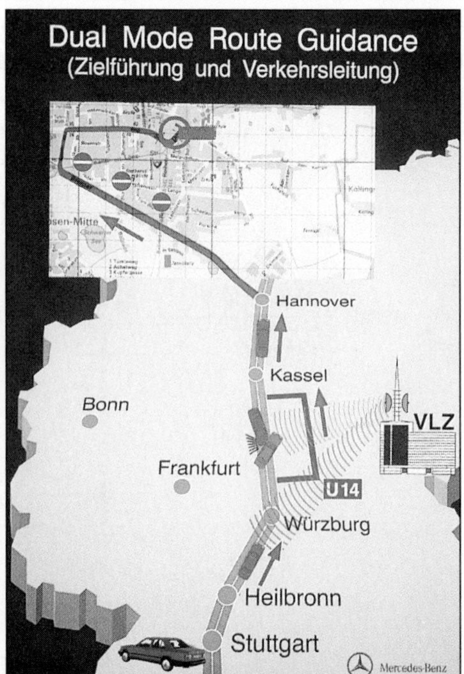

both the carrier and the sidebands of a modulated signal in the frequency domain is the
a. Spectrum analyzer
b. Oscilloscope
c. Digital counter
d. Frequency meter
8. Refer to Fig. 12-7. The carrier input is 1.5000 MHz, the upper sideband (USB) input is 1.5025 MHz, and the lower sideband (LSB) is 1.4975 MHz. The frequency of the detected output is
a. 1.5 MHz
b. 5.0 kHz
c. 2.5 kHz
d. 0.5 kHz
9. Diodes make good AM detectors because
a. They rectify the carrier.
b. They rectify the upper sideband.
c. They rectify the lower sideband.
d. They are nonlinear and produce difference frequencies.

12-2 Simple Receivers

Figure 12-10 shows the most basic form of an AM radio receiver. An antenna is necessary to intercept the radio signal and change it back into an electric signal. The diode detector mixes the sidebands with the carrier and produces the audio information. The headphones convert the audio signal into sound. The ground completes the circuit and allows the currents to flow.

Obviously, a receiver as simple as the one shown in Fig. 12-10 must have shortcomings. Such receivers do work but are not practical.

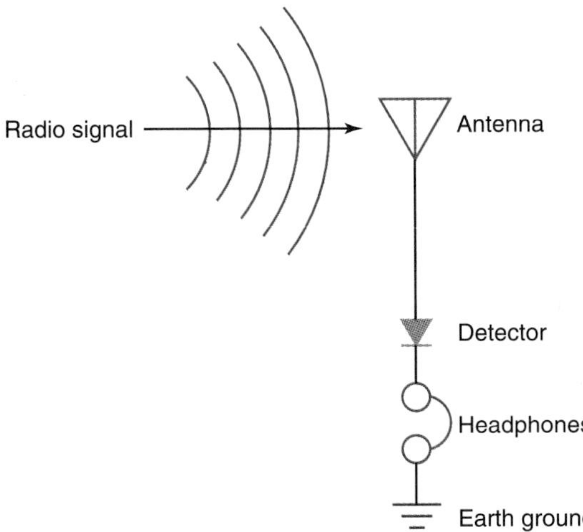

Fig. 12-10 A simple radio receiver.

They cannot receive weak signals (they have poor *sensitivity*). They cannot separate one carrier frequency from another (they have no *selectivity*). They are inconvenient because they require a long antenna, an earth ground, and headphones.

Sensitivity

Selectivity

Before we leave the simple circuit of Fig. 12-10, one thing should be mentioned. You have, no doubt, become used to the idea that electronic circuits require some sort of power supply. This is still the case. A radio signal is a wave of pure energy. Thus, the signal is the source of energy for this simple circuit.

For information on ham Radio, contact the American Radio Relay League or visit their Web site.

The problem of poor sensitivity can be overcome with gain. We can add some amplifiers to the receiver to make weak signals detectable. Of course, the amplifiers will have to be energized. A power supply, other than the weak signal itself, will be required. As the gain is increased, the need for a long antenna is decreased. A small antenna is not as efficient, but gain can overcome this deficiency. Gain can also do away with the need for the headphones. Audio amplification after the detector can make it possible to drive a loudspeaker. This makes the receiver much more convenient to use.

What about the lack of selectivity? Radio stations operate at different frequencies in any given location. This makes it possible to use band-pass filters to select one out of the many that are transmitting. The resonant point of the filter may be adjusted to agree with the desired station frequency.

Figure 12-11 on page 324 shows a receiver that overcomes some of the problems of the simple receiver. A two-stage audio amplifier has been added to allow loudspeaker operation. A *tuned circuit* has been added to allow selection of one station at a time. This receiver will perform better.

Tuned circuit

The circuit of Fig. 12-11 is an improvement, but it is still not practical for most applications. One tuned circuit will not give enough selectivity. For example, if there is a very strong station in the area, it will not be possible to reject it. The strong station will be heard at all settings of the variable capacitor.

Selectivity can be improved by using more tuned circuits. Figure 12-12 on page 324 compares the selectivity curves for one, two, and three tuned circuits. Note that more tuned circuits give a sharper curve (less bandwidth). This improves the ability to reject unwanted

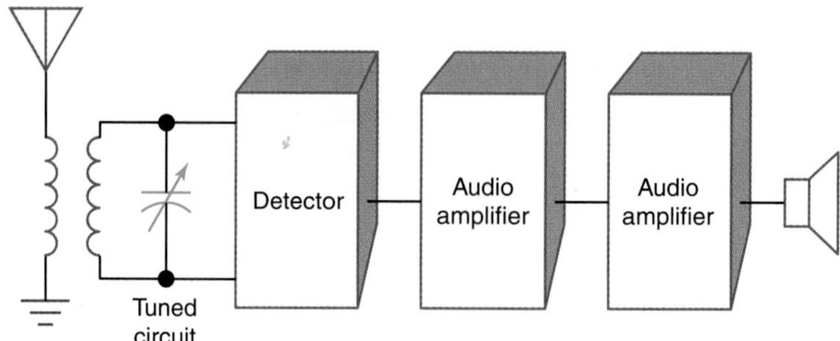

Fig. 12-11 An improved radio receiver.

frequencies. Figure 12-12 also shows that bandwidth is measured 3 dB down from the point of maximum gain. An AM receiver should have a bandwidth just wide enough to pass the carrier and both sidebands. A bandwidth of about 20 kHz is typical in an ordinary AM broadcast receiver. Too much bandwidth means poor selectivity and possible interference. Too little bandwidth means loss of transmitted information (with high-frequency audio affected the most).

TRF receiver

A tuned radio-frequency *(TRF) receiver* can provide reasonably good selectivity and sensitivity (Fig. 12-13). Four amplifiers—two at radio frequencies and two at audio frequencies—give the required gain.

Tracking

The TRF receiver has some disadvantages. Note in Fig. 12-13 that all three tuned circuits are gang-tuned. In practice, it is difficult to achieve perfect tracking. *Tracking* refers to how closely the resonant points will be matched for all settings of the tuning control. A second problem is in bandwidth. The tuned circuits will not have the same bandwidth for all frequencies. Both of these disadvantages have been eliminated in the superheterodyne receiver design that is discussed in the next section.

TEST

Choose the letter that best answers each question.

10. What energizes the radio receiver shown in Fig. 12-10?
    a. The earth ground
    b. The diode
    c. The headphones
    d. The incoming radio signal
11. To improve selectivity, the bandwidth of a receiver can be reduced by which of the following methods?
    a. Using more tuned circuits
    b. Using fewer tuned circuits
    c. Adding more gain
    d. Using a loudspeaker

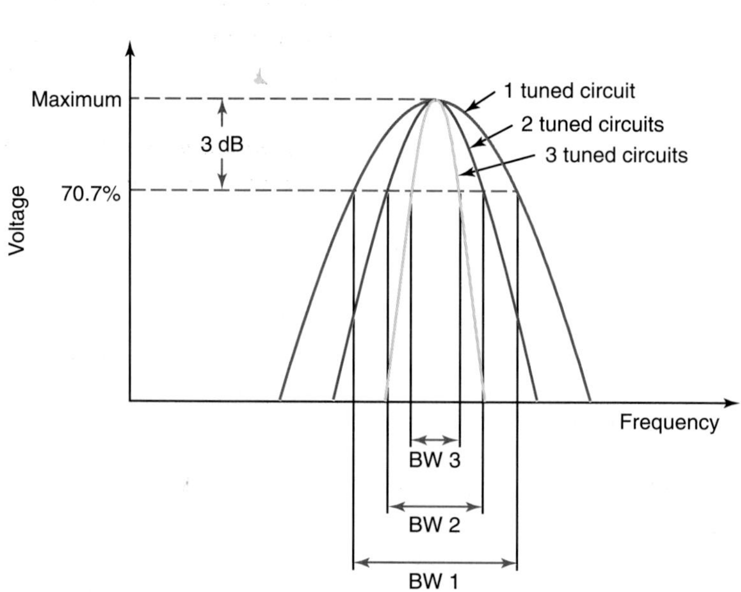

Fig. 12-12 Selectivity can be improved with more tuned circuits.

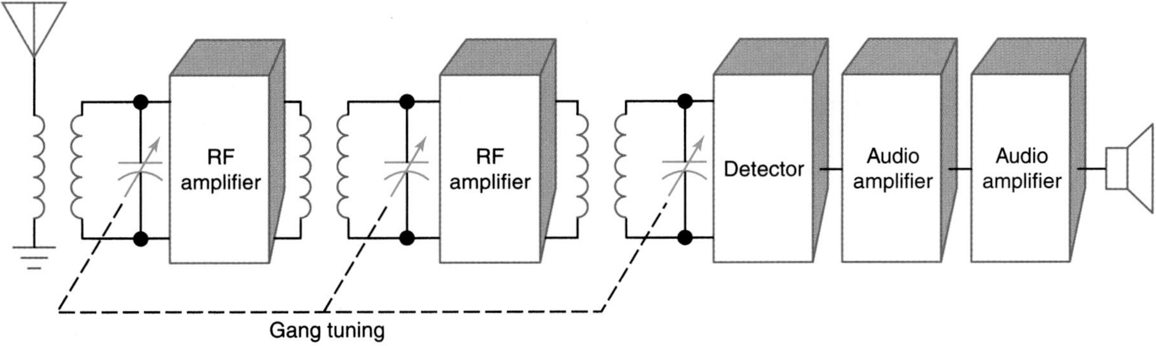

Fig. 12-13 A TRF receiver.

12. A 250-kHz tuned circuit is supposed to have a bandwidth of 5 kHz. It is noted that the gain of the circuit is maximum at 250 kHz and it drops about 30 percent (3 dB) at 252.5 kHz and at 247.5 kHz. What can be concluded about the tuned circuit?
    a. It is not as selective as it should be.
    b. It is more selective than it should be.
    c. It is not working properly.
    d. None of the above.
13. Refer to Fig. 12-13. How is selectivity achieved in this receiver?
    a. In the detector stage
    b. In the gang-tuned circuits
    c. In the audio amplifier
    d. All of the above

12-3 Superheterodyne Receivers

The major difficulties with the TRF receiver design can be eliminated by fixing some of the tuned circuits to a single frequency. This will eliminate the tracking problem and the changing-bandwidth problem. This fixed frequency is called the *intermediate frequency,* or simply IF. It must lie outside the band to be received. Then, any signal that is to be received must be converted to the intermediate frequency. The conversion process is called *mixing* or *heterodyning.* A *superheterodyne receiver* converts the received frequency to the intermediate frequency.

Figure 12-14 shows the basic operation of a heterodyne converter. Here, the letters *A* and *B* represent frequencies. When signals at two different frequencies are applied, new frequencies are produced. The output of the converter contains the *sum and the difference frequencies* in addition to the original frequencies. Any nonlinear device (such as a diode) can be used to heterodyne or mix two signals. The process is the same as AM detection. However, the *purpose* is different. Detection is the proper term to use when information is being recovered from a signal. The terms heterodyning or mixing are used when a signal is being converted to another frequency such as an intermediate frequency.

Most superheterodyne circuits use a transistor mixer rather than a diode. This is because the transistor provides gain. In some cases it can also supply one of the two signals needed for mixing.

Figure 12-15 on page 326 shows a block diagram of a superheterodyne receiver. An *oscillator* provides a signal to mix with signals coming from the antenna. The mixer output contains sum and difference frequencies. If any of the signals present at the mixer output is at or very near the intermediate frequency, then that signal will reach the detector. All other frequencies will be rejected because of the selectivity of the IF amplifiers. The standard intermediate frequencies are

Oscillator

Intermediate frequency

Mixing or heterodyning

Sum and difference frequencies

1. *Amplitude modulation (AM) broadcast band:* 455 kHz (or 262 kHz for some automotive receivers)

Signal *A* → Heterodyne converter (nonlinear device) → *A*, *B*, *A* + *B*, *A* − *B*; Signal *B* ↑

Fig. 12-14 Operation of a heterodyne converter.

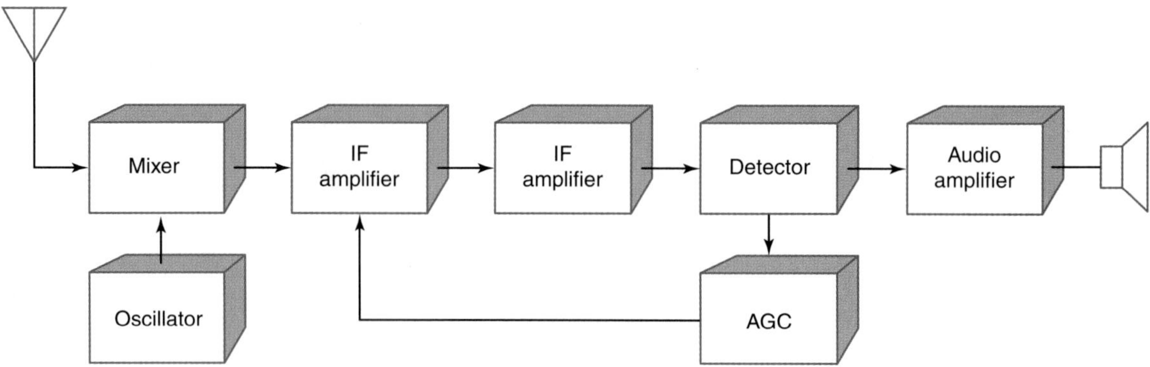

Fig. 12-15 Block diagram of a superheterodyne receiver.

2. *Frequency modulation (FM) broadcast band:* 10.7 MHz
3. *Television broadcast band:* 44 MHz

Shortwave and communication receivers may use various intermediate frequencies, for example, 455 kHz, 1.6 MHz, 3.35 MHz, 9 MHz, 10.7 MHz, 40 MHz, and others.

The oscillator in a superheterodyne receiver is usually set to run above the received frequency by an amount equal to the IF. To receive a station at 1020 kHz, for example, on a standard AM broadcast receiver:

Image interference

$$\text{Oscillator frequency} = 1020 \text{ kHz} + 455 \text{ kHz}$$
$$= 1475 \text{ kHz}$$

The oscillator signal at 1475 kHz and the station signal at 1020 kHz will mix to produce sum and difference frequencies. The difference signal will be in the IF passband (those frequencies that the IF will allow to pass through) and will reach the detector. Another station operating at 970 kHz can be rejected by this process. Its difference frequency will be

Image rejection

$$1475 \text{ kHz} - 970 \text{ kHz} = 505 \text{ kHz}$$

Since 505 kHz is not in the passband of the 455-kHz IF stages, the station transmitting at 970 kHz is rejected.

Trimmer capacitor

Example 12-2

A communication receiver has an IF of 9 MHz. What is the frequency of its oscillator when it is tuned to 15 MHz? Since the oscillator usually runs above the IF:

$$f_{osc} = f_{receive} + \text{IF} = 15 \text{ MHz} + 9 \text{ MHz}$$
$$= 24 \text{ MHz}$$

It should be clear that adjacent channels are rejected by the selectivity in the IF stages. However, there is a possibility of interference from a signal not even in the broadcast band. To receive 1020 kHz, the oscillator in the receiver must be adjusted 455 kHz higher. What will happen if a shortwave signal at a frequency of 1930 kHz reaches the antenna? Remember, the oscillator is at 1475 kHz. Subtraction shows that

$$1930 \text{ kHz} - 1475 \text{ kHz} = 455 \text{ kHz}$$

This means that the shortwave signal at 1930 kHz will mix with the oscillator signal and reach the detector. This is called *image interference*.

The only way to reject image interference is to use selective circuits *before* the mixer. In any superheterodyne receiver, there are always two frequencies that can mix with the oscillator frequency and produce the intermediate frequency. One is the desired frequency, and the other is the image frequency. The image must not be allowed to reach the mixer.

Figure 12-16 shows how *image rejection* is achieved. The antenna signal is transformer-coupled to a tuned circuit before the mixer. This circuit is tuned to resonate at the station frequency. Its selectivity will reject the image. A dual or ganged capacitor is used to simultaneously adjust the oscillator and the mixer-tuned circuit. *Trimmer capacitors* are also included so that the two circuits can track each other. These trimmers are adjusted only once. They are set at the factory and usually will never need to be readjusted.

The mixer-tuned circuit is not highly selective. Its purpose has nothing to do with the adjacent channel selectivity of the receiver. This selectivity is provided in the IF stages. The

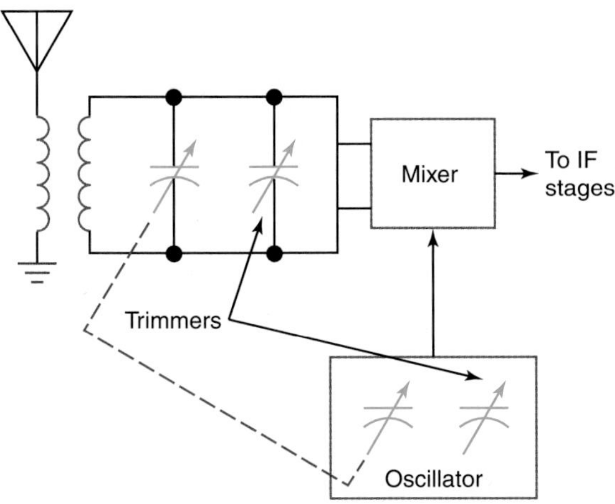

Fig. 12-16 A tuned circuit before the mixer rejects the image.

purpose of the mixer-tuned circuit is to reject the image frequency which is twice the intermediate frequency above the desired station frequency. Thus the desired station and the image are separated by 910 kHz (2 × 455 kHz) in a standard AM broadcast receiver. A signal this far away from the desired one is easier to reject so one tuned circuit before the mixer is often all that is needed.

Example 12-3

An FM receiver is tuned to receive a station at 91.9 MHz. Find the image frequency. We can assume an IF of 10.7 MHz and that the local oscillator runs above the desired frequency. The image frequency is found by:

$$\begin{aligned} f_{\text{image}} &= f_{\text{station}} + 2 \times \text{IF} \\ &= 91.9 \text{ MHz} + 2 \times 10.7 \text{ MHz} \\ &= 113.3 \text{ MHz} \end{aligned}$$

The block diagram of Fig. 12-15 shows an *automatic gain control (AGC)* stage. It may also be called *automatic volume control (AVC)*. This stage develops a control voltage based on the stength of the signal reaching the detector. The control voltage, in turn, adjusts the gain of the first IF amplifier. The purpose of AGC is to maintain a relatively constant output from the receiver. Signal strengths can vary quite a bit as the receiver is tuned across the band. The AGC action keeps the volume from the speaker reasonably constant.

Automatic gain control can be applied to more than one IF amplifier. It can also be applied to an RF amplifier before the mixer if a receiver has one. The control voltage is used to vary the gain of the amplifying device. If the device is a BJT, two options exist. The graph of Fig. 12-17 shows that maximum gain occurs at one value of collector current. If the bias is increased and current increases, the gain tends to drop. This is called *forward* AGC. The bias can be reduced, the current decreases, and so does the gain. This is called reverse AGC. Both types of AGC are used with bipolar transistors.

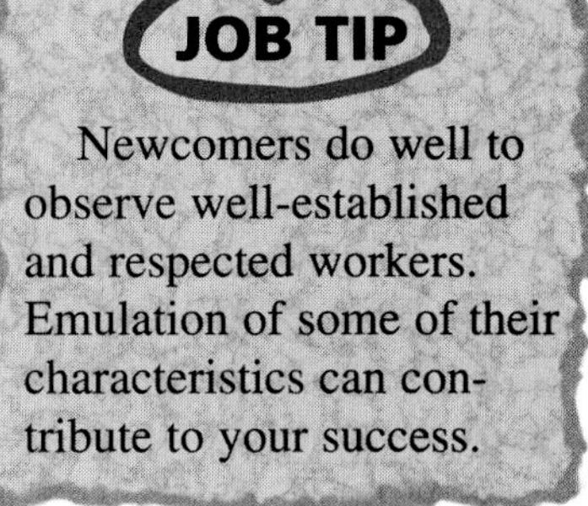

Different transistors vary quite a bit in their AGC bias characteristics. This is an important consideration when replacing an RF or IF transistor in a receiver. If AGC is applied to that stage, an exact replacement is highly desirable. A substitute transistor may cause poor AGC performance and the receiver performance can be seriously degraded.

Dual-gate MOSFETs are often used when AGC is desired. These transistors have excellent AGC characteristics. The control voltage is usually applied to the second gate. You may wish to refer to Sec. 7-3 on field-effect transistor amplifiers.

Integrated circuits with excellent AGC characteristics are also available. These are widely applied in receiver design, especially as IF amplifiers.

AGC

AVC

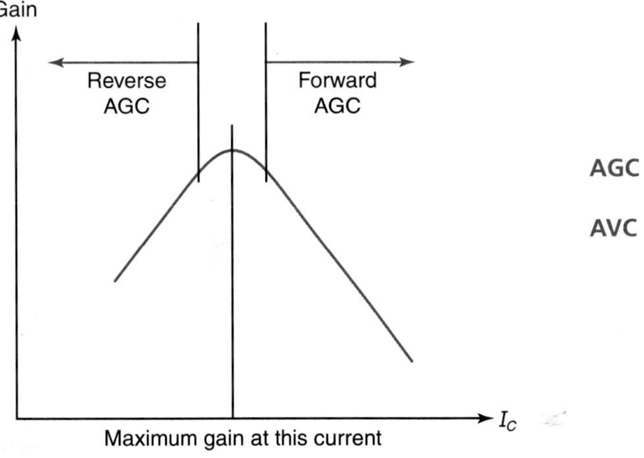

Fig. 12-17 The AGC characteristics of a transistor.

TEST

Choose the letter that best answers each question.

14. It is desired to receive a station at 1160 kHz on a standard AM receiver. What must the frequency of the local oscillator be?
    a. 455 kHz
    b. 590 kHz
    c. 1615 kHz
    d. 2000 kHz
15. A standard AM receiver is tuned to 1420 kHz. Interference is heard from a shortwave transmitter operating at 2330 kHz. What is the problem?
    a. Poor image rejection
    b. Poor AGC action
    c. Inadequate IF selectivity
    d. Poor sensitivity
16. Which of the following statements about the oscillator in a standard AM receiver is true?
    a. It is fixed at 455 kHz.
    b. It oscillates 455 kHz above the dial setting.
    c. It is controlled by the AGC circuit.
    d. It oscillates at the dial frequency.
17. Refer to Fig. 12-15. The receiver is properly tuned to a station at 1020 kHz that is modulated by a 1-kHz audio test signal. What frequency or frequencies are present at the input of the detector stage?
    a. 1 kHz
    b. 454, 455, and 456 kHz
    c. 1020 kHz
    d. 1020 and 1475 kHz
18. In question 17 what frequency or frequencies are present at the input of the audio amplifier?
    a. 1 kHz
    b. 454, 455, and 456 kHz
    c. 1020 kHz
    d. 1020 and 1475 kHz
19. It is noted that a receiver uses an NPN transistor in the first IF stage. Tuning a strong station causes the base voltage to become more positive. The stage
    a. Is defective
    b. Uses reverse AGC
    c. Uses forward AGC
    d. Is not AGC-controlled

Did You Know?

Communications Signals

The ionosphere refracts radio waves, making long-distance communications possible for some frequencies at certain times. Secondary information may be encoded onto radio and television signals, and extra circuits are needed to take advantage of them. A "screen room" may be required for certain RF measurements. EMI tests are one example.

12-4 Frequency Modulation and Single Sideband

FM

Frequency modulation, or *FM,* is an alternative to amplitude modulation. Frequency modulation has some advantages that make it attractive for some commercial broadcasting and two-way radio work. One problem with AM is its sensitivity to noise. Lightning, automotive ignition, and sparking electric circuits all produce radio interference. This interference is spread over a wide frequency range. It is not easy to prevent such interference from reaching the detector in an AM receiver. An FM receiver can be made insensitive to noise interference. This noisefree performance is highly desirable.

Figure 12-18 shows how frequency modulation can be realized. Transistor Q_1 and its associated parts make up a series-tuned Colpitts oscillator. Capacitor C_3 and coil L_1 have the greatest effect in determining the frequency of oscillation. Diode D_1 is a varicap diode. It is connected in parallel with C_3. This means that as the capacitance of D_1 changes, so will the resonant frequency of the tank circuit. Resistors R_1 and R_2 form a voltage divider to bias the varicap diode. Some positive voltage (a portion of V_{DD}) is applied to the cathode of D_1. Thus, D_1 is in reverse bias.

A varicap diode uses its depletion region as the dielectric. More reverse bias means a wider depletion region and less capacitance. Therefore, as an audio signal goes positive, D_1 will reduce in capacitance. This will shift the frequency of the oscillator up. A negative-going audio input will reduce the reverse bias across

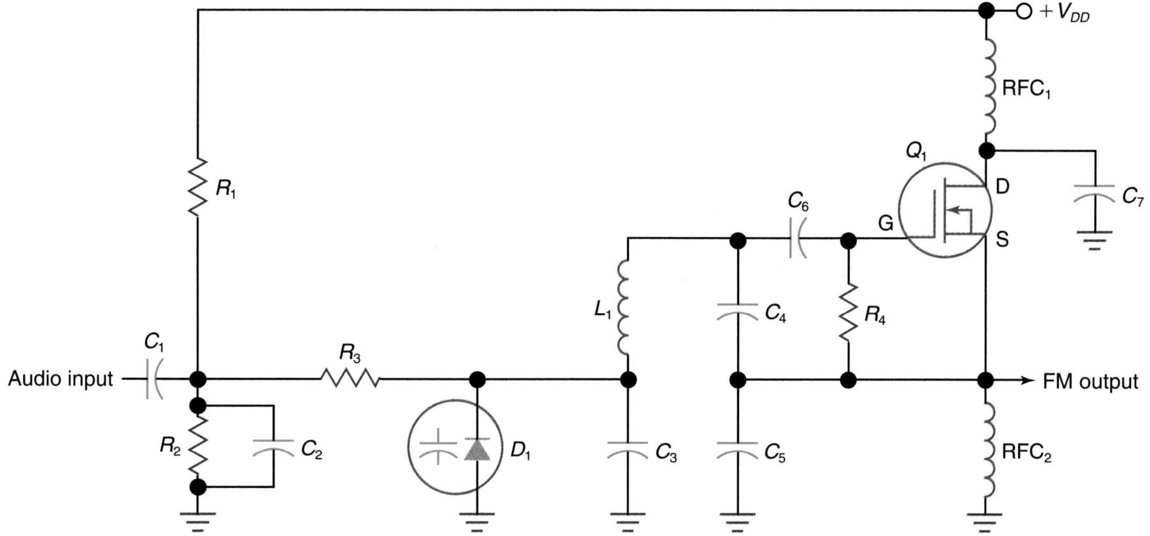

Fig. 12-18 A frequency modulator.

the diode. This will increase its capacitance and shift the oscillator to some lower frequency. The audio signal is *modulating the frequency* of the oscillator.

The relationship between the modulating waveform and the RF oscillator signal can be seen in Fig. 12-19. Note that the amplitude of the modulated RF waveform is constant. Compare this with the AM waveform in Fig. 12-6.

Amplitude modulation produces sidebands, and so does FM (Fig. 12-20). Suppose an FM transmitter is being modulated with a steady 10-kHz (0.01-MHz) tone. This transmitter has an operating (carrier) frequency of 100 MHz. The frequency domain graph shows that *several* sidebands appear. These sidebands are spaced 10 kHz apart. They appear above and below the carrier frequency. This is one of the major differences between AM and FM. An FM signal generally requires more bandwidth than an AM signal.

The block diagram for an FM superheterodyne receiver (Fig. 12-21) on page 330 is quite similar to that for the AM receiver. However, you will notice that a *limiter* stage appears after the IF stage and before the detector stage. This is one way that an FM receiver can reject noise. Figure 12-22 on page 330 shows what happens in a limiter stage. The input signal is very noisy. The output signal is noisefree. By limiting or by amplitude clipping, the noise spikes have been eliminated. Some FM receivers use two stages of limiting to eliminate most noise interference.

Limiter

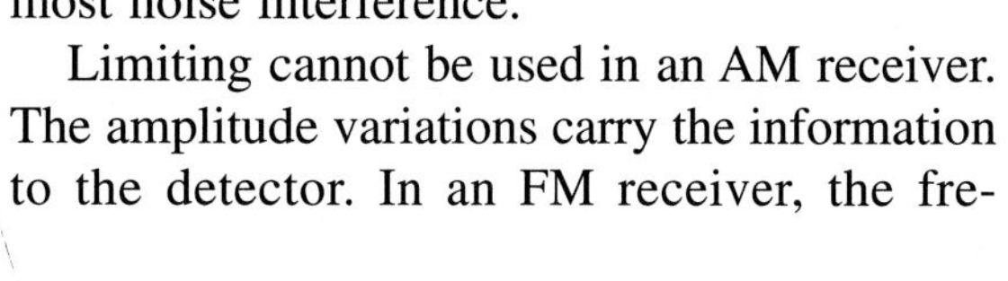

Limiting cannot be used in an AM receiver. The amplitude variations carry the information to the detector. In an FM receiver, the fre-

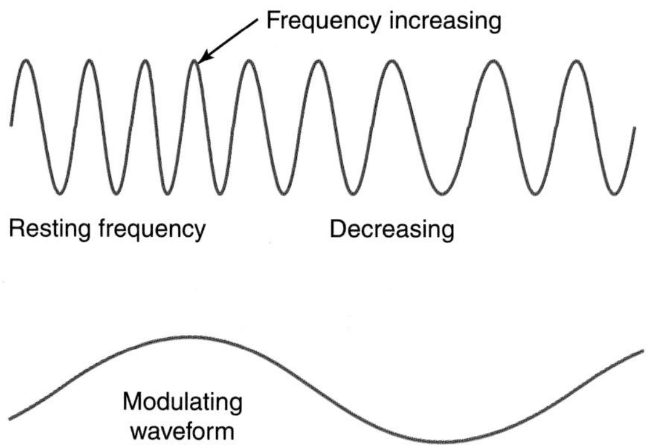

Fig. 12-19 Frequency modulation waveforms.

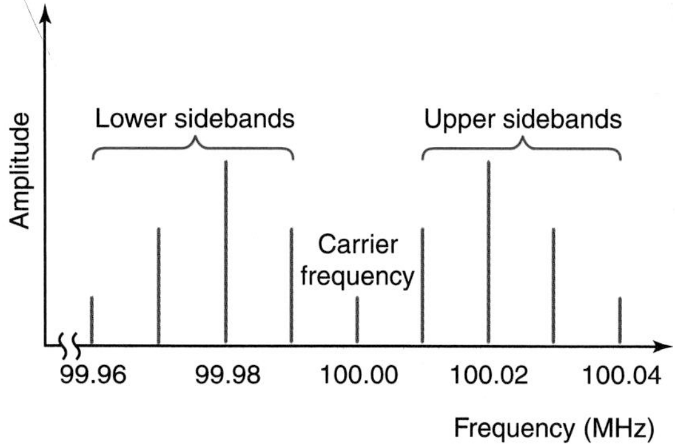

Fig. 12-20 Frequency modulation produces sidebands.

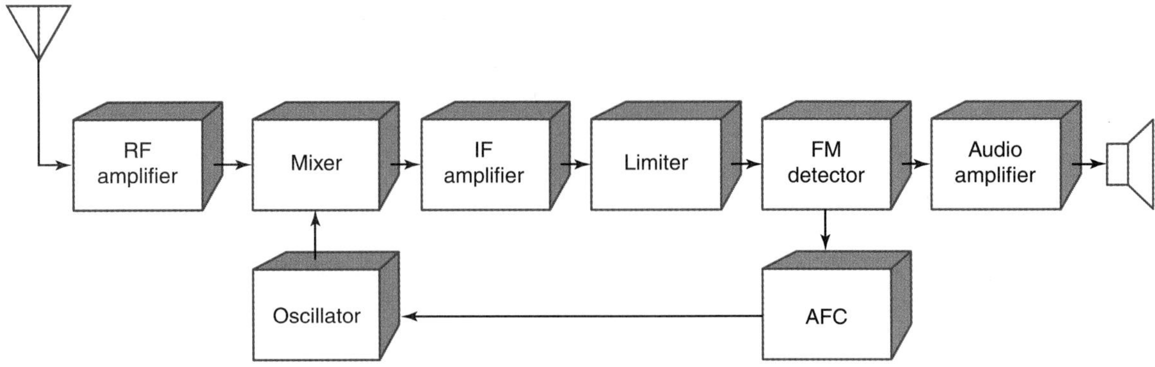

Fig. 12-21 Block diagram of an FM receiver.

quency variations contain the information. Amplitude clipping in an FM receiver will not remove the information, just the noise.

Detection of FM is more complicated than for AM. Since FM contains several sidebands above and below the carrier, a simple nonlinear detector will not demodulate the signal. A double-tuned *discriminator* circuit is shown in Fig. 12-23. It serves as an FM detector. The discriminator works by having two resonant points. One is above the carrier frequency, and one is below the carrier frequency.

Discriminator

In the frequency response curves for the discriminator circuit (Fig. 12-24), f_o represents the correct point on the curves for the carrier. In a superheterodyne receiver, the station's carrier frequency will be converted to f_o. This represents a frequency of 10.7 MHz for broadcast FM receivers. The heterodyning process allows one discriminator circuit to demodulate any signal over the entire FM band.

Refer to Figs. 12-23 and 12-24. When the carrier is unmodulated, D_1 and D_2 will conduct an equal amount. This is because the circuit is operating where the frequency-response curves cross. The amplitude is equal for both tuned circuits at this point. The current through R_1 will equal the current through R_2. If R_1 and R_2 are equal in resistance, the voltage drops will also be equal. Since the two voltages are series-opposing, the output voltage will be zero. When the carrier is unmodulated, the discriminator output is zero.

Suppose the carrier shifts higher in frequency because of modulation. This will increase the amplitude of the signal in L_2C_2 and decrease the amplitude in L_1C_1. Now there will be more voltage across R_2 and less across R_1. The output goes positive.

What happens when the carrier shifts below f_o? This moves the signal closer to the resonant point of L_1C_1. More voltage will drop across R_1, and less will drop across R_2. The output goes negative. The output from the discriminator circuit is zero when the carrier is at rest, positive when the carrier moves higher, and negative when the carrier moves lower in frequency. The output is a function of the carrier frequency.

AFC

The output from the discriminator can also be used to correct for drift in the receiver oscillator. Note in Fig. 12-21 that the FM detector feeds a signal to the audio amplifier and to a stage marked AFC. The letters *AFC* stand for *automatic frequency control.* If the oscillator

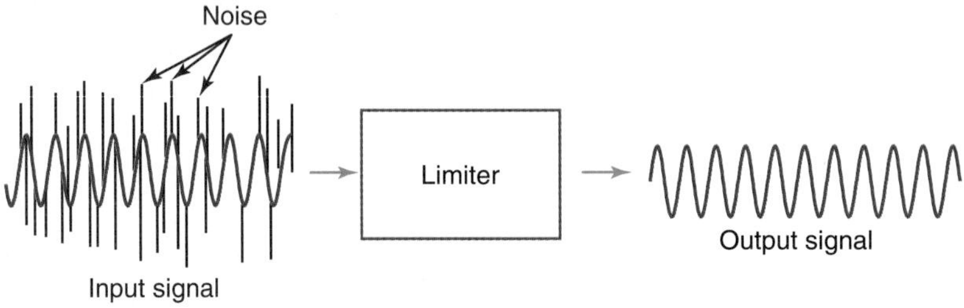

Fig. 12-22 Operation of a limiter.

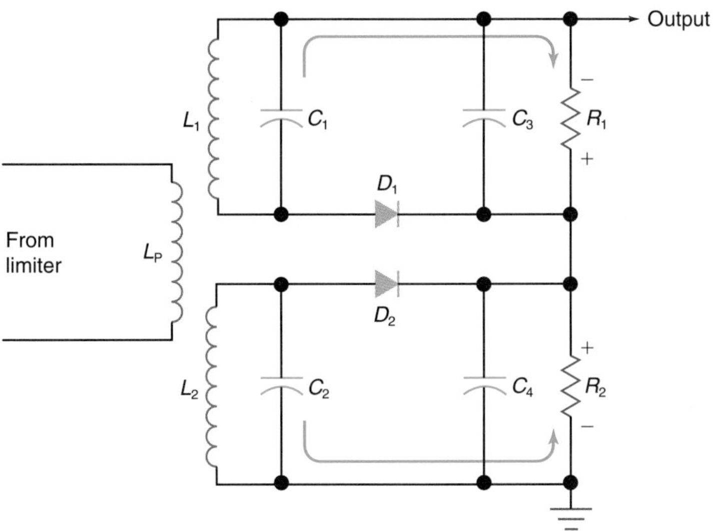

Fig. 12-23 A discriminator.

drifts, f_o will not be exactly 10.7 MHz. There will be a steady dc output voltage from the discriminator. This dc voltage can be used as a control voltage to correct the oscillator frequency and eliminate the drift. Some receivers use the discriminator output to drive a tuning meter as well. A zero-center meter shows the correct tuning point. Any tuning error will cause the meter to deflect to the left or to the right of zero.

Frequency-modulation discriminator circuits work well, but they are sensitive to amplitude. This is why one or two limiters are needed for noisefree reception. The *ratio detector* provides a simplified system; it is not nearly as sensitive to the amplitude of the signal. This makes it possible to build receivers without limiters and still provide good noise rejection.

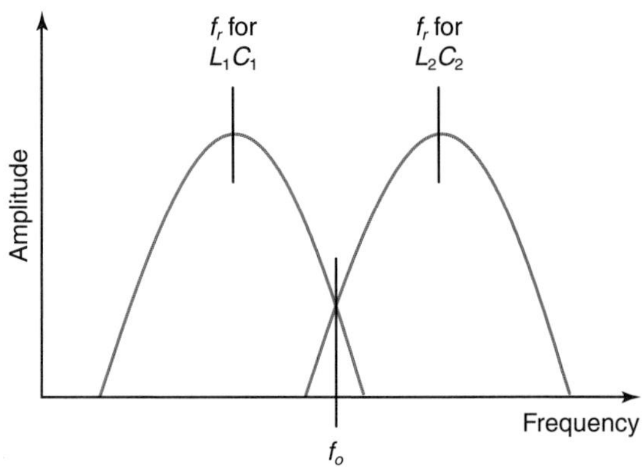

Fig. 12-24 Discriminator response curves.

Figure 12-25 on page 332 shows a typical ratio detector circuit. Its design is based on the idea of dividing a signal voltage into a ratio. This ratio is equal to the ratio of the voltages on either half of L_2. With frequency modulation, the ratio shifts and an audio output signal is available at the center tap of L_2. Since the circuit is ratio-sensitive, the input signal amplitude may vary over a wide range without causing any change in output. This makes the detector insensitive to amplitude variations such as noise.

Ratio detector

There are several other FM detector circuits. Some of the more popular ones are the *quadrature detector,* the *phase-locked-loop detector,* and the *pulse-width detector.* These circuits are likely to be used in conjunction with integrated circuits. They usually have the advantage of requiring no alignment or only one adjustment. Alignment for discriminators and ratio detectors is more time-consuming.

SSB

Single sideband (*SSB*) is another alternative to amplitude modulation. Single sideband is a subclass of AM. It is based on the idea that both sidebands in an AM signal carry the same information. Therefore, one of them can be eliminated in the transmitter with no loss of information at the receiver. The carrier can also be eliminated at the transmitter. Therefore, an SSB transmitter sends only one sideband and no carrier.

Energy is saved by not sending the carrier and the other sideband. Also, the signal will occupy only half the original bandwidth. Single sideband is much more efficient than AM. It has an effective gain of 9 dB. This is equiva-

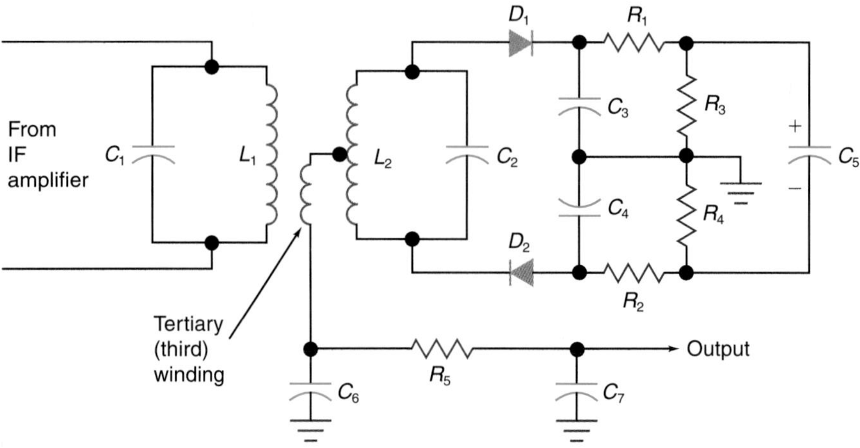

Fig. 12-25 A ratio detector.

lent to increasing the transmitter power eight times!

Balanced modulator

DSBSC

The carrier is eliminated in an SSB transmitter by using a *balanced modulator* (Fig. 12-26). The result is a double-sideband suppressed carrier (*DSBSC*) signal. Note that a balanced modulator produces only the product of the RF and AF signals. Take time to compare Fig. 12-26 to Figs. 12-3 and 12-5. Also note that the oscilloscope display of Fig. 12-26 is not the same as shown in Fig. 12-6(*b*).

Figure 12-27(*a*) shows a diode balanced modulator. The diodes are connected so that no carrier can reach the output. However, when audio is applied, the circuit balance is upset and sidebands appear at the output. There is no carrier in the output. All the energy is in the sidebands.

A band-pass filter can be used to eliminate the unwanted sideband. Figure 12-27(*b*) shows that only the upper sideband reaches the output of the transmitter. The carrier is shown as a broken line since it has already been eliminated by the balanced modulator circuit.

A receiver designed to receive SSB signals is only a little different from an ordinary AM receiver. However, the cost can be quite a bit more. There are two important differences in the SSB receiver: (1) The bandwidth in the IF amplifier will be narrower and (2) the missing carrier must be replaced by a second (local) oscillator so detection can occur. You should recall that the carrier is needed to mix with the sidebands (or sideband) to produce the difference (audio) frequencies.

Single-sideband receivers usually achieve the narrow IF bandwidth with crystal or mechanical filters. These are more costly than inductor-capacitor filters. An SSB receiver must be very stable. Even a small drift in any of the receiver oscillators will change the quality of the received audio. A moderate drift, say 500 Hz, will not be very noticeable in an ordinary AM receiver. This much drift in an SSB receiver will make the recovered audio very unnatural sounding or unintelligible. Stable oscillators are more expensive. This, along with filter costs, makes an SSB receiver more expensive.

Product detector

BFO

Notice the *product detector* in the block diagram for the SSB receiver (Fig. 12-28 on page 334). This name is used since the audio output from the detector is the difference product between the IF signal and the beat-frequency oscillator (*BFO*) signal. Actually, all AM detectors are product detectors. They all use the difference frequency product as their useful output. An ordinary diode detector can be used to demodulate an SSB signal if it is supplied a BFO signal to replace the missing carrier.

The BFO in an SSB receiver can be fixed at one frequency. In fact, it is often crystal-controlled for the best stability. A small error between the BFO frequency and the carrier frequency of the transmitted signal can be corrected by adjusting the main tuning control. The main difference between tuning an AM receiver and an SSB receiver is the need for critical tuning in an SSB receiver. Even a slight tuning error of 50 Hz will make the received audio sound unnatural.

The critical tuning of the SSB makes it undesirable for most radio work. It is useful when maximum communication effectiveness is needed. Since it is so efficient, in terms of both power and bandwidth, it is popular in citizens'

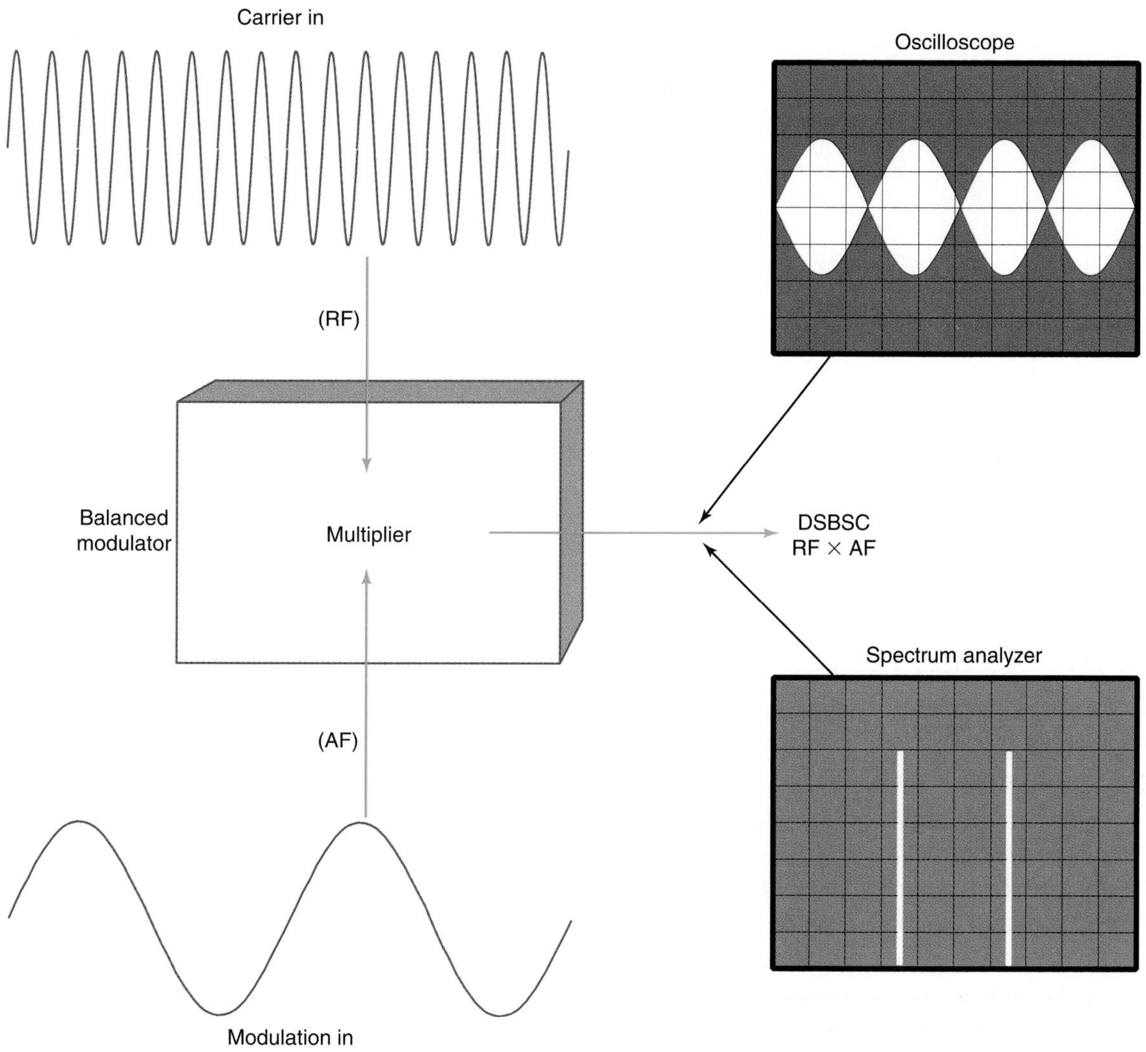

Fig. 12-26 A balanced modulator produces no carrier.

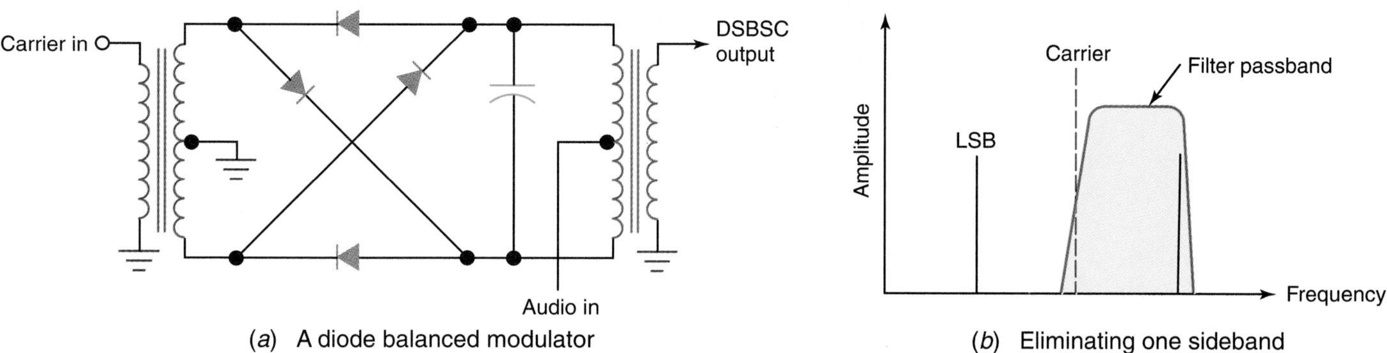

Fig. 12-27 Single sideband.

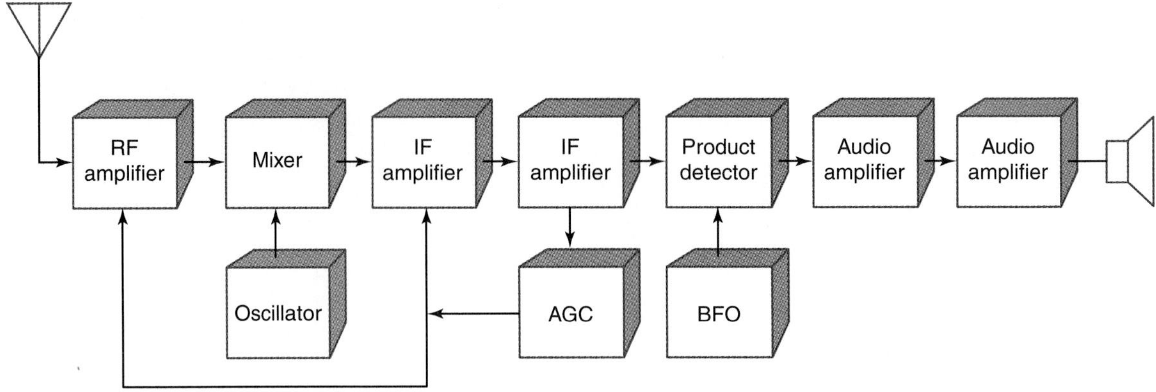

Fig. 12-28 Block diagram of an SSB receiver.

band radio, amateur radio, and for some military communications.

TEST

Choose the letter that best completes each statement.

20. Refer to Fig. 12-18. Resistors R_1 and R_2
    a. Form a voltage divider for the audio input
    b. Set the gate voltage for Q_1
    c. Divide V_{DD} to forward-bias D_1
    d. Divide V_{DD} to reverse-bias D_1
21. Refer to Fig. 12-18. A positive-going signal at the audio input will
    a. Increase the capacitance of D_1 and raise the frequency
    b. Increase the capacitance of D_1 and lower the frequency
    c. Decrease the capacitance of D_1 and raise the frequency
    d. Decrease the capacitance of D_1 and lower the frequency
22. Frequency modulation as compared to amplitude modulation
    a. Can provide better noise rejection
    b. Requires more bandwidth
    c. Requires complex detector circuits
    d. All of the above
23. The function of the limiter stage in Fig. 12-21 is to
    a. Reduce amplitude noise
    b. Prevent overdeviation of the signal
    c. Limit the frequency response
    d. Compensate for tuning error
24. The purpose of the AFC stage in Fig. 12-21 is to
    a. Reduce noise
    b. Maintain a constant audio output (volume)
    c. Compensate for tuning error and drift
    d. Provide stereo reception
25. Refer to Figs. 12-23 and 12-24. Assume that f_o is 10.7 MHz. If the signal from the limiter is at 10.65 MHz.
    a. The output voltage will be 0 V.
    b. The output voltage will be negative.
    c. The output voltage will be positive.
    d. Resistor R_1 will conduct more current than R_2.
26. Refer to question 25. If the signal from the limiter is at 10.7 MHz, then
    a. Diode D_1 will conduct the most current.
    b. Diode D_2 will conduct the most current.
    c. The output voltage will be zero.
    d. None of the above.
27. Refer to Fig. 12-25. The advantage of this FM detector as compared to a discriminator is that it
    a. Is less expensive (uses fewer parts)
    b. Can drive a tuning meter
    c. Can provide AFC
    d. Rejects amplitude variations
28. Refer to Fig. 12-26. The carrier input is 455 kHz, and the audio input is a 2-kHz sine wave. The output frequency or frequencies are
    a. 2 kHz
    b. 455 kHz
    c. 453, 455, and 457 kHz
    d. 453 and 457 kHz
29. Single sideband as compared to amplitude modulation is
    a. More efficient in terms of bandwidth

b. More efficient in terms of power
c. More critical to tune
d. All of the above

30. Refer to Fig. 12-28. The purpose of the BFO circuit is
    a. To correct for tuning error
    b. To replace the missing carrier so detection can occur
    c. To provide noise rejection
    d. All of the above
31. Refer to Fig. 12-28. The IF bandwidth of this receiver, compared with an ordinary AM receiver, is
    a. Narrower
    b. The same
    c. Wider
    d. Indeterminate

12-5 Receiver Troubleshooting

Radio receiver troubleshooting is very similar to amplifier troubleshooting. Most circuits in a receiver are amplifiers. The material covered in Chap. 10 on amplifier troubleshooting is relevant to receiver troubleshooting. For example, Sec. 10-1 on preliminary checks should be followed in exactly the same way.

You should view a receiver as a signal chain. If the receiver is dead, the problem is to find the broken link in the chain. Signal injection should begin at the output (speaker) end of the chain. However, a receiver involves gain at different frequencies. Several signal-generator frequencies will be involved. You must use both an audio generator and an RF generator. Figure 12-29 shows the general scheme of signal injection in a superheterodyne receiver.

It is also possible to make a click test in most receivers. Use the same procedure discussed in Chap. 10 on amplifier troubleshooting. This will work in the audio and IF stages. The noise generated by the sudden shift in transistor bias should reach the speaker. It is also possible to test the mixer with the click test. The oscillator may respond to the click test, but the results would not be conclusive. It is possible that the oscillator is not oscillating, or it may be oscillating at the wrong frequency.

If we assume that the signal chain is intact from the first IF to the speaker, then the problem must be in the mixer or the oscillator. Checking the oscillator is not too difficult. An oscilloscope or frequency counter could be used. A voltmeter with an RF probe is another possibility, but there would be no way to tell whether the frequency was correct. Some technicians prefer to tune for the oscillator signal by using a second receiver. Place the second receiver very close to the receiver being tested. Set the dial on the second receiver above the dial frequency on the receiver under test. The difference should equal the IF of the receiver under test. This is based on the fact that the oscillator is supposed to run above the dial setting by an amount equal to the IF. Now, rock one of the dials back and forth a little. You should hear a carrier (no modulation). This tells you that the oscillator is working, and it also indicates whether the frequency is nearly correct.

If the receiver sounds distorted on strong stations, the problem could be in the AGC circuit. This can be checked with a voltmeter. Monitor the control voltage as the receiver is tuned across the band. You should find a change in the control voltage from no station (clear frequency) to a strong station. The service notes for the receiver usually will indicate the normal AGC range.

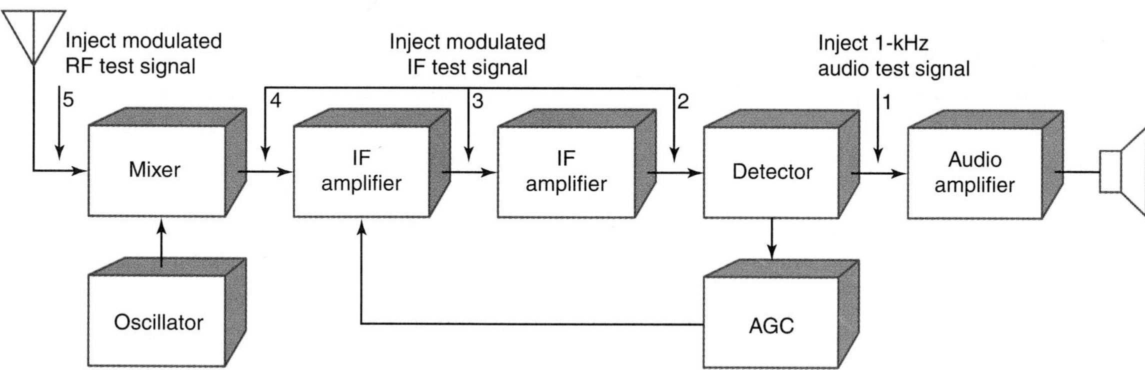

Fig. 12-29 Signal injection in a superheterodyne receiver.

If the receiver has poor sensitivity, again it is possible that the AGC circuit is defective. Since AGC can produce several symptoms, it is recommended that it be checked early in the troubleshooting process.

Poor sensitivity can be difficult to troubleshoot. A dead stage is usually easier to find than a weak stage. Signal injection may work. It is normal to expect less injection for a given speaker volume as the injection point moves toward the antenna. Some technicians disable the AGC circuit when making this test. This can be done by clamping the AGC control line with a fixed voltage from a power supply. A current-limiting resistor around 10 kΩ should be connected in series with the supply to avoid damaging the receiver.

Poor sensitivity can be caused by a leaky detector diode. Disconnect one end of the diode from the circuit and check its forward and reverse resistance with an ohmmeter. Diode testing was covered in Sec. 3-3 of Chap. 3.

Alignment

Improper *alignment* is another possible cause of low gain and poor sensitivity. All the IF stages must be adjusted to the correct frequency. Also, the oscillator and mixer tuned circuits must track for good performance across the band. If the receiver has a tuned RF stage, then three tuned circuits must track across the band.

Alignment is usually good for the life of the receiver. However, someone may have tampered with the tuned circuits, or a part may have been replaced that upsets the alignment. Do not attempt alignment unless the service notes and the proper equipment are available.

Intermittent receivers and noisy receivers should be approached by using the same techniques described in Chap. 10 for amplifier troubleshooting. In addition, you should realize that receiver noise may be due to some problem outside the receiver itself. Some locations are very noisy, and poor receiver performance is typical. Compare performance with a known receiver to verify the source of noise.

It should also be mentioned that receiver performance can vary considerably from one model to another. Many complaints of poor performance cannot be resolved with simple repairs. Some receivers simply do not work as well as others.

A superheterodyne receiver may have a total gain in excess of 100 dB. Unwanted feedback paths or coupling of circuits may cause oscillations. If the receiver squeals only when a station is tuned in, the problem is likely to be in the IF amplifier. If the receiver squeals or motorboats constantly, a bypass capacitor or AGC filter capacitor may be open. Always check to be sure that all grounds are good and that all shields are in place. In some cases, improper alignment can also cause oscillation.

Interference

Interference from nearby transmitters is becoming an increasingly complex problem. When a transmitting antenna is located close to other receiving equipment, problems are likely to occur. Some interference problems can be difficult to solve. Figure 12-30 shows a few techniques that may be successful. Solving interference problems is often a process of trying various things until progress is noted. Try the easiest and least expensive cures first.

Receiver interference can often be traced to nontransmitting equipment. Computers, computer peripherals, light dimmers, touch-controlled lamps, and even power lines are known sources of radio and television interference. The best way to verify if a device is causing a problem is to turn it off. Touch-controlled lamps should be disconnected from the wall

Motorola's DSP56305
24-bit Digital Signal Processor

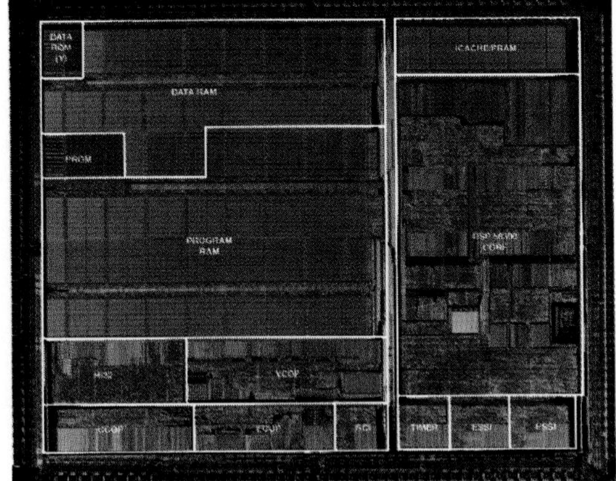

The DSP56305 is intended for telecommunications. It is an 80-MHz device with three coprocessors: The FCOP (filter coprocessor), the VCOP (Viterbi coprocessor), and the CCOP (cyclic code coprocessor). Devices like this support development in integrated digital wireless systems for phones, intercom, alphanumeric paging, and modem, data, and fax services.

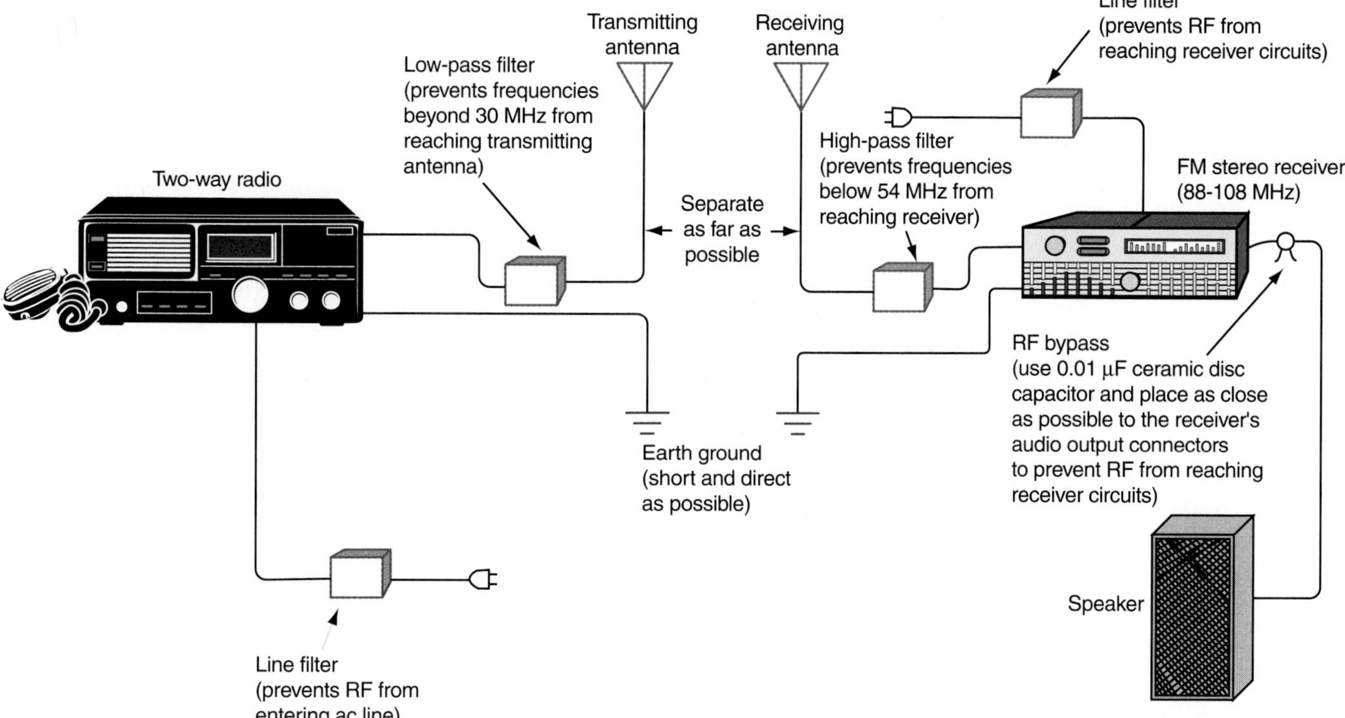

Fig. 12-30 Steps to prevent radio interference.

outlet to determine if they are causing the interference.

TEST

Choose the letter that best answers each question.

32. A 1-kHz test signal can be used for testing which stage of a superheterodyne receiver?
    a. Mixer
    b. IF
    c. Detector
    d. Audio
33. It is desired to check the oscillator of a superheterodyne receiver by using a second receiver. If the dial is set at 980 kHz, where should the oscillator be heard on the second receiver?
    a. 525 kHz
    b. 980 kHz
    c. 1435 kHz
    d. 1610 kHz
34. A receiver sounds distorted only on the strongest signals. Where would the fault likely be found?
    a. In the AGC system
    b. In the loudspeaker
    c. In the audio amplifier
    d. In the volume control
35. What would cause poor sensitivity in a receiver?
    a. A defective mixer
    b. A defective IF amplifier
    c. A weak detector
    d. Any of the above
36. What results from improper alignment?
    a. Poor sensitivity
    b. Dial error
    c. Oscillation
    d. Any of the above

Chapter 12 Review

Summary

1. A high-frequency oscillator signal becomes a radio wave at the transmitting antenna.
2. Modulation is the process of putting information on the radio signal.
3. Turning the signal on and off with a key is called CW modulation.
4. When AM is used, the signal has three frequency components: a carrier, a lower sideband, and an upper sideband.
5. The total bandwidth of an AM signal is twice the highest modulating frequency.
6. Demodulation is usually called detection.
7. A diode makes a good AM detector.
8. Other nonlinear devices, such as transistors, can also be used as AM detectors.
9. A simple AM receiver can be built from an antenna, a detector, headphones, and a ground.
10. Sensitivity is the ability to receive weak signals.
11. Selectivity is the ability to receive one range of frequencies and reject others.
12. Gain provides sensitivity.
13. Tuned circuits provide selectivity.
14. The optimum bandwidth for an ordinary AM receiver is about 15 kHz.
15. A superheterodyne receiver converts the received frequency to an intermediate frequency.
16. Tuning a radio receiver to different stations does not change the passband of the IF amplifiers.
17. The mixer output will contain several frequencies. Only those in the IF passband will reach the detector.
18. The standard IF for the AM broadcast band is 455 kHz.
19. The receiver oscillator will usually run above the received frequency by an amount equal to the intermediate frequency.
20. Two frequencies will always mix with the oscillator frequency and produce the IF: the desired frequency and the image frequency.
21. Adjacent-channel interference is rejected by the selectivity of the IF stages. Image interference is rejected by one or more tuned circuits before the mixer.
22. The AGC circuit compensates for different signal strengths.
23. In an FM transmitter, the audio information modulates the frequency of the oscillator.
24. Frequency modulation produces several sidebands above the carrier and several sidebands below the carrier.
25. Frequency-modulation detection can be achieved by a discriminator circuit.
26. Discriminators are sensitive to amplitude; thus, limiting must be used before the detector.
27. A ratio detector has the advantage of not requiring a limiter circuit for noise rejection.
28. Single sideband (SSB) is a subclass of AM.
29. Receiver troubleshooting is similar to amplifier troubleshooting.
30. The signal chain can be checked stage by stage by using signal injection.
31. A leaky detector can cause poor sensitivity.
32. Good alignment is necessary for proper receiver performance.

Chapter Review Questions

Choose the letter that best answers each question.

12-1. Which portion of a transmitting station converts the high-frequency signal into a radio wave?
a. The modulator
b. The oscillator
c. The antenna
d. The power supply

12-2. What is the modulation used in radio telegraphy called?
a. CW
b. AM

c. FM
d. SSB

12-3. An AM transmitter is fed audio as high as 3.5 kHz. What is the bandwidth required for its signal?
a. It is dependent on the carrier frequency
b. 3.5 kHz
c. 7.0 kHz
d. 455 kHz

12-4. An AM demodulator uses the difference frequency between what two frequencies?
a. USB and LSB
b. Sidebands and the carrier
c. IF and the detector
d. All of the above

12-5. Which of the following components is useful for AM detection?
a. A tank circuit
b. A resistor
c. A capacitor
d. A diode

12-6. Refer to Fig. 12-4. Assume the audio input is zero. The carrier output to the antenna will
a. Fluctuate in frequency
b. Fluctuate in amplitude
c. Be zero
d. None of the above

12-7. Refer to Fig. 12-10. What serves as the energy source for this receiver?
a. The radio signal
b. The detector
c. The headphones
d. There is none

12-8. Refer to Fig. 12-11. How may the selectivity of this receiver be improved?
a. Add more audio gain.
b. Add more tuned circuits.
c. Use a bigger antenna.
d. All of the above.

12-9. A tuned circuit has a center frequency of 455 kHz and a bandwidth of 20 kHz. At what frequency or frequencies will the response of the circuit drop to 70 percent?
a. 475 kHz
b. 435 kHz
c. 435 and 475 kHz
d. 445 and 465 kHz

12-10. An AM receiver has an IF amplifier with a bandwidth that is too narrow. What will the symptom be?
a. Loss of high-frequency audio
b. Poor selectivity
c. Poor sensitivity
d. All of the above

12-11. The major advantage of the superheterodyne design over the TRF receiver design is that
a. It eliminates the image problem.
b. It eliminates tuned circuits.
c. It eliminates the need for an oscillator.
d. The fixed IF eliminates tracking problems and bandwidth changes.

12-12. An AM superheterodyne receiver is tuned to 1140 kHz. Where is the image?
a. 865 kHz
b. 1315 kHz
c. 2050 kHz
d. 2850 kHz

12-13. Refer to Fig. 12-16. The dial of the receiver is set at 1190 kHz. Which statement is true?
a. The mixer tuned circuit should resonate at 1190 kHz.
b. The oscillator circuit should resonate at 1645 kHz.
c. The difference mixer output should be at 455 kHz.
d. All of the above.

12-14. An FM receiver is set at 93 MHz. Interference is received from a station transmitting at 114.4 MHz. What is the problem caused by?
a. Poor selectivity in the RF and mixer tuned circuits
b. Poor selectivity in the IF stages
c. Poor limiter performance
d. Poor ratio detector performance

12-15. What FM receiver circuit is used to correct for frequency drift in the oscillator?
a. AGC
b. AVC
c. AFC
d. All of the above

12-16. A transistor in an FM receiver is controlled by decreasing its current as the received signal becomes stronger. What is this an example of?
a. Forward AGC
b. Reverse AGC
c. Stereo reception
d. None of the above

12-17. How does frequency modulation compare to amplitude modulation with regard to the number of sidebands produced?
a. Frequency modulation produces the same number of sidebands.

b. Frequency modulation produces fewer sidebands.
c. Frequency modulation produces more sidebands.
d. Frequency modulation produces no sidebands.

12-18. What is the function of a limiter stage in an FM receiver?
a. It rejects adjacent-channel interference.
b. It rejects image interference.
c. It rejects noise.
d. It rejects drift.

12-19. The output of Fig. 12-23 is connected to a zero-center tuning meter. How will the meter respond when a station is correctly tuned?
a. It will indicate in the center of its scale.
b. It will deflect maximum to the right.
c. It will deflect to the left.
d. It depends on the station.

12-20. Which of the following circuits is not used for FM demodulation?
a. Diode detector
b. Discriminator
c. Ratio detector
d. Quadrature detector

12-21. The output of a balanced modulator is called
a. SSB
b. DSBSC
c. FM
d. None of the above

12-22. An SSB transmitter runs 100 W. What power will be required in an AM transmitter to achieve the same range?
a. 5 W
b. 20 W
c. 800 W
d. 1200 W

12-23. What is the bandwidth of an SSB signal as compared to that of an AM signal?
a. About 2 times
b. About the same
c. About half
d. About 10 percent

12-24. What must be done to demodulate an SSB signal?
a. Replace the missing carrier
b. Use two diodes
c. Use a phase-locked-loop detector
d. Convert it to an FM signal

12-25. Which of the following test signals would be the least useful for troubleshooting an AM broadcast receiver?
a. 1-kHz audio
b. 455-kHz modulated RF
c. 1-MHz modulated RF
d. 10.7-MHz frequency-modulated RF

12-26. An FM receiver works well, but the dial accuracy is poor. The problem is most likely in the
a. Detector
b. Oscillator
c. IF amplifiers
d. Limiter

Critical Thinking Questions

12-1. Can you identify some ways that radio frequencies are used for purposes other than communication?

12-2. A shortwave listener tells you that some stations can be received at two different frequencies. Are these stations transmitting on two frequencies or is there another explanation? How could you find out?

12-3. Federal Aviation Agency (FAA) rules prohibit passengers on commercial flights from using radio receivers. Why?

12-4. How can a personal computer interfere with radio and television reception?

12-5. Can you think of any significant difference between vehicular cellular telephones and vehicular CB radios?

12-6. The AM broadcast band ranges from 540 to 1600 kHz, for a total bandwidth of a little over 1 MHz. A single television channel is allocated 6 MHz. Why is one television channel wider in bandwidth than the entire AM band?

Answers to Tests

1. d
2. b
3. b
4. a
5. d
6. b
7. a
8. c
9. d
10. d
11. a
12. d
13. b
14. c
15. a
16. b
17. b
18. a
19. c
20. d
21. c
22. d
23. a
24. c
25. b
26. c
27. d
28. d
29. d
30. b
31. a
32. d
33. c
34. a
35. d
36. d

Integrated Circuits

Chapter Objectives

This chapter will help you to:

1. *Compare* integrated circuit (IC) technology to discrete technology.
2. *Explain* the photolithographic process used to make ICs.
3. *Make* calculations for 555 timer circuits.
4. *Recognize* some common linear ICs and their symbols.
5. *Explain* digital signal processing.
6. *Troubleshoot* circuits with ICs.

An integrated circuit (IC) can be the equivalent of dozens or hundreds of separate electronic parts. Digital ICs, such as microprocessors, can equal millions of parts. Now, digital ICs are finding more applications in analog systems. A prime example is in digital signal processing.

13-1 Introduction

The integrated circuit was introduced in 1958. It has been called the most significant technological development of the twentieth century. Integrated circuits have allowed electronics to expand at an amazing rate. Most of the growth has been in the area of digital electronics. Developments in linear integrated circuits lagged behind those of digital ICs for the first 10 years or so. Lately, linear ICs have received more attention, and a broad variety of these types of devices is now available and more analog functions are now achieved using digital technology.

Electronics is growing rapidly for several reasons. One major reason is that electronics continues to advance in performance while the cost remains stable and even decreases from time to time. Another reason for the growth in electronics is that circuits and systems have become increasingly reliable over the years. Integrated circuits have had much to do with these gains.

Discrete circuit

Discrete circuits use individual resistors, capacitors, diodes, transistors, and other devices to achieve the circuit function. These individual or discrete parts must be interconnected. The usual approach is to use a circuit board. This method, however, increases the cost of the circuit. The board, assembly, soldering, and testing all make up a part of the cost.

Integrated circuits do not eliminate the need for circuit boards, assembly, soldering, and testing. However, with ICs the number of discrete parts can be reduced. This means that the circuit boards can be smaller, often use less power, and that they will cost less to produce. It may also be possible to reduce the overall size of the equipment by using integrated circuits, which can reduce costs in the chassis and cabinet.

Integrated circuits may lead to circuits that require fewer alignment steps at the factory. This is especially true with digital devices. Alignment is expensive, and fewer steps mean lower costs. Also, variable components are more expensive than fixed components, and if some components can be eliminated, savings are realized.

Integrated circuits may also increase performance. Certain ICs work better than equivalent discrete circuits. A good example is a modern integrated voltage regulator. A typical unit may offer 0.03 percent regulation, excellent ripple and noise suppression, automatic current limiting, and thermal shutdown. An equivalent discrete regulator may contain dozens of parts, cost six times as much, and still not work as well!

Reliability is related indirectly to the number of parts in the equipment. As the number of parts goes up, the reliability comes down. Integrated circuits make it possible to reduce the number of discrete parts in a piece of equipment. Thus, electronic equipment can be made more reliable by the use of more ICs and fewer discrete components.

Integrated circuits are available in a variety of *package styles.* Some of them are designed for surface-mount technology. The plastic quad flat package and the flat-pack shown in Fig. 13-1 are examples. The pin grid array package and the leadless package can be used with sockets.

DIP

Reliability

The dual in-line package (*DIP*) shown in Fig. 13-1 is popular. It may have 14 or 16 pins. The mini-DIP is a shorter version of the dual in-line package. It has 8 pins. The TO-5 package is available with 8, 10, or 12 pins.

The TO-3 and TO-220 packages are used mainly for voltage regulator ICs. Their appearance can be identical to that of packages used for power transistors. This is a good example of how valuable service literature and

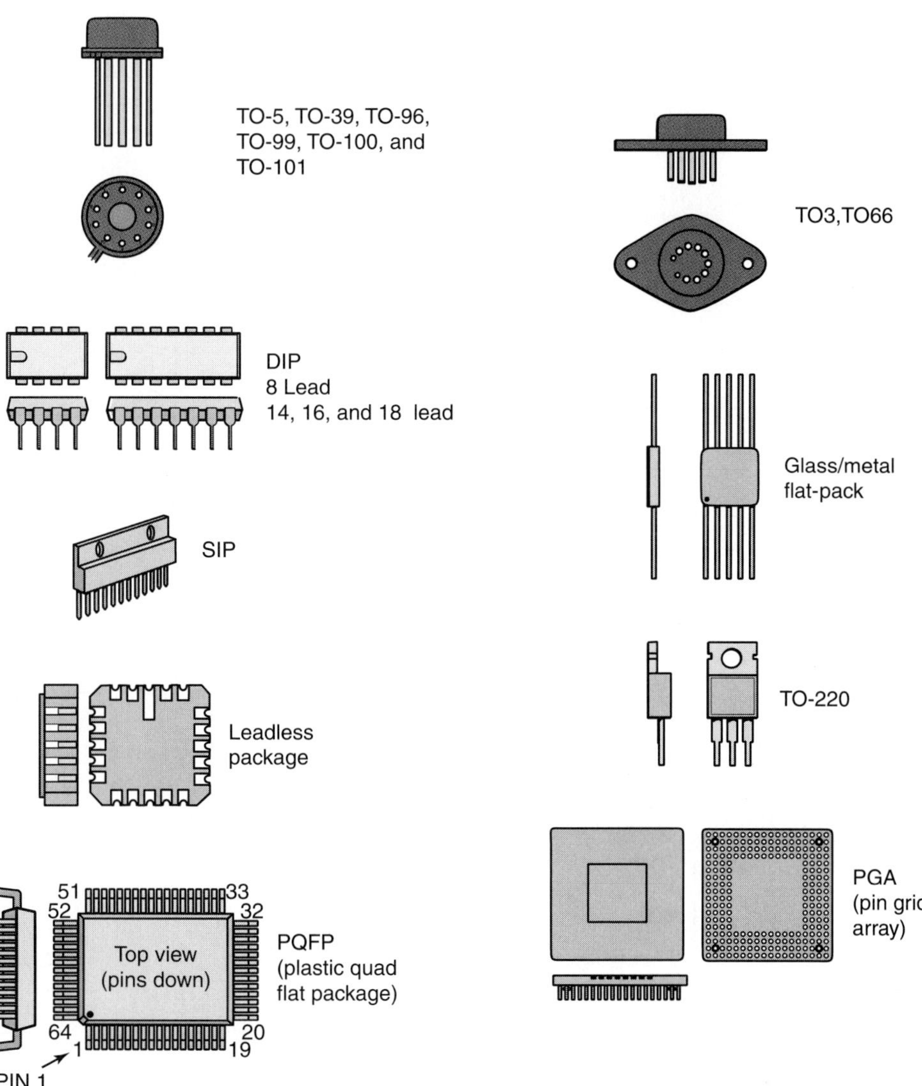

Fig. 13-1 Integrated circuit package styles.

part numbers are when a technician troubleshoots equipment for the first time. Positive component identification cannot be based on a visual check alone.

Schematics seldom show any of the internal features for integrated circuits. A technician usually does not need to know circuit details for the inside of an IC. It is more important to know what the IC is supposed to do and how it functions as a part of the overall circuit. Figure 13-2 shows the internal schematic for a 7812 IC voltage regulator. Most diagrams will show this IC as in Fig. 13-3. Note the simplicity. The voltage regulator function is simple and straightforward. Figure 13-3 plus a few voltage specifications is all that a technician would normally have to check to verify proper operation of the IC.

TEST

Choose the letter that best answers each question.

1. When was the integrated circuit developed?
   a. 1920
   b. 1944
   c. 1958
   d. 1983
2. What is an electronic circuit that is constructed of individual components such as resistors, capacitors, transistors, diodes, and the like called?
   a. An integrated circuit
   b. A chassis
   c. A circuit board
   d. A discrete circuit

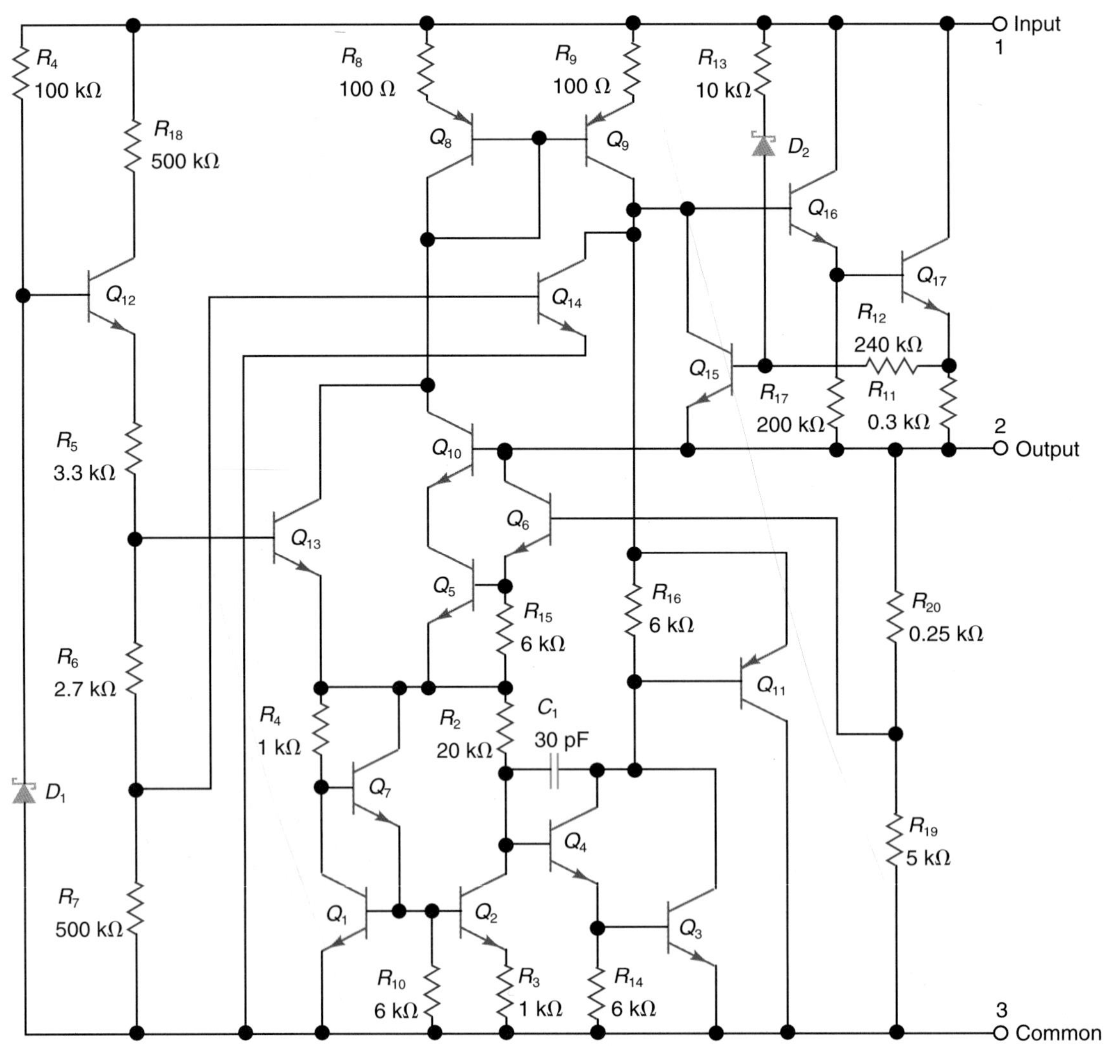

Fig. 13-2 Schematic for an IC voltage regulator.

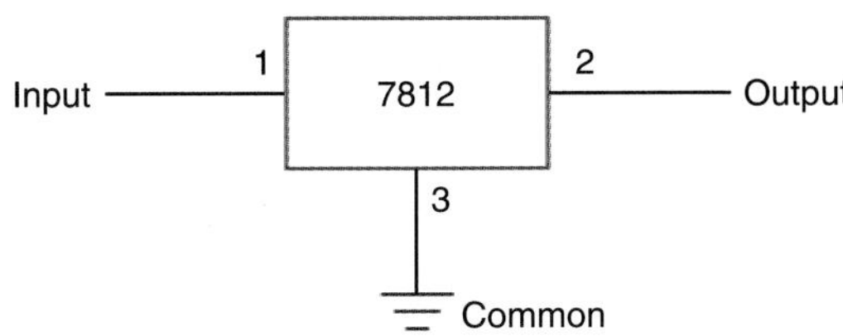

Fig. 13-3 Normal way of showing an IC.

3. The use of ICs in a design can
   a. Decrease the number and size of parts
   b. Lower cost
   c. Increase reliability
   d. All of the above
4. What is the only sure way to identify a part as an integrated circuit?
   a. Look at the package style.
   b. See how it is connected to other parts.
   c. Check the schematic or part number.
   d. Count the pins.
5. When will a technician need the internal schematic for an integrated circuit?
   a. Very seldom
   b. When troubleshooting
   c. When making circuit adjustments
   d. When taking voltage and waveform readings

13-2 Fabrication

The fabrication of integrated circuits begins in a radio-frequency furnace. Silicon that has been doped with a P-type impurity is melted in a quartz crucible. A large crystal of P-type silicon is then pulled from the molten silicon (Fig. 13-4). The crystal is then sliced into wafers 10 mils thick (0.025 cm or 0.01 in.).

The P-type silicon wafers are processed using *photolithography,* as shown in Fig. 13-5 on page 346. The wafer is called the *substrate.* When it is exposed to heat and water vapor the surface is oxidized. Then the oxide surface is coated with *photoresist,* a sensitive material that hardens when exposed to ultraviolet light. The exposure is made through a *photomask.* After exposure, the unhardened areas wash away in the developing process. Now the wafer is etched to expose areas of the substrate. These exposed areas are then penetrated with N-type impurity atoms. Thus, PN junctions are formed in the silicon wafer. The wafer can be reoxidized, and the process repeated several times to make the desired P- and N-type zones in the substrate.

Photolithography

Substrate

Photoresist

Photomask

Integrated circuits are *batch-processed.* As shown in Fig. 13-6 on page 346, one silicon wafer will produce a batch of chips. This keeps the costs down. Also, the wafer is much easier to handle and process since it is large compared to a single IC.

Some companies invest heavily in research and development, and others do not. The level of investment often influences the types of positions open to electronic technicians.

Not all the chips in the batch will be good. They are tested wth needle-sharp probes that contact each chip on the wafer (Fig. 13-7 on page 347). Bad chips are marked with a dot. Next, the wafer is scribed and broken apart. The good chips are mounted on a metal header. Figure 13-8 on page 347 shows how this is done. The process is called *ball bonding* since a ball forms on the end of the wire when the wire is cut by the gas flame. A newer method, called ultrasonic bonding, may also be used to connect the chip pads to the header tabs.

Bonding

It is interesting to examine how some of the specific circuit functions are achieved in the IC. The steps used to form an NPN-junction transistor are seen in Fig. 13-9 on page 348. It begins with the P-type substrate. An N^+ diffusion layer is added (N^+ indicates that it is heavily doped with an N-type impurity). This N^+ layer will improve the collector characteristics of the transistor. Next, an N-type layer is grown over the substrate. This *epitaxial* process produces a

Epitaxial

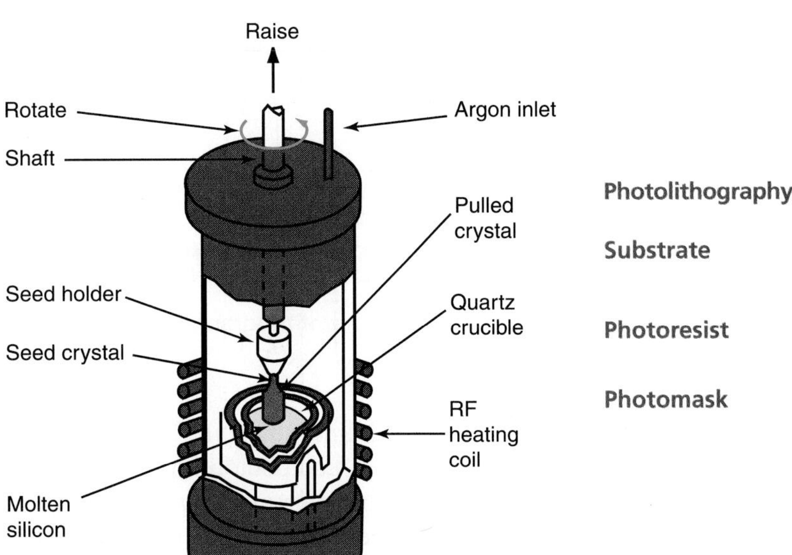

Fig. 13-4 Forming the crystal.

(*a*) Crystalline silicon

Silicon dioxide

(*b*) Oxidize surface of substrate

Photoresist

(*c*) Coat oxide with photoresist

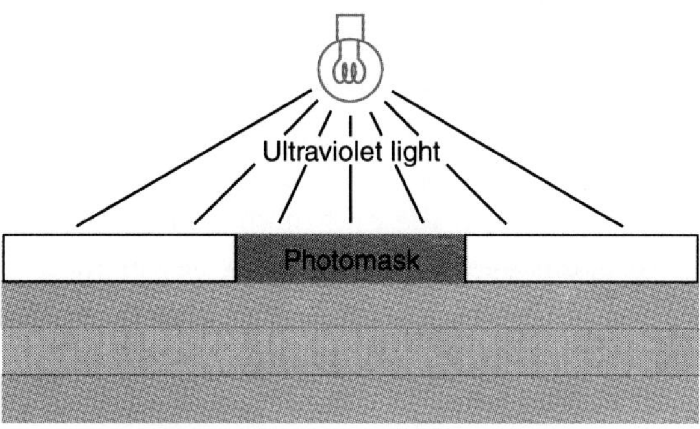

(*d*) Expose through positive photomask

(*e*) Develop, removing the unexposed photoresist

(*f*) Etch through oxide (silicon dioxide)

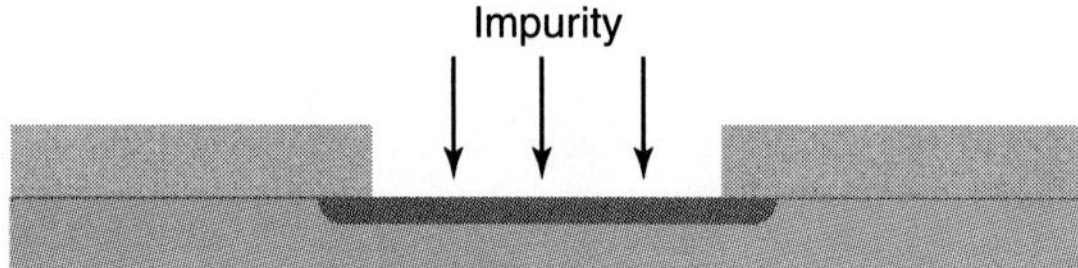

(*g*) Impurity penetrates substrate and a PN junction is formed

Fig. 13-5 The photolithographic process.

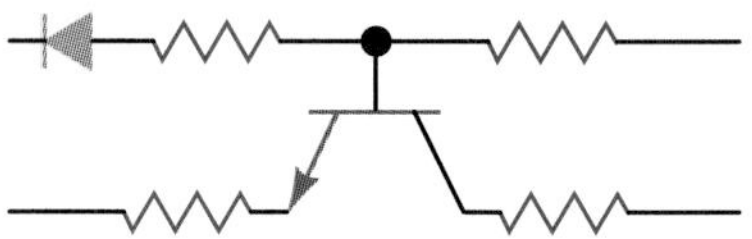

1. Design the circuit

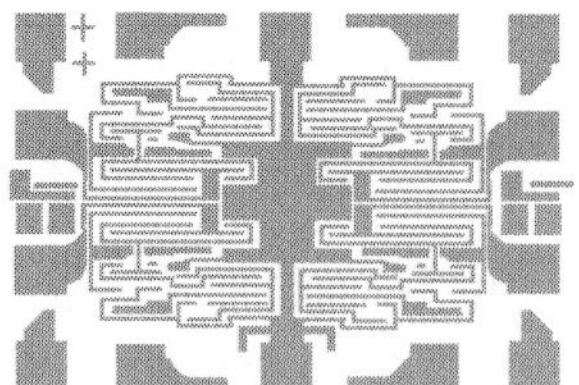

2. Design the layout

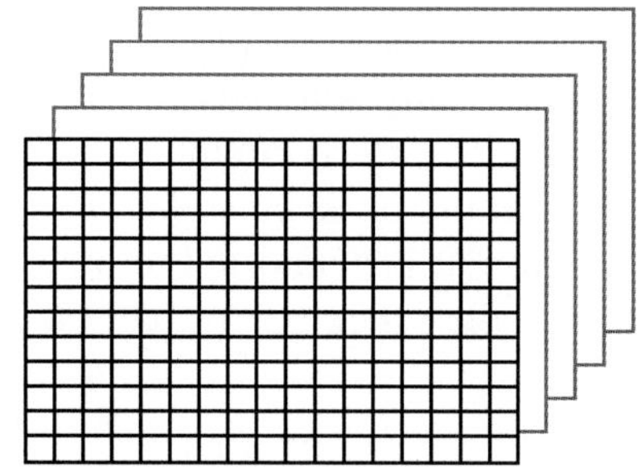

3. Prepare the photomasks—five or more will be required

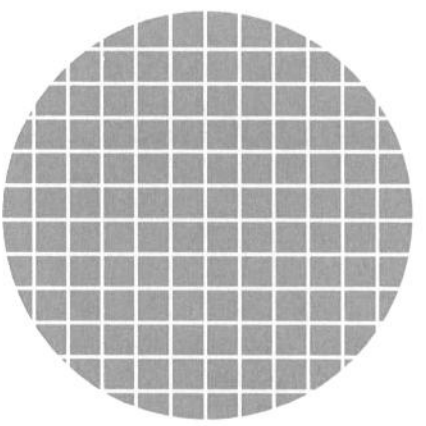

4. Expose the silicon wafer using each photomask

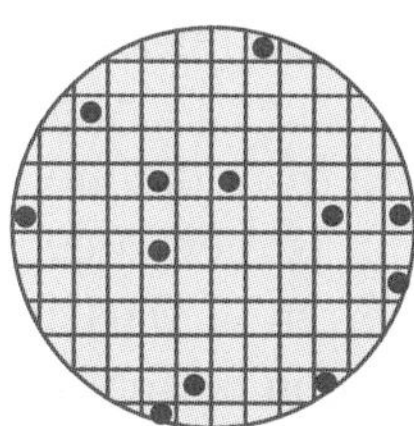

5. Run probe test and scribe the wafer

6. Break into individual chips

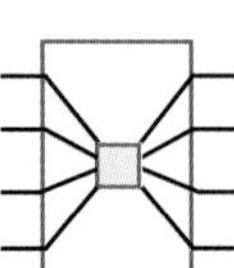

7. Mount chip into package—bond and seal

Fig. 13-6 The major steps in making an IC.

uniform crystalline structure. The epitaxial layer is oxidized, and photolithography is used to expose an area surrounding the transistor. Boron, a P-type impurity, is diffused through the opening until the substrate is reached. This *isolates* a region of the N-type epitaxial layer. The various circuit functions in the IC are electrically insulated from each other by *isolation diffusion.*

Isolation diffusion

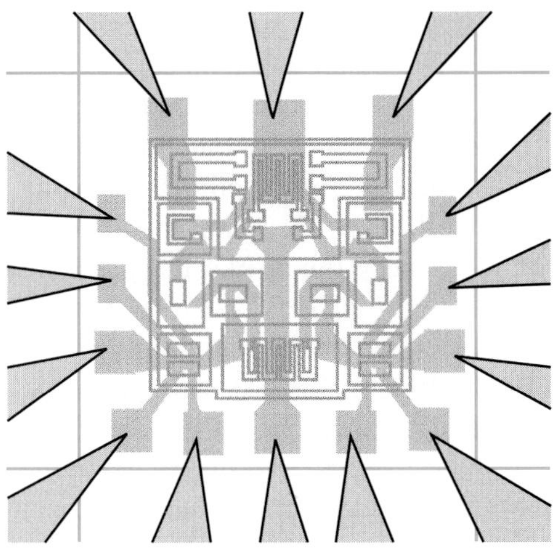

Fig. 13-7 The probe test.

Refer once more to Fig. 13-9. Again, photolithography opens up the desired area in the oxide layer. A P-type impurity penetrates, and the base region is formed. The last diffusion will use an N-type impurity to form the emitter of the transistor. Polarity reversal by diffusion will eventually saturate the crystal. A maximum number is three diffusions. The collector region is formed with the first diffusion and is lightly doped, and the emitter is formed with the third diffusion, so it is heavily doped. This is as it should be for BJTs.

The transistor has been formed. Now it must be connected, as shown in Fig. 13-10 on page 348. Again, the wafer is oxidized, and

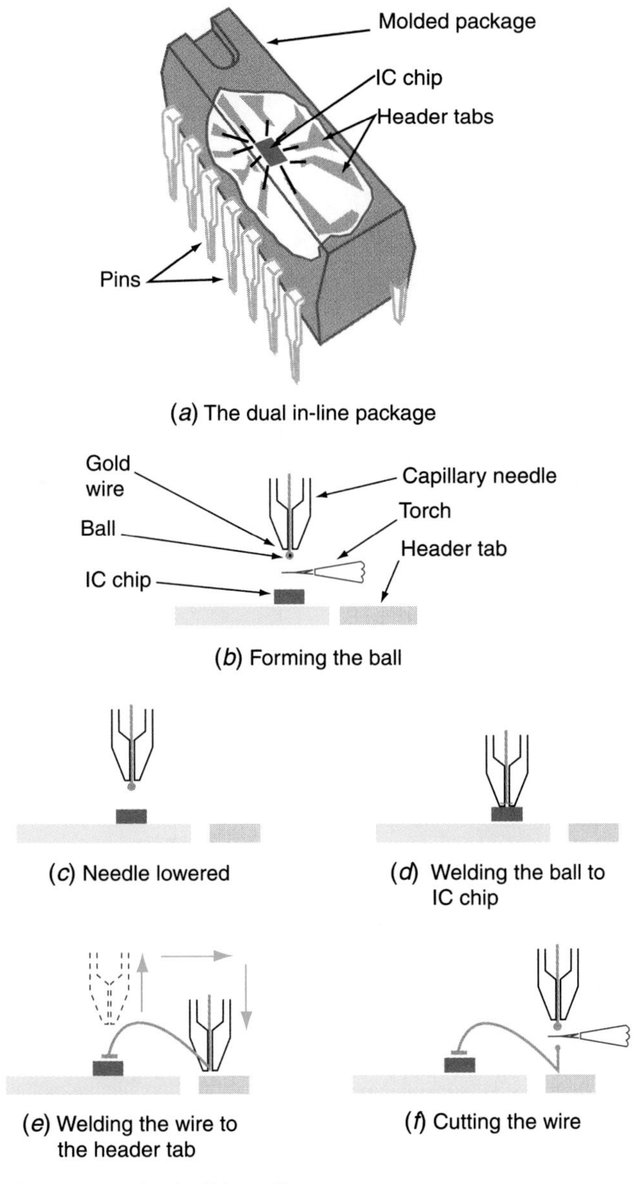

Fig. 13-8 The ball-bonding process.

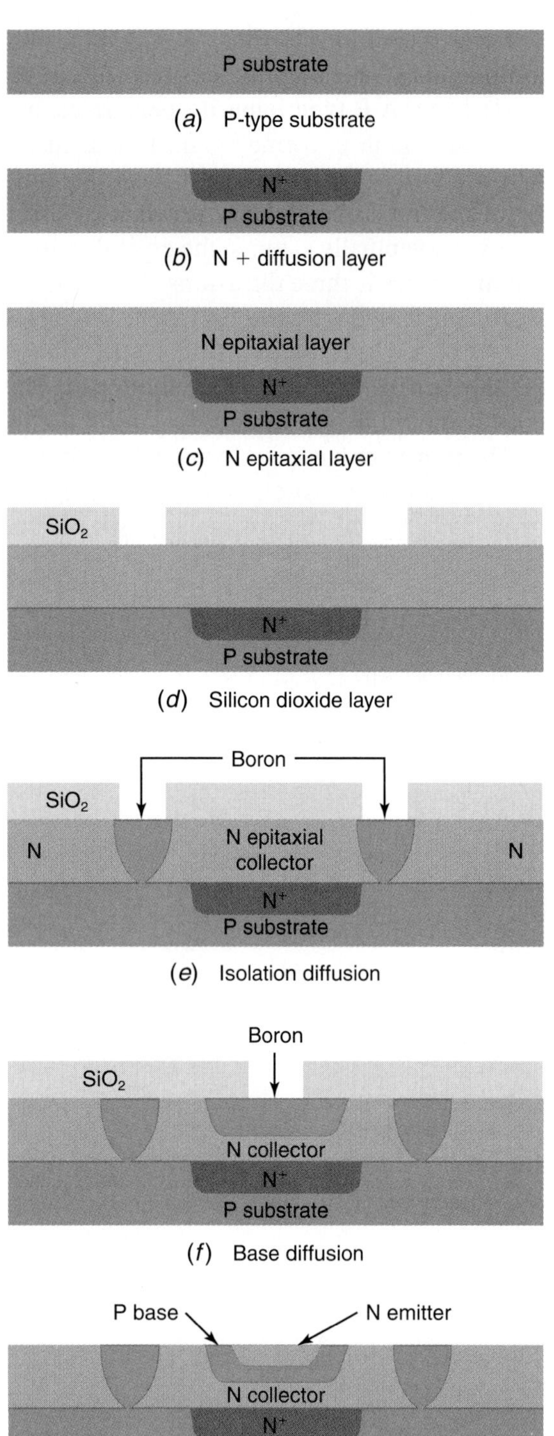

(*a*) P-type substrate

(*b*) N + diffusion layer

(*c*) N epitaxial layer

(*d*) Silicon dioxide layer

(*e*) Isolation diffusion

(*f*) Base diffusion

(*g*) Emitter diffusion

Fig. 13-9 Forming an NPN-junction transistor.

photolithography is used to produce openings that expose parts of the wafer. This time, the openings expose the collector, base, and emitter of the transistor. Aluminum is evaporated onto the surface of the wafer and makes contact with the transistor. The unwanted aluminum is later etched away leaving a separate metal path for the collector, base, and emitter of the transistor. The paths will connect the transistor to other parts of the integrated circuit.

At the same time as transistors are being formed, other circuit functions are being created as well. The PN-junction diode of Fig. 13-11 looks very much like the junction transistor. The emitter diffusion has been eliminated. The collector-base junction is the diode.

The diode (Fig. 13-11) can also serve as a capacitor if it is reverse-biased. This approach is currently used for some IC capacitors. A MOS (metal oxide semiconductor) capacitor is another possibility. Figure 13-12 shows this approach. The N-type region is one plate, the

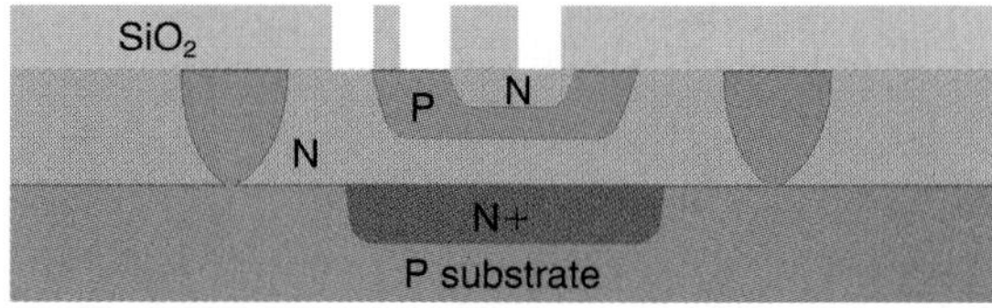

(*a*) Oxide layer with openings

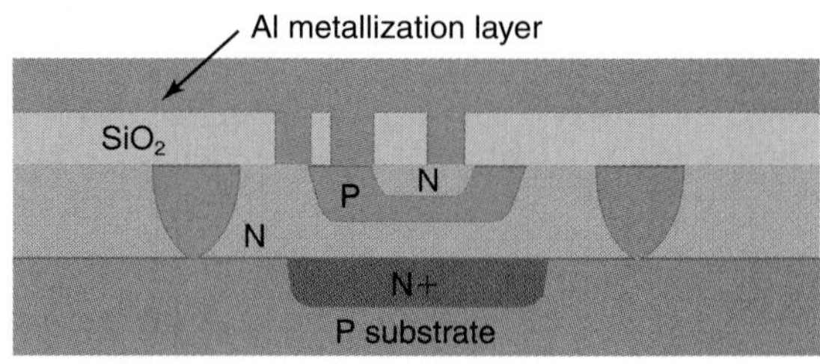

(*b*) Aluminum is evaporated onto the wafer

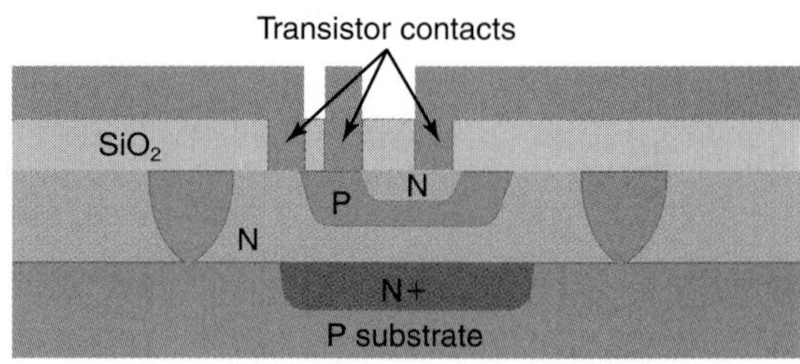

(*c*) The unwanted aluminum is etched away

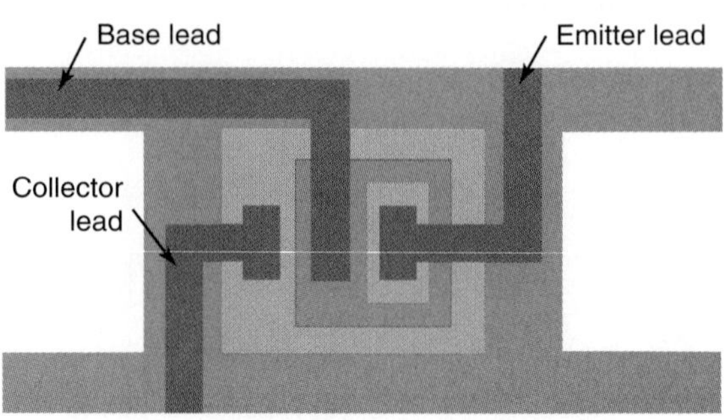

(*d*) Top view showing the remaining aluminum

Fig. 13-10 Connecting the transistor.

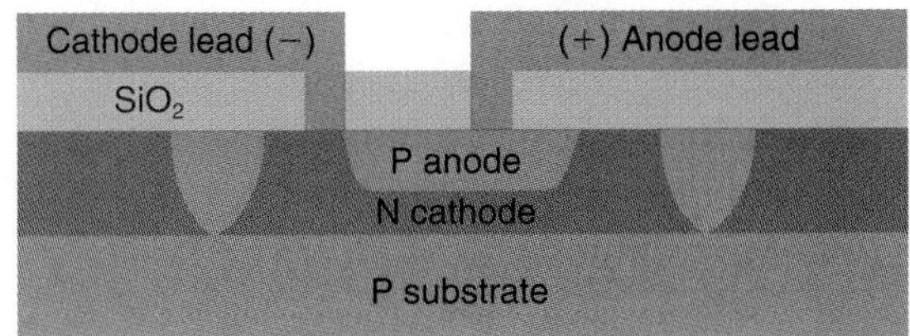

Fig. 13-11 Forming a junction diode.

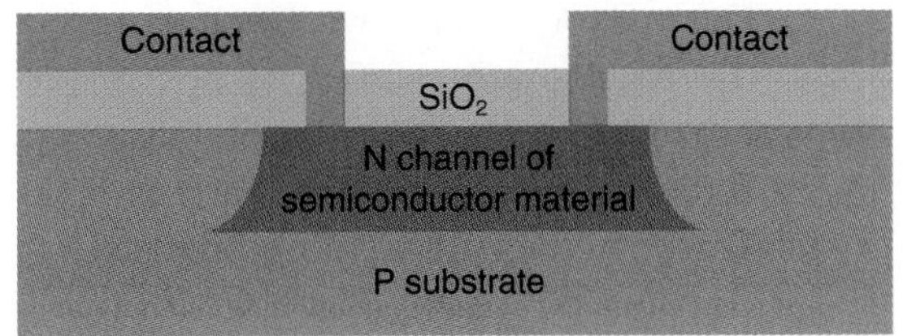

Fig. 13-13 Forming a resistor.

aluminum layer is another, and the silicon dioxide layer forms the dielectric.

The formation of a resistor in the integrated circuit is shown in Fig. 13-13. It is possible to control the size of the N channel and the amount of impurities in order to achieve different values of resistance.

Note the silicon dioxide layer between the gate lead and the channel in the MOS transistor of Fig. 13-14. A MOS transistor has the advantage of using less space in the IC chip than a junction transistor.

Integrated-circuit components have certain disadvantages compared to discrete components:

1. Resistor accuracy is limited.
2. Very low and very high values of resistors are not feasible.
3. Inductors are not practical.
4. Only small values of capacitance can be realized.
5. PNP transistors tend to have poor performance.
6. Power dissipation is limited.

On the other hand, IC components have advantages, too:

1. Since all components are formed at the same time, matched characteristics can be easily achieved.
2. Since all components exist in the same structure, excellent thermal tracking can be achieved.

Thus far, the discussion has been limited to *monolithic* integrated circuits. A monolith (single-stone) type of structure finds all the components in a single chip of silicon. *Hybrid* integrated circuits are also available. These use another substrate such as ceramic to combine various types of components. For example, a thin ceramic wafer may hold several silicon ICs, some film resistors, a few chip-type capacitors, and a power transistor. The hybrid approach is more costly, but it can eliminate some of the disadvantages cited for IC components. Hybrid ICs are available with power ratings of several hundred watts.

Monolithic IC

Hybrid IC

TEST

Choose the letter that best answers each question.

6. How are monolithic ICs made?
   a. On ceramic wafers
   b. By batch processing on silicon wafers
   c. As miniature assemblies of discrete parts
   d. None of the above
7. What is the basic process used in making monolithic ICs called?
   a. Photolithography
   b. Wave soldering
   c. Electron-beam fusion
   d. Acid etching
8. Refer to Fig. 13-5(*e*). How was the window produced?
   a. By boron diffusion
   b. With electron-beam milling

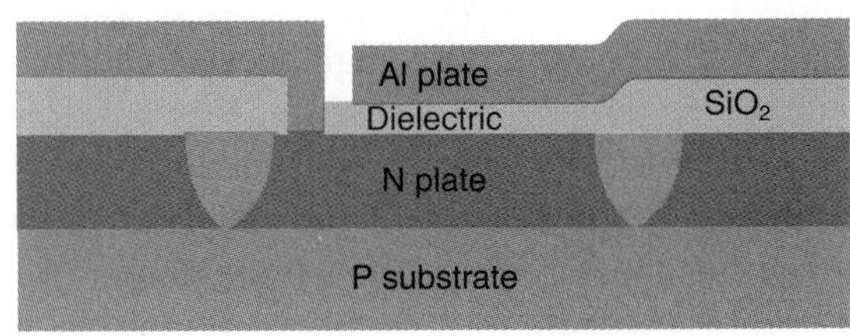

Fig. 13-12 Forming a MOS capacitor.

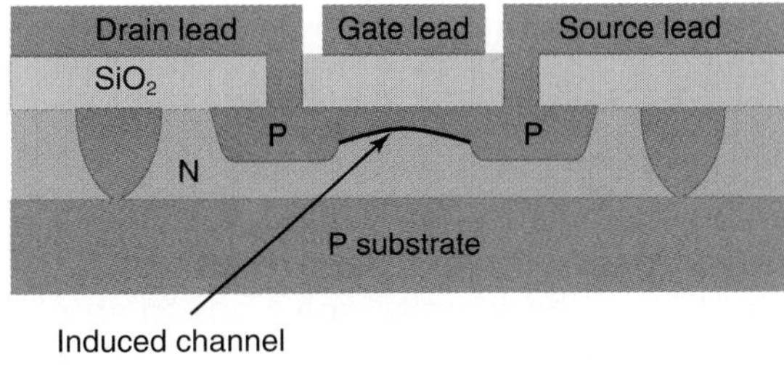

Fig. 13-14 Forming a MOS transistor.

c. The unexposed area washes away
d. By stencil cutting
9. What is the purpose of the probe test?
a. To check the photoresist coatings
b. To verify the ball-bonding process
c. To count how many ICs have been processed
d. To eliminate bad IC chips before packaging
10. What process is used to wire the chip pads to the header tabs?
a. Ball bonding
b. Soldering
c. Epoxy
d. Aluminum evaporation
11. Refer to Fig. 13-9. Why is this called a monolithic IC?
a. A hybrid structure is used.
b. Everything is formed in a single slab of silicon.
c. The base of the transistor is in the collector.
d. All of the above.
12. Refer to Fig. 13-9. Which step prevents the transistor from shorting to other components being formed at the same time?
a. The N^+ diffusion layer
b. The epitaxial layer
c. The silicon dioxide layer
d. The isolation diffusion
13. Refer to Fig. 13-10(*d*). What prevents the remaining aluminum paths from contacting unwanted regions of the IC?
a. The N^+ diffusion layer
b. The P-type substrate
c. The silicon dioxide layer
d. The Teflon spacers
14. How are capacitors formed in monolithic integrated circuits?
a. By forming PN junctions and reverse-biasing them
b. By using the MOS approach
c. Both of the above
d. Capacitors cannot be formed in ICs
15. Which type of IC combines several types of components on a substrate?
a. Monolithic
b. Silicon
c. Digital
d. Hybrid

About Electronics

High Quality and Versatile Signals

- DDS and PLL can be combined to provide signals over a very wide frequency range with excellent resolution.
- DSP is a prime example of how digital solutions to analog problems are on the increase.

13-3 The 555 Timer

The NE555 IC timer has become very popular with circuit designers because of its low cost and versatility. It is available in the 14-pin dual in-line package and the 8-pin mini-DIP. There is also an NE556, which contains two timers in a 14-pin DIP and the NE558 which is a quad timer in a 16-pin DIP. The pin numbers shown in this section are for the NE555 timer in the 8-pin package.

The 555 provides stable time delays or free-running oscillation. The time-delay mode is *RC*-controlled by two external components. Timing from microseconds to hours is possible. The oscillator mode requires three or more external components, depending on the desired output waveform. Frequencies from less than 1 Hz to 500 kHz with duty cycles from 1 to 99 percent can be attained.

Figure 13-15 shows the major sections of the 555 timer IC. It contains two voltage comparators, a bistable flip-flop, a discharge transistor, a resistor divider network, and an output amplifier with up to 200-mA current capability. There are three divider resistors, and each is 5 kΩ. This divider network sets the *threshold* comparator trip point at ⅔ of V_{CC} and the *trigger* comparator at ⅓ of V_{CC}. V_{CC} may range from 4.5 to 16 V.

Suppose that $V_{CC} = 9$ V in Fig. 13-15. In this case the trigger point will be 3 V ($\frac{1}{3} \times 9$ V) and the threshold point 6 V ($\frac{2}{3} \times 9$ V). When pin 2 goes below 3 V, the trigger comparator output switches states and sets the flip-flop to the high state and output pin 3 goes high. If pin 2 returns to some value greater than 3 V, the output stays high because the flip-flop "remembers" that it was set. Now, if pin 6 goes above 6 V, the threshold comparator switches states and resets the flip-flop to its low state. This does two things: the output (pin 3) goes low and the discharge transistor is turned on. Note that the output of the 555 timer

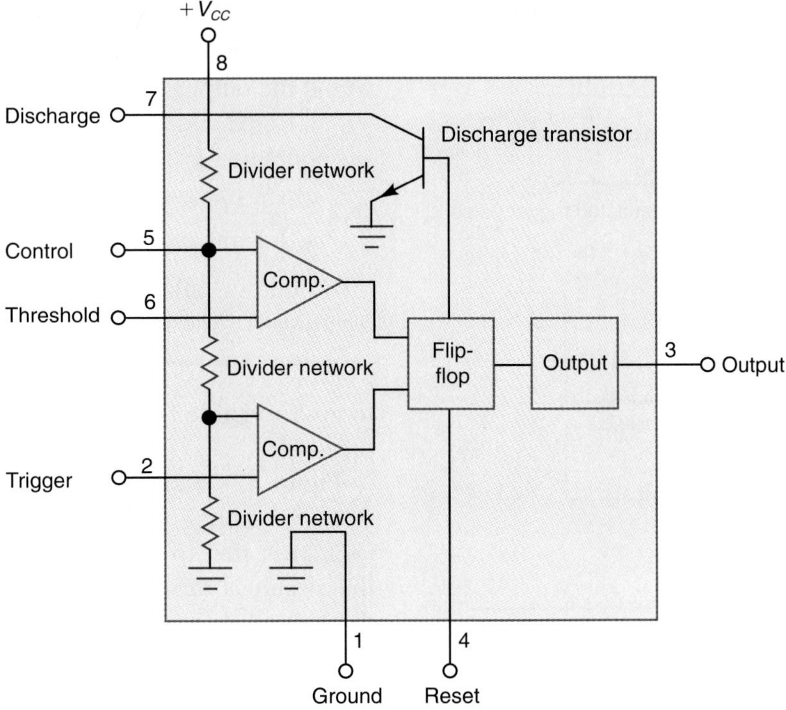

Fig. 13-15 Functional block diagram of the NE555 IC timer.

is *digital;* it is either high or low. When it is high it is close to V_{CC}, and when it is low it is near ground potential.

Pin 6 in Fig. 13-15 is normally connected to a capacitor which is part of an external *RC* timing network. When the capacitor voltage exceeds ⅔ V_{CC}, the threshold comparator resets the flip-flop to the low state. This turns on the discharge transistor which can be used to discharge the external capacitor in preparation for another timing cycle. Pin 4, the reset, gives direct access to the flip-flop. This pin overrides the other timer functions and pins. It is a digital input and when it is taken low (to ground potential) it resets the flip-flop, turns on the discharge transistor, and drives output pin 3 low. Reset may be used to halt a timing cycle. The reset function is ordinarily not needed so pin 4 is typically tied to V_{CC}. Once the 555 is triggered and the timing capacitor is charging, additional triggering (pin 2) will not begin a new timing cycle.

Figure 13-16 shows the IC timer connected for the *one-shot mode* or *monostable mode.* This mode produces an *RC*-controlled output pulse that goes high when the device is triggered. The timer is *negative-edge* triggered. The timing cycle begins at t_1 when the trigger input falls below ⅓ V_{CC}. The trigger input must return to some voltage greater than ⅓ V_{CC} before the time-out period. In other words, the trigger pulse cannot be wider than the output pulse. In those cases where it is, the trigger input must be *ac-coupled* as shown in Fig. 13-17 on page 352. The 0.001-μF coupling capacitor and 10-kΩ resistor *differentiate* the input trigger pulse. The waveforms show that the negative edge of the trigger pulse causes pin 2

Digital output

Differentiate

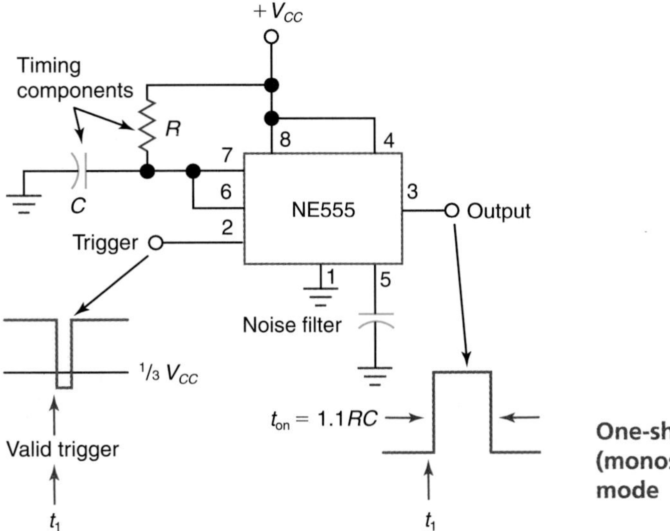

Fig. 13-16 Using the timer in the one-shot mode.

One-shot (monostable) mode

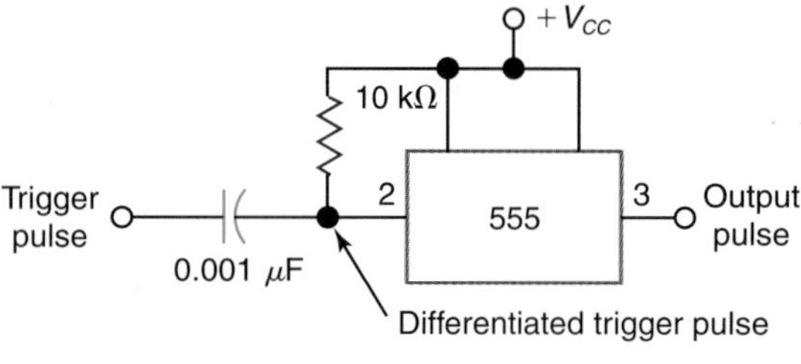

Pulse stretcher

Free-running (astable) mode

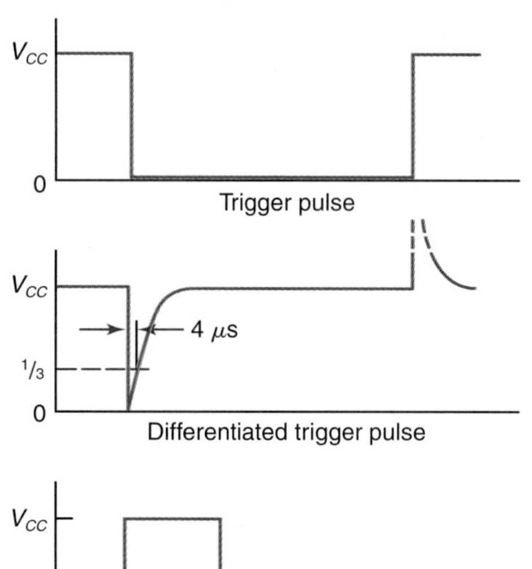

Fig. 13-17 An ac-coupled trigger pulse.

to drop to 0 V. This triggers the timer and the coupling capacitor begins charging through the 10-kΩ resistor. In approximately 0.4 time constants, the voltage at pin 2 will exceed ⅓ V_{CC}, releasing the trigger condition:

$$\begin{aligned}\text{Trigger} &= 0.4 \times R \times C \\ &= 0.4 \times 10 \times 10^3 \times 0.001 \times 10^{-6} \\ &= 0.4 \times 10^{-5} = 4\ \mu\text{s}\end{aligned}$$

Pulse differentiation (ac coupling) decreases the effective width of the trigger pulse.

Output pulse width

The *width* of the *output pulse* is *RC*-controlled in the one-shot circuit. The timing capacitor begins charging through the timing resistor when the timer is triggered. When the capacitor voltage reaches ⅔ V_{CC} the threshold comparator switches and resets the flip-flop. The discharge transistor is turned on and the capacitor is rapidly discharged in preparation for the next timing cycle. The resulting output pulse width is equal to 1.1 time constants.

One application for the one-shot mode is to use it as a *pulse stretcher.* A pulse stretcher is often handy for troubleshooting digital logic circuits. A very narrow pulse can be stretched

Example 13-1

Find the output pulse width for Fig. 13-16 if $R = 10$ kΩ and $C = 0.1$ μF. The pulse width is equal to 1.1 time constants:

$$\begin{aligned}t_{on} &= 1.1\ RC = 1.1 \times 10 \times 10^3 \times 0.1 \times 10^{-6} \\ &= 1.1\ \text{ms}\end{aligned}$$

The output pulse width will be 1.1 ms regardless of the input pulse width.

to give a visible flash of light from an LED indicator.

Figure 13-18 shows the timer configured for the *free-running* or *astable mode*. The trigger (pin 2) is tied to the threshold (pin 6). When the circuit is turned on, timing capacitor C is discharged. It begins charging through the series combination of R_A and R_B. When the capacitor voltage reaches ⅔ of V_{CC}, the output drops low and the discharge transistor comes on. The capacitor now discharges through R_B. When the capacitor reaches ⅓ V_{CC}, the output switches high and the discharge transistor is turned off. The capacitor now begins charging through R_A and R_B again. The cycle will repeat continuously with the capacitor charging and discharging and the output switching high and low.

The charge path for the astable circuit is through two resistors and the time that the output will be held high is given by

$$t_{high} = 0.69(R_A + R_B)C$$

Assume that both timing resistors in Fig. 13-18 are 10 kΩ and that the timing capacitor is 0.1 μF. The output will remain high for

$$\begin{aligned}t_{high} &= 0.69(10 \times 10^3 + 10 \times 10^3)0.1 \times 10^{-6} \\ &= 1.38\ \text{ms}\end{aligned}$$

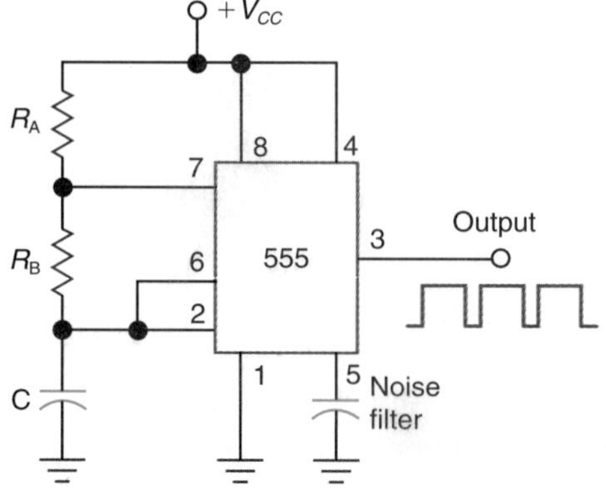

Fig. 13-18 The astable mode.

The discharge path is through only one resistor (R_B) so the time that the output is held low is shorter:

$$t_{low} = 0.69(R_B)C = 0.69(10 \times 10^3)0.1 \times 10^{-6} = 0.69 \text{ ms}$$

The output waveform is nonsymmetrical. The total period can be found by adding t_{high} to t_{low}. The output frequency will be equal to the reciprocal of the total period. Or the output frequency can be found with

$$f_o = \frac{1.45}{(R_A + 2R_B)C} = \frac{1.45}{(10 \times 10^3 + 20 \times 10^3)0.1 \times 10^{-6}} = 483 \text{ Hz}$$

The *duty cycle D* of a rectangular waveform is the percentage of time that the output is high. It can be found by dividing the total period of the waveform into the time that the output is high. For the astable circuit of Fig. 13-18, it can be found from

$$D = \frac{R_A + R_B}{R_A + 2R_B} \times 100\%$$

Assuming two 10-kΩ timing resistors gives

$$D = \frac{10 \times 10^3 + 10 \times 10^3}{10 \times 10^3 + 20 \times 10^3} \times 100\% = 66.7\%$$

Example 13-2

Calculate the output frequency and duty cycle for Fig. 13-18 if $R_A = 1$ kΩ, $R_B = 47$ kΩ, and $C = 1$ μF. Is the output a square wave? The output frequency is given by:

$$f_o = \frac{1.45}{(R_A + 2R_B)C} = \frac{1.45}{(1 \text{ k}\Omega + 2 \times 47 \text{ k}\Omega)1 \,\mu\text{F}} = 15.3 \text{ Hz}$$

The duty cycle:

$$D = \frac{R_A + R_B}{R_A + 2R_B} \times 100\% = \frac{1 \text{ k}\Omega + 47 \text{ k}\Omega}{1 \text{ k}\Omega + 2 \times 47 \text{ k}\Omega} \times 100\% = 50.5\%$$

When R_A is relatively small in value, the output approaches being a square wave.

The circuit shown in Fig. 13-18 cannot be used to produce a *square wave.* A square wave is a rectangular wave with a 50 percent duty cycle. The circuit also cannot provide waveforms with duty cycles smaller than 50 percent. The problem is that the timing capacitor charges through both resistors but discharges only through R_B. The duty-cycle equation shows that making R_A equal to 0 Ω will provide a 50 percent duty cycle. However, this can damage the IC, since there would be no current limiting for the internal discharge transistor.

Square wave

JOB TIP

SMT repairs are routinely done in some manufacturing plants.

Figure 13-19 shows a modification that permits duty cycles of 50 percent or less. A diode has been added in parallel with R_B. This diode bypasses R_B in the charging circuit. Now, the timing capacitor charges through R_A only and discharges through R_B as before. The following equations are appropriate for the modified circuit:

Duty cycle

$$t_{high} = 0.69(R_A)C$$

$$t_{low} = 0.69(R_B)C$$

$$\text{Period} = T = t_{high} + t_{low}$$

$$f_o = \frac{1}{T} = \frac{1.45}{(R_A + R_B)C}$$

$$D = \frac{R_A}{R_A + R_B} \times 100\%$$

Figure 13-20 on page 354 shows the NE555 operating in the *time-delay mode.* This mode calls for the output to change state at some determined time *after* the trigger is received. The time-delay circuit does not use the internal dis-

Time-delay mode

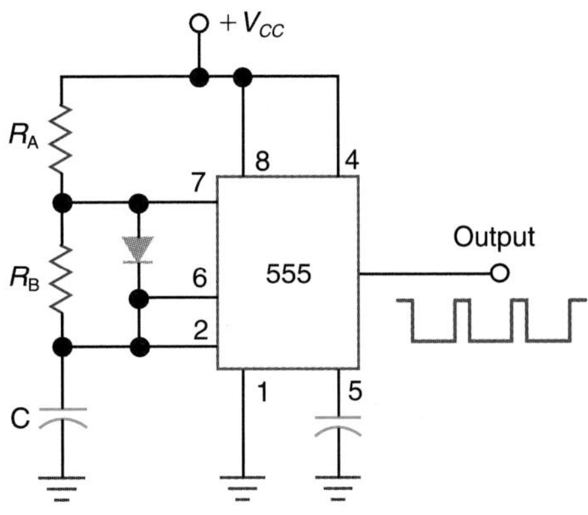

Fig. 13-19 Achieving duty cycles of 50 percent or less.

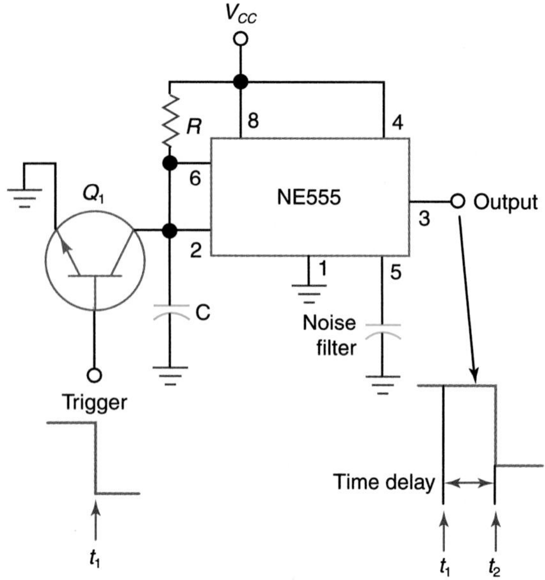

Fig. 13-20 Using the timer in the time-delay mode.

Example 13-3

Select resistor values for Fig. 13-19 that will produce a 1-kHz square wave when the timing capacitor is 0.01 μF. Beginning with the frequency equation:

$$f_o = \frac{1.45}{(R_A + R_B)C}$$

Rearranging gives:

$$R_A + R_B = \frac{1.45}{f_o \times C}$$

$$= \frac{1.45}{1 \times 10^3 \text{ Hz} \times 0.01 \times 10^{-6} \text{ F}}$$

$$= 145 \text{ k}\Omega$$

A square wave has a 50 percent duty cycle, so the resistors should be equal in value. Each resistor must be half of 145 kΩ:

$$R_A = R_B = \frac{145 \text{ k}\Omega}{2} = 72.5 \text{ k}\Omega$$

charge transistor. Operation begins with Q_1 on, which keeps the timing capacitor discharged. Timing begins when the trigger signal goes low, turning Q_1 off. This allows timing capacitor C to begin charging through resistor R. When the capacitor reaches the threshold, the output switches to a low state. If R = 47 kΩ and C = 0.5 μF, the time delay can be found by

$$t_{\text{delay}} = 1.1 \times R \times C$$
$$= 1.1 \times 47 \times 10^3 \times 0.5 \times 10^{-6}$$
$$= 2.59 \times 10^{-2} \text{ s} = 25.9 \text{ ms}$$

If the trigger signal goes high again before the IC times out, the output will not go low. This feature is useful in circuits such as security alarms where some time must be provided to exit an area before arming the alarm circuit.

Example 13-4

Determine how much time is available to leave a protected area after arming an alarm that uses the circuit of Fig. 13-20 assuming R = 470 kΩ and C = 50 μF. The time delay is equal to 1.1 time constants:

$$t_{\text{delay}} = 1.1 \times 470 \text{ k}\Omega \times 50 \ \mu\text{F} = 25.9 \text{ s}$$

For the 555 timer applications discussed so far, the control input (pin 5) has not been used. This input has been bypassed to ground with a noise capacitor (typically 0.01 μF) to prevent erratic operation. By applying a voltage at this pin, it is possible to vary the threshold comparator's trip point above or below the ⅔ V_{CC} value. This feature opens other possibilities and allows the timer IC to function as a voltage-controlled oscillator or as a pulse-width modulator.

TEST

Choose the letter that best answers each question.

16. Refer to Fig. 13-16 and assume that V_{CC} is equal to 12 V. The IC will trigger when pin 2 drops below:
    a. 2 V
    b. 4 V
    c. 6 V
    d. 8 V
17. Refer to Fig. 13-16 and again assume that V_{CC} is 12 V. The timing capacitor will begin to discharge when it reaches:
    a. 2 V
    b. 4 V
    c. 6 V
    d. 8 V

18. Refer to Fig. 13-16 and assume that $C =$ 0.5 μF and $R = 100$ kΩ. For a valid trigger, the output pulse width will be
 a. 220 μs
 b. 1.58 ms
 c. 55 ms
 d. 1.5 s
19. If the trigger input pulse width to a 555 operating in the one-shot mode is greater than the desired output pulse width, the trigger must be
 a. AC-coupled
 b. Inverted
 c. Amplified
 d. All of the above
20. Refer to Fig. 13-18 and assume that $R_A =$ 4.7 kΩ, $R_B = 10$ kΩ, and $C = 0.01$ μF. The output signal will be a
 a. Square wave
 b. Single rectangular pulse
 c. Rectangular wave
 d. Ramp wave
21. For the conditions of question 20, find the output frequency:
 a. 898 Hz
 b. 5.87 kHz
 c. 18.9 kHz
 d. 155 kHz
22. For the conditions of question 20, find the duty cycle:
 a. 59.5 percent
 b. 45.3 percent
 c. 33.7 percent
 d. 21.1 percent
23. Refer to Fig. 13-19 and assume that $R_A =$ $R_B = 22$ kΩ and that $C = 0.005$ μF. The output frequency will be
 a. 567 Hz
 b. 1.06 kHz
 c. 2.22 kHz
 d. 6.59 kHz
24. For the conditions of question 23, find the duty cycle of the output:
 a. 76.6 percent
 b. 50.0 percent
 c. 45.8 percent
 d. 25.0 percent
25. Refer to Fig. 13-20 and assume that $R =$ 1.5 MΩ and $C = 220$ μF. How long will it take the output to go low after Q_1 is turned off?
 a. 134 ms
 b. 1.39 s
 c. 4.98 s
 d. 6.05 min

13-4 Additional ICs

So far, you have been exposed to some of the most widely applied linear integrated circuits. Chapter 9 covered the operational amplifier, which is one of the most important types. The previous section covered the IC timer. Chapter 8 showed an example of an IC audio power amplifier. Chapter 15 covers IC voltage regulators, which are also very popular. This section introduces a few additional types.

An integrated circuit can be used to replace transistor stages. Figure 13-21 shows an IC commonly used as an IF amplifier. You may wonder why an IC would be used when a single transistor works in such a stage. The an-

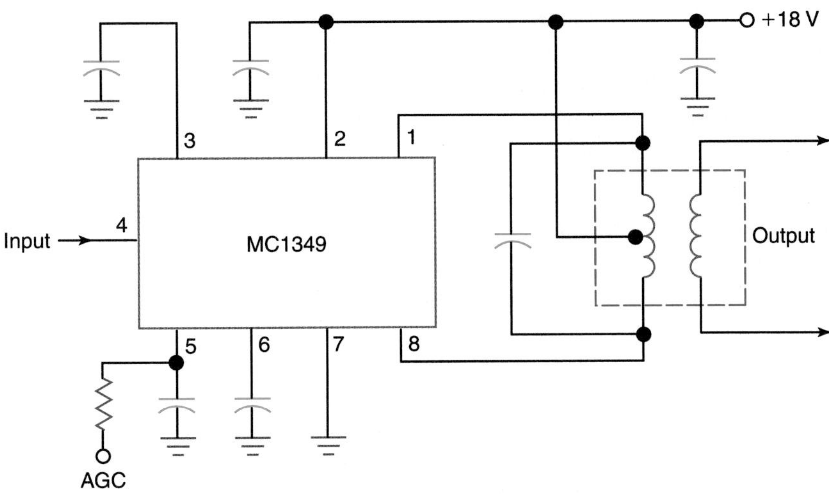

Fig. 13-21 An integrated-circuit IF amplifier.

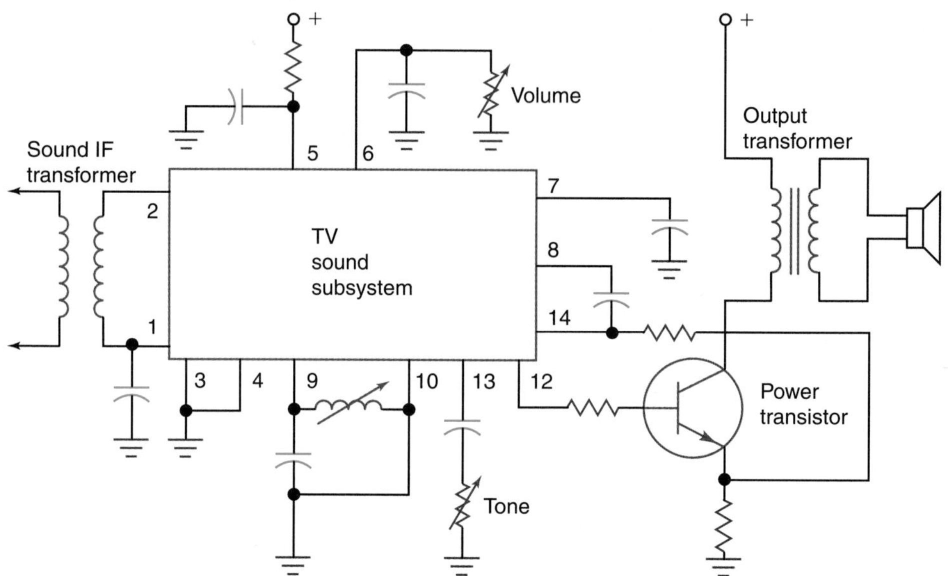

Fig. 13-22 A television sound subsystem.

swer is performance. The MC1349 integrated circuit boasts as much as 61-dB gain at 45 MHz with an AGC range of 80 dB. No single-transistor IF stage can approach this kind of performance.

Subsystem ICs

Some linear ICs provide more than one function. These are often called *subsystem integrated circuits* (Fig. 13-22). This IC contains an IF amplifier, a limiter, an FM detector, an audio driver, a regulated power supply, and an electronic volume control. This greatly reduces the parts count in the sound section of the television receiver. Cost is reduced, performance is good, and the reliability is better than in an equivalent discrete design.

Phase-locked loop

Phase-locked loops are interesting circuits, and some are available in integrated form. A phase-locked-loop circuit is shown in block diagram form in Fig. 13-23. The *phase detector* compares an input signal with the signal from a voltage-controlled oscillator. Any phase (or frequency) difference produces an error voltage. This error voltage is filtered and amplified. It is then used to correct the frequency of the voltage-controlled oscillator. Eventually, the voltage-controlled oscillator will lock with the incoming signal. Once lock is acquired, the VCO will track or follow the input signal.

FM detector

If a phase-locked-loop circuit is tracking an FM signal, the error voltage will be set by the deviation of the input signal. Thus, FM detection is realized. Figure 13-24 shows a 560B IC used as an *FM detector*. The variable capacitor is set so that the voltage-controlled oscillator operates at the center frequency of the FM signal. As modulation shifts the signal frequency, an error voltage is produced. This error voltage is the detected audio output. Phase-locked loops make very good FM detectors.

Tone decoding

Phase-locked loops are also used as *tone decoders*. These are useful circuits which can be used for remote control or signaling by selecting different tones. In Fig. 13-25, two phase-locked-loop ICs are used to build a dual-tone

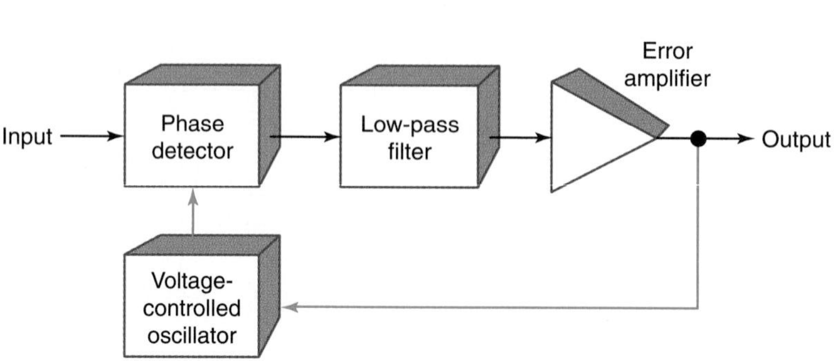

Fig. 13-23 Block diagram for a phase-locked loop (PLL).

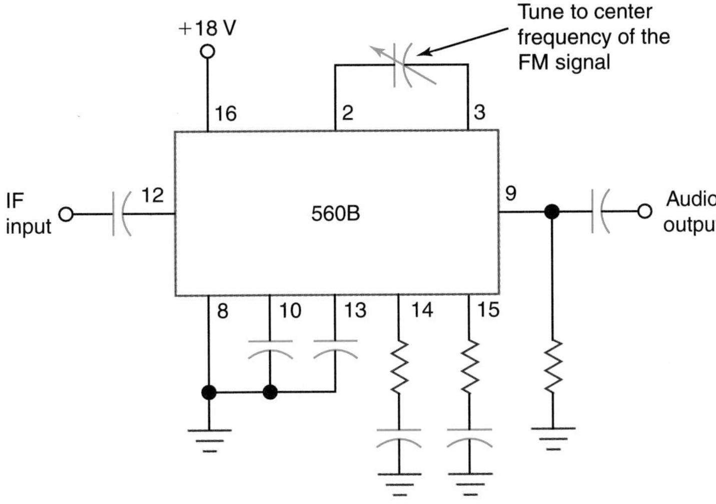

Fig. 13-24 Using the phase-locked loop for FM detection.

decoder. The output will go high *only* when *both* tones are present at the input. This type of approach is less likely to be accidentally tripped by false signals. Telephone touch-tone dialing systems use dual tones for this reason.

Frequency synthesizers have replaced older tuning methods in many electronic communications systems. Some use phase-locked loops combined with digital dividers to provide a range of precisely controlled output frequencies. Figure 13-26 shows a partial block diagram for a synthesized FM receiver. Such a receiver is desirable because it is easy to tune and makes locating a given station easy. Analog dials on FM receivers are often several hundred kilohertz in error, and it may take some time to find a station even when the frequency is known. A synthesized receiver is also very stable—so stable that no automatic frequency control circuit is needed.

Frequency synthesizer

Fig. 13-25 A phase-locked-loop tone decoder.

The FM broadcast band extends from 88 to 108 MHz. The channels are spaced 0.2 MHz apart. The number of channels is found by

$$\text{No. of channels} = \frac{\text{frequency range}}{\text{channel spacing}}$$
$$= \frac{108\ \text{MHz} - 88\ \text{MHz}}{0.2\ \text{MHz}}$$
$$= 100$$

Figure 13-26 on page 358 shows that a phase-locked loop, two crystal oscillators, and a programmable divider can be used to synthesize 100 FM channels. The stability of the output is determined by the stability of the crystals. Since crystal oscillators are among the most stable available, drift is not a prob-

The Long Road from Invention to Sales It is common for there to be a lag as long as several years before a new idea becomes a commercial success.

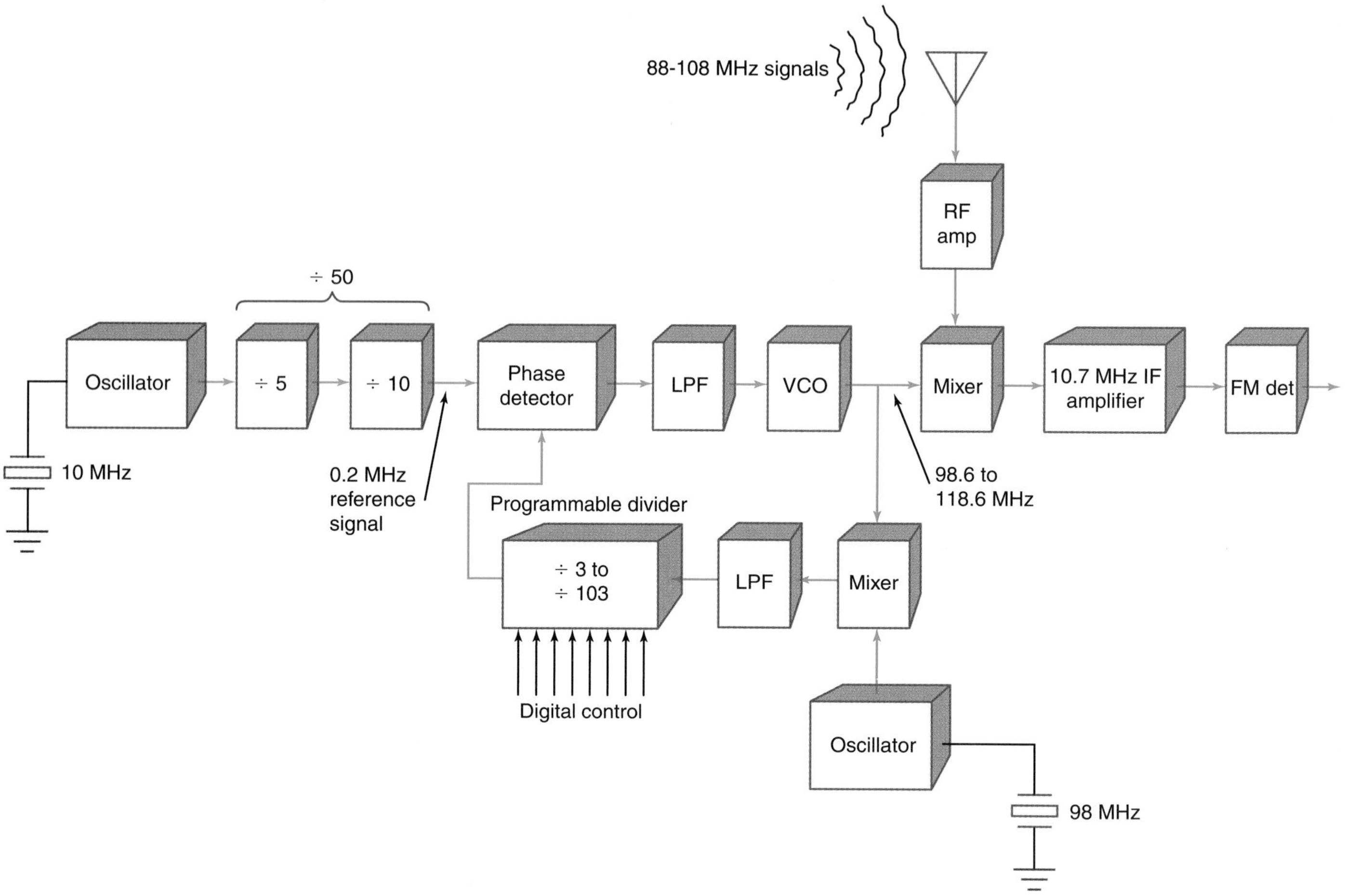

Fig. 13-26 PLL synthesized FM receiver (partial block diagram).

Reference signal

lem. One input to the phase detector is derived by dividing a 10-MHz signal by 50. This produces a signal of 0.2 MHz and is called the *reference signal.* Note that the frequency of the reference signal is equal to the channel spacing. In a PLL synthesizer, the reference frequency is usually equal to the smallest frequency change that must be programmed.

VCO

Figure 13-26 also shows a voltage-controlled oscillator (*VCO*). It feeds the receiver mixer (to the right) and the synthesizer mixer (below). Suppose you wanted to tune in a station that broadcasts at 91.9 MHz. To do so, the VCO would have to produce a signal higher than the station frequency by an amount equal to the IF frequency. The VCO frequency should be 91.9 MHz + 10.7 MHz = 102.6 MHz. The synthesizer mixer would subtract the second crystal-oscillator frequency of 98 MHz from the VCO frequency of 102.6 MHz to produce a difference of 4.6 MHz (102.6 MHz − 98 MHz = 4.6 MHz). This signal would be sent through a low-pass filter to a *programmable divider.* Assume that the divider is currently programmed to divide by 23. Therefore, the second input to the phase detector in Fig. 13-26 is 0.2 MHz (4.6 MHz ÷ 23 = 0.2 MHz), which is the same as the first input. All frequency synthesizers show both inputs to the phase detector to be equal in frequency and phase *when the loop is locked.* The loop corrects for any drift. If the VCO tries to drift low, the signal to the programmable divider becomes slightly less than 4.6 MHz and the output from the divider becomes less than 0.2 MHz. The phase detector will immediately sense the error and produce an output that goes through the low-pass filter and corrects the frequency of the VCO.

Technicians who work with engineers and scientists often make creative contributions to designs and help solve difficult problems.

Now, assume that the programmable divider in Fig. 13-26 is changed to divide by 103. Immediately, the bottom input to the phase detector becomes much less than 0.2 MHz since we are dividing by a much larger number. The phase detector responds to this error and develops a control signal that drives the VCO higher and higher in frequency. As the VCO reaches 118.6 MHz, the system starts to stabilize. This is because 118.6 MHz − 98 MHz = 20.6 MHz and 20.6 MHz ÷ 103 = 0.2 MHz, which is equal to the reference frequency. Any time the divider is programmed to a new number, the phase detector will develop a correction signal that will drive the VCO in the direction that will eliminate the error, and once again both inputs to the phase detector will become equal. Refer to Fig. 13-26 and verify that the entire FM band is covered by the synthesizer and that the channel spacing is equal to the reference frequency of 0.2 MHz.

The digital control signals to the programmable divider in Fig. 13-26 come from a front-panel key pad, a remote control, or perhaps from a scanning circuit controlled by up and down push buttons. The user of the receiver programs the desired station frequency into the receiver. The frequency information is converted to the correct digital code and is sent to the programmable divider. This blend of digital and analog circuits is very common today. Very large scale integration (VLSI) chips that contain most of the synthesizer circuitry in one package are available. It is also worth mentioning that frequency synthesizers open up new areas of performance, such as channel memory, band scanning, and automatic channel change at a prescribed time.

This section is only a sample of modern IC applications. The ones presented here give some indication of the range of circuits available. You should browse through some of the many books and data sheets offered by the IC manufacturers. Their materials often include application notes to show how the ICs can be used in typical circuits. These materials are very informative and interesting.

TEST

Choose the letter that best answers each question.

26. Refer to Fig. 13-23. When is an error voltage produced?
    a. At any time there is an input signal
    b. When the voltage-controlled oscillator is running
    c. Both of the above
    d. When there is a phase/frequency difference between the input and the oscillator
27. Refer to Fig. 13-25. When will the output go high?
    a. When tone 1 is present at the input
    b. When tone 2 is present at the input
    c. When tones 1 and 2 are present at the input
    d. All of the above
28. Refer to Fig. 13-26. The loop is locked. What is the input frequency to the bottom of the phase detector?
    a. 100 MHz
    b. 10 MHz
    c. 1 MHz
    d. 0.2 MHz

13-5 Digital Signal Processing

DSP

Digital signal processing (DSP) involves changing a signal, such as music, into a stream of numbers, performing arithmetic operations on those numbers, and then changing the resulting numbers back into an improved or enhanced signal. DSP can accomplish all sorts of things:

- Remove undesired components from a signal (such as noise or interference)
- Equalize a signal (balance the various frequency components)
- Compress or expand a signal (compression limits the voltage range of signals, and expansion restores the range to normal)
- Enhance a signal to provide special effects (such as dimensional or "surround" sound)
- Eliminate undesired signal characteristics such as echoes
- Demodulate an encoded signal (such as radio or television or in a computer modem)

All of the above can also be solved in other ways. Analog circuits can perform these tasks. The major reason why more designers are opting for DSP is that digital computer technology has become small in size and affordable. A prime example is the computer modem. Early modems were the size of a large hardbound book, whereas current notebook computer modems that use DSP are about the size of a credit card. Here are some additional reasons why DSP might be a better solution than analog circuits:

- Digital designs tend to be more reliable
- Modifications can be made via software changes, thus saving money and time
- The need for circuit alignment is eliminated, and thus manufacturing time and costs are reduced
- Preventive maintenance and calibration are easier and cost less
- Designs are highly reproducible—every manufactured unit works the same way
- Products are not sensitive to environmental changes such as temperature
- Products are not susceptible to aging (analog circuits can deteriorate with time)

DSP solutions and applications have become so widespread that electronics workers need to understand the basics of how they work and how to troubleshoot them.

DSP Block Diagram

Most DSP systems include the stages shown in Fig. 13-27:

- An input amplifier (most signals are too small and need some gain)
- An anti-aliasing filter (a low-pass filter)
- A sample-and-hold circuit (holds the signal voltage constant for digitization)
- A digitizer (an analog-to-digital converter)
- Memory (stores binary values that are representations of the input signal)
- Microprocessor (performs the arithmetic operations according to a software program)
- D/A (digital-to-analog) converter
- LPF (low-pass filter to remove high-frequency noise)

Analog-to-Digital Conversion

The analog signals that come from sources such as sensors and microphones are *continu-*

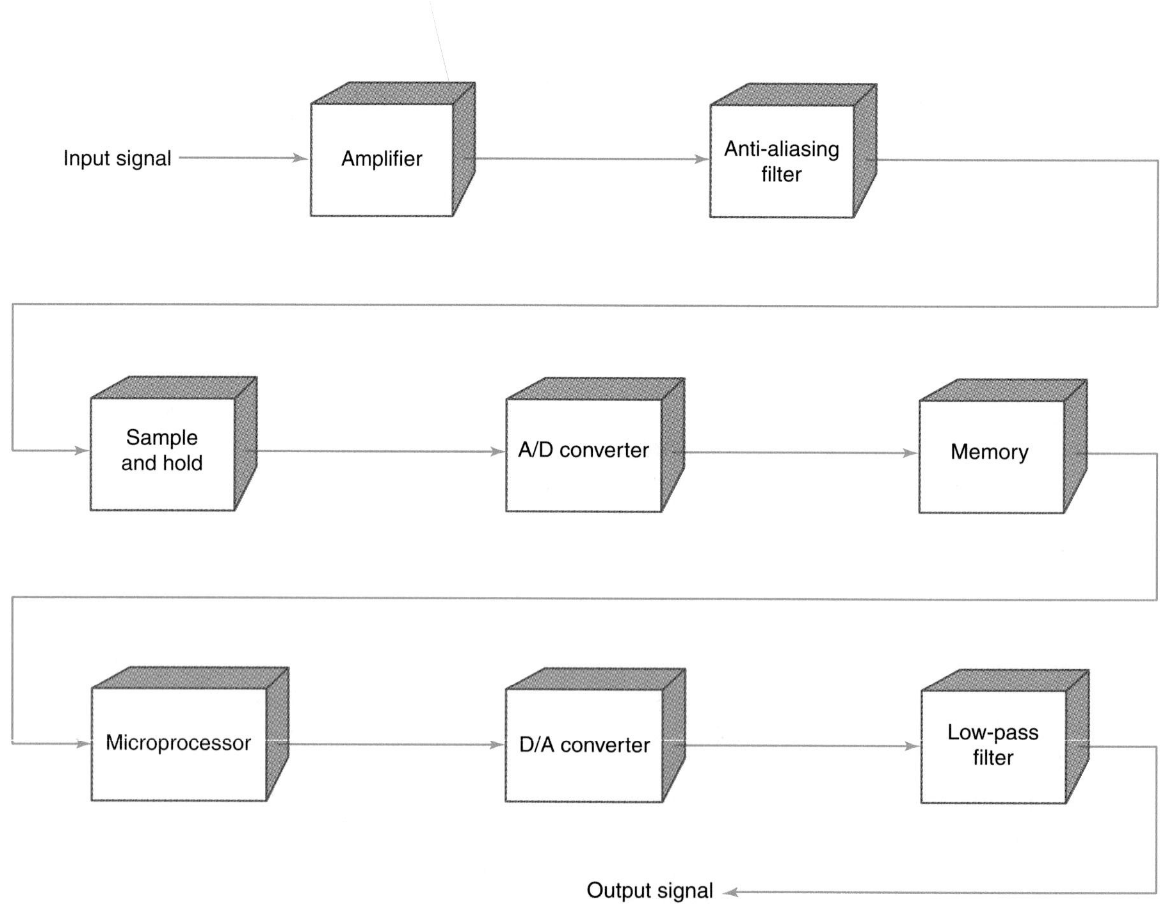

Fig. 13-27 Block diagram of a typical digital signal processing system.

ous. Their voltage value changes smoothly with time. Another type of signal, called *discrete,* can change in voltage value only at specified points in time. A continuous signal must be changed into a discrete signal to make it manageable for digital processing. The process of changing a continuous signal involves *sampling* it at points in time. Usually, a sample-and-hold circuit is employed for the first part of this process. If you will refer again to the block diagram of Fig. 13-27, you will see that the sample-and-hold function precedes the digitizer. In Fig. 13-28(*a*) a continuous signal is shown, and in Fig. 13-28(*b*), its discrete counterpart is shown. Sampling begins at time t_0.

The *Nyquist sampling theorem* states that the sampling rate must be *at least* two times the highest-frequency component of the continuous signal. If intelligible human speech requires frequencies up to 3 kHz, then the minimum sampling rate for speech is twice that, or 6000 samples per second. Often, 6000 samples per second is abbreviated as 6 ksps. Using a minimum rate does not always work out in practice, and signals are often sampled at four times the highest-frequency component of the original signal. Generally, the higher the sampling rate, the better the discrete signal represents the continuous signal. Compact audio disks are prepared by sampling music at a 66-ksps rate, and they are widely regarded as being of excellent quality. If the sampling rate is too low, the overall results will be crude. For example, sound will be distorted or a picture will be grainy or blocklike in appearance.

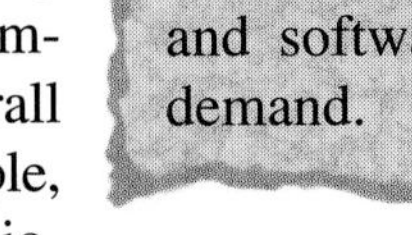

Technicians who have a combination of hardware and software skills are in demand.

Even when the sampling rate is high enough, there is still a potential problem in A/D conversion called *aliasing*. This can be understood by looking at Fig. 13-29 on page 362. Suppose that frequencies up to 1 kHz are to be converted into discrete signals. Also suppose that a 4X sampling rate, or 4 ksps, is selected. This places the sampling intervals at every 0.25 ms:

Nyquist sampling theorem

Aliasing

$$T = \frac{1}{f} = \frac{1}{4\ \text{kHz}} = 0.25\ \text{ms}$$

Look again at Fig. 13-29 and assume that the 1-kHz signal (blue) is 2 V_{p-p}. You can see that

(*a*) Input to the sample and hold amplifier

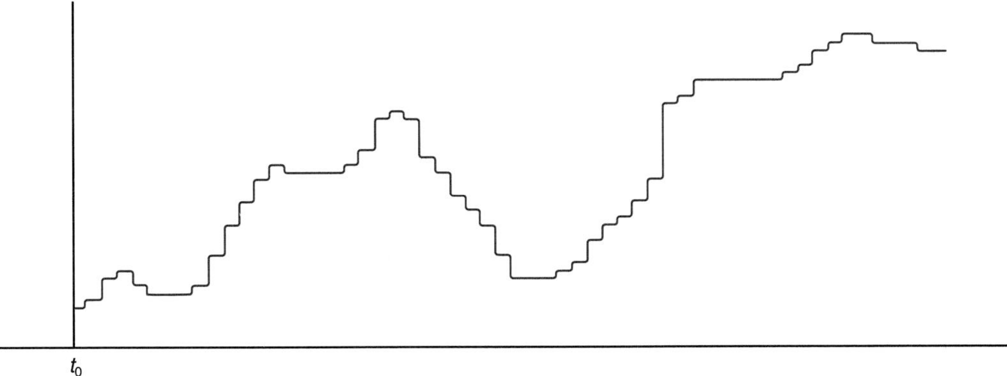

(*b*) Output of the sample and hold amplifier

Fig. 13-28 Changing a continuous signal into a discrete signal.

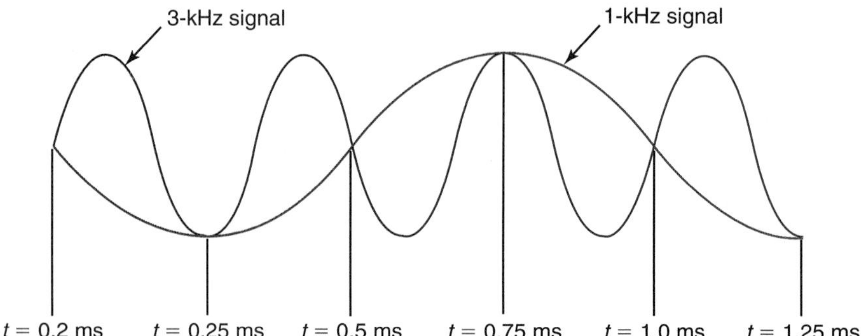

Fig. 13-29 Aliasing in A/D conversion.

its discrete (sampled) values will be 0 V, −1 V, and +1 V. Note that the 3-kHz (red) signal produces *exactly the same discrete values.* So, a 3-kHz signal will show up at the output of the A/D converter disguised (aliased) as a 1-kHz signal. You might be wondering what the 3-kHz signal is doing there in the first place in a system that is supposed to handle signals up to 1 kHz. There are two answers: interference and noise. Both of these can be unpredictable, and designers try to make their products immune to problems and malfunctions.

Luckily, aliasing is usually not a tough problem to solve. Look at Fig. 13-30. By placing an anti-aliasing filter in the signal chain before the sample and hold circuit, high-frequency noise and interference can be reduced to a level at which it will not cause any problems. An anti-aliasing filter is a low-pass filter.

There are various techniques of A/D conversion, but only one will be covered here. Figure 13-31 shows a 3-bit parallel A/D converter. Parallel converters can be very fast and are often called *flash converters*. The problem with flash converters is that they are elaborate when a lot of bits are required. The word *bit* is a contraction of two words: *bi*nary and digi*t*. There are only two binary digits: 0 and 1. Thus, an A/D converter changes each signal sample into some number of bits (0's and 1's). Here are some examples of the output of a 3-bit A/D:

Flash converter

- 000 (this binary number is equal to decimal 0)
- 011 (this binary number is equal to decimal 3)
- 101 (this binary number is equal to decimal 5)
- 111 (this binary number is equal to decimal 7)

Binary 0 is often called *LOW* and binary 1 is often called *HIGH.* If there is any chance of confusion between binary and decimal, binary numbers may be written with a *subscript* of 2, and more familiar decimal numbers with a *subscript* of 10:

$$111_2 = 7_{10}$$

The 3-bit converter in Fig. 13-31 uses 8 resistors in the voltage divider and 7 comparators:

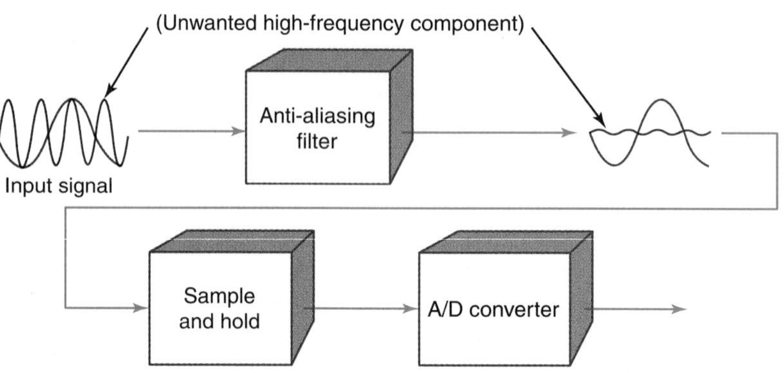

Fig. 13-30 Aliasing can be reduced by using a low-pass filter before A/D conversion.

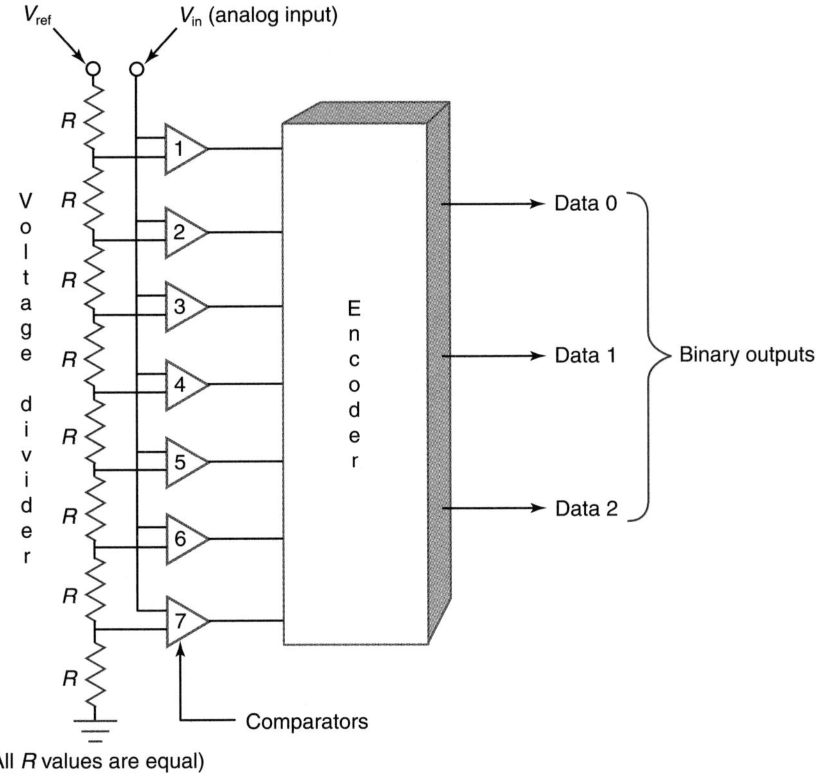

Fig. 13-31 A 3-bit parallel (flash) A/D converter.

$$\text{Number of resistors required} = 2^N = 2^3 = 8$$

$$\text{Number of comparators required} = 2^N - 1 = 2^3 - 1 = 7$$

where N = the number of bits

Flash converters tend to be used when 8 or fewer bits are enough. A flash converter for CD-quality audio would be complicated because 16 bits are required for each sound sample. Even with IC technology, the following calculations show that the circuit would be elaborate:

$$\text{Number of resistors required} = 2^{16} = 65{,}536$$

$$\text{Number of comparators required} = 2^{16} - 1 = 65{,}535$$

Returning to Fig. 13-31, we can see that each comparator has two inputs and that the input signal V_{in} is applied to every comparator. We also see that a reference voltage V_{ref} is applied to a voltage divider. We will assume a 5-V reference. The divider provides a different voltage to each comparator. The input signal will be compared to 7 different voltages so each comparator will trip at a different voltage value, as shown in Table 13-1. Each resistor in the voltage divider of Fig. 13-31 is the same value. You can solve for the voltages shown in Table 13-1 by using the voltage divider equation with $V_{ref} = 5$ V.

Note in Table 13-1 that none of the comparator outputs is high when the input signal is 0. This is because the voltage divider places all comparator inputs at some voltage greater than 0. When the input signal reaches 0.625 V, only the bottom comparator trips since it is located at the lowest point on the voltage divider. At 1.25 V the bottom two comparators are tripped. As the input signal voltage increases, additional comparators go high. Finally, at inputs

Table 13-1

Analog Input, V	Comparator Outputs	Data (Binary) Outputs
0.000	0000000	000
0.625	0000001	001
1.250	0000011	010
1.875	0000111	011
2.500	0001111	100
3.125	0011111	101
3.750	0111111	110
4.375	1111111	111

Thermometer code

of 4.375 V and above, all the comparator outputs are high. This data format is sometimes called the *thermometer code.* The thermometer code is not directly useful for DSP, so the encoder shown in Fig. 13-31 converts it to the binary data output form listed in Table 13-1.

Suppose you are troubleshooting the A/D converter shown in Fig. 13-31. What would you typically expect? If the circuit is normal, $V_{ref} = 5$ V, and when V_{in} is fixed at 2.5 V, data 0 pin will be LOW (close to 0 V), data 1 pin will be the same, and data 2 pin will be HIGH (close to 5 V). Compare this to Table 13-1. Most often, all of Fig. 13-31 is in one IC so the comparator outputs will not be available for your measurements. If V_{in} varies with time, the data output pins will switch back and forth between about 0 and 5 V and an oscilloscope will show a rectangular waveform at these pins.

Look again at Table 13-1. What if the input signal changes from 1.25 to 1.35 V and then to 1.75 V? What happens? The answer is: *nothing.* This particular A/D converter cannot *resolve* these changes. You should commit the following definitions to memory:

- *Resolution:* the ability to distinguish values or the fineness of a measurement.
- *Accuracy:* the conformity of a measured value with an accepted standard.

For more information on integrated circuits, visit National Semiconductor's Internet site.

Now you are ready for some interesting examples. This is necessary since the two words just defined are among those most abused in our language. *High resolution does not guarantee accuracy*—although it does imply it. If you visit a machine shop, you will see devices called *micrometers* that can measure in increments of one ten-thousandth of an inch. In other words, their resolution is 0.0001 in. However, if the micrometer has been dropped and bent, its *accuracy* could be far less. The bent micrometer still resolves 0.0001 in., but maybe it's off by 0.1 in.! The bent micrometer reports 0.5521 in. when measuring a true value of 0.4521 in.

Here is another example of resolution and accuracy: Digital bathroom scales typically have a resolution of 1 lb. If you go on a diet, the scale might report 180 lb on Monday morning and 179 lb on Tuesday morning. On Wednesday, it reports 179 lb, and you start thinking about better resolution. You go out and buy a better(?) scale. You use it, and it reports 181.5 lb. Which of your two scales has better accuracy? The answer is: *There is no way to be absolutely certain without checking both scales with standard weights.* Which of your two scales has better resolution? The new scale does, because it resolves a tenth of a pound.

Moving back to A/D conversion, we find that the *resolution* is simply a function of the number of bits. It is often called *step size* and is found by dividing the signal voltage range by 2^N. The signal range is also called the *span* and is equal to the difference between the lowest and highest signal voltages to be digitized. Two examples for 0- to 5-V signals are

$$\text{For a 3-bit converter, step size} = \frac{5\text{ V}}{2^3} = 0.625\text{ V}$$

$$\text{For a 16-bit converter, step size} = \frac{5\text{ V}}{2^{16}} = 76.3\ \mu\text{V!}$$

We can conclude that a 3-bit A/D converter would not provide the needed resolution for high-quality audio, but a 16-bit converter would.

People appreciate compact disk audio because 16 bits are used for every sound sample. The high resolution makes the reproduction sound as good as the original. But, remember that accuracy is also required in many applications. A scale might resolve a tenth of a pound but be off by 5 lb. Another scale that resolves only 1 lb but is never off by more than 1 lb is a better scale. Unfortunately, most people believe that the scale that resolves a tenth of a pound *has* to be more accurate, and this is not always the case. The accuracy of A/D conversion is a function of the devices used and the reference voltage. Other things, such as temperature, can also affect accuracy. So, a 7-bit converter will *always* have less resolution than an 8-bit converter, but it could have better accuracy.

The Microprocessor

Modern DSP chips can exceed 100 million instructions per second. A DSP chip is actually a special-purpose microprocessor. As can be seen in Fig. 13-27, operations on the analog signal start at the amplifier and continue until it becomes digitized at the A/D converter. It is then placed in memory, and the microprocessor loads each digitized value into one of its registers and performs arithmetic operations on the data. The results of the arithmetic operations are then transferred from the microprocessor to the D/A converter. To understand how the arithmetic operations are performed, look at Table 13-2. This table represents the analog data being processed in a DSP system with a rate of 100 samples per second. The table shows that sample number zero is 0 V and that sample 1 is taken 0.01 s later and happens to be 0.452 V.

As an example of typical DSP terminology, consider the time when the sixth sample point in Table 13-2 is being converted ($n = 6$). The sampled data value $V(6)$ is 1.47 V, and the previous value $(n-1)$ or the fifth sampled point, has a value $V(5)$ of 1.406 V. The point $(n-1)$ represents one time delay, $(n-2)$ represents two time delays, and $(n-k)$ represents k time delays. One time delay represents the sampling rate time period, or the time between two samples. An understanding of this terminology will help you understand how a DSP equation is used to generate output values.

Example 13-5

Refer to Table 13-2. Assume that the sampling occurs at point $n = 8$.

1. Determine the sampled value at point n.
2. Determine the sampled value at point $n-1$ (one time delay).
3. Determine the sampled value at point $n-2$ (two time delays).

The results are:

1. The sampled value at $n = 8$ is V(8) = 1.44 volts
2. The sampled value at $n - 1$ is the seventh sampled point and V(7) = 1.406 volts
3. The sampled value at $n - 2$ is the sixth sampled point and V(6) = 1.47 volts

Filters are among the most popular of all DSP applications. Recall that an electronic filter separates signals of different frequencies. Two types of DSP filters are commonly used. One is called an *FIR* (finite impulse response) filter, and the other an *IIR* (infinite impulse response) filter. The FIR filter will be examined with the aid of Table 13-3 on page 367 and the following *second-order* difference equation:

FIR

IIR

$$V_{out}(n) = a_0 V_{in}(n) + a_1 V_{in}(n-1) + a_2 V_{in}(n-2)$$

where a_0, a_1, and a_2 are called the FIR *coefficients.*

The *order* of a difference equation is determined by the number of time delays in the equation. It is always equal to the order of the filter. Because the above difference equation has two time delays ($n-1$ and $n-2$), this FIR filter is a second-order filter. An example of a third-order FIR difference equation is:

$$V_{out}(n) = a_0 V_{in}(n) + a_1 V_{in}(n-1) + a_2 V_{in}(n-2) + a_3 V_{in}(n-3)$$

In general, a kth-order difference equation looks like:

$$V_{out}(n) = a_0 V_{in}(n) + a_1 V_{in}(n-1) + a_2 V_{in}(n-2) + a_3 V_{in}(n-3) + \dots + a_k V_{in}(n-k)$$

Table 13-2

n = sample number	t = time of sampling, s	$V(n)$ = sampled value, V
0	0	0
1	0.01	0.452
2	0.02	0.81
3	0.03	1.083
4	0.04	1.28
5	0.05	1.406
6	0.06	1.47
7	0.07	1.406
8	0.08	1.44
9	0.09	1.361
10	0.10	1.25

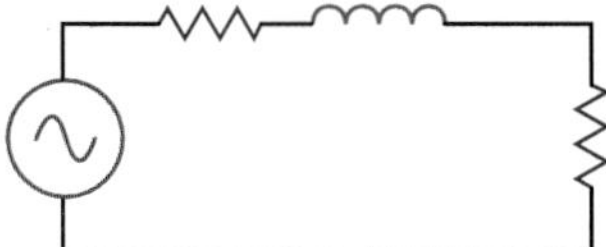

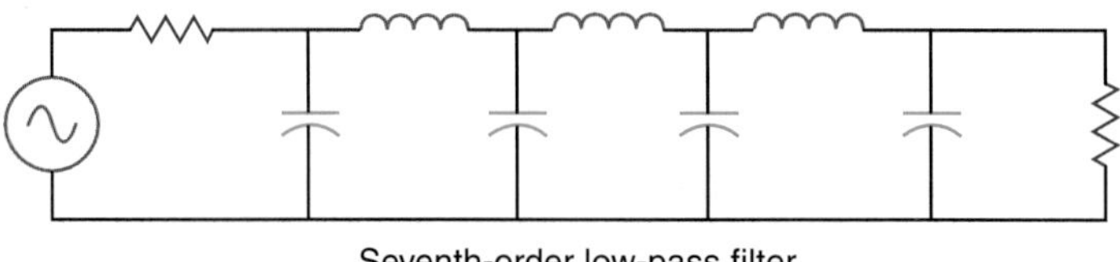

Fig. 13-32 Low-pass filters of different order.

In the case of physical filters (those based on circuits), the order is equal to the number of nonreducible coils and capacitors. Figure 13-32 shows a first-order physical low-pass filter and a seventh-order physical low-pass filter. Note that the first-order filter has a single coil, whereas the seventh-order filter has a total of seven coils and capacitors. Any coils in series or parallel with another coil would not change the order of a filter since the series or parallel combination could be reduced to a single coil. The same is true for capacitors. Figure 13-33 shows that the frequency response curve is much sharper (a fast transition) for the higher-order filter.

Returning now to the world of DSP filters and the second-order difference equation:

$$V_{out}(n) = a_0V_{in}(n) + a_1V_{in}(n-1) + a_2V_{in}(n-2)$$

The data in Table 13-3 assumes that the filter coefficients are $a_0 = 0.5$, $a_1 = -0.2$, and $a_2 = 0.15$.

Example 13-6

Refer to Table 13-3. Determine the output voltage for the fourth sample. Applying the difference equation and the coefficients:

$$V_{out}(n) = a_0V_{in}(n) + a_1V_{in}(n-1) + a_2V_{in}(n-2)$$

$$V_{out}(4) = 0.5V_{in}(4) + (-0.2)\,V_{in}(3) + 0.15V_{in}(2)$$

$$V_{out}(4) = 0.5(1.28) - 0.2(1.083) + 0.15(0.81)$$

$$V_{out}(4) = 0.5449$$

Now that a general understanding of the arithmetic for an FIR filter has been established, the IIR filter will be examined. This discussion will be aided by Table 13-4 and the following second-order difference equation. Note that the IIR filter depends upon the out-

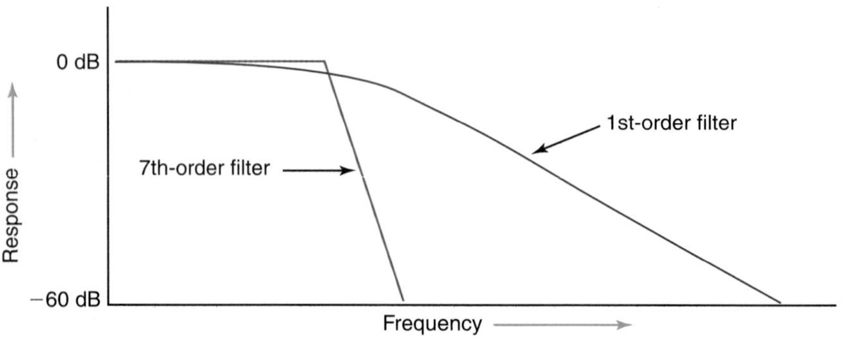

Fig. 13-33 Filter frequency response curves.

TABLE 13-3		
n = sample number	$V_{in}(n)$, input value	$V_{out}(n)$, output value
0	0	0
1	0.452	0.226
2	0.81	0.3146
3	1.083	0.4473
4	1.28	0.5449
5	1.406	0.60945
6	1.47	0.6458
7	1.406	0.6199
8	1.44	0.6593
9	1.361	0.6034
10	1.25	0.5688

put data as well as the input data and that the FIR filter depends solely on the input data:

$$V_{out}(n) = a_0V_{in}(n) + a_1V_{in}(n-1) + a_2V_{in}(n-2) - b_1V_{out}(n-1) - b_2V_{out}(n-2)$$

where a_0, a_1, a_2, b_1, and b_2 are the IIR coefficients. Because the maximum number of time delays in this difference equation is 2, the order of this IIR filter is also 2. An example of an IIR third-order difference equation is:

$$V_{out}(n) = a_0V_{in}(n) + a_1V_{in}(n-1) + a_2V_{in}(n-2) + a_3V_{in}(n-3) - b_1V_{out}(n-1) - b_2V_{out}(n-2) - b_3V_{out}(n-3)$$

The order of an IIR difference equation determines the order of the filter in the same manner that the order of the FIR filter is defined.

The data in Table 13-4 assumes that the filter coefficients are $a_0 = 0.15$, $a_1 = -0.2$, $a_2 = 0.15$, $b_1 = -0.2$, and $b_2 = 0.25$.

Example 13-7

Refer to Table 13-4. Determine the output voltage for the third sample. Applying the difference equation and the coefficients:

$$V_{out}(n) = a_0V_{in}(n) + a_1V_{in}(n-1) + a_2V_{in}(n-2) - b_1V_{out}(n-1) - b_2V_{out}(n-2)$$

$$V_{out}(3) = 0.15V_{in}(3) + (-0.2)\,V_{in}(2) + 0.15V_{in}(1) - (-0.2)V_{out}(2) - 0.25V_{out}(1)$$

$$V_{out}(3) = 0.15(1.083) - 0.2(0.81) + 0.15(0.452) + 0.2(0.04466) - 0.25(0.0678)$$

$$V_{out}(3) = 0.060232$$

The IIR and FIR coefficients are dependent upon the sampling rate and can be determined with the use of a DSP software design program. Since IIR filters depend upon the output data as well as the input data, they are said to

TABLE 13-4		
n = sample number	$V_{in}(n)$, input value	$V_{out}(n)$, output value
0	0	0
1	0.452	0.0678
2	0.81	0.04466
3	1.083	0.060232
4	1.28	0.0977814
5	1.406	0.12184828
6	1.47	0.131224306
7	1.406	0.123582791
8	1.44	0.147210482
9	1.361	0.125596399
10	1.25	0.119616659

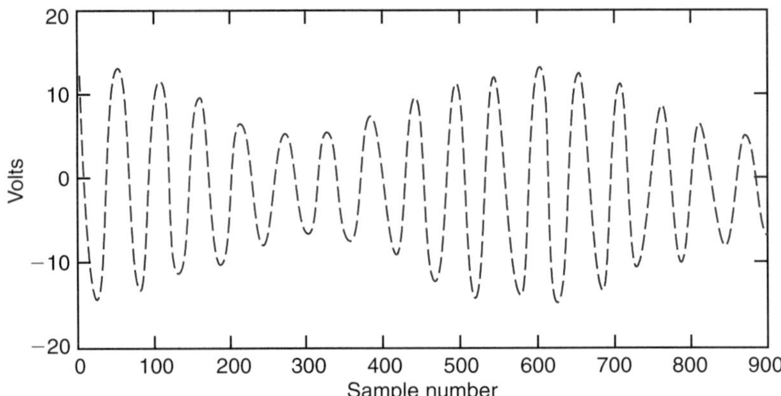

Fig. 13-34 Signal plus noise.

have *feedback properties.* Feedback makes a DSP system very sensitive to any inaccuracies that may arise. Accuracy regarding coefficient approximations and truncation errors from the arithmetic operations can cause serious problems in IIR filters but are not as critical in FIR filters. For an example of a truncation error, consider the fraction 25⁄26, which is equal to 0.961538461. But a truncation of the fraction providing five significant figures would be 0.96153. This truncation error would probably have little effect in an FIR system but would most likely lead to serious problems in a high-order IIR system. A low-order IIR filter would probably behave in an effective manner if the five significant figures of the fraction were rounded to 0.96154.

High-resolution A/D devices are used when implementing IIR filters but are not as important in FIR filters. It is not uncommon to see 16- or 24-bit A/D converters when using IIR filters. Although it is a rare occurrence, 8- or 10-bit A/D converters can be used with FIR filters. It is quite common to use 12-bit A/D converters in both FIR and IIR filters. Furthermore, the sampling rate used in the DSP system must be identical to the software value entered in the DSP application software when determining the coefficients. In general, the higher the order of the filter, the more accuracy required.

An example of how IIR filters can remove unwanted noise from a signal is demonstrated in Fig. 13-34, where the first 900 sampled points from a 4-V peak, 60-Hz sinusoidal noise are superimposed on a 10-V peak, 55-Hz sinusoidal signal. The sampling rate is 3000 samples per second. The purpose of the DSP system is to remove the 60-Hz noise. Figure 13-35 illustrates the first 200 samples from the output of a second-order IIR filter. The blue waveform represents the original 10-V peak, 55-Hz signal, and the red waveform represents the output of the IIR filter. Note that the red waveform is shown at one-half scale so that it can be compared to the original signal. Also note that an initial transient response of the DSP system occurs and

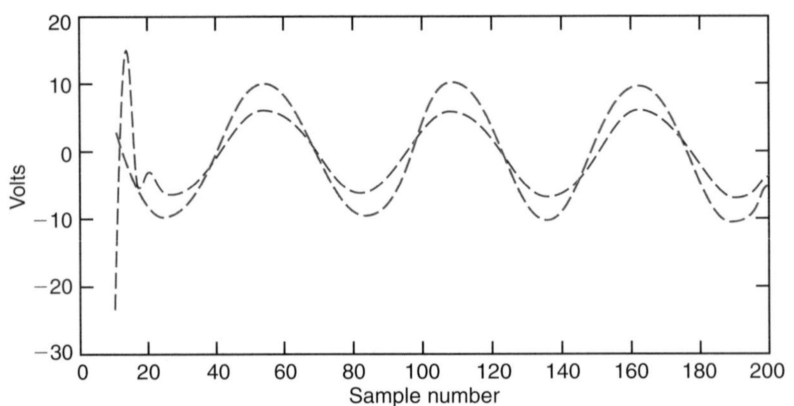

Fig. 13-35 IIR filter output.

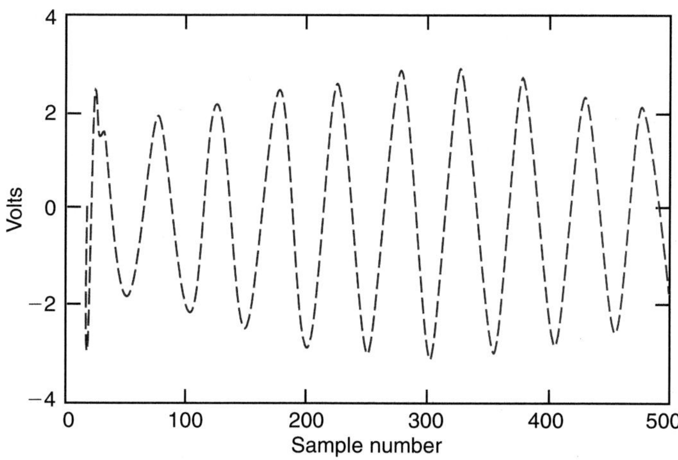

Fig. 13-36 Output due to coefficient error.

that about 20 samples are needed before the desired signal appears. The first 14 sampling points have not been illustrated in Fig. 13-35 since the transient response is so large that it dwarfs the desired output signal.

One of the most common errors that take place in DSP filtering is to enter either a wrong coefficient or an approximation of the coefficient. An example of how output errors can be generated when entering the wrong IIR coefficients is given below. The correct a_0 coefficient is:

$$a_0 = 0.999874856$$

and the incorrect value entered into the microprocessor program is:

$$a_0 = 0.99874856$$

Figure 13-36 shows the first 500 output samples, and it can clearly be seen that there is a problem. The output signal is not 55 Hz at a constant amplitude. Compare Figs. 13-35 and 13-36.

Figure 13-37 shows the output error when the sampling rate is wrong. Instead of the correct rate of 3000 samples per second, the rate has been changed to 3300. The first 500 sampled points make it clear that the output is not the desired 10-V, 55-Hz signal.

Generally speaking, the order of an FIR filter is larger than 20, and the order of an IIR filter is less than 10. In both cases, as the order of the filter increases, more precision is necessary in terms of the:

- Coefficients
- Sampling rates
- A/D converters

Digital-to-Analog Conversion

As shown in the DSP block diagram (Fig. 13-27), a digital-to-analog (D/A) converter fol-

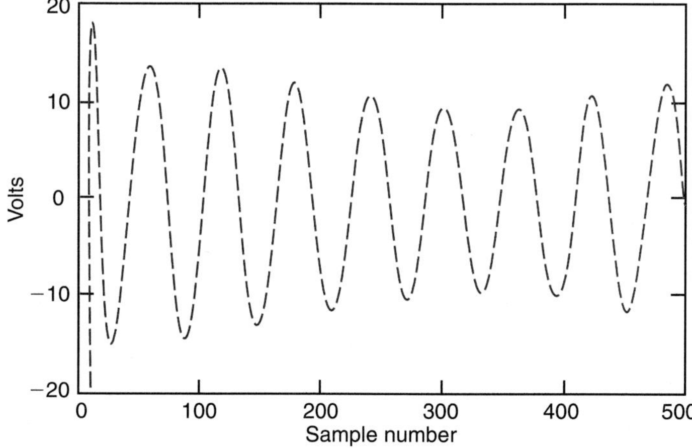

Fig. 13-37 Output due to sampling rate error.

lows the microprocessor. This stage is necessary to convert the digital values to analog form. As is true with A/D converters, D/A converters are most often ICs, and various types are used. Only one type will be presented here.

Figure 13-38(*a*) shows the schematic for a 4-bit D/A converter. Each SPDT switch represents a binary input. When a switch is set to ground, the binary input is LOW or 0; the other switch position is HIGH or 1. As shown in Fig. 13-38(*a*), the 4-bit input happens to be at 1000_2. What will V_{out} be? Using standard op-amp theory:

$$V_{out} = -V_{ref} \times \frac{R_F}{R_{in}} = -5\text{ V} \times \frac{1\text{ k}\Omega}{1\text{ k}\Omega} = -5\text{ V}$$

The resolution (smallest step size) of the D/A converter circuit shown in Fig. 13-38(*a*) can be found by solving for the output voltage

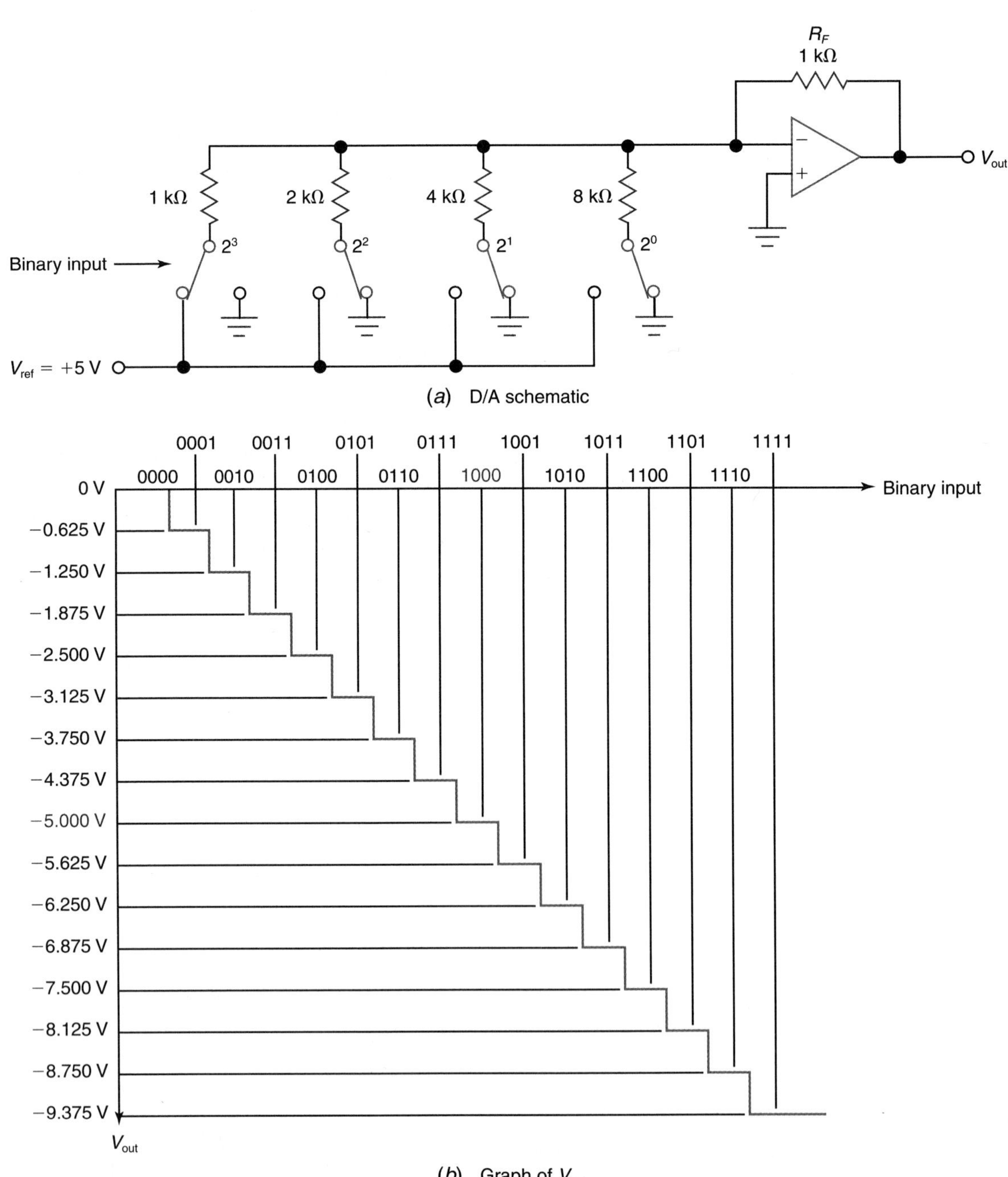

Fig. 13-38 A 4-bit D/A converter.

produced when the least significant bit (2^0) is high:

$$V_{out} = -V_{ref} \times \frac{R_F}{R_{in}} = -5 \text{ V} \times \frac{1 \text{ k}\Omega}{8 \text{ k}\Omega}$$
$$= -0.625 \text{ V}$$

The largest input is 1111_2, which is equal to 15_{10}, and the maximum output voltage can be determined by:

$$V_{out(max)} = -0.625 \text{ V} \times 15 = -9.375 \text{ V}$$

The graph in Fig. 13-38(*b*) shows a staircase waveform with 0.625-V steps from zero to −9.375 V. In cases when positive outputs are needed, V_{ref} can be changed to a negative voltage or the converter can be followed by an inverter. Figure 13-38(*b*) implies that there will be some high-frequency noise inherent in a DSP system. The output cannot change smoothly, as it does in a purely analog system. The staircase shows how the output steps from value to value. This is why the DSP block diagram (Fig. 13-27) shows a low-pass filter in the final stage. The filter eliminates the high-frequency noise due to the stepped output.

Filter Selection

When implementing a DSP filter, several decisions must be made. Should an FIR or an IIR filter be chosen? What filter order is desirable? What type of filter should be selected? The types of DSP filters are numerous. For example, the following FIR types, called *windows,* are available:

- Boxcar
- Bartlet
- Hanning
- Hamming
- Blackman
- Kaiser

IIR filter types include:

- Butterworth
- Chebyshev type I
- Chebyshev type II
- Cauer
- Bessel

With the advent of DSP software, a designer can simultaneously view several digital frequency response curves for a variety of filters. Using these curves and a trial-and-error procedure, the desired filter can be chosen.

As an example of a typical DSP application, consider a case where hospital beds are located near an elevator motor that generates a strong 120-Hz noise that produces interference on patient monitors. Measurements show that the patient monitor amplifiers are receiving about 100-mv peak noise from the motor. The maximum desired signal from the patient is only about 10 mv, so the signal-to-noise ratio is poor. A good differential amplifier along with an analog low-pass filter will remove some but not all of the noise. The decision is made to use an IIR low-pass filter with a passband that ranges from direct current to 100 Hz. This is necessary for a physician to be able to make reliable diagnostic decisions. It is determined that the 100-mv noise must be reduced to 15-mv peak. The required response is:

$$G = 20 \log(V_{out}/V_{in}) = 20 \log(15 \text{ mv}/100 \text{ mv})$$
$$= -16.5 \text{ dB}$$

The DSP filter for this application has the following requirements:

- Between direct current and 100 Hz there should be no gain or attenuation (0 dB)
- At 100 Hz, the response should be between 0 and −3 dB
- At 120 Hz, the response should be equal to or less than −16.5 dB

A DSP software design package shows that the frequency response curves for a fifth-order Butterworth filter fulfill the first two conditions, but the 120-Hz attenuation is only 13 dB. A sixth-order filter not only satisfies the first two conditions but at 120 Hz the response is −17 dB. So a sixth-order Butterworth DSP filter will satisfy the requirements for the patient monitors.

Additional Filters

Adaptive filter

An exceptional DSP filter is the *adaptive filter*. As the conditions surrounding the signal change, the IIR coefficients change in value as a function of time. Law enforcement agencies have been using adaptive filters since the 1960s to foil suspects who try to evade prosecution by playing loud music while discussing unlawful business. Adaptive filters can eliminate the music, and the conversation can be processed and uncovered with clarity.

Adaptive filters are also used in hearing aids. One of the biggest complaints about conventional hearing aids has been annoying background sounds. Many users, especially the elderly, find conventional hearing aids too irritating and refuse to use them. Adaptive filters have eliminated the background noises and have provided the hearing impaired with a beneficial product.

Comb filter

Another type of IIR filter is the *comb filter*. The frequency response graph looks like the teeth of a comb; hence the name. A comb filter can eliminate the fundamental frequency of a noise source plus all of its harmonics. Equipment operators might have to hear almost normally for safe and effective operation. However, some equipment noise can damage hearing and/or cause fatigue. Suppose a rotary machine operates at 3600 rpm. A loud sound is produced at the fundamental frequency of 3600 cycles per minute or 60 cycles per second (60 Hz). Fairly loud second, third, and fourth harmonics are also produced at 120, 180, and 240 Hz. A comb filter can allow an operator to wear a set of headphones and hear all of the important sounds but not the 60-Hz sound and its harmonics.

The medical field has been using two-dimensional IIR filters for several years. Medical imaging devices such as x-rays and ultrasound units have information stored as pixels (picture elements) on the x and the y axes. Gray scale pixels are valued from 0 (white) to 255 (black). IIR arithmetic operations on these pixels, along with the application of pattern recognition techniques, can determine abnormalities smaller than the human eye can detect.

TEST

Choose the letter that best answers each question.

29. Anti-aliasing filters are of which type?
    a. Low-pass
    b. High-pass
    c. Band-pass
    d. Band-stop
30. Analog signals are:
    a. Discrete
    b. Continuous
    c. Digital
    d. Binary
31. An A/D converter must sample:
    a. Every other waveform
    b. Continuously (all points in time)
    c. At the highest frequency to be converted
    d. At twice the highest frequency to be converted
32. Which of the following comes immediately before an A/D converter?
    a. High-pass filter
    b. Microprocessor
    c. Low-pass filter
    d. Sample and hold
33. Which of the following defines the resolution for an N-bit converter?
    a. $2N$-1
    b. $2N$
    c. 2^N
    d. $2N^2$
34. The order of a discrete filter is equal to:
    a. The number of frequencies it rejects
    b. The number of coils and capacitors it uses
    c. The number of parts it uses
    d. Its position in an electronic system
35. What type of waveform can be expected at the output of a D/A converter?
    a. Sine
    b. Discrete
    c. Continuous
    d. Aliased
36. What type of filter follows a D/A converter?
    a. Low-pass
    b. High-pass
    c. Band-pass
    d. Band-stop

13-6 Troubleshooting

Troubleshooting procedures for equipment using integrated circuits are about the same as those covered in Chap. 10. The preliminary checks, signal tracing, and signal injection can all be used to locate the general area of the problem.

The real key to good troubleshooting of complex equipment is a sound knowledge of the overall block diagram. This diagram gives the symptoms meaning. It is usually possible to quickly limit the difficulty to one area when the function of each stage is known. It is really not important if the stage uses ICs or discrete circuits. The function of the stage is what helps to determine if it could be causing the symptom or symptoms.

Figure 13-39 shows a portion of a block di-

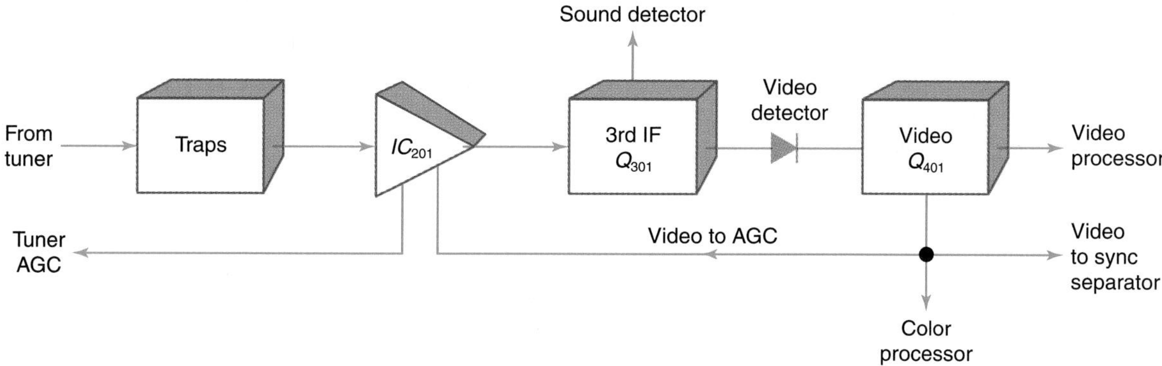

Fig. 13-39 A partial block diagram of a television receiver.

agram for a television receiver. After the preliminary checks, this type of diagram can be used to limit the possibilities. Again, the service literature is very valuable when troubleshooting. Suppose the symptoms indicate the problem could be in IC_{201}. Now it is time to check the schematic diagram.

Figure 13-40 on page 374 shows the schematic diagram for IC_{201}. It shows, in block form, the major functions inside the integrated circuit. It also shows the pin numbers and how the external parts are connected. Note that dc voltages are given. This is very important when trouble-shooting linear ICs. When a particular IC is suspected, the dc voltages should be checked. The dc voltages must be correct if proper operation is to result.

Some schematics, such as that in Fig. 13-40, show many of the dc voltages as two readings. These represent the acceptable voltage range at that particular point. For example, pin 3 is marked

$$\frac{3.2 \text{ V}}{3.6 \text{ V}}$$

This means that any voltage from 3.2 to 3.6 V will be acceptable. A reading outside this range indicates trouble.

The pin voltages tend to be more critical in circuits using ICs. In a normally working discrete circuit, the voltages often vary over a ±20 percent range. Some circuits using ICs will not function properly with an error of 5 percent. Many technicians prefer using a digital voltmeter when working on linear ICs.

Refer again to Fig. 13-40. Note that a waveform is specified for pin 7. This is another valuable servicing aid to be found on some schematics. If the waveform is missing, distorted, or low in amplitude, then a valuable clue has been found. It could indicate a faulty IC. It could also indicate that the signal arriving at the IC is faulty. The schematic usually will have enough sample waveforms that it will be possible to determine where the error is originating.

Pin 15 in Fig. 13-40 is not specified for any dc voltage. This is because pin 15 is grounded. It should, therefore, be at 0 V with respect to ground. Most technicians will take a measurement at this pin in any case. The reason for doing this is that the ground connection could be open. Solder joints can fail. If the IC is in a socket, the socket could be defective. Taking a dc voltage reading at the IC pin will make certain that it is grounded. If an oscilloscope is being used, probing a grounded pin should show a straight line at 0 V.

If a dc voltage error is found, the next step is to determine whether the problem is in the IC or the surrounding circuits. It is not a good idea to immediately change the IC. Integrated circuits are not easy to unsolder, and they can be damaged. If sockets are used, it is easy to try a new IC. However, you will run the risk of damaging the new unit if certain types of faults exist in the external parts. Also, *never* plug or unplug an IC with the power on. This invites circuit damage.

You can see that +12 V is applied to the circuit in Fig. 13-40. If any of the pin voltages is wrong, this 12-V supply circuit should be checked immediately. This must be correct in order for the pin voltages to be correct.

Suppose pin 2 in Fig. 13-40 is reading only 3.5 V. It should not read below 9.1 V. If the 12-V power supply is good, what could be wrong? Notice that there is a resistor in series with pin

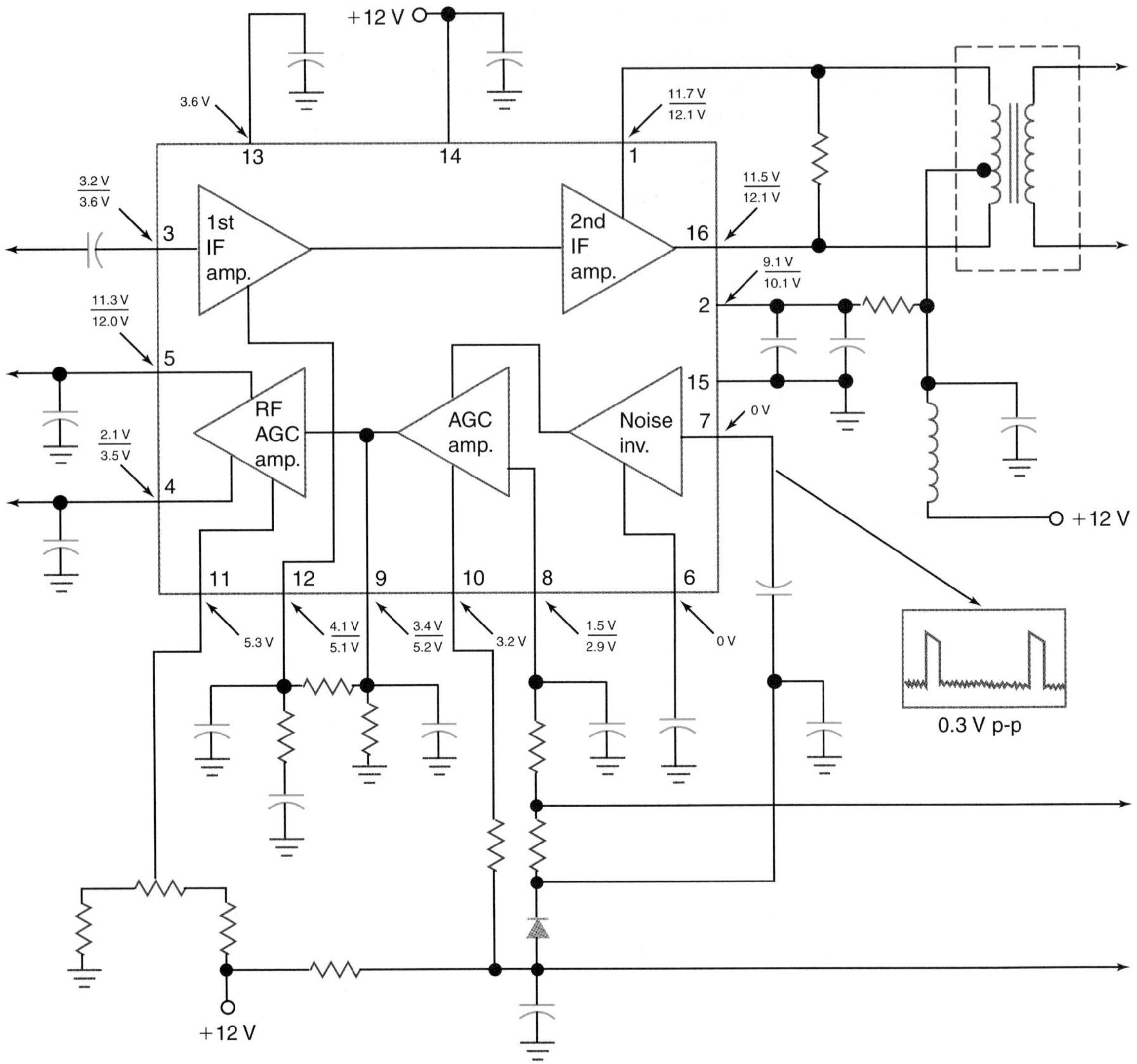

Fig. 13-40 Schematic diagram for IC_{201}.

2. It could be open or high in value. Also note that there are two capacitors from pin 2 to ground. One could be leaky. Since the IC has 16 pins to be unsoldered, it might be a good idea to check these parts first. It is easier and safer to unsolder one lead to run an ohmmeter test on some of the discrete components.

DSP Troubleshooting

When a DSP system is not functioning properly, a technician should test for the following:

- The sampling rate should be measured to ensure that it matches with the desired rate.
- The anti-aliasing filter should be checked to make sure that its cutoff frequency is one-half that of the sampling rate.
- The DSP coefficients entered in the microprocessor program should be examined to see whether they match the values provided by the DSP software application.
- The A/D converter should be tested to ensure that the proper digital output is correct for a given analog input voltage.
- The amplitude of the input analog signal should be tested to see whether it is larger than the maximum allowable to the A/D converter.
- The D/A converter should be tested to ensure that the proper analog output is correct for a given digital input value.

If you reach the conclusion that the fault is in an IC, it must be replaced. Sockets are the exception, not the rule. Thus, a tricky desol-

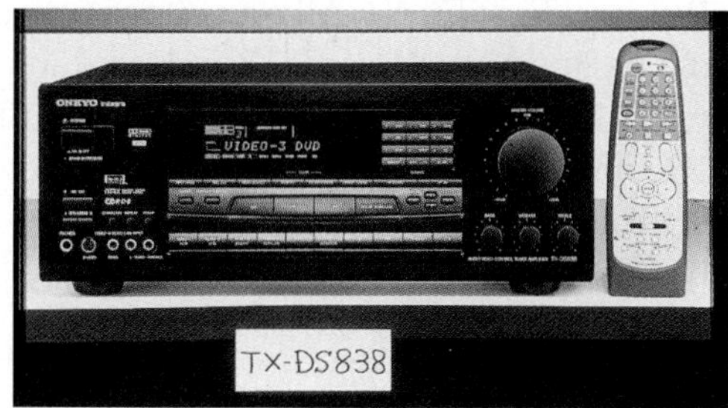

This ONKYO product uses the Motorola DSP 56009 family.

dering job is in store for you. Avoid damaging the circuit board with excess heat and do not apply the heat too long. Use the proper tools and work carefully.

Another important aspect of troubleshooting and repairing equipment that uses integrated circuits is the ability to decipher part numbers. Every IC manufacturer has a unique method of numbering parts. The numbering systems usually include logos, device numbers, prefixes, and suffixes. For example, an IC may have the following part number:

A741AHM

- The prefix A denotes a linear device
- The part number 741 tells which linear device the IC is

Working in the clean room.

- In the suffix AHM, A denotes the electrical class, H signifies that the device is in a metal package, and M signifies the military temperature range of −55° to +125°C

Another example is the following part number:

LF356CN/A+

- The prefix LF denotes the linear (BIFET) family
- The part number is 356
- The optional suffix C denotes the commercial temperature range of 0° to +70°C
- N is the package type and denotes molded dual-in-line epoxy plastic
- The optional suffix /A+ denotes that the IC is a high-reliability type

Information on manufacturers' logos and part numbers may be found in sources such as substitution guides, data books, part vendors' catalogs, and in reference books such as the *IC Master,* published by Hearst Business Communications, Inc.

TEST

Choose the letter that best answers each question.

37. Refer to Fig. 13-40. Pin 1 of the IC measures 11.9 V. What can you conclude?
    a. The IC is defective.
    b. There is a power-supply problem.

This wafer will yield a batch of integrated circuits.

c. The transformer is defective.
d. The voltage is in the normal range.

38. Refer to Fig. 13-40. Pin 1 measures 0 V. What is the next logical step?
    a. Change the IC.
    b. Replace the transformer.
    c. Replace the choke coil in the supply line.
    d. Take measurements at pins 16 and 2.

39. Refer to Fig. 13-40. All dc voltages check out as good. An IF signal is injected into pin 3, and no output is produced from the IC. What is the problem?
    a. The IC is defective.
    b. The capacitor at pin 14 is leaky.
    c. The capacitor at pin 13 is shorted.
    d. None of the above.

Chapter 13 Review

Summary

1. Discrete circuits use individual components to achieve a function.
2. Integrated circuits decrease the number of discrete components and reduce cost.
3. Integrated circuits can reduce the size of equipment, the power required, and eliminate some factory alignment procedures.
4. Integrated circuits often outperform their discrete equivalents.
5. It is possible to increase the reliability of electronic equipment by using more ICs and fewer discrete components.
6. ICs are available in a variety of package styles.
7. Monolithic integrated circuits are batch-processed into 10-mil-thick silicon wafers.
8. The key process in making monolithic ICs is photolithography.
9. Photoresist is the light-sensitive material used to coat the wafer.
10. Aluminum is evaporated onto the wafer to interconnect the various components.
11. A monolithic IC uses a single-stone type of structure.
12. A hybrid IC combines several types of components on a common substrate.
13. The 555 timer can be used in the monostable mode, the astable mode, and the time-delay mode.
14. The output of a 555 timer IC is a digital signal.
15. The 555 timer uses three identical internal resistors in its voltage divider.
16. The internal divider sets trip points at one-third and two-thirds of the supply voltage.
17. The pulse width of a timer IC is controlled by external parts.
18. Applying a voltage to the control pin of the 555 timer allows it to be used as a VCO or as a variable-pulse-width modulator.
19. A linear IC may be used to replace several transistor stages.
20. Subsystem ICs may replace more than six separate discrete stages.
21. A phase-locked loop compares an incoming signal with a reference signal and produces an error voltage proportional to any phase (or frequency) difference.
22. Phase-locked loops are used as FM detectors, as tone decoders, and as a part of frequency synthesizers.
23. The major sections of a DSP system are an A/D converter, a microprocessor, and a D/A converter.
24. Check the power-supply voltages first when troubleshooting IC stages.
25. When troubleshooting ICs, check the dc voltages at all of the pins.
26. Always remove and insert socketed ICs with the power turned off.

Chapter Review Questions

Choose the letter that best completes each statement.

13-1. A monolithic integrated circuit contains all of its components
   a. On a ceramic substrate
   b. In a single chip of silicon
   c. On a miniature printed circuit board
   d. On an epitaxial substrate

13-2. A discrete circuit uses
   a. Hybrid technology
   b. Integrated technology
   c. Individual electronic components
   d. None of the above

13-3. Refer to Fig. 13-1. When troubleshooting ICs, one may find a pin by
   a. Counting counterclockwise from pin 1 (top view)
   b. Counting clockwise from pin 1 (bottom view)
   c. Both of the above
   d. None of the above

13-4. Refer to Fig. 13-1. One may find pin 1 on an IC by
   a. Looking for the long pin
   b. Looking for the short pin
   c. Looking for the wide pin
   d. Looking for package markings and/or using data sheets

13-5. When electronic equipment is inspected, a positive identification of ICs can be made by
   a. Using service literature and part numbers
   b. Counting the package pins
   c. Finding all TO-3 packages
   d. All of the above

13-6. Refer to Fig. 13-2. A technician needs this information
   a. Seldom
   b. For choosing a replacement
   c. For troubleshooting
   d. To determine how to insert the replacement IC

13-7. The major semiconductor material used in making ICs is
   a. Silicon
   b. Plastic
   c. Aluminum
   d. Gold

13-8. When monolithic ICs are made, the following is exposed to ultraviolet light:
   a. Silicon dioxide
   b. Aluminum
   c. Photomask
   d. Photoresist

13-9. Which type of IC is capable of operating at the highest power level?
   a. Discrete
   b. Hybrid
   c. Monolithic
   d. MOS

13-10. The pads on the IC chip are wired to the header tabs
   a. By plastic conductors
   b. With photoresist
   c. By the ball-bonding process
   d. In a diffusion furnace

13-11. Refer to Fig. 13-9. Assume that the last boron diffusion (step *f*) was not performed. The component available is
   a. An inductor
   b. A diode
   c. A resistor
   d. A MOS transistor

13-12. The function of the isolation diffusion is
   a. To insulate the transistors from the substrate
   b. To insulate the various components from one another
   c. To improve the collector characteristics
   d. To form PNP transistors

13-13. The various components in a monolithic IC are interconnected to form a complete circuit by
   a. The aluminum layer
   b. Ball bonding
   c. Printed wiring
   d. Tiny gold wires

13-14. Refer to Fig. 13-11. If this structure is to be used as a capacitor, the dielectric will be
   a. The isolation diffusion
   b. The silicon dioxide
   c. The substrate
   d. The depletion region

13-15. Refer to Fig. 13-21. Assume that signal-injection tests show that no IF signal will pass through the stage. The problem could be
   a. A defective IC
   b. Improper AGC voltage
   c. One of the discrete components has failed
   d. Any of the above

13-16. Refer to Fig. 13-21. Assume that the schematic shows that pin 1 should be 18 V but it measures 0 V. Also assume that all other pin voltages are normal. The trouble is
   a. A shorted capacitor across the transformer
   b. A defective IC
   c. An open in the primary of the transformer
   d. Any of the above

13-17. Refer to Fig. 13-16. A check with an accurate oscilloscope shows that the output pulse is only half as long as it should be. The problem is in
   a. The timing resistor
   b. The timing capacitor
   c. The IC
   d. Any of the above

13-18. Refer to Fig. 13-16. It is desired to make the output pulse 1 s long. A 1-μF capacitor is already in the circuit. The value of the timing resistor should be
   a. 1 kΩ
   b. 90 kΩ
   c. 220 kΩ
   d. 0.909 MΩ

13-19. Refer to Fig. 13-19. It is desired to build a square-wave oscillator with an output frequency of 38 kHz. Assume that a 0.01-μF capacitor is already in the circuit. The values for R_A and R_B are
   a. $R_A = R_B = 1899\ \Omega$

b. $R_A = R_B = 3798\ \Omega$
c. $R_A = 1899\ \Omega$ and $R_B = 3798\ \Omega$
d. None of the above

13-20. In Fig. 13-20, $R = 18\ \text{k}\Omega$ and $C = 4.7\ \mu\text{F}$. The output will switch low, after the trigger, in
a. 18.2 ms
b. 93.1 ms
c. 188 ms
d. 0.82 s

13-21. A phase-locked-loop IC makes an excellent tone decoder or
a. Voltage regulator
b. FM demodulator
c. Television IF amplifier
d. Power amplifier

13-22. The reference frequency in a synthesizer is usually equal to
a. The VCO frequency
b. The output frequency
c. The crystal frequency
d. The channel spacing

Critical Thinking Questions

13-1. The photolithographic process used to make ICs is based on ultraviolet light. There is also a related process called *x-ray lithography.* Can you think of any reason for using x-rays to make ICs?

13-2. Several companies are experimenting with *fault-tolerant* ICs that are capable of repairing themselves. What kinds of applications will they be used for?

13-3. Some ICs combine linear and digital functions. What are some examples?

13-4. IC manufacturers often license their designs to other manufacturers. This gives other corporations the right to make and sell their designs. Why would the original manufacturer do this?

13-5. Some electronic equipment contains ICs with part numbers that cannot be referenced in catalogs, data manuals, substitution guides, or reference books. Why would this be?

Answers to Tests

1. c
2. d
3. d
4. c
5. a
6. b
7. a
8. c
9. d
10. a
11. b
12. d
13. c
14. c
15. d
16. b
17. d
18. c
19. a
20. c
21. b
22. a
23. d
24. b
25. d
26. d
27. c
28. d
29. a
30. b
31. d
32. d
33. c
34. b
35. b
36. a
37. d
38 d
39. a

Electronic Control Devices and Circuits

Chapter Objectives

This chapter will help you to:

1. *Calculate* efficiency in control circuits.
2. *Identify* the schematic symbols for thyristors.
3. *Explain* the operation of thyristors.
4. *Define* conduction angle in thyristor circuits.
5. *Explain* commutation in thyristor circuits.
6. *Troubleshoot* control circuits.

Control of loads is an important application area. For example, a control circuit may be used to accurately set and maintain the speed of a motor. Lights and heating elements can also be regulated with control circuits. The adjustable resistor, or rheostat, can be used to control loads. This chapter describes solid-state control devices and circuits that work much more efficiently than rheostats. It also shows how feedback can be used in control circuits.

14-1 Introduction

Figure 14-1 shows the use of a rheostat to control the brightness of an incandescent lamp. It is obvious that as the rheostat is adjusted for more *resistance,* the circuit current will decrease and the lamp will dim. The rheostat gets the job done but wastes energy. Solving a typical circuit will show why. In order to dim the lamp in Fig. 14-2, the rheostat has been set for a resistance of 120 Ω. This makes the total circuit resistance:

Resistance control

$$R_T = 120\ \Omega + 120\ \Omega = 240\ \Omega$$

The circuit current can now be found by using Ohm's law:

Efficiency

$$I = \frac{V}{R} = \frac{120\ \text{V}}{240\ \Omega} = 0.5\ \text{A}$$

This, of course, is less current than when the rheostat is set for no resistance:

$$I = \frac{120\ \text{V}}{120\ \Omega} = 1\ \text{A}$$

Ohm's law has shown that setting the resistance of the rheostat equal to the load resistance halves the current flow. Now, let's investigate the power dissipated in the load. The current flow is 0.5 A, and the load resistance is 120 Ω:

$$P = I^2R = (0.5\ \text{A})^2 \times 120\ \Omega = 30\ \text{W}$$

When the rheostat is set at no resistance, the power is

$$P = (1\ \text{A})^2 \times 120\ \Omega = 120\ \text{W}$$

The rheostat controls the power dissipated in the load. It has been shown that the power dissipated in the load drops to one-fourth when the current is halved. This is to be expected since power varies as the square of the current.

It is time to look at the *efficiency* of the rheostat control circuit. At full power, the rheostat is set for no resistance. Therefore, no power will be dissipated in the rheostat:

$$P = (1\ \text{A})^2 \times 0\ \Omega = 0\ \text{W}$$

At one-fourth power, the rheostat dissipation is

$$P = (0.5\ \text{A})^2 \times 120\ \Omega = 30\ \text{W}$$

This is not an efficient circuit. Half of the total power is dissipated in the control device

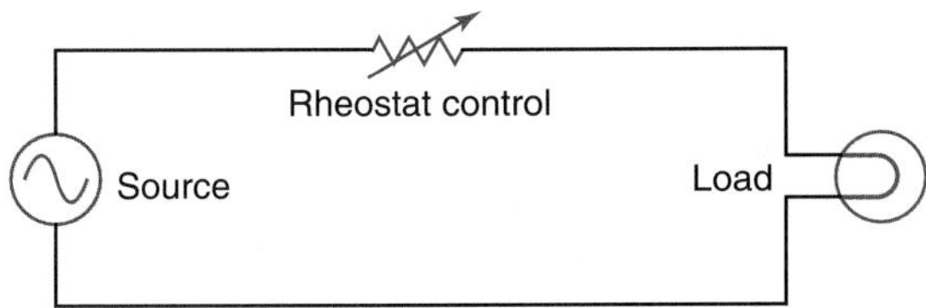

Fig. 14-1 A simple rheostat control circuit.

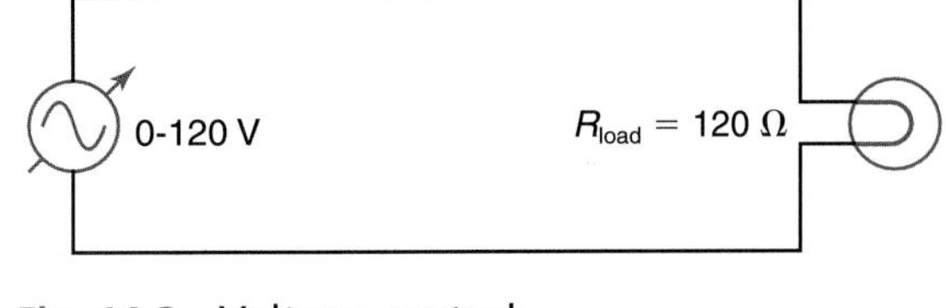

Fig. 14-3 Voltage control.

when the current is halved for an efficiency of only 50 percent. As the resistance of the rheostat is increased, the circuit efficiency decreases. In a high-power circuit, the poor efficiency will produce a high cost of operation. The rheostat will have to be physically large to safely dissipate the heat.

The previous analysis was simplified. It assumed that the resistance of the incandescent lamp remains constant. It does not. However, the conclusions are correct. Rheostat control is inefficient.

What are the alternatives? One is *voltage control.* Figure 14-3 shows such a circuit. As the voltage of the source is adjusted from 0 to 120 V, the power dissipated in the load will vary from 0 to 120 W. This method is much more efficient than the rheostat control circuit. Since there is only one resistance in the circuit of Fig. 14-3, there is only one place to dissipate power. The efficiency of the circuit will always be 100 percent.

Unfortunately, voltage control is not easy to obtain. There is no simple and inexpensive way to control line voltage. A variable transformer is a possibility, but would be a large and expensive item for a high-power circuit.

To be efficient, a control device should have very low resistance. A *switch* is an example of an efficient control device. When the switch in Fig. 14-4 is closed, 1 A of current flows. The power dissipated in the load is 120 W. If the switch has very low resistance, then very little power is dissipated in the switch. When the switch is open, no current flows. With no current, there cannot be any dissipation in the switch. Thus, there is never any significant power dissipation in a switch.

You may be wondering what the circuit of Fig. 14-4 has to do with dimming a lamp or controlling the speed of a motor. It seems that only on-off control is available. This is usually the case with ordinary mechanical switches. However, think for a moment about a very fast switch. Suppose this fast switch can open and close 60 times per second and is closed only half the time. What do you think the condition of the lamp will be? Since the lamp will be connected to the source only half the time, it will operate at reduced intensity and the control device (the fast switch) will run cool.

Voltage control

Mechanical switches cannot serve in this capacity. Even if they could be made to operate quickly, they would wear out in a short time. An *electronic* (solid-state) *switch* is needed. Fast operation will allow the lamp to dim without any noticeable flicker and the electronic switch will run cool. The next section covers such a control device.

Electronic switch

TEST

Choose the letter that best answers each question.

Switch control

1. A load has a constant resistance of 60 Ω. A rheostat is connected in series with the load and set for 0 Ω. How much power is dissipated in the load if the line voltage is 120 V?
   a. 0 W
   b. 60 W

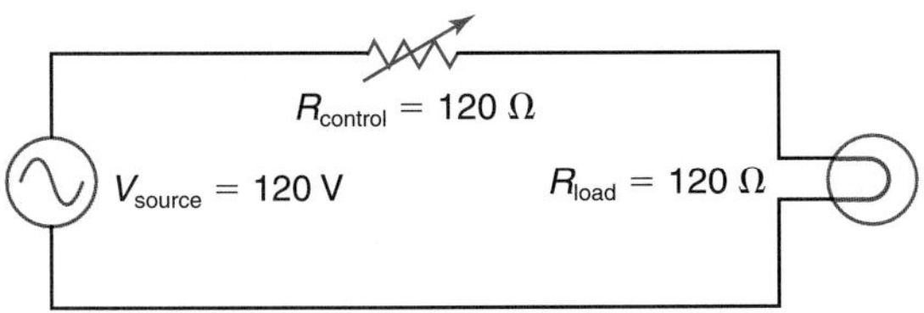

Fig. 14-2 Analyzing a rheostat control circuit.

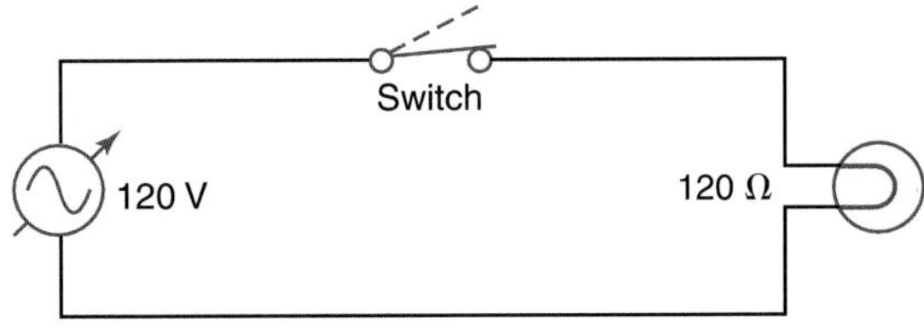

Fig. 14-4 Switch control.

c. 120 W
d. 240 W

2. What is the circuit efficiency in question 1?
   a. 0 percent
   b. 25 percent
   c. 50 percent
   d. 100 percent
3. Refer to question 1. Everything is the same except the rheostat is set for a resistance of 30 Ω. How much power is dissipated in the load?
   a. 18 W
   b. 36 W
   c. 107 W
   d. 120 W
4. What is the circuit efficiency in question 3?
   a. 11 percent
   b. 67 percent
   c. 72 percent
   d. 100 percent
5. The resistance of a certain load is constant. The current through the load is doubled. By what factor will the power dissipated in the load increase?
   a. 1.25 times
   b. 2.00 times
   c. 4.00 times
   d. Not enough information is given
6. Why is it not efficient to use a control resistor or a rheostat to vary load dissipation?
   a. Much of the total power is dissipated in the control.
   b. Power is set by voltage, not by circuit resistance.
   c. Power is set by current, not by circuit resistance.
   d. Loads do not show constant resistance.
7. Why is there no power dissipation in a perfect switch?
   a. When the switch is closed, its resistance is zero.
   b. When the switch is open, the current is zero.
   c. Both of the above.
   d. None of the above.

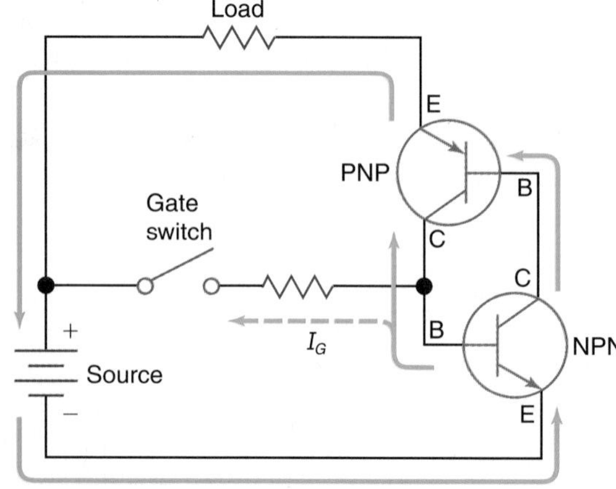

Fig. 14-5 A two-transistor switch.

14-2 The Silicon-Controlled Rectifier

One of the most popular electronic switches is the *silicon-controlled rectifier,* or SCR. This device is easier to understand if we first examine the two-transistor equivalent circuit shown in Fig. 14-5. The circuit shows two directly connected transistors, one an NPN and the other a PNP. The key to understanding this circuit is to recall that BJTs do not conduct until base current is applied. It can be seen in Fig. 14-5 that each transistor must be on to supply the other with base current.

Gate switch

How does the circuit of Fig. 14-5 turn on? Notice that a *gate switch* has been included. When the source is first connected, no current will flow through the load because both transistors are off. When the gate switch is closed, the positive side of the supply is applied to the base of the NPN transistor. This forward-biases the base-emitter junction and the NPN transistor turns on. This applies base current to the PNP transistor and it turns on. With both transistors on, current flows through the load.

Latch

What happens in Fig. 14-5 when the gate switch opens? Will the transistors shut off and stop the load current? No, because once the transistors are on, they supply each other with base current. Once triggered by the gate circuit, the transistors in Fig. 14-5 continue to conduct until the source is removed or the load circuit is opened. The two-transistor switch can be turned on by a gating current, but removing the gating current will not turn the switch back off. Such a circuit is often called a *latch.* Once triggered, a latch stays on.

The two-transistor switch of Fig. 14-5 is efficient. When the transistors are off, they show very high resistance and the current and power dissipation approach zero. When the transistors are on, they are in saturation (turned on hard) and show low resistance. This means low power dissipation in the switch.

Figure 14-6 shows a way to simplify a two-transistor switch. A single four-layer device will do the same job. Study Fig. 14-6 and verify that it is the equivalent of the two transistors shown in Fig. 14-5. The *four-layer diode,* as shown in Fig. 14-6, is an important electronic control device. It is called a diode because it conducts in one direction and blocks in the other. It is usually called a silicon-controlled rectifier (*SCR*).

Figure 14-7 shows the schematic symbol for an SCR. The electron current flow is the same as in an ordinary diode, from cathode to anode. The symbol is that of a solid-state diode with the addition of a gate lead.

Figure 14-8 is a volt-ampere characteristic curve for an SCR. It shows device behavior for both forward bias ($+V$) and reverse bias ($-V$). As in ordinary diodes, very little current flows when the device is reverse-biased until the reverse breakover voltage is reached. Reverse breakover is avoided by using SCRs with ratings greater than the circuit voltages. The forward-bias portion of the volt-ampere curve is very different when compared to an ordinary diode. The SCR stays in the off state until the *forward breakover* voltage is reached. Then, the diode switches to the on state. The drop across the diode decreases rapidly, and the current increases. The *holding current* is the minimum flow that will keep the SCR latched on.

Figure 14-8 is only part of the story because it does not show how gate current affects the characteristics of the SCR. Refer to Fig. 14-9 on page 384. Gate current I_{G_1} represents the

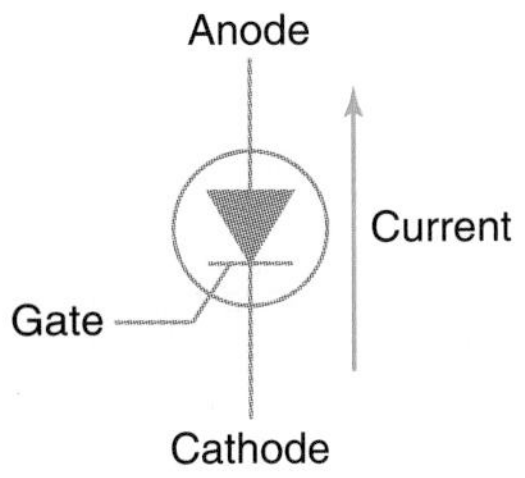

Fig. 14-7 Schematic symbol for the SCR.

smallest of the three values of gate current. You can see that when gate current is low, a high forward-bias voltage is required to turn on the SCR. Gate current I_{G_2} is greater than I_{G_1}. Note that less forward voltage is needed to turn on the SCR when the gate current is increased. Finally, I_{G_3} is the highest of the three gate currents shown. It requires the least forward bias to turn on the SCR.

SCR

In ordinary operation, SCRs are not subjected to voltages high enough to reach forward breakover. They are switched to the on-state with a *gate pulse* large enough to guarantee turn-on even with relatively low values of forward-bias voltage. Once triggered on by gate current, the device remains on until the current flow is reduced to a value lower than the holding current.

Gate pulse

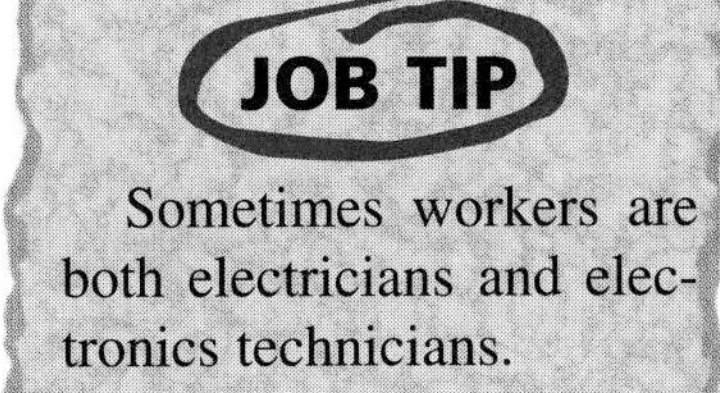

Now that we know something about SCR characteristics, we can better understand some applications. Figure 14-10 on page 384 shows the basic use of an SCR to control power in an ac circuit. The load could be a lamp, a heating element, or a motor. The SCR will con-

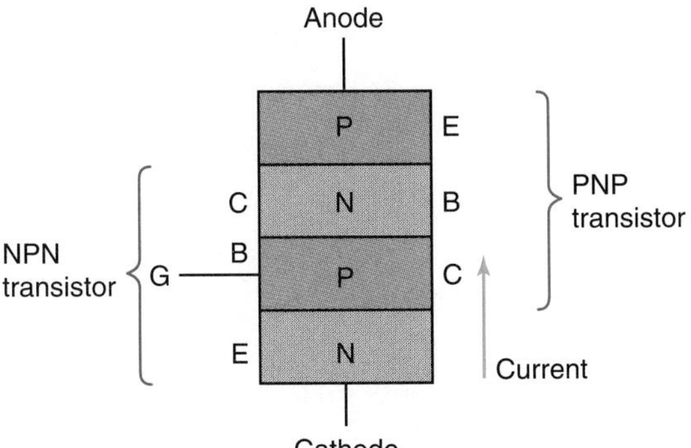

Fig. 14-6 A four-layer diode, or silicon-controlled rectifier.

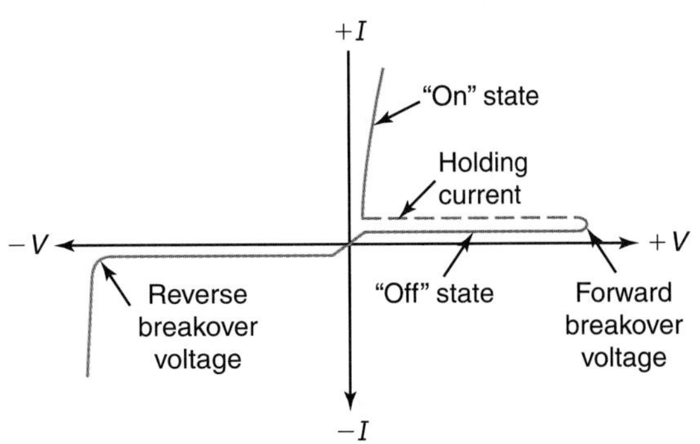

Fig. 14-8 An SCR volt-ampere characteristic curve.

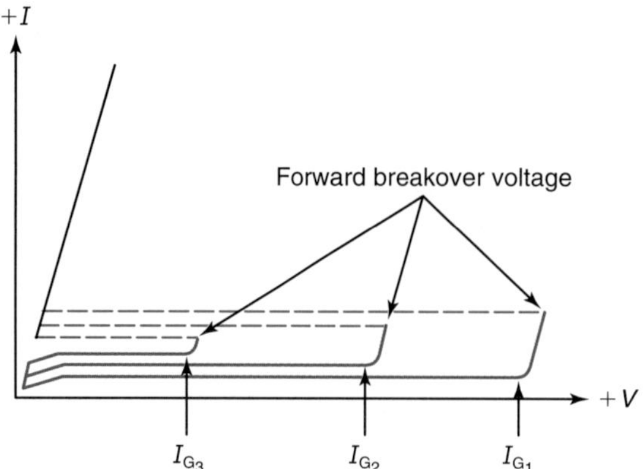

Fig. 14-9 The effect of gate current on breakover voltage.

duct in only the direction shown so this is a *half-wave* circuit. The adjustable gate control determines when the SCR is turned on. Turnoff is automatic and occurs when the ac source changes polarity and reverse-biases the SCR.

Figure 14-11 illustrates how an SCR can control load dissipation in an ac circuit. If the SCR is gated on very late in the positive alternation, the power dissipation will be very low. Since the SCR is off most of the time, the load is effectively disconnected from the source for most of the time. Gating the SCR on earlier increases the load dissipation. Finally, at the bottom of Fig. 14-11, full power is shown. Note that the SCR is turned on at the beginning of the positive alternation. However, "full power" in a simple SCR circuit may be a misleading term. Since the SCR blocks the negative alternation completely, half power is really the most that can be achieved.

Conduction angle

Figure 14-11 is an example of *conduction angle control.* A small conduction angle means the circuit is on for a small portion of the ac cycle. A large conduction angle means the circuit is on for a large portion of the ac cycle.

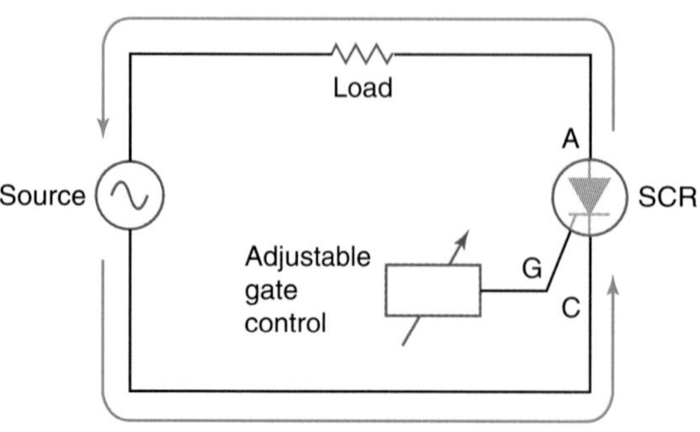

Fig. 14-10 Using an SCR to control ac power.

The largest conduction angle shown in Fig. 14-11 is at the bottom and is equal to 180°. This can be considered full power in a half-wave circuit. In a half-wave circuit, half power would be a conduction angle of 90°. This method of controlling ac power is very efficient. Most of the power will be dissipated in the load device, and only a small fraction in the control device (the SCR).

Silicon-controlled rectifiers serve well in power-switching and power-control applications. They are available with voltage ratings ranging from 6 to around 5000 V and with current ratings ranging from 0.25 to around 2000 A. Even higher voltage and current ratings are possible by using series and parallel combinations of SCRs. Silicon-controlled rectifiers with

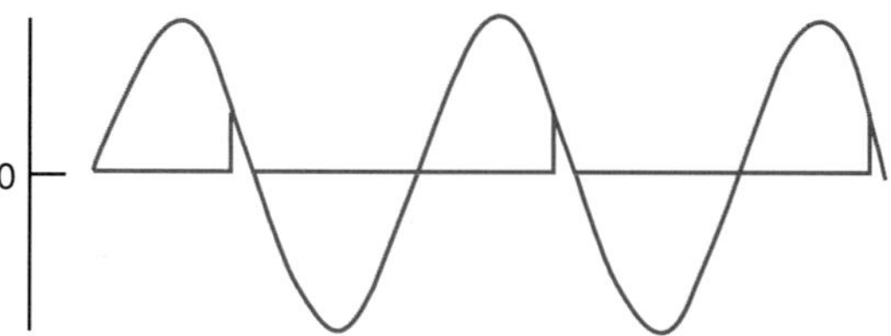

Very low power: SCR is gated on very late

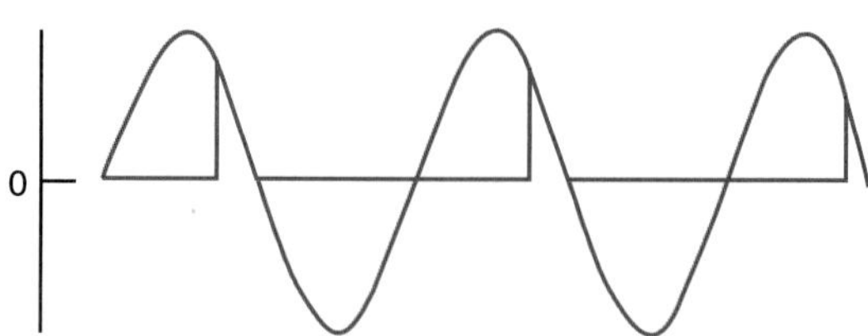

Low power: SCR is gated on late

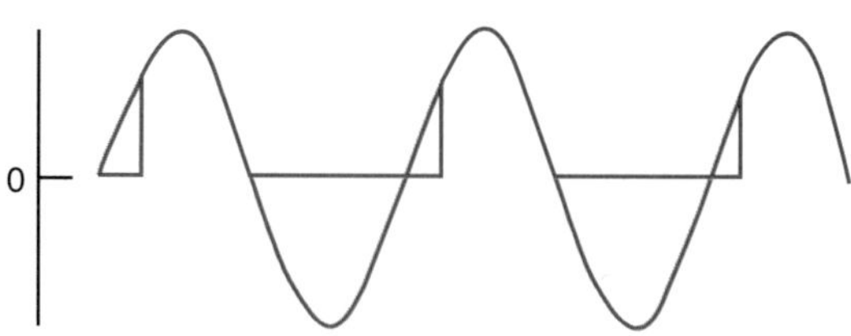

High power: SCR is gated on early

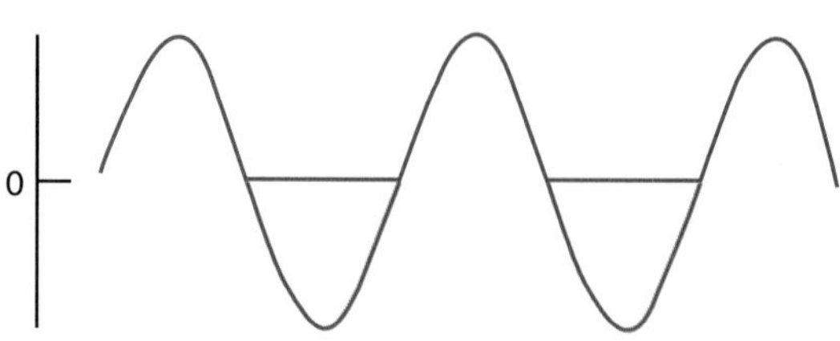

Full power: SCR is gated on at start of cycle

Fig. 14-11 Conduction angles in an SCR power-control circuit.

moderate ratings can switch a load of hundreds of watts with a gate pulse of a few microwatts that lasts for a few microseconds. This performance represents a power gain of over 10 million and makes the SCR one of the most sensitive control devices available today.

Turning off an SCR requires that the anode-cathode circuit be zero-biased or reverse-biased. Reverse bias achieves the fastest possible turnoff. In either case, the turnoff will not be complete until all of the current carriers in the center junction of the device are able to *recombine.* Recombination is a process of free electrons filling holes to eliminate both types of carriers. Recombination takes time. The time that elapses after current flow stops and before forward bias can be applied without turning the device on is the "turnoff" time. It can range from several microseconds to several hundred microseconds, depending on the construction of the SCR.

An SCR can be shut down by interrupting current flow with a series switch. Another possibility is to close a parallel switch, which would reduce the forward bias across the SCR to zero. In ac circuits, turnoff is usually automatic because the source periodically changes polarity. Whatever method is used, the process of shutting down an SCR is called *commutation.* Mechanical switches are seldom suitable for commutation of SCRs. A third approach is called forced commutation and includes six classes or categories of operation:

Class A: Self-commutated by resonating the load. A coil and capacitor effectively form a series resonant circuit with the load. Induced oscillations reverse-bias the SCR.

Class B: Self-commutated by an *LC* circuit. A coil and capacitor form a resonant circuit across the SCR. Induced oscillations reverse-bias the SCR.

Class C: C- or *LC*-switched by a second load-carrying SCR. A second SCR turns on and provides a discharge path for a capacitor or inductor-capacitor combination that reverse-biases the first SCR. The second SCR also provides load current when it is on.

Class D: C- or *LC*-switched by an auxiliary SCR. The auxiliary SCR does not support the flow of load current.

Class E: An external pulse source is used to reverse-bias the SCR.

About Electronics

Popularity of PLCs is on the Rise

- Programmable logic controllers (PLCs) are used in many companies to automate manufacturing and processing.
- PLCs are designed to easily implement the control functions and sequences portrayed by ladder logic diagrams.

Class F: Alternating-current line commutation. The SCR is reverse-biased when the line reverses polarity.

Figure 14-12 shows an example of a class D commutation circuit. No load current can flow until SCR_1 is gated on. The load current flows as shown, and the left-hand portion of L_1 is in the load circuit. As the load current increases through L_1, a magnetic field expands and induces a positive voltage at the right-hand terminal of L_1. This positive voltage charges capacitor *C*, as shown in Fig. 14-12. Diode *D* prevents the capacitor from discharging through the load, the source, and the inductor. When SCR_2 is gated on, the capacitor is effectively connected across SCR_1. Note that the positive plate of the capacitor is applied through SCR_2 to the cathode of SCR_1. Also note that the negative plate of the capacitor is applied to the anode of SCR_1. The capacitor voltage reverse-biases SCR_1 and it turns off.

Commutation

In Fig. 14-12 SCR_1 supports the flow of load current. When it is gated on, load current begins to flow. SCR_2 is used to turn SCR_1 off. When it is gated on, load current stops. Load

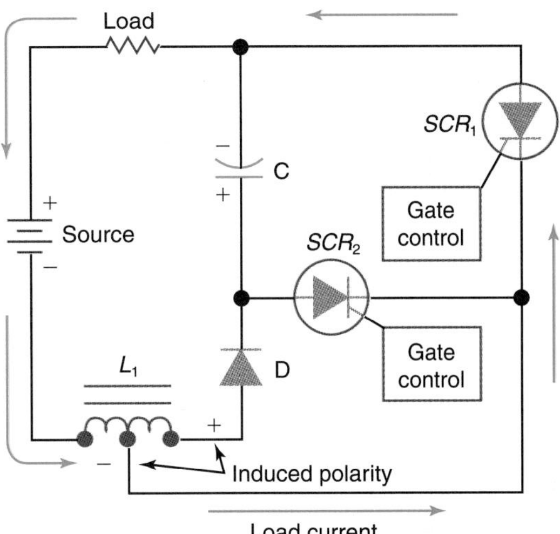

Fig. 14-12 Class D SCR commutation circuit.

power can be controlled by the relationship between the gate timing pulses to the two SCRs. Look at Fig. 14-13(*a*). It shows that SCR_2 is gated on soon after SCR_1. The load current pulses are short in duration since the turnoff comes so soon after the turn-on. Now look at Fig. 14-13(*b*). It shows more delay for the gate pulses to SCR_2. The load current pulses are longer in time and more power dissipates in the load.

Full-wave control

It is possible to achieve *full-wave control* with an SCR by combining it with a full-wave rectifier circuit. Figure 14-14 shows full-wave pulsating direct current. If an SCR is used in this type of circuit, it will no longer be forward-biased at those times when the waveform drops to 0 V. The current in the SCR will drop to some value less than the holding current, and the SCR will turn off.

Figure 14-15 shows a battery charger that uses full-wave, pulsating direct current. Diodes D_1 and D_2, along with the center-tapped transformer, provide full-wave rectification. SCR_1 is in series with the battery under charge. It will be turned on early in each alternation by gate current applied through D_4 and R_4. Commutation is automatic in this circuit as shown earlier in Fig. 14-14.

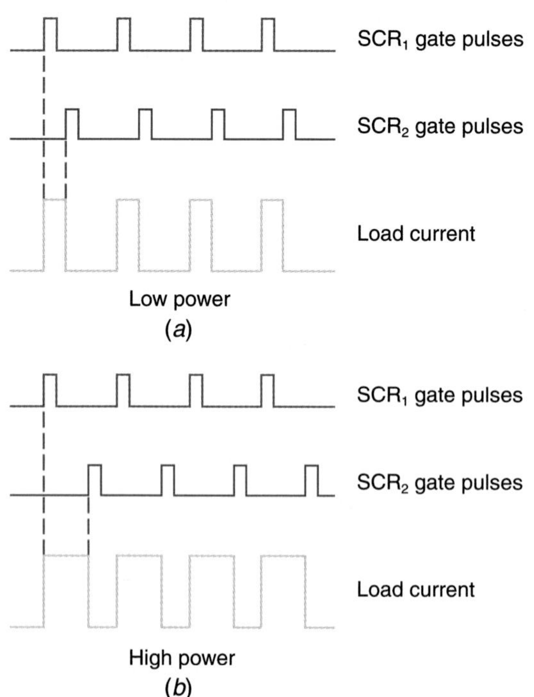

Fig. 14-13 Silicon-controlled rectifier commutation waveforms.

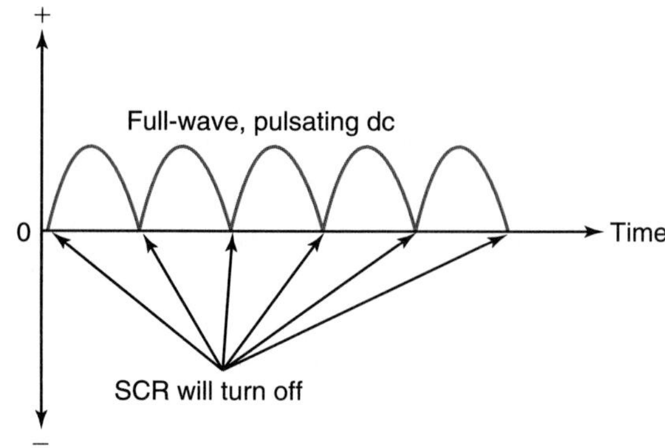

Fig. 14-14 Commutation with full-wave, pulsating direct current.

The battery charger in Fig. 14-15 also features automatic shutdown when full charge is reached. As the battery voltage increases with charge, the voltage across R_2 also increases. Eventually, at the peak of the line, D_5 starts breaking down, and SCR_2 is gated on. As the battery voltage climbs even higher, the angle of SCR_2 keeps advancing (it is now coming on before the line peaks) until SCR_2 is eventually triggering before the input alternation is large enough to trigger SCR_1. With SCR_2 on, the voltage-divider action of R_4 and R_5 cannot supply enough voltage to forward-bias D_4 and gate SCR_1 on. The heavy charging has ceased. The battery is now trickle-charged through D_3 and the lamp (which lights to signal that the battery has reached full charge). The cutout voltage can be adjusted by R_2. Diode D_3 prevents battery discharge through SCR_2 in the event of a power failure.

Silicon-controlled rectifiers can also be used to prevent electrical damage. Some loads can be ruined by excess voltage or current. Figure 14-16 shows a schematic diagram of a circuit that offers dual load protection. In Fig. 14-16 R_1, D_1, and C_1 form a regulated and filtered power supply for the unijunction transistors. The emitter voltage of Q_1 is derived from the load voltage by the divider action of R_2 and R_3. If the load voltage goes too high, the emitter of Q_1 reaches the firing voltage. This causes Q_1 to switch on and rapidly discharge C_2 through R_8. The resulting pulse across R_8 gates on the SCR. The SCR energizes relay K_1, and the armature of the relay opens the contact points. The SCR is now latched on. The lamp is in parallel with the relay, and it lights to indicate that a fault has occurred and that the load has

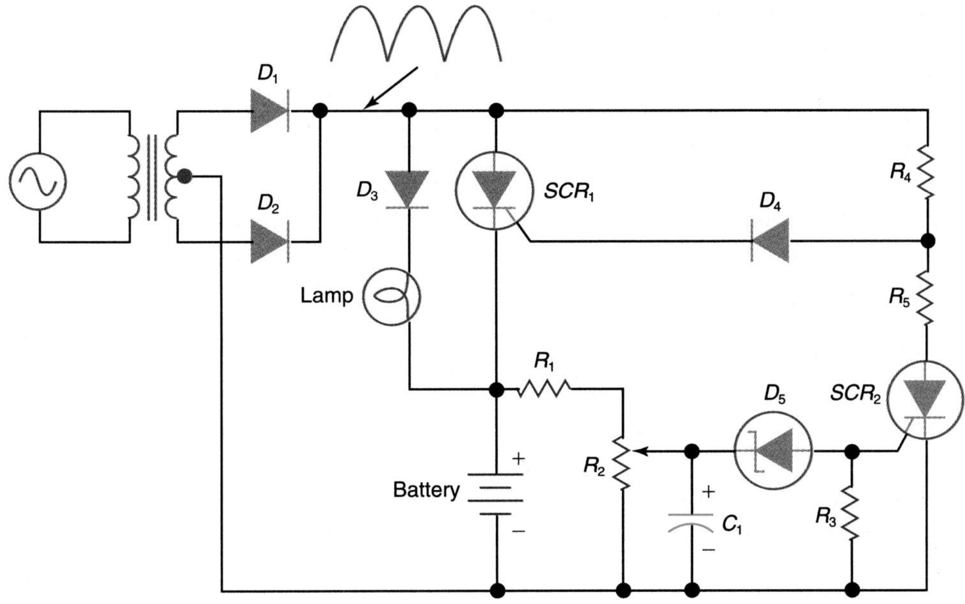

Fig. 14-15 An SCR-controlled battery charger.

been shut down. The load remains off until the circuit is reset by turning S_1 off and then on again.

Too much load current will also cause a shutdown of the circuit in Fig. 14-16. The load current is sensed by R_9. If too much current flows, the drop across R_9 increases and drives the emitter of Q_2 in a positive direction until it fires. Then C_3 is rapidly discharged through R_8, and the resulting pulse gates on the SCR. So, if either UJT fires, the SCR is gated, opening the relay contacts and shutting down the load. The voltage trip point is set at R_3, and the current trip point is set at R_7.

TEST

Choose the letter that best answers each question.

8. Refer to Fig. 14-5. Assume that the source voltage has just been applied and the gate switch has not been closed. What can you conclude about the load current?
   a. The load current will equal zero.
   b. The load current will gradually increase.
   c. It will be mainly determined by V_{source} and the load resistance.

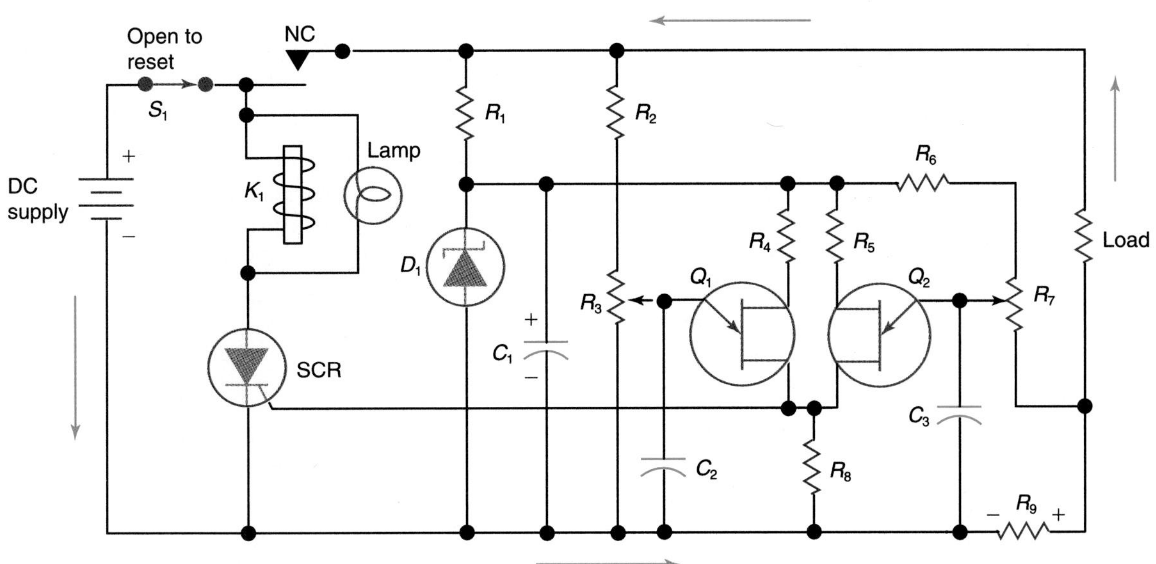

Fig. 14-16 Silicon-controlled rectifier load protection.

d. The load current will flow until the gate switch is closed.

9. Refer to Fig. 14-5. Assume the gate switch has been closed and then opened again. What can you conclude about the load current?
   a. It will go off and then on.
   b. It will go on and then off.
   c. It will come on and stay on.
   d. None of the above.

Triac

10. How are SCRs normally turned on?
    a. By applying a reverse breakover voltage
    b. By applying a forward breakover voltage
    c. By a separate commutation circuit
    d. By applying gate current
11. How can an SCR be turned off in the shortest possible time?
    a. By zero-biasing it
    b. By reverse-biasing it
    c. By reverse-biasing its gate lead
    d. None of the above
12. What happens to the value of forward breakover voltage required to turn on an SCR as more gate current is applied?
    a. It is not changed.
    b. It increases.
    c. It decreases.
    d. None of the above.
13. Refer to Fig. 14-10. Assume that the adjustable gate control is set for maximum power dissipation in the load. What should the load waveform look like?
    a. Half-wave, pulsating direct current
    b. Full-wave, pulsating direct current
    c. Pure direct current
    d. Sinusoidal alternating current (same as the source)
14. How does an SCR control load dissipation in a circuit such as that shown in Fig. 14-10?
    a. The resistance of the SCR is adjustable.
    b. The source voltage is adjustable.
    c. The load resistance is adjustable.
    d. The conduction angle is adjustable.
15. Refer to Fig. 14-15. What is the function of SCR_2?
    a. It provides class D commutation for SCR_1.
    b. It limits the charging current to some safe value.
    c. It prevents SCR_1 from coming on when full charge is reached.
    d. It controls the conduction angle of SCR_1 when charging is started.

14-3 Full-Wave Devices

The SCR is a *unidirectional* device. It conducts in one direction only. It is possible to combine the function of two SCRs in a single structure to obtain *bidirectional* conduction. The device in Fig. 14-17 is called a triac (triode ac semiconductor switch). The *triac* may be considered as two SCRs connected in inverse parallel. When one of the SCRs is in its reverse-blocking mode, the other will support the flow of load current. Triacs are full-wave devices. They have limited ratings as compared with SCRs. They are available with current ratings up to about 40 A and voltage ratings to about 600 V. SCRs are capable of handling much more power, but triacs are more convenient for many low- and medium-power ac applications.

Figure 14-17 shows that the three triac connections are called main terminal 1, main terminal 2, and gate. The gate polarity usually is measured from gate to main terminal 1. A triac may be triggered by a gate pulse that is either positive or negative with respect to main terminal 1. Also, main terminal 2 can be either positive or negative with respect to main terminal 1 when triggering occurs. There are a total of four possible combinations or triggering modes for a triac. Table 14-1 summarizes the four modes for triac triggering. Note that mode

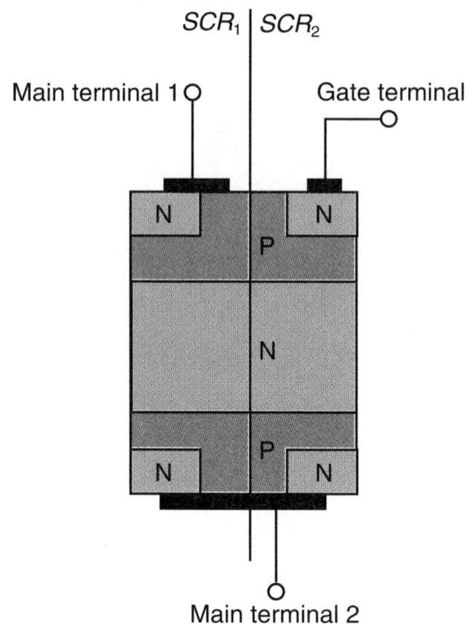

Fig. 14-17 The structure of a triac.

Table 14-1 Triac Triggering Mode Summary			
Mode	**Gate to Terminal 1**	**Terminal 1 to Terminal 2**	**Gate Sensitivity**
1	Positive	Positive	High
2	Negative	Positive	Moderate
3	Positive	Negative	Moderate
4	Negative	Negative	Moderate

1 is the most sensitive. Mode 1 compares with ordinary SCR triggering. The other three modes require more gate current.

Figure 14-18 shows the schematic symbol for a triac. The arrows show that the triac is bidirectional—load current can flow in both directions. Triacs are convenient for controlling (or switching) ac power. Silicon-controlled rectifiers are used when high power levels are encountered. Both devices are in the *thyristor* family. The term thyristor can refer to either an SCR or a triac. Thyristors may be used to perform static switching of ac loads. A static switch is one with no moving parts. Switches with moving parts are subject to wear, corrosion, contact bounce, arcing, and the generation of interference. Static switching eliminates these problems. Most triacs are designed for 50 to 400 Hz and make good static switches over this frequency range. SCRs can operate to approximately 30 kHz.

Figure 14-19 shows a schematic diagram for a simple three-position *static switch.* In position 1, there is no gate signal and the triac remains off. In position 2, the triac is gated at every other alternation of the source, and the load receives half power. In position 3, the triac is gated at every alternation, and the load receives full power. The three-position switch is mechanical, but it operates in a low-current part of the circuit where arcing is not a problem.

Thyristors are also used in applications known as *zero-voltage switching.* They can be gated on when the ac line is near a crossing, or zero-voltage, point. This avoids energizing a load at the moment that the line is at or near one of its maximum points. The idea is to prevent a sudden surge of current that can cause damage and radio-frequency interference. Any circuit that has a very sudden increase in current will generate high frequencies that can interfere with other electronic equipment such as radio and television receivers. Figure 14-20 on page 390 shows a zero-crossing static switch that can be used to control a load such as a heating element. Most of the control circuitry is contained in the TDA1024 IC. It contains a zener regulator that is driven by the ac line at pin 7. It also contains a zero-crossing detector that is driven by the ac line at pin 6. It has comparator inputs at pins 4 and 5, and an output amplifier which sends a gating signal to pin 2. In Fig. 14-20 R_1 and R_2 form a voltage divider for the dc power supply that is available across pins 8 (positive) and 1 (ground). R_2 is a nonlinear resistor that senses temperature. Therefore, the output of the voltage divider is a function of temperature and is sent to one of the comparator inputs at pin 5. The other comparator input is adjustable by setting poten-

Thyristor

Static switch

Zero-voltage switching

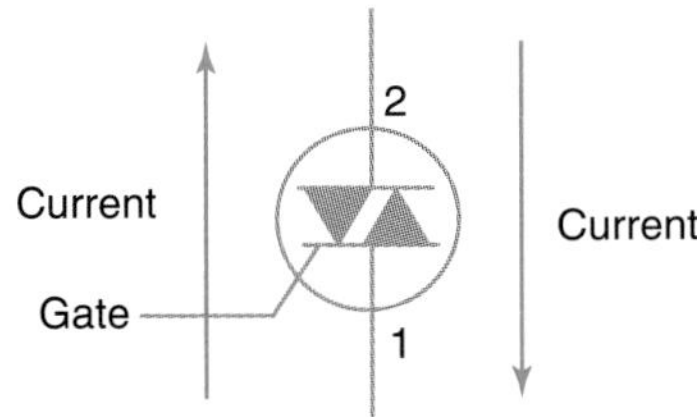

Fig. 14-18 Schematic symbol for the triac.

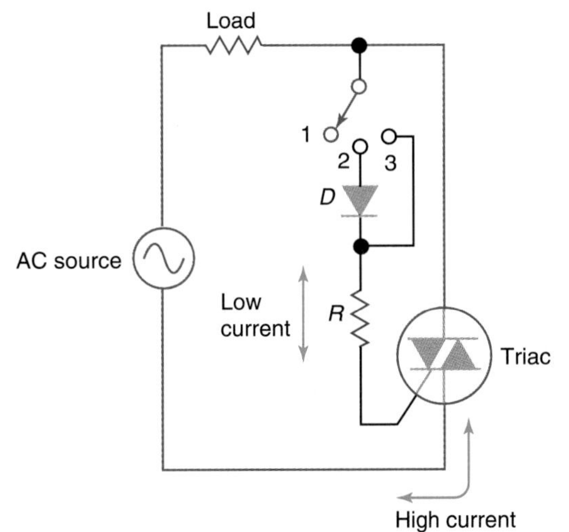

Fig. 14-19 A three-position static switch.

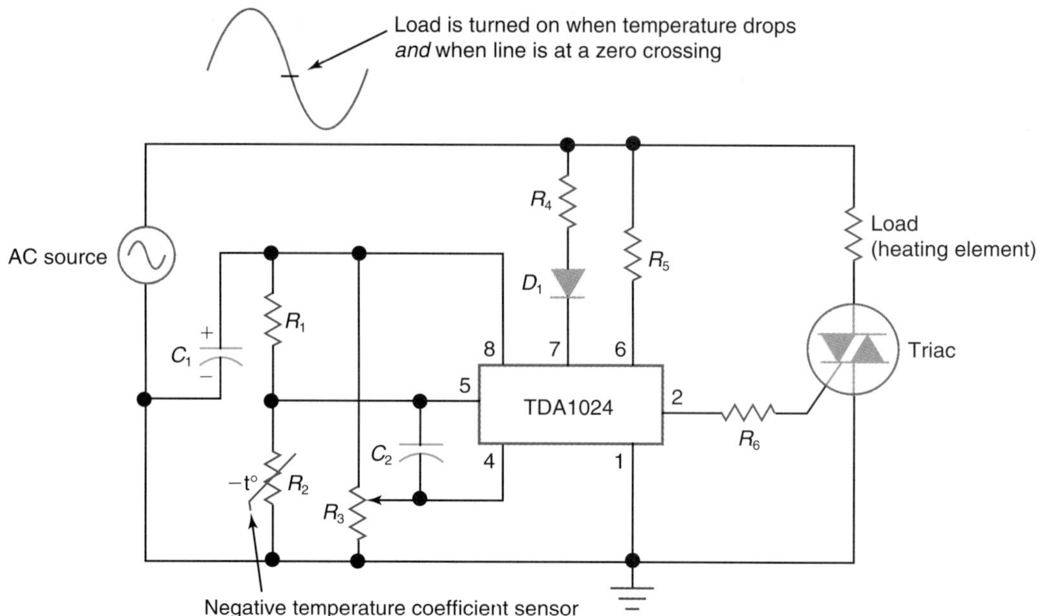

Fig. 14-20 Zero-crossing switch.

tiometer R_3. If the temperature drops, the negative coefficient of R_2 will make its resistance increase. The voltage at pin 5 will go up. The comparator output will switch, and at the zero crossing of the ac line, the output amplifier will supply a signal at pin 2, which will gate on the triac. The triac will be turned on when heat is called for and when the line is at a zero crossing. There will never be a current surge, and RF interference will not be a problem.

JOB TIP

Some companies recruit electronic technicians for positions called *quality technician* or *calibration technician.*

Figure 14-21 shows the general use of a triac to control load dissipation in an ac circuit. The adjustable gate control allows the triac to be gated on for either alternation of the ac source.

Fig. 14-21 Using a triac to control ac power.

Figure 14-22 shows some typical waveforms. As with the SCR, a triac sets load dissipation by controlling conduction angle. However, the triac is a full-wave device and the SCR is a half-wave device.

Commutation may be different in triac circuits. In ac power control, the triac should switch to its off state at each zero-power point. These points occur twice each cycle. If the triac fails to turn off, control is lost. Commutation is automatic when the load is resistive. When the load is inductive (a motor is an example), the current lags the voltage. Thus, when the current goes to zero, a voltage is applied across the triac. This voltage can turn the triac back on. Commutation and power control may be lost with inductive loads.

Line transients can affect thyristors. Transients produce a large voltage change in a short time. A rapid change in voltage can switch a thyristor to its on state. Recall that a PN junction, when not conducting, has a depletion region. Also recall that the depletion region acts as the dielectric of a capacitor. This means that a thyristor in its off state has several internal capacitances. A sudden voltage change across the thyristor terminals will cause the internal capacitances to draw charging currents. These charging currents can act as a gating current and switch the device on.

Inductive loads and transients are both seen as problem areas in triac control. These prob-

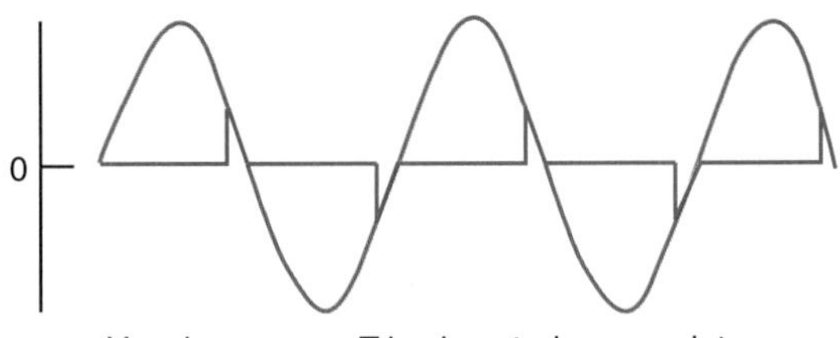

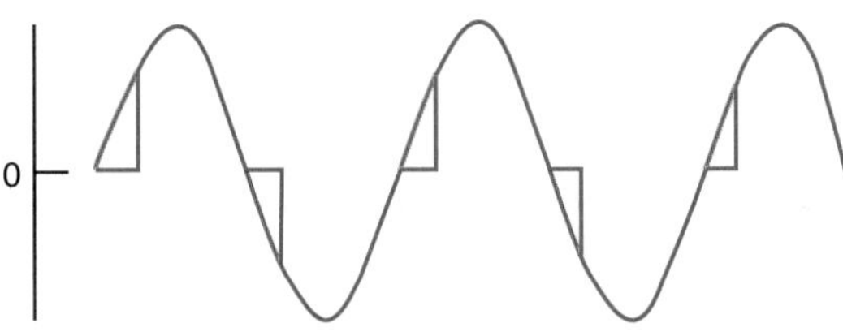

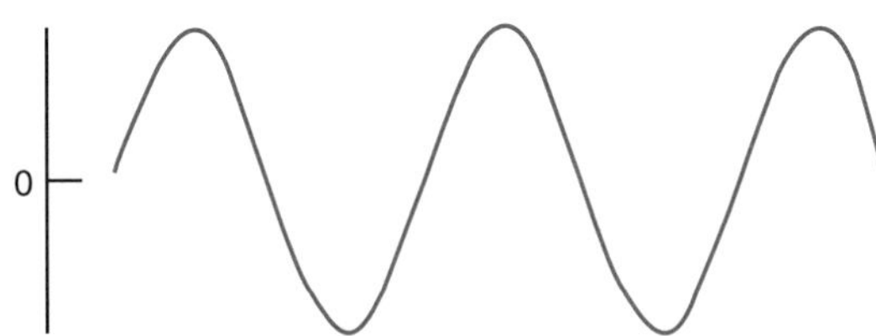

Fig. 14-22 Conduction angles in a triac power-control circuit.

lems can be reduced by special networks that limit the rate of voltage change across the triac. An *RC snubber network* has been added in Fig. 14-23. Snubber networks divert the charging current from the thyristor and help prevent unwanted turn-on.

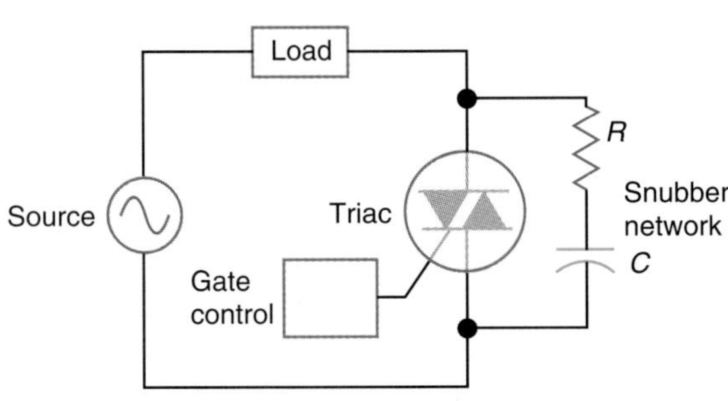

Fig. 14-23 A snubber network.

Triac gating circuits vary from application to application. A triac may simply be switched on or off. Or, it may be gated for various conduction angles. There are many gating circuits in use, and they range from simple to complex. Figure 14-24 shows two simple triac gating circuits. Figure 14-24(*a*) uses a variable-resistor in series with the gate lead. As *R* is set for less resistance, the triac will gate on sooner and the conduction angle increases. This will result in an increase in load power. This approach does not provide control over the entire 360° range and has poor symmetry. The positive alternations will have a different conduction angle than the negative alternations. The circuit is also temperature-sensitive. Figure 14-24(*b*) shows improved operation. This circuit has the advantage of providing a broader range of control. The setting of R_1 will control how rapidly C_1 and C_2 charge. Decreasing R_1 will advance the firing point and increase the load power.

The better gate-trigger circuits use a negative-resistance device to turn on the triac. These devices show a rapid decrease in resistance after some critical turn-on voltage is reached. *Triggering devices* with this negative-resistance quality include neon lamps, unijunction transistors, two-transistor switches, and *diacs*.

Trigger device

Diac

Snubber network

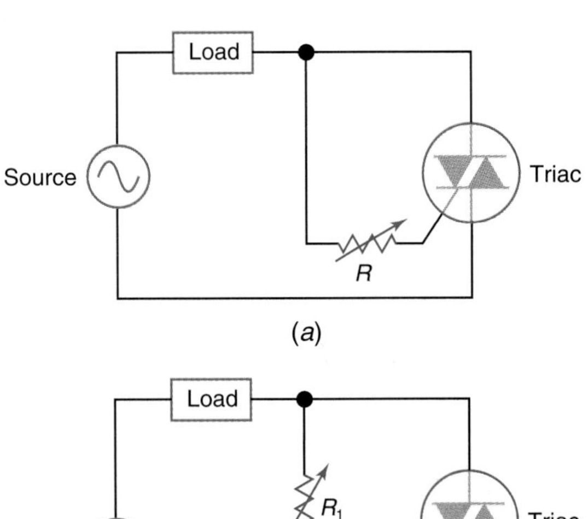

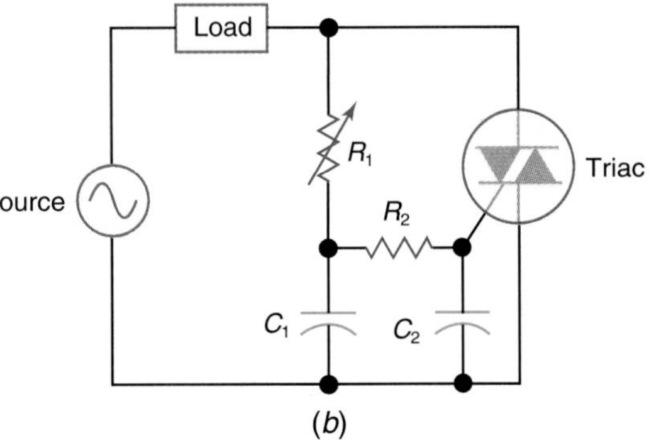

Fig. 14-24 Simple triac gate circuits.

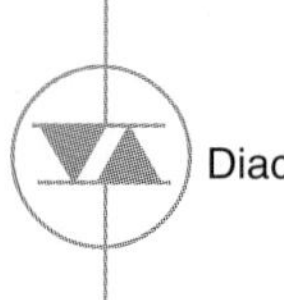

Fig. 14-25 Schematic symbol for the diac.

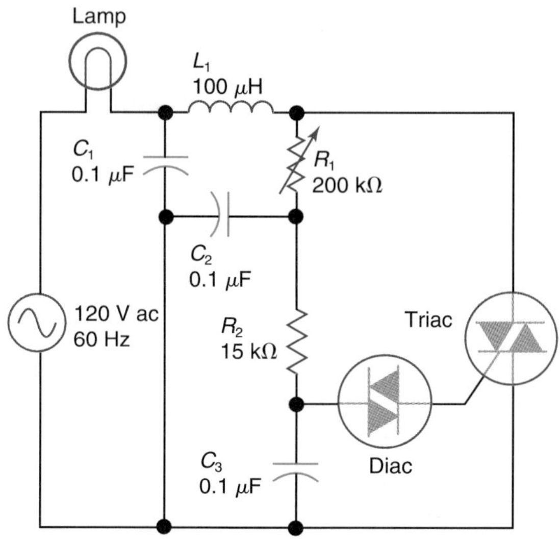

Fig. 14-27 A diac-triac control circuit.

The schematic symbol for a diac is shown in Fig. 14-25. The diac is a bidirectional device and it is well suited for gating triacs. The characteristic curve for a diac is shown in Fig. 14-26. The device shows two breakover points V_P+ and V_P-. If either a positive or negative voltage reaches the breakover value, the diac rapidly switches from a high-resistance state to a low-resistance state.

Figure 14-27 shows a popular circuit that combines a diac and a triac to give smooth power control. Resistors R_1 and R_2 determine how rapidly C_3 will charge. When the voltage across C_3 reaches the diac breakover point, the diac fires. This provides a complete path for C_3 to discharge into the gate circuit of the triac. The discharge of C_3 gates the triac on.

RFI

Figure 14-27 also includes two components to suppress radio-frequency interference (*RFI*). Triacs switch from the off state to the on state in 1 or 2 microseconds (μs). This produces an extremely rapid increase in load current. Such a current step contains many harmonics. A harmonic is a higher multiple of some frequency. For example, the third harmonic of 1 kHz is 3 kHz. The harmonic energy in triac control circuits extends to several megahertz and can produce severe interference to AM radio reception. The energy level of the harmonics falls off as the frequency increases. Interference from thyristors is more of a problem at the lower radio frequencies. Capacitor C_1 and inductor L_1 in Fig. 14-27 form a low-pass filter to prevent the harmonic energy from reaching the load wiring and radiating. This will reduce the interference to a nearby AM radio receiver.

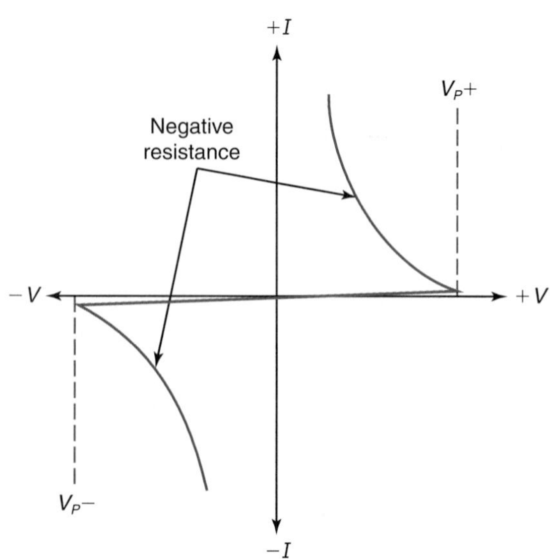

Fig. 14-26 Diac volt-ampere characteristic curve.

TEST

Choose the letter that best answers each question.

16. Which of the following devices was specifically developed to control ac power by varying the circuit conduction angle?
    a. The SCR
    b. The triac
    c. The diac
    d. The two-transistor, negative-resistance switch
17. In a triac circuit, how may the load dissipation be maximized?
    a. Hold the conduction angle to 0°.
    b. Hold the conduction angle to 180°.
    c. Hold the conduction angle to 270°.
    d. None of the above.
18. Suppose a triac is used to control the speed of a motor. Also assume that the motor is highly inductive and causes loss of commutation. What is the likely result?
    a. The triac will short and be ruined.
    b. The triac will open and be ruined.

c. Power control will be lost.
d. The motor will stop.

19. Which of the following events will turn on a triac?
a. A rapid increase in voltage across the main terminals
b. A positive gate pulse (with respect to terminal 1)
c. A negative gate pulse (with respect to terminal 1)
d. Any of the above

20. Why may a voltage transient cause unwanted turn-on in thyristor circuitry?
a. Because of internal capacitances in the thyristor
b. Because of arc-over
c. Because of a surge current in the snubber network
d. All of the above

21. Which of the following is a solid-state, bidirectional, negative-resistance device?
a. Neon lamp
b. Diac
c. UJT
d. Two-transistor switch

22. How many breakover points does the volt-ampere characteristic curve of a diac show?
a. One
b. Two
c. Three
d. Four

23. What is an advantage of a zero-crossing switch?
a. Commutation circuits are never needed.
b. Static switching is eliminated.
c. Snubber networks can be used for greater conduction angle.
d. RFI is eliminated.

14-4 Feedback in Control Circuitry

Electronic control circuits can be made more effective by using *feedback* to automatically adjust operation should some change be sensed. For example, suppose a thyristor is used to control the speed of a motor. After the motor has been set for speed, assume the load on the motor increases. This will tend to slow down the motor. It is possible, by using feedback, to make the speed of the motor constant even though the mechanical load is changing.

Figure 14-28 shows the diagram for a motor-speed control that uses feedback to improve performance. R_1, R_2, D_1, and C_1 form an adjustable dc power supply. Diode D_1 rectifies the ac line. The SCR will fire earlier during a positive alternation if the wiper arm of R_2 is moved up toward R_1. This is because V_1 will be more positive and D_2 will be forward-biased sooner. This increases motor speed. The circuit cannot achieve full speed with a 120-V motor, however. This is because the SCR will conduct only on positive alternations. Sometimes a circuit of this type will be used with special universal motors rated at 80 V to allow full-speed operation on 120 V. Universal motors are so named since they can be energized on alternating current or direct current. They may be identified by their construction, which uses a segmented brass commutator at

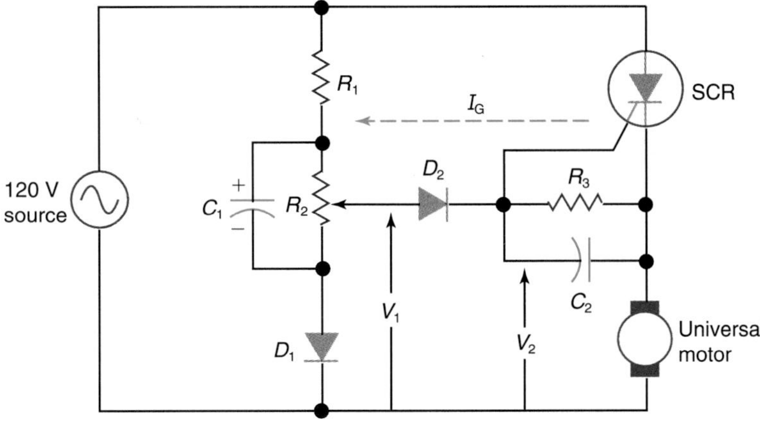

Fig. 14-28 Motor-speed control with feedback.

one end of the armature. Brushes are used to make electrical contact with the rotating commutator.

Positive alternations in Fig. 14-28 will gate the SCR on when D_2 is forward-biased. This occurs when V_1 is more positive than V_2 by about 0.6 V. V_1 is determined by the setting of R_2 and the instantaneous line voltage; V_2 is determined by the counter-electromotive force (*cemf*) of the *motor*. The residual magnetism in a universal motor gives it some of the characteristics of a generator. Therefore, the cemf is determined by the motor's magnetic structure, its iron characteristics, and its *speed.* If the mechanical load on the motor is increased, the motor tends to slow down. The drop in speed will decrease the cemf V_2. This means that V_1 will now exceed V_2 at some earlier point in the alternation. Thus, the SCR is gated on sooner, and the motor speed is stabilized. On the other hand, if the mechanical load is decreased, the motor tries to speed up. This increases V_2, and now V_1 will exceed V_2 later in the alternation. The SCR is on for a shorter period of time, and again the motor speed is stabilized.

Motor cemf

The performance of the motor-speed control circuit shown in Fig. 14-28 is adequate for some applications. However, many motors do not develop a cemf signal that can be used to stabilize speed. It may be necessary to arrange for other types of feedback to make motor speed independent of mechanical load. In some systems, the feedback may relate to the angular position of a shaft rather than its speed. Feedback systems that sense and control position are called *servomechanisms.* Feedback systems that control speed are called *servos.* However, today the distinction is not as important as it once was, and you may find systems that control quantities other than position being classified as servomechanisms. In general terms, a servomechanism is a controller that involves some mechanical action and provides automatic error correction. A servo or servomechanism in its most elementary form consists of an amplifier, a motor, and a feedback element. Servos and servomechanisms can be classified into four basic types:

Servomechanism

1. Velocity control
2. Torque control
3. Position control
4. Hybrid

Figure 14-29 shows the basic arrangement for a *velocity servo.* The motor is mechanically coupled to a *tachometer.* A tachometer is a small generator, and its output voltage is proportional to its shaft speed. The faster the motor in Fig. 14-29 runs, the greater the output voltage from the tachometer. The *error amplifier* compares the voltage from the velocity-set potentiometer with the feedback voltage from the tachometer. If the load on the motor increases, the motor tends to slow down. This causes the output from the tachometer to decrease. Now the error amplifier sees less voltage at its inverting input. The amplifier responds by increasing the positive output voltage to the motor. The motor torque (twisting force) increases, and the speed error is greatly reduced. Changing the position of the velocity-set potentiometer will make the motor

Velocity servo

Tachometer

Error amplifier

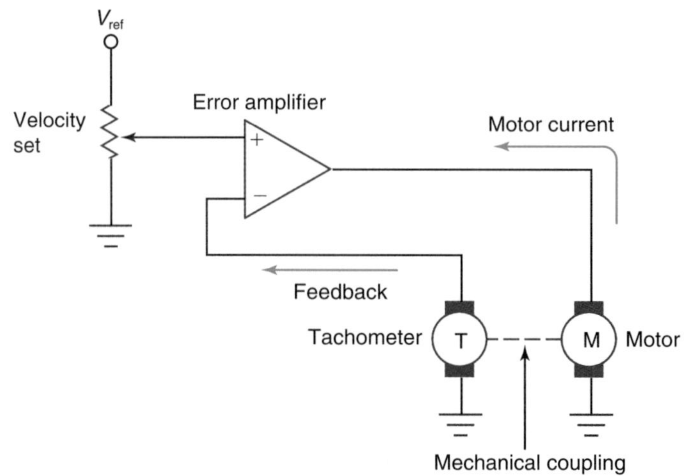

Fig. 14-29 Velocity servo.

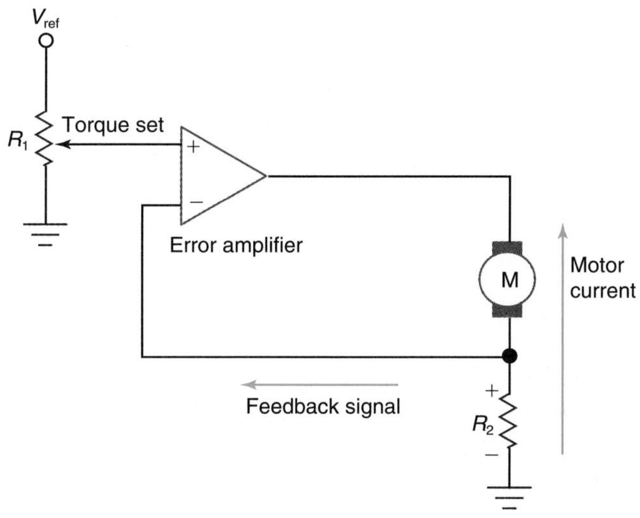

Fig. 14-30 Motor-torque control system.

operate at a different speed. Therefore, the velocity servo provides both speed regulation and speed control.

Figure 14-30 shows a motor-*torque control* system. The torque of the motor is controlled by the current that flows through it. Resistor R_2 provides a feedback voltage that is proportional to motor current. This feedback voltage is compared to the reference voltage that is divided by R_1. Suppose that the load on the motor increases its torque output. The motor will draw more current, and this increased current will increase the voltage drop across R_2. The inverting input of the error amplifier is going in a positive direction. This will make the output of the amplifier go less positive. The motor current will decrease, and the torque output will be held constant.

A positioning servomechanism is shown in Fig. 14-31. The motor drives a potentiometer through a mechanical reduction system (gear train). Many turns of the motor will result in one turn of the potentiometer shaft. The angle of the potentiometer shaft determines the voltage at the wiper arm. This voltage is fed back to the error amplifier. The motor is a dc type that reverses rotation when its supply voltage reverses. Any error between the two potentiometer settings in Fig. 14-31 will cause the amplifier output to drive the motor in the direction that will reduce the error. Therefore, the *position* of the gear train can be *controlled* by adjusting the position-set potentiometer.

Torque control

Position control

A hybrid servomechanism controls two parameters, usually velocity and position. When the position-set control in Fig. 14-32 is changed, the error amplifier develops an output signal to run the motor in the direction that reduces the error voltage between the two position potentiometers. When the motor starts to

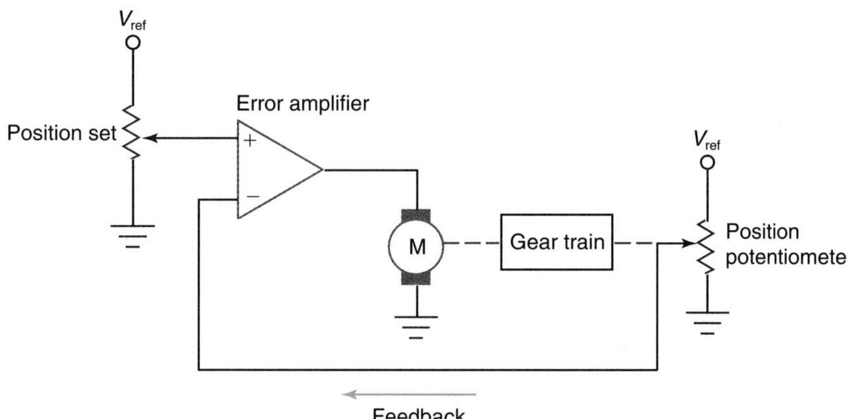

Fig. 14-31 Positioning servomechanism.

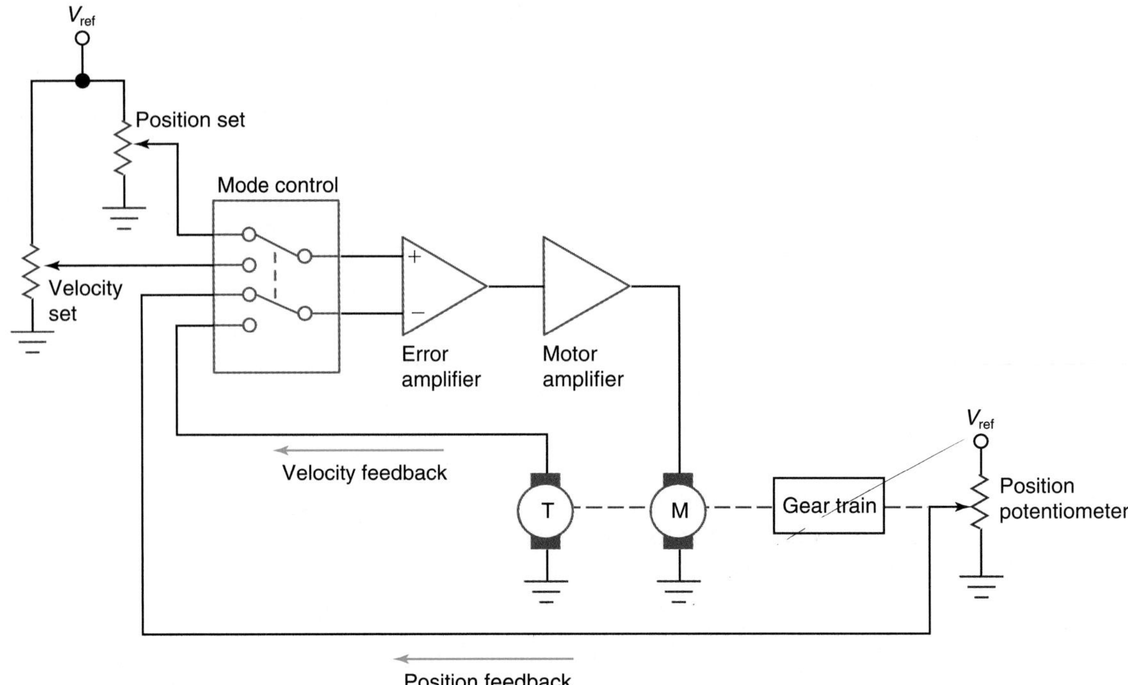

Fig. 14-32 Hybrid servomechanism.

Stiffness

Critically damped

Gain

run, the mode control switches over to compare the tachometer feedback with the setting of the velocity-set control. The motor runs at a controlled velocity until the positioning error is eliminated. As the position error reaches zero, the mode control is switched to position feedback. The mode-control circuitry is shown as a simple double-pole, double-throw (DPDT) switch in Fig. 14-32. In practice it will be more complicated.

The response and accuracy of a servomechanism are a function of *gain.* The more gain the error amplifier has, the greater the positioning accuracy. This is often referred to as the *stiffness* of a servomechanism. Stiffness is usually desirable for fast response to commands and for high positioning accuracy. However, too much gain causes problems. For example, suppose the position-set potentiometer in Fig. 14-31 is suddenly changed. This introduces an abrupt error or transient into the system. Figure 14-33 shows three ways a servomechanism can respond to a transient. The *critically damped* response is the best. It provides the best change from A_1 (the old angle) to A_2 (the new angle). Raising the gain will

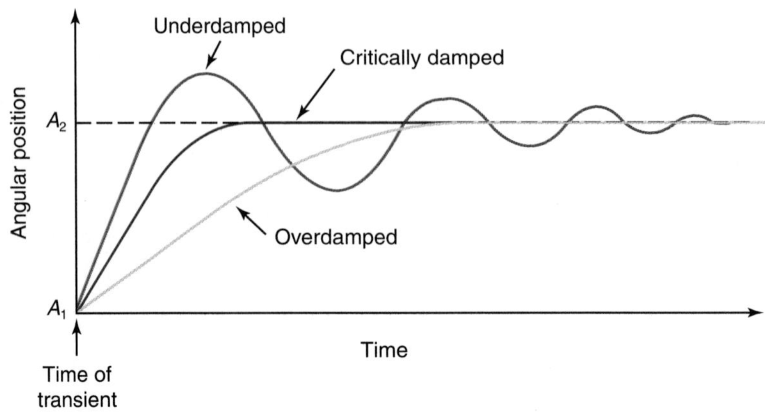

Fig. 14-33 Transient response of a servomechanism.

cause the transient response to follow the *underdamped response* curve. Notice that the servomechanism overshoots A_2 and then undershoots it. This continues until it finally damps out. Too little loop gain provides the *overdamped response.* Here there is no overshoot, but the servomechanism takes too long to reach the new position, A_2. Also, it will not position as accurately as in the critically damped case.

Gain is *critical* in a servomechanism. Too little gain makes the response sluggish and the accuracy poor. Too much gain causes damped oscillations when a transient is introduced into the system. In fact, a servomechanism may oscillate violently and continuously if the gain is too high. The gain of most servomechanisms is adjustable for best stiffness and transient response.

Oscillations will occur in any feedback system when the gain is greater than the loss and when the feedback is positive. It is usually possible to increase gain and still avoid oscillations by controlling the phase angle of the feedback loop. *Phase-compensation* networks are used in most servo systems to improve performance.

Underdamped

Phase compensation

Overdamped

Figure 14-34 shows a computer simulation of a servo system with and without phase compensation. A 0.01-μF compensation capacitor greatly improves the response. With no phase compensation (switch open), the response is underdamped. The simulation shows that the circuit settles in about 10 ms. With compensation, the response is almost ideal and the circuit settles in about 2 ms. You can imagine the problems that would be caused by a positioning system such as a robot arm with an underdamped response.

JOB TIP

Industry trade journals are a good source for fresh information about technology and jobs.

Figure 14-34 has two *time delays* or *lags.* These are due to the feedback capacitors found

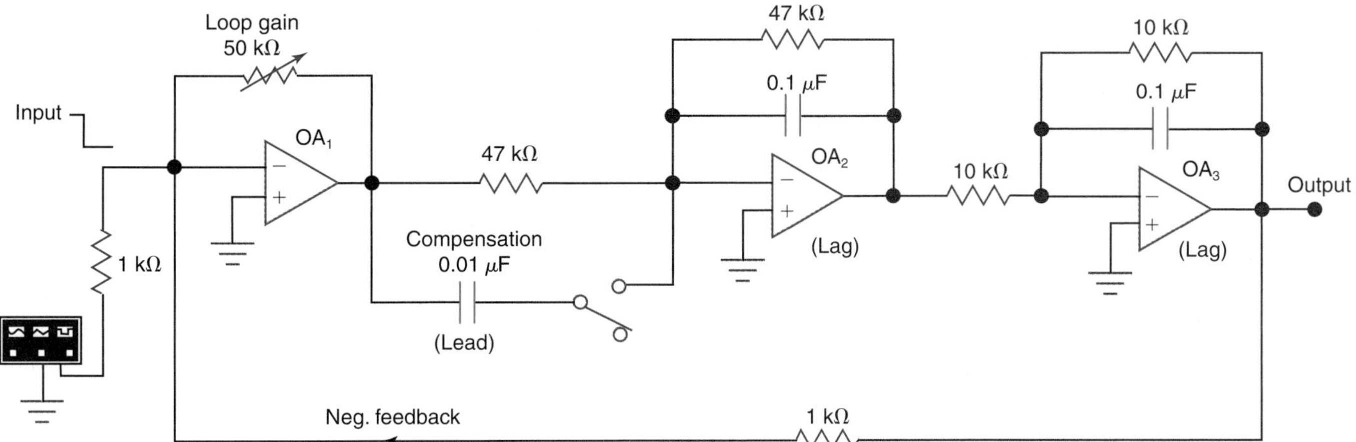

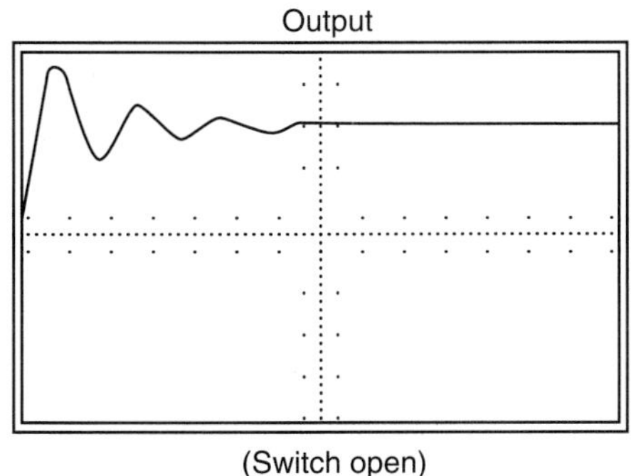

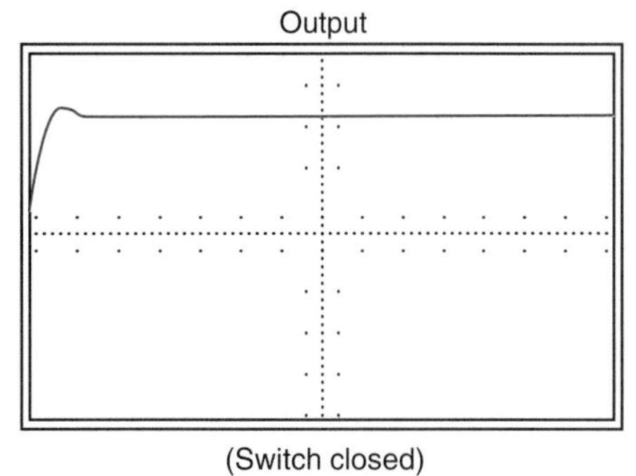

Fig. 14-34 Simulation of a servo.

in op amp 2 and op amp 3, which have outputs that lag behind their inputs. Op-amp integrators were covered in Chap. 9. In a typical servo, an electric motor produces two lags. One is mechanical, and the other is electrical. Multiple lags or delays cause negative feedback to become positive feedback as the frequency goes up. In Fig. 14-34, the feedback is positive at a frequency of about 500 Hz. This circuit will oscillate at 500 Hz if the gain is increased. Likewise, some servos will physically oscillate if the gain is increased.

Lag circuits

Lag circuits are also called integrators, time-delays, or low-pass filters. Which name is used has a lot to do with the particular application. The important idea here is that multiple lags, such as those found in any motor, cause negative feedback to become positive, and positive feedback can cause an underdamped response or continuous oscillation.

Lead circuits

Lead circuits are also called differentiators, anticipators, or high-pass filters. They are used in servos to compensate for the unavoidable lags found in motors and other mechanisms. A lead circuit has a phase angle opposite to that of a lag circuit; therefore, a lead circuit can cancel a lag. Lead circuits are used to compensate for delays in applications such as temperature control systems.

Figure 14-35 shows a servomechanism used to adjust a television lens. Suppose the camera is part of a security system and views a parking lot. Because of changes in light level, it will be necessary to adjust the lens opening from time to time. Part of the signal from the camera is sent to a video processor. This circuit examines the signal and produces an output relative to the brightness of the parking lot. This signal is fed to the error detector along with a reference voltage. Any difference will be amplified and will run the lens motor to adjust for brightness.

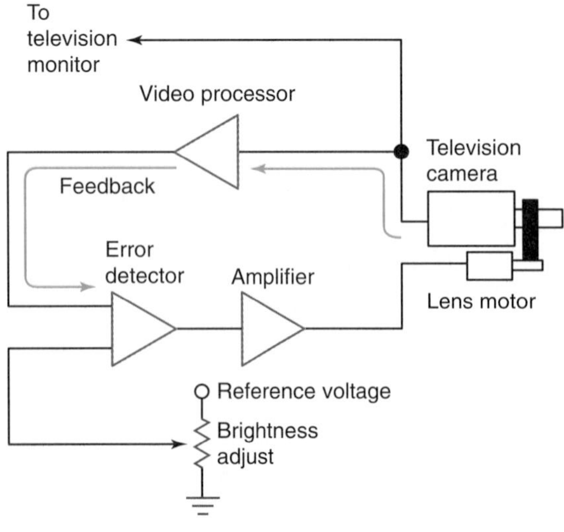

Fig. 14-35 A television lens servomechanism.

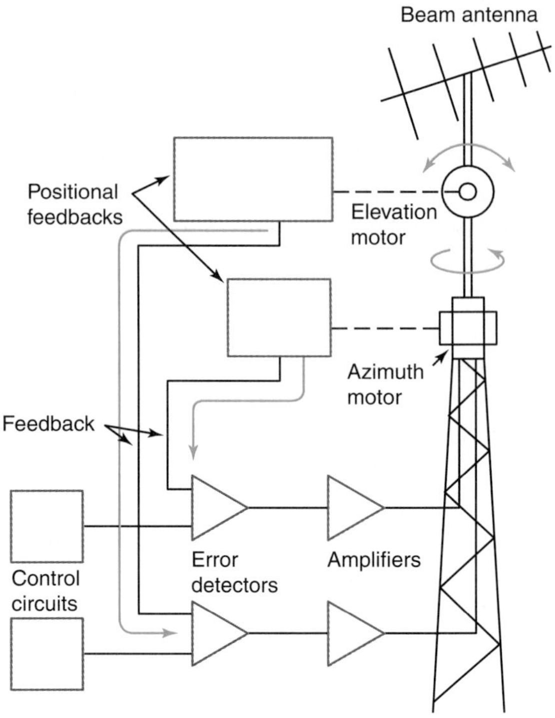

Fig. 14-36 A beam antenna servomechanism.

Examine Fig. 14-36. A beam antenna must be pointed in the proper direction (azimuth) and tilted at the proper angle (elevation) to receive earth satellite signals. Two motors are used to control the proper positioning of the beam antenna. The control circuits in such a system can be computerized to provide automatic tracking of the satellite.

Integrated circuits may be used to reduce the cost and complexity of servomechanisms. Figure 14-37 shows that most of a motor-speed servo can be placed on one chip. The performance and characteristics for this IC are

1. 1 percent speed accuracy
2. −30 to +85°C temperature range
3. 10- to 16-V dc supply range
4. Supplies motor currents up to 2 A
5. Supplies starting currents up to 3 A
6. Very few external parts required
7. Protection for the IC built in
8. Protection for the motor built in
9. Protects against overvoltage, excessive temperature, and motor stalls

Did You Know?

Industrial Computers Computer-controlled lathes and milling machines are common in many machine shops and manufacturing plants.

10. Suited for use in tape players, recorders, and various industrial controls

A complete positioning servomechanism using two integrated circuits is shown in Fig. 14-38 on page 400. A 555 timer IC acts as the reference oscillator. A 543 servo IC contains the error circuits and motor amplifiers required to complete the system.

TEST

Choose the letter that best answers each question.

24. Refer to Fig. 14-28. What would cause V_2 to increase?
    a. A shorted SCR
    b. The motor slowing down
    c. The motor speeding up
    d. A decrease in source voltage

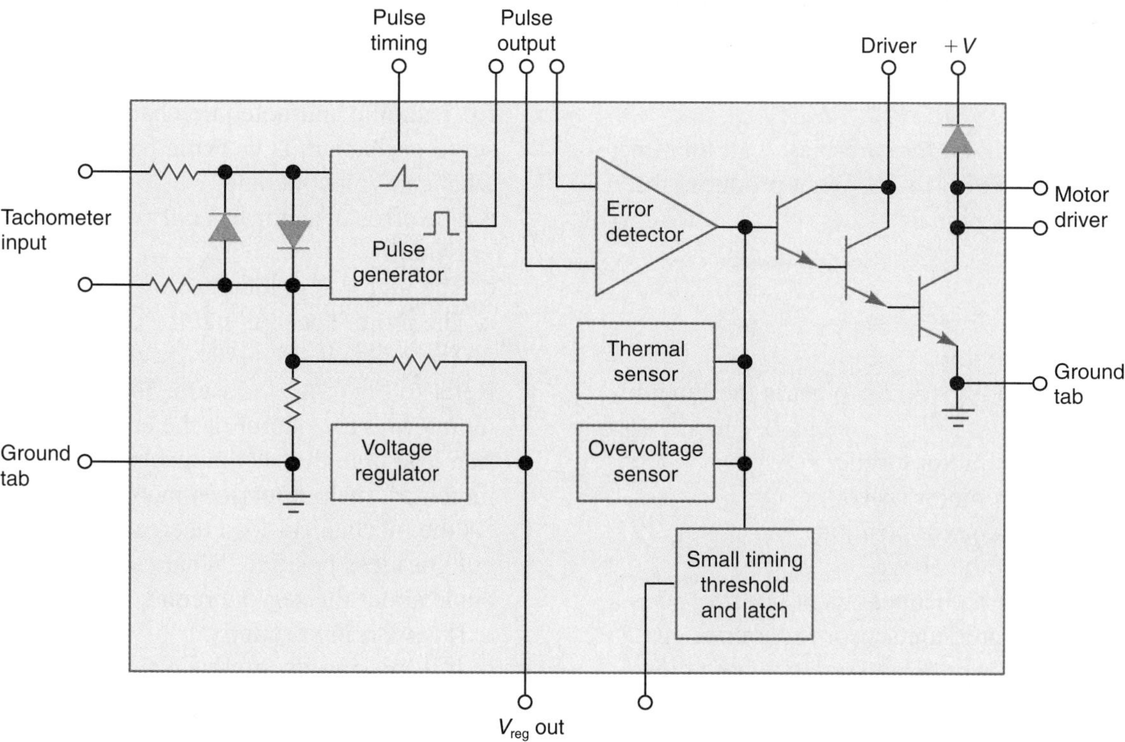

Fig. 14-37 An IC for motor speed control.

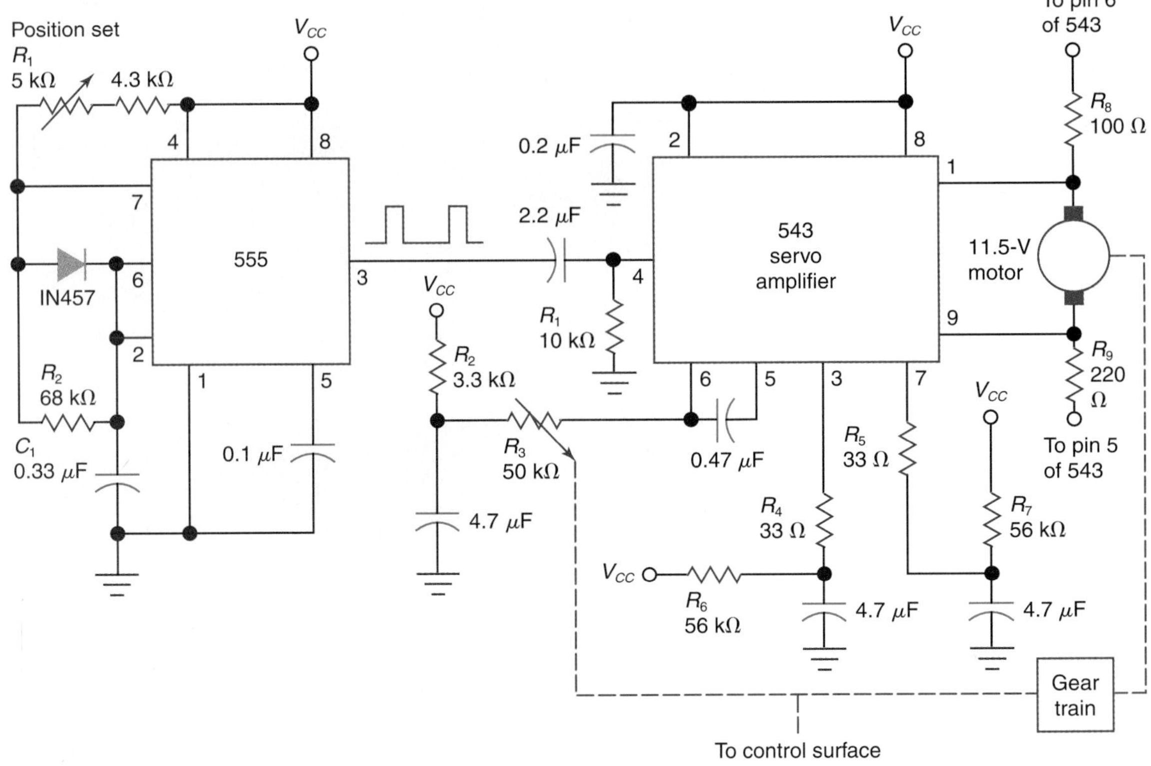

Fig. 14-38 A positioning servomechanism.

25. Refer to Fig. 14-28. What should happen when V_2 increases?
    a. The SCR should gate earlier in the alternation.
    b. The SCR should gate later in the alternation.
    c. There will be no change in conduction angle.
    d. D_2 will be forward-biased all the time.
26. Refer to Fig. 14-28. What produces the feedback signal?
    a D_2
    b. C_1
    c. R_4
    d. The motor
27. Refer to Fig. 14-28. What is the function of R_2?
    a. It sets motor torque.
    b. It sets motor speed.
    c. It sets motor position.
    d. All of the above.
28. Suppose a chemical plant operator uses a remote pressure sensor to monitor gas flow. When the pressure goes too high, the operator closes a circuit which runs a motor and controls a valve. Why would this *not* qualify as a servomechanism?
    a. The system is not automatic.
    b. No mechanical action is involved.
    c. Gas pressure has nothing to do with servomechanisms.
    d. All of the above.
29. Refer to Fig. 14-29. Assume that you have measured the output of the error amplifier for 1 minute and noted no change. If the servomechanism is working properly, what can you conclude?
    a. The error detector has only one input signal.
    b. The motor is gradually slowing down.
    c. The motor speed is stable.
    d. The motor is gradually speeding up.
30. Refer to Fig. 14-29. Assume that you are monitoring the output of the error amplifier. You note that as the mechanical load increases, the output goes more positive. As the mechanical load decreases, the output goes less positive. What can you conclude about the servo circuit?
    a. It is working properly.
    b. It is missing its reference signal.
    c. It is not working at all.
    d. It is connected to a defective power supply.

14-5 Troubleshooting Electronic Control Circuits

Technicians who troubleshoot thyristor control circuits must be aware of their limitations and know safe procedures. The thyristor control circuits covered in this chapter can be used with only certain kinds of loads. Severe damage to the load and the control circuit may result if improper connections are attempted. The general rule for SCR and triac control circuits is this: *Never* attempt to use them with *ac-only* equipment. This type of equipment includes:

AC-only load

1. Fluorescent lamps (unless specially designed for thyristor control)
2. Radios
3. Television receivers
4. Induction motors (including those on fans, record players, tape players, washing machines, large equipment such as air compressors, and so on)
5. Transformer-operated devices (such as soldering guns, model-train power supplies, battery chargers, and so on)

In general, it is safe to use thyristor control circuits with *resistive loads.* These include incandescent lamps, soldering pencils, heating elements, and so on. It is also safe to use thyristor control circuits with universal (ac/dc) motors. These motors usually are found in portable power tools such as drills, saber saws, and sanders. When in doubt, check the manufacturer's specifications. Also be sure that the wattage rating of the load does not exceed the wattage rating of the control circuit.

Resistive load

The general safety rules for analyzing and troubleshooting electronic control circuits are the same as those for any line-operated circuit. It is dangerous to connect line-operated test equipment to thyristor control circuits. A ground loop is likely to cause damage and perhaps severe electric shock. Even if the test equipment is battery-operated, danger still exists. A battery-operated oscilloscope may seem safe, but remember that the cabinet and the probe grounds may reach a dangerous potential when directly connected into power circuits.

If a control circuit is for light duty, it may be possible to use an isolation transformer. Then it is safe to use test instruments for analyzing the circuit. Be sure the wattage rating on the isolation transformer is adequate before attempting this approach.

Suppose you are troubleshooting a thyristor motor-speed control. You notice that the motor always runs at top speed. The speed control has no effect. Assume that you have already completed the usual preliminary checks and have found nothing wrong. What is the next step? Ask yourself what kinds of problems could cause the motor to always run at top speed. Could the thyristor be open? No, because that should stop the motor. Could the thyristor be shorted? Yes, that is a definite possibility. Is it time to change the thyristor? No, the analysis is not over yet. Are there any other causes for the observed symptoms? What if the gating circuit is defective? Could this make the motor run at top speed? Yes, it could.

The last part of the troubleshooting process is to limit the possibilities. How can this be done? One way is to shut off the power. Then disconnect the gate lead of the thyristor. Turn the power back on. Does the motor run at top speed again? If it does, the thyristor is, no doubt, shorted. It should be replaced. What if the motor will not run at all? This means the thyristor is good. With its gate lead open, it will not come on. The problem is in the gate circuit.

There are many types of gate circuits. It will be necessary to study the circuit and determine its principle of operation. If the circuit uses a unijunction transistor, it will be necessary to determine whether the UJT pulse generator is working as it should. If the circuit uses a diac, it will be necessary to determine whether the diac is operating properly. It may be possible to use resistance analysis (with the power off) to find a defect. A resistor may be open. A capacitor may be shorted. Some solid-state device may have failed.

At this point, you should begin to realize that the answers are usually not in the manuals or the textbooks. A good troubleshooter understands the basic principles of electronic devices and circuits. This knowledge will allow a logical and analytic process to flow. It is not always easy. Highly skilled technicians "get stuck" from time to time. However, usually they do not keep retracing the same steps over and over. Once a particular fact is confirmed, it is noted on paper or mentally. Using paper is best because another job or quitting time can interfere. It is too easy to forget what has and has not been checked.

Different technicians use somewhat different approaches to troubleshooting. All good technicians have these things in common, however:

1. They work safely and use a system view.
2. They follow the manufacturer's recommendations.
3. They find and use the proper service literature.
4. They use a logical and orderly process.
5. They observe, analyze, and limit the possibilities.
6. They keep abreast of technology.
7. They understand how devices and circuits work in general terms. They understand what each major stage is supposed to do.
8. They are skilled in the use of test equipment and tools.
9. They are neat, use the proper replacement parts, and put all the shields, covers, and fasteners back where they belong.
10. They check their work carefully to make sure nothing was overlooked.
11. They never consider modifying a piece of equipment or defeating a safety feature just because it is convenient at the time.

Technicians who have developed these skills and habits are in demand and they always will be.

TEST

Choose the letter that best answers each question.

31. Which of the following loads should never be connected to a thyristor control circuit?
    a. Incandescent lamps
    b. Soldering irons
    c. Soldering guns
    d. Soldering pencils
32. Why is it not safe to connect a line-operated oscilloscope across a triac in a light dimmer?
    a. A ground loop could cause damage.
    b. The cabinet and controls of the scope could assume line potential.
    c. Both of the above.
    d. None of the above.
33. Refer to Fig. 14-28. Assume that the SCR is open. What is the most likely symptom?
    a. The motor will run at top speed (no control).
    b. The motor will speed up and slow down.
    c. Diode D_1 will be burned out by the overload.
    d. The motor will not run.
34. In Fig. 14-28, assume D_2 is open. What is the most likely symptom?
    a. The motor will run at top speed (no control).
    b. The motor will be damaged.
    c. The motor will not run at all.
    d. Resistor R_4 will burn up.
35. Refer to Fig. 14-34. With the switch open, the response is:
    a. Overdamped
    b. Underdamped
    c. Critically damped
    d. None of the above.
36. Refer to Fig. 14-35. When the power is turned on, the motor runs the lens wide open and then stops. Light level and the brightness-adjust control seem to have no effect. What is wrong?
    a. Defective error detector
    b. Defective amplifier
    c. Defective reference signal
    d. Any of the above could be defective
37. Refer to Fig. 14-35. Suppose the reference voltage is unstable because of a faulty regulator in the power supply. What is the most likely symptom?
    a. The brightness will change.
    b. The motor will run in only one direction.
    c. The motor will not run at all.
    d. The feedback voltage from the video processor will refuse to change.
38. Refer to Fig. 14-36. The azimuth motor oscillates continuously and the beam antenna is rocking back and forth. The elevation is steady. The problem may be
    a. Too much azimuth gain
    b. Too much elevation gain
    c. The azimuth motor is defective
    d. Both gains are too low

Chapter 14 Review

Summary

1. A rheostat can be used to control circuit current.
2. Rheostat control is not efficient since much of the total circuit power is dissipated in the rheostat.
3. Voltage control is much more efficient than resistance control.
4. Switches dissipate little power when open or when closed.
5. A fast switch can control power in a circuit without producing undesired effects such as flicker.
6. Switch control is much more efficient than resistance control.
7. A latch circuit can be formed from two transistors: one an NPN and the other a PNP.
8. A latch circuit is normally off. It can be turned on with a gating current.
9. Once the latch is on, it cannot be turned off by removing the gate current.
10. A latch can be turned off by interrupting the load circuit or by applying reverse bias.
11. A four-layer diode or silicon-controlled rectifier is equivalent to the NPN-PNP latch.
12. An SCR, like an ordinary diode, conducts from cathode to anode.
13. An SCR, unlike an ordinary diode, does not conduct until turned on by a breakover voltage or by gate current.
14. In ordinary operation, SCRs are gated on and not operated by breakover voltage.
15. The SCR is a half-wave device.
16. Commutation refers to turning off an SCR.
17. The SCR is a unidirectional device since it conducts in only one direction.
18. The triac is a bidirectional device since it conducts in both directions.
19. Triacs are capable of full-wave ac power control.
20. Triacs are useful as static switches in low- and medium-power ac circuits.
21. The term "thyristor" is general and can be used in referring to SCRs or triacs.
22. A snubber network may be needed when triacs are used with inductive loads or when line transients are expected.
23. Negative-resistance devices are often used to trigger thyristors.
24. A diac is a bidirectional, negative-resistance device.
25. Diacs are often used to gate triacs.
26. Feedback can be used in control circuits to provide automatic correction for any error.
27. A load such as a motor may provide its own feedback signal.
28. A separate sensor such as a tachometer may be required to provide the necessary feedback signal.
29. A servomechanism is any control system using feedback that represents mechanical action.
30. Servomechanisms provide automatic control.
31. Servomechanism loop gain determines positional accuracy (stiffness) and transient response.
32. Too much loop gain may cause oscillations in a servomechanism.
33. Thyristor control circuits may be safely used with universal (ac/dc) motors.
34. The wattage rating of a thyristor control circuit must be greater than its load dissipation.
35. Some problems in a thyristor control circuit may be isolated by opening the gate lead.

Chapter Review Questions

Choose the letter that best answers each question.

14-1. Refer to Fig. 14-1. Suppose the load resistance is constant at 80 Ω and the source voltage is 240 V. What will the load dissipation be if the rheostat is set for 160 Ω?
a. 80 W
b. 168 W
c. 235 W
d. 411 W

14-2. What is the dissipation in the rheostat in question 14-1?
a. 62 W
b. 160 W
c. 345 W
d. 590 W

14-3. What is the efficiency of the circuit in question 14-1?
a. 33 percent
b. 68 percent
c. 72 percent
d. 83 percent

14-4. Suppose the resistance of a load is constant. What will happen to the power dissipation in the load if the current is increased three times its original value?
a. The power will drop to one-third its original value.
b. The power will remain constant.
c. The power will increase 3 times.
d. The power will increase 9 times.

14-5. Why is resistance control so inefficient?
a. Resistors are very expensive.
b. The control range is too restricted.
c. Much of the circuit power dissipates in the control device.
d. None of the above.

14-6. Refer to Fig. 14-5. What is the purpose of the gate switch?
a. To turn the transistor switch on and off
b. To commutate the NPN transistor
c. To provide an emergency shutdown feature (safety)
d. To turn on the transistor switch

14-7. Refer to Fig. 14-5. The transistors are on. How can they be shut off?
a. Open the gate switch.
b. Close the gate switch.
c. Open the load circuit.
d. Increase the source voltage.

14-8. Refer to Fig. 14-5. Which of the following terms best describes the way the circuit works?
a. Latch
b. Resistance controller
c. Rheostat controller
d. Linear amplifier

14-9. How is a silicon-controlled rectifier similar to a diode rectifier?
a. Both can be classed as thyristors.
b. Both support only one direction of current flow.
c. Both are used to change alternating current to pulsating direct current (rectify).
d. Both have one PN junction.

14-10. What is the effect of increasing the gate current in an SCR?
a. The reverse breakover voltage is improved.
b. The forward breakover voltage is increased.
c. The forward breakover voltage is decreased.
d. The internal resistance of the SCR increases.

14-11. Refer to Fig. 14-10. What is the maximum conduction angle of this circuit?
a. 45°
b. 90°
c. 180°
d. 360°

14-12. Refer to Fig. 14-10. If the load is a motor, what should the motor do if the conduction angle is increased?
a. Slow down
b. Stop
c. Gradually slow down
d. Speed up

14-13. Why is thyristor control more efficient than resistance control?
a. Thyristors are less expensive.
b. Thyristors are easier to mount on a heat sink.
c. Thyristors vary their resistance automatically.
d. Thyristors are solid-state switches.

14-14. Refer to Fig. 14-12. What happens when SCR_1 is gated on?
a. The load comes on.
b. The load goes off.
c. The capacitor turns off SCR_2.
d. SCR_2 comes on.

14-15. Refer to Fig. 14-12. What happens when SCR_2 is gated on?
a. The load comes on.

b. The load goes off.
c. The capacitor turns off SCR_2.
d. SCR_1 comes on.

14-16. Turning off a thyristor is known as
a. Gating
b. Commutating
c. Forward-biasing
d. Interrupting

14-17. Which of the following devices was developed specifically for the control of ac power?
a. The SCR
b. The UJT
c. The snubber
d. The triac

14-18. Refer to Fig. 14-21. Suppose the load is an incandescent lamp and the conduction angle of the circuit is decreased. What will happen to the lamp?
a. Nothing.
b. It will dim.
c. It will produce more light.
d. It will flicker violently.

14-19. Refer to Fig. 14-21. The load is operating at full power. What is the conduction angle of the circuit?
a. 45°
b. 90°
c. 180°
d. None of the above

14-20. What is the chief advantage of a triac as compared with a silicon-controlled rectifier?
a. It costs less to buy.
b. It runs much cooler.
c. The triac is bidirectional.
d. All the above.

14-21. Refer to Fig. 14-23. What is the function of the snubber network?
a. It prevents false commutation.
b. It reduces television interference.
c. It helps reduce unwanted turn-on.
d. It helps the gate control circuit work sooner.

14-22. Refer to Fig. 14-27. Which component turns on and then gates the triac?
a. Capacitor C_3
b. Capacitor C_1
c. Resistor R_2
d. The diac

14-23. Refer to Fig. 14-27. Which component or components have been added to reduce radio interference?
a. Inductor L_1 and capacitor C_1
b. Capacitor C_3
c. The diac
d. Capacitor C_2 and resistor R_2

14-24. Some devices exhibit a rapid decrease in resistance after some turn-on voltage is reached. What are they called?
a. Negative-resistance devices
b. FETs
c. Linear resistive elements
d. Voltage-dependent resistors (VDRs)

14-25. Refer to Fig. 14-28. What will happen if the SCR shorts from anode to cathode?
a. The motor will stall.
b. The motor will run above its top normal speed.
c. V_2 will fall to zero.
d. None of the above.

14-26. Refer to Fig. 14-28. What will happen if D_2 burns out (opens)?
a. The motor will burn out.
b. The motor will run at above half speed.
c. The motor will run at below half speed.
d. The motor will not run.

14-27. Refer to Fig. 14-29. What symptom would appear if the tachometer coupling is loose and is slipping on its shaft?
a. None because the speed is regulated.
b. The motor will slow down and stop.
c. The motor will run fast.
d. The reference signal will become unstable.

14-28. Which of the following devices should never be operated from a thyristor power-control device?
a. A washing machine motor
b. A heater
c. Christmas tree lights
d. A soldering iron

Critical Thinking Questions

14-1. Which of the power control circuits presented in this chapter qualify as linear circuits. Why?

14-2. Could a BJT be used as a linear dc power controller? Would there be any disadvantage to such an application?

14-3. Is there a way to connect two SCRs so that they will provide full-wave control?

14-4. Some companies manufacture optically coupled triac drivers. They consist of infrared LEDs optically coupled to photodetectors with triac outputs. Can you think of any application for these components?

14-5. What technical term can be used to describe the "cruise control" feature that is found on some vehicles?

Answers to Tests

1. d
2. d
3. c
4. b
5. c
6. a
7. c
8. a
9. c
10. d
11. b
12. c
13. a
14. d
15. c
16. b
17. d
18. c
19. d
20. a
21. b
22. b
23. d
24. c
25. b
26. d
27. b
28. a
29. c
30. a
31. c
32. c
33. d
34. c
35. b
36. d
37. a
38. a

Regulated Power Supplies

Chapter Objectives

This chapter will help you to:

1. *Perform* basic calculations for power supply regulator circuits.
2. *Explain* the use of feedback in voltage regulator circuits.
3. *Identify* the types of current regulation.
4. *Identify* crowbar circuits.
5. *Identify* switch-mode regulators and their characteristics.
6. *Troubleshoot* regulated power supplies.

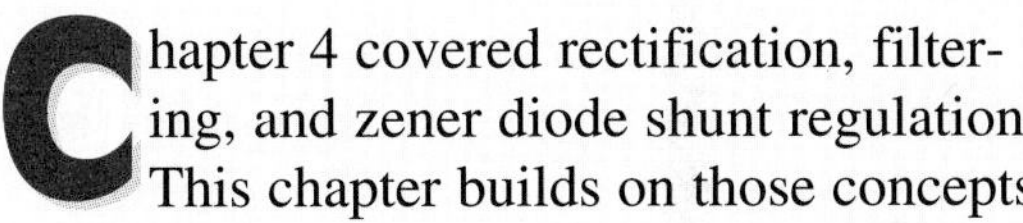

Chapter 4 covered rectification, filtering, and zener diode shunt regulation. This chapter builds on those concepts and shows how basic power-supply performance is enhanced to meet the needs of modern electronic systems.

15-1 Open-Loop Voltage Regulation

Voltage regulation is one of the most important power-supply characteristics. It is the measure of a supply's ability to maintain a constant output voltage. *Open loop* means that feedback is not used to hold the output constant. The next section of this chapter examines the use of feedback (closed-loop) regulator circuits.

Consider Fig. 15-1 on page 408. It is a graph of the performance of a typical nonregulated power supply. You can see that the output drops 6 V (ΔV) as the load on the power supply is increased from 0 to 5 A. Also note that the power supply delivers its rated 12 V only when it is fully loaded. When less of a load is taken, the output is greater than 12 V.

Now examine Fig. 15-2 on page 408. It illustrates the *line regulation* curve for the same power supply. It shows that the output voltage drops as the line voltage falls below its nominal 120 V value. It also shows that high line voltage will increase the output above normal. Line voltage does change. In fact, the word *"brownout"* refers to a condition of low line voltage caused by heavy use of electrical power. Brownouts are common in cities during very hot weather. Power companies are often forced to reduce line voltage under severe load conditions to prevent equipment failure.

When the conditions of low line voltage and high load current are combined, a rather low output voltage will result in a nonregulated supply. Conversely, if high line voltage occurs when the load current is low, a rather high output voltage will occur. Thus, it can be seen that load changes and line changes have significant effects on nonregulated power-supply output voltages.

One answer to this problem is to use a special power transformer. Figure 15-3 on page 408 shows the construction of an ordinary (linear) power transformer. There are two major flux (magnetic flow) paths and the primary and secondary coils are both wound around the center of the laminated core. The core in such a transformer is designed to be linear. That is, the core will not saturate. Now look at the *ferroresonant transformer* in Fig. 15-4 on page 409. It differs in several important ways. There are separate windows for the primary and secondary windings. There are air gaps in the shunt flux path. Finally, there is a *resonating capacitor* across the secondary.

Load regulation

Line regulation

Ferroresonant transformer

Brownout

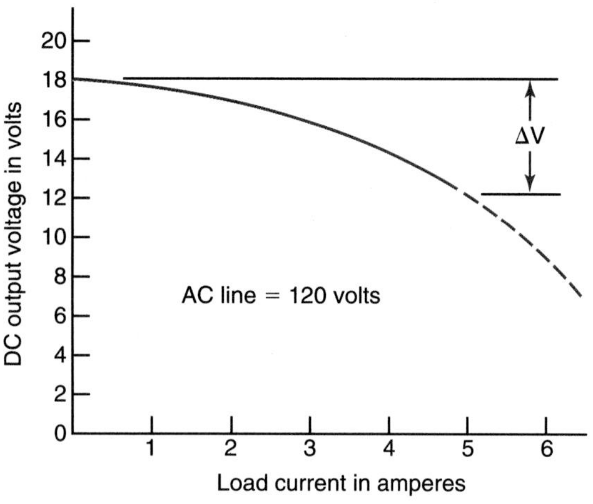

Fig. 15-1 Load regulation curve for a 12-V, 5-A power supply.

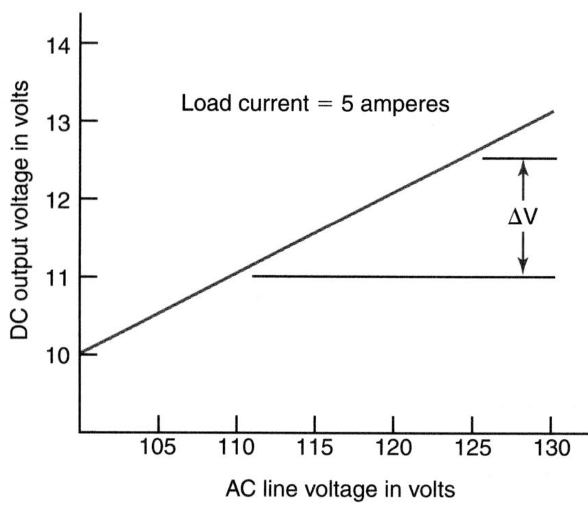

Fig. 15-2 Line regulation curve for a 12-V, 5-A power supply.

Core saturation

The transformer shown in Fig. 15-4 can be used to build a power supply with a much more stable output voltage than the typical unregulated supply. As line voltage is applied to the primary, the main magnetic path excites the secondary. Part, or all, of the secondary winding is tuned by a resonating capacitor (usually several microfarads). As the secondary goes into resonance, large currents flow in the capacitor and the resonant part of the secondary. The circulating current in a parallel resonant circuit is much greater than the line current when the circuit Q is high. The high circulating current drives the main flux path into saturation. This *core saturation* provides *line regulation.*

Core saturation occurs in a magnetic circuit when an increase in magnetizing force is not accompanied by a corresponding increase in flux density. A simple analogy is a saturated transistor circuit where more base current will not produce any more collector current. In a saturated transformer, an increase in primary voltage will not increase the secondary voltage. Similarly, a decrease in primary voltage will not affect the secondary voltage, provid-

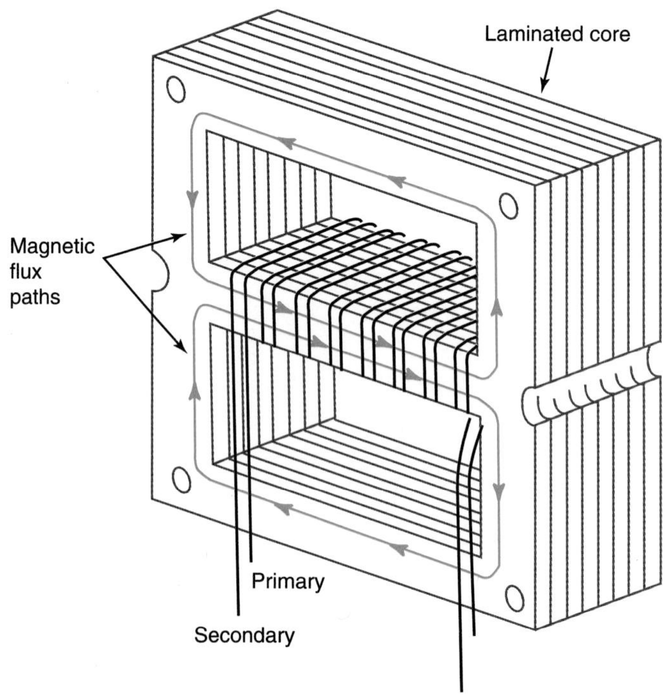

Fig. 15-3 Linear power transformer construction.

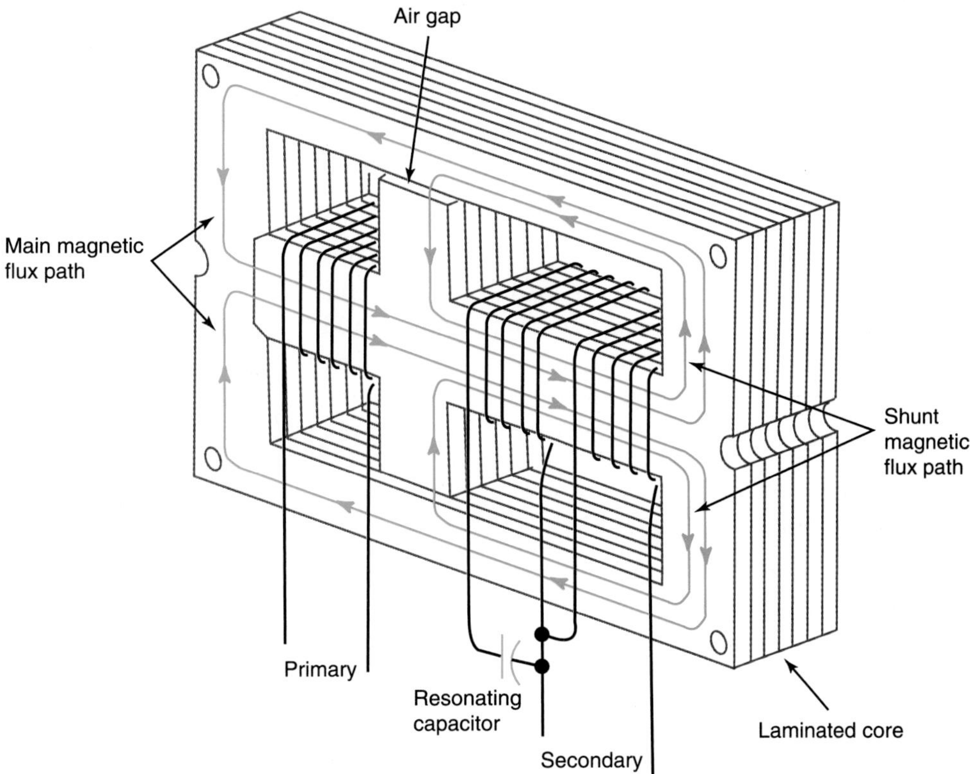

Fig. 15-4 Ferroresonant transformer construction.

ing that the core stays in saturation. Saturated transformers produce a reasonably constant secondary output over some range of primary voltage (typically 90 to 140 V).

Another feature of the ferroresonant transformer of Fig. 15-4 is that the air gaps prevent core saturation for the shunt magnetic flux path. Air has much more reluctance (magnetic resistance) than transformer steel. The air gaps are the equivalent of series resistors and limit flux in the shunt path. Limiting flux prevents saturation and provides a linear response for the

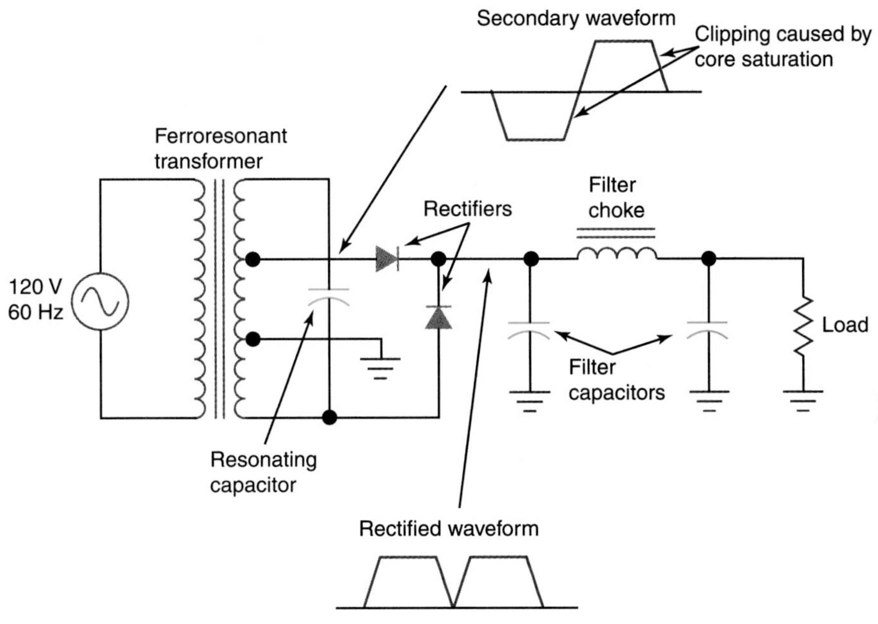

Fig. 15-5 Ferroresonant supply schematic.

About Electronics

No More Power Outages The typical UPS (uninterruptable power supply) uses lead-acid batteries and a 60-Hz oscillator to replace the line voltage in the case of a power failure.

shunt portion of the magnetic circuit. If load current in the secondary is increased, the circuit Q drops and the circulating current decreases. The shunt flux will decrease, allowing an increase in the main flux path. An increase in main flux transfers extra energy from primary to secondary and compensates for the increased load current. This provides *load regulation.* Therefore, ferroresonant transformers provide both load and line regulation.

Figure 15-5 shows the use of a ferroresonant transformer in a dc power supply. Notice that in this case the resonating capacitor is across the entire secondary winding. Also notice that the secondary waveform is clipped. This is caused by core saturation. The clipped sine wave has several advantages. It is easier to filter because the resulting rectified waveform has less ripple content than ordinary full-wave, pulsating direct current. A second advantage is that the clipped waveform has a lower peak voltage which is easier for the rectifiers to handle. These types of power supplies are *very* reliable and provide good voltage regulation for both changing line voltage and changing load current. They are also noted for their efficiency. Unfortunately, the ferroresonant transformer is a large, heavy, and expensive component.

Amplified zener regulator

Zener regulator

Series pass transistor

Figure 15-6 shows another answer to the regulation problem. This circuit was presented in Chap. 4. It uses a zener *diode* connected in parallel with the load. A zener diode will drop a relatively constant voltage when operating in reverse breakover. Therefore, the load will also see a relatively constant voltage. You can see from Fig. 15-6 that the zener current and load current add in resistor R_Z.

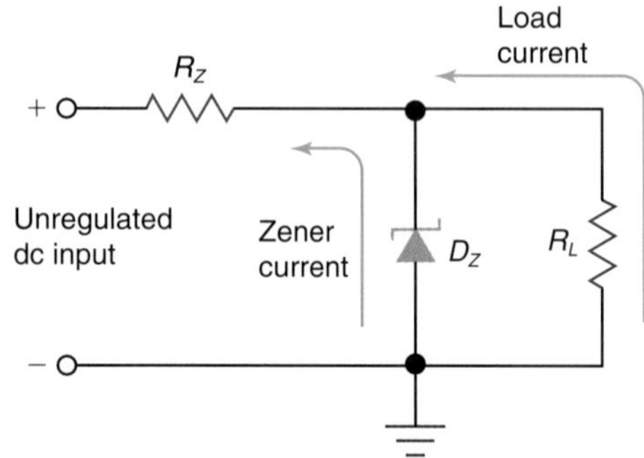

Fig. 15-6 Zener shunt regulator.

A problem with zener shunt regulators is that the diode dissipation is too large in some applications. For example, if the regulator shown in Fig. 15-6 is used to supply 12 V at 1 A, a high-wattage zener will be required. Assume the unregulated dc input to be 18 V. Resistor R_Z will have to drop 6 V (18 V − 12 V = 6 V). If the desired zener current is 0.5 A, R_Z can be found using Ohm's law:

$$R_Z = \frac{V}{I}$$

$$= \frac{6\text{ V}}{1\text{ A} + 0.5\text{ A}} = 4\ \Omega$$

Next, the power dissipation in the diode is

$$P_D = V \times I$$

$$= 12\text{ V} \times 0.5\text{ A} = 6\text{ W}$$

However, if the load is removed from the regulator, all of the current will flow through the zener diode, and its dissipation increases to

$$P_D = V \times I$$

$$= 12\text{ V} \times 1.5\text{ A} = 18\text{ W}$$

For good reliability, a zener diode rated for a power dissipation of at least 2 × 18 W (36 W) would be required. High-power zener diodes are too expensive for most applications.

Figure 15-7 shows a way to reduce the zener dissipation. The circuit is often called an *amplified zener regulator.* The zener diode is used to regulate the base voltage of a power transistor called a *series pass transistor.* If there is a reasonably constant 0.7-V drop across the base-emitter junction of the pass transistor, the emitter voltage and the load voltage will also be reasonably constant.

The current through R_Z in Fig. 15-7 is the sum of the base current (I_B) and the zener current. Assuming a load current of 1 A and a transistor β of 49, the base current is

$$I_B = \frac{I_E}{\beta + 1}$$

$$= \frac{1\text{ A}}{50} = 0.02\text{ A}$$

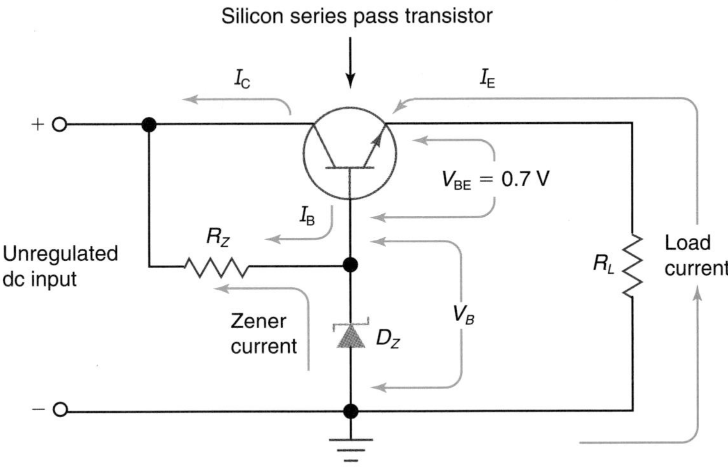

Fig. 15-7 Amplified zener regulator.

Because of the base-emitter drop, the zener voltage should be 0.7 V greater than the load voltage. A 12.7-V zener diode will provide a 12-V regulated output. The zener current should be about half of the base current or 10 mA in this example. Assuming that an unregulated input of 18 V, Ohm's law is used to calculate R_Z:

$$R_Z = \frac{V}{I}$$
$$= \frac{18\text{ V} - 12.7\text{ V}}{0.02\text{ A} + 0.01\text{ A}} = 177\ \Omega$$

Compare Fig. 15-7 with Fig. 15-6 using identical input and output conditions. The worst-case zener dissipation occurs at zero load current. In Fig. 15-7, 30 mA will flow in the zener diode if the load is removed from the regulator. The zener current increases because with no load current, the base current drops to zero. Therefore, all 30 mA must flow in the zener. The zener dissipation will therefore increase:

$$P_D = V \times I$$
$$= 12.7\text{ V} \times 0.03\text{ A} = 0.381\text{ W}$$

A 1-W zener will be safe under all operating conditions in Fig. 15-7. It should now be obvious why the amplified zener regulator is preferred for high-current applications. The circuit does require a series pass transistor, but this is a less expensive component than a high-wattage zener diode.

Figure 15-8 on page 412 shows a *negative* amplified *regulator*. The pass transistor is PNP, and the circuit regulates a negative voltage referenced to the ground terminal. Note that the zener diode cathode is grounded. Compare this connection with that shown in Fig. 15-7.

Example 15-1

Select a value for R_Z in Fig. 15-8 if D_Z is a 5.7-V zener, the load current is 2 A, $\beta = 25$, the unregulated input is 9 V, and the desired zener current is 10 mA. Also, determine the worst-case zener dissipation. Begin by finding the base current:

$$I_B = \frac{I_E}{\beta + 1} = \frac{2\text{ A}}{25 + 1} = 76.9\text{ mA}$$

The total current in R_Z is the sum of the base current and the zener current:

$$I_{RZ} = I_B + I_{ZD}$$
$$= 76.9\text{ mA} + 10\text{ mA} = 86.9\text{ mA}$$

The drop across R_Z is the unregulated input voltage minus the zener voltage:

$$V_{RZ} = 9\text{ V} - 5.7\text{ V} = 3.3\text{ V}$$

Ohm's law can now be used to find R_Z:

$$R_Z = \frac{V_{RZ}}{I_{RZ}} = \frac{3.3\text{ V}}{86.9\text{ mA}} = 38.0\ \Omega$$

The worst-case zener dissipation is:

$$P_D = V \times I$$
$$= 5.7\text{ V} \times 86.9\text{ mA} = 0.495\text{ W}$$

Negative regulator

Figure 15-8 shows another component that is often found in amplified zener regulators. An electrolytic capacitor bypasses the base of

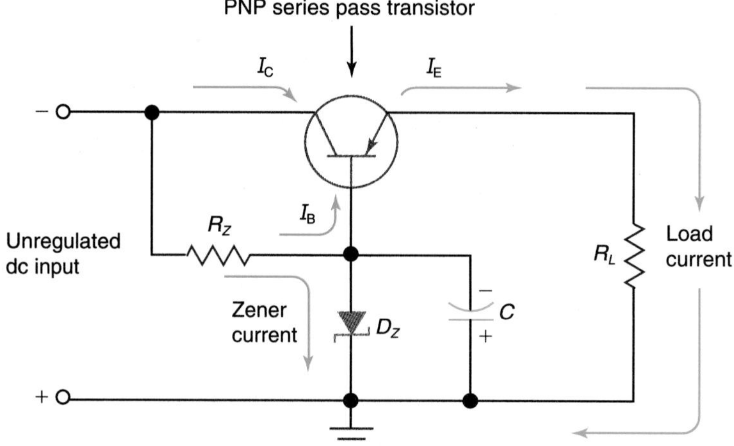

Fig. 15-8 Negative amplified regulator.

the transistor to ground. This capacitor is typically around 50 μF and, in conjunction with R_Z, forms a low-pass filter. This filter helps to remove noise and ripple present at the unregulated dc input. Also, zener diodes generate noise and the capacitor is useful for eliminating it from the output of the regulator. Most amplified zener-regulated power supplies use this capacitor.

Bipolar supply

Figure 15-9 shows a dual-polarity (bipolar) power supply. This circuit provides both a positive and a negative regulated voltage with respect to the ground terminal. Notice that transformer T_1 has two secondary windings. Each secondary is center-tapped and supplies a full-wave rectifier circuit. Capacitors C_1 and C_2 filter the rectifier outputs. Q_1 and Q_2 are series pass transistors.

TEST

Choose the letter that best answers each question.

1. Load regulation is a measure of a power supply's ability to keep a constant output under conditions of changing
   a. Line voltage
   b. Current demand
   c. Temperature
   d. Oscillator frequency
2. Why must a capacitor be connected across the secondary of a ferroresonant transformer?
   a. To filter out ac ripple
   b. To change dc to ac
   c. To cause core saturation
   d. To eliminate radio-frequency interference

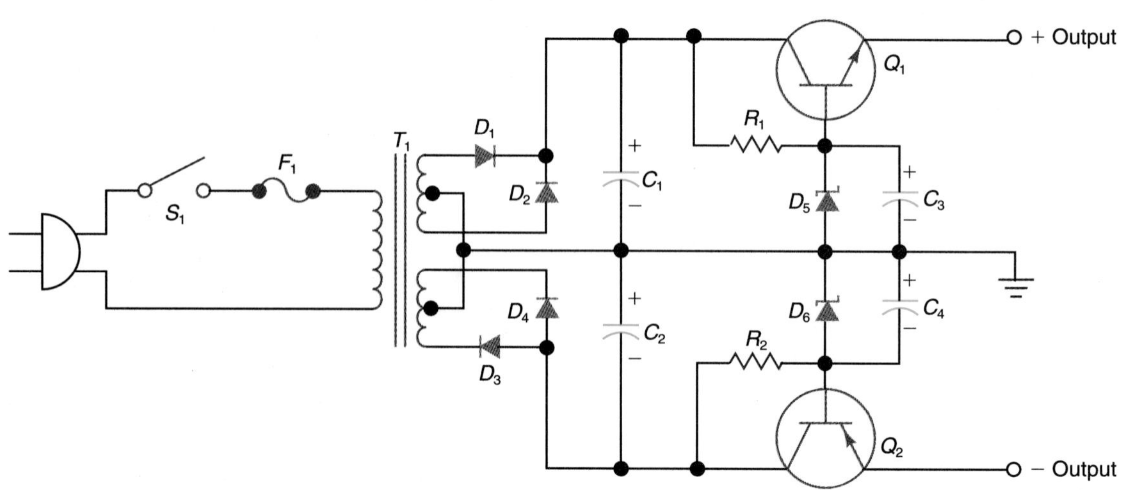

Fig. 15-9 Dual-polarity regulated supply.

3. Refer to Fig. 15-6. If $R_Z = 15\ \Omega$, the unregulated input is 12 V, and the zener operates at 6 V, what is the diode dissipation when the load current is 0 A?
   a. 4.8 W
   b. 2.4 W
   c. 1.2 W
   d. 0 W
4. Refer to Fig. 15-8. Assume a dc input of −20 V, a zener voltage of 14.4 V, and a silicon pass transistor. What is the load voltage?
   a. −20 V
   b. −15.1 V
   c. −14.4 V
   d. −13.7 V
5. Refer to Fig. 15-9. Both zeners are rated at 6.8 V, and both transistors are silicon. What is the voltage from the + output terminal to the − output terminal?
   a. 12.2 V
   b. 10.6 V
   c. 6.8 V
   d. 6.1 V
6. Refer to Fig. 15-9. Resistor R_1 is open (infinite resistance). What symptom can be expected?
   a. The − output will be zero.
   b. Both outputs will be zero.
   c. Both outputs will be high.
   d. The + output will be zero.

15-2 Closed-Loop Voltage Regulation

The amplified zener regulators discussed in the previous section depend on a constant base-emitter voltage drop. As long as this drop and the zener drop do not change, the output voltage will remain constant. However, the base-emitter drop does change when the output current is high. For example, a pass transistor that is conducting 5 A may show a base-emitter voltage of 1.7 V. In other words, as the pass transistor is called upon to conduct higher and higher load currents, its base-emitter voltage increases. This increasing drop will subtract from the zener voltage and cause the regulated output to go down. It is normal to expect about a 1-V decrease in output when a load current of several amperes is taken from an amplified zener regulator.

Open-loop regulators cannot provide highly stable output voltages, especially when large changes in load current are expected. *Feedback* can be used to improve regulation. Examine Fig. 15-10. It shows the basic concept of *closed-loop* regulation. A control device is available to adjust the load voltage. Assume that the unregulated power supply develops 18 V and that the control device drops 6 of these volts. This leaves 12 V (18 V − 6 V) for the load. If the control device can be turned on harder (have its resistance decreased), it will drop less voltage and make more available for the load. Similarly, this control device can be adjusted for a higher resistance to decrease the load voltage. By adjusting the resistance of the control device, the output voltage is controlled.

Feedback

Figure 15-10 also shows a *reference voltage* (V_{ref}). This voltage is stable and is applied to one of the inputs of an *error amplifier.* The other input to the error amplifier is feedback from the load. This feedback allows the amplifier to compare the load voltage with the reference voltage. Any change in load voltage will create a differential signal at the input of the error amplifier. This difference represents an error, and the amplifier adjusts the drive to the control device to decrease the error. If the output voltage tends to drop because of an increased load, the error is sensed and the control device is turned on harder to eliminate the drop in output. The feedback and error amplifier stabilize the output voltage.

Reference voltage

Error amplifier

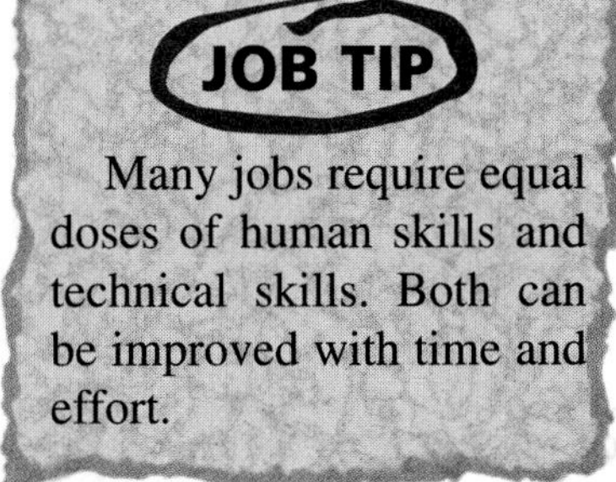

Figure 15-11 on page 414 shows a schematic diagram for a feedback (closed-loop) regulator. Transistor Q_1 is a series pass transistor and serves as the control device. The

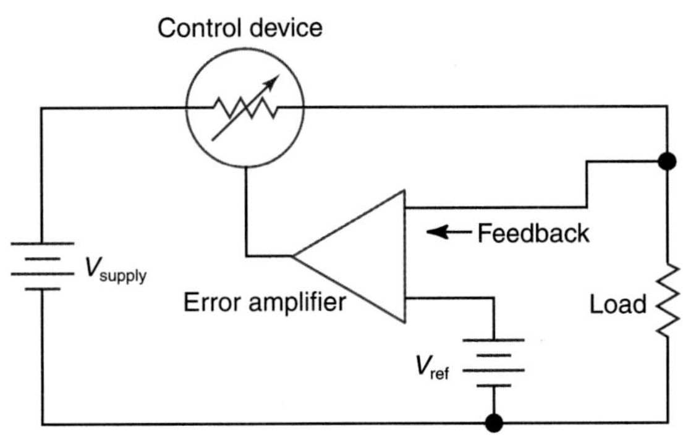

Fig. 15-10 Closed-loop regulation.

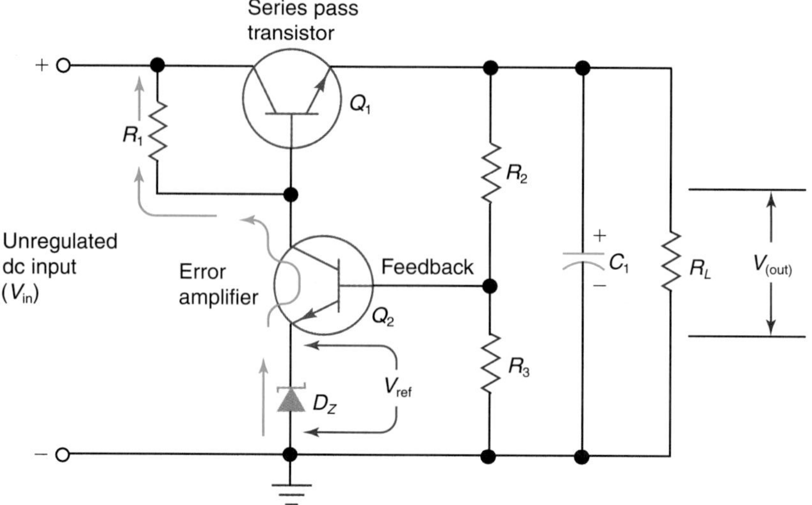

Fig. 15-11 Feedback-regulated power supply.

zener diode produces the reference voltage. Transistor Q_2 is the error amplifier. R_1 is the load resistor for Q_2. R_2 and R_3 form a voltage divider for the output voltage and provide feedback to Q_2. The emitter voltage of Q_2 is zener-regulated, and its base voltage is proportional to the output voltage. This allows Q_2 to amplify any error between the reference and the output.

Assume, in Fig. 15-11, that the load demands more current causing a decrease in output voltage. The divider now sends less voltage to the base of Q_2. Transistor Q_2 responds by conducting less current and less voltage will drop across R_1. The base voltage of Q_1 goes up, and Q_1 is turned on harder, which increases the output voltage. If you trace all of the changes, you will see that output voltage change is reduced by the feedback and the error amplifier.

Example 15-2

Calculate the zener diode current in Fig. 15-11 when the unregulated input is 16 V, the zener is 5.1 V, $\beta_{Q1} = 35$, $R_1 = 47\ \Omega$, $R_2 = 1\ \text{k}\Omega$, $R_3 = 1\ \text{k}\Omega$, and $R_L = 5\ \Omega$. Also, find the zener current when the load is disconnected. This problem takes several steps to solve. The voltage at the base of the error amplifier is found first:

$$V_{B(Q_2)} = V_{DZ} + 0.7\ \text{V} = 5.1\ \text{V} + 0.7\ \text{V}$$
$$= 5.8\ \text{V}$$

V_{out} is determined by the voltage divider and $V_{B(Q_2)}$:

$$V_{B(Q_2)} = V_{out} \times \frac{R_3}{R_3 + R_2}$$

$$5.8\ \text{V} = V_{out} \times \frac{1\ \text{k}\Omega}{2\ \text{k}\Omega}$$

Solving for V_{out}:

$$V_{out} = 2 \times 5.8\ \text{V} = 11.6\ \text{V}$$

With V_{out} known, the base voltage of the series pass transistor is readily determined by adding 0.7 V:

$$V_{B(Q1)} = V_{out} + 0.7\ \text{V}$$
$$= 11.6\ \text{V} + 0.7\ \text{V} = 12.3\ \text{V}$$

The drop across R_1 is calculated with:

$$V_{R1} = V_{in} - V_{B(Q1)}$$
$$= 16\ \text{V} - 12.3\ \text{V}$$
$$= 3.7\ \text{V}$$

Ohm's law will give us the current in R_1:

$$I_{R_1} = \frac{V_{R_1}}{R_1} = \frac{3.7\ \text{V}}{47\ \Omega} = 78.7\ \text{mA}$$

We now have to determine how much of this current flows through the error amplifier and the zener diode. The load current is calculated using Ohm's law:

$$I_L = \frac{11.6\ \text{V}}{5\ \Omega} = 2.32\ \text{A}$$

The pass transistor base current is found next:

$$I_{B(Q_1)} = \frac{I_E}{\beta + 1} = \frac{2.32 \text{ A}}{36} = 64.4 \text{ mA}$$

Therefore, of the total 78.7 mA that flows in R_1, 64.4 mA comes from the base of the pass transistor and the rest comes from the zener and the error amplifier:

$$I_{DZ} = 78.7 \text{ mA} - 64.4 \text{ mA} = 14.3 \text{ mA}$$

When the load is disconnected, we can ignore the small current in the voltage divider network and assume that the pass transistor needs no base current. The error amplifier and zener diode will conduct all the current:

$$I_{DZ\,(\text{no load})} = 78.7 \text{ mA}$$

The ability of a feedback power supply to stabilize output voltage is related to the gain of the error amplifier. A high-gain amplifier will respond to very small changes in output voltage and will provide excellent voltage regulation. Examine Fig. 15-12. An op amp is used as the error amplifier. Op amps are capable of very high gain. Resistor R_1 and the zener diode form a reference voltage for the noninverting input of the op amp. Resistors R_2, R_3, and R_4 form a voltage divider. If the output voltage goes down, there will be a decrease in voltage at the inverting input of the op amp. This decreasing voltage is negative-going and will cause the output of the op amp to go in a positive direction. The positive-going output is applied to the pass transistor and turns it on harder. This tends to increase the output and eliminate the change. The op amps's high gain means that the circuit of Fig. 15-12 can hold the output to within several millivolts, so the voltage regulation is excellent.

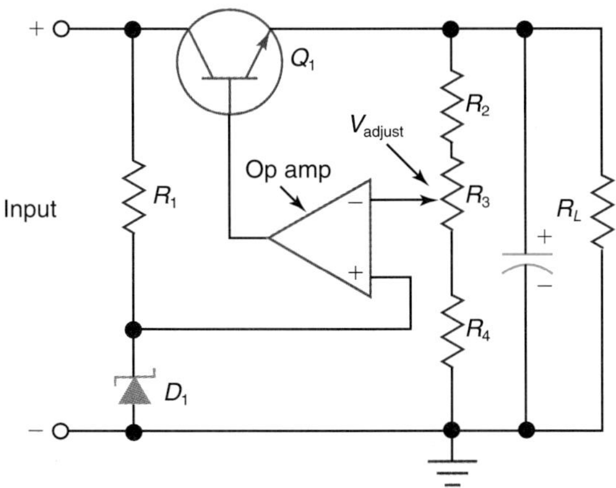

Fig. 15-12 Using an op amp in a feedback-regulated power supply.

The circuit of Fig. 15-12 is adjustable. R_3 is used to set the output voltage. As the wiper arm of the potentiometer is moved toward R_4, less voltage is fed back. This increases the output voltage. As the wiper arm is moved toward R_2, the output is decreased. Adjustable outputs of this type are common in feedback regulators. In practice, the voltage-adjust potentiometer may be a front-panel control, a rear-panel control, or a small trimmer potentiometer mounted on a printed circuit board.

The trend in electronics is to integrate as many circuit functions as is practical onto a single chip of silicon. Regulators have not escaped this trend. Refer to Fig. 15-13 on page 416. It shows an *integrated-circuit voltage regulator.* It is supplied in the TO-220 case and has three leads. The pass transistor, error amplifier, reference circuit, and protection circuitry are all on one chip. The 7812 IC provides 12 V at load currents up to 1.5 A. Typically, it will hold the output within 12 mV over the full range of load currents.

IC voltage regulator

Capacitor C_1 in Fig. 15-13 is required if the IC regulator is located more than several inches from the main power-supply filter capacitor. These ICs are often used as "on-card" regulators. In this configuration, each circuit board in a system has its own voltage regulator so they can be some distance from the main power-supply filter. Capacitor C_2 is optional and can be used to improve the way the regulator responds to rapidly changing load currents.

Some IC regulators, such as the one shown in Fig. 15-13, operate at a fixed output voltage. The 78*XX* series of regulators is typical of this type. The 7805 provides 5 V, the 7812 pro-

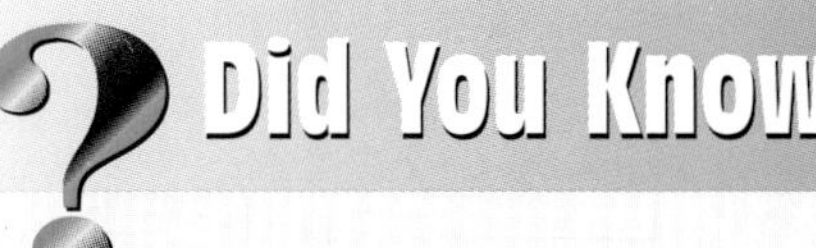

Power and Measurement

- Some modern and portable test equipment can perform the function of a strip-chart recorder. This function is invaluable for investigating power source fluctuations.
- Technicians who troubleshoot power supplies often use dummy loads.

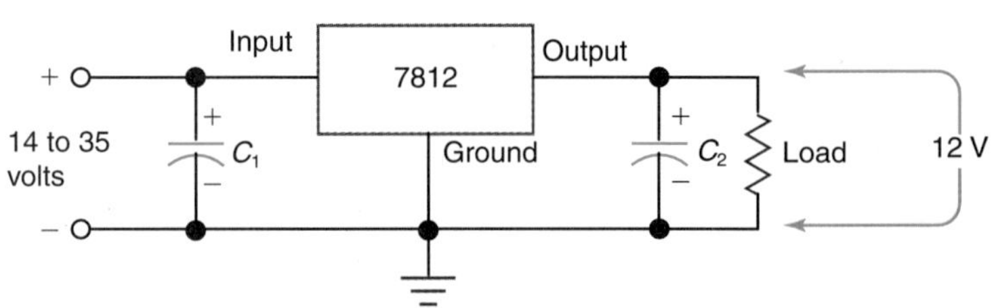

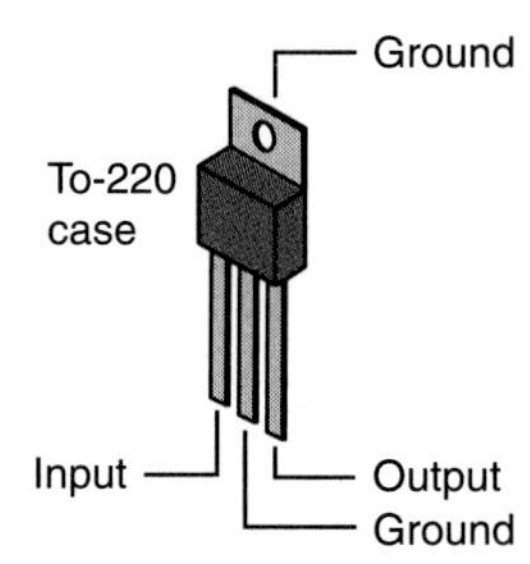

Fig. 15-13 Integrated-circuit voltage regulator.

vides 12 V, and the 7815 provides 15 V. This series is also available in the larger TO-3 case for higher current applications. They all provide a simple and inexpensive alternative to discrete regulators and are widely applied.

Adjustable output

Figure 15-14 shows a way of obtaining an *adjustable output* from a 5-V IC regulator. R_1 and R_2 form a voltage divider. Notice that the ground lead of the regulator is connected to the center of the divider rather than to the circuit ground. By adjusting R_2, the output voltage can be varied from 5 V to some higher voltage.

Two currents flow through R_2 in Fig. 15-14. One is the divider current through R_1. Since the voltage across R_1 must always be 5 V, Ohm's law may be used to find the divider current. The second current through R_2 is I_Q, the quiescent current of the 7805 IC. It is typically 6 mA. The load voltage can be found by adding the regulator voltage (5 V) to the drop across R_2. For example, if we assume R_1 and R_2 to each be 250 Ω, the output voltage is

$$V_{out} = 5\text{ V} + R_2\left(I_Q + \frac{5\text{ V}}{R_1}\right)$$
$$= 5\text{ V} + (250\ \Omega)\left(0.006\text{ A} + \frac{5\text{ V}}{250\ \Omega}\right)$$
$$= 11.5\text{ V}$$

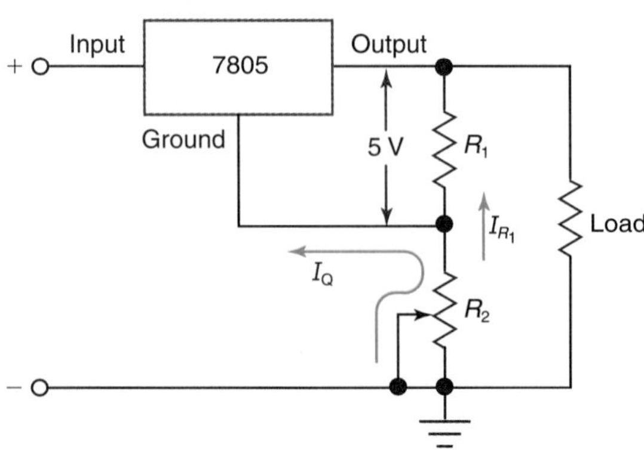

Fig. 15-14 Adjustable output from a fixed regulator.

The output has been adjusted to 11.5 V even though a fixed 5-V regulator is used.

Any increase or decrease in the quiescent current in Fig. 15-14 will cause a change in the drop across R_2 which will affect the output voltage. I_Q is sensitive to the unregulated input, the load current, and the temperature. For example, a 1-mA increase in I_Q would not be unusual and its effect on the output voltage is

$$V_{out} = 5\text{ V} + R_2\left(I_Q + \frac{5\text{ V}}{R_1}\right)$$
$$= 5\text{ V} + 250\ \Omega\left(0.007\text{ A} + \frac{5\text{ V}}{250\ \Omega}\right)$$
$$= 11.75\text{ V}$$

So, the output shift due to the increase in I_Q is 11.75 V − 11.5 V or 0.25 V (250 mV). This shows that adjusting a fixed regulator with a divider degrades its regulation. Since the output is normally held within 12 mV, a 250-mV change is relatively large. Resistor R_2 in Fig. 15-14 should not be too large or the regulation will suffer even more. Values around 100 Ω are practical.

Current-boost circuit

Fixed regulators, such as the 7805, can supply 1.5 A. If more current is required, the *current-boost circuit* of Fig. 15-15 can be used. Transistor Q_1 is used to supply the extra load current. Resistor R_1 determines when Q_1 will turn on and begin sharing the load current. As the IC regulator current increases, the voltage drop across R_1 will increase. This drop is applied to the base-emitter junction of Q_1 and forward-biases it.

If Q_1 in Fig. 15-15 is silicon, it will turn on when its base-emitter voltage reaches 0.7 V. Assume R_1 to be 4.7 Ω. The current required to turn on Q_1 is found by

$$I = \frac{V}{R} = \frac{0.7\text{ V}}{4.7\ \Omega} = 0.149\text{ A}$$

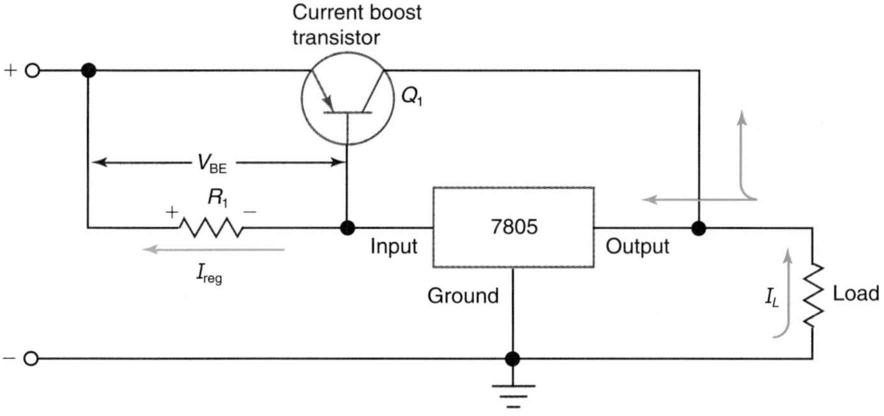

Fig. 15-15 Current-boost circuit.

The IC regulator will conduct all of the load current up to 149 mA. As the load demand exceeds this value, the drop across R_1 turns on Q_1, and it will assist the IC to supply the load. A current-boost circuit, such as the one shown in Fig. 15-15, can provide as much as 10 A by using a high-current transistor to share the load current.

Operational-amplifier circuits often require bipolar power supplies. These power supplies provide both a positive and a negative output voltage with respect to ground. Sometimes these power supplies are adjustable and must *track* one another. A tracking power supply is one in which one or more outputs are *slaved* to a *master*. If the master output changes, so must the output of the slaves change.

Figure 15-16 shows a dual-tracking regulator. The 7805 is a fixed 5-V regulator. The 7905 is also a fixed 5-V device, but it regulates a negative voltage with respect to ground. Neither regulator is directly grounded in Fig. 15-16. The ground leads are driven by operational amplifiers OA_1 and OA_2. This provides adjustable output voltage in a fashion related to the circuit studied in Fig. 15-14. However, the very low output impedance of the op amps ensures that any quiescent current change in the IC regulators will have only a small effect on output voltage.

Tracking

R_4 and R_5 in Fig. 15-16 divide the negative output and apply the result to both op amps. OA_2 is connected in the noninverting mode. As the wiper of R_4 is adjusted toward R_5, the out-

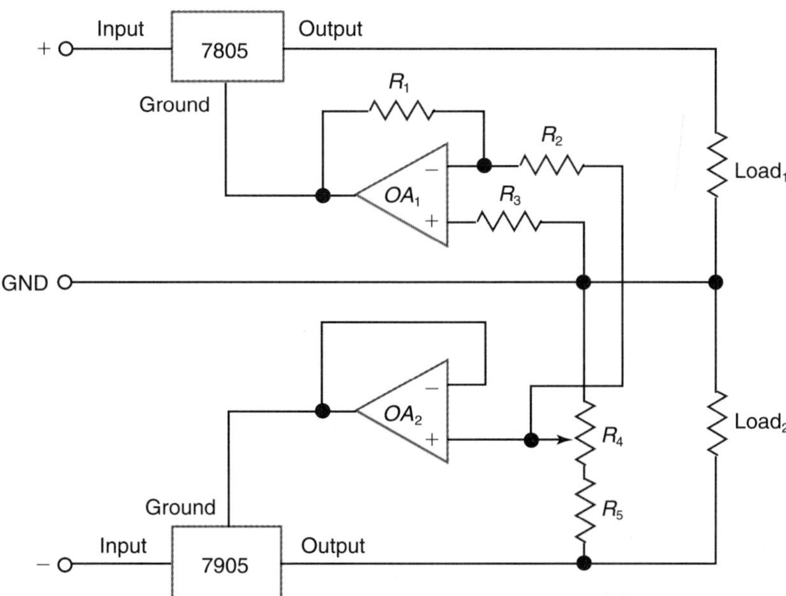

Fig. 15-16 Dual-tracking regulator.

interNET CONNECTION

For more information on power supplies, find the Internet site for Lambda Corp.

put of OA_2 drives the ground lead of the 7905 in a negative direction. This increases the negative output across load 2. At the same time, OA_1 acts as an inverting amplifier. The negative-going signal at the wiper of R_4 becomes a positive-going signal for the ground lead of the 7805. This increases the positive output voltage across load 1. Resistor R_4, therefore, controls both outputs in this tracking regulator. The positive output is slaved to the negative output, and any change in negative output will be tracked by the positive output.

TEST

Choose the letter that best answers each question.

7. Refer to Fig. 15-11. If the unregulated dc input is 18 V and the collector-to-emitter voltage of Q_1 is 12 V, then the voltage across the load resistor will be
   a. 18 V
   b. 12 V
   c. 6 V
   d. 0 V
8. Refer to Fig. 15-11. If the zener diode drops 4.7 V and if the collector-to-emitter voltage of Q_2 is 4 V, what is the voltage across the load resistor? (Assume a silicon pass transistor and a light load current.)
   a. 4 V
   b. 8 V
   c. 12 V
   d. 16 V
9. Refer to Fig. 15-12. Which component provides the reference voltage?
   a. D_1
   b. Q_1
   c. R_3
   d. The op amp
10. Refer to Fig. 15-14. The quiescent current is 5 mA. Resistor R_1 is 100 Ω and R_2 is 200 Ω. What is the output voltage?
    a. 2 V
    b. 5 V
    c. 12 V
    d. 16 V
11. Refer to Fig. 15-15. R_1 is 3.3 Ω and the load current is 150 mA. Transistor Q_1 is silicon. Transistor Q_1 will conduct
    a. No load current
    b. About half the load current
    c. All of the load current
    d. All of the load current in excess of 100 mA
12. Refer to Fig. 15-15. The load current is 4 A. The regulator current is 0.5 A. What is the collector current in Q_1?
    a. 0.5 A
    b. 3.5 A
    c. 4.0 A
    d. 4.5 A

15-3 Current and Voltage Limiting

Current limiting

Some of the regulated power-supply circuits discussed so far are not well protected from damage caused by overloads. The line fuse may not blow quickly enough to protect diodes, transistors, and integrated circuits from damage in the event the power-supply output is short-circuited. A short circuit demands very high current flow from the regulator. This high current will flow through the series pass transistor and other components in the power supply. If there is no *current limiting,* the pass transistor is very likely to be destroyed.

Sometimes, it is just as important to protect the circuits fed by the regulated power supply. Current limiting can prevent serious damage to other circuits. For example, a component in a direct-coupled amplifier may short. The shorted component may overbias an expensive transistor or integrated circuit. The overbiased device can take enough current from the power supply to destroy itself. If the power supply is current-limited, the expensive component may be protected from burnout.

Voltage limiting

Current limiting is helpful, but circuits can also be damaged by too much voltage. A fault in the regulator can cause the output voltage to go up to the nonregulated value. For example, the input to a 5-V regulator may be 10 V. If the pass transistor shorts, the output will go up to 10 V instead of the normal 5 V. This abnormally high voltage will be applied to all devices in the system that are connected to the 5-V supply. Many of them could be destroyed. Therefore, it may be necessary to prevent a power-supply voltage from going beyond some safe value. This is known as *voltage limiting.*

Figure 15-17 shows an example of a circuit that limits current. Much of this circuit was covered in the previous section. The 7812 is a fixed 12-V regulator. Transistor Q_1 boosts the

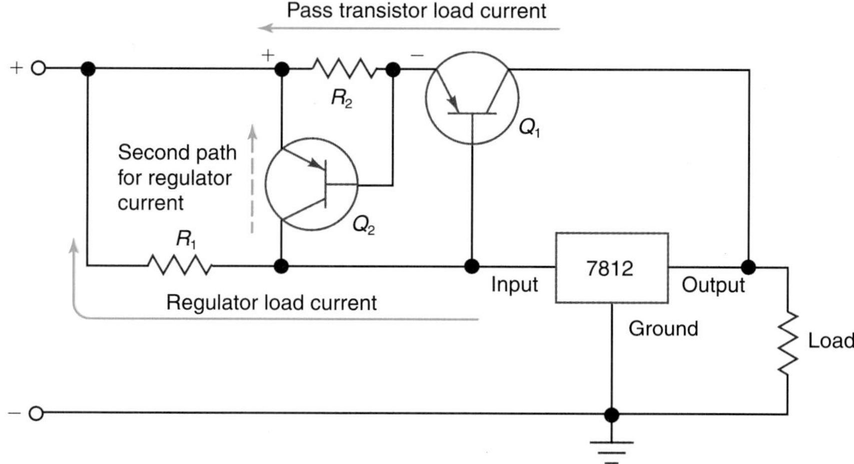

Fig. 15-17 Current-limit circuit.

current output to several amperes. The 78*XX* series of IC regulators is *internally* current limited. A 7812 IC will supply no more than 1.5 A if its output is short-circuited. However, this will not protect transistor Q_1 in Fig. 15-17. Without additional current limiting, it could be destroyed by a short circuit. Transistor Q_2 provides additional current limiting to protect the pass transistor Q_1 and the load.

Most of the load current in Fig. 15-17 flows through Q_1 and R_2. Remember that the 7812 will conduct enough current to produce base-emitter bias for the pass transistor. This bias is produced by the drop across R_1. Now, suppose that the load demands too much current. This current will cause enough voltage to drop across R_2 to turn on Q_2. In this application, R_2 serves as a *current-sensing* resistor. With Q_2 on, there is now a second path for the regulator current. It will flow from the collector of Q_2 to the emitter of Q_2 and on to the + terminal of the input. This second path will reduce the current through R_1. If the current through R_1 is reduced, the voltage drop across R_1 must also be reduced. This reduced voltage means less forward bias for the pass transistor Q_1. With less bias the pass transistor will not conduct as much current, and the circuit goes into current limiting. Even if the load is a short circuit, the current will be limited to some predetermined value.

The maximum current permitted by the limiting action of the circuit shown in Fig. 15-17 is determined by R_2. Q_2 requires about 0.7 V of base-emitter bias to turn on and begin the current limiting action. If R_2 is a 0.1-Ω resistor, it will drop 0.7 V when it conducts 7 A ($V = I \times R$). Some of the load current flows through the 7812 IC (perhaps about 0.5 A). Thus, when the load demands more than 7.5 A, Q_2 will come on and the current will be limited from increasing much beyond this value. Making R_2 larger limits the current to less than 7.5 A. Remember, without the current-limiting action a short circuit will often destroy the pass transistor. Also, the IC regulator must have internal current limiting or it may be damaged by an overload.

Example 15-3

For Fig. 15-17, $R_1 = 4.7\ \Omega$ and $R_2 = 0.22\ \Omega$. Determine the values of load current required to turn on Q_1 and Q_2. Q_1 will turn on when R_1 drops 0.7 V:

$$I_{\text{load}} = \frac{0.7\text{ V}}{R_1} = \frac{0.7\text{ V}}{4.7\ \Omega} = 149\text{ mA}$$

Q_2 comes on when R_2 drops 0.7 V:

$$I_{\text{load}} = \frac{0.7\text{ V}}{R_2} = \frac{0.7\text{ V}}{0.22\ \Omega} = 3.18\text{ A}$$

The design employed in Fig. 15-17 is known as *conventional current limiting*. Figure 15-18 on page 420 shows a graph of circuit performance when conventional current limiting is used. The graph shows that the output voltage remains constant at 12 V as the load current increases from 0 to 5 A. As the load increases beyond 5 A, the output voltage begins to drop

Conventional current limiting

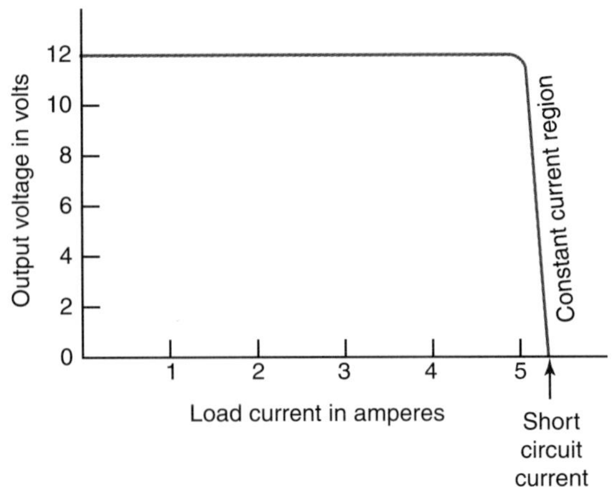

Fig. 15-18 Conventional current-limiting performance.

rapidly. A short circuit will be limited to a little more than 5 A. As the output voltage drops from 12 to 0 V, the curve is in the constant-current region. This is where the circuit operates when it is in current-limiting.

Conventional current limiting may not completely protect the pass transistor. Even if the transistor is rated to conduct the amount of current in the constant-current region, it may overdissipate and be damaged or destroyed if the short persists. For example, a type 2N3055 transistor is rated at 15 A and 117 W. Therefore, it may appear to be safe if operated in the constant-current region as shown in Fig. 15-18. However, this may not be true. Even though 5 A is only one-third the rating of a 2N3055, the transistor can still be destroyed by too much collector dissipation. Suppose the output is shorted. Zero volts will appear across the load, and all the unregulated power supply must drop across the pass transistor. For a 12-V power supply, the unregulated input will probably be around 18 V. The transistor dissipation will be

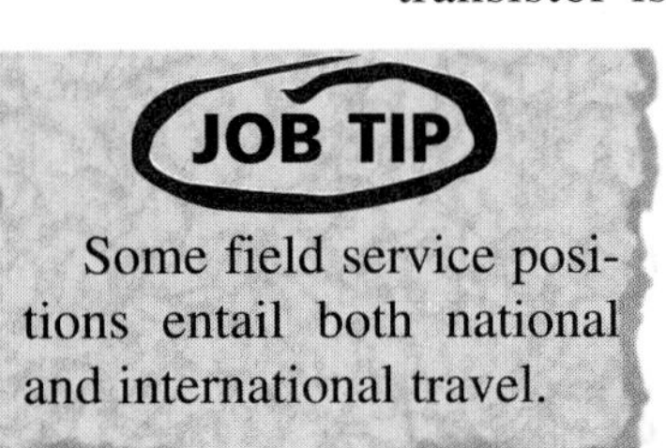

$$P_C = V_{CE} \times I_C$$
$$= 18\text{ V} \times 5\text{ A} = 90\text{ W}$$

Since 90 W is less than 117 W, the transistor is operating within its limits. But the 117-W rating is based on a junction temperature of 25°C (77°F). When a transistor is dissipating 90 W, it is going to get *very* hot. A large heat sink will help, but it is likely that the junction temperature will exceed 65°C. At this temperature, the maximum collector dissipation is less than 90 W. Power transistors must be *derated* for temperatures over 25°C. Thus, the 2N3055 will be damaged or destroyed if the short circuit lasts long enough for the transistor temperature to exceed 65°C. Conventional current limiting may provide protection only when short circuits are momentary.

Foldback current limiting

Figure 15-19(*a*) shows a voltage regulator that uses *foldback current limiting.* An analysis of this circuit will help you understand how this improved type of current limiting works. There are actually two important current limits that can be calculated, and one of them is partly determined by the output voltage. V_{out} is established by the zener reference voltage and the $R_5 - R_6$ voltage divider. We can assume that V_{out} will turn on the error amplifier (Q_4). Allowing for a 0.7-V drop from base to emitter, the base voltage of Q_4 will be:

$$V_{B(Q_4)} = 5.1\text{ V} + 0.7\text{ V} = 5.8\text{ V}$$

This voltage must be equal to some fraction of V_{out}, as determined by the voltage divider:

$$V_{B(Q_4)} = V_{out} \times \frac{R_6}{R_6 + R_5}$$
$$5.8\text{ V} = V_{out} \times \frac{620\ \Omega}{620\ \Omega + 660\ \Omega}$$
$$V_{out} = 12.0\text{ V}$$

The short-circuit current in Fig. 15-19(*a*) is set by R_2, R_3, and R_4. With the output shorted, V_{out} is zero and the drop across R_2 will be high enough to turn on Q_3, the current limit transistor. With Q_3 on, drive current is diverted from Q_2 and the pass transistor starts to shut down. Since R_3 and R_4 form a voltage divider, Q_3 does not see the full drop across R_2:

$$V_{BE(Q_3)} = 0.7\text{ V}$$
$$= V_{R_2} \times \frac{120\ \Omega}{120\ \Omega + 12\ \Omega}$$
$$0.7\text{ V} = V_{R_2} \times 0.909$$
$$V_{R_2} = 0.770\text{ V}$$

Now Ohm's law is used to find the current in R_2. This is the current flow when the output is shorted:

$$I_{SC} = \frac{V_{R_2}}{R_2} = \frac{0.770\text{ V}}{0.38\ \Omega} = 2.03\text{ A}$$

The maximum load current in Fig. 15-19(*a*) is greater than the short-circuit current. Its

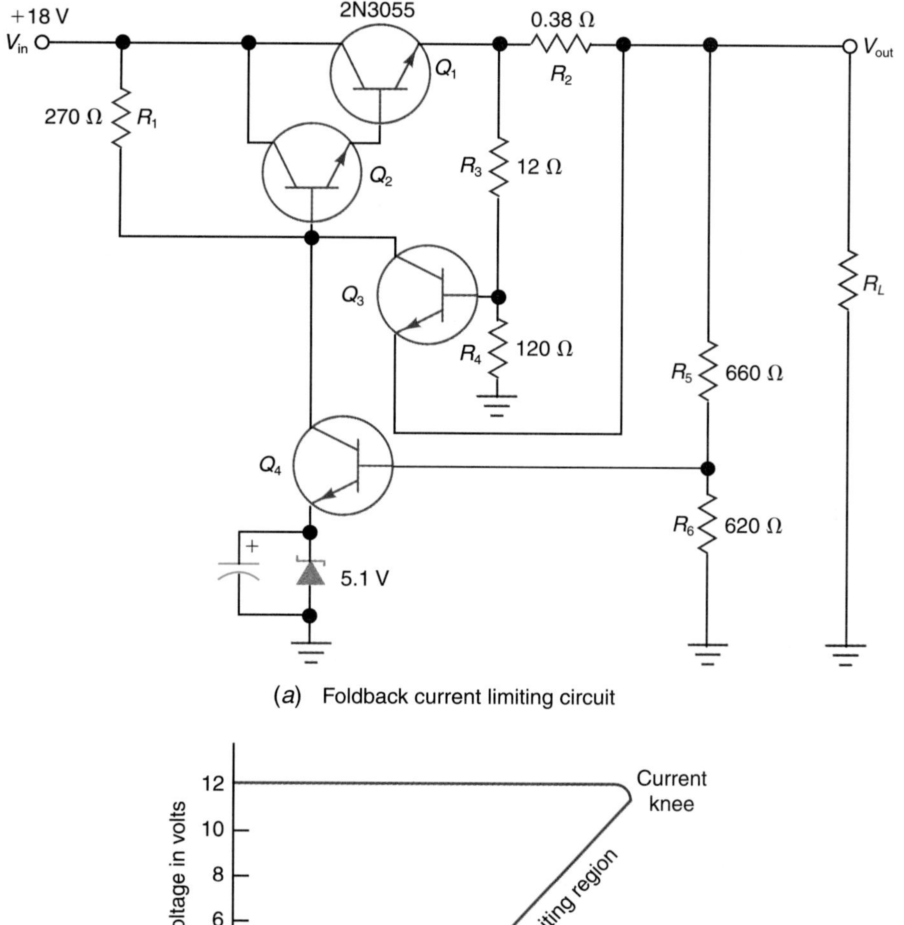

(*a*) Foldback current limiting circuit

(*b*) Circuit performance

Fig. 15-19 Foldback current limiting.

value is calculated by assuming that the output voltage is normal. V_{out} establishes V_E for Q_3, and V_B for Q_3 is 0.7 V higher, or 12.7 V:

$$V_{B(Q_3)} = (V_{R_2} + V_{out}) \times \frac{R_4}{R_4 + R_3}$$

$$12.7 \text{ V} = (V_{R_2} + 12 \text{ V}) \times \frac{120\ \Omega}{120\ \Omega + 12\ \Omega}$$

$$12.7 \text{ V} = 0.909\ V_{R_2} + 10.9 \text{ V}$$

$$V_{R_2} = 1.97 \text{ V}$$

The current in R_2 is found with Ohm's law:

$$I_{R_2} = I_{max} = \frac{1.97 \text{ V}}{0.38\ \Omega} = 5.18 \text{ A}$$

This demonstrates that the maximum load current is significantly higher than the short-circuit current in regulators that use foldback current limiting.

Figure 15-19(*b*) shows the performance graph for Fig. 15-19(*a*). This type of protection folds back the current flow once some preset limit is reached. Note that 5 A is the limiting point. However, in this case, the current begins to decrease instead of remaining constant near 5 A. If the overload is a short circuit, the current folds back to a value near 2 A. This greatly limits the dissipation in the pass transistor for short circuits. If we again assume an unregulated input of 18 V, the collector dissipation will be

$$P_C = V_{CE} \times I_C$$
$$= 18 \text{ V} \times 2 \text{ A} = 36 \text{ W}$$

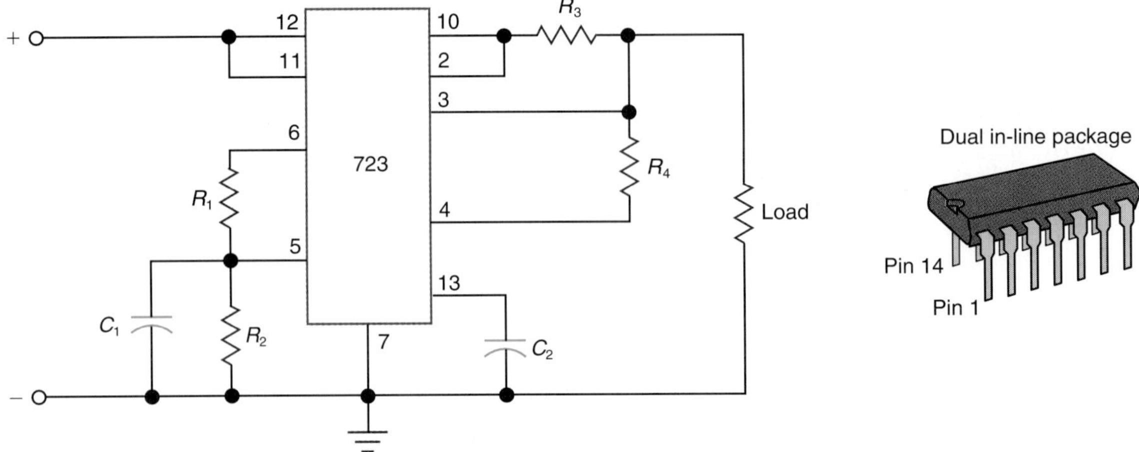

Fig. 15-20 An IC regulator configured for conventional current limiting.

A dissipation of 36 W is much more reasonable for a 2N3055 transistor. The transistor will now be safe up to a junction temperature of 150°C. A good heat sink will be able to maintain the junction below this temperature. With foldback current limiting and a good heat sink, the pass transistor will be able to withstand a short circuit for an indefinite period of time.

We have already learned that the 78*XX* series of IC regulators features internal current limiting. This current limiting is of the conventional type. Another popular IC regulator, the 723, is capable of both types of current limiting. This IC is available in the dual-in-line package. Figure 15-20 shows it connected in a circuit to provide conventional current limiting. In this circuit, R_1 and R_2 divide an internal reference voltage to set the output voltage between 2 and 7 V. R_3 is the current-sensing resistor. When the drop across this resistor reaches about 0.7 V, the regulator goes into conventional current limiting.

Figure 15-21 shows the 723 regulator configured for an output greater than 7 V and for foldback current limiting. R_4 and R_5 determine the output voltage. R_1, R_2, and R_3 determine the current knee and the short-circuit current [refer to Fig. 15-19(*b*)].

Overcurrent circuits protect systems from damage. Sometimes, though, a power supply will fail and destroy other circuits even though current-limiting circuitry is included. We have learned that the series pass transistor is used to drop the unregulated voltage to the desired value. If the pass transistor shorts from emitter to collector, the entire unregulated voltage will be applied to all loads connected to the power supply. When this happens, many circuits may be damaged. Some form of *overvoltage* protection may be needed to prevent this from happening.

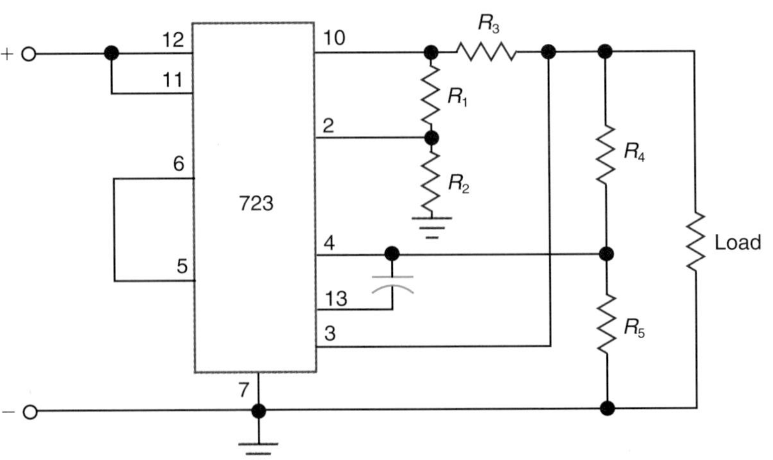

Fig. 15-21 An IC regulator configured for foldback current limiting.

Figure 15-22 shows the schematic diagram for a high-current power supply with *crowbar protection.* A crowbar is a circuit that shorts the power supply when some voltage limit is exceeded. Zener diode D_1 is part of the crowbar circuit. Normally it will not conduct. However, if the output voltage goes too high, D_1 will turn on and the resulting current through R_9 will create a voltage drop which is applied to the gate of the SCR. This voltage will gate the SCR on. The SCR will then "crowbar" (short) the power supply and blow the fuse. A blown fuse is far more desirable than damaged load circuitry.

Crowbar

Another interesting feature of the power supply shown in Fig. 15-22 is the high current capability provided by parallel pass transistors. Power supplies of this type can provide currents in excess of 25 A. Transistors Q_3 through Q_6 share the load current. Resistors R_5 through R_8 ensure current sharing among the parallel transistors. They are called *swamping resistors* and are typically 0.1 Ω in value. The swamping resistors ensure that one or two high-gain transistors will not "hog" more than their share of the load current. Suppose, for example, that Q_5 has a higher β than the other three pass transistors. This would tend to make it conduct more than its share of the load current and that would make it run hotter than the other transistors. Since β increases with temperature, it would then conduct more of the load current.

Swamping resistor

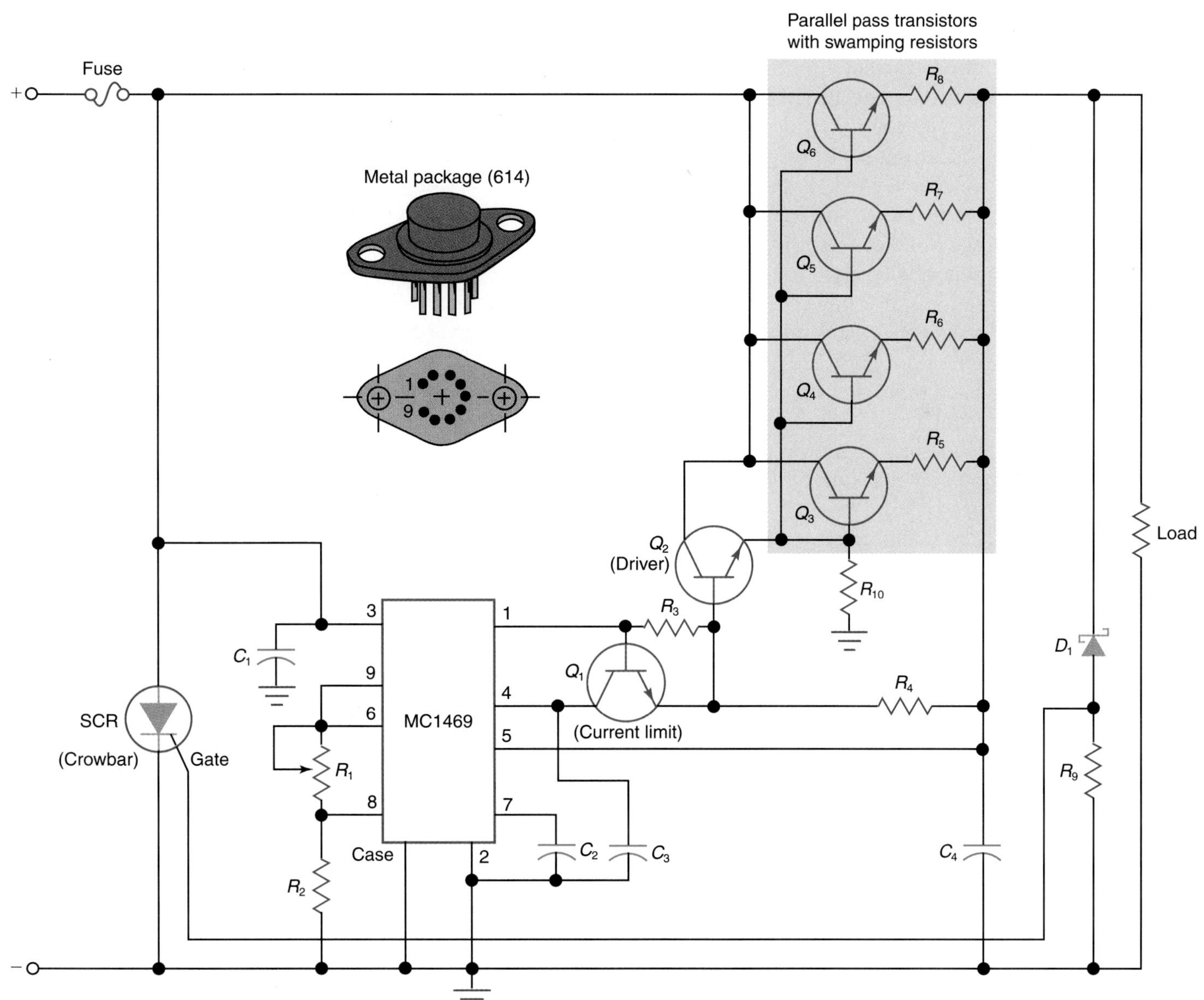

Fig. 15-22 High-current supply with crowbar protection.

Thermal runaway

MOV

Driver transistor

Line transient

Varistor

It would again increase in temperature and so on. This condition is called *thermal runaway* and could destroy Q_5. The swamping resistor decreases the chance of thermal runaway because it drops more voltage if the current in Q_5 increases. This drop subtracts from the transistor's forward bias and reduces the current in Q_5. Therefore, the swamping resistors in Fig. 15-22 help ensure current sharing among the four pass transistors.

In Fig. 15-22 Q_2 is called a *driver transistor.* The MC1469 regulator cannot supply enough current for four transistors and Q_2 boosts the drive from the IC regulator. R_3 and Q_1 form a current-sensing circuit. If the current supplied to Q_2 causes a 0.7-V drop across R_3, Q_1 comes on and activates pin 4 of the IC. This limits the drive current and the output current to a safe value. This circuit provides conventional current limiting, and long-term shorts may damage the pass transistors if the fuse does not blow; R_1 adjusts the output voltage.

Current-limiting circuits and crowbar circuits do a good job of protecting electronic circuitry. However, line transients may still damage solid-state devices. A line transient is an abnormally high voltage, usually of short duration, on a power-supply line. For example, transients of several thousand volts may occur on an ordinary 120-V ac circuit in a building. Such transients are caused by lightning, equipment failures, and by switching of inductive loads such as motors and transformers. Studies predict that one 5000-V transient can be expected each year on every 120-V service circuit in this country. More occurrences of lower voltage transients can be expected. Many electronic equipment failures are caused by *line transients.*

Transients lasting several microseconds are capable of damaging circuits, contacts, and insulation. Protection devices such as crowbars and spark gaps (ionizing breakdown protection devices) are too slow-acting. Also, these devices may not be self-clearing. That is, they may remain in conduction after the transient has passed, and this characteristic can cause further problems. Voltage clipping devices are considered better choices for protecting electronic circuitry from transients. These devices include selenium cells, zener diodes, and *varistors.*

Varistors are voltage-dependent resistors. Their resistance is not constant as it is with ordinary resistors. As the voltage across a varistor increases, its resistance decreases. This feature makes them valuable for clipping transients. Varistors are made from silicon carbide or, more recently, zinc oxide. Zinc oxide varistors are usually called *MOVs* (for metal oxide varistors) and are widely applied for protecting electronic equipment from line transients.

Figure 15-23 shows the structure of an MOV device. It is made up of a wafer of granular zinc oxide. A silver film is deposited on both sides of the wafer. Leads are soldered to the silver electrodes. When a normal voltage is applied across the leads, very little current flows. This is because of the boundaries between the zinc oxide grains. These boundaries act as semiconductor junctions and require about 3 V for turn-on. The boundaries act in series. Therefore, more than 3 V will be required to turn the entire wafer on. Designing an MOV device for a given varistor voltage is a matter of controlling wafer thickness. A thick wafer will have more series boundaries and turn on at a higher voltage. Figure 15-24 shows a typical MOV volt-ampere characteristic curve. Notice that the current is about zero over the normal line-voltage range. Also note that if the line voltage exceeds normal, a very sharp increase in current can be expected.

Figure 15-25 shows four package styles for MOV devices made by General Electric. The small axial devices can safely absorb 2 joules (J) of energy and conduct 100 A during a transient. A joule is equal to a watt-second (1 W × 1 s). Suppose that an MOV device absorbs a

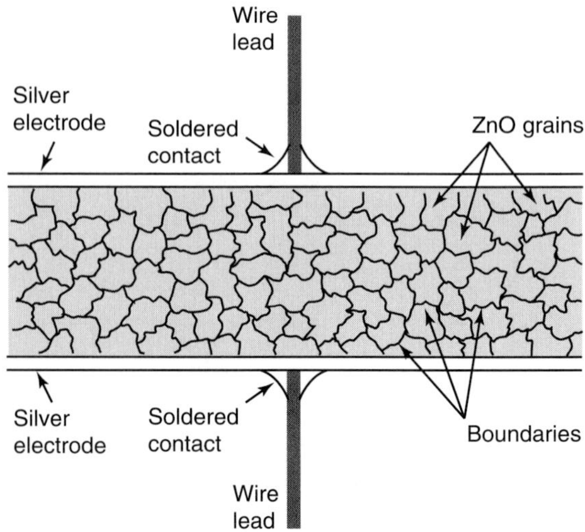

Fig. 15-23 Metal oxide varistor structure.

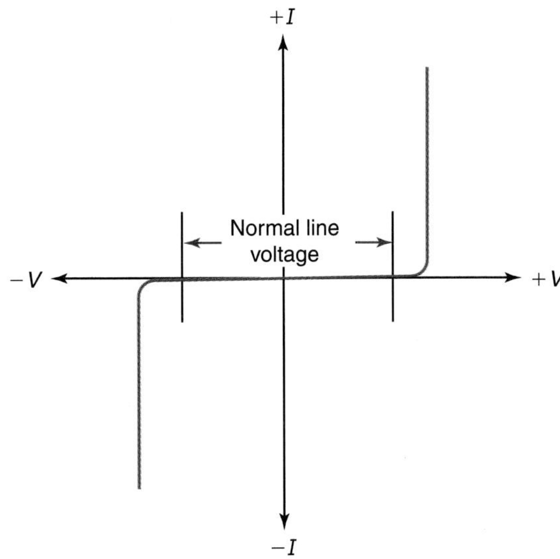

Fig. 15-24 Typical MOV volt-ampere characteristic curve.

1000-V, 100-A transient that lasts 20 μs. The energy E dissipated in joules is

$$\begin{aligned} E &= V \times I \times t \\ &= 1000 \text{ V} \times 100 \text{ A} \times 20 \times 10^{-6} \text{ s} \\ &= 2 \text{ J} \end{aligned}$$

The high-energy devices are rated as high as 6500 J and 50,000 A. The response time of the MOVs is measured in nanoseconds (ns). They are effective in safely absorbing transient energy to protect electronic equipment.

Figure 15-26 on page 426 shows a power-supply circuit protected with an MOV. The varistor is connected in parallel with the power transformer. Normally, it will conduct very little current. A transient will turn the varistor on and much of the transient energy will be absorbed. After the transient passes, the MOV will return to its high-resistance state and the circuit will

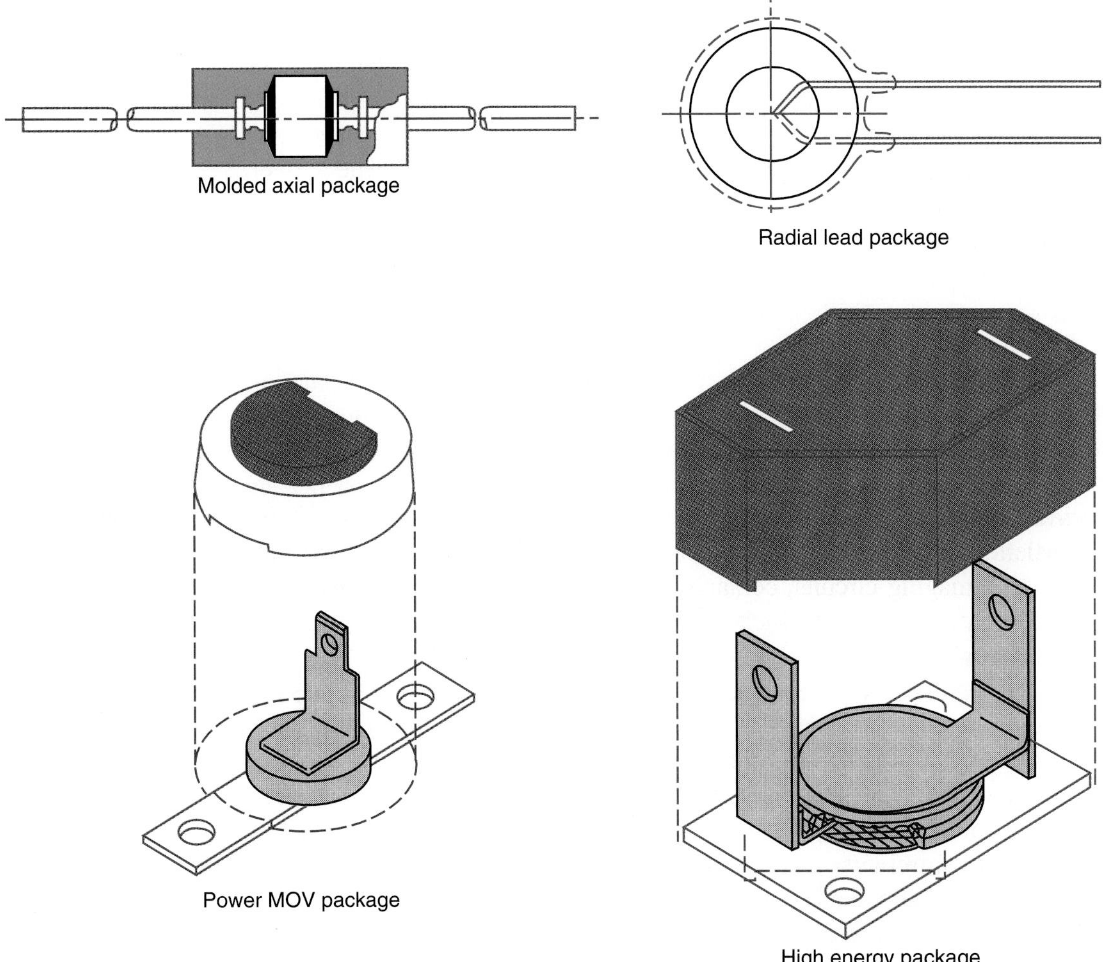

Fig. 15-25 General Electric MOV package styles.

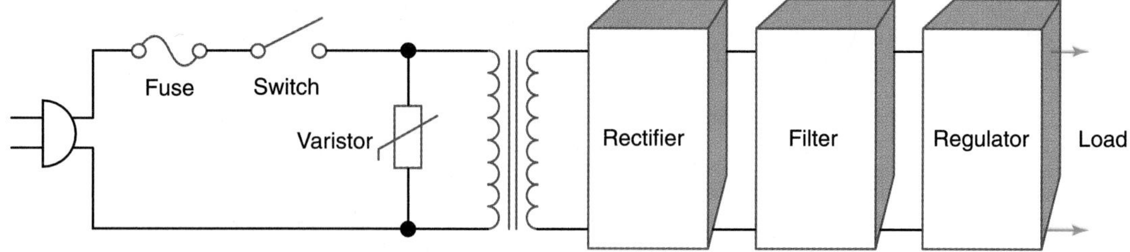

Fig. 15-26 Varistor-protected power supply.

resume normal operation. A long-term transient will cause the fuse to blow and it will have to be replaced. Note the schematic symbol for the varistor in Fig. 15-26. The line drawn through it shows a nonlinear resistance characteristic.

TEST

Choose the letter that best answers each question.

13. Current-limited power supplies can prevent damage to
    a. Rectifier diodes and power transformers
    b. Pass transistors
    c. Other circuits in the system
    d. All of the above
14. Refer to Fig. 15-17. Both transistors are silicon. R_1 is 10 Ω and R_2 is 0.2 Ω. At what load will Q_1 begin to provide current?
    a. 0 A
    b. 0.07 A
    c. 3.57 A
    d. 7.07 A
15. Refer to Fig. 15-17. Both transistors are silicon; R_1 is 10 Ω and R_2 is 0.2 Ω. At what load will current limiting begin?
    a. 0 A
    b. 0.07 A
    c. 3.57 A
    d. 7.07 A
16. Foldback current limiting has the advantage of
    a. Better pass transistor protection for long-term overloads
    b. A defined turn-on point
    c. Circuit simplicity
    d. All of the above
17. A crowbar is a power-supply circuit that provides
    a. Conventional current limiting
    b. Foldback current limiting
    c. Temperature control
    d. Voltage limiting
18. Refer to Fig. 15-22. What is the function of resistors R_5 through R_8?
    a. To ensure current sharing among Q_3 through Q_6
    b. To adjust the crowbar trip point
    c. To provide current sensing to shut down Q_2
    d. To improve the voltage regulation
19. In general, MOV devices are used to protect electronic equipment from dangerous operating conditions such as
    a. High temperatures
    b. Line transients
    c. Overcurrent
    d. All of the above

15-4 Switch-Mode Regulators

Linear regulation

The regulator circuits discussed up to this point are of the linear variety. They work by using a series pass transistor to drop more or less of the unregulated input voltage to maintain a stable output voltage. The circuits are considered *linear regulators* because the series pass transistor operates in the active (linear) region. There is a serious disadvantage to using linear regulation and that is poor efficiency. For example, assume that a 12-V power supply must deliver 5 A of load current. Also assume that the unregulated input voltage is 18 V. This means that the pass transistor will have to drop the extra 6 V. The power dissipated in the pass transistor will be 30 W (6 V × 5 A). This dissipation is wasteful and requires a large heat sink.

The efficiency of the linear regulator can be calculated by comparing the useful output power to the input power. The useful output power is 60 W (12 V × 5 A). The input power

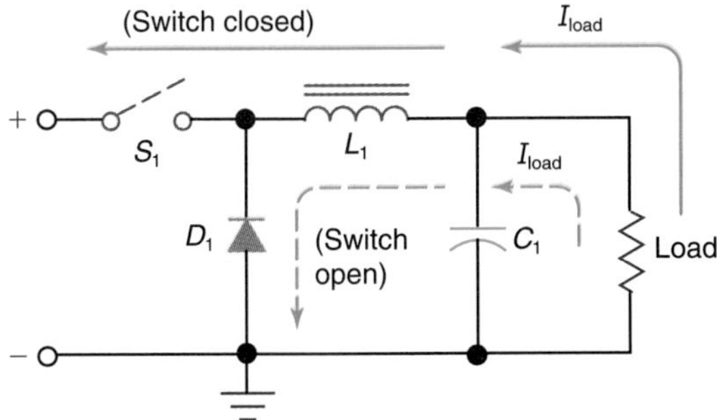

Fig. 15-27 Step-down configuration.

is 90 W (18 V × 5 A). Efficiency is given by

$$\eta = \frac{P_{out}}{P_{in}} \times 100\%$$

$$= \frac{60\text{ W}}{90\text{ W}} \times 100\% = 66.7\%$$

The overall efficiency of the power supply will be less than 66.7 percent. This is because of additional losses in the transformer, the rectifiers, and other parts of the circuit. Linear power supplies usually have overall efficiencies of less than 50 percent. This means that much of the electrical energy will be wasted in the form of heat.

Another approach to power-supply design replaces the linear regulator with a *switching transistor.* A switching transistor operates in either of two modes: cutoff or saturation. Remember, a saturated transistor drops very little voltage and therefore has low power dissipation. When the switching transistor is in cutoff, its current is zero, and the power dissipation is also zero. Therefore, a switching regulator will dissipate much less energy than a linear regulator. Smaller devices and smaller heat sinks can be used. A compact, cool-running power supply is the result. In fact, a switching power supply can be less than one-third the weight and volume of an equivalent linear power supply and it will cost less to operate.

Figure 15-27 shows the basic configuration for a *step-down* switching *regulator.* When S_1 is closed, load current flows through L_1 and through the switch into the unregulated input. The current through L_1 creates a magnetic field, and energy is stored there. When S_1 opens, the magnetic field in L_1 begins to collapse. This generates a voltage across L_1, which forward-biases D_1. Load current is now supplied by energy that was stored in inductor L_1. After a short period of time, S_1 is closed again and the inductor is recharged. L_1 acts as a smoothing filter to maintain load current during those periods of time when S_1 is open. C_1 helps to filter the load voltage. The overall result is reasonably pure dc at the load even though the switch is opening and closing. The step-down configuration of Fig. 15-27 supplies less load voltage than the unregulated input voltage. As you will see later, it is also possible to have a step-up configuration in switch-mode power supplies.

Step-down regulator

Switching power supplies regulate output voltage by using *pulse-width modulation.* Examine Fig. 15-28. The waveform in Fig. 15-28(*a*) of the illustration shows a rectangular wave with a *duty cycle* of 50 percent. Notice that the average value of the waveform is half of the peak value. Now look at Fig. 15-28(*b*). This rectangular wave has a duty cycle of much less than 50 percent, and the average value is much less than half of the peak value. Rectangular waves are used to drive the switch-

Pulse-width modulation

Duty cycle

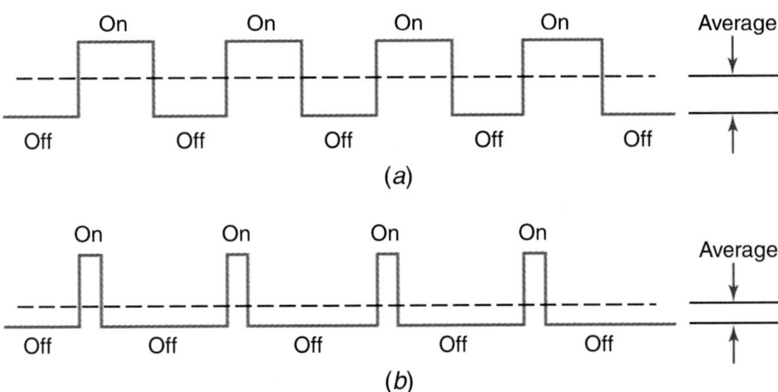

Fig. 15-28 Using pulse-width modulation to control average voltage.

mode regulators. By modulating (controlling) the duty cycle of the rectangular wave, the average load voltage can be controlled. The load voltage is smoothed by the filter action of inductors and capacitors to provide a low-ripple direct current.

Figure 15-29 shows a more complete circuit for a step-down switch-mode regulator. Q_1 is the switch, and it is driven by a rectangular wave with a varying duty cycle. An error amplifier compares a portion of the output voltage with a reference voltage. If the load on the power supply increases, the output tends to drop. This error is amplified, and a control signal is applied to the pulse generator, which increases the duty cycle of its output. The switching transistor is now turned on for longer periods of time. This increased duty cycle produces a higher average dc voltage, and the output voltage goes back toward normal. L_1 and C_1 eliminate the ripple; D_2 turns on when the switching transistor cuts off and allows the inductor to discharge through the load.

JOB TIP

Some of the happiest people to be found are the ones who love their jobs.

Switch-mode voltage regulators tend to be more complicated than linear voltage regulators. However, ICs can help to simplify designs. Look at Fig. 15-30. It shows a 78S40 IC that contains much of the circuitry needed for switch-mode operation. The oscillator is built-in (integrated) and can be set to the desired frequency of operation by C_1, an external component. The typical switch-mode regulator operates at 20 kHz or above. Higher frequencies mean smaller magnetic cores in transformers and inductors. Smaller filter capacitors can also be used. Remember that capacitive reactance goes down as frequency goes up. This means that far fewer microfarads are required to filter 20-kHz ripple than 60-Hz ripple. Therefore, many of the components can be much smaller and lighter than they would be in a 60-Hz power supply.

Figure 15-30 shows that pin 14 of the IC provides another input to the oscillator. R_1 is connected to pin 14 and acts as a current-sensing resistor. If too much load current flows, the voltage drop across R_1 will reach 0.3 V and the oscillator duty cycle will be reduced. This will protect the IC and other components from damage. The oscillator output is combined with the output of a comparator (error amplifier) in a logical AND gate. An AND gate will allow the oscillator signal to go positive for the period of time that the comparator output is also high. Thus, this gate controls the pulse width supplied to the latch. A latch is a digital storage circuit. In this application, it will produce a positive-going signal at its Q output until it is reset by the negative-going oscillator signal. The latch drives Q_1 and Q_2, which form a Darlington switch. Load current will flow through L_1 and through the switch when the Darlington pair is turned on. When the Darlington switch turns off, D_1 turns on and allows L_1 to discharge through the load.

The reference voltage is also integrated in Fig. 15-30. A voltage of 1.3 V is fed from pin

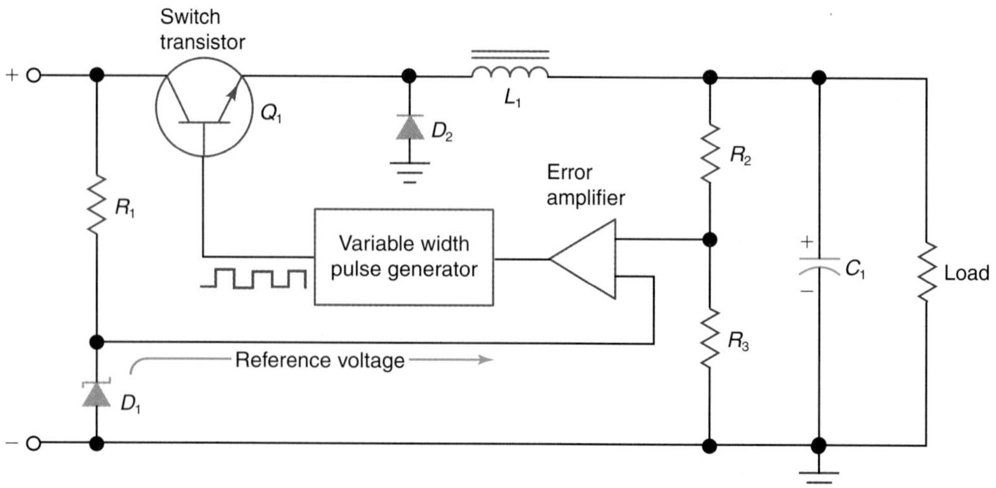

Fig. 15-29 Step-down switching voltage regulator.

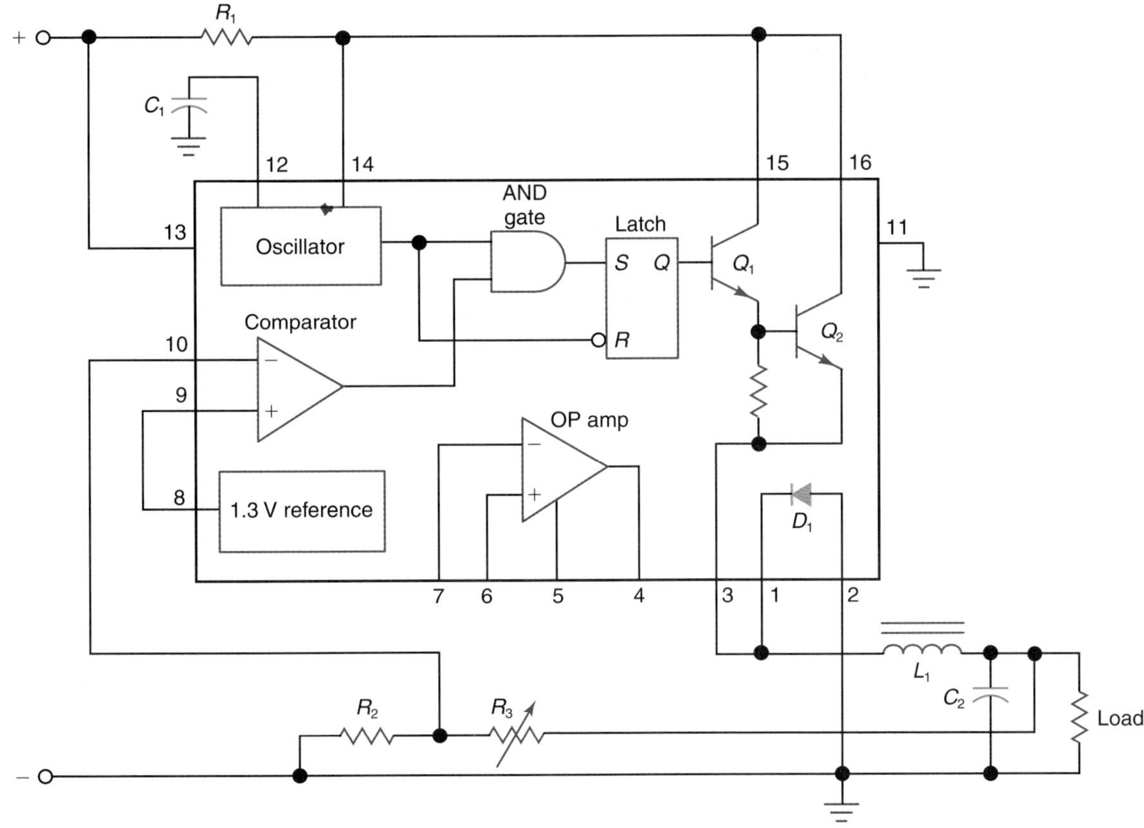

Fig. 15-30 Using the 78S40 IC as a step-down regulator.

8 to one of the inputs of the comparator (pin 9). The other comparator input (pin 10) comes from the voltage divider formed by R_2 and R_3; R_3 is used to adjust the load voltage. If the output voltage drops, the comparator will invert this drop and send a positive-going signal to the gate. The gate will then allow more of the positive-going oscillator signal to reach the latch. The latch will provide a higher duty-cycle drive to Q_1. This raises the average output voltage and eliminates much of the error.

There is also an op amp on the chip in Fig. 15-30. It is not used in this application. The IC manufacturer includes it to make other designs easier by eliminating as many external components as possible. Sometimes other components must be added. The built-in switching transistor and diode in the 78S40 are capable of handling 40 V and 1.5 A peak current. If the regulator must handle more, external components must be added. Pin 3 can be used to drive the base of an external transistor, and an external switching diode can be used.

Ordinary rectifiers and transistors will not work in switch-mode power supplies. The high frequencies of operation demand very fast components. For example, Q_1, Q_2, and D_1 in Fig. 15-30 have switching times of around 400 ns. Special switching transistors and fast-recovery rectifiers are used in switch-mode power supplies. A fast-recovery rectifier is specially designed to recover (turn off) as quickly as possible when reverse-biased. Ordinary silicon rectifiers take too long to turn off to be used in high-frequency applications.

A switch-mode power supply can also be connected in the *step-up configuration.* Refer to Fig. 15-31 on page 430. The inductor is now connected in series with the unregulated input, and the switching transistor is connected to ground. When the transistor is turned on by the positive-going part of the rectangular wave, a charging current flows through the transistor and through the inductor. This charging current stores energy in the inductor's magnetic field. When the rectangular wave goes negative, the transistor turns off. The field in the inductor begins to collapse. This induces a voltage across the inductor. The polarity of the induced voltage is shown in Fig. 15-31. Note that it is series-aiding with the polarity of the unregulated input. Therefore, the load circuit sees two

Step-up configuration

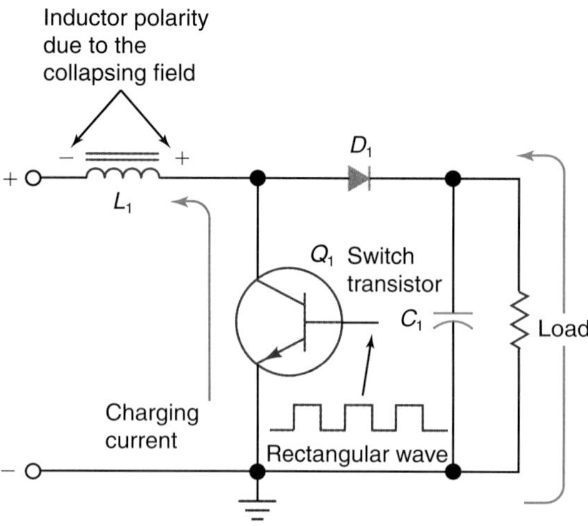

Fig. 15-31 Step-up configuration.

voltages in series, and a step-up action is achieved. D_1 prevents filter capacitor C_1 from being discharged when the switching transistor is turned on again. A complete step-up switcher would have a reference supply, an error amplifier, an oscillator, and a pulse-width modulator to regulate the output voltage. The 78S40 integrated circuit studied earlier can be used in the step-up configuration.

Inverting configuration

The *inverting configuration* is shown in Fig. 15-32. Here, the switching transistor is in series and the inductor is connected to ground. When the transistor is turned on, current flows through L_1, as shown, and charges it. When the transistor is turned off, the field collapses and the induced voltage at the top of the inductor is negative with respect to ground. D_1 is forward-biased by this induced voltage, and the current flows through L_1, through D_1, and down through the load. The top of the load resistor is negative with respect to ground. Inverting regulators are useful in systems where a positive power supply energizes most of the circuits and one negative voltage is needed. The 78S40 can be used in the inverting configuration.

Converter

Figure 15-33 illustrates a *converter.* A converter is a circuit that changes direct current to alternating current and then changes the alternating current back to direct current again. Converters can be considered dc transformers and are used for step-up and step-down action and for isolation. Q_1 and Q_2 are driven by out-of-phase rectangular waves. They will never be on at the same time. The collector current of each transistor flows through the primary of T_1. Alternating voltage is induced across the secondary of T_1. D_1 and D_2 form the familiar full-wave rectifier arrangement. D_3 serves the same purpose as it did in the step-down configuration (Fig. 15-27). There are periods of time when both transistors are off and L_1 will discharge to maintain load current. D_3 is forward-biased by the discharge of L_1 and completes the circuit. The circuit will work without D_3, but then the discharge current will flow through rectifiers D_1 and D_2 and through the secondary of T_1. This discharge path is not desirable since it will increase the dissipation in the rectifiers and the transformer.

Regulation is provided by pulse-width modulation in Fig. 15-33. R_1 and R_2 provide a sample of the output voltage for the inverting input of the op amp. The other input to the op amp is a reference voltage. Any error is amplified and controls the pulse width of the rectangular wave supplied to the two switching transistors.

A converter circuit such as the one shown in Fig. 15-33 will often work off the ac line. The line voltage must first be rectified, filtered, and then applied to the converter. This may seem too complicated, but in practice it still produces a more efficient and compact power supply. It is more compact because the 60-Hz power transformer has been eliminated, and it is more efficient because the linear regulator has been eliminated. Transformer T_1 in Fig. 15-33 operates at 20 kHz or higher. Its magnetic core is only a small fraction of the size and weight required for a 60-Hz transformer with comparable ratings. It also uses much less copper than a comparable 60-Hz transformer. Therefore, a transformerless 60-Hz dc rectifier and filter circuit are used to change the ac line power to dc power. Then, that dc power is

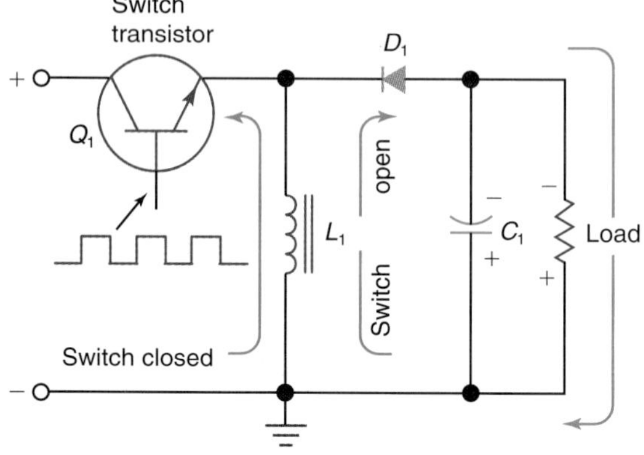

Fig. 15-32 Inverting configuration.

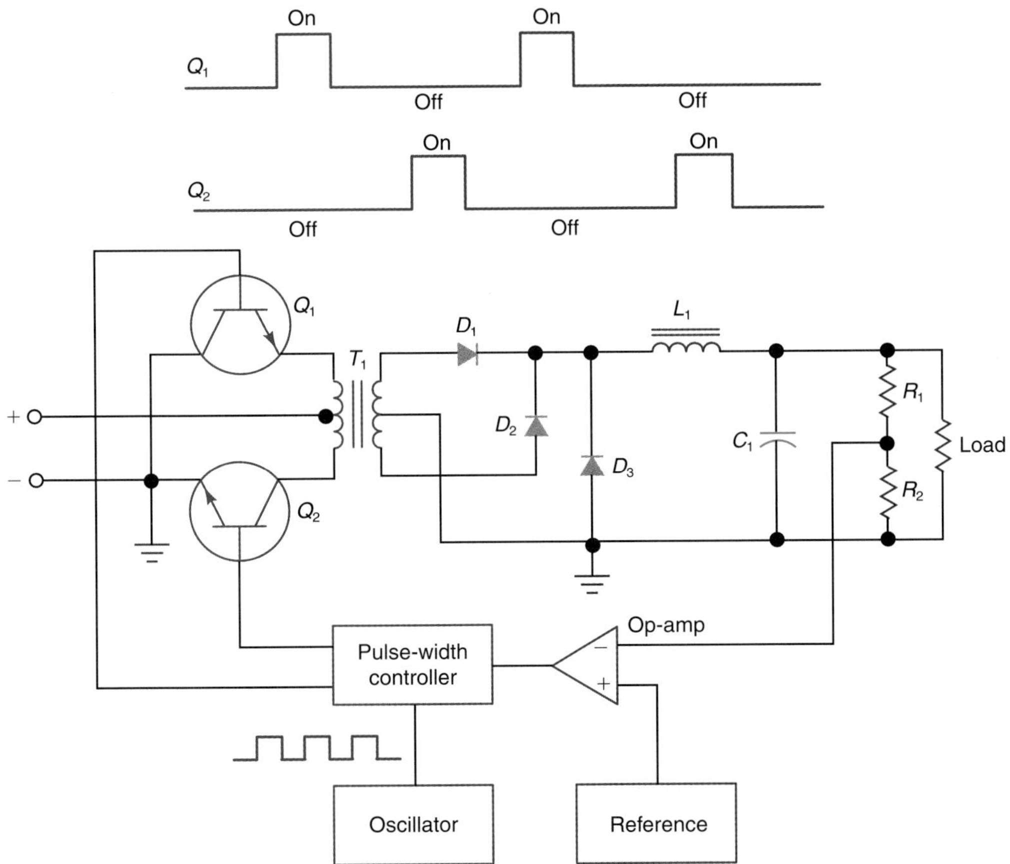

Fig. 15-33 Width-controlled converter/regulator.

transformed to the desired voltage level in a high-frequency switching converter. Line isolation can be achieved in the high-frequency switching transformer; therefore, many of the ground loop and shock hazards associated with line-operated (transformerless) power supplies are eliminated.

Switch-mode power supplies are more efficient, lighter, and more compact than linear power supplies. However, they are also *noisier.* Rectangular waves have high-frequency components that can cause interference. Certain products must meet electromagnetic interference (*EMI*) standards to prevent interference with communications and other electronic equipment. Sine waves have no high-frequency components and are therefore preferred when interference is a problem.

A *sine-wave converter* design is shown in Fig. 15-34 on page 432. It uses power field-effect transistors and frequency control of the dc output voltage. The power FETs do not have problems of storage time associated with bipolar transistors. Storage time in a bipolar transistor is caused by carriers (holes and electrons) stored in the crystal when the device is saturated. The stored carriers keep current flowing for a period of time after the base emitter forward bias is removed. Field-effect transistors do not store carriers and can be turned off much faster. The circuit shown in Fig. 15-34 uses power FETs (Q_1 and Q_2) for switching. The FETs are driven with out-of-phase square waves and are operated around 200 kHz.

The square wave is converted to a sine wave in Fig. 15-34 by resonating L_1 with C_3. T_1 effectively couples the tuning components to form a *tank circuit.* This tuned circuit provides voltage control. When the voltage-controlled oscillator (*VCO*) is tuned to the resonant frequency, maximum voltage appears across C_3. When the VCO is tuned above resonance, the tank circuit voltage drops 12 dB per octave. A 12-dB drop amounts to the voltage decreasing to one-fourth its value. An octave frequency change is twice the original value. Thus, if the tank voltage was 20 V at 150 kHz, it would drop to 5 V at 300 kHz. The VCO is controlled by comparing a sample of the output voltage with a reference voltage. Any error produces a

Noise

EMI

Tank circuit

VCO

Sine-wave converter

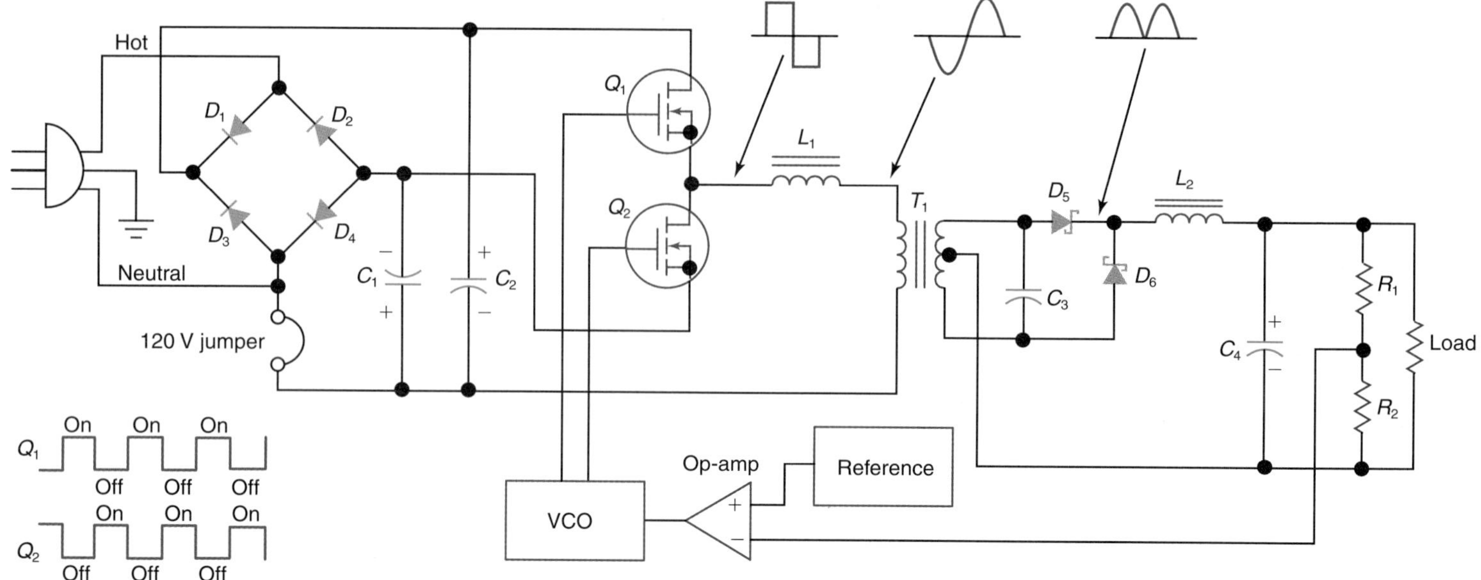

Fig. 15-34 Frequency-controlled sine-wave converter.

frequency change, and the output voltage is adjusted up or down to reduce the error.

Schottky rectifier

In Fig. 15-34, D_5 and D_6 are *Schottky rectifiers.* These diodes can be turned off very quickly and therefore make good high-frequency rectifiers. The diodes used in the 200-kHz power supply have a turn-off time of about 50 ns.

Bridge doubler circuit

In Fig. 15-34, D_1 through D_4 and C_1 and C_2 form a *bridge doubler circuit.* With the 120-V jumper installed, the 120-V ac line is doubled to about 240 V dc. In order to operate the power supply from the 240-V ac line, the jumper is removed, and the circuit acts as a bridge rectifier and again provides about 240 V dc. Line isolation and voltage transformation take place in T_1. Since it operates around 200 kHz, it is tiny compared to a 60-Hz power transformer.

TEST

Choose the letter that best answers each question.

20. What is the function of D_1 in Fig. 15-30?
    a. It provides overvoltage protection.
    b. It prevents C_2 from discharging when Q_2 switches on.
    c. It rectifies the square wave into smooth direct current.
    d. It allows L_1 to discharge when Q_2 is switched off.
21. How is voltage regulation achieved in Fig. 15-30?
    a. Q_1 and Q_2 act as a variable resistor to drop excess voltage.
    b. The 1.3-V reference is pulse-modulated.
    c. By pulse-width modulation at the base of Q_1 and Q_2
    d. By frequency-modulating the oscillator
22. Refer to Fig. 15-31. Assume that the input voltage is 5 V and that the average voltage induced across L_1 is 7 V. What is the load voltage?
    a. −2 V
    b. 5 V
    c. 7 V
    d. 12 V
23. Why is the inverting configuration used in switch-mode power supplies?
    a. To produce an opposite-polarity power supply voltage
    b. To isolate circuits from the ac line
    c. To step up direct current
    d. To step down direct current
24. Which of the following is not considered an advantage of pulse-width-modulated power supplies?
    a. Small size
    b. Low EMI
    c. Cool running
    d. High efficiency
25. How does the circuit of Fig. 15-34 achieve voltage regulation?

a. By pulse-width modulation
b. By zener clamping
c. By frequency modulation
d. All of the above

26. D_5 and D_6 in Fig. 15-34 are
a. Zener regulators
b. Metal oxide varistors
c. Schottky rectifiers
d. Fast-recovery field-effect diodes

15-5 Troubleshooting Regulated Power Supplies

The first and foremost consideration when you are troubleshooting any power supply is *safety.* In general, high-voltage power supplies are the most dangerous. However, it must be emphasized that all electronic circuits must be treated with care and respect. A switch-mode power supply designed to deliver 5 V may develop several hundred volts in an earlier stage (the circuit of Fig. 15-34 is an example). Safe workers always use good procedures. They know circuit principles, they have and use all relevant literature, and they use the correct tools and test equipment. They never defeat safety features such as interlocks unless it is recommended by the equipment manufacturer as a valid servicing technique. They never modify a piece of equipment so that it could become a fire hazard or a shock hazard. They make every effort to work safely and to restore circuits and equipment to meet original specifications.

Some power-supply circuits are transformerless. Special caution must be used when working on circuits without power transformers. Many pieces of test equipment, such as oscilloscopes, are provided with three-wire power cords. This equipment is automatically grounded when plugged into the ac outlet. This is a safety feature and prevents the case and ground leads from reaching a dangerous voltage. Unfortunately, the grounded leads can create a ground loop and a short circuit when connected to transformerless equipment. A switch-mode power supply may use a high-frequency transformer for isolation. However, a ground loop is still possible when you are analyzing the rectifier and filter circuits that precede the switch and transformer section. Examine Fig. 15-34. There is no line isolation for any of the components to the left of T_1. Connecting test equipment to any of those components may create a *ground loop.* If possible, use an isolation transformer when troubleshooting electronic equipment. The isolation transformer will prevent ground loops. Ground loops are covered in more detail in Chap. 4.

Ground loop

Safety

Some of the symptoms that can be observed in a regulated power supply are

1. No output
2. Low output
3. High output
4. Poor regulation and/or instability
5. Excessive ripple or noise
6. High temperature and possibly a burning odor
7. A clicking and/or squealing sound

Many of the symptoms have already been discussed in Chap. 4. It is recommended that you review the earlier section on power-supply troubleshooting. The difficulty could be in a rectifier or filter circuit that precedes the regulator. The troubleshooting information presented here assumes normal operation of the circuits preceding the regulator. It is not sensible to troubleshoot a regulator until you have verified that the regulator has the correct input voltage.

The symptom of no output in a linear power supply often means that the series pass transistor is open or has no drive. With no drive, it will not support the flow of any load current and the output voltage will be zero. It is possible to determine if the pass transistor or the drive circuit is at fault by measuring the base voltage of the transistor. It is usually expected to be near the value of the output (emitter) terminal. If the base voltage is zero, then the drive circuit is probably at fault. A normal base voltage will indicate an open transistor or possibly a shorted output. Refer to Fig. 15-7. If R_Z opens, the base voltage will be zero and the pass transistor will be in cutoff. The symptom is no output. If the base voltage is normal and there is no output, the pass transistor is probably open. A shorted output can also cause the voltage to be zero; however, in a circuit such as Fig. 15-7, this would cause other symptoms. The other symptoms would include low base voltage and a hot pass transistor.

The symptom of no output in a switch-mode power supply could be caused by a short circuit, a defective switch transistor, a defective pulse-width modulator, or a defective oscillator. If the output is shorted, finding and removing the short should restore normal power-supply operation. An oscilloscope can be used to determine if the base or bases of the switching transistors are driven with the correct signal. With no drive, the transistors will not come on and the output will be zero. Signal tracing with an oscilloscope will allow limiting the defect to the oscillator or the modulator. If the drive is normal, the switching transistors may be defective. The fault may also be in the high-frequency transformer, inductor, rectifiers, or filter capacitors.

If a power supply is current-limited, a short circuit may not cause extreme symptoms. This is especially true if the circuit uses foldback current limiting. It may be necessary to disconnect the power supply from the other circuits to determine if the fault is in the power supply or somewhere else in the system. Another technique is to measure the output current. This involves breaking the circuit and measuring the current drain on the power supply. Remember to start with a high range on the ammeter since the current could be above normal. Since it is usually not convenient to break into circuits for current measurements, try to use Ohm's law instead. Many power supplies use a resistor for current sensing. If you know that the power-supply current flows through the resistor and if you know the value of the resistor, it is possible to measure the voltage drop across the resistor and use Ohm's law to calculate the current. If the resistor is in tolerance, this method will provide enough accuracy for troubleshooting purposes.

Troubleshooting for low output is similar to troubleshooting for no output. Again, the base voltage at the series pass transistor should be investigated. If it is low, then the power supply may be overloaded. Check for an overload by measuring current, as discussed before. If the power supply is not overloaded, investigate the reference circuit and error amplifier to determine why the base voltage is low. For example, refer to Fig. 15-8. Three things could go wrong to make the reference voltage low: R_Z might be high in value, D_Z could be defective, or the capacitor could be leaky. Refer to Fig. 15-12. The problems here could include R_1, D_1, the op amp, or the divider network. Finally, if the power supply you are troubleshooting has more than one output, check to see if it is a tracking power supply. If it is a tracking power supply, an overloaded or defective master power supply will cause errors in the slaves.

High output in a regulated power supply is often caused by a shorted pass transistor. Pass transistors are hardworking parts and are therefore prone to fail. When they fail, they often short from emitter to collector. When shorted, they drop no voltage and the output goes up to the value of the unregulated input. An ohmmeter test with the transistor out of the circuit will provide conclusive evidence. You may also test with an ohmmeter while the transistor is in the circuit. Be sure that the power supply is unplugged and all the capacitors have discharged. Check from emitter to collector. Reverse the ohmmeter leads and check again. Zero ohms in both directions usually indicates that the pass transistor is shorted.

A shorted switching transistor can cause various symptoms, depending on the circuit configuration. Refer to Fig. 15-29. If Q_1 shorts, the output voltage will be too high. Now refer to Fig. 15-31. Here, if Q_1 shorts, the output will be zero, and the unregulated power supply will be overloaded. A line fuse may blow in this case.

The output voltage in most regulated power supplies should be quite stable. Changes indicate that something is wrong. If the power-supply voltage varies from normal to some voltage less than normal, there may be an intermittent overload on the power supply. As before, the load current must be checked to determine if it is too high. If the power-supply voltage varies above normal, the power supply itself is unstable. Check the reference voltage. It must be stable. Any change in reference voltage will cause the output to change. Check the base of the pass transistor. With a steady load on the power supply, the base voltage should be constant. An intermittent may be found in the pass transistor itself, the error amplifier, or the voltage divider. If the power supply is located near a source of radio-frequency energy such as a transmitter, the source could be causing instability. This is usually easy to diagnose, since turning off the source would make the power supply return to normal operation. Extra shielding and bypassing may be required if a power supply must operate in a strong RF field.

Regulated linear power supplies may go into

oscillation. It is normal for switchers to oscillate but not for linear regulators. A capacitor can open, and the power supply may oscillate under some load conditions. If the power-supply voltage seems unstable, use an oscilloscope and view the output waveform. It should look like pure direct current (a straight line on the scope). Any ac content may be a result of oscillation in the regulator. Check the output capacitors and especially the bypass capacitors on any ICs in the power supply. Check for bad solder joints. Refer to Fig. 15-22. Capacitors C_1 through C_3 are very important for stability. A defect in one of these or an associated solder joint could cause oscillations.

Excessive ripple or noise on the output of a regulated power supply is usually due to the failure of a filter or a bypass capacitor. Electrolytic capacitors are widely applied in power-supply circuits. These capacitors may have a shorter life than most other electronic components. They can slowly dry out, and their effective series resistance may increase. Their capacitance may also drop. They will not be nearly as effective for filtering and bypassing. Integrated circuits and transistors can also develop noise problems. If the capacitors are all good, then the IC voltage regulator could be defective. An oscilloscope can be used to probe for the source of the noise.

Power transistors and transformers can safely run hot in some equipment. A device can be too hot to touch, yet be operating normally. Probes are available for measuring the temperature of heat sinks, solid-state devices, and transformers. If a power supply seems too hot, check for an overload first. If the current and voltage are normal, the power supply may be safe. Check the manufacturer's specifications. If there is an odor of burning parts, the power supply is probably *not* safe. Troubleshoot the power supply using voltage readings. Sometimes it is necessary to turn the power supply off between readings to allow the part or parts to cool. Minimize the damage as much as possible. As always, make sure the power supply is not overloaded since this is the most frequent cause of hot and burning components.

A clicking or squealing sound may be heard in switch-mode power supplies. If they are defective or if they are overloaded, they can make sounds. The sounds are caused by the oscillator circuits operating at the wrong frequency. One of the reasons for running a switcher above 20 kHz is to keep it above the range of human hearing. If you can hear a switcher, then it may be running abnormally low in frequency because of an overload or a defect in the power supply. A clicking sound may mean that the power supply is overloaded and is shutting down. Every time it tries to start up again, it makes a click. The first step will be to reduce the load on the power supply. If readings are normal and the sounds stop, the circuits fed by the switcher are probably overloading it. (Completely unloading a switcher is not a good idea since many of them do not produce normal outputs under this condition.)

Troubleshooting switch-mode power supplies demands safe work habits and proper test equipment. The first section of a switcher is a line rectifier and filter. Voltage doublers may be used. Therefore, lethal dc voltages should be expected even in 5-V switch-mode power supplies. The frequencies and waveforms found in switchers are beyond the capabilities of many meters. You have learned that pulse-width modulation is used to control the output voltage in many switchers. Since the duty cycle is changing and the peak voltage is not, an ordinary ac meter may not properly indicate circuit action. A true root-mean-square meter with a frequency rating at least as high as the power-supply operating frequency will be required for accurate testing. True rms meters indicate the correct rms (effective) value for all ac waveforms. Most meters indicate the correct rms value for sinusoidal ac only. Since waveforms are so important in switchers, most technicians prefer the oscilloscope for troubleshooting. If the power supply uses frequency control, such as the one in Fig. 15-34, a frequency meter may be useful in testing. The VCO must operate near or above the resonant frequency of the tank circuit. As the load on the power supply is increased, the VCO frequency should drop to come closer to resonance. This can be seen on an oscilloscope as an increase in the period of the waveform. Period and frequency are reciprocals.

The final step in the repair process is replacing the defective part or parts. An exact replacement is usually the best. One exception is an upgraded part that is recommended by the manufacturer. Substitutions may affect the performance, reliability, and safety of a system.

Special components

Some *components* are *special.* The Schottky rectifiers (D_5 and D_6) in Fig. 15-34 are a good example. There is no way that ordinary rectifiers will work in this circuit. The high frequency would cause tremendous dissipation in ordinary rectifiers. They would probably burn up in a short period of time and could cause damage to other components in the power supply. Capacitor C_3 in Fig. 15-34 is another example of a special component. It is a special high-current capacitor. An ordinary capacitor, even though it meets the voltage and capacity requirements, would overheat and be destroyed in this circuit.

Lead dress

Lead dress is important when replacing components. Lead dress refers to the length and position of the leads on a part. Leads that are too long can make some circuits unstable. It was mentioned before that linear IC voltage regulators can become unstable and oscillate. It is absolutely necessary that some bypass capacitors have very short leads. Always install replacement parts with the same lead dress as the originals.

TEST

Choose the letter that best answers each question.

27. An open series pass transistor produces the symptom of
    a. Low output voltage
    b. High output voltage
    c. Unstable output voltage
    d. No output voltage
28. A series pass transistor with a collector-to-emitter short produces the symptom of
    a. Low output voltage
    b. High output voltage
    c. Unstable output voltage
    d. No output voltage
29. Refer to Fig. 15-17. The output voltage is zero. The input is on the high end of normal. There is no short or overload in the load circuit; in fact, the load current is zero. The defect is in
    a. The 7812 IC
    b. Q_1
    c. Q_2
    d. R_2
30. Refer to Fig. 15-22. D_1 is shorted. What is the symptom?
    a. High output
    b. Excessive ripple voltage
    c. Low output
    d. A blown fuse
31. Why should an isolation transformer be used when troubleshooting?
    a. To prevent shock
    b. To prevent circuit damage
    c. To prevent a ground loop
    d. All of the above
32. Refer to Fig. 15-30. Q_2 is shorted from collector to emitter. What is the symptom?
    a. No output
    b. High output
    c. The reference voltage on pin 8 will be over 1.3 V
    d. None of the above
33. Refer to Fig. 15-26. The varistor is open. What is the symptom?
    a. No output
    b. Low output
    c. High output
    d. No symptom, but transient protection has been lost

Chapter 15 Review

Summary

1. In a nonregulated power supply, output voltage varies with the line voltage and with the load current.
2. The output voltage tends to drop as the load on a power supply is increased.
3. Open-loop voltage regulators do not use feedback to control the output voltage.
4. A ferroresonant transformer with a saturated core can be used to regulate voltage.
5. Ferroresonant transformers use a resonating capacitor as part of their secondary circuit.
6. A zener diode shunt regulator is not practical in high-current applications because a high-power zener is required.
7. A series pass transistor can be used in conjunction with a zener diode to form a practical high-current power supply.
8. Negative regulators often use PNP pass transistors while positive regulators use NPN pass transistors.
9. Dual-polarity (bipolar) power supplies provide both negative and positive voltages with respect to ground.
10. Better voltage regulation is obtained with feedback (closed loop) power-supply operation.
11. Feedback power supplies use an error amplifier to compare the output voltage to a reference voltage.
12. Zener diodes are often used to provide a reference voltage in feedback-operated power supplies.
13. Op amps can be used as error amplifiers in regulated power supplies.
14. Integrated circuit voltage regulators provide fixed or variable output voltages in an easy-to-use package.
15. Adjusting a fixed IC voltage regulator with a resistive divider somewhat degrades its voltage regulation.
16. A current-boost transistor can be used with IC voltage regulators to provide more load current.
17. Tracking power supplies have a master output and one or more slave outputs. Any change in the master will be tracked by the slaves.
18. Shorting the output of a regulated power supply may damage the series pass transistor and other components in the power supply.
19. Current-limited power supplies protect themselves and also protect the load circuits connected to them.
20. Foldback current limiting is superior to conventional current limiting for preventing damage caused by long-term overloads.
21. Some IC voltage regulators can be configured for either type of current limiting.
22. A crowbar circuit provides voltage limiting by shorting the supply.
23. Swamping resistors can be used to ensure current sharing among parallel pass transistors.
24. Line transients can be clipped by varistors.
25. Metal oxide varistors turn on in nanoseconds and can safely handle hundreds or thousands of amperes.
26. Switch-mode regulators are more efficient than linear regulators and result in smaller and lighter power supplies.
27. Switch-mode power supplies operate at very high frequencies, allowing smaller transformers and filter components to be used.
28. Pulse-width modulation can be used to control the output voltage in switch-mode supplies.
29. Increasing the duty cycle of a waveform increases its average voltage.
30. Switch-mode power supplies use high-speed transistors, fast-recovery rectifiers, or Schottky rectifiers.
31. A converter is a circuit that changes direct current to alternating current and then back to direct current again.
32. Switch-mode supplies are noisier than linear types and can cause electromagnetic interference.
33. Sine-wave converters solve the noise and EMI problems associated with switchers.
34. An isolation transformer should be used when servicing or troubleshooting electronic equipment to avoid ground loops.
35. An open pass transistor (or no drive to the tran-

sistor) will cause the symptom of no output in a linear regulator.

36. A shorted pass transistor will cause the output to be abnormally high.
37. No output, low output, overheating, or a blown fuse are indications of an overloaded power supply.
38. An error in the reference voltage will cause an error in output voltage.
39. Switchers generate waveforms and frequencies beyond the capabilities of many meters.
40. When replacing parts, use exact replacements when possible and pay attention to lead dress.

Chapter Review Questions

Choose the letter that best answers each question.

15-1. Electrical brownouts are
   a. Caused by lightning and accidents
   b. Periods of low line voltage
   c. Periods of high line voltage
   d. Line transients

15-2. Refer to Fig. 15-5. What is the function of the resonating capacitor?
   a. To prevent damage from line transients
   b. To change pulsating direct current to pure direct current
   c. To cause high circulating currents in the secondary
   d. None of the above

15-3. Refer to Fig. 15-6. What happens to the zener diode dissipation if the load is disconnected?
   a. It stays the same.
   b. It decreases.
   c. It increases.
   d. It goes to zero.

15-4. Refer to Fig. 15-8. C is open. What is the most likely symptom?
   a. Excessive noise and ripple across the load
   b. Low output voltage
   c. No output voltage
   d. High output voltage

15-5. Refer to Fig. 15-9. What is the function of Q_1 and Q_2?
   a. They are error amplifiers.
   b. They establish the reference voltage.
   c. They provide overcurrent protection.
   d. They are series pass transistors.

15-6. What happens to the base-emitter voltage in a transistor as that transistor is called upon to support more current flow?
   a. It drops.
   b. It increases.
   c. It remains constant.
   d. It approaches 0 V at high current.

15-7. Refer to Fig. 15-11. Assuming that the circuit is working normally, what will happen to the series pass transistor when the load demands more current?
   a. It is driven toward cutoff.
   b. It dissipates less power.
   c. It is turned on harder.
   d. None of the above.

15-8. Refer to Fig. 15-12. How will the output of the op amp be affected if the load suddenly demands less current?
   a. It will go more positive.
   b. It will go less positive.
   c. It will not change.
   d. It will shut down.

15-9. Refer to Fig. 15-12. What is the purpose of R_3?
   a. To adjust the output voltage
   b. To adjust the voltage gain of the op amp
   c. To adjust the reference voltage
   d. To adjust the voltage limiting point

15-10. Linear IC voltage regulators, such as the 78*XX* series, are useful
   a. For decreasing costs in power supply designs
   b. As on-card regulators
   c. In decreasing the number of discrete parts in supplies
   d. All of the above

15-11. Refer to Fig. 15-14. What is the disadvantage of this circuit?
   a. The voltage regulation is somewhat degraded.
   b. It is too costly.
   c. It is difficult to troubleshoot.
   d. All of the above.

15-12. Refer to Fig. 15-14. Assume that the quiescent IC current is 6 mA, R_1 is 220 Ω, and R_2 is 100 Ω. What is the load voltage?
   a. 4.35 V
   b. 5.00 V
   c. 7.87 V
   d. 9.00 V

15-13. Refer to Fig. 15-15. Q_1 is silicon and R_1 is 12 Ω. At what value of current will the external pass transistor turn on and help to supply the load?
a. 0.006 A
b. 0.022 A
c. 0.058 A
d. 1.25 A

15-14. Refer to Fig. 15-16. What is the function of R_4?
a. To adjust the negative output voltage
b. To adjust the positive output voltage
c. To adjust both outputs
d. None of the above

15-15. Refer to Fig. 15-17. Q_2 is open. What is the symptom?
a. No output
b. Low output
c. High output
d. No current limiting

15-16. Refer to Fig. 15-22. Q_2 is open. What is the symptom?
a. No output
b. Low output
c. High output
d. No current limiting

15-17. Refer to Fig. 15-22. What could happen if R_2 is adjusted for too much output voltage?
a. The IC may overheat.
b. The crowbar may blow the fuse.
c. The current limiting may change to foldback.
d. All of the above.

15-18. Refer to Fig. 15-26. What is the function of the varistor?
a. It prevents brownouts from spoiling regulation.
b. It resonates the transformer.
c. It provides overcurrent protection.
d. It suppresses line transients.

15-19. A 5-A-rated power supply is normal but supplies only 2 A when short-circuited. This supply is protected by
a. Foldback current limiting
b. Conventional current limiting
c. An MOV device
d. A slow-blow fuse

15-20. Compared to switchers, linear power supplies with the same ratings are
a. Heavier
b. Larger
c. Less efficient
d. All of the above

15-21. Refer to Fig. 15-29. what is the function of L_1?
a. It takes on a charge when the transistor is on.
b. It dissipates its charge when the transistor is turned off.
c. It helps smooth the load voltage.
d. All of the above.

15-22. Refer to Fig. 15-29. What is the function of D_2?
a. It regulates the output voltage to the error amplifier.
b. It turns on when Q_1 is off to keep load current flowing.
c. It provides overcurrent protection.
d. All of the above.

15-23. Refer to Fig. 15-30. Suppose the load suddenly demands less current. What happens to the signal supplied to the base of Q_1?
a. The peak-to-peak amplitude goes down.
b. The duty cycle increases.
c. The duty cycle decreases.
d. The square wave changes to a sine wave.

15-24. Refer to Fig. 15-30. R_1 is damaged and has increased in value. What is the symptom?
a. Excessive output ripple
b. High output voltage
c. Output drops as the supply is loaded
d. The IC runs hot

15-25. Refer to Fig. 15-31. Diode D_1 is open. What is the symptom?
a. No output voltage
b. High output voltage
c. Reverse output polarity
d. C_1 burns up

15-26. Why are switch-mode power supplies operated at frequencies so much above 60 Hz?
a. To limit dissipation in transistors and diodes
b. To allow smaller transformers and filters
c. So that pulse-width modulators can be used
d. All of the above

15-27. Refer to Fig. 15-33. How could the output voltage be increased?
a. By increasing the oscillator frequency
b. By decreasing the oscillator frequency
c. By removing D_3
d. By keeping Q_1 and Q_2 on longer

15-28. Refer to Fig. 15-34. What is the purpose of C_3?
a. It resonates L_1 and changes the square waves to sine.
b. It changes the frequency of the VCO.
c. It provides transient protection.
d. All of the above.

15-29. Refer to Fig. 15-34. What do Q_1 and Q_2 accomplish?
   a. They control voltage by linear resistance change.
   b. They change the ac line power to pulsating dc power.
   c. They change direct current to alternating current.
   d. They provide conventional current limiting.

15-30. Refer to Fig. 15-34. Where is isolation from the ac line accomplished?
   a. In the bridge rectifier
   b. By C_1 and C_2
   c. In T_1
   d. In the Schottky diodes

Critical Thinking Questions

15-1. You are troubleshooting a power supply with three outputs: one master and two slaves. Which section of the power supply should be verified first? Why?

15-2. It is desired to use a crowbar circuit to protect equipment that is remotely located. How could the basic crowbar design be modified so that the equipment would automatically come back on line after the fault cleared?

15-3. Is there any situation when the modified design of question 15-2 could perform in an undesirable way?

15-4. Can you think of any physical (nonelectrical) problems that could cause intermittent operation in power supplies?

15-5. Why does some battery-operated equipment contain voltage regulators?

15-6. What type of power-supply circuit would you expect to find in a photographer's battery-operated electronic flash unit? Why?

Answers to Tests

1. b
2. c
3. b
4. d
5. a
6. d
7. c
8. b
9. a
10. d
11. a
12. b
13. d
14. b
15. c
16. a
17. d
18. a
19. b
20. d
21. c
22. d
23. a
24. b
25. c
26. c
27. d
28. b
29. a
30. d
31. d
32. b
33. d

Term	Definition	Symbol or Abbreviation
AC component	The fluctuating or changing value of a waveform or of a signal. Pure direct current has no ac component.	
Active	An operating region between saturation and cutoff. The current in an active device is a function of its control bias.	
Active filter	An electronic filter using active gain devices (usually operational amplifiers) to separate one frequency, or a group of frequencies, from all other frequencies.	
Amplifier	A circuit or device designed to increase the level of a signal.	
Amplitude modulation	The process of using a lower frequency signal to control the instantaneous amplitude of a higher frequency signal. Often used to place intelligence (audio) on a radio signal.	AM
Analog	That branch of electronics dealing with infinitely varying quantities. Often referred to as linear electronics.	
Analog to digital	A circuit or device used to convert an analog signal or quantity to digital form (usually binary).	A/D
Anode	That element of an electronic device that receives the flow of electron current.	
Attenuator	A circuit used to decrease the amplitude of a signal.	
Automatic frequency control	A circuit designed to correct the frequency of an oscillator or the tuning of a receiver.	AFC
Automatic gain control	A circuit designed to correct the gain of an amplifier according to the level of the incoming signal.	AGC
Automatic volume control	A circuit designed to provide a constant output volume from an amplifier or radio receiver.	AVC
Avalanche	The sudden reverse conduction of an electronic component caused by excess reverse voltage across the device.	
Balanced modulator	A special amplitude modulator designed to cancel the carrier and leave only the sidebands as outputs. It is used in single sideband transmitters.	
Barrier potential	The potential difference that exists across the depletion region in a PN junction.	
Base	The center region of a bipolar junction transistor that controls the current flow from emitter to collector.	B
Beat frequency oscillator	A radio receiver circuit that supplies a carrier signal for demodulating code or single sideband transmissions.	BFO
Beta	The base-to-collector current gain in a bipolar junction transistor. Also called h_{FE}.	β
Bias	A controlling voltage or current applied to an electronic circuit or device.	
Bipolar	Having two polarities of carriers (holes and electrons).	
Bleeder	A fixed load designed to discharge (bleed off) filters.	
Block diagram	A drawing using a labeled block for every major section of an electronic system.	
Blocking capacitor	A capacitor that eliminates the dc component of the signal.	
Bode plot	A graph showing the gain or phase performance of an electronic circuit at various frequencies.	
Bootstrap	A feedback circuit usually used to increase the input impedance of an amplifier. May also refer to a circuit used to start some action when the power is first applied.	

Term	Definition	Symbol or Abbreviation
Break frequency	A frequency in which the response or gain of a circuit decreases 3 dB from its best response or gain.	f_b
Bypass	A low-pass filter employed to remove high-frequency interference from a power supply line or a component such as a capacitor that provides a low impedance path for high-frequency current.	
Capacitive coupling	A method of signal transfer that uses a series capacitor to block or eliminate the dc component of the signal.	
Capacitive input filter	A filter circuit (often in a power supply) using a capacitor as the first component in the circuit.	
Carrier	A movable charge or particle in an electronic device that supports the flow of current. Also refers to an unmodulated radio or television signal.	
Cathode	That element of an electronic device that provides the flow of electron current.	
Characteristic curves	Graphic plots of the electrical and/or thermal behavior of electronic circuits or components.	
Choke input filter	A filter circuit (often in a power supply) using a choke or an inductor as the first component in the circuit.	
Clamp	A circuit for adding a dc component to an ac signal. Also known as a dc restorer.	
Clapp oscillator	A series-tuned Colpitts configuration noted for its good frequency stability.	
Class	One way to categorize an amplifier based on bias and conduction angle.	
Clipper	A circuit that removes some part of a signal. Clipping may be undesired in a linear amplifier or desired in a circuit such as a limiter.	
Collector	The region of a bipolar junction transistor that receives the flow of current carriers.	
Colpitts oscillator	A circuit with a capacitively tapped tank circuit.	
Common base	An amplifier configuration where the input signal is fed into the emitter terminal and the output signal is taken from the collector terminal.	CB
Common collector	An amplifier configuration where the input signal is fed into the base terminal and the output signal is taken from the emitter terminal. Also called emitter follower.	CC
Common emitter	The most widely applied amplifier configuration where the input signal is fed into the base terminal and the output signal is taken from the collector circuit.	CE
Common-mode rejection ratio	The ratio of differential gain to common-mode gain in an amplifier. It is a measure of the ability to reject a common-mode signal and is usually expressed in decibels.	CMRR
Commutation	The interruption of current flow. In thyristor circuits, it refers to the method of turning the control device off.	
Comparator	A high-gain amplifier that has an output determined by the relative magnitude of two input signals.	
Complementary symmetry	A circuit designed with opposite polarity devices such as NPN and PNP transistors.	
Conduction angle	The number of electrical degrees that a device is on.	
Continuous wave	A type of modulation where the carrier is turned off and on following a pattern such as Morse code.	CW
Converter	A circuit that transforms dc from one voltage level to another. Also refers to a circuit that changes frequency.	
Coupling	The means of transferring electronic signals.	
Crossover distortion	Disturbances to an analog signal that affect only that part of the signal near the zero axis or average axis.	

Term	Definition	Symbol or Abbreviation
Crowbar	A protection circuit used to blow a fuse or otherwise turn a power supply off in the event of excess voltage.	
Crystal	A piezoelectric transducer used to control frequency, change vibrations into electricity, or filter frequencies. Also refers to the physical structure of semiconductors.	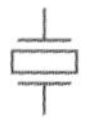
Current gain	The feature of certain electronic components and circuits where a small current controls a large current.	A_I
Current limiter	A circuit or device that prevents current flow from exceeding some predetermined limit.	
Curve tracer	An electronic device for drawing characteristic curves on a cathode-ray tube.	
Cutoff	That bias condition where no current can flow.	
Darlington	A circuit using two direct-coupled bipolar transistors for very high current gain.	
DC component	The average value of a waveform or of a signal. Pure ac averages zero and has no dc component.	
Decibel	One-tenth of a bel. A logarithmic ratio used to measure gain and loss in electronic circuits and systems.	dB
Demodulation	The recovery of intelligence from a modulated radio or television signal. Also called detection.	
Depletion	The condition of no available current carriers in a semiconducting crystal. Also refers to that mode of operation for a field-effect transistor where the channel carriers are reduced by gate voltage.	
Diac	A silicon bilateral device used to gate other devices.	
Differential amplifier	A gain device that responds to the difference between its two input terminals.	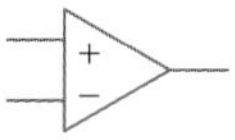
Digital	That branch of electronics dealing with finite and discrete signal levels. Most digital signals are binary: they are either high or low.	
Digital to analog	A circuit or device used to convert a digital signal into its analog equivalent.	D/A
Digital signal processing	A system using A/D and D/A converters plus a microprocessor to alter some characteristic of an analog signal.	DSP
Diode	A two-terminal electronic component. They usually allow current to flow in only one direction. Different types of diodes can be used for rectification, regulation, tuning, triggering, and detection. They can also be used as indicators.	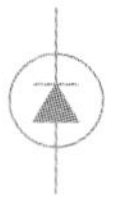
Direct digital synthesis	A method of generating waveforms based on a lookup table and a phase accumulator.	DDS
Discrete circuit	An electronic circuit made up of individual components (transistors, diodes, resistors, capacitors, etc.) interconnected with wires or conducting traces on a printed circuit board.	
Discriminator	A circuit used to detect frequency-modulated signals.	
Distortion	A change (usually unwanted) in some aspect of a signal.	
Doping	A process of adding impurity atoms to semiconductor crystals to change their electrical properties.	
Drain	That terminal of a field-effect transistor that receives the current carriers from the source.	D 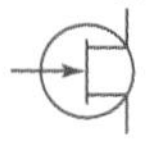
Dual power supply	A power supply that produces positive and negative outputs with reference to ground. Also called a bipolar supply.	
Efficiency	The ratio of useful output from a circuit to the input.	η
Electromagnetic interference	A form of interference to and from electronic circuits resulting from the radiation of high-frequency energy.	EMI
Emitter	That region of a bipolar junction transistor that sends the current carriers on to the collector.	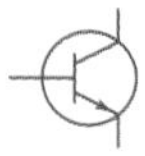 E
Enhancement mode	That operation of a field-effect transistor where the gate voltage is used to create more carriers in the channel.	

Term	Definition	Symbol or Abbreviation
Epitaxial	A thin, deposited crystal layer that forms a portion of the electrical structure of certain semiconductors.	
Error amplifier	A gain device or circuit that responds to the error (difference) between two signals.	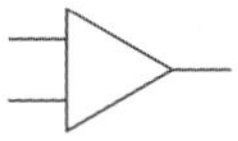
Feedback	The application of a portion of the output signal of a circuit back to the input of the circuit. Any of a number of closed-loop systems where an output is connected to an input.	
Ferroresonant	A special type of power-supply transformer using a resonating capacitor and a saturated core to provide both load and line voltage regulation.	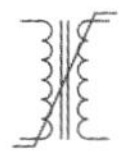
Field-effect transistor	A solid-state device that uses a terminal (gate) voltage to control the resistance of a semiconducting channel.	FET 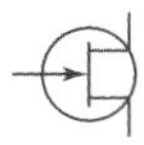
Filter	A circuit designed to separate one frequency, or a group of frequencies, from all other frequencies.	
Flip-flop	An electronic circuit with two states. Also known as a multivibrator. May be free-running (as an oscillator) or exhibit one or two stable states.	
Foldback current limiting	A type of current limiting in which the current decreases beyond the threshold point as the load resistance drops.	
Frequency modulation	The process of using a lower frequency signal to control the instantaneous frequency of a higher frequency signal. Often used to place intelligence (audio) on a radio signal.	FM
Frequency multiplier	A circuit where the output frequency is an integer multiple of the input frequency. Also known as a doubler, tripler, etc.	
Frequency synthesis	A method of generating many accurate frequencies without resorting to multiple crystal-controlled oscillators. Usually based on PLL or DDS technology.	
Gain	A ratio of output to input. May be measured in terms of voltage, current, or power. Also known as amplification.	A or G
Gain-bandwidth product	The high frequency at which the gain of an amplifier is 0 dB (unity).	f_t
Gate	That terminal of a field-effect transistor that controls drain current. Also the terminal of a thyristor used to turn the device on.	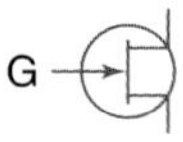 G
Ground loop	A short (or otherwise unwanted) circuit across the ac line caused by grounded test equipment or some other ground path not normally intended to conduct current.	
Hartley oscillator	A circuit distinguished by its inductively tapped tank circuit.	
Heterodyne	The process of mixing two frequencies to create new (sum and difference) frequencies.	
Holes	Positively charged carriers that move opposite in direction to electrons and can be found in semiconducting crystals.	
Hysteresis	A dual threshold effect exhibited by certain circuits.	
House numbers	Nonregistered device numbers peculiar to the manufacturer.	
Image	The second, unwanted, frequency that a heterodyne converter will interact with to produce the intermediate frequency.	
Impedance match	The condition where the impedance of a signal source is equal to the impedance of the signal load. It is desired for best power transfer from source to load.	
Integrated circuit	The combination of many circuit components into a single crystalline structure (monolithic) or onto a supporting substrate (thick-film) or a combination of the two (hybrid).	IC
Integrator	An electronic circuit that provides continuous summation of signals over some period of time.	
Intermediate frequency	A standard frequency in a receiver that all incoming signals are converted to before detection. Most of the gain and	IF

Term	Definition	Symbol or Abbreviation
	selectivity of a receiver are produced in the intermediate-frequency amplifier.	
Intermittent	A fault that only appears from time to time. It may be related to mechanical shock or temperature.	
Intrinsic standoff ratio	In a unijunction transistor, the ratio of the voltage required to fire the transistor to the total voltage applied across the transistor.	η
Inverting	An amplifier where the output signal is 180° out of phase with the input.	
Latch	A device that, once triggered on, tends to stay on. Also a digital circuit for storing one of two conditions.	
Lead dress	The exact position and length of electronic components and their leads. Can affect the way circuits (especially high-frequency) perform.	
Lead-lag network	A circuit that provides maximum amplitude and zero phase shift for one (the resonant) frequency. It produces a leading angle for frequencies below resonance and a lagging angle for frequencies above resonance.	
Leakage	In semiconductors, a temperature-dependent current that flows under conditions of reverse bias.	
Limiter	A circuit that clips off the high-amplitude portions of a signal to reduce noise or to prevent another circuit from being overdriven.	
Line transient	An abnormally high voltage of short duration on the ac power line.	
Linear	A circuit or component where the output is a straight-line function of the input.	
Majority carriers	In an N-type semiconductor, the electrons. In a P-type semiconductor, the holes.	
Metal oxide semiconductor	A discrete or integrated semiconductor device that uses a metal and an oxide (silicon dioxide) as an important part of the device structure.	MOS
Metal oxide varistor	A device used to protect sensitive circuitry and equipment from line transients.	MOV
Minority carriers	In an N-type semiconductor, the holes. In a P-type semiconductor, the electrons.	
Modulation	The process of controlling some aspect of a periodic signal such as amplitude, frequency, or pulse width. Used to place intelligence on a radio or television signal.	
Neutralization	The application of external feedback in an amplifier to cancel the effect of internal feedback (inside the transistor).	
Noise	Any unwanted portion of, or interference to, a signal.	
Noninverting	An amplifier where the output signal is in phase with the input signal.	
Numerically controlled oscillator	Another name for a direct digital synthesizer (DDS).	NCO
Offset	An error in the output of an operational amplifier caused by imbalances in the input circuit.	
Open circuit	A condition of infinite resistance or infinite impedance and zero current flow.	
Operational amplifiers	High-performance amplifiers with inverting and noninverting inputs. They are usually in integrated circuit form and can be connected for a wide variety of functions and gains.	Op amp
Operating point	The average condition of a circuit as determined by some control voltage or current. Also called the quiescent point.	
Opto-isolator	An isolation device that uses light to connect the output to the input. Used where there must be an extremely high electrical resistance between input and output.	

Term	Definition	Symbol or Abbreviation
Oscillator	An electronic circuit for generating various ac waveforms and frequencies from a dc energy source.	
Pass transistor	A transistor connected in series with a load to control the load voltage or the load current.	
Phase-locked loop	An electronic circuit that uses feedback and a phase comparator to control frequency or speed.	PLL
Phase-shift oscillator	An oscillator circuit characterized by an *RC* phase-shift network in its feedback path.	
Pi filter	A low-pass filter using a shunt input capacitor, a series inductor, and a shunt output capacitor.	
Power amplifier	An amplifier designed to deliver a significant level of output voltage, output current, or both. Also known as a large-signal amplifier.	
Power gain	The ratio of output power to input power. Often expressed in decibels.	A_P or G_P
Printed circuit	A lamination of copper foil on an insulating substrate such as fiberglass and epoxy resin. Portions of the foil are removed, leaving circuit paths to interconnect electronic components to form complete circuits.	PC
Product detector	A special detector for receiving suppressed carrier transmissions such as single sideband.	
Programmable	A device or circuit in which the operational characteristics may be modified by changing a programming voltage, current, or some input information.	
Pulsating direct current	Direct current with an ac component (i.e., the output of a rectifier).	
Pulse-width modulation	Controlling the width of rectangular waves for the purpose of adding intelligence or to control the average dc value.	PWM
Pure alternating current	Alternating current with no dc component. It has an average value of zero.	
Pure direct current	Direct current with no ac component. Pure direct current has no ripple or noise and is a straight line on an oscilloscope.	
Push-pull	A circuit using two devices, where each device acts on one-half of the total signal swing.	
Radio-frequency choke	A coil used to block or eliminate radio (high) frequencies.	RFC
Ratio detector	A circuit used to detect frequency-modulated signals.	
Rectification	The process of changing alternating to direct current.	
Regulator	A circuit or device used to hold some quantity constant.	
Relaxation oscillator	Those oscillators characterized by *RC* timing components to control the frequency of the output signal.	
Ripple	The ac component in the output of a dc power supply.	
Saturation	The condition where a device, such as a transistor, is turned on hard. When a device is saturated, its current flow is limited by some external load connected in series with it.	
Schmitt trigger	An amplifier with hysteresis used for signal conditioning in digital circuits.	
Selectivity	The ability of a circuit to select, from a broad range of frequencies, only those frequencies of interest.	
Semiconductors	A category of materials having four valence electrons and electrical properties between conductors and insulators.	
Sensitivity	The ability of a circuit to respond to weak signals.	
Servomechanism	A control circuit that regulates motion or position.	
Sidebands	Frequencies above and below the carrier frequency created by modulation.	
Silicon	An element. The semiconductor material currently used to make almost all solid-state devices such as diodes, transistors, and integrated circuits.	
Silicon-controlled rectifier	A device used to control heat, light, and motor speed. It will	SCR 

Term	Definition	Symbol or Abbreviation
	conduct from cathode to anode when it is gated on.	
Single sideband	A variation of amplitude modulation. The carrier and one of the two sidebands are suppressed.	SSB
Slew rate	The measure of the ability of a circuit to produce a large change in output in a short period of time.	
Small-signal bandwidth	The total frequency range of an amplifier in which its gain for small signals is within 3 dB of its best gain.	
Source	That terminal of a field-effect transistor that sends the current carriers to the drain.	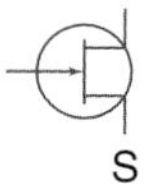 S
Static switch	A switch with no moving parts, generally based on thyristors.	
Superheterodyne	A receiver that uses the heterodyne frequency conversion process to convert the frequency of an incoming signal to an intermediate frequency.	
Surge limiter	A circuit or a component (often a resistor) used to limit turn-on surges to some safe value.	
Surface-mount technology	A method of printed circuit fabrication in which the component leads are soldered on the component side of the board and do not pass through holes in the boards.	SMT
Swamping resistor	A resistor used to swamp out (make insignificant) individual component characteristics. Can be used to ensure current sharing in parallel devices.	
Switch mode	A circuit where the control element switches on and off to achieve high efficiency.	
Tank circuit	A parallel *LC* circuit.	
Temperature coefficient	The number of units change, per degree Celsius change, from a specified temperature.	
Thermal runaway	A condition in a circuit where temperature and current are mutually interdependent and both increase out of control.	
Thyristor	The generic term referring to control devices such as silicon-controlled rectifiers and triacs.	
Transistor	Any of a group of solid-state amplifying or controlling devices that usually have three leads.	
Triac	A full-wave, bidirectional control device that is equivalent to two silicon-controlled rectifiers (*tri*ode *ac* switch).	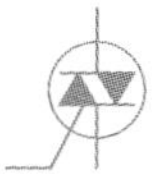
Troubleshooting	A logical and orderly process to determine the fault or faults in a circuit, a piece of equipment, or a system.	
Twin-T network	A circuit containing two branches, each arranged in the form of the letter *T*, that can be used as a notch filter or to control the frequency of an oscillator.	
Unijunction transistor	A transistor used in control and timing applications. It turns on suddenly when its emitter voltage reaches the firing voltage.	UJT 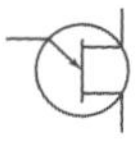
Variable-frequency oscillator	An oscillator with an adjustable output frequency.	VFO
Varistor	A nonlinear resistor. Its resistance is a function of the voltage across it.	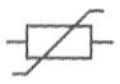
Virtual ground	An ungrounded point in a circuit that acts as a ground as far as signals are concerned.	
Voltage-controlled oscillator	An oscillator circuit where the output frequency is a function of a dc control voltage.	VCO
Voltage gain	The ratio of amplifier output voltage to input voltage. Often expressed in decibels.	A_V or G_V
Voltage multipliers	Direct current power-supply circuits used to provide transformerless step-up of ac line voltage.	
Voltage regulator	A circuit used to stabilize voltage.	
Zener diode	A diode designed to operate in reverse breakover with a stable voltage drop. It is useful as a voltage regulator.	
Zero-crossing detector	A comparator that changes states when its input crosses the zero volt point.	

Appendix B Thermionic Devices

Thermionic devices (vacuum tubes) dominated electronics until the early 1950s. Since that time, solid-state devices have all but completely taken over. Today, vacuum tubes are used only in special applications such as high-power RF amplifiers, cathode-ray tubes (including television picture tubes), and some microwave devices.

Thermionic emission involves the use of heat to liberate electrons from an element called a *cathode.* The heat is produced by energizing a filament or heater circuit within the tube. A second element, called the *anode,* can be used to attract the liberated electrons. Since unlike charges attract, the anode is made positive with respect to the cathode.

A third electrode can be placed between the cathode and the anode. This third electrode can exert control over the movement of electrons from cathode to anode. Thus, it is called the *control grid.* The control grid is often negative with respect to the cathode. This negative charge repels the cathode electrons and prevents them all from reaching the anode. In fact, the tube can be cut off by a high negative grid potential. Figure B-1 shows the schematic symbol for a three-electrode vacuum tube and the polarities involved.

The vacuum tube shown in Fig. B-1 is an amplifier. The signal to be amplified can be applied to the control grid. As the signal goes in a positive direction, more plate current will flow. As the signal goes in a negative direction, less plate current will flow. Thus, the plate current is a function of the signal applied to the grid. The signal power in the grid circuit is much less than the signal power in the plate circuit. The vacuum tube is capable of good power gain.

Vacuum tubes may use extra grids located between the control grid and the plate to provide better operation. The extra grids improve gain and high-frequency performance. The tube in Fig. B-1 is called a *triode* vacuum tube (the heater is not counted as an element). If a screen grid is added, it becomes a *tetrode* (four electrodes). If a screen grid and a suppressor grid are added, it becomes a *pentode* (five electrodes).

Vacuum tubes make excellent high-power amplifiers. It is possible to run some vacuum tubes with plate potentials measured in thousands of volts and plate currents measured in amperes. These tubes offer output powers of several thousand watts. It is even possible to develop 2,000,000-W amplitude-modulated RF output by using four special tetrodes. This is an example of the outstanding power capacity of vacuum tubes.

The cathode-ray tube is a vacuum tube used for the display of graphs, pictures, or data. Figure B-2 shows the basic structure. The cathode is heated and produces thermionic emission. A positive potential is applied to the first anode, the second anode, and the aquadag coating. This positive field accelerates the electrons toward the screen. The inside of the screen is coated with a chemical phosphor that emits light when hit by a stream of electrons.

As shown in Fig. B-2, the electrons are focused into a narrow beam. This makes it possible to produce a small dot of light on the screen. The deflecting plates can move the beam vertically and horizontally. For example, a positive voltage applied to the top vertical deflecting plate will attract the beam and move it up. The dot of light can be positioned anywhere on the screen.

The grid shown in Fig. B-2 makes it possible to control the intensity of the beam. A negative voltage applied to the grid will repel the cathode electrons and prevent them all from reaching the screen. A high negative voltage will completely stop the electrons, and the dot of light will go out.

By controlling the position and the intensity of the dot, any type of picture information can be presented on the screen. Because the phosphor will retain its brightness momentarily and because the eye will retain the image for a brief period, the effect of the moving dot is to produce what seems to be a complete picture on the screen. If this is done repeatedly, a movie effect is produced. This is how a television picture tube works. Colors can be shown by using several different chemical phosphors.

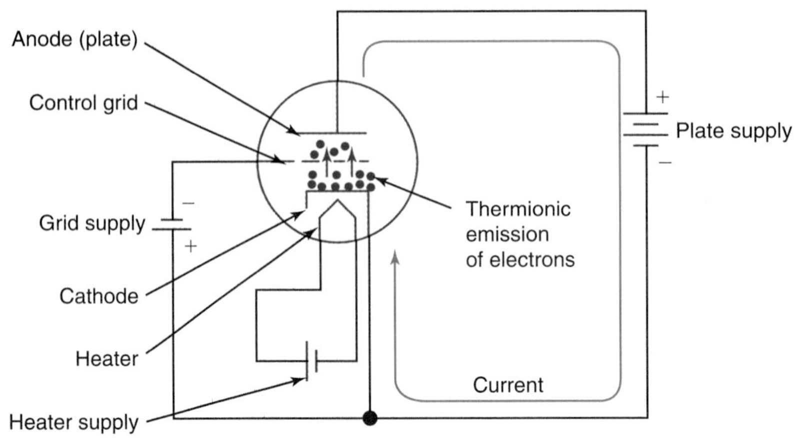

Fig. B-1

The deflection system may be different from that shown in Fig. B-2. *Magnetic deflection* uses coils around the neck of the cathode-ray tube. When a current flows through the coil, the resulting magnetic field will deflect the electron beam. Television picture tubes generally use magnetic deflection. Oscilloscopes generally use electrostatic deflection.

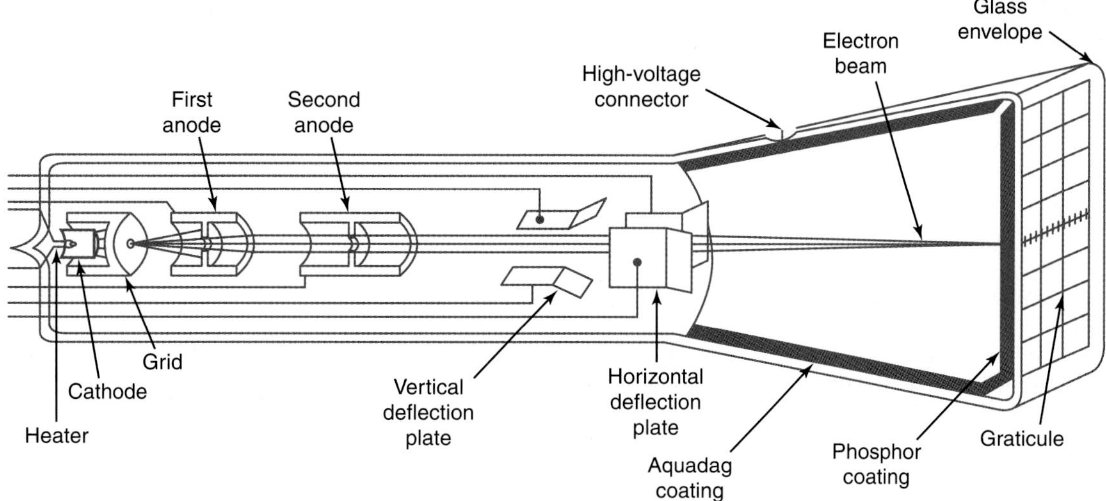

Fig. B-2 A cathode-ray tube using electrostatic deflection.